RNATIONAL ENCYCLOPAEDIA OF FOOD AND NUTRITION

VOLUME 17

NUTRITION OF ANIMALS OF AGRICULTURAL IMPORTANCE

VOLUME EDITOR

SIR DAVID CUTHBERTSON, C.B.E., F.R.S.E., *Glasgow*

PART 1

INTERNATIONAL ENCYCLOPAEDIA OF FOOD AND NUTRITION

NUTRITION OF ANIMALS
OF AGRICULTURAL IMPORTANCE

PART 1

THE SCIENCE OF NUTRITION OF FARM LIVESTOCK

VOLUME EDITOR

SIR DAVID CUTHBERTSON, C.B.E., F.R.S.E.

Department of Pathological Biochemistry, Glasgow University and Royal Infirmary, Glasgow C.4

1966
THE QUEEN'S AWARD TO INDUSTRY 1966

PERGAMON PRESS

OXFORD · LONDON · EDINBURGH · NEW YORK

TORONTO · SYDNEY · PARIS · BRAUNSCHWEIG

Pergamon Press Ltd., Headington Hill Hall, Oxford
4 & 5 Fitzroy Square, London W.1

Pergamon Press (Scotland) Ltd., 2 & 3 Teviot Place, Edinburgh 1

Pergamon Press Inc., Maxwell House, Fairview Park, Elmsford, New York 10523

Pergamon of Canada Ltd., 207 Queen's Quay West, Toronto 1

Pergamon Press (Aust.) Pty. Ltd., 19a Boundary Street, Rushcutters Bay, N.S.W. 2011, Australia

Pergamon Press S.A.R.L., 24 rue des Écoles, Paris 5e

Vieweg & Sohn GmbH, Burgplatz 1, Braunschweig

First edition 1969

Library of Congress Catalog Card No. 69-14229

PRINTED IN GREAT BRITAIN BY THE ANCHOR PRESS LTD., TIPTREE, ESSEX

08 003583 3

Contents

PART 1

Contents

PART 2

*These chapters are reprinted from Agricultural Research Council Publications, published by H.M.S.O.

List of Contributors

Bolton, W., *Agricultural Research Council, Poultry Research Centre, Edinburgh*

Care, A. D., *The Rowett Research Institute, Bucksburn, Aberdeen*

Corbett, J. L., *C.S.I.R.O., Pastoral Research Laboratory, Armidale, New South Wales*

Coates, M. E., *National Institute for Research in Dairying, Shinfield, Reading*

Cuthbertson, D. P., *Department of Pathological Biochemistry, Royal Infirmary, Glasgow C.4*

Dodsworth, T. L., *The North of Scotland College of Agriculture, Aberdeen*

Fell, B. F., *The Rowett Research Institute, Bucksburn, Aberdeen*

Flatt, W., *United States Department of Agriculture, Beltsville, Maryland*

Garton, G. A., *The Rowett Research Institute, Bucksburn, Aberdeen*

Greenhalgh, J. F. D., *The Rowett Research Institute, Bucksburn, Aberdeen*

Hill, F. W., *Department of Nutrition, University of California, Davis, California*

Hobson, P. N., *The Rowett Research Institute, Bucksburn, Aberdeen*

Jakobsen, P. E., *Department of Animal Physiology, Rolighedsvej 25, Copenhagen*

Jayne-Williams, D. J., *National Institute for Research in Dairying, Shinfield, Reading*

Jones, A. S., *The Rowett Research Institute, Bucksburn, Aberdeen*

Lamming, G. E., *University of Nottingham, School of Agriculture, Sutton Bonington, Loughborough*

Lodge, G. A., *University of Nottingham, School of Agriculture, Sutton Bonington, Loughborough*

McDonald, I. W., *Division of Animal Physiology, Ian Clunies Ross Animal Research Laboratory, C.S.I.R.O., Parramatta, New South Wales*

Mills, C. F., *The Rowett Research Institute, Bucksburn, Aberdeen*

Olsson, N., *Verberöd, Sweden*

Payne, W. J. A., *F.A.O./U.N.D.P., Dairy Training and Research Institute, Philippines*

Phillipson, A. T., *Department of Veterinary Clinical Studies, The University of Cambridge*

Preston, T. R., *Instituto de Ciencia Animal, Catalina de Güines, Habana, Cuba*

Pullar, J. D., *The Rowett Research Institute, Bucksburn, Aberdeen*

Riis, P. M., *Department of Animal Physiology, Rolighedsvej 25, Copenhagen*

Robinson, K. L., *Agricultural Chemistry Department, The Queen's University of Belfast*

Rolfe, E., *National College of Food Technology, Weybridge*

Roy, J. H. B., *National Institute for Research in Dairying, Shinfield, Reading*

Scott, M. L., *Department of Poultry Science and Graduate School of Nutrition, Cornell University, Ithaca, New York*

Slen, S. B., *Canada Department of Agriculture, Research Station, Lethbridge, Alberta*

Van Soest, P. J., *United States Department of Agriculture, Beltsville, Maryland*

Woodham, A. A., *The Rowett Research Institute, Bucksburn, Aberdeen*

Wragg, S. R., *Department of Economics (Agricultural Economics), the University, Bristol*

Preface

THE purpose of Volume 17 is to provide first a general background to the science of nutrition of animals of agricultural importance and then to assess their requirements in the light of existing knowledge. Thus the contents are presented in two parts.

Part 1 provides first an introduction to the study of nutrition of animals of agricultural importance and there is a chapter on the newer knowledge of the composition and methods of analysis of feeding stuffs. Then follow chapters on the physiology, biochemistry and microbiology of digestion and absorption in the main classes of livestock. These are followed by details of energy metabolism and the metabolism of specific groups of nutrients and on the relation of nutrition to reproduction and the endocrine control of growth and metabolism. The final chapter in this first part is on the preservation of animal products for human consumption.

Part 2 is concerned with the special nutrition of the different classes of farm livestock, including the horse, and the assessment of their requirements. 'Requirement' has been defined as the amount of a nutrient that must be supplied in the diet to meet the needs of the normal healthy animal given a completely adequate diet in an environment compatible with good health. In general, the requirements include no margins of safety and are estimated requirements, but some discussion on allowances is provided. The methods of estimating these requirements have differed for the different nutrients and are further expounded in the Technical Reviews of the Nutrient Requirements of Farm Livestock published for 'Poultry', 'Ruminants' and 'Pigs' by the U.K. Agricultural Research Council, London. By kind permission of the Agricultural Research Council, Summaries of these Technical Reviews are included in these volumes as separate chapters, and the comparable publications on Nutrient Requirements of Domestic Animals by the Committee on Animal Nutrition of the U.S. National Academy of Sciences National Research Council also provide much valuable knowledge.

While most of the observations relate to temperate climates, attention has also been paid to the problems of the nutrition of ruminants in the tropics. There is also a chapter on nutritional disorders and the so-called 'metabolic' diseases.

Recent developments in our basic understanding of ruminant nutrition, in particular, and in the evolution of a new system of measuring nutrient requirements and feeding values have a very special significance in agricultural economics. Economists will regard current progress in this field of animal science as a means of making the direct econometric approach more effective through a better understanding of the underlying principles of the input/output relationships which this new knowledge provides. Some of the information on metabolizable energy requirements and feed values may allow them to predict, for example, the economic allowances of concentrates for various milk/concentrate price ratios. A number of the basic assumptions in linear programming may be upset by the fact that nutrient affects are not additive. These matters of import are dealt with in the penultimate chapter of Part 2. The final chapter surveys world needs.

As Editor of this volume I should like to say how fortunate I have been in securing as authors many of the leading investigators in the science of nutrition and its application to the animal production industry, and I am deeply conscious of the honour they do us in participating in the writing of this work. I am greatly indebted to my very efficient secretaries Miss R. A. Watson and Mrs. E. Coulter. We should also like to take this opportunity of thanking the publishers for their great help and care.

General Introduction

THE three primitive instincts of all higher animals including man are nutrition, self-preservation and sex. Of these nutrition is the earliest since all animals and plants ingest and metabolize food. Hence the science of nutrition and the practical application of that science interest those engaged in a wide variety of disciplines. For instance, the chemist studies the composition of foodstuffs, the bio-chemist their transformations in the body with consequent growth and energy-production, the physiologist the normal processes of metabolism and the pathologist the changes that result from abnormalities of those processes. The clinician must recognize the disorders produced by faulty or insufficient food and the dietist helps to correct these; agriculturalists and economists are concerned with the production and distribution of food whether animal or vegetable, sociologists and politicians with the national and international aspects of that production and distribution.

There are two factors that have made desirable the collection and integration of existing knowledge of food and nutrition at the present time. First, this knowledge has been advanced very rapidly in the past few years largely by the great discoveries in biochemistry. Secondly, there is no more urgent problem in the world today than the nutritional disasters that are now present amongst us and will increase with the explosive rise in the population of the world. Hundreds of thousands of years have been needed to reach our present population of 3400 million, but another 3400 million persons will probably be produced in the next 35 years. And it is alleged that two-thirds of the present population is underfed. Whatever the proportion, we have a food deficit that is certain to increase and must be met by increased production of food whether by more efficient methods, bringing new lands into cultivation or by the use of new types of foodstuffs, for instance from oil.

Nor need the other third be complacent about its present food and nutrition, for the rapid developments in food technology have created new problems as foods are increasingly processed and sophisticated. We now treat foods with a variety of chemicals. Such additives are essential for two reasons. First, the housewife increasingly demands partially cooked foods in plastic wrappers which she merely warms at home; she no longer prepares and cooks what she gathered from her garden. Secondly, we have to produce the maximum amount of food from the land and then store it and transport it over long distances. Therefore we very properly use chemical fertilizers, pesticides, hormones, antibiotics and improvers (such as vitamins A and D in margarine). All these activities create problems.

An encyclopaedia is a work that contains exhaustive information on a branch of knowledge, arranged systematically and not necessarily alphabetically. The aims, scope and purpose of this one were discussed with Mr. Robert Maxwell, the publisher, at the International Congresses of Nutrition in Washington, Edinburgh and Hamburg in 1960, 1963 and 1966. The plan is to provide an authoritative and comprehensive presentation of knowledge in the whole field of nutrition of man and lower animals, with brief reference to the nutrition of plants and microorganisms. Each volume is written primarily for nutritional scientists, but is of value to others such as chemists, biochemists, medical practitioners, dietists, veterinary surgeons, agriculturalists and economists. Experimental details of an essentially practical nature are not emphasized in the Encyclopaedia; but a thorough background of the theoretical basis of techniques is included so that these may be understood and applied at the fullest extent. The Encyclopaedia provides for investigations in the fundamental biological sciences—biophysics, biochemistry, physiology, pathology, psychology, experimental therapeutics, experimental zoology and related fields—data which will help each in his own field. Further, the advanced student in both the basic medical sciences and clinical specialities will find these volumes an invaluable guide and an authoritative source of reference to literature.

As Editor-in-Chief I have been supported by five Regional Editors and by an Honorary Editorial Advisory Board composed of the world's leading nutritional scientists and under the Chairmanship of the Publisher, Mr. Robert Maxwell. We regret that the authorship of the articles has not been as widely international as is desirable, but while in some volumes there is a preponderance of articles from the English-speaking countries, greater emphasis on international authorship has been achieved in many parts of the Encyclopaedia. Perhaps it is inevitable that in multi-author works delays in one or two chapters should affect a whole volume, but we believe that this has not detracted seriously from the usefulness of the work as a whole. Indeed, the rapid advances in the science of nutrition create a problem. To meet it an annual review will be published entitled 'Progress in the Science of Nutrition and Food Technology' which will report advances in the respective fields covered by the Encyclopaedia and give full bibliographical details of recent works for easy reference.

H. M. SINCLAIR
Oxford

CHAPTER 1

Introduction to the Nutrition of Animals of Agricultural Importance

D. P. CUTHBERTSON

*Department of Pathological Biochemistry, Royal Infirmary, Glasgow, C.4**

INTRODUCTION

IN NO field of agriculture has more outstanding progress been made in recent times than in the nutrition of farm livestock, and knowledge continues to grow as the result of investigations in the many research institutions throughout the world. This activity is but part of the expression of man's concern with the quality of his life. As his economic circumstance improves there is accompanying intensification in the scale of his livestock production and accompanying increased efficiency in feed conversion. Man also finds himself in greater or lesser degree competing with his animals for sources of primary production of food and also some secondary. Where his competition is only for secondary needs it can in certain areas be tough enough, but when it arises from competition for primary needs, then indeed it can be a fight for life. The basis of war and revolution lies in his fight for a larger share of this world's natural resources and of markets to which its surpluses can be sold and this struggle for a better standard of living than other competing groups—and frequently at their expense—still continues. The ideologies are, as Sir William Slater (1964) has expressed it, but merely sophisticated expressions of all living things, namely, the urge first to survive and then to better the conditions of life.

When man congregates in towns he ceases to receive information about the rural basis of his survival and thrusts all recognition of this competition with other species into the background. His anxiety is that he has enough money to pay for his purchases in the local shop or supermarket.

We tend to think of this balance of nature without reference to ourselves as part of it. We tend to think of it as static, but it is a flux of shorter or longer changes which in terms of productivity are generally reversible. The most rapid changes are those induced by rapidly breeding species such as fungi, bacteria and insects. When man takes a hand to control them the balance is upset and the net gain has to be carefully assessed. If there is a

*Formerly Director, Rowett Research Institute, Bucksburn, Aberdeen

succession of unfavourable years famine with its co-horseman pestilence may result. War too may ride in.

Up to the beginning of the nineteenth century there was no great increase in man's numbers. Then things began to change, slowly at first, but faster with each decade, for man was fighting to conquer the conditions which shortened his life and limited his numbers. This was a new feature in evolution, for now it was not necessary to wait for the slow process of evolution to adapt man to his environment. Now man without any material change in his genetic make-up was able to control and change his environment to make it fit his special needs. Rapid changes in environment to fit man overtook the pace of the chance emergence of a genetic structure more suited to it. The plant and animal breeder had already found means of telescoping chance adaptations and culling out his less productive stock. It is in the survival of the young and in the numbers reaching puberty that his control of environment has been most effective and where it differs most from the results of evolution.

The nursing mother provides a balanced diet for her baby and even when her own diet is deficient her milk does not fail proportionately but tends to be maintained. It is when the child is weaned that difficulty arises in many developing countries where there is a lack of an adequate and safe supply of cow's or other ruminant milk, or where other sources of animal proteins or suitable mixtures of plant proteins with comparable amino-acid contents are not available. Milk, in addition, supplies bone-forming minerals and vitamins.

Given ample food, good housing and clothing, and all the aids of modern medicine, there is no reason why the average woman should not produce and rear to puberty not only two children but ten or more. It is obvious that the human population could increase five-fold or more in a generation, but in so doing it would cancel the advantage of the pre-existing supply of food for worldwide famine would then occur. Currently, the world's population is doubling in the space of some 35 years for it is growing at the rate of about 2 per cent per annum.

The very nature of the high nutritional standards, the better housing, the better clothing and generally improved standards of living make it increasingly difficult to sustain a large family in the state it should be kept. In the U.K. a rough balance has been struck between the rates of survival and reproduction allowing a two-fold increase in a little under 70 to 80 years. This will mean at the turn of the century some 80 million people in the U.K., and these will need some 28 per cent more food than is at present consumed.

The old colonial powers attempted to match the growing population which resulted from control of internecine war, infanticide—particularly of females—and other human sacrifices, and treated disease with medicine developed in the ruling country. But it has never been so easy to match the growing population by improving primitive agriculture; doing so is more difficult than providing a medical service. Few of the crops which form the basis of the agricultural systems of the developed world were suitable to the tropics, and almost none of the traditional methods of cultivation. Efforts had to be divided between food production and cash crops which provided the farmers with money to spend and the territory with export earnings. Before the agricultural adviser to these parts could recommend better equipment, seeds, or the use of fertilizers, he had to make certain that the farmers could afford them, that they were available and that the final result would justify the new procedure. A serious mistake resulting in crop failure would blast confidence for years.

The most difficult fight is always against tradition and prejudice, and so in many areas food production has failed to keep up with increasing population. Overcropping and overstocking have increased as pressure of population increased. Years of continuous cultivation, without any adequate return in dung or other fertilizer, has exhausted the soil in many parts of the world. Without added nutrients crop yields were pitiably low and herds of largely unproductive cattle overgrazed what pasture was available and so further reduced their value as food producers.

Thus we see that in the new developing countries three main factors affect the relationship between the population and available food supplies; the expanding health services which affect the survival rate; the growth of literacy and the efforts to raise the standard of living which should, if the pattern of the older countries is followed, lead to the voluntary limitation of pregnancies; and the increase in food production. To maintain the already insufficient food supply for a rapidly growing population will require a great increase in food supplies by 20 to 30 per cent in each decade. The present high level of animal protein in the diets of countries with a high standard of living is produced at the expense of many times the energy value in crops on which the stock is fed (see Table 1 a and b). High-quality animal protein or its equivalent in amino-acid composition in plant proteins is essential to a balanced diet. In many of the developing countries more than twice as much of animal protein is eaten by the adult as the old recommended level of 33g/day. It is, however, the supply of essential amino acids that is important. With animal protein go too its associated fat and vitamins. Its consumption by adults can be done without any danger to health provided there is sufficient of the essential amino acids and vitamin B_{12} in the diet, but at the expense of gastronomic satisfaction, for animal protein is a great asset in the making of attractive dishes.

Just as great as is the difference in level of production of high-quality animal protein between Europe, North America and Australasia on the one hand and the rest of the world on the other so also is the difference in the average yields of the basic staples, for the average yield of rice there is well over 4000 kg/hectare while in Africa and the Far East (which includes Japan) yields are only 1100 kg and 1650 kg respectively. Wheat, the second biggest crop, has yields of 1930 kg/hectare in Europe and only 570 kg in Africa and 840 kg in the Far East.

TABLE 1A

Approximate Efficiencies of European and Tropical Livestock in Converting Animal Feeds to food for Man derived from data of Leitch and Godden (1953) and Holmes (1962)§

Ruminants	European Energy*	European Protein†	Tropical Energy*	Tropical Protein†	Non-ruminants	Energy*	Protein†
Cattle					*Pigs*		
Milk production					Pork	27	19
Lactation	27	40	16	25			
Lifetime	—	—	11	14			
Beef production					*Poultry*		
Fattening period	11	11‡	16	10	Eggs	13-17	18-25
Lifetime	—	—	7	7	Meat	7	24

* Energy expressed as kcal of food produced. Feed as gross digestible energy (4090 kcal/kg) except for eggs where metabolizable energy was used (Bolton, 1958).
† Protein expressed as protein in food and as protein equivalent in feed (after Leitch and Godden, 1953).
‡ Because protein efficiency is lower on a high protein diet.
§ No differentiation made between Europe and Tropics.

TABLE 1B

Food Output—Wye College, Kent, England (Holmes, 1965)

Products	Energy (Mcal/ha.)	Protein (kg/ha.)	Ratio of protein production to lamb production taken as unity
Wheat	15,017	531	24
Barley	12,735	430	20
Peas	3,602	308	14
Milk	4,000	176	8
Beef	1,890	78	3.5
Lamb	462	22	1

We must remember that agriculture is not the only source of animal protein and fat, for the sea and inland waters provide large quantities of fish. Fish farming in shallow waters such as paddy fields and ponds can be rich reservoirs of food. These waters can be fertilized to encourage the vegetation on which small fish feed and so achieve a high yield of these fish. It is claimed that such properly managed fish ponds in the tropics can produce more protein than any collectable crop on land there: 700 kg/hectare have been reported.

The cells of the green parts (leaves) of plants such as grass contain proteins similar in nutritional value to those in foods of animal origin. They are, however, largely inaccessible to animals other than herbivores—ruminants in particular—because of the lack of adequate microbial activity in the alimentary tract of non-herbivorous mammals. Attempts have been made with varying success to break the cell walls and so free the protein which can then be separated and used to feed human beings or lower animals like pigs and poultry. But it is not always easy to persuade people to eat these proteins, which are at first unpalatable. Some countries readily accept greenish looking food concoctions and it should be easier for them.

The recent discovery of microorganisms which can multiply on certain petroleum products and synthesize protein from inorganic nitrogen is promising but careful screening for toxic substances is very necessary.

Although yeast as food and fodder has been known for some time, quite recently there has been a resurgence of interest in it. The cultivation of algae for protein requires a great volume of medium and is costly.

We must recognize that man differs from the other species around him in that he can bring to bear on each problem his scientific knowledge and can exterminate any serious competitors for his food supply and thereby augment his food supply far beyond that of any other mammal. He can preserve, store and distribute his food over long periods and long distances and can overcome his periodic local famines due to crop failures. Britain during the war lifted the net output from its farms to 160 per cent of the pre-war figures and this on a diminishing acreage and with a fall of more than 15 per cent in labour force. What is lacking in many developing countries is the incentive and drive to do comparable things. The term developing country has been used for what were formerly under-developed countries, but any country which is not developing is stagnant and will soon be moribund. It is preferable to consider countries as more or less developed when planning.

THE FOOD OF FARM LIVESTOCK: PROXIMATE PRINCIPLES

Plants constitute the primary source of nutrients and energy for man and his domesticated livestock. It is with their productivity that we are first here concerned. Let us turn to the plant. It is anchored by its roots in the soil and it spreads its upper portion to the air and the sun. From the soil it takes in water and salts, from its aerial system carbon dioxide and solar energy. All these substances and energy source are the raw materials out of which it builds carbohydrates, fats, proteins and other complex substances. The plant materials are then used by animals with varying degrees of efficiency. Ruminants have the greatest capacity to utilize plant material, in particular cellulosic material, other herbivores less so and carnivores least. From the proteins, fats and carbohydrates the animal can derive energy and along with the appropriate minerals and vitamins can grow and reproduce. For a further account of animal nutrition the reader is recommended to read such textbooks as that by Maynard and Loosli (1962) or McDonald, Edwards and Greenhalgh (1966).

The many hundreds of chemical compounds in feeding stuffs are classified quantitatively into four major groups:

1. Mineral or ash constituents.
2. Nitrogen containing constituents.
3. Fatty and waxy materials.
4. Carbohydrates.

This is a general classification and each of these four groups contain a multitude of constituents: the chemical constitution of some is not exactly known. To these groups must also be added the very important constituent—water—which is a significant part of all living things.

WATER

The determination of water is the first step a chemist takes in the analysis of a feeding stuff. He heats it to the boiling point of water, then he determines the amount of the other major classes of material in the dry matter. Obviously any other substance which is volatile at, or below, this temperature will also be lost. Even the so-called 'dry feeds' contain water. Freshly purchased concentrated cubes contain the least water, approximately 10 per cent; hays and straws about 10 to 15 per cent; grain and other green forages between 65 and 85 per cent. Turnips such as swedes often contain 90 per cent.

Since the animal body contains some 65 per cent of water and this is constantly being lost by the skin, lungs, and also in the course of excreting waste metabolic products and unused residues in the urine and faeces, it is in constant need of replacement. During milk secretion there is further loss by this route. Water requirements are met not only by drinking water, but also as we have noted, by the water in the feed, and water also arises from oxidation —metabolic water. In the Spring, on lush succulent pasture, heifers receive sufficient water and do not need to go to the drinking bowl. The highly fed dairy cow will lose daily up to 4 gallons in her milk.

To grow and reproduce its kind an animal requires both materials and energy. It requires some eight to ten essential amino acids, about fifteen vitamins and about the same number of mineral elements. By virtue of the activity of their rumen microflora and microfauna ruminants can dispense with some of these organic nutrients in the food, but biosynthetic activity is needed to replace them. This is mainly in the rumen. Biosynthetic activity in the caecum and large intestine may also reduce to some extent the dietary requirement for some vitamins not only in the ruminant but also in non-ruminants. But of all the requirements that which limits production most quickly, apart from lack of water, is lack of energy. Indeed it has been estimated that the main part of the economic loss that is due to faulty nutrition of farm livestock results from failure to meet the animal's energy requirement.

Energy

For production energy input must exceed energy output. The requirement for maintenance is the amount of energy the animal must expend on the various processes and reactions which are essential to life. When a milking cow is dry or when a steer is not growing or fattening it still requires a source of energy to keep it 'ticking over'. This is its maintenance level. The production requirement is the amount of energy which is needed in addition so that it may support the production of milk, eggs, meat or fat, or meet the energy lost to the body in muscular work.

Energy inputs and outputs are measured in terms of calories, 1 Calorie being the quantity of heat which is required to raise the temperature of 1 kg of water 1°C. The term kilocalorie (kcal) is preferred, sometimes written as Cal, to distinguish it from its sub-unit or calorie which is 1/1000 kcal. A megacalorie (Mcal) is 1000 kcal and is also termed a therm. It is preferable to use kcal to Cal.

Gross Energy

The gross energy of a feed or feeding stuff is the amount of heat liberated when it is burnt completely in a bomb calorimeter, the carbon and hydrogen in it being completely oxidized to CO_2 and H_2O. Gross energy values only state what energy the food contains not what the animal can obtain from it. Obviously only such proteins of the food which can be digested by the animal furnish it with any useful energy. Most of that which appears in the faeces has not been digested. Some substances are excreted by the body before they have been completely oxidized. Thus the bulk of the nitrogen excreted in the urine and which represents essentially wear and tear is not excreted as gaseous nitrogen, but is locked up as urea, hippuric acid and ammonia. Their excretion means a loss of energy. In the ruminant the production of combustible gases, mainly methane and, under some conditions, hydrogen also means a loss of energy (see Chapter 4).

Metabolizable Energy

The metabolizable energy is the gross energy less the sum of the heat of combustion of the faeces, fermentation gases and urine. Where fermentation in the gut is negligible the metabolizable energy can be taken as the absorbed energy less the energy of the urine. For further definitions and discussion on this important but complex subject the reader is referred to Chapter 4 for ruminants and to Chapters 5 and 30 for pigs and poultry respectively. A new feeding system in terms of energy has been devised for ruminants (Blaxter, 1962) for the Starch Equivalent System of assessing the energy needs of ruminants is not fully in accord with experimental evidence, and somewhat similar criticisms can be levied against the Total Digestible Nutrient System of assessing the energy needs of ruminants. In particular the efficiencies with which the metabolizable energy of different diets are used by ruminants to meet energy requirements or to promote retentions in growth and fattening or energy secretion as milk, are not simple multiples one of another. The new system evaluates diets as wholes, and takes into account the effects of level of feeding. In many practical conditions little error may be incurred by using the old Starch Equivalent System of Kellner (1926). (See also Chapter 19.)

As the requirements of energy depend on the concentration of metabolizable energy of the whole diet they can be expressed in terms of the number of megacalories of metabolizable energy of a diet, M, with a particular concentration of M/D, where D is the dry matter. Equally, energy requirements could be expressed in terms of an amount of dry matter, D, in which the metabolizable energy concentration is M/D. Requirements for other nutrients can be expressed in this way, that is in terms of total dry matter: requirements for other nutrients can be expressed as percentages of the dry ration. Not all the metabolizable energy can be used by the animal for productive processes nor for meeting its maintenance requirement as a portion is always lost as heat. This is because a great deal of internal work must be done such as in prehension, mastication, digestion and assimilation. There is a further wastage of energy in the transference of absorbed nutrients to the tissues. The energy expended in the work of digestion and the losses entailed in metabolic transformations of the metabolizable energy is eventually given up as heat which may be of value under certain conditions for keeping the animal warm, but in general, this heat is a waste product and in very hot weather may be a liability causing animals to reduce their dry matter intake.

Critical Temperature

The air temperature at which an animal's normal body processes generate just sufficient heat to maintain its body temperature at the normal level is called the critical temperature. Should the air temperature drop below this temperature the animal has to expend energy by shivering in order to keep the body temperature at the normal level. When chicks huddle together, or pigs crawl under their litter, it may be assumed that the atmosphere temperature has dropped beneath their critical temperature.

The critical temperature varies with age and degree of growth stage reached. Thus with bacon pigs it is about 13°C, and for the adult pig on full feed, about 4·5°C, but for baby pigs the temperature is about 30°C. If the temperature falls below these levels energy is diverted to muscle activity to keep warm which obviously lowers feed efficiency. It is obvious that it is very important to provide young stock with additional heat as their critical temperatures are higher than adults. Young pigs and baby chicks are particularly vulnerable. Ruminants on normal rations have very low critical temperatures and the shaggy breeds like the Highland cattle are particularly well adapted. This is because of the considerable heat of fermentation in their reticulo-rumens: but when fasting the critical temperature may be as much as 10°C above that of one on a normal diet. Thus if the critical temperature of the hill ewe rises through starvation her reserves of energy in the form of body fat are exhausted, and eventually she can no longer equate her heat production with her heat loss, so body temperature will fall and death will eventually result. While a heavy fleece provides much insulation, after shearing the critical temperature may rise to about 21°C.

High environmental temperature. At high air temperatures the animal may have difficulty getting rid of heat. Cattle and sheep have particularly poorly developed sweating mechanisms and have to rely in part on the evaporation of water from the respiratory passages to keep themselves cool, and in consequence pant vigorously, sheep in particular. If air temperature continues to rise still further the animal is unable to get rid of enough heat and the body temperature increases. Between this point of rising temperature and the critical temperature is what is called the range of thermal neutrality. This is the range through which the animal's expenditure of energy is not affected by variation of temperature. Just as a lowered food intake increases the critical temperature so, at the other end of the scale, a lowered intake raises the point of hyperthermal rise. In other words, an animal on a high level of nutrition has a relatively poor heat tolerance. It is obvious that in hot climates, particularly hot humid conditions, high levels of feeding cannot be achieved. Dairy cows imported from temperate regions deteriorate under these hot conditions and give little milk because they cannot maintain the food intake necessary for high milk yield and yet keep their body temperature normal. The tropical breeds of cattle are better equipped to get rid of surplus heat than are those of temperate regions.

Net Energy

We have already noted that the net energy is the energy of the food which the animal actually uses for maintenance of its vital processes and for growth or for deposition as fat or protein in its body or secretion in the milk. The maintenance cost in terms of net energy is, however, a first charge on the total supply. Only the surplus is available for tissue growth, fat deposition or the production of milk or eggs.

Kellner found that the proportion of metabolizable energy which is really of value to the animal as net energy, varies considerably from one feeding stuff to another and the values are not the same for all animals. The pig for example is much more efficient than the steer in utilizing the energy of materials such as starch and simply reflects the differences between the digestive processes of the two species. The pig, with its simpler digestive system, is well adapted to the utilization of concentrated foods while the complex stomach system of the ruminant is specially adapted to predigest food, and fibre in particular, by fermentation through the microorganisms there. The horse digests cellulose microbiologically in its large intestine.

The net energy value of a feed may become a minus quantity if the metabolizable energy it supplies is less than the expenditure involved in its utilization. Straw has very little value for ruminants, and for horses it is worse than useless, since the horse expends more energy in eating and digesting it than the straw yields in metabolizable form.

Starch Equivalents

Kellner in 1905 put forward his concept of starch equivalent before Armsby suggested net energy values, but the two ideas have much in common and, broadly speaking, express the same concept in rather similar ways. Kellner established a series of factors for the fat-producing powers in bullocks of the energy yielding nutrients. He calculated 1 kg starch produced 0·250 kg fat in the animal body and this was equivalent to 2380 kcal of chemical energy. He further calculated the theoretical fat-producing powers of feeding stuffs from their content of digestible nutrients and then determined the actual fat-producing power experimentally. He found good agreement except where the feeding stuffs were high in crude fibre. Kellner suggested for those that agreed a correction factor or 'V value' which should be applied to the theoretical value and this is usually of the order of 80–95 per cent. For those rich in crude fibre a sliding scale in deductions based on the crude fibre was proposed. It is now known that the latter relate to the effect of the ration in relation to the ratio of acetic acid to propionic acid in the fermentation products, as the greater the ratio the lower is the efficiency of fat production. On the other hand, acetic acid production favours milk fat synthesis. The starch equivalent is the amount of pure starch, perfectly digested, which has the same fat-producing power as 100 lb of the feeding stuff. He also expressed his results in kilocalories. For an approximate conversion of starch equivalent to net energy value it is necessary to multiply by a factor 1071/100, the net energy value of starch being 1071 kcal/lb. The division by 100 takes into account that starch equivalents are usually expressed per 100 lb feeding stuffs, whereas net energy values are expressed per lb.

Total Digestible Nutrients

The total digestible nutrients per 100 lb of feeding stuff can be obtained by weighing the ether extract, multiplying by a factor of 2·25, because of its energy content after correcting for apparent digestibility, and adding to this the weights of apparently digestible crude protein, digestible nitrogen-free extract and digestible crude fibre. The analysis of the feed is done by the Weende system.

Minerals

The ash constituents of a feeding stuff or animal tissue or excretion are determined by incinerating (ashing) the dry material at a dull red heat in a muffle furnace to constant weight. If this is extracted with hot dilute acid and filtered, the residue is referred to as insoluble ash but may largely be due to dirt contamination. Certain trace elements may volatilize during the process, e.g. iodine and selenium. The residue from the above filtration consists of silica, and the sulphates, chlorides, phosphates and oxides of calcium, magnesium, potassium, sodium, iron and a large number of other elements not all of which are essential to the animal. The major mineral elements in relation to animals are calcium, phosphorus, potassium, sodium, chlorine and magnesium. These major elements are all of nutritional significance: of the other main mineral elements found in plants little is known but it is generally accepted that while boron, aluminium and nickel are present they do not appear to be essential, although boron is essential for plants. During the last 35 years in particular, a steadily increasing, though still small, number of elements has been found which are essential for animals in trace amounts, the so called 'trace elements'. Some in excess produce toxic effects. That the number known to be essential for higher animals is small could simply be a reflection of the great technical difficulties imposed by the extraordinary low concentrations involved, and it seems probable that, as more highly purified media for plants and more highly purified nutritionally adequate diets for animals are developed, and as physiological and analytical techniques improve, some at least of these 'accidental' trace elements will be found to have essential physiological functions in the tissues.

The trace mineral elements required by higher animals are: iron, copper, molybdenum, cobalt, zinc and manganese; iodine, and probably also fluorine and selenium amongst the non-metals. The metals all belong to the transitional group of the periodic table. Most proteins with side chains containing carboxyl, amino or thiol groups can form very stable associations with these elements. The metals that act in traces may form an indispensable but small part of the molecule of essential substances, like the iron porphyrins of the haemoproteins and the cobaltic complex of vitamin B_{12}, or they may occur in association with enzymes as mediators or integral components of these, and in hormones, for example iodine in thyroxine and triiodothyronin.

Some feedstuffs contain very little ash, e.g. flaked maize which has less than 1 per cent whereas white fish, one of the highest in content, has more than 20 per cent of dry matter in this form.

Part of the sulphur requirement is for the sulphur-containing amino acid methionine either in the food or produced microbiologically in the gut. Cystine which also contains sulphur can spare some of the methionine.

Some minerals are toxic in large amounts, e.g. common salt in excess may easily poison poultry and pigs. One way of limiting intake of supplements of protective nutrients is to increase the simultaneous consumption of sodium chloride. Cattle which are pastured on land recently given heavy doses of phosphatic fertilizer and which have not yet been washed away by rain may suffer scours for a few days. Certain pastures or soils where there is an excess of molybdenum also cause scouring followed by loss of condition. Such pastures are known as 'teart' pastures. The condition increases under alkaline conditions so that the application of lime or basic slag aggravates the position. The symptoms are similar to copper deficiency and indeed copper cake or a copper drench will correct the deficiency. The interesting interrelations of these and other minerals in this and other conditions are discussed in Chapter 11. Suffice to note here that there is evidence that molybdenum is probably required in minute amounts but these are so small that rations are unlikely to be deficient in it. Selenium is present in toxic amounts in certain soils and causes 'alkali disease' or 'blind staggers' with loss of hair and hooves in chronic cases. In this condition there is no antidote to the disorder and, since selenium is present in the soil of these areas and in varying amount appears in the different pasture species, it is impossible to carry on any system of livestock husbandry without running into serious loss. Like molybdenum, selenium is probably also required in small amount and conditions of illthrift in sheep have responded to traces but the risk of overdosage rules it out for general usage. Although independent of vitamin E in certain anti-oxidant roles, selenium may protect against vitamin E deficiency (see Chapter 31) in pigs and poultry.

While the element fluorine occurs in bones and teeth and is probably necessary for their healthy development, this element more often causes trouble, for it occurs in excess in areas of the country where the natural waters are rich in fluoride; in addition proximity to aluminium works, brick works and superphosphate works may cause surface contamination of the plants with ill effects to grazing animals. Rock phosphates contain fluoride, but supplies for feeding purposes are now available in a defluorinated form. Of all toxic elements lead probably causes most trouble, particularly as the result of stock chewing lead-painted premises, fences, gutters, etc.

Though the deficiency or excess of certain trace minerals can cause ill health or even death, much more harm to the productivity of livestock is generally due to a shortage in their daily rations of those major minerals which are essential. The use of more concentrated rations and the intensification of animal production has brought this to light in increasing degree. The deficiency of a mineral may arise in several ways. A soil may be deficient in a particular mineral or the mineral may be present but unavailable to the plants eaten. In either case the plant itself fails to grow if the element is essential to it, or if it does grow, it itself is deficient in this particular element. But as the requirement of the plant may be relatively much smaller than that of the animal, the plant may be able to grow and indeed grow well, yet the stock grazing it be rendered deficient.

It sometimes happens that the diet contains an adequate supply of minerals but if one or more is in an unavailable form the animals will not thrive.

Most of the mineral elements of the body are present in the ionic form, apart from the calcium, phosphorus and magnesium of bone—stable yet subject to flux—the iron of the blood, the organic phosphorus and sulphur compounds, and the minerals present in enzyme systems and some hormones.

About 50 per cent of the total ash of farm animals consists of calcium and phosphorus. Ninety-nine per cent of the calcium of the body is in the bony skeleton and while 15–20 per cent of the phosphorus is present in appreciable amounts in the soft tissues, only 1 per cent of the total body calcium is outside the bones and teeth. Organic bound phosphorus occurs as phosphoprotein, nucleoprotein, nucleotides, phospholipid, phosphocreatine, hexose phosphate and adenosine triphosphate. One of the most vitally important functions is in the part it plays in high energy bonds. Such organic linkages with phosphorus are easily broken and liberate far more energy than that liberated on cleavage of an ester linkage with phosphorus.

The young growing animal and the milking animal have great need of these mineral elements and they must be supplied in the feed. During early lactation the milking cow draws part of the requirements from her own bones and, later in lactation, rebuilds her depleted bones in preparation for her next lactation. An imbalance of these two elements generally brought about by inadequate supply, coupled with the deficiency of the vitamin which controls their absorption and deposition—vitamin D—will cause rickets and other bone abnormalities.

The soils of some large areas, especially in South Africa, are by nature grossly deficient in phosphorus, and the natural herbage is so deficient that cattle grazing it develop the disease aphosphorosis which is characterized by stunting of growth and depraved appetite.

Most concentrates are low in calcium and some starchy foods are very poor in phosphorus as well. Indeed, the normal home-grown mixtures for milk production are generally deficient in calcium, phosphorus and common salt, so that a proprietary mineral mixture or—more cheaply—a mixture of two parts common salt, one part ground chalk and one part sterilized bone-flour should be added to the concentrated ration at the rate of 2·7 parts in each 100. For pigs 1.5 per cent ground chalk is generally provided but 1 per cent will suffice if the diet is otherwise reasonably good.

Magnesium is closely allied to calcium and phosphorus in the body and 70 per cent of it is in the bones and teeth. About one-third of this may be mobilized for use by the soft tissues during time of deficiency. It is also concerned with the activity of certain enzymes. Magnesium tetany may occur in calves reared for an extended period on milk and hypomagnesaemia with or without tetany is frequently found on lush spring pastures.

In general, shortage of trace elements is not common, occurring only over a small range of soil types or under highly artificial intensive systems of rearing. Iodine deficiency, which affects the thyroid gland, frequently arises in limestone areas far from the sea. Certain antithyroid substances or goitrogens in brassicas may also affect its metabolism. The iodine content of the soil is generally reflected in the iodine content of the crops grown. Although not as yet identified as essential for the plant, animals suffer from its deficiency in iodine-deficient areas. It is required throughout life to keep the thyroid gland functioning. Provided there is sufficient calcium, phosphorus and common salt present in mixed rations and that a fair variety of feeding stuffs is used throughout the year, there is seldom any danger of deficiencies of the other mineral elements. It is among stock living under natural conditions of grazing on certain soil types that deficiencies may arise. Unless there is a previous history of inability to thrive on such pastures it is almost certain that on traditionally good pastures trace element deficiency will not occur though there are reports that with excess fertilizer treatment such deficiencies may arise, e.g. 'swayback' or 'swingback', an unco-ordinated condition in lambs from excess lime or other calcium fertilizer affecting copper utilization by the animal (see Chapter 11). It is of interest that the effects of cobalt deficiency in the soil are on animals—ruminants in particular—dependent on the pasturage, and not on the pasture plants themselves, whose growth is unaffected; their requirements, if any, being presumably very much less. The bacteria in the rumen synthesize the cobalt-containing cyanocobalamin (vitamin B_{12}). Non-ruminants must receive vitamin B_{12} pre-formed in the diet and it is essentially the 'animal protein factor'.

The body contains about 0·004 per cent iron mostly in the form of haemoglobin in the blood, myoglobin in the muscles, and in various forms in the liver, but it is also present in some oxidizing enzymes. Sty-reared pigs often develop an iron deficiency once their stores at birth are reduced. It is common practice to inject them with iron salts to prevent iron-deficiency anaemia. All milks are low in iron and copper.

Deficiencies of manganese may affect bone and egg development in the bird, but do not occur in animals at pasture. At least two dozen enzyme systems are known to be activated *in vitro* by manganese; but of these only a few have been shown to have a specific requirement for manganese *in vitro*. As most forages contain 30–100 p.p.m. zinc on a dry matter basis there is little chance of grazing animals suffering from zinc deficiency. Zinc deficiency may be found in pigs (parakeratosis) and may be conditioned by excess of certain mineral elements, e.g. calcium and iron.

Net Requirement

The *net requirements* of a mineral element is defined as the sum of:

(a) The endogenous losses of the element in the faeces, urine and from the skin.

(b) The storage of the element in the body in growth and the amount used in producing eggs, milk or wool.

Endogenous losses are determined from metabolic trials. Analyses of carcasses, metabolic balances and analysis of milk, eggs and wool provide the estimate of the second term.

With some mineral elements, particularly those that are structural elements of the skeleton and those sequestered by the liver it is not easy to assess what degree of mineralization is optimal as this may depend on the stage of gestation, lactation or post-lactational pause at the time of assessment. When potentially toxic elements are involved the time relations of continued accumulation relative to expected life span must be given attention.

Availability

Ideally, feeds should be assessed in terms of their *available* mineral content, availability being defined as the percentage of the mineral supplied by the food that can be used to make good the endogenous loss or promote storage. Availability is measured by the ratio:

$$(\text{Mineral retention plus endogenous loss} \div \text{Mineral intake}) \times 100,$$

and must be measured at that level of intake which provides for optimum retention or some level below rather than above this level. Availability varies with the food and with age, for example, calcium in ruminants varies from 99 per cent at birth to 40 per cent at maturity. Availability falls when the level of supply exceeds the requirement. The need is best expressed as actual quantity of dietary mineral element rather than as available.

$$\frac{\text{Net requirement as defined above}}{\text{Average availability of mineral for class of stock in question}} \times 100$$

Food values are expressed in terms of concentration (g/kg). Where a mineral element of a particular feed is less available than average, the value is adjusted by multiplying g/kg by the factor

$$\frac{\text{Determined availability}}{\text{Average availability}}$$

It is only with three major elements (Ca, P and Mg) that availability departs much from 100 per cent. With Na, K and Cl, available mineral element is virtually synonymous with actual content. With Ca, P and Mg the differences in availability which are of physiological origin may be as great as the difference between feeds.

With the exception of iron, data on trace elements enabling net requirements to be estimated scarcely exist. This necessarily means that requirements and feed values have to be expressed in terms of amount of feed. This does not preclude the possibility of correction of feed values for those instances when the minor element of a feed is unavailable. Again the same method of statement of 'adjusted' values should be used.

Nitrogenous Constituents in Plants and Animals

Proteins are fundamental to life for they govern and take part in all phases of the activities of the living cell: they serve as structural elements in all soft tissues and play a part in the metabolism of skeletal tissues. The functions of proteins are diverse and multitudinous. All of the known enzymes, many of the animal hormones, the oxygen-carrying pigment of the blood, collagen, antibodies and the chemical units of hereditary transmission are proteins.

Fundamentally proteins are composed of chains of amino acids. Their molecular weights may range from 6000 in a simple molecule of only fifty-one amino acid residues, such as insulin, to many millions in the complex protein of a virus particle.

The alpha-amino acids, the building blocks of the protein molecule, are characterized by having a carboxyl group (–COOH) and an amino group ($-NH_2$) attached to the same carbon atom. Also attached to this alpha-carbon atom is another group or groups designated R in conventional working of formulae. The R group may contain another carboxyl or amino group, so that some amino acids are acidic like aspartic, some basic, like arginine and lysine. The neutral amino acids may be aliphatic—their R group being a chain of carbon atoms—or, as in the aromatic amino acids phenylalanine and tyrosine, the R group may be a ring structure. The sulphur containing amino acids are important in protein structure because they may provide cross-binding of one polypeptide chain to another or between the parts of the same chain through disulphide bridges.

The main mode of linkage of proteins is the covalent peptide bond in which the alpha-carboxyl group of one amino acid residue is coupled to the alpha-amino group of another amino acid residue.

Some of the nitrogenous constituents are simple, many are proteins. In animal tissues the latter predominate. Examples of proteins in animal tissues are the casein, lact-albumin and lact-globulin of milk; the gliadin and glutelin of wheat are examples of plant proteins.

The term crude protein is applied to the total N content and is made up of protein and non-protein nitrogenous substances. The term "amides" has sometimes been applied to the latter and given half the nutritive value of protein.

Proteins are made up of up to twelve to twenty or more different amino acids. Those which cannot be synthesized in the non-ruminant or calf in its pre-ruminant stage, or at a rate adequate to meet the needs for growth and reproduction, are termed essential and must be present in the diet. Those which the non-ruminant or calf in pre-ruminant stage can synthesize at adequate rates from other nitrogenous sources are termed non-essential. In the case of the ruminant, microbial synthesis can largely overcome the need for preformed protein or essential amino acids in the food for the bacteria and protozoa can synthesize the protein in their bodies from non-protein non-amino acid sources such as urea and ammonium salts and in amounts sufficient to sustain milk production. The microbial protein is subsequently digested and the liberated amino acids supply all those essential for growth and reproduction.

The essential or indispensable amino acids for non-ruminants and calves in the pre-ruminant stage are presumed to be the same as for the rat: valine, methionine, leucine, isoleucine, threonine, phenylalanine, tryptophan, lysine, arginine, and histidine, and for the bird also glycine. Arginine and histidine are not essential in the adult organism as they can be synthesized at a rate sufficient for adult maintenance. All these amino acids except lysine and threonine take part in reversible transfer of amino acids and keto acids. Cystine spares the requirement for methionine, and tyrosine that for phenylalanine.

The proportion of crude protein which is in the form of pure protein varies greatly in feedstuffs. In ripe seeds it is generally well over 90 per cent, whereas in mangels it is only about 10 per cent—the remainder in this latter case includes some nitrate. Non-protein compounds such as the amines, asparagine and glutamine, may constitute 50 per cent of the nitrogenous constituents of rapidly growing green crops.

Deficiency in an essential amino acid leads to failure in growth and eventually may lead to death. Different classes of non-ruminant livestock have different requirements for some of these, and in the bird glycine cannot be synthesized sufficiently rapidly for adequate feathering since the protein keratin of the feathers has a large proportion of glycine in the molecule. Lysine followed by methionine, threonine and tryptophan are the most likely essential amino acids to be limiting growth and reproduction through insufficiency in the proteins of the diet. Their deficiency can be made good by adding the limiting amino acid to the rest of the feed. In general, animal proteins, such as those of blood meal, meat meal and white fish meal are much superior to plant proteins as sources of essential amino acids—their biological value is greater. But it is possible to supplement plant proteins or have them in such an admixture that their individual deficiencies are compensated by the pattern of amino acids in the rest of the feed. Ammonia produced in the digestive tract is absorbed, converted to urea in the liver and mostly excreted. Some will recirculate as urea in transudates and secretions and be reavailable where there is microbial synthesis. This is known as 'N-recycling'. (For ruminants see Chapter 4).

Protein Requirements

While the amino acid composition of the food protein is important in relation to growth of pigs and poultry and calves in the pre-ruminant stage, by contrast the importance of biological value for older ruminants is much less. Thus dairy cows show very little response in milk yield to major changes in the make-up of their food protein, and from this it would seem that, for ruminants, the non-protein nitrogen fraction has a value not far short of true protein. The protein requirements of ruminants are best expressed in terms of digestible crude protein rather than as pure protein. A convention which has been used in the past and needs further investigation is the practice of defining the sum of the digested pure protein plus half the digested non-protein as the 'protein equivalent'.

The 'protein requirement' of an animal is thus in reality simply a convenient phrase to use in assessing the animal's need for about ten essential amino acids (depending on species) released from intact proteins in the gut at a particular rate and in a correct proportion, together with the need for some non-essential amino acids, or a source of nitrogen from which the animal can synthesize those amino acids that need not be supplied pre-formed in the diet. The need is determined by the rate of wastage in the turnover of the nitrogen containing compounds of cells and body fluids and the rate of synthesis of protein in growth of tissues or in production of milk or eggs.

In simple stomached species it may be possible to express requirements in terms of minimum quantities of essential amino acids with a proviso that the total N supply must provide the non-essential acids or the bases for their synthesis. Owing to the very considerable variation in the results of experiments to determine amino acid requirements and the analytical problems of the diets themselves, this approach may be impracticable, but recently greater definition has been possible. The alternative has generally been to express requirements in terms of total N or total protein N. This entails the necessity for adjustments to the biological value of the protein sources. For ruminants an approximation can be made since the degree to which proteins vary one from another in their biological value for them is relatively small, and requirements can be stated in terms of "average protein" of biological value 70. But this approach is inappropriate when applied to non-ruminants and particular sources of feed proteins should be expressed in terms of biologically effective protein or net protein by multiplication of the protein content by the true digestibility and the biological value, for such values are not additive because of supplementary effects between protein sources referable to compensating differences in amino acid composition.

A reasonable approach is to express requirements for the non-ruminant as digestible crude protein, but in addition to stipulate that the N of the ration must contain *X* per cent from lysine, *Y* per cent from methionine, etc. This is indeed a compromise as there is evidence that the need for a particular essential amino acid expressed as a percentage of the total N absorbed may change with rate and stage of growth and with breed.

The calculation of minimal requirements involves:

(a) Assessment of endogenous urinary N loss and the metabolic faecal N loss.

(b) Assessment of N retention in growth and fattening and of N used in production of milk, eggs, hair, fleece, or feathers.

The sum of these two multiplied by 6·25 is the net protein requirement. Division by an average biological value expressed as a decimal provides an estimate of the amount of the 'truly' digestible crude protein needed. This need is met by the 'truly' digestive crude protein of the feed. To express it in terms of apparently digestible crude protein necessitates making an estimate of the metabolic faecal loss of N for each feed. However, as the metabolic faecal N is in part determined by the nature and the amount of feed given, so no single estimate will be correct for all intakes, but only with ruminants is the error large. Requirements can be stated as available protein (Blaxter and Mitchell, 1948).

The Lipids

Included in the standard method of analysis are not only the true fats or glycerides, but resins, organic acids, essential oils, sterols and plant pigments such as chlorophyll and the carotenoids. All these are extracted from the dry, finely ground feeding stuffs by prolonged extraction with hot ether, and hence they are more correctly known as 'ether extractives'. In some foods, e.g. the oil seeds, the glycerides account for the larger part of the ether extractives, whereas in others—notably the straws and hays— the non-glycerides are the more abundant. In general the glycerides have a much greater feeding value than the non-glycerides, but the latter include vitamins A, D, E, and K, which are essential nutrients.

Lipolysis is mainly in the upper part of the small intestine of non-ruminants. Recent work has shown that, contrary to fairly widespread belief, very little if any triglyceride fat is absorbed intact in particulate form except when the gut is grossly overloaded with fat. The 'normal' processes are as follows:

(i) Triglycerides $\xrightarrow[\text{lipase}]{\text{Pancreatic}}$ 2-monoglycerides + fatty acids.

$$\text{i.e.}\quad \begin{array}{l} CH_2OR_1 \\ | \\ CHOR_2 \\ | \\ CH_2OR_3 \end{array} \longrightarrow \begin{array}{l} CH_2OH \\ | \\ CHOR_2 + R_1 + R_3 \\ | \\ CH_2OH \end{array}$$

(ii) Bile 'salts' (i.e. taurine and glycine conjugates of bile acids) form mixed micelles with mono-glycerides and free fatty acids. These micelles are *completely water-soluble*.

(iii) The jejunal mucosal cells absorb the monoglycerides and free fatty acids; the bile salts are not absorbed and return to the lumen to act as vehicle for more monoglycerides and fatty acids. Re-synthesis of triglycerides in the mucosal cells takes place by the reverse of the process.

Bile salts are absorbed in the ileum.

In ruminants due to hydrolysis of esterified fatty acids in the rumen little or no glycerol reaches the jejunum. Bacterial hydrolysis of triglycerides or diglycerides is non-specific with respect to fatty acids and so no monoglyceride remains, as is the case with pancreatic lipase. Dehydrogenation of unsaturated long chain acids occurs in the rumen and glycerol is attacked and converted largely to propionic acid where it is absorbed (see Chapter 10).

It is known that, in the absence of monoglyceride, bile salts do not form mixed micelles with free fatty acids. It is likely that, in the ruminant small intestine, the place of monoglyceride in mixed micelle formation may be taken by lysolecithin derived from the action of pancreatic lecithinases on biliary lecithin of which the ruminant produces a copious supply. It is likely that pH plays an important part in relation to mixed micelle formation with long-chain fatty acids. In the non-ruminant the pH of jejunal contents is usually about 6·3–6·5 and at this pH *in the presence of bile salts* about 50 per cent of the fatty acids is ionized and is readily incorporated into micelles. However, at lower pH's such as obtain in the jejunum of the ruminant, very little fatty acid is apparently present in micellar solution and it seems likely therefore that absorption from a micellar solution takes place in the upper ileal region where the pH is about 6·0–6·5.

Essential Fatty Acids

It is well established that when either linoleic acid, linolenic or arachidonic acid is given to small laboratory animals subsisting on a fat-free diet the dermal symptoms of essential fatty acid deficiency are readily cured and in the case of young animals growth is also restored. Though no absolute need for dietary lipids has, as far as is known, been proved for any adult herbivores it has been shown by two groups of workers that linoleic acid is probably a specific dietary requirement for the very young ruminant as linolenic and arachidonic can be formed from it.

THE CARBOHYDRATES

The carbohydrates of feeding stuffs are generally classified according to their complexity as:

1. Monosaccharides or simple five- or six-carbon sugars, with the general formula $C_5H_{10}O_5$ for the pentoses, e.g. ribose, and $C_6H_{12}O_6$ for the hexoses, e.g. glucose and fructose.
2. Disaccharides, consisting of two hexoses with general formula $C_{12}H_{22}O_{11}$ and which include cane and beet sugar (fructose and glucose), milk sugar (galactose and glucose), and malt sugar (2 glucose mols and glucose).
3. Polysaccharides, all of high molecular weight and made up of condensations of simple sugars. They include:
   (a) The starches which are the main reserve carbohydrates in plants, and the dextrins which arise during the build-up or breakdown of starch.

In the pig and bird nearly 100 per cent of the starches, dextrins and disaccharides are hydrolysed to simple sugars by enzymes (carbohydrases) secreted in the digestive tract. These simple sugars are then absorbed into the blood.

   (b) The pectins, cellulose, lignin, hemicelluloses, pentosans and polyuronides or 'cell wall carbohydrates'. Of these, cellulose and lignin are the most important, and they are the most abundant organic substances in the world. The former is a polymerized glucose yielding it on hydrolysis. However, none of the enzymes secreted into the digestive tract of higher animals can carry out this hydrolysis. For their digestion to occur microbial action is necessary.

In simple stomached animals little such activity occurs until the food residues reach the caecum and rest of the large intestine. In those animals with a complex or multi-chambered stomach like the ruminant, great microbial activity is normal and there cellulose yields to attack by bacteria in particular, but also by protozoa—mainly ciliates. In the horse the large intestine is the main seat of microbial activity. (See Chapter 25.)

The products of attack, other than those used directly for further microbial synthesis, are not glucose but the lower steam volatile fatty acids (VFA), acetic, propionic and butyric acids. Methane is also produced by reduction of carbon dioxide. Some of the cellulose is converted to bacterial and protozoal starch and some of the energy locked up in cellulose is liberated as heat—the heat of fermentation. The ruminant does not obtain the whole of the energy of the cellulose which the bacteria break down.

The steam volatile fatty acids, which account for some 70 per cent of the energy-yielding compounds the ruminant absorbs, are used for maintenance and for the synthesis of body fat and milk constituents. Acetic acid is incorporated into the milk fat, and propionic acid is used in the synthesis of lactose (see Chapter 4). The other main acid is butyric acid.

Lignin is not appreciably attacked by microorganisms in the rumen and there is no secreted digestive juice which attacks it. The process of lignification makes the plant less nutritious as the cellulose and the cell constituents are protected from the action of bacteria by the coating of lignin. This process accounts for the decline in the digestibility of hay when cutting is delayed. The residue of lignified cellulose (woody fibre) which is insoluble first in dilute acid and then in dilute alkali is called "crude fibre". But as lignin is partially soluble in alkali and cellulose partially in the acid, the crude fibre of a late cut hay or a straw is very different in terms of nutritive value from the crude fibre of young spring grass.

Fat-soluble Vitamins—Vitamins A, D, E and K

Vitamin A

Vitamin A does not occur in plants as such but is formed in the animal body largely from the orange-yellow β-carotene, one of the group of yellow pigments called carotenoids which are associated in the plant with the green colouring matter chlorophyll. β-Carotene is not the only precursor of vitamin A, but is the most common. In yellow maize grain it is crypto-xanthine.

While all the animals have the capacity to convert carotene to vitamin A, some do it more efficiently than others. Thus the Guernsey cow produces a yellow milk fat because she passes into the milk a considerable proportion of the carotene she ingests. The Ayrshire, on the other hand, given the same ration, converts relatively more to vitamin A before secreting it. The goat is an even better converter. The values of such milks in preventing disease due to vitamin A deficiency can be the same. The vitamin A potency of eggs and milk is directly proportional to the amount of carotene obtained from the food.

In terms of content, carotene is largely found in green, leafy plants; their storage organs contain but little—even yellow maize grain is a rather poor source. Carrot roots are very rich in carotene but mangels, swedes, turnips, sugar beet pulp and potatoes supply virtually none. Some fish meals contain a little pre-formed vitamin A but this may be largely lost in processing. As a leafy plant matures its carotene content falls markedly. It is largely destroyed in the process of hay making and it is only if the hay retains a green colour that it can be regarded as a valid source. Artificial grass drying is the best method of conserving carotene but in bales it is gradually lost by oxidation.

Cod-liver oil which supplies vitamin A as such is one of the best sources but when mixed with feeding stuffs its vitamin A value falls off very rapidly. Attempts to stabilize it in fish liver oils have been made.It is probably best to rely on young green material for meeting the requirements for this vitamin. Some 80 g/day of really young leafy grass will meet the daily requirements of a growing pig and 1 kg that of a milking cow. During the summer and autumn when green stuff is plentiful, animals are able to store both vitamin A and carotene in their livers and are able to fall back on those stores during periods when their diet is deficient. In milking animals there is a steady drain on the reserves, so they become more rapidly exhausted. A cow may give birth to a weak, dead or blind calf before she herself has begun to show signs of deficiency.

In general, young animals are born with very small reserves of vitamin A in their livers. They acquire a large amount from the colostrum of their dams which is much richer than full lactation milk. Vitamin A deficiency in young cattle is not uncommon in this country and mostly occurs on rations of sugar beet pulp and poor straw with carotene-poor concentrates. The main signs are failure to thrive, night blindness and, later, total and incurable blindness. In pigs, paralysis of the hind quarters with a staggering gait results. In all animals secondary bacterial infections occur where the deficiency is prolonged.

Vitamin D (the so-called anti-rachitic factor)

In all, there are at least ten different forms of vitamin D, the most important being vitamin D_2 (the form found in plant products such as sun-cured hay and irradiated yeast) and vitamin D_3 which is found in animal products, notably fish oils and is, itself, found in the animal body during exposure to ultraviolet light. While the D_2 and D_3 forms have the same antirachitic potency for the laboratory rat, vitamin D_3 is far more effective than D_2 for poultry. Fish oils and irradiated animal sterols are thus better sources of vitamin D for poultry than are irradiated plant materials.

While vitamin D is very sparsely distributed in animal feeding stuffs, the sterols are fairly widely distributed, particularly in the dead areas of leaf which have been irradiated by the sun. Artificially dried grass, not having been exposed to the sun, has a very low vitamin D content. On the other hand, vitamin D is synthesized in the skins of animals thriving out of doors when irradiated by sun and skyshine. Vitamin D deficiency tends to arise in animals kept indoors and not receiving sun-cured foods. This vitamin does not decline during storage nor when mixed with other feeds.

Vitamin D deficiency leads to a failure to lay down calcium and phosphorus in cartilage and the bones in consequence bend and malformed limbs result. It may also arise from a deficiency of either calcium or phosphorus. When the ratio of Ca/P is abnormal, vitamin D will often permit normal growth, whereas if the ratio is normal it may be very difficult to demonstrate any requirement for vitamin D. This vitamin enables growth to take place under dietary conditions which would otherwise soon result in deformity. Fewer eggs of lowered hatchability result in hens on rations low in vitamin D, and the few eggs which are laid have very thin shells. Rickets is very uncommon in the bird but can be recognized by the peculiar attitudes affected by the birds in balancing their bodies and by their stiff-legged gait. The breast-bone is frequently bent and the beak becomes rubbery and is easily deformed.

Vitamin E

The first discovery indicated that this vitamin was an essential nutrient for reproduction in the rat. In its absence males were sterile and in pregnant females the embryos degenerated and were absorbed. While pigs and poultry may require it for normal reproduction, and adult birds if deprived of it may lay eggs of low hatchability, there is no clear evidence that vitamin E in the diet is essential for reproduction in the ruminant. It is concerned with the normal functioning of most tissues. Vitamin E is important for with rations low in vitamin E, cattle and sheep develop a muscular dystrophy, and birds develop severe nervous and circulatory disturbances. Many of these muscular and nervous disturbances due to deficiency of vitamin E can be prevented by selenium. The amounts of the latter to maintain normality are very small—parts per thousand million in the diet. Deficiencies of selenium can also be reversed by giving vitamin E. Selenium and vitamin E are mutually independent in the performance of several anti-oxidant roles one of which is intimately concerned with sulphur metabolism in the prevention of muscular dystrophy (See Chapter 31.) The administration of cod-liver oil with its unsaturated fatty acids markedly increases the animal's vitamin E requirements due to oxidative destruction. Small amounts of most proprietary concentrates of vitamin A and D do not have this effect.

Vitamin E is found in large amounts in green foods, and, as with vitamin A, it is the young animals born in the later part of winter to dams which have received poor rations which are most vulnerable to deficiency.

Vitamins K_1 and K_2

These vitamins which are concerned with the maintenance of the clotting time of the blood were first discovered in experiments with chicks. K_1 is found in alfalfa: K_2 in a bacterial product in fish meal. The vitamins occur in the intestinal tract, arising from the food and through bacterial synthesis, but they are only absorbed in the presence of bile. There is no body store. Requirements are greatly increased by an attack of coccidiosis.

Poisoning by the substance dicoumarol, which is present in mouldy lespedeza hay, can be cured by giving vitamin K. It has not been reported in this country.

Water-soluble Vitamins

Vitamin C

Farm animals can synthesize this adequately in their bodies.

Vitamin B Group

No member of the B-group vitamins is required in the feed by adult ruminants as the microorganisms, amongst their other activities, synthesize the components of this vitamin B complex. Before this state exists in the young calf, that is, in its pre-ruminant stage, sources of the B-vitamins are required. Pigs and poultry and the farm dog are the only farm animals which are likely to suffer from major deficiencies of the vitamin B complex.

Thiamine (Vitamin B_1)

Its deficiency in pigs leads to loss of weight and in the bird there occurs the nervous disease, polyneuritis. Deficiency is not likely to occur in poultry particularly since cereals and miller's offals are rich in this vitamin. However, thiamine deficiency due to thiaminase in the diet may arise in non-ruminants receiving certain types of uncooked fish, e.g. carp. Bracken also contains a destructive agent and if horses eat too much of it symptoms of paralysis arise. These can be counteracted by injections of thiamine. In ruminants excess bracken causes damage to the blood-forming bone marrow. It is also well known that increased intakes of carbohydrate necessitate increased intakes of thiamine.

Riboflavine (Vitamin B_2)

Its deficiency causes limb stiffness in pigs and curled-toe paralysis in chicks. Baby chicks should receive a supplement rich in riboflavine such as dried whey or dried skimmed milk.

Nicotinic Acid (Niacin: the pellagra preventive factor)

In pigs severe diarrhoea may result with loss of appetite and in poultry there is inflammation of the mouth and darkening of the tongue. There have been reports of deficiency where the rations are restricted to maize and maize products, but lack of the vitamin is unlikely in normal farm practice. It is synthesized from tryptophan, so when tryptophan is in short supply and needed for protein synthesis, the synthesis of nicotinic acid may suffer.

Pyridoxine (Vitamin B_6)

Its deficiency may lead to dermatitis, oedema, anaemia and fits in pigs, and in birds to convulsions, but such is not likely to occur in ordinary rations.

Pantothenic Acid

Its deficiency in pigs leads to slow growth, nervous disturbance and 'goose-stepping', and in poultry to failure to grow, but again such deficiency is very unlikely to exist in common farm rations.

Biotin

In pigs cracked hooves may result from its deficiency and in poultry the signs are poor growth and skin disorders. Its deficiency is very unlikely to occur.

Choline

A hock disorder called 'perosis' occurs in poultry where choline is deficient, but this is very unlikely to occur except where the diets are predominantly maize.

Folic Acid

Its deficiency in poultry feeds results in slowing of growth, anaemia, and also in poor feather development and depigmentation; but in practical rations is unlikely to occur. In the pig, intestinal synthesis is apparently sufficient to meet the requirements, thus only poultry require a supply in the diet.

Cobalamin (Vitamin B_{12})

Its deficiency leads to anaemia and poor growth and this is liable to occur on all vegetable rations. Cyanocobalamin is now regarded as the major factor in the animal-protein factor complex. As already noted a deficiency can arise in ruminants when cobalt is lacking, because the rumen bacteria cannot then synthesize this vitamin. Non-ruminants require it pre-formed in the feed. Diets of non-ruminants supplied with animal or fish protein will not lack the vitamin and fermentation residues will also provide it.

With more refined techniques it is possible to be more accurate in defining criterion of response to different levels of intake. Concentrations of the vitamin B-complex are best stated as concentrations per kg feed *or* per kcal and can be 'adjusted' for known low availability.

NUTRIENT REQUIREMENTS

The nutrient requirements of the various classes of farm livestock are dealt with in separate chapters by experts in the subject. Also included by kind permission of the Agricultural Research Council, though not necessarily expressing the views of the Council itself, are the summaries of estimated requirements as set out by the Council's Technical Committee on Nutrient Requirements of Farm Livestock and its working parties. These have been published by the Agricultural Research Council and by mutual arrangement are set out again here. (See Chapters 24, 29, 32.)

We recognize that most of our present knowledge of nutrient requirements has been gained by comparatively short-term studies. As higher and higher productive performance is sought of our farm animals, we recognize the limitations of our estimates of adequate nutrition from the standpoint of productive life. The desirability of rapid growth, not only to bring animals into production as soon as possible, but also to develop the highest producers has been a dominant principle in feeding practice. At the same time an increasing appreciation has developed that a large number of animals which are promising from the standpoint of growth performance, fail to come into profitable production or have an abnormally short productive life. There is a real need for a reconsideration of the nutrition of the animal from birth.

The requirements of livestock should be considered in relation to the optimum of a completely adequate diet given to normal healthy animals in an environment compatible with good health. The net requirement contains an element corresponding to the true variation from animal to animal in need which is independent of the food source and the error of the method. Provided there is available a number of such estimates or alternatively the between animal variances and co-variances of the components which go to make these estimates are available, the standard deviation of the estimated net requirement can be ascertained. This allows a measure of dispersion of the needs of the individual. If an average value is assumed for the availability of an element in the diet, this factor can be regarded as constant and the dispersion of actual requirements calculated.

For energy supply no margin of safety need be allowed. The input–output relationship has been shown to act as a guide to optimal amount to be given. When this relationship is flat over a wide range of intake, and the nutrient acts as a determinant of production only when intake is very much reduced or a toxic amount is given, then the requirement is stated with a margin of safety.

The margins of safety for the non-determinant nutrients take into account variation in the needs of the animal only, together with experimental error in measuring them. They do not include error that can be attributed to variation in the feed supply. The overall error in meeting the animals' need thus depends also on the accuracy with which feed values can be assessed.

Direct analysis of a sample of the feed in use with a particular animal obviously results in least error and in experimental work is the only correct method; the identification of the feed given with a table entry may well result in considerable error. For a fuller discussion on requirements the reader is referred to the chapters specifically dealing with the different classes and ages of livestock and to Mitchell (1962).

In Chapter 2 a scheme of analysis of feedstuffs is presented. It might be argued that a standard summative scheme of analysis such as this scheme or any modern alternative based on similar principle is not necessarily the best, as the desirable end is to relate chemical analytical data to measures of nutritive value as is being done by the Laboratory for Soil and Crop Testing in the Netherlands (1958). Joint United States/Canada Tables of Feed Composition have been published (1964). A compilation is being undertaken by the Agricultural Research Council.

The nature and relative values of the feedstuffs currently in use in the United Kingdom are set out by the United Kingdom Ministry of Agriculture (Evans, 1960) and Watson and More (1962).

The Food of Animals

Forage includes any crop which can be eaten off by grazing, or be cut and carted to the stock, either in some other field or indoors. In addition to grass there is a whole host of crops like lucerne, sainfoin, vetches, clover, cabbages, kales, rape and mustard. By appropriate sowing and management various types of green stuff can be made available at different times of the year in temperate climates. Grasslands in the temperate zones range from heaths, moors and more temporary leys. There is obviously a great range of quality even in the best grass which is usually to be found in more permanent pastures.

Of the products obtained from grass by various methods of conservation the best known is hay. Indeed, in dry lands the stalky grass is generally in this condition. Hay is extremely variable in composition as its quality depends greatly on the time of cutting and the weather during hay making. Grass may also be cut at the leafy stage and dried artificially. Thus, a product rich in protein is obtained and can be classed as a concentrate rather than as a roughage. The simplest method is to ensile. For this process grass, or other crops such as oats and vetches grown specially for the purpose, is cut and in either fresh or slightly wilting condition is filled into a tower or pit or stacked and compacted in such a manner as to keep the bulk of the material in relatively anaerobic condition. In the original method the crop was allowed to ferment with production of a lactic-acetic type of fermentation and without any addition its reaction came down to pH 4 to 4·5 and a whole range of silages could be produced varying with the tightness of the packing in the silo. By mixing in molasses in the case of highly nitrogenous material such as clover, vetches and beans the production of a butyric acid type of fermentation can be restrained and the pH lowered more quickly to about 4, but a large addition is generally necessary. In some countries mineral acids are often used, the best known of such is that called the A.I.V. process, called after Professor A. I. Virtanen, its inventor; sometimes formic acid or sulphur dioxide. The losses in carbohydrate and in digestibility of the protein through fermentation at the higher temperatures, the former in particular, can be considerable. Nevertheless, well-made silage is greatly relished. The temperature of the process should be about 38°C, serious overheating with caramelization occurs at 49°C.

For a fuller discussion see Chapters 17 and 20 and Watson and More (1962).

Cereals and their By-products

During the manufacture of wheat flour for human consumption a number of by-products of great value as animal feeding stuffs arise; the best known of these are bran and wheatings. In the preparation of the flour the wheat is ground and passed through a series of silk cloths which retain the coarser particles of wheat offals. Normally about 72 per cent of the wheat becomes flour, at higher rates of extraction for human consumption the proportion of wheat offals becomes lower. Wheat flour is a starchy food containing some protein, whereas the offals contain less carbohydrate but more fibre and protein.

Oats are normally fed in large quantities to stock, and the oatmeal is prepared for human consumption. The preparation of oatmeal does not give rise to such a large supply of by-products as does wheat.

Barley of the non-malting varieties is now being grown in greater amount for beef production. It is normally bruised or ground to a meal before being fed to stock, in addition there is a small supply of offals from the manufacture of pearl barley for human consumption. Brewers' grains are a by-product of barley in the brewing industry. The material has little if any starch, but still contains much of the protein, fibre and ash of the original grain.

Maize after grounding may be fed direct to stock but often it is cooked and rolled before use. The manufacture of starch and cornflour gives rise to a variety of maize by-products. Rye and rice are also used.

Cereal straws vary greatly in palatability and nutritional value.

Legumes

The seeds of the legumes such as soya beans are valuable sources of meals which are rich in protein. If precautions are taken to prevent the meal going mouldy and if the seeds are mature and aged before grinding, they are safe and provide excellent feeding stuffs. Their proteins are rather low in methionine content. Legumes are also used as forage.

Oil Seeds, Cakes and Meals

Many seeds are rich in oil, but this is removed by squeezing the seed or dissolving the oil out with a solvent. It is then purified for use in the manufacture of soap and margarine. The residue contains a little oil, but is rich in protein. These residues are valuable animal feeds and are known as oil cakes and oil cake meals. Solvent extraction being more efficient leaves a by-product containing less oil.

Of the oil seed cakes, the best known are cotton seed, ground nut, coconut, soya bean, palm kernel and linseed. Certain seeds such as cotton seed and ground nut, have either a fibrous seed coat or pod. If this is left on when the oil is removed from the seed then the resulting product is said to be undecorticated; if it is removed the product is said to be decorticated. The undecorticated cake will naturally be the richer in fibre. Fungal attack may produce toxins e.g. aflatoxin from *Aspergillus flavus*. (See Chapter 8.)

Roots and Tubers

Roots have a much greater percentage of water and starch than tubers such as the potato. Potatoes are indeed a useful source of starch and a very safe food, particularly if boiled. It is customary either to feed roots indoors to stock or to let the stock consume the roots actually out in the fields and so enrich the land with their droppings.

Animal and Fish By-products

A number of by-products result from the slaughter house. The various waste parts resulting in the preparation of carcasses for human consumption and often the carcasses of diseased animals are treated in order to obtain the fat from them and the residue, after the removal of fat, can be dried and ground. If there is no bone it is referred to as meat meal and is very rich in protein whereas if the residue contains bone as well it is described as meat and bone meal and obviously is richer in calcium and phosphorus than the meat meal. Dried blood is also rich in protein but is normally fed in very small quantities.

Similarly, the by-products of the whaling industry, which is mainly centred on the oil obtained from the whales, yields useful protein-rich meat meals.

The wastage from the preparation of white fish, which are much less oily than herring, can be turned into white fish meal after drying and grinding. Actually its colour is yellow. It is rich in protein, calcium and phosphorus and is excellent for growing stock. If the best is used it is unlikely to taint bacon and pork particularly if omitted in the last fortnight before the pigs are killed. When the wastage from the preparation of oily fish like the herring are similarly treated, a second grade of fish meal is produced, containing a higher percentage of oil and this is called herring meal. It can be used successfully where the products are not to be consumed direct, that is to breeding stock other than dairy cows and laying hens. There is some evidence that when fed to poultry, fish meal affects hatchability of the eggs.

Separated milk, a by-product from the production of cream, is an excellent feeding stuff for stock, being rich in protein and also in minerals and certain water-soluble vitamins. Similarly, when cream is churned for butter the by-product is buttermilk which differs from separated milk in that the lactose of the former is replaced by lactic acid. In the making of cheese from the curd the by-product whey which separates out has lost not only fat, but also protein and some minerals; hence it is not so good as either separated milk or buttermilk, nevertheless it is a useful additive. Frequently these fluids are dried.

For further details on the nature and value of feedstuffs the reader is recommended to consult Morrison (1956) and Watson and More (1962).

FOOD ADDITIVES

Many feeds consumed on the farm at the present time are bought as compound feeds for a special growth or reproductive phase in the various classes of livestock. The compounder often adds materials other than feed-stuffs, in order to improve his products. Under this heading can be included vitamin-mineral supplements and individual amino acids likely to be lacking in the rest of the diet and therefore limiting for growth and reproduction. Of the amino acids, lysine is frequently the most likely to be lacking for pigs and poultry. Methionine, threonine and tryptophan may also be lacking. The addition of antibiotics and hormones is described in the next sections.

Antibiotics

The widespread and often indiscriminate use of antibiotics by dairy farmers to eradicate mastitis from their milking herds has resulted in relatively high residual levels in the milk. They are undesirable, not only because they can have adverse effects on the cheese-making process, but also from the public health angle. Mastitis would appear to be due to the enhanced activity of microorganisms normally associated with the udder and their flare-up brings about the inflammatory reaction. Stress, or some other triggering mechanism, would appear to be responsible for this change in the host's constitution. We need to know much more about this major disease problem of our dairy stock.

Antibiotics in the feed of fattening calves on concentrate rations such as the 'barley-beef' system for up to about 12 weeks reduce the incidence of liver abscesses and may be of economic value in farms consistently below the expected average in growth and feed utilization performance.

The practice of feeding antibiotics under the conditions permitted in some countries has been of economic benefit to livestock producers. In young pigs and poultry, feeding antibiotics at the permitted levels significantly lowers the total costs of production through either improvement in rate of growth or improvement in the efficiency of food conversion, or both. The physiological mechanisms are not precisely known but under laboratory conditions feeding antibiotics to germ-free chicks has no such effects. It would seem, therefore, that the effects in farm animals are often due to control of subclinical infection. After slaughter there are no more than traces of antibiotics in the carcasses and resulting animal products.

By inducing the emergence of resistant strains of bacteria, continued routine field usage of antibiotics may lead to a reduction in their efficacy as feed additives in so far as this efficacy is dependent on the reduction of subclinical infection, or as agents for the therapy of manifest infection. In view of the limited contribution antibiotics can make to egg production, there should be no relaxing in requirements for labelling of these substances.

The observation of a rise in multiple resistance to antibiotics in Shigellae by Japanese workers in 1959, and their demonstration that this resistance could be transferred to other Enterobacteriaceae by conjugation (Watanabe, 1963), opened a new phase both in bacterial genetics and in the study of drug resistance in this group of organisms. Briefly, this type of resistance is transferred from cell to cell by resistance factors (R-factors), which have been postulated to consist of the genetic determinants for drug resistance (R-determinants), and resistance transfer factors (RTFs). It has been assumed that the R-factors are cytoplasmic in location and that the communicability of the complex is dependent on the RTFs, which are episomal in nature.

The drug resistance is usually multiple and is directed against the anti-bacterial drugs commonly used in human and veterinary medicine—for example, ampicillin, chloramphenicol, neomycin, kanamycin, streptomycin, sulphonamides, and tetracycline.

Meantime this mode of acquisition of antibiotic resistance depends on the transfer of cytoplasmic elements between two species existing together in the alimentary tract. Multiple resistance to antibiotics and other anti-bacterial drugs can thus be transferred from a pathogen such as *Salmonella* to *Escherichia coli*, and thence to other pathogens.

It is now established (Anderson, 1965) that two elements are concerned in the process of transfer, each of which may exist alone in any given culture. One is the resistance determinant, a genetic element situated in the cytoplasm, although perhaps originally of chromosomal origin. This confers resistance on the bacterium possessing it, but it is not itself transmissible. The second is the resistance transfer factor, which it is now proposed to call the transfer factor, since it is concerned in the transmission of other properties than drug resistance. Only when these two come together can resistance be transferred to another organism.

The donor of the transfer factor to a culture of another species possessing the resistance determinant can receive back from it the complete resistance factor derived from the separate elements in the two cultures.

These observations explain why drug resistance in intestinal bacteria is sometimes transferable and sometimes not. When it is not, the organism possesses the resistance determinant only, and it has to acquire the transfer factor from another before its resistance becomes transferable. The likelihood of this occurrence is increased by the presence of an antibiotic in the bowel favouring the selective multiplication of organisms possessing the R-determinant. This brings us again to the conclusion reached earlier by Anderson and Lewis (1965): "The time has clearly come for a re-examination of the whole question of the use of antibiotics and other drugs in the rearing of livestock."

Until we know how often and in what doses certain antibiotics and other drugs are given to calves, no one can lay blame for what is happening on any particular mode of use of antibiotics in these animals. Since, as Anderson says, antibacterial drugs are being given too freely to try to control infections, the prevalence of which is due to "bad conditions of hygiene in animal husbandry, especially in the intensive farming, transport and marketing of calves", does not the only ultimate remedy lie in a reform of farming methods?

Meantime, the use of antibiotics for feeding should probably be restricted to the narrow spectrum antibiotics, leaving the broad spectrum antibiotics for therapeutic purposes. It is currently wise not to use antibiotics in feeds compounded for adult stock. Antibiotics should not be allowed for laying poultry. See also Report of the Joint Committee on Antibiotics in Animal Feeding, 1962, by the United Kingdom Agricultural Research Council and Medical Research Council.

Synthetic Oestrogens

The careful use of synthetic oestrogens to promote growth and induce caponisation presents little danger, but it should be realized that since they largely pass through animals unchanged, particular care is necessary when stock is kept under intensive condition. For ruminants implantation in the ear with subsequent destruction of the ear at slaughter is apparently more certain and distinctly safer than feeding, and is more economical.

The United Kingdom Ministry of Health's Committee on Medical and Nutritional Aspects of Food Policy has stated that there seems no reason for the time being to believe that the present use of oestrogens in poultry offers any risk to those consuming the meat. Continuous vigilance is necessary in the public interest.

Other Food Additives

Arsenic compounds are widely used as an aid in prevention of blackhead in turkeys and coccidosis in chickens. They are also helpful in stimulating the growth of chicks and pigs but dairy calves have not shown responses. The results are more favourable under stress conditions when the chicks and pigs are somewhat unthrifty or exposed to a low 'disease level'. When conditions are ideal, healthy animals may show no growth response. Of the organic arsenicals available, arsanilic acid and 3-nitro-4-hydroxyphenyl-arsonic acid have been most widely used.

It is stated that provided the arsenical is eliminated from the ration a few days before slaughter, no residue is present in the meat. It is very necessary to ensure that no food residues remain in the vicinity which could be ingested.

TOXIC CHEMICALS

Concerning the toxic chemicals used in agriculture and food storage there is no doubt that further research is needed to the development of more selective and safer pesticides of limited persistence, to the evaluation of the toxicological significance of pesticides in the human body tissues and for toxicological and biochemical studies for pesticides in mammals. The Agricultural Research Council in its report of the Research Committee on Toxic Chemicals (1964) has stressed the need to ensure that every suspected case of pesticide poisoning in man be fully investigated and that the information be made known. These have only occasionally happened and have almost always been due to faulty handling of the pesticide or its container. Suggestions that there are more widespread effects have not been substantiated (see also Hodges, 1965). "Safety is a negative condition, the absence of hazard and no amount of research or study will ever provide absolute assurance of safety" (Smith, 1964).

Pesticides have been defined as 'materials which kill unwanted organisms and include insecticides, rodenticides, niticides, fungicides and herbicides'.

It is important to know whether the presence of traces of pesticides in body fats or other tissues has any toxicological significance. It is also necessary to examine for contamination of fresh waters by pesticides and to study their accumulation in animals.

The rate of advance in pesticide development is such that existing knowledge and the methods at present available for the study of possible effects of pesticides on human health, on the welfare of domestic animals, and on wild life, are not adequate. The demonstration of small amounts of pesticide residue in the tissues of man and his domesticated animals, or in wild animals, is of little value until more information is available on their toxicological significance. Ecological studies involving also wild life population studies on chosen species is necessary before the effects of these chemicals can be measured. Every new pesticide must therefore be adequately studied. One of the difficulties is what to look for.

There has been much discussion of the way in which the use of chemical insecticides upsets 'the balance of nature' although it seems to have been forgotten that, from the very beginning of cultivation, building towns, developing forests, treating disease with chemicals, etc., man has been continuously upsetting this balance. It is not the use of chemical insecticides which should be condemned, but their misuses. The correct application of the terms 'use' and 'misuse' require consideration of the general eco-system. The modern trend should be to use them only when this is absolutely necessary as an emergency measure and where other methods fail, trying to avoid as far as possible the deleterious side effects associated with such applications.

Biological control is being more strongly emphasized today, but it is no panacea for all pest problems. Some use of pesticides is not optional but mandatory.

PLANT HORMONES

Vigilance is also needed to safeguard man and his livestock against the possibility of a toxic chemical being produced in plants by the action of a plant hormone itself non-toxic to animals.

REFERENCES

Agricultural Research Council (1964) *Report of the Research Committee on Toxic Chemicals*. H.M.S.O., London.

Agricultural Research Council and Medical Research Council (1962) *Report of the Joint Committee on Antibiotics in Animal Feeding*. H.M.S.O., London.

Anderson, E. S. (1965) *Brit. Med. J.*, **2**, 1289.

Anderson, E. S. and Lewis, M. J. (1965) *Nature, Lond.*, **206**, 579.

Bedryslaboratorium Voor Crond—en Gewasonderzock Mariendaal (1958) Oosterbrook, The Netherlands.

BLAXTER, K. L. (1962) *The Energy Metabolism of Ruminants*. Hutchison, London.
BLAXTER, K. L., and MITCHELL, H. H. (1948) *J. Anim. Sci.* **7**, 351.
BOLTON, W. (1958) *J. Agric. Sci.* **50**, 97.
EVANS, R. E. (1960) *Rations for Livestock*. 15th ed. H.M.S.O. Ministry of Agric. Fish and Food Bull. No. 48.
HODGES, R. E. (1965) *Nutr. Rev.* **23**, 225.
HOLMES, W. (1962) *Nutritio et Dieta*, **4**, 179.
HOLMES, W. (1966) *J. Agric. Prog.* **41**, 60.
JOINT UNITED STATES/CANADA TABLES OF FEED COMPOSITION, *A Joint Report of the Committee on Animal Nutrition, U.S.A., and the National Committee on Animal Nutrition, Canada* (1964) **Publication 1232,** National Academy of Sciences—National Research Council, Washington, D.C.
KELLNER, O. (1926) *The Scientific Feeding of Animals*. Translated by Goodwin, W. 2nd ed. Duckworth, London.
LEITCH, I. and GODDEN, W. (1953) *The Efficiency of Farm Animals in the Conversion of Feeding-stuffs to Food for Man*. Comm. Bur. Nutrit. Tech. Comm. No. 14.
MCDONALD, EDWARDS and GREENHALGH (1966).
MAYNARD, L. A. and LOOSLI, J. K. (1962) *Animal Nutrition*. McGraw-Hill, New York.
MITCHELL, H. H. (1962) *Comparative Nutrition*, Vols. 1 and 2. Academic Press, New York.
MORRISON, F. B. (1956) *Feeds and Feeding*. 22nd ed. The Morrison Publishing Co., Ithaca, New York.
SLATER, W. (1964) *Man Must Eat*. University of Chicago Press, Chicago and London.
SMITH, R. B., Jr. (1964) *Arch. Environ. Health*, **9**, 634.
WATANABE, T. (1963) *Bact. Rev.* **27**, 87.
WATSON, J. A. S. and More, J. A. (1962) *Agriculture: The Science and Practice of Farming* 11th ed. Oliver and Boyd, Edinburgh.

CHAPTER 2

Newer Knowledge on the Composition and Methods of Analysis of Feeding Stuffs

PETER J. VAN SOEST

United States Department of Agriculture, Beltsville, Maryland

CHEMICAL analysis of feedstuffs serves two basic purposes. In conjunction with feeding trials, it helps the investigator to elaborate and assess nutritional phenomena. It also provides a practical means of evaluating feedstuffs in the laboratory, without the time and expense associated with feeding experiments.

In the nineteenth century, the proximate or Weende system of analysis was invented to serve these purposes at a time when chemical understanding was poor. The original concepts that attended promotion of this system are archaic and now inaccurately remembered but no satisfactory replacement has been found for it in the 100 years since its inception. The system continues to be used because of the difficulties involved in bridging various scientific gaps as well as in overcoming a kind of investigative inertia that has settled over this branch of research in the course of the last century.

In an attempt to resolve some of these problems, a brief historical survey will first be made of the chemistry of feedstuffs. This will be followed by a discussion of the structural components of plants and the more easily digested portions of feedstuffs. A final section is devoted to the estimation of nutritive value from compositional data.

HISTORY OF THE WEENDE SYSTEM OF ANALYSIS

Since the development of the so-called proximate system of feed analysis of Henneberg and Stohmann (1860, 1864) at the Weende Agricultural Experiment Station in Germany, this often criticized system has nevertheless achieved universal usage. While the methodology is often attributed to them, the origin of the crude fibre procedure which is the basis of the system considerably antedates Henneberg and Stohmann.

Einhof had earlier devised a method which involved the successive extraction of plant material with alcohol, water, dilute acid, and dilute alkali to yield a residue of 'woody fibre' regarded as being indigestible (Einhof, 1808; Thaer, 1809). This analysis became the basis of Thaer's hay value system of comparing feedstuffs. Hay values for different feeds were calculated by subtracting fibre percentage from 100 and dividing the result with the correspondingly determined value for meadow hay. The ratios obtained were finally adjusted to a scale in which meadow hay was given the value of 100.

Implicit in the minds of the early agricultural chemists was the belief that nutritional availability was a function of solubility, a notion still important in modern nutrition. Sir Humphrey Davy (1813) prepared fibre by macerating plant tissue with cold and hot water, thereby removing most of the extractives, starch and protein. The residue was regarded as indigestible woody fibre. This method was used by Voelcker (1852), who noted the large increment of carbohydrate substances (now known to be chiefly hemicellulose) dissolved when Davy's residue was extracted with acid and alkali in the Einhof procedure. The dissolved carbohydrates were considered to be equal to starch in food value (Voelcker, 1861; Anderson, 1861).

Use of the hay value system became widespread in the first half of the nineteenth century, and extensive summaries of feedstuff analyses were compiled by Hemming (1852) and Wolff (1856). Discovery of the importance of protein in the diet led Boussingault (1844) to rate feeds on the basis of their nitrogen content, and Wolff (1856) attempted to improve the hay value system by adjusting the equivalents according to nitrogen content as well as fibre, but the results were severely criticized by Henneberg (1859).

Criticism first arose because of the considerable digestibility of crude fibre, discovered by Haubner (1855), and because of the discrepancies between hay values and observed performance of animals. Henneberg and Stohmann (1860, 1864) performed many digestion trials replete with chemical analyses. Their system of analysis, which was a practical adaptation of existing methods, combined ideas of Liebig regarding division of foods into carbohydrates, proteins and fats with those of Einhof regarding fibre as a special kind of carbohydrate. Different classes of feedstuffs were characterized by proximate analysis and empirical digestion coefficients for the different chemical fractions. Their feeding tables were the first published in a form still in use today. The superiority of Henneberg and Stohmann's contribution over systems in use prior to their time lay in the greater precision and practical usefulness of their data. The fame of their achievement rapidly promoted their systems and methods to the point that earlier work was eclipsed.

The Weende experiment station initiated extensive chemical studies on the carbohydrates comprising what are now called the cellulose and hemicellulose fractions of plants. Henneberg and Stohmann (1864) found that lignin was indigestible and was dissolved into the nitrogen-free extract (NFE) by the use of sodium hydroxide. That Henneberg did not eliminate the fibre determination, but rather standardized the Einhof method and made it an integral part of his system, may be accounted for by his recognition that high-fibre feeds are less effective than low-fibre ones in promoting growth and animal production.

The attempts to characterize the NFE led to the discovery and characterization of xylan, araban, and uronide hemicellulose (Tollens, 1897). These developments contributed to the field of organic chemistry but did not feed back really useful information on the chemical evaluation of feedstuffs. Meanwhile, the Weende system of proximate analysis became widely adopted, irrespective of its limitations.

In the period from 1880 to 1920, the proximate system became the official method of analysis for all human and animal foods. Foods of animal origin were analysed by procedures originally conceived for plant material. Von Knierem (1885) compared fibre and cellulose procedures, using horn powder to see how much residue remained. Criterion of judgement was that materials of animal origin should leave no residue. Therefore crude fibre was recommended as being superior. From a modern point of view, the chosen method, which was supposed to measure the difficult-to-digest fraction of a feed, was imperfect with regard to both plant and animal residues by its failure to retain either lignin or keratin.

Eventually the progress in standardizing feeding and energy-balance trials led to a realization of the need for improved analyses that would more truly reflect the value of feeds (Crampton and Whiting, 1943). A search for suitable chemical entities during the period from 1930 to 1940 was based on the ideal of a pure fibrous carbohydrate fraction that would be common to all plant feeds. Although this led to the recommendation of methods for cellulose and lignin (Crampton and Maynard, 1938; Matrone *et al.*, 1946), the fractions obtained from various plant species and stages of maturity were very different in both carbohydrate composition and digestibility. The hemicellulose tended to be ignored because of the difficulties in handling this fraction in the laboratory and because of its heterogeneity (Phillips, 1940b; Yanovsky, 1940). An explanation for the variability in digestibility of cellulose and other entities was sought and found in the degree of lignification.

The analytical systems emphasizing cellulose and lignin which were developed at Cornell (Crampton and Maynard, 1938; Crampton and Whiting, 1943) did not replace crude fibre, although cellulose is much used by researchers. From a theoretical standpoint, the best was the lignin method, but the procedure was slow, expensive, and difficult, and the purity of the product being analysed was open to question. Furthermore, the systems proposed by Crampton did not characterize the NFE, which was likely to contain much hemicellulose along with the more available carbohydrates.

The problem of isolating the carbohydrates of the cell wall has been attacked by various workers. The researches of Harwood (1954), Gaillard (1958), Waite and Gorrod (1959), and Paloheimo *et al.* (1961) have provided valuable information. Ely and Moore (1954) adapted a rapid method for these carbohydrates using $NaClO_2$ and acetic acid, and termed the residue holocellulose. The procedure has been little used because none of the carbohydrates or fibrous residues, with the exception of lignin, are capable of predicting digestibility more reliably than is crude fibre (Sullivan, 1964).

Another approach has been to prepare fibre residues in which lignin has been retained. The use of lignocellulose should have a sounder nutritional basis than the use of cellulose alone because the most indigestible chemical fraction is included in the residue. The earliest attempt to prepare lignocellulose was the normal-acid fibre procedure developed in Great Britain (Norman, 1935; Raymond *et al*, 1955). The factor limiting this method was the difficulty in removing protein from the residue (Walker, 1959). Residual protein gave high results for the fibre determination and interfered with the estimation of lignin. The problem of nitrogen removal in the determination of lignin has been the principal hindrance to development of a rapid method of lignin analysis (Van Soest, 1963a). The preparation of lignocellulose residues without resort to the tedious expedient of enzyme digestions has now been accomplished by the use of detergents (Van Soest, 1963b).

In recent years, alternative approaches to the evaluation of forages have included the development of *in vitro* fermentation and enzyme digestion techniques. The thought behind this kind of approach—which places emphasis on the solubility of foods in the medium—has much in common with that of chemists in the early nineteenth century, who believed that they could determine availability by the solubility of foods in acids and alkalis.

Methods of forage analysis using an *in vitro* rumen fermentation estimate the digestion of a fibrous fraction, usually cellulose (Quicke *et al.*, 1959). Apart from poor analytical precision, the difficulties with this system are that cellulose is only a part of the cell wall and a small part of the forage dry matter, and that factors affecting the availability of cellulose may not be the same as those affecting other constituents. For these reasons, *in vitro* fermentation is not a suitable means of evaluating concentrate feeds. Recognition of its faults has led to procedures in which the total cell-wall material is measured (Tilley and Terry, 1963).

The search for the combination of chemical entities in the plant cell wall that determines the nutritive availability of its components has been a difficult task. Many researchers have been prone to evade the troublesome basic questions, and to be satisfied with limiting their scope to the modification of existing procedures or to the practical end of predicting nutritive quality. Many scientists aware of this state of affairs have cast scorn on the pragmatic use of obsolete methods, yet find the prospect of further research in this area dreary. For even when a better understanding is achieved and proper procedures are invented, there is still the overwhelming work of changing official procedures, rewriting textbooks in all nutritional fields, and achieving adoption of the new methods.

THE STRUCTURAL COMPONENTS OF PLANTS

It is hardly necessary to point out that crude fibre does not measure the total fibrous matter of plants (Paloheimo, 1962). The total woody tissue of forages is composed of lignin, cellulose, hemicellulose, pectin, insoluble mineral constituents, and nitrogen compounds that may be in combination with the lignin and hemicellulose. In addition, particularly at the surfaces of leaves and stems, there may be waxes and other ether-insoluble cutin material. Of the foregoing parts of the cell wall, cellulose and small but variable portions of lignin and hemicellulose will be retained in crude fibre (Tollens and Glaubitz, 1897; Norman, 1935). The realization that total fibre, which is properly the cell-wall material of plants, is composed of a complex variety of substances physically and chemically present as a unit, demonstrates the weakness of expressing estimates of fibre on the basis of supposedly chemically uniform fractions such as cellulose. Nutritionally, the structural carbohydrates of the cell wall are of major importance, but to stress these at the expense of non-carbohydrate components and physico-chemical structure does not lead to understanding of factors controlling fibre digestibility.

The physico-chemical structure of the cell wall is important in determining the bio-degradability of the component parts of the structure. There are three possible modes of action of lignin upon the availability of plant carbohydrates: (1) lignin is toxic to digesting microorganisms; (2) lignin physically incrusts the carbohydrates, rendering parts of them inaccessible to digestive organisms; or (3) lignin is linked to carbohydrates by covalent and hydrogen bonds which cellulytic organisms do not possess appropriate enzymes to hydrolyse.

Of these possibilities, the first has been recognized but largely ignored (Crampton and Maynard, 1938). The theory of incrustation has been supported recently by evidence that ball milling of lignified tissue greatly increases the *in vitro* digestibility of cellulose (Dehority *et al.*, 1962). Explanation of this phenomenon on the basis that milling breaks away incrusting lignin does not allow for the very probable degradation of carbohydrate by ball milling (Halliwell, 1960). Ball milling of cotton, which contains no lignin, greatly increases the rate of digestion by cellulytic organisms. As regards the third possibility, involving lignin–carbohydrate bonds, there has long been a controversy as to whether such bonds exist (Phillips, 1940a; Head, 1960). The negativeness of the evidence is related to the fact that chemical isolation of lignin is so drastic that any evidence of bonding is usually lost.

However, in view of recent studies on the biosynthesis of lignin and milder means of isolation, strong support can be found for an intimate combination between lignin and cell-wall carbohydrates (Wardrop and Bland, 1958; Siegel, 1962; Freudenberg, 1965). Lignin-carbohydrate complexes have been isolated (Bjorkman, 1957), and the existence of covalent linkage is probable (Bolker, 1963).

The present view of the chemical structure of the cell wall is that cellulose forms the fibrous structural backbone, the fibrils of cellulose being embedded in a matrix of hemicellulose and lignin in chemical combination (Siegel, 1962). It has become clear from electron micrographs (Hepler, 1964) that lignin is always laid down with carbohydrates, and the proportion of the lignin varies with the site in the cell wall, as shown in Fig. 1. Thus, the lignin content of forages—which is observed to increase with plant maturity—varies not only with respect to such different parts of the plant as leaf and stem, but also within the structure of an individual fibre or cell wall.

In this connection, it is not surprising to note the presence of cell-wall carbohydrates of differing nutritive availabilities within the same plant. Jarrige (1960) and Gaillard (1962) have found that xylose has a low digestibility relative to glucose, whereas arabinose has just the reverse relationship. Pectin, obtained by extraction using hot water and ammonium oxalate (Jermyn, 1955), is in Gaillard's work (1962) completely digestible.

Variability in degree of lignification is a possible factor in rendering some parts of the cell wall more digestible than others. For example, in studies of cellulose-digestion rates in single forages, the rates of cellulose disappearance are more rapid in the initial 12-hour period than in subsequent periods, particularly in legumes (Donefer *et al.*, 1960; Reid and Jung, 1965). A hypothetical point that deserves experimental examination is that as lignified tissues digest in a fermenting medium, the lignin content of the undigested residue rises as cell-wall carbohydrates are hydrolysed. The effect of this increase in lignin upon the fermentation rate of the remaining carbohydrates would be of interest.

That the digestibility of fibre fractions might be related to the proportion of lignin associated with each fraction was first examined by Naumann (1940), who found a very high degree of relationship between the digestibility and lignin content of crude fibre in different forages. Saarinen *et al.* (1956) showed that digestibilities of crude fibre and hemicellulose in delignified wood pulps were best explained by the ratios which the respective fractions formed with the amounts of lignin in the pulp. Maymone (1962) has noted that the ratio of lignin to crude fibre varies with forage species, particularly between grasses and legumes, the latter having a much higher proportion of lignin and a generally lower fibre digestibility than the former (Archibald *et al.*, 1962). Jarrige and Minson (1964) have explained differences in the digestibilities of ryegrass and orchard grass on this basis.

Lignin as a percentage of lignocellulose (acid-detergent fibre) is highly correlated with the digestibility of cell-wall fractions, as is the lignin content

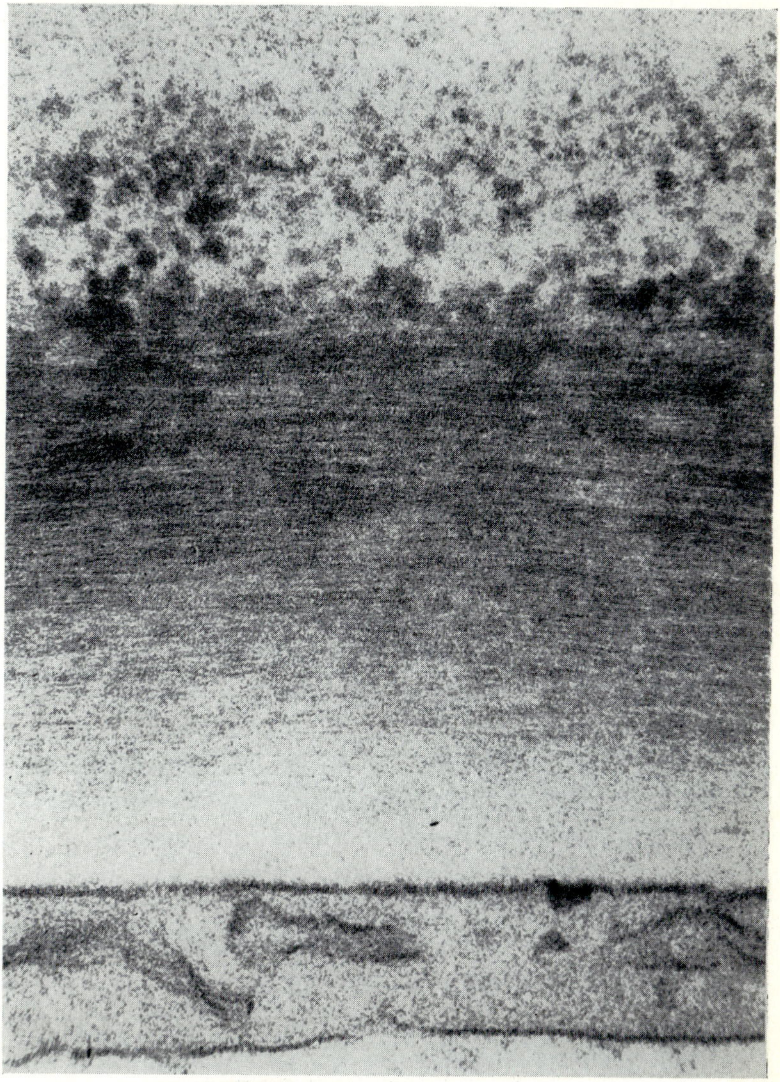

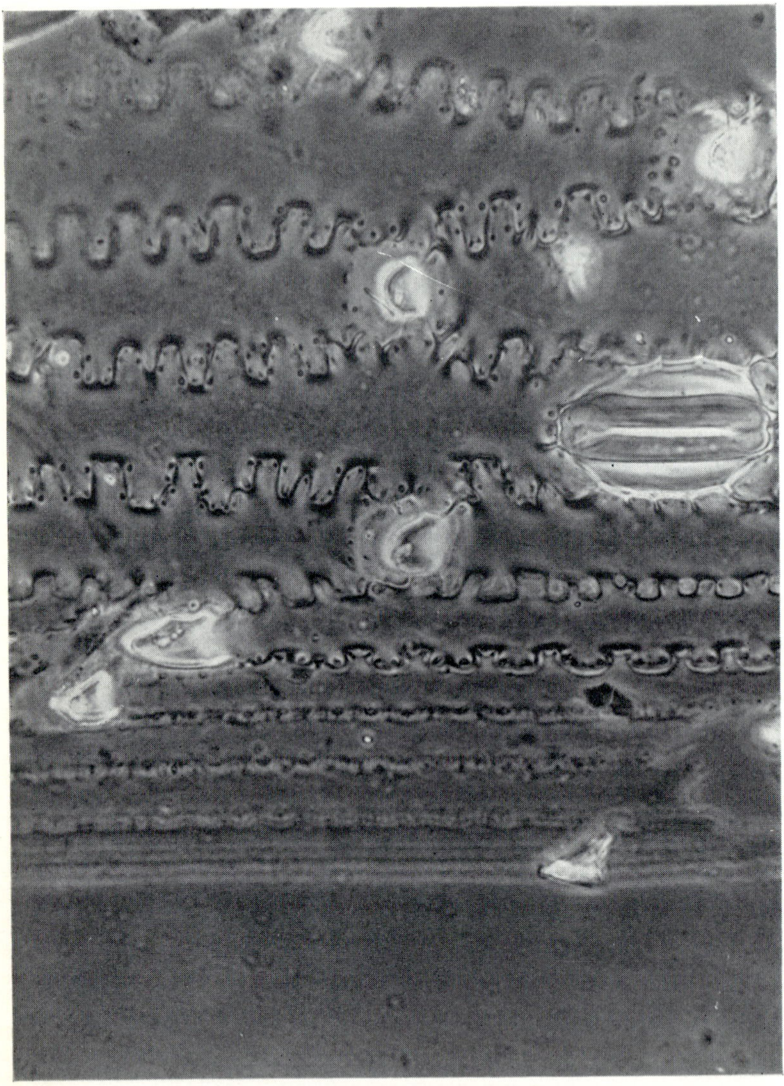

of the cell walls (Van Soest and Moore, 1965). Correlations in Table 1 show that digestibilities of fractions not contained in the cell wall (cell contents) are not significantly correlated with lignin content expressed as a percentage of a cell-wall fraction. Similar results have been reported for digestibility of crude fibre (Van Soest, 1964). The cellular contents of forages are, for the most part, completely digestible.

TABLE 1
Correlations between Digestibility of Forage Fractions and the Lignin Content of 19 Grass and Legume Forages (Van Soest and Moore, 1965)

Forage fraction	Correlations with lignin expressed as concentration in:		
	Dry matter	Cell walls	Lignocellulose
Cell walls	−0·87†	−0·90†	−0·95†
Hemicellulose	−0·92†	−0·85†	−0.86†
Lignocellulose	−0·82†	−0·86†	−0·93†
Cellular contents	−0·51*	−0·14	−0·15
Nitrogen	−0·65†	−0·14	−0·15

* Significant at the 5 per cent level of probability.
† Significant at the 1 per cent level of probability.

From these and other studies it is evident that hemicellulose, which is not ordinarily included in the fibre determination but rather as a part of the NFE, is a lignified fraction with nutritive availability similar to that of cellulose (Ely *et al.*, 1953). Microbial fermentation is the means of utilizing hemicellulose, and digestive secretions cannot hydrolyse it. Evidence that lignin is the principal factor affecting the availability of cell-wall carbohydrates lies in the increase in digestibility obtained when plant fibres are delignified (Lehman, 1891; Hvidsten and Simonsen, 1952). The very low digestibility of wood can be increased to about 95 per cent for ruminants by paper-making processes (Saarinen *et al.*, 1959).

While most fibre-digestibility studies have been made with ruminants, the factors affecting the nutritive availability of cellulose and hemicellulose are probably similar for non-ruminants, although expected digestibilities will be of a much lower order (Mangold, 1934). Fibre of a very low lignin content might have a significant digestibility while the highly lignified fibre of legumes might appear indigestible. Furthermore, it is the cell-wall content of a feed and not the crude-fibre content (as is commonly used) that should be used as representing the difficult-to-digest fraction of feeds for non-ruminants; also, it is the cell-wall content or perhaps the amount of digestible cell-wall contents that should be related to efficiency and the heat increment, as well as to voluntary intake (Van Soest, 1965b). The last point may be true for both ruminants and non-ruminants and should be investigated. However, the inclusion of lignin and hemicellulose in the NFE fraction may be a more serious matter for non-ruminants than for ruminants.

While lignification is usually presented as the factor influencing digestibility of fibre, several other possibilities should also be mentioned. For instance, it should be kept in mind that stage of maturity and other plant environment factors constitute the main variables of experiments in which nutritive value is measured. In such experiments it is reasonable to expect lignin to emerge as the chief factor. Other suggestions have included cutinization, which was recognized by the German chemists of the nineteenth century. Baker and Harriss (1947) have shown that cutinized surfaces of leaves and stems provide a barrier to digesting microorganisms, and Sakurai (1963) has suggested that this phenomenon may be relatively more important in forage species having more heavily cutinized leaf and stem surfaces. Gaillard (1961), in her studies showing the differential digestibility of hemicellulose fractions, has suggested that differences in the availability of hemicellulose polysaccharides may result from the kind of linkage that exists between the sugar residues and the degree of branching of the polymer. The structure has a marked influence on the water-solubility of these carbohydrates, and it is not unreasonable to suppose that rate of digestion is also affected.

Grinding and pelleting of hays will also affect the digestibility of fibre fractions. While grinding increases digestibility of fibre *in vitro*, *in vivo* digestibility is decreased. Advantages of pelleting are gained by increased feed efficiency largely through increased consumption (L. A. Moore, 1964). Particle size distribution of feedstuffs is obviously an important attribute, but few reported measurements have been made (Campling *et al.*, 1963).

Silica is another possible factor affecting fibre digestibility (Jones and Handreck, 1965), and it has received little attention. Relatively large amounts of silica can be metabolized by some plants, particularly cereals, and deposited within the matrix of the cell walls (see Fig. 2). Like lignin, silica is an integral part of the matrix of the cell wall. The common practice of determining fibre fractions on an ash-free basis has little to support it if one considers that fibre is used to determine a residue fundamentally non-nutritive in character. If a feedstuff contains silica, even as sand or dirt, it would be more appropriate to include silica with other insoluble ash as a component of the fibre.

Finally, certain general basic aspects of the biological degradation of woody tissue should be pointed out. The carbohydrates of the cell wall represent the more available parts, so that lignin and other indigestible parts tend to accumulate in the residue. This is generally true whether it be during the rotting of wood, the decomposition of cellulosic material in soil (Flaig, 1958; Sternhill, 1964), or the digestive processes of an herbivore. An exception exists in the case of certain fungi that possess ligninase activity (Sundman *et al.*, 1964). It appears that degradation of lignin by these organisms involves oxidative cleavage of the aromatic rings. The fermentative processes of herbivores by which cell walls of plants are utilized are obligately anaerobic. Therefore the conditions by which lignin is biologically degraded are not satisfied. Although some degradation of lignin in the digestive process may occur (Pigden and Stone, 1952), it is probably restricted to terminal and substituent groups such as methoxyl. This likelihood is supported by the high recoveries of lignin in faeces of ruminants when the digestion trials and the lignin analyses are carefully done (Journet and Jarrige, 1962; Van Soest, 1963).

THE HIGHLY DIGESTIBLE COMPONENTS OF FEEDS

Substances in this class form the main source of energy in feedstuffs and represent the readily available nutrients. Feedstuffs high in these components are classed as concentrates. Feedstuffs of animal origin are of value principally as protein concentrates—lipids and proteins forming the bulk constituents of the organic matter of these materials. In feedstuffs of plant origin, concentrates usually mean seeds of commercially grown crops in which reserve carbohydrates, proteins and fats form the bulk of the dry matter. However, the cell contents of roughages or forages are composed of carbohydrates, proteins and organic acids—substances that are also fundamentally concentrate in nature. The justification for regarding coarse forages as being partially concentrate is exemplified in the preparation of protein concentrates by isolating appropriate fractions from forage (Duckworth and Woodham, 1961; Gerloff *et al.*, 1965).

A concentrate food is defined as being readily digestible by the usual processes in the simple-stomached animal. The difference between concentrate feed and fibrous feed is that fibre is digestible with difficulty and only by microbial fermentation, because the digestive tracts of animals do not secrete the appropriate enzymes. In the lower digestive tracts of herbivores the principal undigested food residues are made up almost entirely of cell-wall material. The highly digestible portion of forages corresponds very nearly to that of the cell contents, which are almost completely digestible (Van Soest, 1965; Jarrige, 1965). The faeces of forage-fed ruminants are qualitatively devoid of soluble carbohydrates. In concentrate feeding, some starch may escape digestion, but this is not a large amount (Karr *et al.*, 1965). The faeces of herbivores are composed chiefly of undigested plant cell walls, excretion products, sloughing from the gut, and bacterial residues. The acid- and alkali-soluble parts of these components are included in the NFE fraction in the analysis of faecal material by the proximate system, a matter further confounding the digestibility coefficients of this so-called crude carbohydrate fraction.

The principal limitations to the full utilization of concentrate feedstuffs probably lie in the rate at which they can be digested. The rate in turn is affected by the solubility of the constituents, their susceptibility to enzymic degradation, and the physical modifications induced in the feedstuffs by such processing techniques as fine grinding or steaming.

Carbohydrate components. In the Weende system of feed analysis the fraction calculated by difference, the NFE, is supposed to represent available carbohydrates. In reality, however, many non-carbohydrate constituents may exist under this heading (Browne, 1940). In addition to solubilized lignin and hemicellulose, there may be organic acids (Buch, 1960), tannins and miscellaneous compounds (Denny, 1940). In forages little or no starch may be present, and the largest carbohydrate fraction will consist of the soluble carbohydrates (Waite and Boyd, 1953; Whelan, 1955). The soluble carbohydrates are composed of sugars (glucose, fructose, sucrose, and miscellaneous others); oligosaccharides; and either fructosans or starch or both, depending on whether the forage is a grass or a legume (Percival, 1952; Bell, 1955; Rendig *et al.*, 1964). In silages, volatile fatty acids and lactic acid have formed at the expense of sugar and other readily available substances, and very little or no soluble carbohydrate may be present (Langston *et al.*, 1962). Preservation with mineral acid will enable more of the original forage composition to be retained (Virtanen, 1938).

The pectins form a class somewhat intermediate between the soluble carbohydrates and the hemicelluloses. While formally a part of the cell membrane, pectin may be readily water-soluble (Jermyn, 1955; Bath, 1960). The material obtained by extraction with dilute ammonium oxalate is essentially completely digestible in ruminants (Gaillard, 1962), but is less so in poultry and perhaps other non-ruminants (Becker, 1964). Digestibility may depend on fermentation due to the absence of pectinases in the digestive tract. The so-called pectin fraction obtained by acid extraction or estimated by acid decarboxylation is not identical to the above pectin because of including lignified polyuronides. In grasses the uronide hemicelluloses form a large fraction, and soluble pectin is a very minor constituent. In legumes the ammonium oxalate pectin is the more abundant fraction.

In cereal concentrates, the NFE is composed of starches and hemicelluloses (Thurber, 1940). The starches are divided into the branched amylopectin and the linear amylose (Fraser *et al.*, 1956). Starches of different sources vary in composition, grain size, solubility and rates of digestion or fermentation. Undoubtedly the physical structure of starch grains is an important factor affecting these rates. Cooking, steaming or fine grinding destroys the physical structure of starch grains and greatly increases the rates at which they can be assimilated (Baker *et al.*, 1950).

Nitrogenous constituents. The crude protein of feedstuffs is commonly calculated as total Kjeldahl nitrogen times the factor 6·25. However, not all of the nitrogen of feeds is in the form of protein, since non-protein nitrogen compounds such as ammonium salts, urea, nitrates, nitrites, free amino acids, peptides, heterocyclic compounds (including nucleic acids), and lignified nitrogenous substances may be present in varying proportions (Synge, 1955; Butler, 1959). The hot-water-soluble, non-protein nitrogen fraction of forages comprises roughly one-third of the total nitrogen and is made up largely of non-essential amino acids (Kolousek and Coulson, 1954; Ferguson and Terry, 1954). The remaining two-thirds of the nitrogen is not all true protein, as is commonly assumed, because about 5 to 10 per cent of the total nitrogen is combined with the lignin fraction and is essentially indigestible (Whitehead and Quicke, 1960; Van Soest, 1965a). The nitrogenous portion of the lignin fraction is extraordinarily difficult to dissolve (Fraser and Holmes, 1958) and should not be regarded as protein (Bondi and Meyer, 1948). A protein fraction appears to be combined with carbohydrates of the cell wall (Ely *et al.*, 1956; Preston, 1961), and probably has a high digestibility. The true protein of leaf tissue has a high quality (Pirie, 1955; Gerloff *et al.*, 1965), although it may appear to be lower if the protein has been determined as total Kjeldahl nitrogen (Robertson *et al.*, 1940; Garry, 1951) or if the forage has been dried at too high a temperature (Woodham, 1963).

In the case of concentrates of plant origin, reserve proteins of a lower quality comprise most of the nitrogen (Pace, 1955). In these concentrates and also in feeds of animal origin, where distinction between cell walls and contents has no relevance, the presence of heat-damaged protein and indigestible residues such as keratins may be of concern (Duckworth, 1955). Also, there may be inhibitory factors present in seed proteins (Pace, 1955). In these instances, available true protein can be estimated by means of enzyme digestion procedures (Elmslie, 1960; Lovern, 1965). The need for distinguishing rate as opposed to extent of digestion may also exist (Lovern, 1965).

In ruminants, the quantity of available nitrogen is of primary importance, although too fast a rate of breakdown in the rumen may lead to urinary losses (Chalmers, 1960; Tagari *et al.*, 1962). In non-ruminants, a slow rate of *in vitro* digestion may be related to poor animal performance (Lovern, 1965), but usually the primary factors are problems of protein quality and availability of such critical amino acids as lysine (Woodham, 1963). Protein quality can be estimated by amino-acid analyses, but because these do not measure availability, the use of biological techniques may be necessary (Longenecker, 1963). In some cases, animal feeding trials by nitrogen balance or growth may be indicated.

Alteration of feedstuffs by processing. Heating or cooking of feeds will denature proteins and enzymes and also induce other reactions. Some of these may be beneficial to feed quality, as in the inactivation of an inhibitory factor by the heating of soybean products (Pace, 1955; Hohls, 1964) or the improvement of protein solubility by the pressure-cooking of keratins with alkali. However, the heating of proteins is usually detrimental to quality. Because steaming, pelleting, drying or dehydrating is commonly used to render feedstuffs less bulky, more stable for storage, or more likely to ensure greater intake, quality control becomes a problem.

The principal chemical change causing damage to feedstuffs upon heating is the non-enzymic browning or Maillard reaction (Hodge, 1953). This reaction is catalyzed by water and consists of the degradation of carbohydrates to dehydration products bearing a reactive carbonyl group which condenses with amino groups. Rearrangement of the double-bond structure and polymerization yield a dark-coloured substance containing about 11 per cent of nitrogen. Apart from the nitrogen content, this material has some of the properties of lignin, including indigestibility (Donoso *et al.*, 1962) and acid-insolubility, and interferes with the determination of lignin (Van Soest, 1965a).

Different carbohydrates vary in their rate of and susceptibility to thermal degradation, and amino acids also vary in their reactivity, lysine bearing a free amino group being one of the most active (Hodge, 1953). Phenols and lignin itself are capable of participating in the general reaction (Burton *et al.*, 1963), which can occur in dry silages (Wieringa, 1960; Hill and Noller, 1963) and in improperly cured hays. The Maillard reaction is undoubtedly an important factor in the increase in temperature and development of spontaneous combustion in damp hay.

The fermentative changes that occur upon ensiling usually cause more or less degradation of the protein fraction to non-protein nitrogen compounds—in particular, ammonia and volatile amines (Barnett, 1954). Large amounts of the protein fraction may be degraded in poor-quality silages. A high initial soluble-carbohydrate content ensures a rapid drop in pH in the development of lactic acid and tends to reduce the protein breakdown. Naturally fermented grass and legume silages appear to contain a factor depressing consumption (Thomas *et al.*, 1961). The use of additives such as acids or carbohydrates allows control of the fermentation (Watson and Nash, 1960). Fish and whale meat have been satisfactorily ensiled by controlled fermentation as an alternative to direct heat-drying (Bertullo, 1962).

Lipids. Although the lipids represent a broad class of substances, all of which are likely to be found in some feedstuff, by far the most important in terms of quantity and distribution are the triglyceride fats and oils. The properties of fats and oils may be characterized by saponification number, iodine number, Reichert–Meissl number, or Polenske number—all of which involve factors of chain length, unsaturation, presence of volatile acids, and so on (Hilditch and Williams, 1964). Recently it has become possible to determine the distribution of fatty-acid residues by gas chromatography, which provides much more precise information. Nutritionally, the degree of unsaturation is of considerable importance. For example, feeding of unsaturated oils depresses the output of milk fat in milking cattle, whereas relatively more saturated fats tend to have the opposite effect (Van Soest, 1963c). Some degree of hydrogenation of unsaturated fatty acids goes on in the rumen (Chapter 3). The presence of the unsaturated fatty acid—linoleic acid—is essential in the feeds of non-ruminants, and other unusual acids may also be of importance in animal nutrition.

Whereas triglyceride oils constitute the principal fraction of seed-concentrate lipids, a great variety of ether-soluble substances make up the principal fraction of green-forage lipids. Some of these substances, such as the waxes, chlorophylls, and essential oils, may have little nutritional value (Meara, 1955). The most valuable of them are the galactosyllipids, which contain relatively unsaturated fatty acids (Weenink, 1961). The fractions containing esterified fatty acids are highly digestible, as is true of the triglycerides in general (Garton, 1965).

One of the problems of unsaturated lipids is that they spontaneously oxidize and polymerize, although these reactions may be inhibited with suitable antioxidants. The products are probably not digestible and do not appear in the ordinary determination of lipids as ether extract. Thus, the concentration of ether extract in hays, for example, is not constant but declines with storage (Couchman, 1959). The practical importance of this effect is seen in the carotene and vitamin A activity of leaf meals, which declines by virtue of the same mechanism.

Energy content. In the proximate analysis of feedstuffs, ether extract is often determined for the sole purpose of presenting digestible nutrients on an equal-energy basis. Usually the empirical factor 2·25 is multiplied with the estimated amount of digestible ether extract in order to place the latter on a comparable basis with carbohydrate. However, there are inaccuracies in this system due to the presence of undigested lipid in an ether-insoluble form in the excreta (Roberts *et al.*, 1963) and the questionable universality of the factor 2·25—particularly when applied to forages in which the lipids contain little triglyceride. It is likely that the American system of calculating total digestible nutrients involves a multiplication of errors. It would be less work and just as informative in low-fat feeds to present evaluations in the form of digestible organic matter. In cases where high precision is needed, the direct approach of determining digestible energy by bomb calorimetry is much to be recommended (Lofgreen, 1951).

Vitamins and minerals. These extremely important nutritional substances comprise *in toto* but a small part of the dry weight of feedstuffs. Because of the chemical diversity of both vitamins and minerals, detailed treatment here is not possible.

They are discussed briefly in Chapter 1, but reference should be made to Chapter 11 on Minerals, and to the Chapters dealing with the different classes of livestock.

Comprehensive summaries of the more critical vitamin and mineral contents of foods and feeds are given by the National Research Council (1956, 1958, 1964) and Watt and Merrill (1963). Discussions of various mineral elements and their distribution can be found in Comar and Bronner (1962), Schütte (1964), Beaton and McHenry (1964a), and Rosenfeld and Beath (1964). For vitamins see T. Moore (1957), Hartman and Dryden (1965), and Beaton and McHenry (1964b).

The best compilation of analytical procedures for vitamins and minerals occurring in foods and feeds is found in *Methods of Analysis*, published by the Association of Official Agricultural Chemists (1965). All methods in this text have been studied and approved for reliability, and therefore do not necessarily represent the most recent analytical methodology. Some recent publications on specific subjects deserve mention: trace minerals (Ward and Johnston, 1960); chloride (Cotlove, 1964); fat-soluble vitamins (Analytical Methods Committee, 1964B); vitamin B_6 (Storvick *et al.*, 1964); and flavins (Yagi, 1962). Determination of many chemical elements by modern instrumental techniques will have increasing application in nutritional studies. These include atomic absorption spectroscopy (Willis, 1963), X-ray emission spectroscopy (Natelson and Whitford, 1964), and radioactivation analysis (Bowen and Gibbons, 1963).

Miscellaneous constituents. There are many constituents of feedstuffs that do not ordinarily comprise more than a small part of the dry matter, but are of nutritional importance and may have either positive or negative effects on the utilization of the diet. Minerals and vitamins are examples of this type of constituent.

Other examples are found in a great class of aromatic plant substances, including organic acids (Buch, 1960), tannins (Burns, 1963), pigments (Davidson, 1954), anthocyanins, phenols (Meara, 1955), and saponins (Coulson, 1958). The tannins combine with soluble proteins to form insoluble leather-like adducts, and some forages high in tannin—such as *Lespedeza cuneata* or the vetches—may have a low protein digestibility for this reason (Hawkins, 1959). Tall fescue (*Festuca* sp.) has been shown to contain a toxic factor of unknown constitution (Jacobson *et al.*, 1962). Legumes can contain an oestrogenic factor, such as coumestrol, which has been isolated from alfalfa; and apparently other similar compounds exist (Bickoff *et al.*, 1964). A factor reducing palatability has been isolated from reed canary grass (*Phalaris arundinacea*) (Roe and Mottershead, 1962). Undoubtedly, many unknown compounds of this sort remain to be discovered and elucidated.

There is another group of substances having nothing else in common than that they are added to feedstuffs either purposefully or as accidental contaminants. These include antibiotics; pesticides; drugs with hormone activity, such as thyroid-active iodinated proteins or diethylstilbestrol; and many other substances. Assays for materials in this class are often called for by regulatory agencies and are needed in the very rapidly developing field of feedstuff formulation and processing.

SOME PROBLEMS IN SAMPLE PREPARATION AND METHODOLOGY

The mechanical means by which compositions are obtained must be simple, reproducible and economical. These are factors of utmost importance in the acceptance of new methods, regardless of how much more meaningful or significant the innovations may be. The difficulty involved in replacing the crude-fibre method illustrates the extent to which an outmoded but economical technique can become entrenched. Methods of sample preparation offer many examples of situations in which convenience overrides other considerations. For instance, it frequently happens that no thought is given to the consequences of preliminary steps by which biological samples are commonly dried with heat and then ground in a mill. Among the determinations in which the mode of preparing and drying the sample is known to affect results are sugar, soluble carbohydrates (Jones, 1962), *in vitro* cellulose digestion (Johnson *et al.*, 1962), available protein and available lysine, crude fibre (Harberts and Immink, 1959), acid-detergent fibre (Van Soest, 1965a), lignin (MacDougall and Delong, 1942), and volatile constituents generally.

The non-enzymic browning reaction is generally responsible for increased yields in fibre and lignin determinations because the products of the browning have solubility characteristics similar to that of lignin. Results of sugar methods are lowered because of degradation of the sugar residues in the browning reaction. Drying at temperatures below 50°C would avoid problems in the foregoing instances, but in determination of volatile constituents prior drying must be avoided. The latter caution also applies to determinations of caloric content in samples containing organic volatiles.

Standard procedures for the determination of water usually involve volatile loss either *in vacuo* or in forced draft at 100°C. Under such conditions all volatile matter will be measured as water. This defect is serious when such methods are used on fermented products such as silages, where a variety of volatile acids, bases, alcohols, aldehydes and other compounds may be present. The toluene distillation affords some improvement, but leaves much to be desired (McDonald and Dewar, 1960). The volatile substances of silages have a high caloric value, and their loss in drying by conventional sample preparation may be the reason why the digestible energy of silages sometimes appears to be more efficiently utilized than those of other forages (Ekern and Reid, 1963). The true energy contents of silages can be determined on fresh matter with bomb calorimetry, provided a suitable primer such as dimethylformamide or butyl cellosolve is used (Coppock and Van Soest, 1959; Fenner and Archibald, 1959).

Another case of error is the loss of nitrogen when excreta from balance or digestion trials are dried, an effect which tends to make the nitrogen balance appear more favourable than it really is. In this case, as in others, error can be avoided by performing the analysis directly on the wet sample.

ESTIMATION OF NUTRITIVE VALUE FROM COMPOSITIONAL DATA

The proximate system of analysis serves many purposes. Besides being the basis for feedstuff evaluation, it is also the basis for quality control in the feed industry. Regulatory laboratories utilize the proximate system, and legal definitions of feedstuffs are related to specified limits of their content of crude fibre, crude protein, ether extract, and so on.

As has been shown, the proximate system has many failings in regard to the characterization of nutritive value. The system was accepted early in the development of nutritional science on the basis that it *ought* to estimate nutritive value because crude fibre *ought* to be indigestible. Similar criticism could be made of some of the methods proposed more recently.

TABLE 2

Composition of Some Feedstuffs (Van Soest, 1966)

Feedstuff	Crude fibre	Cell wall constituents	Acid det. fibre	Lignin	Dry matter digest.
Maize–soybean meal mixture	5·0	11·0	7·1	0·6	79
Cooked feathers	1·5	—	20·0	—	83*
Raw feathers	1·3	—	68·2	—	17*
Alfalfa-hay (*Medicago sativa*)	23·5	40·2	25·1	5·3	62
Alfalfa hay late cut	38·6	55·2	39·5	8·7	53
Orchard grass (*Dactylis glomerata*)	24·1	52·3	27·1	2·7	72
Orchard grass late cut	35·0	70·4	40·1	4·7	57
Wheat straw (*Triticum vulgare*)	42·4	81·8	53·3	7·6	21

* Digestibility by AOAC pepsin digestion (Elmslie, 1960).

It is a relatively recent development, and so far largely confined to the ruminant field, that proposed measures of nutritive value are compared on feedstuffs fed in digestion trials. When making elaborate comparisons of this sort, it is not satisfactory merely to correlate various chemical components with digestibility and to demonstrate significant correlations and regressions, for these relationships are too often only apparent. Moreover, they are not reproducible because basic cause-and-effect relationships are not under test.

Table 2 illustrates the inconsistencies that can occur in comparing analyses for crude fibre and lignin with digestibility. It also illustrates that no one measure can adequately classify the entire range of feedstuffs.

As will be seen, the feather samples are properly classified by the value for acid-detergent fibre, whereas grasses and straw are classified by the value for cell-wall constituents. The low cell-wall contents of alfalfa are offset by the high lignin contents. Since lignification affects the digestibility of only the cell walls, it is unimportant in evaluating concentrates; hence the total fibre or cell-wall content is needed as a measure of quality.

It is important that any prospective replacement for the proximate system be equally comprehensive in scope—that is, feedstuffs of both plant and animal origin must be evaluated for all species of livestock, both ruminant

TABLE 3

Division of Forage Organic Matter by System (Van Soest, 1966) of Analysis Using Detergents

Fraction	Components	Nutritional available		
		Ruminant	Non-ruminant	Category
Cell contents (soluble in neutral detergent)	Lipids	Virtually complete	Highly available	A
	Sugars, organic acids, and water-soluble matter	,,	,,	
	Starch	,,	,,	
	Non-protein nitrogen	,,	,,	
	Soluble protein	,,	,,	
	Pectin	,,	,,	
Cell-wall constituents (fibre insoluble in neutral detergent)				
(1) Soluble in acid detergent	Hemicellulose	Partial	Very low	B
(2) Insoluble in acid detergent (acid-detergent fibre)	Cellulose	,,	,,	
	Lignin	Indigestible	Indigestible	
	Lignified nitrogen compounds	,,	,,	
	Heat-damaged protein	,,	,,	
	Keratin	,,	,,	
	Silica	,,	,,	

and non-ruminant. The general problem outlined above is capable of solution if feed nutrients are classified according to the factors affecting their availability. These must be determined from digestibility data obtained by careful analysis of feeds and corresponding faeces for characteristic entities—solubles, starch, proteins, cellulose, and so on. The principles outlined here have guided the system of dividing the dry matter of feedstuffs using detergents, developed at Beltsville.

It should be noted that the division of forage dry matter into cell-wall or membrane substances follows botanical lines. The importance of this distribution in nutrition has been pointed out by a number of investigators, and a number of procedures for cell-wall isolation and determination have been proposed (Gaillard, 1958; Harwood, 1954; Paloheimo *et al.*, 1961; Van Soest, 1963a).

On the basis of such studies, two general categories of constituents can be established, as shown in Table 3.

Category A includes substances for which digestive enzymes exist in the digestive tracts of all animals. Category B includes substances that are digested to a significant extent only in those species possessing an important gastrointestinal fermentation capability. Category B also includes indigestible fractions, among them lignin, which affect the availability of partially digestible carbohydrates. The indigestible residues of keratin proteins and the products of the non-enzymic browning reaction remain insoluble in the acid-detergent fibre determination, as shown in Table 3, but can be distinguished from lignin on the basis of the nitrogen content (Van Soest, 1965).

If one considers the residues of acid-detergent fibre or lignin as representing not specific chemical constituents but difficult-to-digest or non-nutritive residues, then a common boundary having meaning for ruminants and non-ruminants alike can be drawn around roughages, forages and concentrates of both plant and animal origin.

REFERENCES

ANDERSON, T. (1861) *Trans. Highland Soc. Scotland*, **10**, 249.
ANALYTICAL METHODS COMMITTEE (1964a) *Analyst*, **89**, 1.
ANALYTICAL METHODS COMMITTEE (1964b) *Analyst*, **89**, 7.
ARCHIBALD, J. G., BARNES, H. D., FENNER, H., and GERSTEN, B. (1962) *J. Dairy Sci.* **45**, 858.
ASSOCIATION OF OFFICIAL AGRICULTURAL CHEMISTS (1965) *Official Methods of Analysis*. 10th ed., 957 pp.
BAKER, F., and HARRISS, S. T. (1947) *Nutr. Abstr. Rev.* **17**, 1.
BAKER, F., NASR, H., MORRICE, F., and BRUCE, J. (1950) *J. Path. Bact.* **62**, 617
BARNETT, A. J. G. (1954) *Silage Fermentation*. Academic Press, New York.
BATH, I. H. (1960) *J. Sci. Food Agr.* **11**, 560.
BECKER, M., HARNISCH, H., and HEILMAN, D. (1964) *Arch. Tierernähr.* **14**, 291.
BELL, D. J. (1955) In *Moderne Methode der Pflanzenanalyse*, Vol. II, p. 1. Ed. K. Paech and M. V. Tracey. Springer Verlag, Berlin.
BERTULLO, V. H. (1962) *Rev. Inst. Invest. Pesqueras*, **1**, 7.
BICKOFF, E. M., GUGGOLZ, G., LIVINGSTON, A. L., SPENCER, R. R., KNUCKLES, B. E., WITT, S. C., TAYLOR, K. W., and KOHLER, G. O. (1964) *J. Agr. Food Chem.* **12**, 537.
BJÖRKMAN, A. (1957) *Ind. Eng. Chem.* **49**, 1395.
BOLKER, H. L. (1963) *Nature*, **197**, 489.
BONDI, A., and MEYER, H. (1948) *Biochem. J.* **43**, 248.
BOUSSINGAULT, J. B. (1844) *Economie Rurale*, Vol. II, p. 438. Paris.
BOWEN, H. J. M., and GIBBONS, D. (1963) *Radioactivation Analysis*. Oxford University Press, Oxford.
BROWNE, C. A. (1940) *J. Assoc. Off. Agr. Chem.* **23**, 102.
BUCH, M. L. (1960) *Agriculture Handbook No.* **64**. U.S. Dept. Agr., Washington, D.C.
BURNS, R. E. (1963) *Ga. Agr. Exp. Sta. Tech. Bull.* **32**, 1963.
BURTON, H. S., MCWEENY, D. J., and PANDHI, P. N. (1963) *Nature*, **199**, 659.
BUTLER, G. W. (1959) *Proc. N.Z. Soc. Animal Prod.* **19**, 99.

Campling, R. C., Freer, M., and Balch, C. C. (1963) *Brit. J. Nutr.* **17**, 263.
Chalmers, Margaret (1960) In *Digestive Physiology and Nutrition of the Ruminant*, p. 205. Ed. D. Lewis. Butterworths, London.
Comar, C. L., and Bronner, F. (1962) *Mineral Metabolism.* Vols. I and II. Academic Press, New York.
Coppock, C. E., and Van Soest, P. J. (1959) ARS-44-55. Agriculture Research Service, U.S. Dept. Agr., Washington, D.C.
Cotlove, E. (1964) *Methods of Biochemical Analysis*, **12**, 277. Ed. D. Glick.
Couchman, J. F. (1959) *J. Sci. Food Agr.* **10**, 519.
Coulson, C. B. (1958) *J. Sci. Food Agr.* **9**, 281.
Crampton, E. W., and Maynard, L. A. (1938) *J. Nutr.* **15**, 383.
Crampton, E. W., and Whiting, F. (1943) *J. Animal Sci.* **2**, 278.
Davidson, J. (1954) *J. Sci. Food Agr.* **5**, 1.
Davy, Sir Humphrey (1813) *Elements of Agricultural Chemistry*. London.
Dehority, B. A., Johnson, R. R., and Conrad, H. R. (1962) *J. Dairy Sci.* **45**, 508.
Denny, F. E. (1940) *J. Assoc. Off. Agr. Chem.* **23**, 151.
Donefer, E., Crampton, E. W., and Lloyd, L. E. (1960) *J. Animal Sci.* **19**, 545.
Donoso, G., Lewis, O. A. M., Miller, D. S., and Payne, P. R. (1962) *J. Sci. Food Agr.* **13**. 192.
Duckworth, J. (1955) In *Moderne Methode der Pflanzenanalyse*, Vol. IV, p. 106. Ed. K. Paech and M. V. Tracey. Springer Verlag, Berlin.
Duckworth, J., and Woodham, A. A. (1961) *J. Sci. Food Agr.* **12**, 5.
Einhof, H. (1808) *Grundriss der Chemie für Landwirthe.* Cited by Henneberg and Stohmann (1860) p.6, and Thaer (1809).
Ekern, A., and Reid, J. T. (1963) *J. Dairy Sci.* **46**, 522.
Ellis, G. H., Matrone, G., and Maynard, L. A. (1946) *J. Animal Sci.* **5**, 285.
Elmslie, W. P. (1960) *J. Assoc. Off. Agr. Chem.* **43**, 320.
Ely, R. E., Kane, E. A., Jacobson, W. C., and Moore, L. A. (1953) *J. Dairy Sci.* **36**, 334.
Ely, R. E., Melin, C. G., and Moore, L. A. (1956) *J. Dairy Sci.* **39**, 1742.
Ely, R. E., and Moore, L. A. (1954) *J. Agr. Food Chem.* **2**, 826.
Fenner, H., and Archibald, J. G. (1959) *J. Dairy Sci.* **42**, 1995.
Ferguson, W. S., and Terry, R. A. (1954) *J. Sci. Food Agr.* **5**, 515.
Flaig, W. (1958) *Proc. 4th Int. Congr. Biochem., Vienna*, p. 227.
Fraser, J. R., Brandon-Bravo, M., and Holmes, D. C. (1956) *J. Sci. Food Agr.* **7**, 577.
Fraser, J. R., and Holmes, D. C. (1958) *J. Sci. Food Agr.* **9**, 511.
Freudenberg, K. (1965) *Science*, **148**, 595.
Gaillard, Blanche D. E. (1958a) *J. Sci. Food Agr.* **9**, 170.
Gaillard, Blanche D. E. (1958b) *J. Sci. Food Agr.* **9**, 353.
Gaillard, Blanche D. E. (1961) *Nature*, **191**, 1295.
Gaillard, Blanche D. E. (1962) *J. Agr. Sci.* **59**, 369.
Garry, R. O. (1951) *Brit. J. Nutr.* **5**, 243.
Garton, G. A. (1965) In *Physiology of Digestion in the Ruminant*, p. 390. Ed. R. W. Dougherty. Butterworths, Washington.
Gerloff, E. D., Lima, I. H., and Stahmann, M. A. (1965) *J. Agr. Food Chem.* **13**, 139
Halliwell, G. (1960) In *Digestive Physiology and Nutrition of the Ruminant*, p. 119. Ed. D. Lewis. Butterworths, London.
Harberts, C. L., and Immink, H. J. (1959) *Mededeling* **87**, *I. B. S. Jaarboek.*, p. 159. Instituut voor Biologisch en Scheikundig Onderzoek van Landbouwgewassen, Wageningen.
Hartman, A. M., and Dryden, L. P. (1965) In *Fundamentals of Dairy Chemistry*, p. 261. Ed. B. H. Webb and A. H. Johnson. Avi Publ. Co., Westport, Conn.
Harwood, V. D. (1954) *J. Sci. Food Agr.* **5**, 270.
Haubner, G. C. (1855) *Z. Deut. Landw.*, p. 177. Cited in Henneberg (1859) and von Knierem (1885).
Hawkins, G. E. (1959) *J. Animal Sci.* **18**, 763.
Head, M. J. (1960) In *Digestive Physiology and Nutrition of the Ruminant*, p. 263. Ed. D. Lewis. Butterworths, London.

HEMMING, E. T. (1852) *J. Royal Agr. Soc.* **13**, 409.
HENNEBERG, W. (1859) *J. Landw.* **7**, B. III, p. 299.
HENNEBERG, W., and STOHMANN, F. (1860, 1864) *Bergründung einer rationellen Futterung der Wiederkäuer*, Vols. I (1860) and II (1864). Braunschweig.
HEPLER, P. K. (1964) Ph.D. Thesis, University of Wisconsin.
HILDITCH, T. P., and WILLIAMS, P. N. (1964) *The Chemical Constitution of the Natural Fats*. John Wiley, New York.
HILL, D. L., and NOLLER, C. H. (1963) *J. Animal Sci.* **22**, 850.
HODGE, J. E. (1953) *J. Agr. Food Chem.* **1**, 928.
HOHLS, H. W. (1964) *Z. Tierfyziol. Tierernähr. Futtermittelk.*, Sonderheft No. 18.
HVIDSTEN, H., and SIMONSEN, H. (1952) *Tidsskr. Norske Landbruk.* **59**, 85.
JACOBSON, D. R., MILLER, W. M., YATES, S. G., TOOKEY, H. L., and WOLFF, I. A. (1962) *J. Dairy Sci.* **45**, 663.
JARRIGE, R. (1960) *Proc. 8th Int. Grassl. Congr., Reading*, p. 628.
JARRIGE, R. (1965) *Proc. 9th Int. Grassl. Congr., Sao Paulo*, p. 809.
JARRIGE, R., and MINSON, D. J. (1964) *Ann. Zootech.* **13**, 117.
JERMYN, M. A. (1955) In *Moderne Methode der Pflanzenanalyse*, Vol. II, p. 197. Ed. K. Paech and M. V. Tracey. Springer Verlag, Berlin.
JOHNSON, R. R., DEHORITY, B. A., CONRAD, H. R., and DAVID, R. R. (1962) *J. Dairy Sci.* **45**, 250.
JONES, D. I. H. (1962) *J. Sci. Food Agr.* **13**, 83. 1962.
JONES, L. H. P., MILNE, A. A., and WADHAM, S. M. (1963) *Plant Soil*, **18**, 358.
JONES, L. H. P., and HANDRECK, K. A. (1965) *J. Agric. Sci.* **65**, 129.
JOURNET, M., and JARRIGE, R. (1962) *Ann. Biol. Animale Biochim. Biophys.* **2**, 223.
KARR, M. R., LITTLE, C. O., and MITCHELL, G. E. (1965) *J. Animal Sci.* **24**, 890.
KNIEREM, W. VON (1885) *Z. Biol.* **21**, 67.
KOLOUSEK, J., and COULSON, C. B. (1954) *J. Sci. Food Agr.* **5**, 126.
LANGSTON, C. W., WISEMAN, H. G., GORDON, C. H., JACOBSON, W. C., MELIN, C. G., and MOORE, L. A. *J. Dairy Sci.* **45**, 396.
LEHMAN, F. (1891) *Landw. Versuchstat.* **38**, 337.
LOFGREEN, G. P. (1951) *J. Animal Sci.* **10**, 344.
LONGENECKER, J. B. (1963) In *Newer Methods of Nutritional Biochemistry*, p. 113. Ed. A. Albanese. Academic Press, New York.
LOVERN, J. A. (1965) *J. Assoc. Off. Agr. Chem.* **48**, 60.
MACDOUGALL, D., and DELONG, W. A. (1942) *Can. J. Res.* **20**, 20.
MANGOLD, E. (1934) *Nutr. Abstr. Rev.* **3**, 647.
MATRONE, G., ELLIS, G. H., and MAYNARD, L. A. (1946) *J. Animal Sci.* **5**, 306.
MAYMONE, B. (1962) *Aliment. Animale*, **6**, 371.
MCDONALD, P., and DEWAR, W. A. (1962) *J. Sci. Food Agr.* **10**, 566.
MEARA, M. L. (1955) In *Moderne Methode der Pflanzenanalyse*, Vol. II, p. 317. Ed. K. Paech and M. V. Tracey. Springer Verlag, Berlin.
MOORE, L. A. (1964) *J. Animal Sci.* **23**, 230.
MOORE, T. (1957) *Vitamin A*. Elsivier Publ. Co., Amsterdam.
NATELSON, S., and WHITFORD, W. R. (1964) *Methods of Biochemical Analysis*, **12**, 1. Ed. D. Glick.
NATIONAL RESEARCH COUNCIL (1956) *Composition of Concentrate By-product Feeding Stuffs*. Publ. 449. Nat. Acad. Sci., Washington, D.C.
NATIONAL RESEARCH COUNCIL (1958) *Composition of Cereal Grains and Forages*. Publ. 585. Nat. Acad. Sci., Washington, D.C.
NATIONAL RESEARCH COUNCIL (1964) *Joint United States-Canadian Tables of Feed Composition*. Publ. 1232. Nat. Acad. Sci., Washington, D.C.
NAUMANN, K. (1940) *Z. Tierernähr. Futtermittelk.* **3**, 193.
NORMAN, A. G. (1935) *J. Agr. Sci.* **25**, 529.
PACE, J. (1955) In *Moderne Methode der Pflanzenanalyse*, Vol. IV, p. 69. Ed. K. Paech and M. V. Tracey. Springer Verlag, Berlin.
PALOHEIMO, L. (1962) *Nordisk Jordbrugsforskning*, **2**, 1.

PALOHEIMO, L., VAINIO, K. A., KERO, M. L., and HERKOLA, E. (1961) *Maataloustieteellinen Aikakauskirja*, **33**, 51.

PERCIVAL, E. G. V. (1952) *Proc. Nutr. Soc.* **6**, 105.

PHILLIPS, M. (1940a) *J. Assoc. Off. Agr. Chem.* **23**, 108.

PHILLIPS, M. (1940b) *J. Assoc. Off. Agr. Chem.* **23**, 119.

PIRIE, N. M. (1955) In *Moderne Methode der Pflanzenanalyse*, Vol. IV, p. 23. Ed. K. Paech and M. V. Tracey. Springer Verlag, Berlin.

PIGDEN, W. J., and STONE, J. E. (1952) *Sci. Agr.* **32**, 502.

PRESTON, R. D. (1961) In *Macromolecular Complexes*, p. 229. Ed. M. V. Edds. Ronald Press, New York.

QUICKE, G. V., BENTLEY, O. G., SCOTT, H. W., and MOXON, A. L. (1959) *J. Animal Sci.* **18**, 275.

RAYMOND, W. F., JONES, E. C., and HAINS, C. E. (1955) *Agr. Prog.* **30**, **(II)** 120.

REID, R. L., and JUNG, G. A. (1965) *J. Animal Sci.* **24**, 615.

RENDIG, V. V., MCCOMB, E. A., and HU, C. L. (1964) *J. Agr. Food Chem.* **12**, 421.

ROBERTS, W. K., MCKIRDY, J. A., and STRINGAM, E. W. (1963) *J. Animal Sci.* **22**, 846.

ROBERTSON, E. I., CARVER, J. S., and COOK, J. W. (1940) *Wash. State Agr. Exp. Sta. Bull.* 388.

ROE, R., and MOTTERSHEAD, B. E. (1962) *Nature*, **193**, 255.

ROSENFELD, I., and BEATH, O. A. (1964) *Selenium.* Academic Press, New York.

SAARINEN, P., JENSEN, W., and ALHOJÄRVI, J. (1959) *Acta Agral. Fennica*, **94**, 41.

SAKURAI, M. (1963) *Research Report No. 15.* Grassland Division Kanto-Tosan Agr. Exp. Sta.

SCHÜTTE, K. H. (1964) *The Biology of the Trace Elements.* Lippincott Co., Philadelphia.

SIEGEL, S. M. (1962) *The Plant Cell Wall.* Pergamon, New York.

STERNHILL (1964) *Australian J. Chem.* **17**, 1236.

STORVICK, C. A., BENSON, E. M., EDWARDS, M. A., and WOODRING, M. J. (1964) *Methods of Biochemical Analysis*, **12**, 183. Ed. D. Glick.

SULLIVAN, J. T. (1964) ARS 34-62. Agriculture Research Service, U.S. Dept. Agr., Washington, D.C.

SUNDMAN, VERONICA, KUUSI, T., KUHANEN, S., CARLBERG, G. (1964) *Acta Agr. Scand.* **14**, 229.

SYNGE, R. L. M. (1955) In *Moderne Methode der Pflanzenanalyse*, Vol. IV, p. 1. Ed. K. Paech and M. V. Tracey. Springer Verlag, Berlin.

TAGARI, H., ASCARELLI, I., and BONDI, A. (1962) *Brit. J. Nutr.* **16**, 237.

THAER, A. VON (1809) *Grundsätze der rationellen Landwirtschaft.*, Sect. 275.

THOMAS, J. W., MOORE, L. A., OKAMOTO, M., and SYKES, J. F. (1961) *J. Dairy Sci.* **44**, 1471.

THURBER, F. H. (1940) *J. Assoc. Off. Agr. Chem.* **23**, 126.

TILLEY, J. M. A., and TERRY, R. A. (1963) *J. Brit. Grassl. Soc.* **18**, 104.

TOLLENS, B. (1897) *J. Landw.* **45**, 295.

TOLLENS, B., and GLAUBITZ, H. (1897) *J. Landw.* **45**, 97.

VAN SOEST, P. J. (1963a) *J. Assoc. Off. Agr. Chem.* **46**, 825.

VAN SOEST, P. J. (1963b) *J. Assoc. Off. Agr. Chem.* **46**, 829.

VAN SOEST, P. J. (1963c) *J. Dairy Sci.* **46**, 204.

VAN SOEST, P. J. (1964) *J. Animal Sci.* **23**, 838.

VAN SOEST, P. J. (1965a) *J. Assoc. Off. Agr. Chem.* **48**, 785.

VAN SOEST, P. J. (1965b) *J. Animal Sci.* **24**, 834.

VAN SOEST, P. J. (1966) *J. Assoc. Off. Anal. Chem.* **49**, 546.

VAN SOEST, P. J., and MOORE, L. A. (1965) *Proc. 9th Int. Grassl. Congr. Sao Paulo*, p. 783.

VIRTANEN, A. I. (1938) *Cattle Fodder and Human Nutrition.* Cambridge University Press.

VOELCKER, A. (1852) *J. Royal Agr. Soc.* **13**, 385.

VOELCKER, A. (1861) *J. Royal Agr. Soc.* **22**, 382.

WAITE, R., and BOYD, J. (1953) *J. Sci. Food Agr.* **4**, 257.

WAITE, R., and GORROD, A. R. N. (1959) *J. Sci. Food Agr.* **10**, 317.

WALKER, D. C. (1959) *J. Sci. Food Agr.* **10**, 415.

WARD, G. M., and JOHNSTON, F. B. (1960) *Chemical Methods of Plant Analysis.* Publ. 1064. Res. Branch Canada Dept. Agric. Ottawa.

WARDROP, A. B., and BLAND, D. E. (1958) *Proc. 4th Int. Biochem. Congr., Vienna*, p. 92.
WATT, B. K., and MERRILL, A. L. (1963) *Composition of Foods*. U.S. Dept. Agric., Agriculture Handbook No. 8. Washington, D.C.
WATSON, S. J., and NASH, M. J. (1960) *The Conservation of Grass and Forage Crops*. Oliver and Boyd, Edinburgh.
WEENINK, R. O. (1961) *J. Sci. Food Agr.* **12**, 34.
WHELAN, W. (1955) In *Moderne Methode der Pflanzenanalyse*, Vol. II, p. 145. Ed. K. Paech and M. V. Tracey. Springer Verlag, Berlin.
WHITEHEAD, D. L., and QUICKE, G. V. (1960) *J. Sci. Food Agr.* **11**, 151.
WIERINGA, G. W. (1960) *Proc. 8th Int. Grassl. Congr., Reading*, p. 497.
WILLIS, J. B. (1963) *Methods of Biochemical Analysis*, **11**, 1. Ed. D. Glick.
WOLFF, E. (1856) *Die naturgesetzlichen Grundlagen des Ackerbaues*. 3rd. ed. Leipzig.
WOODHAM, A. A. (1963) In *Progress in Nutrition and Allied Sciences*, p. 105. Ed. D. P. Cuthbertson. Oliver and Boyd, Edinburgh.
YAGI, K. (1962) *Methods of Biochemical Analysis*, **10**, 319. Ed. D. Glick.
YANOVSKY, E. (1940) *J. Assoc. Off. Agr. Chem.* **23**, 131.

CHAPTER 3

Microbiology of Digestion in Ruminants and its Nutritional Significance

P. N. HOBSON
Rowett Research Institute, Bucksburn, Aberdeen

THE subject of microbial metabolism in the rumen has been reviewed in general a number of times over the last few years, and reviews on specialized aspects have also been quite numerous. There thus seemed to be little point in repeating all the details given in these reviews and so here an overall picture of rumen metabolism has been attempted. A minimum number of references is given, as the reviews quoted give detailed lists, and many statements made are based on general knowledge of rumen metabolism derived from the results of many workers, or from unpublished results obtained in the author's laboratories.

GENERAL INTRODUCTION

All animals harbour microorganisms in their gut, but the importance of these microorganisms in the digestive processes of the animal depends on the diet and the construction of the gut of the animal. In most carnivorous and omnivorous animals intestinal microorganisms play a small part in the digestion of food, although they may be of some importance as producers of vitamins or other nutritional factors for the host. The commensal organisms may also help to control the growth of pathogenic, or potentially pathogenic, microorganisms in the intestines. However, in herbivorous animals the microorganisms of the gut have a vital part to play in the digestive processes, especially those related to the utilization of the cellulosic materials of plants, as cellulose cannot be digested by any of the normal intestinal enzymes. Since microbial digestion of materials such as cellulose is a comparatively slow process, for optimal microbial action the passage of food must be slowed down in some portion of the gut by an enlargement of the lumen, and for optimal utilization of the products of microbial digestion the site of microbial action must be such that these products can be absorbed by the host either from this site or from some distal portion of the intestine as the ingesta pass on. In some herbivorous animals the principal site of microbial action is in the large intestine and thus the enzymic and absorptive capacities of the stomach and small intestine cannot be brought to bear on the products of microbial activity unless coprophagy takes place. In the ruminant, however, the main microbial activity takes place in the first part of the gut and the normal digestive processes can be brought to bear on the microbial products as they pass to the abomasum and intestines. While it might be argued that the digestion of the otherwise resistant cellulose is of primary importance in the rumen and that feedstuffs such as starch, sugars and proteins would be more efficiently utilized if subject only to the normal mammalian digestive processes, the microbial action is not selective and the whole of the ruminant's feed is subject to radical alteration in the rumen. The processes occurring in the rumen, then, are of fundamental importance to the economy of the animal, and thus to the economy of mankind.

The study of rumen microorganisms and their metabolism has increased radically over the past 15 years and a very large number of papers with a bearing on the subject has been published in this time. In a short review only the broad outlines of the subject can be given; the reader is referred to original papers and reviews (for instance, on general aspects of rumen microbiology: Bryant, 1959; Cuthbertson and Hobson, 1961; Hobson, 1960, 1961, 1963; Hungate, 1963; Hungate, Bryant and Mah, 1965) to fill in the details, and other papers in this series will deal with the overall metabolism of the ruminant. An exhaustive list of references supporting all the statements made is not given in this review. The reviews quoted above and references given in the papers quoted in the text will enable the reader to explore the subject further.

We are concerned with the microbial digestion occurring in the rumen. Microbial digestion occurs in the large intestine of the ruminant, but, because of the comparative volumes of portions of gut, intestinal microbial action is insignificant compared with that of the rumen in the adult animal. In the young milk-fed animal the position may be reversed.

Because of the economic importance of cattle and sheep most of the investigations of rumen metabolism have been carried out using these animals. However, the rumen microorganisms and reactions are essentially the same in each animal, and in ruminants as a whole, and observations on one may be generally applied to the others. Thus, in general, in this article, no mention will be made of the source of the rumen organisms.

The Development of the Rumen and Its Microorganisms and Factors influencing the Established Population

The rumen is small in comparison with the rest of the digestive system in the newly-born sheep or cow and its development is caused both by the physical action of the solid feed ingested as the animal grows and by the action of the products of microbial fermentation of this feed. Kay (1963) has recently described the physiological development of the rumen and related functions such as the salivary flow. Rumen microorganisms are picked up by the developing animal either entirely through physical contact with adult animals, as in the case of the larger protozoa, which appear to be transmitted only by mouth contact, or by a mixture of physical contact and airborne contamination in the case of the smaller organisms. Transmission of rumen bacteria in airborne-droplets or in communal drinking water or feed has been shown to be a definite possibility. Thus the young ruminant in its normal surroundings has the possibility at any age of developing the normal adult rumen flora. What flora will develop at any time depends mainly on the feed ingested, and it is impossible to generalize and say that a ruminant of such and such an age will have a rumen flora of this or that type. The initial population of the rumen is probably derived mainly from back-flow of the intestinal and abomasal contents and contains large numbers of lactobacilli, streptococci and coliforms (Mackay and Oxford, 1954; Mann and Oxford, 1955). As more solid feed is consumed this predominantly lactic acid-producing flora changes to a more varied population, producing a larger proportion of volatile fatty acids. Bryant *et al.* (1958) found large numbers of lactic acid bacteria in calves of from 1 to 6 weeks of age fed on milk, but with access to hay and concentrates, and conversely large numbers of lactic acid-utilizing bacteria are found in young animals. Also, in young animals the proportion of facultative anaerobes in the bacterial population is higher than in adult animals. The continued ingestion of large amounts of starchy concentrates after weaning tends to perpetuate the lactic acid-producing, lactic acid-fermenting flora of the young animal and the pH of the rumen remains rather low. This low pH inhibits the growth of protozoa (Eadie, 1962a) and probably some of the anaerobic bacteria normally associated with the adult rumen. However, in an animal weaned on to a diet containing a normal amount of roughage a predominantly adult, cellulolytic flora will have developed when about 13 weeks old.

Since the rumen population is affected so much by diet, and under normal conditions the young ruminant is surrounded by sources of rumen organisms, the value of inocula said to contain rumen bacteria would seem to be marginal in aiding the development of an adult rumen.

The rumen population, once established on whatever dietary régime, is resistant to change from outside sources. The animal ingests thousands or millions of organisms with its feed. Some of these are aerobes and cannot proliferate in the rumen, others may be capable of life in the rumen environment but are kept to very small numbers, or entirely suppressed, by the growth of the established flora. Also true rumen organisms which may flourish under some conditions cannot be induced to proliferate under other conditions by artificial inoculation into the established flora (Hobson and

Mann, 1961). In the case of the rumen protozoa the established population may be antagonistic to other types which alone may proliferate readily in the rumen (Eadie, 1962b).

The rumen population, once established, then may remain essentially the same throughout the life of the ruminant if the diet remains the same. Gradual changes in diet may be followed by gradual changes in the balance of the flora and if the diet is greatly changed in composition a new flora wherein different organisms predominate may appear, but this flora will produce essentially the same metabolic products of benefit to the animal. An abrupt change in diet, however, if it is to a diet high in easily-fermented carbohydrate, may lead to rapid changes in the balance or amount of fermentation products and this may have grave effects on the well-being of the animal.

Apart from influences such as these, the established rumen population is subject to short-term factors which bring about changes, of which the inflow to the rumen of feedstuffs is the greatest. In animals fed once or twice daily the rumen organisms are given a large amount of substrate over a short period. This may cause an initial fall in microbial activity due to the diluting effect of the feed and saliva and the passage of some organisms out of the rumen to the abomasum. This may then be followed by a burst of fermentative activity in which the end products may so alter the rumen conditions (such as by a fall in pH) that the microbial activity may again be diminished for a period until the normal conditions are re-established. The overall utilization of the feed may thus be not as efficient as if the same amount were added to the rumen in small portions over a long period, and practical experience seems to bear out the point that feeding 'little and often' gives better results than does the same amount of feed given once or twice daily. Other factors which will affect the rumen microbial population are the secretions of the host animal, of which the main one is the saliva which will cause effects by its volume, its buffering capacity and its ionic content and surface tension. The addition of urea to the rumen either by the saliva or through the rumen epithelium alters the substrates available to the bacteria. The rates of absorption from the rumen of microbial fermentation products and the removal of fermentation products and microbial cells by passage to the abomasum will also affect the microbial metabolism. The ruminant's feed will not only affect the microbial population by its gross composition but also by its content of trace elements and other minor constituents needed for the growth of the microorganisms. It will also affect the rumen population by its physical state which affects its rate of passage through the rumen and the length of time it is available for microbial action. The physical state of the feed may also alter the whole rate of flow of digesta through the rumen and, for instance, it has recently been suggested that a finely-ground diet may make the rumen turnover rate faster than the division rate of the larger protozoa and thus cause these organisms to 'wash out' of the rumen and the population to disappear (Christiansen, Woods and Burroughs, 1964).

The rates of growth of the microbes will be influenced by the availability of substrates. These may be introduced in a fermentable form, such as oligosaccharides, in the feed and may be immediately available, or they may need preliminary hydrolysis (e.g. cellulose) and thus their fermentation may be delayed. Or they may be the products (such as lactate) of some preliminary fermentation. Thus, Warner (1962) showed that different morphological types of organisms attained maximum numbers at different times after feeding, as might be expected if they were fermenting different substrates.

The rates of growth and enzyme production, and the fermentation products of the microorganisms may all be affected by the pH in the rumen and this in turn will be influenced by the amount and composition of the feed and will vary with time after feeding (Hobson, 1965).

The rumen contents, then, may be regarded as a type of continuous culture but one in which a true 'steady state' is never attained in practice and in which the microbial activity may be influenced by many factors.

MICROBIAL METABOLISM OF FEEDSTUFFS IN THE RUMEN

Introduction

As was mentioned at the beginning of this chapter, the metabolism of feedstuffs by microorganisms is not a unique feature of ruminant digestion and there is evidence that the processes of the microbial fermentation of plant feedstuffs are similar in all animals. The reactions occurring in the rumen have, however, been the subject of particular study partly because the position and size of the rumen makes it a relatively easy organ to sample, but mainly because of the importance of ruminants to the world economy.

Early workers suggested that the main nutritional significance of the rumen organisms lay in their ability to digest the cellulose of plant structures and thus make the remainder of the material, the proteins and oligosaccharides, more easily digestible in the abomasum and intestines. One of the first microscopic observations to be made on the rumen bacteria was that the cells surrounding and digesting plant fibres often contain an iodophilic polysaccharide. This led to the suggestion that the bacteria converted the indigestible cellulose to a starch-type polysaccharide which was digested by intestinal amylases as the cells passed down the gut. The volatile fatty acids known to be produced by microbial fermentation in the rumen were thought to be insignificant nutritionally. Later work by a number of people has shown that on most, if not all, diets the amount of microbial 'starch' passing to the intestines is nutritionally insignificant and the work begun by Elsden, Phillipson and their collaborators in the early 1940's and extended in many laboratories has shown the overriding importance of the fermentation acids as sources of energy and body constituents for the ruminant. On the other hand, the microbial cells have been shown to be the major source of digestible protein for the ruminant.

The rumen provides an anaerobic environment for microbial proliferation and the major reactions occurring there are fermentations in the strict sense of the word. Because the carbohydrates of the diet are fermented and the products absorbed and then (in some cases) resynthesized in the tissues to glucose which the host can metabolize, and because the dietary protein is broken down and resynthesized by the microbes before reaching its site of digestion in the intestines, the overall utilization of feed by the ruminant may be wasteful when compared with normal digestive processes. However, the microbial digestion allows the host to utilize carbohydrates not susceptible to normal digestive processes, and the processes of microbial protein synthesis allow the utilization as feed of non-protein nitrogen and the recycling of feed nitrogen in the animal body in times of stress. Microbial synthesis also produces vitamins and essential amino acids which may be lacking in the feed. So overall microbial digestion is of benefit to the ruminant under natural feeding régimes. The position of the rumen, before the rest of the digestive system, results in a much better utilization of microbial products than in the case of other herbivores, where the principal site of microbial digestion is the large intestine. Here coprophagy may be the only way in which some products of microbial synthesis can be utilized.

The overall picture of microbial digestion in the rumen is thus as follows. Carbohydrates are hydrolysed, if necessary, to mono- and di-saccharides and these are fermented to produce either directly, or indirectly, acetic, propionic and butyric acids. These are absorbed through the rumen epithelium and metabolized by the ruminant. Carbon dioxide and methane are also fermentation products, some of the former may be anabolized by the microbes, but most of it and all the methane are excreted by mouth and the carbon is lost to the animal. Proteins of the feed are largely broken down to peptides, amino acids, volatile fatty acids and ammonia, which can then be synthesized into microbial cells and later digested in the intestines. Or, in the case of ammonia absorbed from the rumen, converted to urea, and either excreted by the animal or recycled via the saliva to the rumen where ammonia is again formed. Lipids are hydrolysed and glycerol fermented and the fatty acids passed on to the intestines for eventual absorption. Unsaturated fatty acids are hydrogenated in the rumen before passing to the intestines. In addition the microbes form B vitamins. They form sulphur amino acids using inorganic sulphur and can utilize non-protein nitrogen, including inorganic nitrogen such as nitrate, for synthesis of protein.

These processes will be considered in detail in the next sections. However, it must be emphasized that what is being discussed here are the capabilities of rumen microbes. It is not suggested that these processes occur to the same extent in animals under all conditions. The predominating reactions will be governed by the feeding and management of the animal and there will be diurnal and other variations in the patterns of microbial metabolism.

The Types and Numbers of Microorganisms in the Rumen

Rumen contents of adult animals under normal conditions of feeding contain about 10^{10} bacteria per ml. This number can be determined fairly readily, but because of their larger, and very variable, size and consequent sampling problems the concentration of protozoa is less readily determined and quoted figures have varied widely. There may be up to something like 200,000 of the smaller species/ml, but perhaps not more than 4,000/ml of the larger species. These bacteria and protozoa are living in a symbiotic association both with themselves and their host, and experiments have suggested that in animals on the same diet a lack of protozoa is compensated for by an increase in bacterial concentration (e.g. Eadie and Hobson, 1962). The rumen protozoa are strict anaerobes and obtain energy by fermentation of carbohydrates, the fermentation products being the same as the bacteria, i.e. various mixtures of lower fatty acids, lactic acid, carbon dioxide and hydrogen. The bacteria mainly responsible for the rumen fermentations are also strict anaerobes and their isolation has been affected only in the last fifteen years or so. However, application of the techniques used in rumen bacteriology (Hungate, 1950) to other sites of microbial digestion has shown that similar bacteria may be present in large numbers in the intestines of many animals, including man. The bacteria usually found in the intestines of animals, coliforms, streptococci, lactobacilli and clostridia, are absent from, or present in relatively small numbers in, the adult rumen under conditions where the diet is not exceptionally high in carbohydrate. However, small numbers of facultative anaerobes may be responsible for the oxygen uptake which keeps the rumen anaerobic in spite of the air which the animal must swallow at times, although there is evidence that some anaerobic bacteria have a capacity for taking up oxygen. Yeasts can be isolated from rumen contents, but the numbers in which they occur are small and they are nutritionally of no significance to the animal. The rumen bacteria are thus adapted to growth at a temperature of 39°C in an atmosphere of carbon dioxide and methane in a well-reduced medium and at a pH normally about 6 to 6·5, and any great departure from these conditions will inhibit growth. The most important types of rumen bacteria have probably now been isolated and a description of most of them, many classified in new species or genera, is given in the review of Bryant (1959). It is not proposed to give a detailed list here of types of rumen bacteria, but some mention will be made of the properties of different bacteria when the reactions in the rumen are being discussed. However, there is still some confusion over the classification of rumen bacteria and there seem to be many strain differences, so that although a species may be mentioned in connection with a reaction it cannot be taken that all strains will have this particular property.

The Fermentation of Carbohydrates in the Rumen

The digestion of cellulose has always been considered to be the most important function of the rumen microorganisms, and in ruminants feeding on their natural diet it undoubtedly is. However, all rumen microorganisms known obtain energy through fermentation of carbohydrates or organic acids, and mixed organisms have a very wide range of fermentative powers, so that all carbohydrate which enters the rumen is subject to fermentation. Besides cellulose, forage plants contain other polysaccharides and oligosaccharides which are easily fermented, and most farm rations contain a proportion of starch-containing concentrates. Under the intensive-feeding conditions now coming into favour, the greater part of the ration carbohydrate may be in the form of starch. The extent of fermentation of a particular carbohydrate will depend on the ease with which it can be attacked by microbial enzymes and its rate of passage through the rumen. The latter may depend on the physical nature of the diet (ground, pelleted, chopped, etc.) and the way in which it is fed. The former will depend on the solubility of the carbohydrate, obviously, soluble polysaccharides and oligosaccharides will be readily fermented, but in plant materials the availability of even soluble sugars may depend on their position in the plant material and the degree to which they are 'protected' by cellulose and lignin. The ease of hydrolysis and solubilization of polysaccharides will also depend on the treatment of the feed. For instance, starch granules which have been crushed in the manufacture of concentrates will be more easily attacked than whole starch granules, and a finely-chopped hay will present a bigger surface area for cellulolysis than will a long-fibre hay.

The overall picture is, then, that polysaccharides are hydrolysed to mono- or di-saccharides by certain organisms. The sugars may be utilized by these organisms, or may become available to organisms lacking the initial hydrolytic power, and are finally fermented. The fermentation process is very rapid, and fermentable sugars are present only in very low concentration in the rumen liquid, so that the rate-governing factor of the overall breakdown of carbohydrates is probably the initial hydrolytic step.

A lot of data is now accumulating on the various pathways of, and enzymes involved in, rumen metabolism of carbohydrates, but these will not be discussed in this chapter, which is mainly concerned with the value of the fermentations to the economy of the ruminant.

Cellulose. The results of work on cellulases from different sources, and experiments such as those of Halliwell and Bryant (1963) using rumen bacteria, suggest that microorganisms such as the rumen bacteria may possess a number of β-glucosidases of different specificities and that a distinction might be made between organisms which can breakdown native cellulose fibres and those which can only hydrolyse poly- β-glucosides of a lesser degree of polymerization. However, mixed rumen organisms can degrade all forms of cellulose *in vivo* and some pure strains of bacteria have been shown to degrade native cotton fibres *in vitro*. In a sheep feeding on hay some 70 per cent of the cellulose decomposed in the gut was degraded in the rumen, the rest being degraded in the intestines (Gray, 1947). Cellulose is never completely utilized by the ruminant, however, and from about 20 to 70 per cent of that eaten may be found in the faeces.

A number of types of rumen bacteria have been shown to be cellulolytic, but cellulolysis is a character that seems to be strain-, rather than species-, specific. The predominant types of cellulolytic bacteria in the rumen may depend, amongst other things, on the cellulose-containing fibre consumed. For instance, Gouws and Kistner (1965) found that the predominant cellulolytic bacteria in the rumens of sheep on a diet of lucerne hay were strains of *Ruminococcus albus*, while the cellulolytic bacteria in sheep on a teff hay diet were mainly Gram-negative rods (*Butyrivibrio*).

Cellulose is hydrolysed and the hydrolysis products are fermented by mixed rumen bacteria mainly to acetic and propionic acids. However, the fermentation products produced by the initial fermentation of cellulose are more complex than this. The most important cellulolytic rumen bacteria are all strict anaerobes of the genera *Ruminococcus*, *Bacteroides*, and *Butyrivibrio*. The ruminococci have been divided into two species, one of which, *R. albus*, ferments cellulose principally to acetic and formic acids, ethanol and hydrogen (with, probably, CO_2). *R. flavifaciens* produces principally acetic, formic, succinic and lactic acids. The formic, lactic and succinic acids and ethanol and hydrogen are then the substrates for further bacterial action, as will be described later. *Butyrivibrio* ferment carbohydrates, including cellulose, mainly to butyric acid, with smaller amounts of formic acid, carbon dioxide and hydrogen.

The cellulolytic ruminococci and *Bacteroides succinogenes* (which produces acetic and succinic acids) are examples of the interdependence of rumen bacteria. The ruminococci require mixtures of straight and branched-chain lower fatty acids as growth factors, and the latter acids are produced by deamination of amino acids by other bacteria (see later). Both organisms are stimulated by, or have an absolute requirement for, B vitamins, including B_{12} (Scott and Dehority, 1965), and these vitamins are produced by other rumen bacteria.

Early microscopic observations (Baker and Harriss, 1947) showed that many bacteria were in close proximity to plant fibres in the rumen and were apparently digesting the cellulose to form pits in which the bacteria lay. Later cultural work has shown that cellulase is generally secreted by the bacteria into the surrounding medium and that cellulose is hydrolysed at some distance from the cell. The products of cellulolysis then diffuse into the medium and are available to the cellulolytic bacteria and any other nearby bacteria. In a crowded microbial environment such as the rumen much of the products of cellulolysis may be fermented by non-cellulolytic bacteria, and Hungate (1963) has suggested that lack of success in competing for the products of cellulose hydrolysis may be a factor in keeping the numbers of cellulolytic bacteria to a low level (1 – 5 per cent of the total) although up to one-third of the fermentable carbohydrate of the ration may be cellulose.

The larger ophryoscolecid rumen protozoa appear to be truly cellulolytic in that they can be seen to swallow and digest plant fibres. However, the cellulosic substrates which have been shown to be hydrolysed *in vitro* by whole cells or extracts of protozoa have all been chemically treated or ground cellulose fibres (Hungate, 1942, 1943; Abou Akkada, Eadie and Howard, 1963).

Pentosans. Pentosans are found in herbage in association with cellulose and their digestion usually parallels that of cellulose. Mixed rumen bacteria *in vitro* ferment pentosans to mainly acetic and propionic acids, but again, the pentosan-fermenting bacteria in pure culture produce a more varied selection of fermentation products. Some of the cellulolytic bacteria also ferment pentosans as do non-cellulolytic species, but again this appears to be a strain, rather than a species, property. The ability to ferment the products of pentosan hydrolysis such as xylose is more widely distributed and may occur without accompanying xylan-hydrolysing ability. However, many enzymes hydrolysing starch, cellulose and xylans produce mainly the constituent disaccharide and it is a common property of rumen bacteria that they will ferment the polymer and the disaccharide, but not the monosaccharide.

Some of the rumen protozoa (*Polyplastron multivesiculatum* and *Epidinium ecaudatum*) possess enzymes capable of hydrolysing pentosans, but it is not known to what extent they can ferment the products of hydrolysis (Abou Akkada *et al.*, 1963; Bailey, Clarke and Wright, 1962).

Starch. Starch is not a major polysaccharide constituent of herbages, so that under natural feeding conditions the rumen microorganisms have very little of this as substrate. However, many concentrates added to rations for domestic ruminants contain appreciable amounts of starch, and in the latest intensive feeding systems the principal ration carbohydrate is starch. However, because of the very different rumen conditions of conventionally and intensively-fed animals it is impossible to generalize about the microorganisms and products of starch fermentation in animals. The rumen pH is generally lower in the intensively-fed than the conventionally-fed animals and this may preclude the growth of some bacteria capable of fermenting starch in conventional ruminants, and protozoa are almost always missing from the rumens of intensively-fed animals. Certain protozoa may be the principal starch-fermenting organisms in animals on conventional diets.

Streptococcus bovis, a facultative anaerobe, can usually be cultured from rumen fluid and it was thought for some time that this was the most important starch-fermenting rumen bacterium. Later work has shown that, except under certain conditions, other starch-fermenting organisms are of more importance than *S. bovis*. *Bacteroides amylophilus* is a strict anaerobe which ferments only starch and maltose to give acetic, formic and succinic acids and it seems to be widely distributed (Hamlin and Hungate, 1956; Blackburn and Hobson, 1962). Recent studies in the author's laboratories on cattle intensively fed on barley suggest that *Bacteroides* spp. may be amongst the most important rumen bacteria. *S. bovis* seems to occur in low numbers only. *Butyrivibrio* strains often ferment starch, giving rise principally to butyric acid. Succinic acid is a major end product of starch fermentation by a number of important rumen bacteria such as *Bacteroides succinogenes*, *B. rumincola*, and *Succinimonas amylolytica*, but this is subjected to a secondary fermentation by other bacteria to give propionic acid. Some strains of *Selenomonas ruminantium* are starch-fermenting, and here propionic acid is a direct fermentation product.

A large number of species of rumen protozoa can ferment starch, but in all cases the starch is ingested in the granular form, so that the type of starch fermented depends to some extent on the size of the protozoon. For instance, the small holotrich *Dasytricha ruminantium* is unable to swallow any starch granules, while the larger *Isotricha* species can swallow small rice starch granules. Many of the ophryoscolecid protozoa can ingest and ferment large starch granules. The products of fermentation of starch by all types of protozoa are very similar. Acetic and butyric acids predominate and propionic acid is a minor product, but hydrogen and carbon dioxide are also produced. The protozoa also convert a large proportion of ingested starch into a reserve amylopectin-type polysaccharide which can later be fermented to the same products as feed starch. However, as mentioned before, in animals on diets containing mainly starch the low rumen pH precludes the growth of protozoa and starch digestion is carried out by the bacteria. Although very little starch passes undigested through the rumen in animals fed normal diets (Weller and Gray, 1954), it seems likely that in intensively-fed animals a much greater proportion of starch passes undigested to the abomasum and intestines and intestinal digestion may play a much greater role in the metabolism of these animals than in ruminants on more conventional diets.

Pectin. Pectin is readily fermented by rumen organisms *in vivo* (Pressey *et al.*, 1960) and *in vitro*, and suspensions of rumen bacteria *in vitro* ferment pectin to mainly acetic acid, with a little propionic acid. A number of strains of strictly anaerobic pectinolytic bacteria have been isolated from the rumen and found to be of two types, of which one, Gram-positive diplococci, could attack native pectin in potato slices, whilst the others, Gram-negative rods and cocci, could attack commercial pectin but not native pectin. It thus seems that a situation analogous to that amongst the 'cellulolytic' bacteria exists amongst the 'pectinolytic' bacteria. The primary products of pectin fermentation produced by pure cultures of rumen bacteria are more complex than the products formed by mixed rumen organisms, emphasizing again the role of secondary fermentations in the overall rumen metabolism. Some of the rumen protozoa can decompose pectic materials, but the holotrichs and *Ophryoscolex purkynei* are unable to ferment the products of pectin hydrolysis. It might be assumed that the ability to hydrolyse pectin would be of value to the microorganism itself, and to the rumen metabolism as a whole, because it would assist in the fragmentation of plant particles. Pectin esterase is found in rumen microorganisms and methanol is rapidly released from pectin in rumen contents and then converted to unknown products (Howard, 1961). It is possible that the methanol is fermented to methane.

Soluble sugars. The concentration of soluble sugars (in which group might be included fructosans) in herbage can rise to very high values at certain seasons. In grasses just before flowering soluble sugars can be as much as 25 per cent of the dry weight (Waite and Boyd, 1953). Soluble sugars may, then, contribute a large amount of the carbohydrate fermented in the rumens of animals at pasture, and rumen metabolism of these sugars is almost complete. Few of the known types of rumen bacteria are unable to ferment soluble sugars and the fermentation products are the same as those already mentioned for polysaccharides. The ophryoscolecid protozoa seem to be unique amongst carbohydrate-fermenting organisms in that they can ferment only particulate carbohydrates and not carbohydrates in solution. The holotrich protozoa are, however, active fermenters of soluble sugars and most of the invertase activity of the rumen, which causes the primary hydrolysis in the fermentation of fructosans, resides in the holotrich protozoa (Thomas, 1960). The greater part of the soluble sugar fermented by the holotrich protozoa is converted to lactic acid with smaller amounts of acetic and butyric acids, carbon dioxide and hydrogen (Heald and Oxford, 1953).

Secondary Reactions during Carbohydrate Fermentation in the Rumen

It was mentioned earlier that carbohydrate fermentation by rumen microorganisms produces initially a varied mixture of acids and gases, but only acetic, propionic and butyric acids, carbon dioxide and methane are found in any quantity in the rumen. The initial products of fermentation are subject to secondary fermentations which yield additional energy for microbial growth.

Lactate and succinate. These are primary fermentation products that are very rapidly metabolized under normal conditions. Lactate is found in the rumen only transitorily and lactic acid introduced into the rumen artificially, or in silage, also disappears rapidly. At one time the principal product of fermentation of lactate was thought to be propionic acid, but later experiments using ^{14}C-labelled lactic acid have shown that acetate is usually the main product (Jayasuriya and Hungate, 1959; Bruno and Moore, 1962; Baldwin, Wood and Emery, 1962). However, the relative proportions of acetate and propionate formed probably vary with the diet, and a diet rich in carbohydrate may lead to more propionate being formed (Baldwin *et al.*, 1962). The role of lactic acid in digestive disturbances is considered later.

Succinic acid is rapidly metabolized by rumen microorganisms with the formation of propionic acid, and recent experiments (Blackburn and Hungate, 1963) have suggested that succinate and not lactate may be the main precursor of ruminal propionic acid.

Formic acid and hydrogen. Formic acid is a primary product of carbohydrate fermentation by a number of rumen bacteria, but it is found in the rumen to the extent of only about 1 per cent of the total volatile fatty acids (Annison, 1954). Formic acid is rapidly decomposed in the rumen to hydrogen and carbon dioxide and these products then react further to produce methane. A bacterium (*Methanobacterium ruminantium*) able to convert formate, or hydrogen and carbon dioxide, to methane has been isolated from the rumen and shown to be present in numbers sufficient to account for the methane production in the rumen (Smith and Hungate, 1958). Hydrogen is also a primary fermentation product of carbohydrates, but it is found to compose only about 0·05 per cent of the rumen gas, and hydrogen added to rumen contents rapidly disappears (Hungate, 1963). The potential of the rumen contents to metabolize hydrogen is thus greater than its normal rate of production. As well as to the production of methane, hydrogen is diverted to other reducing reactions in the rumen, notably the hydrogenation of unsaturated fatty acids (see later).

The rumen gas usually contains about 30 per cent methane, and methane production can lose the ruminant about 10 per cent of the energy value of the feed consumed, but on the other hand the production of methane is a reaction which provides energy for microbial growth and thus provides protein of use to the ruminant. Little, if any, methane arises from the metabolism of volatile fatty acids, such as acetic, in the rumen, although it does so in other environments. Bacteria capable of such reactions seem to be present in the rumen in very low numbers (Opperman, Nelson and Brown, 1961), but their proliferation is normally suppressed for some reason, perhaps because of a low growth rate. The possibility of methane production from methanol derived from pectic substances has already been mentioned.

Volatile fatty acids. Although acetic, propionic and butyric acids are the main acids found in the rumen, these are not inert products in a microbiological sense. The formation of the higher volatile fatty acids butyric and valeric has been demonstrated to occur through condensation of acetate with itself or with propionate (Gray, *et al.*, 1952; Van Campden and Matrone, 1960), Carbon dioxide can also be incorporated into the carboxyl group of volatile fatty acids (Van Campden and Matrone, 1960; Otagaki *et al.*, 1955). In these experiments mixed rumen organisms were used, but studies on pure cultures have shown that *Butyrivibrio* takes up acetate from the medium, presumably in the formation of butyrate, and the lower volatile fatty acids can all act as growth factors, probably in all cases by incorporation into cell material, for a number of rumen bacteria.

Ethanol. Ethanol is produced by some rumen bacteria in pure culture fermentation of carbohydrate, but ethanol is found only in traces in the rumen and is decomposed slowly in rumen contents. It is possible that the ethanol is not produced *in vivo* as it is *in vitro* (Hungate, 1963). However, small concentrations of ethanol have been shown to be decomposed to acetic acid in rumen contents *in vitro* and the molar proportion of acetate in rumen volatile fatty acids was increased 4 hours after feeding ethanol (Chalupa, Evans and Stillions, 1964; Moomaw and Hungate, 1963). A bacterium capable of fermenting ethanol and acetic acid to give butyric acid has been isolated from rumen contents (Gray, 1958), but it is not known how widespread is this organism. However, under certain conditions ethanol production in the rumen may be increased. Acid culture conditions tend to favour ethanol formation *in vitro*, so might be expected to do so *in vivo*, and it is of interest that Allison *et al.* (1964) found ethanol in rumen contents of sheep fed large amounts of readily-fermented carbohydrate. The presence of ethanol was consistently associated with acid indigestion due to overfeeding, in these experiments.

General. A wide range of products can be formed from fermentation of carbohydrates by the many and varied types of rumen microorganisms. These products are then subjected to further metabolism to give three main acid end-products, acetic, propionic and butyric acids. However, the proportions of these end-products are by no means constant. They can vary with the type of diet and in one animal can be different at different times. These variations are possibly due to similar causes in many cases. For instance, a higher ratio of propionic to acetic acids is found as the rumen pH decreases in a particular animal at different times. A higher ratio of propionic to acetic acids is found in starch concentrate-fed animals than in roughage-fed animals, and feeding starchy concentrates usually leads to a comparatively low rumen pH. In a mixed population like the rumen, changes in fermentation products may be due to changes in the predominant bacteria metabolizing a particular substrate or to an 'inactivation' of a particular link in a chain composed of a number of bacterial types, but even in pure cultures changes in culture pH can lead to changes in fermentation products. The number of stages present in the rumen fermentation means that there is always the possibility that some stage may be disturbed and intermediary products may accumulate. This may have disastrous results to the whole process and cause digestive upsets in the animal, but this aspect will be dealt with later.

THE SYNTHESIS OF CARBOHYDRATES IN THE RUMEN

As was mentioned before, the formation of iodophilic materials in the rumen microorganisms was one of the first observations made on rumen contents, but later work has shown that this iodophilic material, although starch-like, is generally of little nutritive value to the ruminant. However, formation of starch-like reserve polysaccharides may have a part in the metabolism of individual organisms and in this way be of ultimate benefit to the host. The protozoa, most strikingly the holotrichs, convert part of the sugars or polysaccharides ingested and metabolized to storage amylopectin-type material which can be later metabolized when exogenous substrate is lacking. The products of this endogenous metabolism are the same as those of normal fermentations, and this storage mechanism could tend to even out the rumen fermentation after feeding. (This topic has been reviewed a number of times; see for instance, Howard, 1962.) The formation and subsequent utilization of a starch-type polysaccharide has been noted in mixed rumen bacteria, but has only been studied in one pure culture (Doetsch *et al.*, 1957). Starch formation in bacteria, though, may be indicative of an interruption in normal metabolic processes and in this case the starch may not be further utilized (Hobson and Mann, 1955).

Besides starch-type intracellular polysaccharides many rumen bacteria produce capsules or slimes which are probably in the main more complex polysaccharides. However, although production of these materials may seem large on a microbial scale when preparations of rumen contents are examined, except under exceptional circumstances (see later) they are comparatively small in quantity so far as the ruminant is concerned and it seems unlikely that even the small quantities produced could be digested lower down the digestive tract.

Lignin. Lignin is highly resistant to microbial attack, and it appears to be microbially decomposed by an oxidative process. Thus, the only known microbes able to decompose lignin are all aerobes, and one would not expect lignin decomposition to take place in the rumen. This, in fact, is the case and lignin in feeds is generally regarded as inert and suitable for use as a 'marker' compound in digestibility studies. Some workers, however, have reported losses of lignin in the gut, but this may be more a reflection of the experimental difficulties in determining this rather ill-defined substance than a true digestion. The main importance of lignin in the digestive processes in the rumen is that it slows down the digestion of plant constituents by the microorganisms. Highly lignified plant tissue is usually poorly digested in the rumen, and chemical delignification can make an otherwise indigestible plant material readily susceptible to microbial decomposition.

NITROGEN METABOLISM IN THE RUMEN

This topic has been reviewed a number of times recently (e.g. Hobson, 1963; McLaren, 1964; Blackburn, 1965) and the reader should consult these reviews for detailed references.

It might be argued that the breakdown of protein in the rumen is essentially a wasteful process, and protein could be more efficiently utilized if it were presented directly to the normal digestive enzymes. There is, in fact, some evidence that casein added to the duodenum is more efficiently utilized than casein added to the rumen (Cuthbertson and Chalmers, 1950). However, the rumen microbes need a nitrogen source for growth and production of cellulases, etc., and this must be supplied in the feed. The greatest value to the ruminant of its rumen organisms, considered solely from the point of view of nitrogen metabolism, is the synthetic ability of the organisms. The ability of the organisms to synthesize amino acids means that amino-acid deficiencies in the feed may be made good and a large proportion of the feed nitrogen may be provided as non-protein material such as urea. Further, the ability to use urea for protein formation gives the ruminant an ability to 'recycle' nitrogen, and this may be a definite advantage in times of low feed supply.

Feed protein entering the rumen is hydrolysed. The speed of this hydrolysis depends on the availability of the protein. The action of microbial enzymes in breaking down the structural polysaccharides of plants may help in exposing the protein constituents to microbial attack, and so the availability of protein in herbage may depend on the age and degree of lignification of the plant. In concentrate meals the processing of the meal changes the availability of the protein; e.g. heat treatment may render the protein hard and difficult to hydrolyse.

The proteolytic activity of mixed rumen microorganisms has been known for some time, but it is only recently that the rumen bacteria responsible for proteolysis have been isolated and identified. Blackburn and Hobson (1962), Fulghum and Moore (1963) and Giesecke (1962), came to the same conclusions in investigations of cattle and sheep. These were, that proteolytic activity is a strain property of many types of anaerobic rumen bacteria and that proteolytic bacteria are present in similar numbers in animals on a wide variety of diets. Amongst the proteolytic rumen bacteria are strains of *Bacteroides amylophilus*, *Selenomonas ruminantium*, *Bacteroides ruminicola*, *Lachnospira multiparus*, *Butyrivibrio* spp., *Borrelia* spp., and *Succinivibrio* spp. Two rumen protozoa have been shown to be proteolytic, *Entodinium caudatum* (Abou Akkada and Howard, 1962) and *Ophryoscolex purkynei* (Mah and Hungate, 1965), but other types no doubt have this property as well. The majority of the proteolytic bacteria examined by Abou Akkada and Blackburn (1963) possessed both endo- and exo-peptidases and produced hydrolysates with a large amount of amino acids and small peptides.

However, free amino acids and peptides are present in the rumen contents in extremely small amounts and these primary products of protein hydrolysis are very rapidly metabolized. There is evidence that amino acids can be absorbed by the rumen epithelium (Cook, Brown and Davis, 1965), but it seems likely that this is not nutritionally significant and the greater part of the amino acids are metabolized by the rumen contents. Peptides have been found to be growth stimulants for some rumen bacteria and recent work has shown that *Bacteroides ruminicola* can absorb and utilize only peptides and not free amino acids (Pittman, Bryant and Lakshmanan, 1965), and some rumen bacteria have a requirement for certain amino acids or a mixture of amino acids, but many of the most important rumen bacteria can satisfy their nitrogen requirements with ammonia, and ammonia is the final product of protein catabolism in the rumen. Amino acids are rapidly deaminated by mixed rumen organisms to give volatile fatty acids and ammonia. Studies with pure cultures (Bladen, Bryant and Doetsch, 1961; Abou Akkada and Blackburn, 1963) have shown that deaminative ability is possessed by certain strains of different species of rumen bacteria, amongst which *B. ruminicola* may be most important, but deaminative activity is not as widely distributed as is proteolytic activity. Ammonia production seems to be correlated with active growth of the bacteria, which agrees with the increase in ammonia production in the rumen after feeding when the increase in bacterial numbers is most rapid. Deamidation, a property of some strains of rumen bacteria (Abou Akkada and Blackburn, 1963), would also release ammonia from amides naturally occurring in feedstuffs (Warner, 1964.) Urease activity is also found in mixed rumen organisms and this will produce ammonia from urea in the saliva, but a high urease activity has not been demonstrated in the most predominant rumen bacteria. It may be that more actively ureolytic bacteria are still to be isolated, or that the urease activity found *in vitro* is not a true reflection of the activity of the organisms *in vivo*. Ammonia is also an end product of endogenous metabolism of the protozoa.

Ammonia is absorbed by the rumen epithelium and converted to urea in the animal body. It may then be excreted in the urine, and a rapid ammonia production in the rumen often causes a net loss of nitrogen to the animal by this pathway, or the urea may be recycled via the saliva or diffusion back into the rumen and again become available for microbial action. However, if a suitable source of carbohydrate as energy-producer for the bacteria is available the greater part of the ammonia is used for synthesis of bacterial protein. Experiments have shown that up to 80 or 90 per cent of feedstuff proteins, whether naturally-occurring in plants, or added as, say, casein to the feed, is converted to microbial protein in the rumen (McDonald and Hall, 1957; Weller, Gray and Pilgrim, 1958), and this microbial protein is of good digestibility and has an amino-acid content similar to that of plant protein (Weller, 1957; Abdo, King and Engel, 1964). There is, however, some evidence that when urea is the sole source of nitrogen for the bacteria, a condition never met with in practice, that histidine is not formed, but the rumen organisms seem to be able to synthesize all other amino acids (Duncan *et al.*, 1953; Land and Virtanen, 1959). The ability of rumen bacteria to utilize ammonia for protein synthesis has been turned to practical use in that a proportion of the protein of ruminant diets may be replaced by non-protein nitrogen in the form of urea, ammonia or amides and diets low in protein (such as poor pasture) may be supplemented by the addition of urea. (Many papers and reviews on this topic have been published and significant references may be found in the reviews quoted at the beginning of this chapter. A recent paper on urea metabolism is that by Chalupa *et al.*, 1964.)

Experiments with pure and mixed cultures of rumen bacteria have shown the activities responsible for this aspect of rumen metabolism.

The deamination of amino acids produces ammonia, carbon dioxide and volatile fatty acids, including branched-chain butyric and valeric acids. These latter acids are not produced by carbohydrate fermentation, although the straight-chain isomers can be. Mixtures of straight- and branched-chain fatty acids have been known for some time to be growth factors for a number of rumen bacteria and to stimulate cellulolysis by mixed rumen bacteria. More recent work has shown that branched-chain acids serve as carbon skeletons for amino-acid synthesis by some bacteria. Work with ^{14}C-labelled fatty acids has shown this incorporation into cell protein. For instance, ^{14}C isovaleric acid is the precursor of leucine in *Ruminococcus flavifaciens* protein (Allison, Bryant and Doetsch, 1962). The carboxyl group of leucine and other synthesized amino acids in rumen bacteria comes from carbon dioxide (Allison and Bryant, 1963; Wright, 1960; Huhtanen, Carleton and Roberts, 1954). A large number of rumen bacteria have been shown to need carbon dioxide for growth. More recent studies have shown that carbon from volatile fatty acids (acetic to valeric) and glucose (so presumably other carbohydrates) are used for protein synthesis by mixed rumen bacteria (Hoover, Kesler and Flipse, 1963).

Sulphur is needed for the formation of sulphur-containing amino acids and this sulphur can be supplied to the rumen bacteria in the form of inorganic sulphate, sulphide or elemental sulphur. Sulphate is reduced by the rumen microorganisms to sulphide and this is utilized by the bacteria. Sulphide is also produced from the sulphur amino acids of feed proteins by the rumen bacteria and this sulphide can then be incorporated again into amino acids. It would seem from the evidence of sulphide utilization that the protozoa have little, if any, ability to form amino acids and must be dependent on ingested bacteria or preformed feed amino acids for a source of protein (Müller and Krampitz, 1955). Studies on sulphur utilization by rumen microorganisms have also provided evidence that the protein of rumen bacteria is utilized by the ruminant to form body proteins such as wool and milk (Hale and Garrigus, 1953; Block, Stekol and Loosli, 1951).

Non-protein nitrogen other than ammonia can be utilized by rumen bacteria through formation of ammonia. Purines are fermented and ammonia is formed, nitrate is reduced to ammonia and then utilized and the possibility exists that atmospheric nitrogen swallowed by the ruminant could be 'fixed' to form ammonia in the rumen.

However, besides forming protein, the rumen bacteria and protozoa must form nucleic acids and nucleic acid synthesis represents a source of loss of nitrogen to the host animal. Ellis and Pfander (1965) have recently reported experiments which suggest that at least 5 to 13 per cent of the ruminants' dietary nitrogen can be converted to microbial polynucleotide nitrogen and this is largely lost to the animal. Purine nitrogen excretion in ruminants is considerably greater than in non-ruminants, and this excretion decreases more on starving the ruminant than the non-ruminant, indicating that this purine nitrogen is probably of microbial origin (Morris and Ray, 1939).

LIPID METABOLISM IN THE RUMEN

A detailed list of references to lipid metabolism will be found in the chapter by Garton (Chapter 10) so only a brief summary will be given here.

Lipids form a part of the diet of ruminants on all rations and these lipids undergo various changes in the rumen. Glyceryl esters, or the galactolipids of herbage, are hydrolysed by rumen organisms, as are lecithin and lysolicethin. Glycerol and galactose formed in these reactions are fermented by rumen bacteria, the former largely to propionic acid. The fatty acids formed pass from the rumen to the intestines, where they can be absorbed and converted to depot fats of the animal, but the unsaturated fatty acids are first hydrogenated to more saturated ones. The hydrogenation in the rumen of unsaturated fatty acids is the cause of the predominantly saturated form of ruminant depot fats. The bacteria are mainly responsible for glyceride hydrolysis in the rumen and an anaerobic lipolytic bacterium has been isolated from sheep on a number of diets. Although hydrogenation of unsaturated acids can be demonstrated with suspensions of mixed rumen bacteria, the actual bacteria responsible for this action have not yet been defined, although there is some evidence that *Butyrivibrio* may play a part. Some types of protozoa have been shown, in mixed suspension, to hydrogenate unsaturated lipids.

The lipids leaving the rumen also contain the lipids of microbial cells. Experiments have shown that straight- and branched-chain volatile fatty acids can be incorporated into bacterial lipids, as well as into protein. Microbial lipid, however formed, can form a considerable amount of the lipid material finally incorporated into ruminant tissue or milk fats.

VITAMIN FORMATION AND DESTRUCTION IN THE RUMEN

The need of various rumen bacteria for vitamins has been mentioned before, as has the fact that B vitamins are synthesized in the rumen. The synthesis of vitamins in the rumen before the sites of vitamin absorption in the intestines also means that the ruminant is almost immune to the various vitamin-deficiency diseases of monogastric animals. Biotin, folic acid, nicotinic acid, pantothenic acid, pyridoxine, riboflavin, thiamine and vitamin B_{12} are all synthesized by microbial action in the rumen (Kon and Porter, 1956), as is vitamin K (McElroy and Goss, 1940). However, the young ruminant, whose rumen does not contain an active adult flora, is as dependent on extraneous vitamins as the non-ruminant. Vitamin B_{12} contains cobalt, so that synthesis of this vitamin depends on an adequate supply of cobalt to the rumen. A deficiency of cobalt in the diet leads to a vitamin B_{12} deficiency. In a recent survey (Dryden *et al.*, 1962) *Selenomonas ruminantium* and *Peptostreptococcus elsdenii* were shown to produce vitamin B_{12}, and factors A and B, in pure culture and these may be responsible for the B_{12} production *in vivo* as *S. ruminantium* is found in most rumen contents in high members. *Peptostreptococcus elsdenii* occurs in most rumen contents, but except in young animals or those on high-grain diets, its numbers are low. Some bacteria which synthesize other B vitamins and vitamin K in pure culture have been isolated from the rumen, but whether they are the main, or only, producers of these vitamins *in vivo* is not known.

Loss of vitamins synthesized in the rumen can take place by uptake into bacterial cells (as growth factors) or by other types of microbial action, but the rate of synthesis is greater than the rate of destruction. The requirements of rumen protozoa for vitamins are unknown, but they may be responsible for some vitamin loss in the rumen.

Vitamin C in the feed is destroyed in the rumen and there is evidence that extensive vitamin A destruction can occur there (King, 1963; Klatte *et al.*, 1963).

MISCELLANEOUS REACTIONS IN THE RUMEN

Ruminant feeds can contain a large number of minor substances of varied composition. Some of these are naturally occurring, others may be residues of herbicides or pesticides and so on. The very varied flora of the rumen can change these substances in many ways that are beneficial to the ruminant. Activities of this kind include the destruction of oxalic acid and hepatotoxic alkaloids (Morris and Garcia-Rivera, 1955; Dick *et al.*, 1963), the inactivation of digitalis (Westermarck, 1959) and the fermentation of plant acids such as citric and malic to volatile fatty acids (Packett and Fordham, 1965). The rumen population may also adapt to substances which are normally antimicrobial. For instance, white cedar oils inhibited fermentation by the rumen microorganisms of a steer on an alfalfa corn-concentrate diet, but did not inhibit the microorganisms from a deer adapted to a diet of cedar leaf sprays (Short, 1963). There has been much discussion recently about the dangers to farm animals of insecticide and pesticide residues and their products used for human consumption, and also to wild animals. It would appear that the rumen microorganisms can destroy certain insecticides and thus they decrease the possibility of these substances appearing in the ruminant's tissues. Robbins and Kastelic (1961) found that ruminants could tolerate considerable amounts of tetramethylthiuram disulphide (TMTD)-treated corn without toxic effects, due to the action of the rumen microorganisms in degrading TMTD. Williams *et al.* (1963) showed that some rumen protozoa degraded Thiodan and Diazinon.

DIGESTIVE DISTURBANCES CAUSED BY THE RUMEN MICROORGANISMS

The ability of the rumen microorganisms to destroy potentially toxic substances in the ruminant's feed has been mentioned above, but there also exists the possibility that the rumen organisms may produce toxic substances. However, the term 'toxic' must be interpreted rather widely as most of the digestive disturbances of ruminants which can be attributed to the rumen microorganisms are caused by changes in magnitude of normal fermentation patterns rather than by the introduction of completely new metabolic pathways. Some examples of these disturbances are given below.

Glycosides are normally attacked by rumen organisms and the degradation of these can liberate hydrogen cyanide. Small amounts of cyanide are converted to thiocyanate, which is non-toxic, but if the feed contained large amounts of certain glycosides it is possible that toxic amounts of cyanide could be liberated. Similarly, nitrate contained in feedstuffs is reduced by the rumen organisms to nitrite and then ammonia, but if the nitrate concentration in the diet is high then the reaction nitrate/nitrite takes place faster than nitrite/ammonia, and toxic amounts of nitrite can accumulate. Nitrite appears to be the toxic principle in, for instance, oat–hay poisoning and in the digestive disturbances caused by fresh mangolds and certain South African plants. Sulphate is reduced to sulphide in the rumen and this sulphide is used by the bacteria to form sulphur amino acids, as mentioned earlier. If the feed were to contain large amounts of sulphate then hydrogen sulphide production could become greater than assimilation and toxic amounts of hydrogen sulphide could accumulate. Hydrogen sulphide has been suggested as a cause of inhibition of rumen motility in bloat. Decarboxylation of amino acids can take place in rumen contents and the rate of such reactions increases under acid conditions. Production of toxic amounts of histamine by decarboxylation of histidine under the acid rumen conditions caused by high-carbohydrate feeding may play a part in production of bloat, and also in some degenerative changes in the tissues observed in animals on high-carbohydrate diets.

Ammonia production is a normal metabolic process in the rumen and ammonia produced is either converted by the bacteria to amino acids or is absorbed into the blood stream and later converted to urea which is excreted or in part recycled through the secretions into the digestive tract. However, if an easily digested protein or urea is fed in large amounts then large quantities of ammonia can be produced in the rumen and the animal can become ill, or die, from ammonia poisoning. This type of reaction would most likely occur when the diet is rapidly changed to one high in ammonia precursors. A large ammonia production can be avoided to some extent by feeding carbohydrate which acts as an energy source for rumen bacterial protein synthesis, or by limiting the intake of the ammonia precursors. This latter is done in various types of urea supplements by combining the urea with salts and other materials into a hard block which the animal can assimilate only slowly by licking. A sudden feeding of ruminants, either intentionally or accidentally, with large amounts of easily-fermented carbohydrate can cause an acidosis which can prove fatal. This type of acidosis seems to be caused by an excess production of lactic acid in the rumen, and the microbial changes have been studied by a number of workers. Krogh (1963) has compared the acidosis produced by experimental feeding of large amounts of carbohydrate with that produced in cattle which had accidentally eaten large quantities of carbohydrate-rich feed. These results show that the rumen microorganisms can utilize quite large amounts of carbohydrate without appreciably changing. However, increasing the carbohydrate beyond a certain level results in a low rumen pH due to the excess fermentation and the rumen flora rapidly changes. A more aciduric flora composed initially of streptococci then becomes dominant. If carbohydrate ingestion is continued the streptococci are replaced by lactobacilli and the rumen pH falls to very low values, 3·8–4·5. The rumen contains large numbers of yeasts and cells from the epithelium at this time and the animal is seriously ill. The streptococci and lactobacilli produce lactic acid in much greater quantities than is usual in the rumen fermentation. The breakdown of the normal rumen flora occurs very quickly, over a day or even less, once the process has been initiated. However, if the change in rumen conditions has not proceeded too far, the flora returns to normal conditions and the animal recovers in a few days after cessation of carbohydrate feeding and the introduction of a diet higher in roughage.

It would seem that when easily-fermented carbohydrate (starch or sugars) is fed to cattle or sheep then if it is introduced into the feed gradually then the balance of the rumen population, or fermentation rates, change so as to metabolize the carbohydrate and its fermentation products. However, although the rumen population is a mixed one in animals accustomed to high-starch diets it contains more lactobacilli than are usually found in animals on roughage diets and the pH often fluctuates to a greater extent and reaches lower values (*ca.* 5) than in a normally fed animal. The low pH of the rumen usually precludes the growth of protozoa. The population is then an unstable one in that any stress such as slight overfeeding may cause the pH to fall even lower and this will encourage the growth of more aciduric streptococci and lactobacilli which produce more lactic acid, leading to a lower pH and the acidosis described above. Ethanol production has also been associated with acute indigestion caused by over-feeding of carbohydrate to sheep and may be another cause of toxic symptoms (Allison *et al.*, 1964).

Perhaps the most important of the digestive disturbances on a world-wide scale in ruminants is bloat. A voluminous literature exists on the subject of bloat and many different theories of its cause and treatment have been suggested. It seems likely that there are a number of causes of bloat, but in all cases the animal is unable to release the gas produced by rumen fermentations and so a pressure develops in the rumen and this organ is distended. Pressure from the distended rumen may then impair the functions of other organs, with results which can be fatal to the animal. The physiological reasons for cessation of the normal eructation mechanisms are outside the scope of this Chapter, but some of the microbiological processes causing the production of gas will be briefly considered here.

Bloat has been divided into two main types, one, where the gas is in free form above the rumen contents but is not eructated, and the other where the gas is entrapped in a mass of foam filling the rumen, the 'frothy' bloat. The gas is produced by metabolism of carbohydrate either directly or by the neutralization of fermentation acids by the bicarbonate of the saliva. The carbohydrate can be starch-type, as in 'feed-lot' bloat, or soluble sugars as in 'pasture' bloat. Feed-lot bloat is slower in developing than pasture bloat and seems to be connected with changes in the rumen microbial population. Gutierrez *et al.* (1959) found that *Peptostreptococcus elsdenii* and *Streptococcus bovis* increased in numbers in the rumen on the onset of feed-lot bloat in cattle. They suggested that capsular polysaccharides from *S. bovis* and long chains of *P. elsdenii* together with fermentation gases produced a stable foam in the rumen. This may happen in some cases, but Bryant, Robinson and Lindahl (1961) could not correlate any changes in the rumen flora with the onset of feed-lot bloat, and this indicates the difficulties of ascribing the signs of bloat to any one cause. Bacterial slimes have been implicated as foam stabilizers in bloat by a number of workers. Apart from the work of Gutierrez *et al.* mentioned above, a correlation was found by Jacobson *et al.* (1957) between bloat and the number of capsulated bacteria in the rumen, and strains of *Butyrivibrio* and *Bacteroides* from the rumen produced slimes when grown on lucerne saponins, and bloat is found commonly in animals feeding on lucerne (Gutierrez, Davis and Lindahl, 1959). However, the composition of the slime isolated from feed-lot-bloating animals by Gutierrez *et al.* (1961) suggested that it contained autolysed cell debris as well as capsular polysaccharides. Plant materials from the feed, such as pectic materials, saponins and proteins, have also been implicated as foam stabilizers in bloat. Salivary mucins have also been suggested as foam stabilizers and foam destroyers and in the latter case bloating was ascribed to the destruction of antifoaming mucin by mucinolytic bacteria (Fina *et al.*, 1961). The foam in bloat can be destroyed by addition to the rumen of such substances as oils, tallow, medicinal paraffin, commercial detergents and silicone antifoams, either in drinking water or sprayed on the pasture (e.g. Reid, Johns and Vlieg, 1961). The lipids of chloroplasts also have antifoaming properties and destruction of these lipids by rumen micro-organisms has been suggested as a cause of pasture bloat (Mangan, Johns and Bailey, 1959).

It is still not clear whether in the slower developing feed-lot bloat there are changes in the microbial system which produce a greater fermentation rate and so more gas, or whether changes in the microbial population produce substances which stabilize a foam trapping the gas which can normally be eructated in a free form. However, it would seem that in pasture bloat which is sudden in its onset, the main factor producing the bloat is the increase in fermentation produced by the increased soluble sugar intake of animals grazing lush spring pastures. The production of a stable foam may be a secondary consequence of increased gas production. Antibiotics have had a limited success as bloat-preventing agents. Even if they do prevent bloat, their action is of limited duration, a few hours or days, for if the antibiotic dose is not repeated the antibiotic is rapidly destroyed in the rumen or lost by passage of digesta from the rumen. If the dose is repeated then the bacteria rapidly become resistant to the antibiotics so far tried and the antibiotic has no further effect.

REFERENCES

ABDO, K. M., KING, K. W. and ENGEL, R. W. (1964) *J. Anim. Sci.* **23**, 734.
ABOU AKKADA, A. R. and BLACKBURN, T. H. (1963) *J. Gen. Microbiol.* **31**, 461.
ABOU AKKADA, A. R., EADIE, J. M. and HOWARD, B. H. (1963) *Biochem. J.* **89**, 268.
ABOU AKKADA, A. R. and HOWARD, B. H. (1962) *Biochem. J.* **82**, 313.
ALLISON, M. J. and BRYANT, M. P. (1963) *Arch. Biochem. Biophys.* **101**, 269.
ALLISON, M. J., BRYANT, M. P. and DOETSCH, R. N. (1962) *J. Bact.* **83**, 523.
ALLISON, M. J., DOUGHERTY, R. W., BUCKLIN, J.A. and SNYDER, E. E. (1964) *Science* **144**, 54.
ANNISON, E. F. (1954) *Biochem. J.* **58**, 670.
BAILEY, R. W., CLARKE, R. T. J. and WRIGHT, D. E. (1962) *Biochem. J.* **83**, 517.
BAKER, F. and HARRISS, S. T. (1947) *Nut. Abs. and Revs.* **17**, 3.
BALDWIN, R. L., WOOD, W. A. and EMERY, R. S. (1962) *J. Bact.* **83**, 907.
BLACKBURN, T. H. (1965) In *Physiology of Digestion in the Ruminant*. Butterworths, London.
BLACKBURN, T. H. and HOBSON, P. N. (1962) *J. Gen. Microbiol.* **29**, 69.
BLACKBURN, T. H. and HUNGATE, R. E. (1963) *Appl. Microbiol.* **11**, 132.
BLADEN, H. A., BRYANT, M. P. and DOETSCH, R. N. (1961) *Appl. Microbiol.* **9**, 175.
BLOCK, R. J., STEKOL, J. A. and LOOSLI, J. K. (1951) *Arch. Biochem. Biophys.* **33**, 353.
BRUNO, C. F. and MOORE, W. E. C. (1962) *J. Dairy Sci.* **45**, 109.
BRYANT, M. P. (1959) *Bact. Rev.* **23**, 125.
BRYANT, M. P., ROBINSON, I. M. and LINDAHL, I. L. (1961) *Appl. Microbiol.* **9**, 511.
BRYANT, M. P., SMALL, N., BOUMA, C. and ROBINSON, I. M. (1958) *J. Dairy Sci.* **41**, 1747.
CHALUPA, W., EVANS, J. L. and STILLIONS, M. C. (1964a) *J. Anim. Sci.* **23**, 802.
CHALUPA, W., EVANS, J. L. and STILLIONS, M. C. (1964b) *J. Nut.* **77**, 1964.
CHRISTIANSEN, W. C., WOODS, W. and BURROUGHS, W. (1964) *J. Anim. Sci.* **23**, 984.
COOK, R. M., BROWN, R. E. and DAVIS, C. L. (1965) *J. Dairy Sci.* **48**, 475.
CUTHBERTSON, D. P. and CHALMERS, M. I. (1950) *Biochem. J.* **46**, 17.
CUTHBERTSON, D. P. and HOBSON, P. N. (1961) *World Rev. Nutr. Diet.* **2**, 67.
DICK, A. T., DONN, A. T., BULL, L. B. and CULVENOR, C. C. J. (1963) *Nature, Lond.* **197**, 207.
DOETSCH, R. N., HOWARD, B. H., MANN, S. O. and OXFORD, A. E. (1957) *J. Gen. Microbiol.* **16**, 156.
DUNCAN, C. W., AGRAWALA, I. P., HUFFMAN, C. F. and LUECKE, R. W. (1953) *J. Nut.* **49**, 41.
DRYDEN, L. V., HARTMAN, A. M., BRYANT, M. P., ROBINSON, I. M. and MOORE, L. A. (1962) *Nature, Lond.* **195**, 201.
EADIE, J. M. (1962a) *J. Gen. Microbiol.* **29**, 565.
EADIE, J. M. (1962b) *J. Gen. Microbiol.* **29**, 579.
EADIE, J. M. and HOBSON, P. N. (1962) *Nature, Lond.* **193**, 503.
ELLIS, W. C. and PFANDER, W. H. (1965) *Nature, Lond.* **205**, 974.

FINA, L. R., HAY, C. A., BARTLEY, E. E. and MISHRA, B. (1961) *J. Anim. Sci.* **20**, 654.
FULGHUM, R. S. and MOORE, W. E. C. (1963) *J. Bact.* **85**, 808.
GIESECKE, D. (1962) *Zbl. Bakt.* **186**, 170
GOUWS, L. and KISTNER, A. (1965) *J. Agric. Sci.* **64**, 51.
GRAY, F. V. (1958) *J. Bact.* **76**, 535.
GRAY, F. V. (1947) *J. Exp. Biol.* **24**, 15.
GRAY, F. V., PILGRIM, A. F., RODDA, H. J. and WELLER, R. A. (1952) *J. Exp. Biol.* **29**, 57.
GUTIERREZ, J., DAVIS, R. E., LINDAHL, I. L. and WARWICK, E. J. (1959) *Appl. Microbiol.* **7**, 16.
GUTIERREZ, J., DAVIS, R. E. and LINDAHL, I. L. (1959) *Appl. Microbiol.* **7**, 304.
GUTIERREZ, J., DAVIS, R. E. and LINDAHL, I. L. (1961) *Appl. Microbiol.* **9**, 209.
HALE, W. H. and GARRIGUS, V. S. (1953) *J. Anim. Sci.* **12**, 492.
HALLIWELL, G. and BRYANT, M. P. (1963) *J. Gen. Microbiol.* **32**, 441.
HAMLIN, L. J. and HUNGATE, R. E. (1956) *J. Bact.* **72**, 548.
HEALD, P. J. and OXFORD, A. E. (1953) *Biochem. J.* **53**, 506.
HOBSON, P. N. (1960) In *Digestive Physiology and Nutrition of the Ruminant*, p. 107. Butterworths, London.
HOBSON, P. N. (1961) In *Rowett Institute Collected Papers*. Rowett Institute, Aberdeen.
HOBSON, P. N. (1963) *J. Dairy Res.* **30**, 288.
HOBSON, P. N. (1965) *J. Gen. Microbiol.* **38**, 167.
HOBSON, P. N. and MANN, S. O. (1955) *J. Gen. Microbiol.* **13**, 420.
HOBSON, P. N. and MANN, S. O. (1961) *J. Gen. Microbiol.* **24**, i.
HOOVER, W. H., KESLER, E. M. and FLIPSE, R. J. (1963) *J. Dairy Sci.* **46**, 733.
HOWARD, B. H. (1961) *Proc. Nut. Soc.* **20**, xxix.
HOWARD, B. H. (1962) *Proc. N.Z. Soc. Anim. Prod.* **22**, 109.
HUHTANEN, C. N., CARLETON, T. S. and ROBERTS, H. R. (1954) *J. Bact.* **68**, 749.
HUNGATE, R. E. (1942) *Biol. Bull. Wood's Hole* **83**, 303.
HUNGATE, R. E. (1943) *Biol. Bull. Wood's Hole* **84**, 157.
HUNGATE, R. E. (1950) *Bact. Rev.* **14**, 1.
HUNGATE, R. E. (1963) In *Symbiotic Associations*, p. 266. 13th Symp. Soc. Gen. Microbiol., University Press, Cambridge.
HUNGATE, R. E., BRYANT, M. P. and MAH, R. (1964) In *Annual Review of Microbiology*. Annual Reviews Inc., Palo Alto, U.S.A.
JACOBSON, D. R., LINDAHL, I. L., MCNEILL, J. J., SHAW, J. C., DOETSCH, R. N. and DAVIS, R. E. (1957) *J. Anim. Sci.* **16**, 515.
JAYASURYA, G. C. N. and HUNGATE, R. E. (1959) *Arch. Biochem. Biophys.* **82**, 274.
KAY, R. N. B. (1963) *J. Dairy Res.* **30**, 261.
KING, T. B. (1963) *Dissert. Abstracts* **24**, 454.
KLATTE, F. J., HUBER, T. L., LITTLE, C. O. and MITCHELL, G. E. (1963) *J. Anim. Sci.* **22**, 239.
KON, S. K. and PORTER, J. W. G. (1954) *Vit. & Horm.* **12**, 53.
KROGH, N. (1963) *Acta Vet. Scand.* **4**, 27.
LAND, H. and VIRTANEN, A. I. (1959) *Acta Chem. Scand.* **13**, 489.
MCDONALD, I. W. and HALL, R. J. (1957) *Biochem. J.* **67**, 400.
MCELROY, L. W. and GOSS, H. (1940) *J. Nut.* **20**, 527.
MACKAY, E. S. M. and OXFORD, A. E. (1954) *J. Gen. Microbiol.* **11**, 472.
MCLAREN, G. A. (1964) *J. Anim. Sci.* **23**, 577.
MAH, R. A. and HUNGATE, R. E. (1965) *J. Protozool.* **12**, 131.
MANGAN, J. L., JOHNS, A. T. and BAILEY, R. W. (1959) *N.Z. J. Agric. Res.* **2**, 342.
MANN, S. O. and OXFORD, A. E. (1955) *J. Gen. Microbiol.* **12**, 140.
MOOMAW, C. R. and HUNGATE, R. E. (1963) *J. Bact.* **85**, 721.
MORRIS, S. and RAY, S. C. (1939) *Biochem. J.* **33**, 1217.
MORRIS, M. P. and GARCIA-RIVERA, J. (1955) *J. Dairy Sci.* **38**, 1169.
MÜLLER, R. and KRAMPITZ, G. (1955) *Z. Tierg. Zuchtbiol.* **65**, 187.
OPPERMAN, R. A., NELSON, W. O. and BROWN, K. E. (1957) *J. Dairy Sci.* **40**, 779.
OTAGAKI, K. K., BLACK, A. L., GOSS, H. and KLEIBER, M. (1955) *J. Agric. Food Chem.* **3**, 948.
PACKETT, L. V. and FORDHAM, J. R. (1965) *J. Anim. Sci.* **24**, 488.

PITTMAN, K. A., BRYANT, M. P. and LAKSHMANAN, S. (1965) *Bact. Proc.* p. 75.
PRESSEY, R., SYNHORST, S. H., ALLEN, R. S. and JACOBSON, N. L. (1960) *J. Anim. Sci.* **19,** 1311.
REID, C. S. W., JOHNS, A. T. and VLIEG, P. (1961) *N.Z. J. Agric. Res.* **4,** 476.
ROBBINS, R. C. and KASTELIC, J. (1961) *J. Agr. Food Chem.* **9**, 256.
SCOTT, H. W. and DEHORITY, B. A. (1965) *J. Bact.* **89**, 1169.
SHORT, H. C. (1963) *J. Wildlife Management* **27,** 184.
SMITH, P. H. and HUNGATE, R. E. (1958) *J. Bact.* **75**, 713.
THOMAS, G. (1960) *J. Agric. Sci.* **54**, 360.
VAN CAMPDEN, D. R. and MATRONE, G. (1960) *J. Nut.* **72**, 277.
WAITE, R. and BOYD, J. (1953) *J. Sci. Fd. Agric.* **4**, 197.
WARNER, A. C. I. (1962) *J. Gen. Microbiol.* **28**, 129.
WARNER, A. C. I. (1964) *Aust. J. Biol. Sci.* **17**, 170.
WELLER, R. A. (1957) *Aust. J. Biol. Sci.* **10**, 384.
WELLER, R. A. and GRAY, F. V. (1954) *J. Exp. Biol.* **31**, 40.
WELLER, R. A., GRAY, F. V. and PILGRIM, A. F. (1958) *Brit. J. Nut.* **12**, 421.
WESTERMARCK, H. (1959) *Acta Vet. Scand.* **1**, 67.
WILLIAMS, P. P., ROBBINS, J. D., GUTIERREZ, J. and DAVIS, R. E. (1963) *Appl. Microbiol.* **11**, 517.
WRIGHT, D. E. (1960) *J. Gen. Microbiol.* **22**, 713.

CHAPTER 4

Physiology of Digestion, Absorption and Metabolism in the Ruminant

I. W. McDONALD

Division of Animal Physiology,
Ian Clunies Ross Animal Research Laboratory, C.S.I.R.O., Parramatta,
New South Wales, Australia

INTRODUCTION

THE most characteristic feature of digestion in the ruminant is the modification of structure and function enabling the utilization of cellulose. Throughout the animal kingdom (with the exception of some protozoa and herbivorous molluscs) there is a conspicuous lack of capacity to degrade cellulose by enzymic secretions comparable with other carbohydrases, proteinases, lipases, etc. This is an extraordinary feature of animal evolution, as many species of animals in many phyla are herbivorous or omnivorous in their feeding habits. In those species capable of using cellulose as a significant contributor of energy, it is usual to find some modification of the digestive tract designed to harbour symbiotic microorganisms that can degrade cellulose to products useful in the animal's economy. In this chapter, special attention will be given to the modifications found in ruminants and to their consequences in respect to the digestion of other nutrients. Much of the ruminant digestive tract is strictly analogous in structure to that of the monogastric* animals and in consequence has attracted little attention from the research scientist: it has been taken for granted that function is also analogous to the corresponding mechanisms in the animals most studied—especially man, dog and rat. Mention will be made of the few features that are known to be different in the ruminant.

In brief, the special features of ruminant digestion are associated with the proventriculus** (the so-called forestomach). Here a milieu is provided for a remarkably dense population of bacteria and protozoa which ferment carbohydrates and produce volatile fatty acids as end-products that can serve

* Non-ruminant animals are often referred to as 'monogastric' animals, though anatomically this usage is quite erroneous. The ruminant also has only one stomach. During embryonic development, the cardiac region of the stomach becomes modified and greatly enlarged to form the three compartments of the 'forestomach'.

** For anatomical terminology, the recommendations of Habel (1965) have been followed.

as energy sources for the host animal. In respect to cellulose and related substances of plant cell walls, this is a gain to the host; the same fermentative processes occur with starches and sugars that could be digested and absorbed by the small intestine, and this represents a net loss to the animal's nutritional economy because the end-products of fermentation are less valuable than the sugars. The transactions in respect to nitrogenous compounds are more complex and, depending on the dietary circumstances, the microbial activities may provide a gain or a loss to the protein economy of the animal. In the diets of ruminants, lipids occur in lower concentrations than in the usual foods of carnivorous and omnivorous animals; there are, however, important aspects of lipid digestion to be considered.

Of outstanding nutritional importance is the influence of microbial activity on the ruminant's vitamin requirement. The synthetic abilities of the microbes provide the host with a full complement of the B-vitamins.

Some of the special features of digestion, and their metabolic consequences, are dealt with in other chapters in this volume. Particular attention is drawn to microbiology of digestion (Chapter 3), morphology of the absorptive surfaces of the alimentary tract (Chapter 9), lipid metabolism (Chapter 10), mineral nutrition (Chapter 11), and metabolic disorders (Chapter 33).

In this chapter, a brief account is given of our present knowledge of the main features of ruminant digestion. The subject is highly complex and is attracting a great deal of attention in research laboratories in many countries. This work is proving very fruitful, not only in general biology, but also in many practical applications to animal husbandry and veterinary medicine.

Most of the experimental work on ruminant digestion has been done with sheep, cattle and goats. The basic features show no important species difference and so it may be assumed that a similar pattern will occur in the numerous other species of ruminants, domestic and feral. Very little is known of the closely related species in the genus Camelidae, which is usually not included in the order Ruminantia. Recently, it has been found that analogous gastric digestion occurs in some species of the Marsupialia, and similar mechanisms occur in other classes of mammals (Moir, 1964); these are interesting examples of convergent evolution and deserve more intensive investigation. For simplicity of presentation it has been assumed that observations on one ruminant species are applicable to ruminants generally, though it is realized that some specific differences have been recorded, and no doubt others will be found as a wider range of species is studied.

Anatomical Modifications of the Digestive Tract

Brief mention of the anatomical form of the ruminant stomach is essential to provide a basis for the description of digestive function. (See p. 136, Fig. 1.)

The ruminants are strictly herbivorous and have evolved a unique dental apparatus for prehension of the plant fodder. The maxillary incisors are lacking and are replaced by a tough pad on which the mandibular incisors, with sharp edges, can bite. A distinct interdental space separates the incisors from the molars which are well adapted for the grinding necessary to comminute coarse plant fragments. The mouth is supplied with many salivary glands, which are capable of high secretion rates. The copious saliva assists the animal to eat the dry, coarse fodder that is probably the portion of most ruminants, in the natural state, during some part of every year. However, even when consuming lush pasture, salivary secretion is copious (Wilson, 1963). Both serous and mucous glands are present; the parotid glands are serous and secrete continuously, though the rate of secretion is enhanced during chewing.

During grazing, the animal collects food quite rapidly, chews only sufficiently to prepare a suitable bolus, and then swallows. The oesophagus is notable in having striated muscle throughout, although nervous control is mediated via the vagus nerve as usual in mammals. The oesophagus is guarded by three powerful sphincters—the cranial oesophageal sphincter at the oral end, and the diaphragmatic and cardiac sphincters at the gastric end (Dougherty and Habel, 1955). The cardia is located where the oesophagus enters the stomach at the dorsal junction of the rumen and reticulum. These sphincters play critical roles during the functions of rumination and eructation as well as during deglutition.

The stomach has evolved into a complex of four compartments. The so-called true stomach, abomasum, is analogous to the conventional mammalian stomach and consists of the usual fundic and pyloric regions, with typical secretory glands producing digestive enzymes and hydrochloric acid. The fundus has the mucosa thrown into numerous soft folds which greatly increase the surface area of secretory tissue, and which may influence the passage of solids through the viscus and facilitate the mixing of digesta with gastric juice. The cardiac region of the primitive stomach has become modified into three compartments, the rumen, reticulum and omasum. The rumen and reticulum form a single functional unit, though the form of the two compartments and the appearance of the mucous membranes make anatomical differentiation quite easy; these two compartments communicate by a large orifice which permits digesta to be easily transferred from one to the other.

From the cardia a strongly muscled groove, the reticular groove (commonly, and erroneously (Habel, 1965), called the oesophageal groove), passes down the medial wall of the reticulum to terminate at the reticulo-omasal orifice. In the young ruminant, the lips of the reticular groove close reflexly during sucking, and the swallowed milk passes directly from the oesophagus, down the groove, through the reticulo-omasal orifice, along the base of the omasum, and thus directly into the abomasum (Watson, 1941). In the adult, the reticular groove is still functional, but does not play any essential role. A curious species difference may be noted here: in adult sheep, the groove closes in response to stimulation of oral mucosa by solutions of copper salts (Clunies Ross, 1934), while in cattle the groove responds to solutions of sodium bicarbonate but only erratically to copper (Riek, 1954).

Both the rumen and reticulum are lined with a tough stratified squamous epithelium which is formed into characteristic patterns of papillae and ridges. The papillae, which are large and very numerous, require for their development the stimulation provided by the volatile fatty acids (especially propionic and butyric acids) produced by ruminal fermentation (Warner and Flatt, 1965). The weight and capacity of the ruminoreticulum are fully developed only when the animal consumes an adequate amount of roughage in its diet. The mucous membranes* have no secreting glands—not even mucous glands.

The reticulum is a simple sac joined dorsally to the rumen, has a long free ventral part, and lies against the diaphragm. The ventral extremity is close to the pericardium, with only the thin diaphragm separating the two organs; this juxtaposition occasionally has disastrous consequences, since sharp metallic objects, not uncommonly ingested accidentally, tend to lodge in the reticulum and may penetrate the viscus and diaphragm to produce a fatal pericarditis.

The rumen is the largest of the four compartments and has a complex sacular structure. Most of the ingested food is held in the rumen until sufficiently fragmented to pass with the fluid digesta into the reticulum, from which passage to omasum or return to the rumen can occur. The rumen and reticulum function as a single unit; hence the term 'rumen' will often be used to refer to the ruminoreticulum.

The omasum is the most highly modified of the compartments and is found only in the Ruminantia; an analogous structure occurs in the Camelidae (Hansen and Schmidt-Nielsen, 1957; Vallenas, 1965). The mucous membrane again has a stratified squamous epithelium and is devoid of glands; it is thrown into a series of thin, papillated leaves, resembling a book, which almost fill the organ. At the base of the omasum, a firm smooth groove passes from the reticulo-omasal orifice to the omaso-abomasal orifice. These two orifices are effectively controlled by circular muscle fibres so that digesta move only in one direction, although some reverse movement is possible. So definite is this pathway that vomition, so readily achieved by most non-ruminant animals, cannot proceed from the abomasum; it is remarkable too, that in spite of the efficiency of regurgitation from the ruminoreticulum during rumination, noxious material cannot be vomited from these compartments. It has been reported that vomition can be induced by the mixed alkaloids of *Veratrum* (Mullenax *et al.*, 1966), and that sheep with advanced adrenal insufficiency

* While it appears a contradiction in terms to refer to a mucous membrane with no mucous glands, this tissue, and the related mucosa of the cardiac region in the equine stomach, are strictly analogous, embryologically, to all the lining membranes of the alimentary tract.

will readily vomit rumen contents when subjected to the stimulus of gastric intubation (Goding, personal communication).

The small intestine of ruminants and the associated secretory glands are analogous to those of non-ruminants. The large intestine in general resembles that of the omnivores, having a rather large caecum and a long but simple colon, which is coiled into a characteristic double spiral. The complex structures found in the Equidae have no parallel in the Ruminantia.

THE DIGESTIVE SECRETIONS

The copious saliva secreted by ruminants seems to be essentially devoid of digestive enzymes (Somers, 1951; Otterby *et al.*, 1964), though an esterase is probably secreted by oral glands of calves (Wise *et al.*, 1940); the function of saliva in the digestive process is discussed in a later section (a detailed review is given by Kay, 1966). There is no secretion, in the accepted sense, from the rumen, reticulum or omasum, though water and solutes can pass into the digesta from the blood.

The secretion of gastric juice by the abomasum resembles that of the stomach in non-ruminants (Hill, 1960). There are, however, very important differences in detail, probably influenced principally by three factors that will be discussed in more detail later: (a) the continuous flow of digesta into the abomasum, in contrast with the periodic filling in monogastric animals; (b) the fluid nature of digesta arriving from the omasum; and (c) the high buffering capacity of the digesta.

The glandular structure of the abomasal mucosa is similar to that of monogastric animals, and the secretion of hydrochloric acid and pepsin presents no unique features—indeed, much of the research on the properties of pepsin has been done with ruminant preparations. However, the gastric juice also contains rennin, which seems to occur only in the secretion of young ruminants (Henschel *et al.*, 1961). Rennin is secreted as a zymogen, prorennin, and is activated very rapidly at pH 2–4; it is a proteolytic enzyme and produces clotting of milk (Rand and Ernstrom, 1964).

Ruminant pancreatic juice is known to contain trypsinogen, chymotrypsinogen, procarboxypeptidase, carboxypeptidase, amylase and lipase (Keller *et al.*, 1958; Taylor, 1962). As secreted, the proteolytic activity of pancreatic juice is negligible and the proenzymes need to be activated by the enterokinase secreted by the small intestine. The enzyme components differ in no important way from those in monogastric animals; this is in conformity with the finding that the function of the small intestine seems to be conventional. Pancreatic secretion is continuous, shows no direct response to feeding, and is controlled by vagal stimulation, the passage of acid digesta into the duodenum, and the secretin mechanism, in a manner similar to that of non-ruminants (Hill, 1961; Taylor, 1962). The volume and bicarbonate concentration tend to be lower in ruminants (Magee, 1961).

Bile is also continuously secreted, in conformity with the continuous flow of digesta from the stomach to the small intestine (Harrison, 1962). No digestive enzymes occur in the bile; the principal digestive function is expressed by the lipids and bile salts as in non-ruminants.

Little study has been made of the secretions of the small intestine, but the usual enzymes appear to be present. The large intestine contributes no digestive secretions.

THE ENVIRONMENT OF THE RUMEN

The biological success of the ruminant, as an evolutionary development, seems to be due to the physiological regulation of the ruminal environment as a microbial habitat. This has enabled the evolution of a characteristic microbial population of quite remarkable density. The special features of the bacteria and protozoa inhabiting this environment are discussed in Chapter 3. Here it is necessary to consider only the mechanisms by which the ruminant controls the ruminal milieu.

In Nature, the ruminants obtain food by grazing, a slow process occupying several hours a day. Thus the substrate for microbial metabolism is provided gradually, in contrast with the rapid consumption of a meal by a carnivore. Under domestication, ruminants are often given concentrated rations, usually based on grains and oil-seed meals, so that rapid eating can occur, but unless the animals are trained to eat rapidly they spread their intake over some hours every day.

Salivary Secretion

The grazed plants are partially chewed and mixed with a large volume of saliva which provides the principal fluid medium for microbial support. The volume of saliva secreted is so great that the animal does not need to consume water while eating even the driest of food over several hours; water is consumed as an entirely separate function, and this makes it easy for ruminants, in Nature, to travel long distances between feed and water when necessary.

Although the parotid and sublingual glands secrete continuously, the acts of feeding and rumination stimulate a great increase in the rate of secretion. The submaxillary glands function only during feeding. Dry food stimulates secretion more than succulent food, and the glands on the chewing side of the mouth secrete faster than their opposite members (Coats *et al.*, 1956; Somers, 1957; Kay, 1958).

The volume of saliva secreted per day is chiefly dependent on the type of ration fed (Kay, 1966). The average daily secretion tends to be consistent, but the rate of secretion over short intervals can vary greatly with changing stimuli. The large volume of saliva maintains the fluid state in the rumen so that the dry matter content of the ruminal digesta is usually about 10 per cent; this facilitates mixing of the digesta, provides suitable conditions for regurgitation and rumination, and gives a suspension for transport of digesta to the omasum. The fluid conditions also favour the movements of microorganisms within the digesta.

The volume of fluid taken into the rumen each day has a critical influence on the microbial population in the rumen. Saliva contributes the major part of this volume; the quantity of water drunk is very much influenced by the water content of feed (varying from 10–90 per cent of the feed), as well as the ambient temperatures. It is probable that relatively little water is absorbed from the rumen, so that outflow of fluid from the rumen is approximately equal to inflow (Stacy and Warner, 1966). It will be evident that the ratio of fluid outflow to rumen volume will determine the number of generations per day that each species of rumen microbe will have to achieve in order to maintain an approximately uniform concentration in the rumen contents. The rate of salivary secretion is therefore a primary factor in determining growth-rates of the rumen microorganisms (Warner, 1966).

A feature in the chemical composition of ruminant saliva is the high concentration of sodium, phosphate and bicarbonate. The phosphate is present chiefly in inorganic form and has a concentration that may exceed 10 times that in the blood (Watson, 1933; Bailey and Balch, 1961); the functional significance of the high phosphate concentration is not clear, though it has been suggested that this is favourable to bacterial growth (Watson, 1933), as well as aiding the buffering capacity of the rumen fluid (Turner and Hodgetts, 1955). As the natural diets of ruminants are commonly very low in phosphorus, this mechanism may be very important to ruminal bacteria. The very large amounts of phosphate secreted daily in the saliva are efficiently absorbed lower in the alimentary tract, so that the ruminant can maintain its phosphorus status on very low intakes. (The phosphorus deficiency diseases will be discussed in Chapter 11.)

The bicarbonate of saliva is usually about decinormal in concentration with sodium as the principal base. The pH is about 8. Although Na is the principal base, this is not physiologically obligatory. The diets of ruminants are usually rather low in Na and high in K; the Na status is very efficiently maintained, but under extreme conditions of Na deprivation or Na loss, K can replace Na in the saliva to a high degree (Denton, 1956). Under ordinary dietary conditions, the salivary concentrations of Na and K vary inversely.

Buffering of Rumen Contents

The major products of microbial fermentation in the rumen are the volatile fatty acids (VFA): acetic, propionic and butyric acids. In an unbuffered solution, the average concentration of these acids, as found in the rumen, would give pH values of about 2·5–3·0; in practice, the rumen pH usually lies in the range of pH 6–7 (Turner and Hodgetts, 1955b). Much of the VFA is neutralized by the $NaHCO_3$ of saliva with the liberation of CO_2; additional CO_2 is derived from microbial metabolism, so that a high pCO_2 is maintained in the rumen. The dissolved CO_2 contributes in a major degree to maintaining the pH of the rumen; for example, mere exposure of ruminal fluid to air can lead to a marked rise in pH (Turner and Hodgetts, 1955a). A high concentration of dissolved CO_2 in the rumen contents is maintained by equilibration with the ruminal gas that is always present in the dorsal sac of the rumen; the concentration of CO_2 in rumen gas varies with the time after feeding and averages about 50 per cent—thus the pCO_2 of rumen fluid is some 10 times the level in arterial blood. The CO_2 content of rumen gas ranges from about 20–80 per cent; this range can account for much of the change in rumen pH found with ordinary feeding regimes, i.e., about pH 5·8–7·0.

In normal circumstances, the buffering of the VFA formed in fermentation is principally by phosphate-bicarbonate (note that the pK_2 of phosphoric acid is about 6·85 and the pK_1 of bicarbonate is about 6·1). The rumen fluid is rather poorly buffered against alkali, so that a deleterious rise in pH can occur when NH_3 is formed from dietary urea in the rumen (Turner and Hodgetts, 1955b).

At a normal pH (say pH 6·5), the VFA are almost completely neutralized by salivary base and, as mentioned previously, the high concentration of CO_2 has a major role in determining pH. With a rapid fermentation, for example, of starch or sugar, VFA accumulate and the pH may fall to about 5·0; bicarbonate ceases to have any effect as a buffer at about pH 5·8, and then the buffering action is mainly dependent on phosphate.

The only stronger acid normally occurring in rumen fermentation is lactic acid (pK_a 3·7; Lockwood *et al.* 1965); at ordinary concentrations, this is effectively buffered by the VFA which occur in relatively higher concentration, with pK about 4·8. Only in abnormal fermentations (see Chapter 33) does the lactic acid accumulate to high concentration and thus produce very low pH values.

The rumen pH is also controlled by the absorption of the VFA (to be discussed later). Although only a small fraction of VFA occurs in the rumen fluid in the unionized form (less than 0·5 per cent at pH 7), this form accounts for about half the total VFA absorbed (Ash and Dobson, 1963). It is probable that a supply of hydrogen ions from the conversion of dissolved CO_2 to bicarbonate allows the transformation of fatty acid anion to the unionized form which is then absorbed.

The Osmolality of the Rumen Contents

The osmolality of the medium is probably of considerable importance to the growth of the microbes, particularly the protozoa (Hungate, 1942). Until recently, it had been considered that the rumen fluid and blood tend to become isosmotic, but it is now known that the tendency is for the rumen fluid to be hypotonic (Warner and Stacy, 1965). During feeding, osmolality rises and greatly exceeds plasma values which, in the sheep, fall chiefly in the range 290–310 mosm/kg. After feeding, the osmolality steadily declines to become hypotonic even if water is not drunk; probably hypotonic conditions would prevail for part of each day, even with grazing animals. The rise in osmolality during feeding is due to the solutes of the fodder consumed; by contrast, drinking of water may be sufficiently copious to produce a marked hypotonic state which, however, is steadily changed by the addition of saliva and by water absorption from the rumen. On fasting, the rumen contents remain consistently hypotonic. In general, then, it appears that in spite of the great changes due to feeding and drinking, the rumen environment tends to return to within fairly narrow limits of osmolality.

The major components contributing to the osmotic pressure are the salts of sodium and potassium (Munch-Petersen, 1964). There is an inverse relation between the concentrations of these cations, and there are definite differences in concentrations exhibited by individual animals on a given diet; the physiological mechanisms involved in the control of these concentrations are not yet clear. The post-prandial rise in rumen osmotic pressure does not always provide a stimulus to drinking.

Removal of metabolites

The microbial density in the rumen is exceptionally high for a natural environment. Clearly, this density and the growth required to maintain it, could not be sustained unless the metabolic products of the microbes were removed at a sufficient rate. Three main mechanisms provide for the removal: fluid flow to the omasum, absorption from the ruminoreticulum, and mutual use of metabolites by the various species of organisms (see Chapter 3). The volume of flow of fluid passing through the rumen per day is subject to a great deal of variation; on restricted diets, values range from 1–4 times the average volume of the rumen contents, while the values for grazing animals may exceed 8 times the rumen volume (Hydén, 1961; Tulloh *et al.*, 1965). The outflow to the omasum is probably much more regular and consistent than inflow via saliva, drinking water and diffusion of water from the blood (Balch *et al.*, 1951). Absorption of metabolites is a major feature and will be discussed in more detail later; quantitatively, the metabolites of greatest importance are the VFA and ammonia.

The Gas Phase

Under normal conditions, there is always a pocket of gas in the dorsal sac of the rumen. As gas accumulates, the excess is removed by eructation. This phenomenon is characteristic of ruminant digestion and is vital to the survival of the animal; the mechanism is described in detail in Chapter 33 where consideration is given to the disease 'bloat', in which abnormal quantities of gas accumulate in the rumen. The principal gas is CO_2, which is derived from salivary bicarbonate when neutralized by the VFA, and is produced during fermentation.

Methane is the second most abundant component of the rumen gas mixture, but appears to be a stable metabolite and plays no apparent role in controlling the environmental conditions that influence microbial activity. Nitrogen in the rumen gas is derived from air swallowed with food and saliva.

Some oxygen enters the rumen with food and saliva, but it is unlikely that any oxygen diffuses from the blood into the rumen. By analogy with gaseous exchanges through the human skin (Fitzgerald, 1957) and into tissue slices *in vitro* (Dixon, 1951), it is most probable that all oxygen diffusing from blood capillaries into the ruminal mucosa is utilized in tissue oxidation. The considerable numbers of aerobic and facultative anaerobic bacteria always present in the rumen remove the oxygen during ordinary feeding conditions, and the uptake of oxygen by anaerobic organisms (Dobson and Bullen, 1964) reduces oxygen concentration to extremely low levels. None of the aerobic bacteria seem to be obligate residents of the rumen (or 'true' rumen organisms) but they are continually added to rumen contents from the external environment; also, none reaches the very high concentration achieved by the principal ruminal bacteria, nearly all of which are strictly anaerobic. The rumen gas contains appreciable amounts of oxygen after fasting.

Hydrogen is produced by some organisms and is always present in rumen gas (McArthur and Miltimore, 1961), but usually does not contribute appreciably except on re-alimentation after a prolonged fast (Pilgrim, 1948); hydrogen produced during the fermentation processes is taken up by other reactions, especially the formation of methane. Traces of hydrogen sulphide are also normally present (McArthur and Miltimore, 1961).

Reducing Conditions in the Rumen

The chemical reactions that induce the highly reducing conditions in the rumen are not well understood. It is conventional to refer to the 'redox potential' in the rumen (Broberg, 1957), but this is perhaps misleading as no equilibrium state is reached; as pointed out by Dobson and Bullen (1964), electron transfer reactions can be very slow and a mixture of redox systems, such as occurs in the rumen, does not possess a unique Eh until equilibrium between them has been reached.

Inorganic Constituents and Growth Factors

The rumen fluid provides a suitable basal inorganic medium for microbial growth (Munch-Petersen, 1964). The various cations and anions are regulated at rather narrow ranges of concentrations by salivary secretion, food intake, absorption from the rumen, and diffusion from blood to the rumen. The organic nutrients for microbial growth are wholly supplied by the food. Some growth factors required by certain microbes are provided by the metabolism of other microbial species; the biota as a whole are independent of an external supply of B-vitamins. (The special role of dietary cobalt in the bacterial synthesis of vitamin B_{12} is described in Chapter 11).

Temperature

The temperature in the rumen tends to be a little higher (*ca.* 1°C) than rectal temperature (Krzywanek, 1929). Feeding leads to a slight rise in rumen temperature. The response to drinking is, of course, dependent on the temperature of the water drunk, though the temperature of the water, from near freezing to a little above body temperature, has no effect on daily water intake (Cunningham *et al.*, 1964). Very cold drinking water affects particularly the ventral rumen, where the temperature may fall by 10°C or more; the temperature equilibrates throughout the rumen contents in about 1 hour, and returns to normal in about 2 hours.

When cattle take all their drinking water at very low temperatures, there is no fall in the digestibility of fodder; this suggests that temporary fluctuations in rumen temperature have no deleterious effect on microbial activity.

Mixing of Rumen Contents

A most important aspect for control of the rumen environment is the mixing of the contents though homogeneity is never achieved. This is a rather formidable mechanical task as the rumen contents commonly comprise up to 20 per cent of the animal's body weight. Mixing is achieved by the strong musculature of the rumen wall. The circular and longitudinal muscle fibres of the primitive gut become greatly modified in disposition in the adult ruminoreticulum to provide pillars and folds that divide the organ into several sacs, as well as thick muscle sheets that cover the entire viscus. The mobility of the rumen and its neural control have been intensively studied (see Titchen and Reid, 1965); only a brief outline can be given here. The contractions of the ruminoreticulum are entirely reflex and are governed by medullary centres with transmission principally through the vagus nerve. The whole gastric complex has a highly integrated pattern of movements on which are superimposed the special mechanical functions of regurgitation and eructation.

Mixing of the digesta is achieved in a churning manner by waves of contractions that pass systematically over the wall of the ruminoreticulum. When a bolus of feed is swallowed, it tends to pass over the ruminoreticular fold into the dorsal sac of the rumen. Although the feedstuffs have a higher specific gravity than the rumen fluid, they tend to float in the dorsal sac by virtue of entrapped gas bubbles, especially the bubbles formed during fermentation; in cattle there is usually a pronounced stratification with coarse forage particles forming a mat at the surface of the digesta; sheep chew their food more thoroughly and the smaller particles do not so readily form a mat. As the feed particles become progressively smaller during digestion, they tend to sink. Thus the digesta in the ventral sac of the rumen are more fluid than in the dorsal sac.

The fluid part of the digesta can move more freely than the solids. Contraction of the ventral sac with simultaneous relaxation of the reticulum enables the more fluid digesta of the ventral sac to be moved forward to the reticulum. Then follows a characteristic double contraction of the reticulum which forces the liquid digesta backward to the dorsal sac, so that the liquid can percolate down through the solid digesta to the ventral sac; thus a continuous fluid cycle is set up (McAnally and Phillipson, 1944).

By these means, newly consumed feedstuffs are mixed with partially digested food from previous meals, and the solids are constantly being washed by the fluids carrying the microbial population. The whole scheme provides a remarkably efficient process that enables ready access of microbes to their substrate and facilitates the operation of the several mechanisms that control the rumen environment.

Only the smallest particles can leave the ruminoreticulum and thus this viscus, in contrast with the stomach of monogastric animals, is never emptied —not even after a prolonged fast. Some particles of a given meal will start leaving the rumen soon after its consumption, while large fragments of more resistant food will remain in the rumen for days. At any given time, the rumen will contain the equivalent of about one day's feed intake. This system is admirably adapted to the microbial digestion of cellulose, a process that is always very much slower than the fermentation of the other digestible components of the feed.

Rumination

The process of rumination is an integral part of the mechanical functioning of the system. Under normal dietary conditions, the animal may spend as much as 8 hours per day ruminating; the actual time spent is a reflection of the coarseness of the ration consumed. When foodstuffs are soft and easily disintegrated, or when the whole ration is finely ground, rumination time is very greatly reduced or even abolished (Gordon, 1958; Freer and Campling, 1965). The mechanical aspects of rumination are governed by a medullary centre (Bell and Lawn, 1955) and are highly stereotyped for a given individual animal; the process can, however, be interrupted quite easily by external stimuli.

The function of rumination is to reduce the size of coarse particles of feed; during eating, the animal chews its food only sufficiently to form a bolus satisfactory for swallowing. With coarse fibrous rations, rumination plays a major role in reducing food particle size and thus becomes an important factor in determining the rate of passage of food, the extent of digestion, and the animal's feed intake (Pearce and Moir, 1964). As noted above, the feed from successive meals is mixed; hence, the actual particles returned to the mouth during rumination are simply those that happen to be in the region of the cardia at the moment of regurgitation—there is no selection of the most recently consumed food. A subsidiary function of rumination is the reflex stimulation of salivary secretion (Denton, 1957) with its sundry effects on the ruminal environment.

Another consequence of rumination must be noted. Most of the true rumen microbes seem to be obligate residents of the rumen; this is almost certainly true of the protozoa and probably true of many bacterial species—they occur nowhere else in Nature. None of the protozoa has been found to form cysts, and few of the bacteria have resistant forms (e.g. spores) that can pass through the intestinal tract and be voided in the faeces (Bryant, 1959). Thus, a young ruminant can easily be raised in an isolated environment and kept free from infection with rumen protozoa (Becker, 1929). The process of rumination ensures that the ruminant's mouth and saliva are continuously inoculated with rumen microbes, and a ready means exists for easy transmission of microbes from animal to animal by aerosol formation, contamination of feed and water, grooming or contact; these mechanisms will be especially important in the transmission of microbes from mother to offspring.

The mechanism of rumination is very complex but, in brief, the process is as follows (Bergman and Dukes, 1926; Stevens and Sellers, 1960). After a regurgitated bolus has been chewed, the glottis closes and swallowing occurs; the glottis remains closed till the act of regurgitation is complete. An extra contraction of the reticulum, aided by contraction of the diaphragm and abdominal wall, floods the region of the cardia with digesta. The sphincters of the cardia relax. An inspiratory effort markedly reduces intrapleural pressure, because the glottis is closed. The thoracic oesophagus is thus dilated and digesta pour in through the cardia, which then contracts. An expiratory effort now raises the intrapleural pressure and aids the swift anti-peristaltic contraction of the oesophagus in forcing the digesta into the mouth. Immediately, the more liquid portion is swallowed, leaving the coarse fragments in the mouth. Now the glottis opens, chewing commences and breathing can be resumed. This sequence takes about 3 seconds and is followed by chewing movements, the number of which is remarkably regular for the individual animal. When the chewing sequence is complete, the food and saliva are returned to the rumen, and not passed on to the omasum via the reticular groove; thus any given fragment of food might be chewed several times.

All these mechanisms, then, serve the dual function of mechanical fragmentation of food particles and the provision of an environment in which microbes can flourish and digest fodder for the host. Mechanical fragmentation of food particles greatly increases the surface area available for microbial attack, which in turn tends to fragment the particles still further. The anatomical arrangement and the control of the rumen environment provide for a continuous culture of a highly complex group of protozoal and bacterial species.

MICROBIAL DIGESTION IN THE RUMEN

In deference to convention, the processes in the rumino reticulum are described as 'digestion', but it must be realized that the microbial digestion bears little relation to the animal's own digestive processes operated by enzymes secreted into the gut. These enzymes simply reduce complex substances to simpler molecular forms that can be absorbed from the gut and metabolized; there is no loss of substance or of nutrient value in the process. By contrast, the rumen microbes are actively seeking nutrients for their own metabolic needs and rapidly denude the rumen fluid of soluble organic substances, such as sugars and amino acids, that could be used by the ruminant as nutrients. There are left for the host the end-products of microbial fermentation and the bodies of the microbes themselves as they are washed out of the rumen into the omasum. The real gain to the ruminant is the energy derived from cellulose for which the animal possesses no digestive enzymes.

CELL WALL CONSTITUENTS

There are three main groups of substances in the plant cell walls: cellulose, lignins and hemicelluloses. Of these, only cellulose can be described as a chemical entity; it comprises chains of glucose molecules joined exclusively by β-(1,4) glycosidic links, the repeating disaccharide unit being cellobiose. Only rarely is cellulose found in virtually pure condition, as for example in the natural fibres of cotton. In ordinary plant cell walls, the cellulose is associated, in chemical relations not yet understood, with lignin. In highly lignified material, as for example in sawdust of hard woods, the cellulose is completely indigestible in the ruminant gut; by contrast, in leaves of pasture legumes, with lignin contents as low as 3 per cent, there is apparently no interference with cellulose digestion.

Digestion of cellulose presents many unsolved problems and is the subject of much current research. The difficulty stems from the fact that the ruminant ingests cellulose in its native form in which the chains are very highly polymerized, with as many as 10,000 glucose units per molecule; in this native form, the highly insoluble cellulose is extremely resistant and few microorganisms possess the capacity to initiate depolymerization. When experimental work is conducted on natural feedstuffs, it is not easy to determine the exact reaction on the cellulose molecules, as so much other material (hemicelluloses, for example) is usually present. However, if the natural materials are subjected to preliminary treatment to 'purify' the cellulose, some degradation of the cellulose is very likely to occur. The cellulose molecules can be degraded by acids or alkalies, or even by grinding, to give much smaller molecules which still, however, retain their great insolubility. Natural alpha-cellulose is extremely insoluble except in strong acid or alkali. When cellulose is degraded by chemical or mechanical means, β- and γ-cellulose are formed, and these are much more readily attacked by enzyme preparations and by a much greater range of microbial species (see Halliwell, 1957).

It is quite definite that the true cellulolytic activity in the rumen is extremely high—possibly the highest of any naturally occurring system. The early stages of cellulose breakdown are not definitely known. It is not certain that the process is hydrolytic; phosphorolytic or other mechanisms could also be involved. Although a considerable number of bacterial, fungal and protozoal species are known to digest cellulose in culture (Halliwell and Bryant, 1963), only one organism, *Trichoderma koningii*, has been shown to produce a culture filtrate that can degrade native cellulose; this cell-free enzyme preparation contains cellobiase and carries the cellulose degradation to glucose (Halliwell, 1965). There is still some doubt whether any rumen protozoa are truly cellulolytic; it is possible that bacteria within the protozoa are responsible for the observed cellulolysis.

It is remarkable that undried cotton fibres in the boll are quite readily hydrolysed by filtrates from cellulolytic organisms, but once the fibres have been dried they become very resistant to enzymic attack; this is likely to prove of interest in ruminant digestion, as the fodder may be either fresh green plant or dried material.

Presumably, the cellulolytic bacteria apply themselves to the fibres and secrete a depolymerizing enzyme; this results in the formation of microscopic pockets in which the bacteria are found (Baker and Harriss, 1947). It is likely that the very close proximity of a bacterium to the point at which soluble products are released, enables the cellulolytic bacterium to absorb the breakdown products before they diffuse out into the general medium and thus become available to the other non-cellulolytic species. The first products are probably oligosaccharides but these are metabolized and do not accumulate in the medium.

The fermentation products of cellulose formed by individual rumen bacteria vary with the different species; the following have been recorded: acetic acid, butyric acid, formic acid, lactic acid, succinic acid, ethanol, CO_2 and H_2 (Oxford, 1964). Of these products, only acetic acid, butyric acid and CO_2 accumulate; the other substances participate in further microbial activities.

Although ethanol has been found in the products of pure cultures, it occurs in negligible amounts in the rumen; also, as ethanol has a low turnover rate in the rumen, it is probably not a metabolite of significance (Moomaw and Hungate, 1963). Propionic acid does not seem to be a significant direct product of cellulose fermentation, but is derived from succinic acid by the decarboxylation reactions of other bacterial species (Johns, 1951).

It has been assumed, without experimental basis, that rumen bacteria utilizing cellulose as an energy and carbon source absorb glucose or cellobiose as end-products of cellulolysis. This is not necessarily so, as it has been shown (Schafer and King, 1965) that *Cellvibrio gilvus* (an organism occurring in soil), could absorb oligosaccharides with as many as 6 glucose units (cellohexaose) and indeed could utilize the larger polymers even more efficiently for growth than the smaller polymers. The absorption appeared to be due to an active process and not simple diffusion. Intracellular phosphorolysis would conserve the energy of the glucosyl bonds in oligosaccharides, and thus be advantageous to the organisms, though it appeared from growth studies that there are other factors which make the larger oligosaccharides more efficient as energy sources for growth. If such a mechanism for absorbing oligosaccharides occurred in rumen cellulolytic organisms, it would no doubt confer an advantage on the organism initiating the degradation of cellulose.

In some cereal grains (barley, oats) there occurs a water-soluble polysaccharide related to cellulose; this is β-glucan, in which the glucose units are combined in β-1,3 and β-1,4 glycosidic links (Templeton and Dyer, 1964). Similar polysaccharides occur in lichens (lichenin) and seaweeds (laminarin), both of which have been used by ruminants as fodders. Little is known of the digestion of these substances in the rumen (McNaught *et al.*, 1954).

The hemicelluloses are wrongly so named and comprise a mixture of chemical compounds that are chiefly polymers of pentoses, hexoses or both. In contrast with the highly insoluble cellulose, they are soluble in cold aqueous alkali. The principal group is the xylans, consisting of polymers of xylose and arabinose, sometimes with glucuronic and related acids. The hexosans are chiefly mannans, glucomannans, galactans, and arabinogalactans. This group of substances is, nutritionally, as important as cellulose and may comprise over 20 per cent of the dry weight of hay (Oxford, 1964).

Information on the fate of hemicelluloses in the rumen is rather meagre—principally due to the lack of specific and accurate analytical procedures. However, tests, *in vitro*, indicate that some forms of hemicellulose are readily digested by several species of rumen organisms (Bryant, 1959). Microbial enzymes degrade the polymers to oligo- and mono-saccharides, and related substances, but these do not accumulate in the rumen as they are rapidly metabolized by the microbes (Bath and Head, 1961). It is likely that the hemicelluloses are, like cellulose, potentially completely digestible in the rumen provided that they are not combined with lignin (Van Soest, 1964).

Naturally-occurring lignins are polymers of phenyl propane compounds in which the aromatic nucleus may carry methoxyl or carboxyl radicals; a great variety of lignins occurs by combinations of the differing aromatic nuclei (Bland and Logan, 1965). Methods of analysis for lignin in foodstuffs and excreta are still rather unsatisfactory, and the accuracy of some current procedures is questionable; this uncertainty has led to a good deal of confusion on the digestibility of lignin. The best assessment of the present situation is that lignin is completely indigestible, and that apparent digestibility, especially in young plant materials, is due to inaccuracy inherent in the standard analytical procedures (Van Soest, 1965). An important modification of lignified fodder (such as cereal straws) can be achieved by treatment with cold dilute alkali (usually about N-NaOH). This reagent removes some lignin and hemicellulose and modifies the linkages between lignin and cellulose-hemicellulose, with the result that the digestibility of the latter is increased (McAnally, 1942).

Pectins also may form a substantial fraction of fodder plants (e.g. up to 12 per cent of the dry matter of lucerne (Smart *et al.*, 1964; Hirst *et al.*, 1959). These substances are polymers composed chiefly of galacturonic acids, and although they are constituents of the plant cell walls, they are readily extracted with aqueous solvents. Several species of rumen bacteria and protozoa are known to digest pectins (Abou Akkada and Howard, 1961; Williams *et al.*, 1961; Wright, 1960). Two types of enzyme are produced: pectin esterases that split off methanol, and polygalacturonases that hydrolyse α-1,4-glycosidic linkages giving galacturonic acid or oligogalacturonides; pectins appear to be completely digestible (Gaillard, 1962).

Carbohydrates of Cell Contents

Within the plant cell a great range of carbohydrates is found. In the cereal grains, starch is the major component; in plant leaves and stems, which form the bulk of the grazing ruminant's diet, starch is a minor component and extremely variable in concentration, both between species and at different times in a single species. Although many rumen organisms store starch (or polysaccharides resembling starch) when fermenting sugars and other substrates (Baker and Harriss, 1947), the actual amount made available to the host animal seems to be quite small (e.g. about 5 g/day to a sheep fed on hay—Heald, 1951). In periods between feeds, the microbes metabolize their stored starch. The microbial polysaccharides are comparable with amylopectin or glycogen but only one has been fully characterized (Forsyth and Hirst, 1953).

A considerable array of microbial enzymes, capable of completely degrading starch, is produced by rumen microbes. These include α-amylase, amylomaltase, maltase and isomaltodextrinase. Although the initial digestion of starch granules must be accomplished by extra-cellular enzymes, bacteria-free filtrates of rumen fluid are found to be inactive. Microscopic examination shows that bacteria apply themselves to starch granules and steadily erode the structure (Baker *et al.*, 1950); some species of protozoa can ingest entire starch granules which are then digested with their own specific amylases or with the aid of attached bacteria.

Several different bacteria and protozoa are known to be active starch digesters and the following fermentation products have been recorded: acetic acid, formic acid, lactic acid, succinic acid and CO_2. Lactic acid may accumulate with harmful effects under conditions discussed in detail in Chapter 33.

Fructosans are important carbohydrate reserves in the leaves and stems of fodder plants and may form a very large part of the dry matter, e.g. up to 36 per cent of grass stems (Waite and Boyd, 1953). The fructosans are polymers of fructose, are readily soluble, and are easily fermented in the rumen.

Sugars form an important group of nutrients in ruminant foodstuffs and a great variety of compounds is known to occur in plant materials. In addition, sugars and related compounds are probably the initial breakdown products of cellulose, hemicelluloses, starch and fructosans. Most of the rumen microbes can ferment one or more sugars, with the production of the metabolites mentioned above with starch and cellulose. The fermentative characteristics of the known microbial species in the rumen have been discussed in Chapter 3.

In addition to the sugars, a wide variety of organic substances occurs in the cell-sap of fodder plants; examples are pyruvate, malate, fumarate, tartrate, oxaloacetate, succinate, citrate and oxalate. Most of these compounds are present in small concentrations and little attention has been paid to their fate in the rumen. It is probable that most of these substances can be utilized in microbial metabolism and become converted principally to VFA and CO_2. A substance of special interest is oxalate which is found in high concentration (up to 15 per cent of D.M.) in the leaves of some pasture plants and weeds; the ruminal catabolism of oxalate protects the animal from toxic effects (Dodson, 1959; Maymone *et al.*, 1961). The recent observation that trans-aconitate sometimes accumulates to a high concentration in pasture plants indicates that a variety of plant metabolites may prove to be potentially harmful to ruminants (Burau and Stout, 1965).

The products of fermentation mentioned above are those recorded from tests with cultures *in vitro*. There is now good evidence that conditions within the rumen are highly competitive for the biota, and that any compound that is capable of providing energy in the highly anaerobic environment is likely to be fermented. Thus the only fermentation end-products of consequence are the volatile fatty acids (acetic, propionic, butyric and small amounts of higher acids), carbon dioxide and methane. Although lactic acid is an important, or even the sole product of some organisms, only traces are present in the rumen of animals fed high-roughage diets (Balch and Rowland, 1957). Lactic acid is known to be a nutrilite for several rumen bacteria (Oxford, 1964) and yields the usual volatile fatty acids and CO_2; when the diet contains much starch, lactic acid is a major precursor in the formation of propionic acid.

Propionic acid appears to be formed in the rumen by two major pathways: the decarboxylation of succinate (Johns, 1951) and the reduction of lactate (Elsden *et al.*, 1956). The former pathway seems to account for most of the propionate when cellulose and hemicellulose are being fermented (Satter *et al.*, 1964). With increasing amounts of the more readily fermented carbohydrates (starches and sugars) in the diet, the formation of propionate from pyruvate via lactate and acrylate becomes more dominant (Jayasuriya and Hungate, 1959).

The relative amounts of end-products of carbohydrate fermentation are not simply determined by the chemical composition of the diet. It has been amply demonstrated that the physical form of the feedstuffs can influence the ratio of acetic to propionic acid in the fermentation products (Balch, 1960); thus when lactating cows are given all their roughage as finely ground hay, the production of acetic acid falls and of propionic acid rises. This phenomenon, whose mechanism is not yet understood, has important consequences, as the deficit of acetate leads to a marked fall in the butterfat content of milk (O'Dell *et al.*, 1963).

Hydrogen is produced by both bacteria and protozoa but does not accumulate in the rumen gas. It appears to provide the main substrate for the formation of methane by the reduction of CO_2 (Smith and Hungate, 1958).

Another common product of fermentation is formate, but this substance is rapidly dissimilated in the rumen and only occasionally is present in detectable amounts in the rumen fluid (Annison, 1954a). Formate may provide an intermediate for the formation of methane, but it is more likely that the formate is first metabolized to H_2 and CO_2 (Smith and Hungate, 1958).

Methane is a consistent end-product of digestion within the rumen; it usually represents some 10–13 per cent of the total digestible energy of the ration consumed, the amount being dependent on the level of feeding and the type of ration fed (Blaxter and Clapperton, 1965). Where sheep or cattle are fed once or twice a day the rate of methane production rises rapidly after feeding and reaches a maximum in about 2 hours; thereafter, the rate steadily declines (Pilgrim, 1948; Washburn and Brody, 1937). It has been definitely shown by ruminal infusion of $Na_2{}^{14}CO_3$ that CO_2 is the principal precursor of methane (Williams *et al.*, 1963). Only one bacterial species has been shown to be a significant methane producer in the rumen—*Methanobacterium ruminantium* (Smith and Hungate, 1958). Hydrogen provides the energy source for growth of the organism and CO_2 reduction to methane.

Lipids

In the ruminal digestion of lipids, special interest attaches to the hydrogenation of unsaturated fatty acids (Shorland *et al.*, 1957). Hydrolysis of esterified fatty acids occurs, and the free glycerol formed is readily fermented.

Most ruminant foods consist of plant leaves and stems, the grains of cereals, and oil-seed meals. The neutral fats in these foodstuffs are characterized by having a high proportion of C_{18}-unsaturated fatty acids. In pasture plants, the predominant fatty acid is linolenic acid, with smaller amounts of linoleic and oleic acids; palmitic acid is the major saturated acid. Similarly, in the oil-seed meals, from which most of the oil is usually removed before feeding to animals, triglycerides containing C_{18}-unsaturated fatty acids comprise the major part of the lipid. A noteworthy feature of plant-leaf neutral lipid is that triglycerides are minor components and their place is taken by galactoglycerides (Weenink, 1961).

Quite independently of the dietary lipid, the ruminal microbes synthesize their own lipids, which thus contribute to the digesta passing into the abomasum and intestine.

The mechanism of hydrogenation of unsaturated fatty acids is not fully understood but has affinities with hydrogenation by a metallic catalyst. Both bacteria and protozoa are known to hydrogenate chloroplast lipids (Wright, 1960). Hydrogen is an end-product of the fermentation of carbohydrates and amino acids in the rumen, and possibly the molecular hydrogen is activated by microbial hydrogenase. Under physiological conditions, *in vitro*, oleic, linoleic and linolenic acids are hydrogenated at similar velocities; linolenic acid is hydrogenated to two types of dienoic acids, subsequently to a monoenoic acid, and finally to stearic acid; linoleic is converted first to a *trans*-monoenoic acid and then to stearic acid (Ward *et al.*, 1964; Wilde and Dawson, 1966). The effect of these changes on the dietary fats is to increase greatly the ratio of C_{18}-saturated to C_{18}-unsaturated fatty acids leaving the rumen. Active hydrogenation of fatty acids also occurs in the caecum and colon of the sheep, so that the faecal fatty acids are mostly saturated (Ward *et al.*, 1964).

No lipase is found in ruminant saliva, except perhaps in young animals (Wise *et al.*, 1940). However, the rumen contents are actively lipolytic, though there appears to be little or no degradation of the long-chain fatty acids (Garton, 1965). The lipolytic activity seems to be chiefly due to bacteria, and a lipolytic species has been isolated from the rumen (Hobson and Mann, 1961). The glycerol moiety released by lipolysis is rapidly fermented with the formation of propionic acid.

The fatty acids of galactoglycerides are also readily released, and the galactose is liberated by the action of galactosidases which are produced by both bacteria and protozoa (Bailey, 1962; Howard, 1963); the galactose is readily susceptible to fermentation.

The phospholipids are also attacked by rumen microbes, giving rise to fatty acids and glycerylphosphorylcholine which is further hydrolysed.

The microbes leaving the ruminoreticulum contribute to the animal's intake of lipid as some 9 per cent of the dry weight of rumen bacteria was found to be lipid, both neutral lipid and phospholipid; the bacterial lipids include no linoleic acid, so that if the ruminant has a specific nutritional requirement for this acid, it must be derived from the fodder and escape hydrogenation in the rumen (Garton and Oxford, 1955). The microbial lipids are noteworthy for the presence of branched-chain fatty acids with 13, 14, 15, 16 and 17 carbon atoms, normal fatty acids with an odd number of carbon atoms, and fatty aldehydes—these may be the source of the ruminant plasmalogens (see Garton, 1964). The branched-chain acids are presumably synthesized from the short-chain acids, *iso*butyric, *iso*valeric and 2-methylbutyric, formed in the rumen by bacterial deamination and decarboxylation of valine, leucine and isoleucine (Annison, 1954). The branched, short-chain fatty acids are known to be essential nutrilites for some rumen bacteria; it may well be that these acids are required for the synthesis of long-chain acids and aldehydes by the bacteria. The C_{13} to C_{17} fatty acids become incorporated in adipose tissue and butterfat of the ruminant.

Nitrogenous Substances

Protein

The ruminal digestion of protein is essentially due to microbial activity as no digestive enzymes are secreted into the rumen, and it is unlikely that autolytic changes due to the release of plant enzymes make a significant contribution (Sym, 1938; McDonald, 1952). Few species of rumen microbes have been shown, in culture, to be strongly proteolytic, though many more are weakly proteolytic; this probably explains the observation that all fractions of the biota are found to be proteolytic (see Blackburn, 1965). Presumably, proteases are secreted into the medium by bacteria, but do not accumulate; the bacteria-free fluid has little or no activity, perhaps due to mutual destruction by the proteases (Blackburn and Hobson, 1960). An active protease has been found in the rumen protozoan, *Entodinium* (Abou Akkada and Howard, 1962).

Proteolytic extracts can readily be made from acetone-dried preparations of bacteria or protozoa. The proteolytic activity of rumen organisms is not dependent on the protein content of the diet—this is in marked contrast to the deaminating capacity (Warner, 1956). Soluble proteins are much more readily attacked than insoluble proteins (Chalmers *et al.*, 1954); this is a point of considerable nutritional significance as particles of insoluble proteins have a greater opportunity of leaving the rumen before being degraded; in turn, this will influence the biological value of the fodder protein, depending on whether it has a higher or lower value than the microbial proteins formed in the rumen. In fresh plant feed, much of the protein is soluble in the dilute salt solution of rumen fluid (Steward and Thompson, 1954), and is rapidly released from plant cells as the result of chewing (Reid *et al.*, 1962; Hogan, 1965). Quite high concentrations of leaf protein may be found in the rumen fluid while ruminants are grazing on succulent plants. When plant material is dried, as in hay-making, much of the cellular protein is denatured and becomes highly insoluble; thus in the rumen fluid of hay-fed animals the concentration of soluble protein is quite low. The ease of attack on hay protein will be greatly influenced by the rapidity of degradation of the plant cell walls.

The proteins of cereal grains and seed meals are readily attacked. It should be noted, however, that oil-seed meals may be subjected to high temperatures during processing and this is known to reduce the rate of ruminal proteolysis (Chalmers and Synge, 1954).

Amino Acids

The proteases produced by rumen microbes digest feed proteins to amino acids; *in vitro*, these are the end-products when microbial preparations are kept under toluene. However, in the rumen, amino acids do not accumulate owing to the presence of very active microbial deaminases. Small quantities of free amino acids and diffusible peptides are always present in rumen contents, and the concentrations increase after feeding (Annison, 1956). Within the micro-organisms, there is also a significant concentration of amino acids (*ca.* 2·3 per cent of total N) which are probably in equilibrium with the amino acids in the rumen fluid.

In natural ruminant rations, amino acids and amides (together with some peptides) may form an appreciable fraction of the total nitrogen; this is especially the case with young, rapidly growing plants. In addition to the amino acids that are found in proteins, a large and varied array of other amino acids occurs in fodder plants—and new compounds are frequently being reported (Steward and Pollard, 1957). It is interesting that some of these amino acids (e.g. mimosine) are toxic substances, so that the deaminase and other fermentative reactions provide a first line of defence in detoxication.

The deaminase activity of rumen microbes is strongly influenced by the nitrogen content of the diet (Warner, 1956), in contrast with the proteolytic activity. As deaminases are produced by many species of organisms, and as the rumen fluid will always present a wide range of amino acids derived from protein, it is perhaps not surprising that no striking pattern of deamination has been observed, and that the precise nature of the deamination reactions has not been demonstrated (Lewis, 1955). Only a few amino acids (aspartic acid, glutamic acid, serine, arginine and cystine) are readily deaminated and decarboxylated when added singly to rumen culture *in vitro* (El-Shazly, 1952).

It is likely that some Stickland reactions occur, in which a coupled deamination of the two amino acids occurs with one acting as a hydrogen donor and the other as a hydrogen acceptor; however, the quantitative importance of this reaction is doubtful, as a strong reaction has been found only for alanine and proline. It is probable, too, that the various species of microbes carry out a variety of degradative reactions; for example, *Peptostreptococcus elsdenii* attacks only serine, threonine and cystine, and produces hydrogen gas as well as CO_2, NH_3 and VFA (Lewis and Elsden, 1955). All amino acids of protein are known to be attacked, and the rate of attack seems to be enhanced by increasing complexity of the amino acid mixture presented in test systems.

Amino acids are degraded to ammonia, volatile acids and CO_2 in roughly equimolar proportions, although the products from all amino acids are not known. Special interest attaches to the metabolism of valine, leucine and *iso*leucine which are deaminated and decarboxylated to yield *iso*butyric, *iso*valeric, and 2-methylbutyric acids respectively (El-Shazly, 1952; Van den Hende *et al.*, 1963); these branched-chain acids are required by certain bacteria for their growth (see Chapter 3) and may play an important role in the synthesis of branched long-chain fatty acids and aldehydes in the fats of microbes and of adipose tissue.

Microbial metabolism of the aromatic amino acids (phenylalanine, tyrosine and tryptophan) yield phenyl-substituted fatty acids: phenylacetic, 3-phenylpropionic acid, *p*-hydroxyphenylacetic, *p*-hydroxypropionic, and indolylacetic acids; it is likely that Stickland reactions are involved in the formation of these products (Scott *et al.*, 1964).

The two dicarboxylic amino acids, aspartic and glutamic acids, are prominent in the non-protein nitrogen (NPN) of fodders as well as in plant proteins. Aspartic acid is more readily attacked by rumen microbes than any other amino acid; it appears to be first deaminated to succinic acid and then decarboxylated to propionic acid (Sirotnak *et al.*, 1964). Glutamic acid is less readily attacked, with NH_3, acetic acid and CO_2 as the main products. Specific deaminases for threonine and serine have been detected in rumen bacteria (Walker, 1958). Arginine is degraded via citrulline, ornithine, proline and δ-aminovaleric acid to ammonia and valeric acid (Van den Hende *et al.*, 1964). Other amino acids occurring in proteins are degraded with the usual end-products as CO_2, NH_3 and VFA (see Blackburn, 1965).

The amides of aspartic and glutamic acid are also common components of NPN as well as of proteins. There appear to be distinct enzymes attacking asparagine and glutamine, again with the formation of ammonia (Warner, 1964). It has been known since 1879 (Weiske *et al.*) that asparagine could serve as a source of dietary nitrogen for sheep, and thus support gain in body weight and positive nitrogen balance. These observations stimulated great interest in the unique ability of ruminants to use NPN-substances as protein-sparing nutrients. It is now known that this is achieved through the ruminal formation of NH_3 which can be used by microbes for protein synthesis. The early interest in asparagine led to the study of urea, on which more detail is given below. The capacity of the rumen to deaminate asparagine is so great that the amount likely to be consumed is readily degraded. Other aliphatic amides, e.g., formamide, acetamide, propionamide, do not occur in natural feedstuffs, and are but slowly attacked by rumen microbes; however, after repeated feeding of these synthetic amides, the rate of hydrolysis increases (Repp *et al.*, 1955) but it is not known whether this is due to the formation of adaptive enzymes or to a change in the distribution of microbial species within the rumen.

Decarboxylation as a primary degradation of amino acids seems to be of little importance in the normal rumen; this is to be expected as the rumen pH is rather higher than the acid conditions in which bacterial amino acid decarboxylases are most active. Thus, in contrast to the large concentrations of NH_3 in the rumen, amines are inconspicuous. The only exception is methylamine, but its origin has not been established; methylamine appears in the rumen contents after feeding, reaches a peak and then disappears. As the addition of casein, which is readily deaminated, does not provoke methylamine formation, amino acids are probably not the source (Hill and Mangan, 1964).

Little information is available on the fate in the rumen of the great array of other organic nitrogenous components of fodders, though it is known that purine and pyrimidine bases (guanine, xanthine, hypoxanthine, uric acid, uracil and thymine) are degraded with ammonia formation (Jurtshuk *et al.*, 1958). Since ammonia dominates the soluble NPN fraction in rumen fluid, and since many types of degradations are known to be conducted by the numerous microbial species, it is probable that most of the organic nitrogen compounds are attacked and that ammonia is the principal end-product.

Nitrate

In many natural feedstuffs there is a significant content of nitrogen in the form of nitrate; indeed, high concentrations of nitrate are sometimes found in the leaves and stems of a wide variety of plant species (see Wright and Davison, 1964). It has been noted above that chemical reactions in the rumen can protect the animal from toxic materials; in the case of nitrate, the reverse occurs, as the relatively non-toxic nitrate can be rapidly reduced in the rumen to nitrite, which after absorption produces toxic and even fatal effects (for details see Chapter 33). The nitrite is further reduced in the rumen to ammonia; however, this reaction appears to be slower than the reduction of nitrate so that nitrite can accumulate to dangerous concentrations (Lewis, 1951a, b). This will occur only when the feedstuffs have a high nitrate content, and particularly, when there is also a relative lack of readily fermentable carbohydrate; during active fermentation the reduction of nitrite is accelerated. Under ordinary conditions, the reactions involved must be considered as part of the normal metabolic pattern in the rumen, contributing again to the ammonia produced by so many other pathways.

Cyanide

Many plants, e.g. white clover and linseed, consumed by ruminants, contain cyanogenetic glucosides and related compounds. Sometimes these glucosides are accompanied by glucosidase, so that when the integrity of the plant cells is disturbed by the animal's chewing, HCN is liberated. In other plants the enzyme may be absent. The rumen microbes always have glucosidase activity though the level of activity varies greatly, and is much reduced during active fermentation of carbohydrates. The released HCN seems to be stable in rumen contents and would be present mostly in undissociated form as the pK for HCN is 9·1; absorption from the rumen is rapid, sometimes with fatal results (Coop and Blakley, 1949, 1950). Thus the rumen activity is a potential source of danger to the animal whose detoxication mechanism, mainly through the formation of thiocyanate (Smith and Foulkes, 1966), may be overwhelmed by the high rate of cyanide absorption.

Urea

The interest in amino acid amides as a useful source of nitrogen for ruminants led to nutritional study of urea when this substance became readily available from industrial synthesis early in this century. The enormous number of urea-feeding experiments undertaken served to obscure the physiological role of urea in ruminant digestion until recent years. The degree to which urea can serve as a N-source is discussed in Chapter 21.

As in all other mammals examined, the saliva of ruminants contains urea, at a concentration about 70 per cent of that in the blood plasma (McDonald, 1948). The very large volume of saliva secreted by ruminants makes this addition of urea to the rumen of great significance. Further, it is now clear that urea tends to diffuse from the mucous membrane into the contents of the gastro-intestinal tract; it has been fully demonstrated that, in this respect, the rumen is no exception (Simonnet *et al.*, 1957). Much more urea enters the rumen by diffusion than with the saliva (Houpt, 1959); the significance of this movement of urea to nitrogen metabolism is discussed later. Urea is thus normally ingested and enters the rumen continuously. It is therefore not surprising to find that microbial urease is produced in the rumen; what is rather surprising is that although none of the individual microbial species tested is highly ureolytic, the general level of activity is such that urea is completely degraded and never accumulates in normal rumen contents. The protozoa are not known to produce urease. The urease is associated with the bacteria and is not present in bacteria-free rumen fluid (Schwartz *et al.*, 1964). As with urease from other sources, urea is converted to CO_2 and NH_3; the ammonia is absorbed and is converted into urea by the liver and thus establishes a urea-NH_3 cycle of great significance in metabolism (McDonald, 1948).

The ruminal urease activity is so high that all the urea likely to be fed in ruminant rations is easily converted to ammonia; for example, Schwartz *et al.* (1964) found that 10g urea was degraded in the sheep's rumen in about 30 minutes. This is an important feature, for in circumstances that permit the animal to consume large amounts of urea in a short period, NH_3 is so rapidly produced that the rumen pH rises to an abnormal level and absorption of NH_3 becomes sufficiently fast to produce fatal concentrations in the blood stream (see McDonald, 1958).

In contrast with the ease of degradation of urea, biuret seems to be little, if at all, attacked in the rumen even after some weeks of 'adaptation' to diets containing biuret (Johnson and McClure, 1964). Creatine is degraded by rumen microorganisms to urea and sarcosine, both of which are then further metabolized to yield free ammonia (Campbell *et al.*, 1959).

Nucleic Acids

Although the plant materials used by ruminants as food are not rich in nucleic acids, purines or pyrimidines, special interest attaches to these substances because the ruminal microbes themselves have a high nucleic acid content (Ellis and Pfander, 1965). However, no experimental work on the ruminal digestion of the nucleic acids has been published.

General Comment

It may be noted in passing that the special flora and fauna of the rumen have been able to evolve in a protected and relatively well-controlled environment in which they have experienced contact with an immense range of chemical compounds derived from the wide range of plant species found in natural pastures. This probably accounts for the great biochemical versatility of the rumen biota. In Nature, ruminants tend to graze over large areas of land and select a wide variety of species as food; under modern conditions of agriculture, however, it is common to find that the animal is obliged to subsist on a single species or very few species of plants. One consequence of this is a growing list of metabolic and nutritional disorders produced by toxic substances in plants that have long enjoyed a high reputation as fodder species. This subject is pursued in more detail in Chapter 33.

PRODUCTS OF DIGESTION IN THE RUMEN

It is clear that the extensive dissimilation reactions of the rumen microbes reduce a very complex mixture of food constituents to a small group of metabolic end-products, viz: carbon dioxide, methane, acetic, propionic and butyric acids (with traces of higher fatty acids), long-chain fatty acids, and ammonia (with traces of amino acids and peptides). These few substances, together with the bodies of microorganisms themselves, comprise the bulk of the ruminant's real food. The CO_2 and methane are waste products removed chiefly by eructation. The ammonia leaving the rumen, by absorption or passage to the omasum, probably also is a waste product except in special circumstances, to be discussed later, when the recycling of nitrogen can contribute to microbial growth in the rumen.

In most ruminant foods, fats comprise a relatively small fraction of the total digestible nutrients; hence the long-chain fatty acids of the ruminal digesta make only a small contribution to the animal. The three volatile fatty acids (acetic, propionic and butyric) make the chief contribution to the ruminant's energy intake; for example, Leng and Leonard (1965) found that, in sheep fed lucerne chaff, nearly 80 per cent of the apparently digestible energy of the ration could be accounted for by VFA production.

In addition to these metabolic products, bacteria and protozoa are continuously swept out of the rumen as the digesta pass through the reticulo-omasal orifice. As the organisms pass into the abomasum, they meet the unfavourably low pH of the contents there, disintegration occurs (explosively for the protozoa, less so for the bacteria), and digestion of their bodies starts. The organisms contribute chiefly their proteins, which have a relatively high biological value. Small amounts of carbohydrate and lipid are also present. A significant proportion (perhaps 10 per cent) of the microbial nitrogen is present as nucleic acid, which is, presumably, also a waste product for the animal.

Some fraction of the ingesta consumed passes through the ruminoreticulum unchanged (McDonald, 1954; Weller *et al.*, 1958). As stated previously, freshly consumed fodder is immediately mixed with partially digested food in the rumen; small particles may be carried by the rumen liquid into the reticulum and then either swept back into the rumen or passed into the omasum. The amount of undigested food passing through to the omasum will depend on the rate of reduction to small particle size and the rate of degradation by the microbes. Soluble materials meet no barrier to passage into the omasum, so that the amount escaping digestion will depend on the relative rates of fluid flow and of digestion.

When protein, starch or sugar escapes ruminal digestion, this will be advantageous to the animal as these substances, after intestinal digestion and absorption, can be more efficiently used in metabolism than their fermentation products. On the basis of available data, it is difficult to make a satisfactory statement about the amounts of these constituents reaching the abomasum. The general impression is that very little sugar escapes breakdown; that starch granules, except on high starch rations, are also extensively degraded; that soluble proteins are largely digested in the rumen, but that a considerable proportion of insoluble, granular proteins may pass unchanged into the abomasum. The situation with cellulose is very different. The natural cellulose fibres are only slowly dissolved so that any small particles have a high chance of leaving the rumen undegraded; roughly, it seems that about one third of the potentially digestible cellulose escapes digestion in the rumen. This estimate is based on the finding that more than 80 per cent of cellulose digestion occurs in the rumen and less than 20 per cent in the caecum and colon, and that part of the cellulose excreted in the faeces is still in a form susceptible to ruminal digestion.

Vitamins in the Rumen

Detailed consideration of the vitamin requirements of ruminants is given in Chapter 24. Here it is desirable only to comment briefly on the changes imposed by the rumen.

Retinol (vitamin A) does not occur in the usual ruminant fodders and its place is taken by the precursors, the carotenes. In spite of the highly reducing conditions in the rumen, the conjugated double bond systems in the carotene molecule are resistant to hydrogenation, and most of the carotene passes through the rumen unchanged (Keating *et al.*, 1964); retinol, however, is partially destroyed in the rumen. The grazing animal is entirely dependent on its fodder to provide these vitamin A precursors, which are converted into retinol chiefly by the intestinal epithelial cells and the liver (see Bridge and Spratling, 1966).

The vitamin B group presents a very complex picture in the rumen. Like all living cells, the rumen microbes require B vitamins for their own metabolism and growth. Some ruminal organisms are exacting in their nutritional requirements for members of the B complex, but these requirements can be met by the synthetic activities of other species (Ford *et al.*, 1958). In this way, a flourishing rumen population can be maintained independently of a dietary supply of vitamins. In turn, this synthetic activity ensures the ruminant an adequate supply of all the required B vitamins, regardless of dietary content (Kon and Porter, 1954). The only known failure of this system is that which occurs when ruminants consume a diet deficient in cobalt (Marston, 1952); this prevents the formation of vitamin B_{12} (cobalamin) or its analogues, and leads to a characteristic wasting disease which has been recorded in sheep and cattle in many countries.

Ascorbic acid (vitamin C) is rapidly destroyed in the rumen (Vavich *et al.*, 1945), but the ruminants share with most other mammals the capacity to synthesize their own requirements.

Rumen bacteria also synthesize vitamin K and thus make the ruminant independent of a dietary supply (McElroy and Goss, 1940). No information is available on the stability of vitamins D and E in rumen contents, but neither vitamin is synthesized in the rumen.

THE OMASUM

This organ occurs only in the order Ruminantia and a related structure rs found in the Tylopoda (family Camelidae); no comparable organ has evolved in the variety of other mammals (for example, the Marsupialia) that have developed gastric modifications adapted for microbial digestion. It is, therefore, clear that the functions of the omasum are not essential to successful exploitation of the symbiotic relation of host and gastric microbes. Nevertheless, the high degree of specialized anatomical structure in the Ruminantia suggests that the omasum does play a valuable role in digestive processes; this role is expressed in two activities: control of flow and absorption of nutrients.

Digesta from the reticulum are aspirated into the omasum by a simultaneous relaxation of the omasum and contraction of the reticulum (see Stevens *et al.*, 1960; Balch *et al.*, 1951). Large food particles are blocked at the reticulo-omasal orifice by the omasal leaves, while the fluid and smaller particles fill the body of the organ. As digesta accumulate in the omasum, the reticulo-omasal orifice is closed, the omaso-abomasal orifice relaxes, and contraction of the body of the omasum forces material on into the abomasum; a reversal of this action can also occur to clear coarse particles retained, at the reticulo-omasal orifice, by the grid of the omasal leaves. The effect of this straining is to retain coarse, undigested particles in the ruminoreticulum for more prolonged digestion and re-chewing.

The rates of contraction of reticulum, omasum and rumen are closely co-ordinated, and the reticulo-omasal orifice relaxes with each cycle of reticular contraction; this makes it possible for digesta to flow onward with each cycle of contraction (Stevens *et al.*, 1960). However, the amplitude of omasal contraction is reflexly controlled by the degree of distension of the abomasum and by its own degree of filling; the omasum can regulate the rate of flow of digesta into the abomasum which is thus continuously maintained with a moderate amount of contents though there is a marked rise in the volume of abomasal contents soon after feeding (Boyne *et al.*, 1956). This is in contrast to the stomach of the monogastric animal, in which filling and emptying occur with the episodes of eating.

The rate of flow of digesta through the omasum is highly variable and so is the form of the digesta, varying from solid material to a very fluid material containing little solids (Ash, 1962). Presumably, the filling and emptying of the omasum can be controlled in response to conditions in the abomasum and ruminoreticulum. Filling the abomasum decreases flow from the omasum, while the consumption of food increases omasal flow; rumination has no definite effect. No evidence is yet available on the range of times spent by the digesta within the omasum.

It has long been supposed that the omasal leaves exerted a grinding action on coarse plant fragments in the digesta; this is a misconception derived from the fact that the larger fragments are prevented from entering the omasum from the reticulum.

THE ABOMASUM

A feature of abomasal function is the very copious secretion of gastric juice—some 5 l./day in sheep and 30 l./day in cattle (see Hill, 1965). This high secretion rate is made possible by the very rich blood supply to the mucosa, and the extensive surface area of the mucosa, effected by numerous large folds in the fundic area. The entire surface of the abomasum carries secretory glands as there is no cardiac region—during foetal life this region develops into the rumen, reticulum and omasum.

Functionally, the abomasum differs greatly from the stomach of monogastric animals. In the ruminant, feed is extensively digested before reaching the abomasum, and small quantities of digesta are received at frequent intervals throughout the entire day from the omasum; in conformity with this, abomasal contents are passed to the duodenum in gushes at irregular intervals throughout the day (Phillipson, 1952). Although appreciable volume changes do occur (Ash, 1961a), the abomasal contents are maintained at a fairly steady volume, and the organ is never emptied under ordinary feeding conditions.

In contrast to the powerful, complicated and co-ordinated movements of the rumen, reticulum and omasum, the abomasum exhibits only sluggish contractions which are independent of forestomach activities (McAnally and Phillipson, 1944); no doubt this is related to the fact that the abomasal contents are always very fluid and little effort is needed transporting digesta to the duodenum.

In conformity with the frequently repeated passage of digesta into and out of the abomasum, there is a continuous secretion of gastric juice (Hill, 1960). The act of feeding does not directly stimulate the secretion of gastric juice; the principal stimulus is provided by the flow of digesta from the omasum (Ash, 1961a). Thus, when the animal is frequently fed there is a continuous high level of secretion. When the animal is forced to consume its daily ration during one short feeding period, the act of feeding provokes a rapid passage of digested food from the rumen to the abomasum, and this has a marked stimulatory effect on abomasal secretion. This is not the natural state for ruminants; when grazing or when allowed free access to feed in troughs, ruminants spend many hours each day in feeding.

The control of gastric secretion is highly complex and appears to be directed toward maintaining a steady state of the pH of digesta leaving the abomasum. The secretion of pepsin seems to be controlled by the vagal centre; the normal sources of stimulation and inhibition of the vagal centre have not yet been clarified, though it is likely that activities of the forestomach and the associated act of rumination are chiefly involved; these activities may have direct or indirect effects, as they are themselves under vagal control.

The secretion of hydrochloric acid and water is controlled by several mechanisms. The hormone, gastrin, is produced by both pyloric and fundic regions of the abomasum (Anderson *et al.*, 1962), and stimulates the secretion of acid; stimuli transmitted through the vagus potentiate the effect of gastrin on the parietal cells. The principal factor affecting the release of gastrin is the acidity of the abomasal contents. The digesta coming from the omasum are only slightly acid and highly buffered, and tend to raise the pH of the abomasal contents; this stimulates the release of gastrin and thus the secretion of acid. A low pH inhibits gastrin release, and acid secretion thus declines; this inhibition occurs when the acidity of the digesta reaches about pH 2. The VFA of the digesta have a marked stimulatory effect on secretion, but the mode of action is not known. Also, distension of the duodenum or a low pH in the duodenum tends to inhibit the secretion of acid by the parietal cells (Ash, 1961a).

The rate of flow of digesta into the abomasum is controlled by vagal reflexes which produce an increased rate of flow with increase in rumen volume; distension of the abomasum reduces the flow, and emptying accelerates flow (see Phillipson and Ash, 1965). Similarly, distension and emptying of the duodenum control the flow rate through the pylorus. As in monogastric animals, the presence of fat in the duodenum provokes the release of enterogastrone; this hormone reduces motility of the abomasum and forestomach, and inhibits gastric secretion (Hill, 1965). Under natural conditions, this mechanism is probably of little importance as ruminant diets usually have a low fat content.

The high acidity of the abomasal contents causes instant destruction of the protozoa and a slower disintegration of many bacteria (Masson, 1950). The peptic digestion of protein is initiated, and continues within the duodenum. This peptic activity, together with the maintenance of flow into the duodenum throughout the entire 24 hours, probably constitute the main functions of the abomasum. Little is known concerning the absorptive function of the organ.

THE INTESTINES

The characteristic feature of the ruminant small intestine is its very great length; it seems likely that this evolutionary development is a response to the very large quantity of digesta produced as a result of ruminal and abomasal activities; for example, a sheep weighing some 40 kg may secrete up to 27 litres of fluid per day into the gut (Phillipson and Ash, 1965). Functionally, there are no important differences from the monogastric animals.

Little research on the large intestine of ruminants has been undertaken, due no doubt to resemblance of the organ to those of the monogastric animals. Bacterial activity is high in the caecum and first part of the colon so that appreciable digestion of cellulose occurs; volatile fatty acids are produced and absorbed. Some 12 per cent of the digestible cellulose in the ration is, in fact, digested in the lower gut (Goodall and Kay, 1965).

ABSORPTION OF NUTRIENTS

THE RUMINORETICULUM

The unique structure and digestive functions of the ruminoreticulum are accompanied by important absorptive functions; it is tempting to speculate that these absorption mechanisms are primarily directed toward maintenance of the environmental status within the rumen rather than to the recovery of nutrients for metabolism. Thus, the undue accumulation of microbial products (e.g. VFA and NH_3) is prevented, and the flux of water and inorganic ions (Na^+, K^+, HCO_3^-, Cl^-, $PO_4^{\equiv}$) maintains fluidity, osmolality, pH, ionic strength and inorganic composition of the medium. The absorption of VFA is, nutritionally, the most important aspect.

Although the mucosa has a stratified squamous epithelium, the structure is adapted to absorption; only a thin layer of epithelium (*ca.* 0·05 mm) separates the digesta from a rich network of capillaries, and electron microscopy reveals the presence of intercellular spaces between the epithelial cells (Hydén and Sperber, 1965). Further, the powerful churning movements of the organ ensure a continuous flow of digesta over the epithelial surfaces. Thus the rumen has developed an effective absorptive function while retaining a tough surface structure capable of withstanding abrasion from the coarse forage consumed.

The VFA (principally acetic, propionic and butyric acids) are absorbed by diffusion and the concentration gradient is therefore the major factor influencing rate of absorption (Danielli *et al.*, 1945); there is probably no active mechanism involved. However, absorption is slower than normal in animals deprived of rumen digesta experimentally and in the young animals with undeveloped rumen; in both cases, the lack of VFA (especially butyric acid) for the metabolism of the epithelial cells seems to be the critical factor

(Armstrong *et al.*, 1957; Sutton *et al.*, 1963). The nature of this mechanism is at present not known, but it is possible that stimulation of blood flow through the rumen wall is involved; within the physiological range, in the rumen, butyrate concentration in rumen contents is directly related to rumen arterial blood flow (Sellers, 1965). Acetate and propionate have a less stimulating effect on blood flow. Increased CO_2 tension in the rumen also increases rumen blood flow; during fermentation, productions of CO_2 and the VFA tend to be correlated, so that the simultaneous formation of these products would enhance absorption mechanisms.

The undissociated VFA molecules are absorbed much faster than the ionized forms, and thus the rate of absorption (within the normal range of pH) increases as the pH is reduced (Danielli *et al.*, 1945; Gray, 1948). Conversely, the absorption of VFA tends to maintain the rumen pH at its normal, slightly acid state above pH 6. At this pH, the ratio of ionized to unionized acetic acid is about 200:1, yet the rates of absorption of the two forms are about equal. With the absorption of each unionized acetic acid molecule, a molecule of HCO_3^- appears in the rumen (Masson and Phillipson, 1951). This mechanism leads to a concentration of bicarbonate in rumen fluid greatly in excess of that in plasma, while conversely the concentration of CO_2 is proportionately lowered (Ash and Dobson, 1963). The rapid absorption of unionized acid and the simultaneous formation of bicarbonate represent a major mechanism in the neutralization of ruminal VFA; the effect is of about the same magnitude as the neutralization by the bicarbonate of saliva.

In the arterial blood, the only VFA with an appreciable concentration is acetate, but this is small in proportion to the usual rumen concentration (Annison, 1954a); thus the concentration gradients for the fatty acids depend only on rumen concentration and the rate of blood flow through the mucosa. Acetate, propionate and butyrate, during absorption, stimulate ruminal blood flow but there is, as yet, no evidence that the rate of absorption of VFA is influenced by rate of blood flow.

Butyric acid is absorbed faster than propionic and acetic, and is largely converted to β-hydroxybutyric and acetoacetic acids during passage through the rumen mucosa (Pennington, 1952); as a result, the portal blood carries little butyrate (Annison *et al.*, 1957).

Owing to the great experimental difficulties involved, it is not yet possible to give an accurate estimate of the proportion of the total VFA that is absorbed from the rumen under ordinary feeding conditions, but it is likely (Murray *et al.*, 1962) that only a small proportion passes on to the omasum and abomasum.

The longer-chain fatty acids in the rumen are formed by condensation reactions of short-chain acids and by fermentation of some aliphatic amino acids. The absorption of valeric, *iso*butyric, *iso*valeric and 2-methylbutyric acids has been demonstrated (Bensadoun *et al.*, 1962); their concentration in rumen digesta is always low and hence they make only a small contribution to the absorbed VFA.

Lactic acid is a normal constituent of the digesta but is present in very low concentration when ruminants are given roughage rations (Balch and Rowland, 1957). When the diet contains large amounts of readily fermented carbohydrates (especially sugars and starches), lactic acid may accumulate to significant concentrations (Dunlop and Hammond, 1965). Lactic acid seems to be absorbed chiefly in the undissociated form; since the pK of lactic acid is about 3·7, most of the lactate would be dissociated at normal rumen pH values and hence absorption is slow (Williams and Mackenzie, 1965). However, in animals on rations containing high proportions of easily fermented carbohydrates, lactic acid can accumulate and reduce the rumen pH to levels as low as 4; under such conditions lactic acid is fairly readily absorbed.

It is known that glucose can be absorbed from the rumen, but this is of negligible importance as the concentration of glucose in rumen fluid is normally very low due to the high rate of microbial fermentation.

The ammonia produced by breakdown of amino acids and other nitrogenous dietary constituents is readily absorbed from the rumen by a process of simple diffusion, and the rate of absorption is therefore dependent on concentration in the rumen (McDonald, 1948). The undissociated molecules are probably absorbed more rapidly than the ionized form (Coombe *et al.*, 1960); at normal rumen pH, only a small fraction of the ammonia is present in the unionized form; however, in dietary circumstances which permit rapid accumulation of ammonia (as for example with urea-containing feeds), the pH may rise to high levels, increasing the proportion of undissociated ammonia and thus enhancing the rate of absorption, even to the point of producing toxic manifestations. Absorption of ammonia is complicated by the diffusion of urea into the rumen. Under normal conditions, the rumen digesta contain no urea, and the arterial blood has only a trace of ammonia. Thus the two substances tend to diffuse in opposite directions into areas of negligible concentration. In the rumen, urea is swiftly converted to ammonia, while in the liver ammonia is converted to urea. Thus a nitrogen cycle is set up leading to important metabolic consequences to be discussed in a later section (McDonald, 1952).

In contrast to ammonia, the amino acids are present at only low concentrations in the rumen. Experimentally, it has been shown that amino acids can be absorbed from the rumen (Cook *et al.*, 1965), but this cannot be of more than minor significance in nitrogen metabolism.

Little information is available on the absorption of other organic nutrients (e.g. vitamins, lipids) from the rumen.

During fermentation, methane and carbon dioxide are produced in large quantities; only in exceptional circumstances is hydrogen present in significant amounts. Methane is freely absorbed from the rumen, the rate being dependent chiefly on the partial pressure in the rumen; the absorbed methane is then excreted by the lungs. Carbon dioxide is very readily absorbed, but the rate of absorption is not explained by simple diffusion, since there is an exchange of CO_2, in the form of bicarbonate ions, between the blood and rumen contents (Hörnicke *et al.*, 1965).

Measurement of gas absorption from the rumen is complicated by the fact that eructated gas passes into the trachea and lungs in quite large amounts; the gases are partially absorbed by the lungs then re-excreted in the expired air (Dougherty *et al.*, 1962). This remarkable phenomenon seems to have no nutritional consequence, as the absorbed gases are so swiftly cleared by the lungs.

The absorption of electrolytes from the rumen is of great metabolic importance; the processes involved are complex and at present incompletely understood (see Kay and Hobson, 1963; Hörnicke, 1964). It is now clear that electrolytes move from the rumen digesta to the blood vessels in the mucous membrane and also move in the reverse direction. Thus 'absorption' is really the net flux of the two movements. Water appears to move between blood and rumen digesta depending on the osmotic pressures in the two fluids; the net effect is a tendency for rumen fluid to be maintained slightly hypotonic to blood plasma. The ruminal osmotic pressure rises to a hypertonic state after the consumption of food and slowly declines during the digestive process (Warner and Stacy, 1965).

The ruminal fluid is electro-negative with respect to blood, and the magnitude of the potential difference is positively correlated with the concentration of potassium, the principal cation in the foodstuffs (Dobson and Phillipson, 1958). The electrical potential thus aids the absorption of anions and impedes the absorption of cations; in the absence of active absorption mechanisms, a net flux of principal cations (Na^+ and K^+) would occur only when ruminal concentration is considerably higher than in plasma, while anions (chiefly Cl^-) could be absorbed against a concentration gradient. It follows too, that these electrolytes will exhibit a net flux from blood into the rumen when the electrochemical gradient has this direction. These ionic movements are important nutritionally as well as for the maintenance of ruminal fluid composition.

In spite of the low sodium content of common feedstuffs, very large amounts of sodium are secreted in the saliva, together with bicarbonate and phosphate (Kay, 1966). The rumen wall exhibits an active mechanism for absorption of sodium, thus enabling Na^+ to be driven against the concentration and electrical gradients into plasma (Dobson, 1959); a hypertonic state, as found after feeding, stimulates the absorption of sodium from the rumen (Stacy and Warner, 1966). It appears that about half the salivary Na^+ is reabsorbed from the rumen. Potassium probably moves through the rumen wall by diffusion; absorption requires movement against the potential gradient, and hence K^+ is absorbed only when the ruminal concentration is more than three times the plasma concentration. As feedstuffs usually have a high content of K, net absorption will usually occur.

The potential gradient favours the absorption of Cl^- which can thus be absorbed when the chloride concentration in the rumen is in excess of 35 m–equiv./l.; at lower concentrations, Cl^- enters the rumen (Phillipson, 1955).

Like Na^+, phosphate plays important nutritional and physiological roles. Ruminant feedstuffs often have a low P-content, but large amounts of P are secreted in the saliva. It is likely that a high phosphate concentration in ruminal fluid is required for optimal microbial growth. Thus a highly efficient re-absorption of P is essential to the nutritional economy of the ruminant, but absorption from the rumen seems to be quite small (Wright, 1955), in spite of a strong electro-chemical gradient (inorganic-P commonly has a rumen concentration more than 10 times as great as in plasma—Kay, 1966).

The concentration of the divalent ions, Ca^{++} and Mg^{++}, in rumen fluid, is usually too low to permit absorption against the potential gradient.

Absorption from the Omasum

There is considerable uncertainty on the absorptive functions of the omasum; the technical difficulties in studying the organ have not yet been fully solved. There may also be important differences between species, as the omasum of cattle holds some 12 per cent of the total stomach contents compared with only 2 per cent for the sheep (Warner and Flatt, 1965).

Absorption from the omasum is facilitated by the enormous epithelial surface, presented by the leaves and their papillae, relative to the volume of digesta in the organ. The absorptive functions appear to aid the efficiency of the abomasum in several ways. Water is absorbed, but the extent of absorption in normal conditions is uncertain; it is possible that the dry matter concentration of the digesta may be doubled before passing into the abomasum (Gray *et al.*, 1954; Hydén, 1961); this permits the addition of large volumes of gastric juice without excessive dilution. The absorption of bicarbonate (Ekman and Sperber, 1953) prevents the release of large volumes of gas when hydrochloric acid of the gastric juice is added; this is probably an important protective function, as the abomasum has no mechanism for the expulsion of gas by belching. It may be noted that a gas bubble is always present in the abomasum (Benzie and Phillipson, 1957); this gas is probably absorbed from the abomasum or passed on to the small intestine—it is not known whether gas bubbles ever pass orally into the omasum. Volatile fatty acids are absorbed in large part (Joyner *et al.*, 1963) and thus are prevented from producing undue buffering of the hydrochloric acid secreted in gastric juice. Other soluble constituents, especially NH_3, Na^+ and K^+, are absorbed while chloride ions tend to diffuse into the digesta from the plasma.

Absorption from the Abomasum

It has been shown that several nutrients are absorbed from the abomasal contents, notably VFA (Ash, 1961), NH_3, inorganic P, and B vitamins. However, little is known of the quantitative aspects of absorption; as the digesta remain in the abomasum for only a short time, it is probable that absorption is unimportant relative to that of the forestomach and small intestine (see Hörnicke, 1964).

Absorption from the Intestines

The function of the small intestine appears to be strictly comparable to that in non-ruminants and consequently has attracted little research activity. The very large volumes of water in the digesta received by the small intestine are efficiently absorbed by the time the terminal ileum is reached. No doubt, the very long intestine of ruminants is related to the need for absorption of so much fluid. The few data available indicate that most of the soluble nutrients are readily absorbed (Phillipson and Ash, 1965). One notable difference from non-ruminants is that in adult ruminants the capacity to absorb glucose seems to be appreciably reduced, though the rate of absorption is still adequate for the amounts of glucose received by the small intestine with ordinary diets (Ridges and Singleton, 1962); this may well be associated with the fact that in the mature ruminant, sugars and starches are so extensively degraded to VFA in the rumen.

A special case is also presented by the lipids in the intestine. As mentioned previously, lipolysis and hydrogenation in the rumen result in the dietary fats reaching the small intestine chiefly as palmitic and stearic acids with some C_{18}–monoenoic acids. The absorption of these long-chain acids is facilitated by the addition, in bile, of large amounts of lecithin, with smaller amounts of cholesterol and free fatty acids (Adams and Heath, 1963); the lecithin has about 30 per cent oleic acid and 10 per cent linoleic acid; reaction of the bile lecithin with pancreatic juice yields free fatty acids and lysolecithin which facilitate the absorption of the dietary fatty acids, principally by the ileum. The process of absorption of long-chain fatty acids appears to be the same as in other mammals, with the uptake by mucosal cells of soluble micelles consisting largely of bile salts and free fatty acids (Garton, 1965). After absorption, the fatty acids are transported as chylomicron lipid, chiefly in the form of triglycerides, with small amounts as phospholipids and non-esterified fatty acids, as in the non-ruminant animals (Heath and Morris, 1962).

As with the small intestine, there appear to be no special features in the absorptive function of the ruminant large intestine. In both sheep and cattle, the contents of the caecum and first part of the colon are quite fluid and contain the major electrolytes (Na^+, K^+, and Cl^-) at about the same concentration as plasma. The Na^+ and Cl^- are very efficiently absorbed (Goodall and Kay, 1965). Potassium is usually abundant in ruminant rations and appears in faeces as the major cation (Brouwer, 1961); most of the excess K^+ is, however, excreted in the urine. Active bacterial fermentation occurs, the principal substrates being cellulose and hemicelluloses that escaped attack in the rumen; more than 10 per cent of the total cellulose digested is, in fact, degraded in the large intestine (Goodall and Kay, 1965). The VFA are produced by the fermentation in molar ratios very similar to those of the rumen and are absorbed, together with significant amounts of NH_3 (Williams, 1965; Ward *et al.*, 1961). The absorption of water is completed, though it is noteworthy that water absorption from faecal material is much more effective in sheep and goats than in cattle.

METABOLISM OF RUMINANTS

There is little doubt that the fundamental reactions within the cells of ruminant animals conform with those of other mammalian species. In this section, therefore, special consideration is given to those features involved in the initial processing of absorbed nutrients and their transport to sites of utilization. The peculiarities of ruminant digestion result in quite important quantitative differences from the non-ruminants; for example, the short-chain fatty acids are produced in and absorbed from the large intestine in mammals generally, but whereas this source of nutrients is of little note in most species, it is of major significance to the ruminant because very large amounts of VFA are produced in the rumen.

The Volatile Fatty Acids

The principal end-products of microbial digestion in the forestomach are the VFA, acetic, propionic and butyric acids. Together, they comprise the chief energy source for ruminants. The extent to which they do this is dependent on the type of ration and the level of feeding. On roughage rations and at low levels of feeding with any ration, nearly all of the digestible carbohydrates in the fodder will be absorbed as the VFA. As the amount of cereal grains or other starchy feeds is increased in the ration, an increasingly large fraction of the starch tends to pass unchanged through the forestomach, and can be digested in the conventional manner in the small intestine and absorbed as glucose. The actual amounts of glucose thus obtained cannot yet be stated with confidence as few experiments have been reported. In Nature, such a situation would rarely occur, but under intensive farming conditions, fattening cattle and sheep, and high-producing dairy cows, are frequently given as much grain as they will voluntarily consume.

Acetic acid is the major metabolic end-product of rumen digestion and is also the largest single contributor to the metabolizable energy of the host ruminant—it may provide up to half the total carbon oxidized by ruminants (Sabine and Johnson, 1964). Although this situation confers on the ruminant the capacity to derive energy from otherwise useless feed constituents, it imposes some metabolic restrictions on the host because there are definite limitations to the pathways in which acetate can be utilized. This is in marked contrast to the universality of reactions involving glucose in the energy-yielding pathways. It is true that nearly all tissues (notable exceptions are brain and testis, and to a lesser extent, liver—Mayfield *et al.*, 1965) can use acetate as a source of energy, but the efficiency of utilization of acetate is less than that of glucose.

A small amount of acetate is metabolized by the epithelium of the fore-stomach during absorption, but most passes into the portal vein and finally into the peripheral blood stream. Acetate represents about 90 per cent of the VFA of peripheral blood (Annison, 1954a), though it is important to note that not all the blood acetate is derived from the digestive tract. Acetate also finds its way into the blood from the body tissues; indeed, the concentration of acetate in the blood of ruminants is comparable with that of non-ruminants, in which the blood acetate is chiefly of endogenous origin (see Annison, 1964). In the ruminant, the absorption of exogenous acetate suppresses endogenous acetate formation (Annison and White, 1962a).

There appear to be no metabolic reactions of acetate that are unique to the ruminant. Oxidation occurs via acetyl-CoA and the TCA cycle, and there is no mechanism for net synthesis of glucose. Acetate is taken up by adipose tissue for fat synthesis, but this contribution to synthesis requires the concurrent metabolism of glucose to provide NADPH for reduction reactions; it is likely that this interaction is a factor that limits the metabolic usefulness of acetate in the fattening process.

In the lactating ruminant, acetate makes a major contribution to the synthesis of fatty acids (especially those with chain length up to C_{16}) in the triglycerides secreted in milk (Hardwick *et al.*, 1961). About half of the milk fatty acids is thus derived; the remainder comes from the fat in plasma chylomicrons. This has important practical implications in milk production by dairy cows, as dietary changes that lead to a marked reduction of the acetate/propionate ratio in rumen fermentation may induce a fall in the fat content of milk (Van Soest, 1963). The explanation of the phenomenon is still not certain but seems to involve changes in the rumen that produce an increase in glucogenic metabolites which then induce a reduced concentration of ketones and lipids in the blood, and stimulate deposition of fat in the adipose tissues at the expense of milk fat synthesis.

Propionate is the second most abundant VFA produced in the rumen and plays a key role in ruminant metabolism. Together with the non-essential amino acids, propionate is the main source of glucose. Some propionate is oxidized to lactic acid by the rumen wall during absorption (Pennington and

Sutherland, 1956), and the remainder passes in the portal blood stream to the liver which removes most of the propionate (Annison *et al.*, 1957). The concentration of propionate in the peripheral blood is very low (less than 5 per cent of the total VFA) and is of the same magnitude in ruminants and non-ruminants. Propionate enters the metabolic pathway by the formation of propionyl-CoA and the fixation of CO_2 to form methylmalonyl-CoA (Smith and Osborne-White, 1965); this substance is converted to succinyl-CoA by the enzyme methylmalonyl-CoA isomerase which requires cobalamin (vitamin B_{12}) as a co-factor. In this connection it is interesting to recall that ruminants are remarkably susceptible to deficiency of dietary cobalt, needed for the synthesis of cobalamin by ruminal bacteria. Succinate can then be used for the synthesis of glucose via oxaloacetate, phosphopyruvate, and reversal of the glycolytic pathway. The formation of oxaloacetate also facilitates the oxidation of acetyl-CoA derived from acetate, butyrate or β-hydroxybutyrate.

Butyrate takes a metabolic pathway very different from either acetate or propionate. The epithelium of the rumen, reticulum and omasum extensively metabolizes butyrate so that less than half of the absorbed butyrate passes into the portal blood (Pennington, 1952). This epithelium constitutes a surprisingly large organ—equivalent to about two-thirds the size of the liver. At birth, the rumen tissue of lambs does not produce significant amounts of ketone bodies from butyrate, but this activity rapidly increases with age (Hird and Weidemann, 1964). This change in epithelial metabolism runs *pari passu* with the increase in size of the ruminal papillae. As noted previously, the presence of VFA in the rumen contents stimulates the development of the papillae; it may well be that this stimulation is associated with the metabolic system that oxidizes butyrate. The products of oxidation in the rumen wall are the ketone bodies β-hydroxybutyrate and acetoacetate (some ketone bodies are also produced from the metabolism of acetate by rumen epithelium). The liver effectively removes butyrate from the portal blood, so that only traces appear in the peripheral blood stream (Annison *et al.*, 1957). The oxidation of butyrate to ketone bodies accounts for about 70 per cent of the oxygen consumption of rumen epithelium and up to 40 per cent for liver (Hird and Weidemann, 1964). In many tissues, the ketone bodies can substitute for glucose in oxidative metabolism via the Kreb's cycle, but it appears definite that no net synthesis of glucose can be made from these substances in ruminant tissues (see Annison, 1964). The apparent conversion of butyrate or β-hydroxybutyrate to glucose in some experiments can be accounted for by glucose-sparing actions and by the known function of acetoacetate in stimulating gluconeogenesis from lactate (Krebs *et al.*, 1963).

During the cellular metabolism of fatty acids, little butyrate is released from the cells into the extracellular fluids; this is in marked contrast to β-hydroxybutyrate (or acetoacetate) which is always present in blood, and may rise from the normal low levels to very high concentrations when metabolic patterns are disturbed. In non-ruminants, the blood ketones are chiefly derived from condensation of acetyl-CoA units; by contrast, in ruminants, the chief origin is from exogenous butyrate after oxidation by the epithelium of the forestomach or by the liver. Most ruminant tissues, except liver and brain, can readily utilize ketones (Leng and Annison, 1964); a curious exception is the forestomach epithelium—hence the output of ketones from absorbed butyrate.

The formate present in blood is derived endogenously, as there is usually none in the rumen fermentation products. In the blood, formate may comprise 10–30 per cent of the total VFA in both ruminants and non-ruminants (Annison, 1954b). Although this substance participates in many metabolic reactions, its turnover rate is low (only 1–2 per cent of the rate for acetate in the sheep), and it is therefore clear that formate cannot be an important contributor to energy metabolism (Annison and White, 1962b).

Glucose

Quantitative data are at present inadequate to estimate, with precision, the extent to which propionate can meet the ruminant's need for glucose; the available information indicates that other sources of glucose, especially amino acids, would be needed to meet the requirements of pregnant and lactating animals (Armstrong, 1965).

In spite of the limited supplies of glucose, this substance is just as important a metabolite as in non-ruminants. The chief difference between the two groups is that, in ruminants, glucose is not an important contributor of 2-carbon units for fat synthesis; these are available usually from the absorbed VFA, though, as noted above, glucose oxidation is still needed to provide the reduced coenzyme (NADPH) for the reduction reactions in fatty acid synthesis. Thus it is found that the rate of glucose utilization, expressed per unit of metabolic size ($BW.^{0 \cdot 75}$), in ruminants is quite comparable with that in non-ruminants (Armstrong, 1965).

The peculiar vulnerability of the ruminant to glucose deficit is well exemplified in the occurrence of the so-called pregnancy toxaemia of ewes; in this disease, which is discussed in more detail in Chapter 33, inadequate nutrition (notably, insufficient supplies of glucose precursors) leads to a train of metabolic and hormonal disturbances terminating in failure of the central nervous system and death (Reid, 1963). In the more acute forms of the disease, early administration of glucose or precursors of glucose can reverse the trend and lead to the animal's recovery. The critical factor here is that the rapidly growing foetal lamb, in the late stages of pregnancy, has a high demand for glucose and successfully competes with the maternal tissues for this metabolite. Eventual failure of the central nervous system seems to be due to inhibition of glucose utilization by the brain, as alternative metabolites cannot be used by brain cells.

Although glucose contributes little carbon to the synthesis of fatty acid chains, it remains an essential source for the glycerol in triglycerides. Glucose is needed, too, for lactose synthesis during lactation (Linzell, 1960). Although glucose can serve as a precursor for the synthesis of non-essential amino acids, the reverse pathways of gluconeogenesis are certainly far more critical in ruminant metabolism.

Lactic Acid

Under normal feeding conditions, lactate formed in the rumen is rapidly metabolized to propionate either via CO_2 fixation to form succinate with subsequent decarboxylation (Johns, 1957), or via acrylate (Ladd and Walker, 1959). The lactate remains at a small or negligible concentration and hence makes no significant contribution to the animal.

Even when the diet contains large amounts of starch or sugars, the lactic acid is rapidly removed provided the rumen pH is maintained above 5 (Dunlop and Hammond, 1965). However, if the rate of lactic acid production is sufficiently rapid to bring the pH below 5, the mechanisms for conversion to propionate are inhibited and lactate accumulates (Reid *et al.*, 1957). In these circumstances, absorption of lactate is significant; absorption of the free acid appears to be much more rapid than of the lactate anion.

Lactate formed during fermentation occurs as both the L– and D–isomers, the proportions of which may vary widely, with either isomer being in excess; the ratio is apparently dependent on the species and strains of the bacteria producing lactate (Dunlop and Hammond, 1965). After absorption into the blood stream, the L–lactate is rapidly converted to pyruvate and glucose and hence does not accumulate to high concentrations in the blood. On the contrary, D–lactate is only very slowly metabolized and may accumulate in the blood. The deleterious effects from excessive production of lactate in the rumen are discussed in Chapter 33.

Lipids

The contribution of acetate and butyrate to the formation of the fatty acids of adipose tissue triglycerides has already been mentioned. In addition, long-chain fatty acids, mostly saturated or mono-unsaturated, derived from fats in the fodder, are made available to the tissue after absorption from the small intestine. The metabolic pathways in the synthesis of long-chain fatty acids, in the formation and hydrolysis of triglycerides, and in the oxidation of fatty acids appear to be common to ruminant and non-ruminant. A special feature in the ruminant is the provision, by fermentation, of the branch-chain fatty acids, *iso*butyric, *iso*valeric and 2–methylbutyric acids, derived from valine, leucine and *iso*leucine respectively. The addition of 2–carbon units to these acids in fatty acid synthesis by rumen bacteria results in the formation of a series of unusual branched-chain acids with odd-numbered carbon chains, especially C_{15} and C_{17} acids (Allison *et al.*, 1962). Also, the hydrogenation of unsaturated fatty acids in the rumen has the effect that the ruminant depot fats are relatively little influenced by the composition of the dietary fats. However, dietary and microbial lipids are not the major sources of tissue lipid; acetate, butyrate and β-hydroxybutyrate are the main sources, and hence the gross composition of lipids in the adipose tissue is rather a characteristic of the individual species. The ruminal hydrogenation serves merely to reduce the contribution of highly unsaturated fatty acids in the diet.

Some of the unsaturated fatty acids of the diet escape hydrogenation in the rumen, and these appear particularly in the cholesterol esters of plasma (Duncan and Garton, 1962). The degree to which ruminants are dependent on a dietary source of unsaturated fatty acids has not been determined. The mammary gland is capable of desaturating stearic acid to form oleic acid, but it is not known whether the animal tissues are capable of producing the more highly unsaturated fatty acids.

In the formation of milk fats, about half of the fatty acid is derived by synthesis in the mammary gland from acetate, while the remainder, especially the C_{18} acids, originate from the fat of plasma chylomicrons and low-density lipoproteins (Linzell, 1960; Barry *et al.*, 1963).

The free fatty acids (FFA) of plasma are a major energy source in ruminants, as in non-ruminants, and the level in the circulating blood is dependent on the supply of other substrates available for tissue oxidation.

The relative importance of the major contributors to the ruminant's energy supply may be gauged from experimental data (Annison *et al.*, 1965, and personal communication). In fed sheep, about 80 per cent of the total oxidative metabolism could be accounted for from the following substrates: glucose 7 per cent, acetate 40 per cent, propionate 14 per cent, butyrate 12 per cent, free fatty acids 8 per cent, ketones 5 per cent. The dominant role of acetate is very evident, while the three VFA together account for about 65 per cent of the energy. The FFA account for only a small fraction, but this should not obscure the vital role played by the FFA in the caloric homeostatis of the body; for example, in the fasted animal, it was found that acetate contributed only 20 per cent of the energy, while the FFA now contributed over 25 per cent.

Protein and Amino Acids

Few experimental studies have been published on the intermediary metabolism of protein in ruminants. All the indications are that, qualitatively, the ruminants conform with the patterns observed in the usual experimental animals.

Indirect evidence indicates that the usually accepted 'essential' amino acids are required by ruminants. This conclusion is based on the finding that radioactive tracers, in substances that are not amino acids, are subsequently found only in the non-essential amino acids (Black *et al.*, 1957; Downes, 1961). This subject has attracted less interest in ruminant nutrition, as it has long been realized that microbial protein from the rumen contributes to the host's protein intake.

It has been noted above that many nitrogenous substances are degraded to ammonia in the rumen, and that this substance occupies a central role in ruminal nitrogen metabolism. The absorption of large amounts of ammonia from the forestomach, together with smaller amounts from the intestines, presents the liver with a high concentration of ammonia in the portal blood (Lewis *et al.*, 1957)—in fact, the concentration (e.g. 17 μg NH_3–N/ml in portal blood) may be far above that which would represent a toxic level in the peripheral circulation. As the blood passes through the liver, the ammonia is so completely removed that peripheral blood concentration is maintained at an extremely low value (less than 1 μg N/ml—McDonald, 1948). This detoxication mechanism operates with such efficiency that, under ordinary feeding conditions, blood ammonia concentration is uninfluenced by nitrogen intake. When ruminal activity produces very high ammonia concentrations, the rate of absorption may increase so much that the liver's capacity to remove ammonia is exceeded, and the peripheral blood concentration starts to rise (Lewis *et al.*, 1957). This situation is reached, in experimental sheep, when rumen concentration rises to about 60 mM (i.e. 840 μg NH_3–N/ml). Animals grazing rich pasture consume very large amounts of protein and non-protein N substances, and frequently exhibit very high concentrations of rumen ammonia (Hogan, 1964); nevertheless, there is no evidence that, in natural feeding conditions, the physiological capacity of the normal liver is ever exceeded to the extent that pathological effects ensue.

A different situation exists when ruminants are given diets with urea replacing a substantial part of the dietary protein. The potent urease activity of rumen contents leads to a very rapid increase in ammonia concentration. This increase is, however, accompanied by a rise in ruminal pH as the non-electrolyte urea is converted to the alkaline ammonia; with normal rations, a high ammonia production is always accompanied by an active fermentation with acid production, so that the pH tends toward acidity. With the ingestion of excess urea, the combination of high rumen ammonia concentration and elevated pH leads to such rapid absorption that the liver's protective function is overwhelmed, ammonia concentration in blood rises, and toxic reactions appear; death usually occurs if the blood ammonia concentration reaches about 8 μg/ml (see McDonald, 1958).

It is a general phenomenon that blood urea returns to the alimentary tract in saliva and other secretions, and by diffusion through the mucous membranes. It is also quite general that the bacteria inhabiting the alimentary tract have sufficient urease activity to produce rapid hydrolysis of the urea with ammonia formation. It is still in dispute (Regoeczi *et al.*, 1965) whether there exists a mammalian urease, but this is unlikely as urea is not transformed to ammonia in germ-free animals (Levenson *et al.*, 1959). It has been suggested that ammonia absorbed from the gastro-intestinal tract is a growth inhibitor but the evidence is equivocal (Visek, 1962).

Although urea diffuses into all the body water, the diffusion is not unimpeded, so that for example, the urea concentration in saliva is appreciably less than in plasma; on the other hand, urea concentration in the renal medulla exceeds that of plasma. It has not yet been established whether there is any impediment to the passage of urea into the gut fluids, although there is some evidence that this is the case (Gärtner *et al.*, 1961). The movement of urea into the gut, hydrolysis to NH_3, absorption of NH_3 and its reconversion to urea in the liver, establishes a continuous urea cycle (McDonald, 1948). Experimental work in recent years has shown that the magnitude of this cycle is surprisingly high; for example, in the rabbit 30–70 per cent of all urea synthesized in the body may pass into the gut and be degraded to ammonia (Regoeczi *et al.*, 1965); in man, values of 15–30 per cent have been recorded (Walser and Bodenlos, 1959). In the ruminant, a large part of the body urea passes into the gut (Decker *et al.*, 1961), but the available data do not allow an estimation of the relative amounts entering the ruminoreticulum (via saliva and by diffusion) and the remainder of the gastro-intestinal tract.

There appears to be no evidence that this urea cycle has any influence on the efficiency of nitrogen utilization by most non-ruminants, though it is conceivable that the provision of recycled nitrogen to the amino N pool in the liver could be a factor in determining the required ratio of essential amino acids to the total nitrogen intake. In the ruminant, however, (and probably in some other species with ruminant-like stomachs), the cycling of nitrogen can have a most important influence on the nitrogen economy, through the utilization of ammonia for microbial growth in the rumen (see Reid, 1953). The microbial growth converts the non-protein N of the rumen digesta to protein; in this way, the urea entering the rumen is returned to the host animal as protein. There is no doubt that recycling of nitrogen helps the ruminant to establish nitrogen equilibrium at a lower level of N intake than would otherwise be possible.

When the diet has a high protein content, nitrogen is available to the animal in excess of its requirements for maintenance, growth, lactation, etc.; the carbon skeletons of amino acids are used for energy production and the excess of nitrogen is converted chiefly to urea; blood urea concentration is high, and renal excretion is rapid, with a high ratio of urea N to total nitrogen in the urine. Under these circumstances, the cycling of urea N confers no benefit on the animal. By contrast, when the ruminant has a diet with a low nitrogen content, the train of events is very different. The rumen microbes now have an excess of energy sources relative to the nitrogenous nutrients; rumen ammonia is taken up by bacteria so that the concentration may be reduced to extremely low levels (less than 1 mg N/100 ml). Urea N entering the rumen by diffusion or in the saliva contributes ammonia for microbial growth, and since the rumen ammonia is kept at a low concentration, absorption is reduced to a negligible amount. This situation is associated with a marked fall in blood urea concentration, reduced urinary excretion of urea, and a profound fall in the urea/total N ratio in the urine.

The difference in urea metabolism at high and low N intakes is well shown by injection of urea into the blood stream; at high N intake, the urea is promptly excreted in the urine, but at low N intake, nitrogen is retained in the body, and a negative N balance may be converted to a positive N balance (Somers, 1961). It has been suggested (Schmidt-Nielsen and O'Dell, 1959) that the kidney has an active mechanism for conserving urea when N intake is low, as the tissues of the renal medulla of sheep are found to contain urea at a higher concentration than in urine.

In Nature, when the ruminant's diet has a low nitrogen content, it is usual that the fodder is highly fibrous and rather resistant to fermentation in the rumen, with the result that microbial growth tends to be limited by the energy supply rather than the nitrogen supply. In modern farm practice, however, it is often feasible to provide rations rich in carbohydrates in the form of starches and sugars, which are energy sources readily available to rumen microbes; now, the capacity of the microbes to take up ammonia can far exceed the rate of entry of urea into the rumen; this situation can be exploited in practice by the addition of urea to the ration.

When the diet provides a plethora of nitrogen, a great deal of ammonia is absorbed from the rumen—much more than enters the rumen as urea from the saliva and blood; thus the total nitrogen in the digesta passing out of the ruminoreticulum tends to be less than the amount of nitrogen in the food (Ellinger and Phillipson, 1964). At the other extreme, with low N intake, the entering urea tends to become 'trapped' by the microbial synthetic activity, and the quantity of digesta N leaving the rumen now exceeds that in the diet, and the amount of amino acid N leaving the rumen may be very much more than is contained in the food (Weston and Hogan, 1966). The nitrogen gained in this way enhances not only the animal's protein status but also its energy status; urea N entering the rumen facilitates the microbial digestion of cellulose and enables the animal to increase its intake of low quality roughage. This has been demonstrated experimentally by administration of urea (Campling *et al.*, 1962), and almost certainly operates when urea is made available to ruminants grazing poor quality pastures. The effect on intake, however, is not simply due to the availability of nitrogenous substances for rumen microbes; there appear to be metabolic influences on the animal such that an improved protein status stimulates voluntary food consumption; thus a supplement of protein, given per duodenum, can increase intake without changing the rate of digestion in the rumen, and even when excess of urea N is provided in the diet (Egan, 1965).

Between the two extremes of plethora and deficit lies a continuous spectrum of dietary situations; at the present time adequate criteria do not exist for delineating the limits of useful contribution of recycled N to the nitrogen economy of the ruminant. The same reservation has to be made regarding the addition of urea to ruminant diets; a very large number of dietetic experiments on urea-containing feeds has been made during the past half century, but progress in evaluating the mechanisms involved has been hampered by lack of critical experimental procedures for measuring the rate of microbial protein synthesis in the rumen.

For similar reasons, there has been uncertainty (and some difference of opinion) concerning the usefulness of ammonia N absorbed from the rumen. In non-ruminant animals, the amino acid assemblage available for absorption by the animal is essentially identical with that in the food proteins. This has enabled the general concepts of food protein evaluation to be developed; there is now substantial agreement that the biological value of food proteins is determined primarily by the content of essential amino acids (Block and Mitchell, 1946), and more specifically by the content of that amino acid in shortest supply relative to the animal's needs. The ratios and contents of the non-essential amino acids in proteins appear to be of little biological importance. A corollary to this concept is that if the animal's requirement of essential amino acids is met by food proteins (as distinct from amino acids supplied as such in the food), then the entire nitrogen requirement will also be met. This corollary seems to be substantially correct, though it has been demonstrated (Bender, 1961) that the highest quality protein (e.g. whole egg protein) does in fact have all the essential amino acids present at concentrations higher than the minimal required; in consequence, this protein could be diluted with some 15 per cent of non-essential nitrogen without reducing its biological value for the growing rat. With this reservation, the following argument appears to be valid.

The biological value of a protein is conventionally expressed as the percentage of the absorbed nitrogen that is retained within the body, i.e. the degree to which a protein can serve the body's needs without waste. Few proteins are rated at a biological value approaching 100 per cent; those proteins with a lower value supply an excess of nitrogen which is eventually excreted in the urine. All the evidence then points to the conclusion that if food protein supplies an adequacy of essential amino acids, additional non-protein N in the diet cannot contribute to the N-economy—the animal is already receiving as much nitrogen as it can utilize.

In transferring this argument to ruminants, it is necessary to postulate that the intermediary metabolism of nitrogenous compounds is essentially the same as in the usual experimental animals (e.g. the rat). As mentioned previously, methods for measuring the amount of protein reaching the abomasum are still rather inadequate. However, it is clear that the ruminant receives very little amino acid by absorption from the forestomach; proteolysis is accompanied by microbial uptake of amino acids and by rapid deamination reactions, so that the amino acid concentration in the ruminal digesta is always very small in comparison with the concentration of protein. With little error then, it can be said that the ruminant's ultimate supply of essential amino acids is wholly derived from proteins of the two categories: undegraded food proteins, and microbial proteins synthesized in the rumen. Plant proteins, both of leaves and seeds, usually have biological values appreciably below 100 per cent. Also, the few experiments so far published indicate that rumen microbial 'protein' (usually taken as being equivalent to total nitrogen) also has a biological value much less than 100 per cent (McNaught *et al.*, 1954). The conclusion then is inevitable that additional nitrogen absorbed as non-protein N, notably ammonia, cannot be useful in the nitrogen economy of the animal's tissues.

This conclusion is not at variance with the previous statement on the value of recycled nitrogen to the ruminant. When recycled nitrogen is of value, it must first be incorporated into the proteins of rumen microbes; the non-protein N can contribute to microbial nitrogen economy but not directly to that of the animal's tissues.

Further support for this conclusion is the finding that a readily digested protein, casein, has a higher biological value when administered per duodenum than when given in the feed (and thus subjected to microbial degadration in the rumen); also, when the solubility of the casein was reduced by heating, its biological value was increased (Chalmers *et al.*, 1954). Similarly, it has been shown (Reis and Schinckel, 1964) that wool growth can be markedly increased by abomasal administration of protein as compared with provision of the same amounts of protein in the food. All these observations can be fully accounted for by the microbial conversion of the amino acids in the food protein to non-amino acid compounds that cannot be utilized by the ruminant.

Available information on the biological value of microbial proteins is inadequate for precise assessment. Individual proteins vary greatly in amino acid composition, and particularly is this true for the specialized proteins such as the storage proteins of seeds; thus, for example, the zein of maize lacks lysine but has a very high content of leucine. In ordinary cells, however, there are very many proteins, and the mixture tends to have a rather consistent amino acid pattern; published analyses of the whole proteins of yeasts, bacteria, plant leaves and liver cells show relatively small differences in composition. On general grounds then, it would be expected that the ruminal microbial proteins would have a biological value comparable with the proteins of the pasture plants or mixed rations usually consumed by ruminants. The measurements of biological value of microbial proteins reported are in conformity with this expectation (see McNaught *et al.*, 1954), but the values given are probably too low as corrections for non-protein substances (especially nucleic acids) have not been made. For the same reason it is not clear whether protozoal proteins have a higher biological value than bacterial proteins. Further uncertainty arises due to the lack of information on the digestibility of bacteria, some components of which may be very resistant to gastro-intestinal digestion. The protozoa swiftly disintegrate in the acid digesta of the abomasum, so presumably their proteins are readily digested.

On the basis of amino acid composition (Block and Mitchell, 1946), the proteins of rumen micro-organisms probably have a biological value of about 75 (compared with whole-egg protein of nearly 100), with methionine+cystine and *iso*leucine as the amino acids in greatest deficit (Weller, 1957; Abdo *et al.*, 1964).

It has been concluded that absorbed non-protein N, especially ammonia, is lost to the N economy of the ruminant. There is another source of loss imposed by microbial activity in the rumen, viz., the excess of metabolic faecal N. In non-ruminants, the faecal N has been conventionally described as consisting of undigested N (chiefly protein) and the metabolic faecal N (MFN) derived from intestinal secretions; appropriate corrections for MFN permitted calculation of the so-called 'true digestibility' of food protein. This concept does not give an adequate description of events in the ruminant gut (Blaxter, 1964). The MFN for ruminants is about 0·45g/100g dry matter ingested (Blaxter and Mitchell, 1948) compared with values of 0·2 for the rat, 0·1 for pig and 0·2 for man. Also, since ruminants usually eat a greater mass of less digestible food, relative to body size, than non-ruminants, it is obvious that this large loss of N in the faeces becomes a major factor determining the protein requirements of ruminants. The critical role of rumen function in this respect is well shown by the finding that urea, ammonium carbonate or casein (Blaxter and Martin, 1962) infused into the abomasum caused little increase in faecal N, but urea added to a low protein diet caused a marked increase in faecal N (Campling *et al.*, 1962). The exact nature of this nitrogen loss has not been ascertained, but it seems likely that some constituents of the rumen bacteria are highly resistant to digestion in the abomasum and small intestine.

Thus, in evolving a highly efficient mechanism for obtaining energy from coarse, fibrous feeds, the ruminant has had to sacrifice some efficiency in the digestion of proteins; this loss is at least partly offset by the recycling of urea N, which is probably most helpful in enabling the ruminant to survive

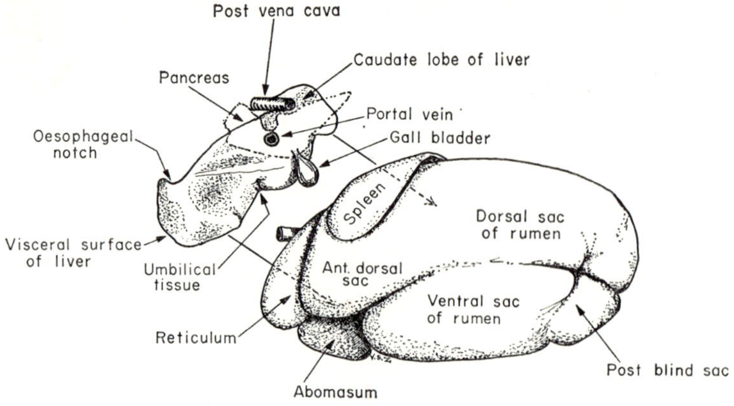

Fig. 1 (a) Digestive tract of ox (left side).

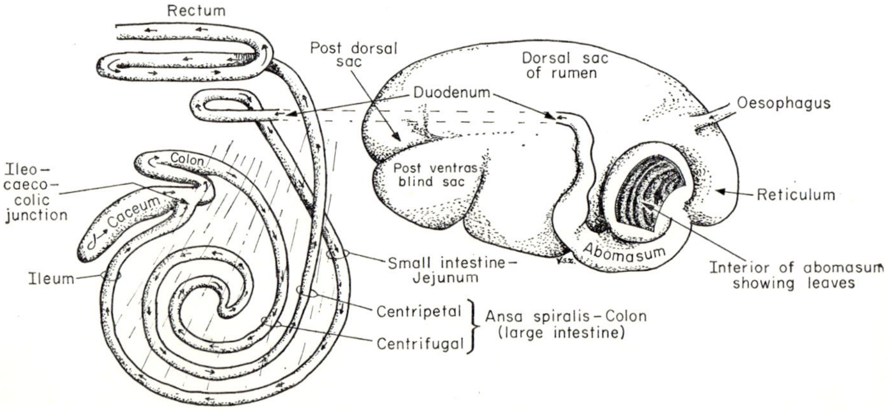

Fig. 1 (b) Intestine of ox (right side).

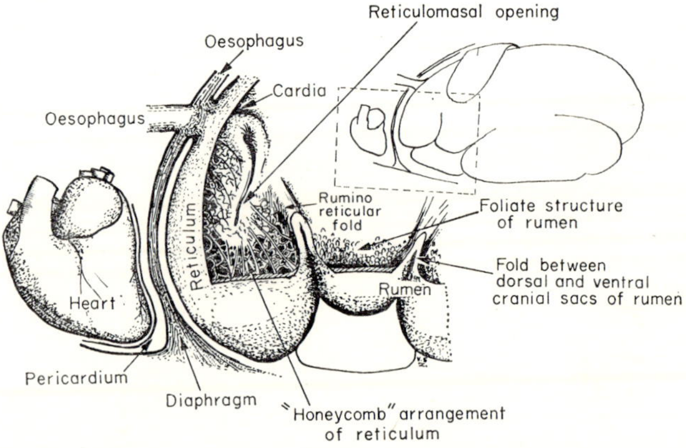

Fig. 1 (c) Detail of rumino reticular area to show relationship of diaphragm and heart to reticulum.

long periods of undernutrition when the fodders have a very low protein content.

It has been noted that a substantial portion of the protein leaving the rumen is composed of microbial 'protein', as measured by conventional procedures. A significant fraction actually consists of the nucleic acids, with rather higher proportions in bacteria than in protozoa; this is in conformity with the known composition of typical micro-organisms grown in pure culture. Published data are inadequate to assess the amount of nucleic acid thus produced in the rumen, but it is likely to be of the order of 10 per cent (expressed as nitrogen) of the apparent microbial protein (Ellis and Pfander, 1965).

The subsequent fate of these nucleic acids in the digestive tract has not been examined, although it is known (Keller *et al.*, 1958) that bovine pancreatic juice contains both ribonuclease and deoxyribonuclease; it is likely that the purine and pyrimidine bases are absorbed and finally excreted in the urine, chiefly as allantoin (Blaxter, 1961). It has been shown experimentally (Blaxter and Martin, 1962) that protein added to the rumen provides an exogenous source of urinary purines. All available evidence indicates that mammalian tissues are capable of synthesizing their own purines and pyrimidines for nucleic acid formation, and the animal is thus independent of an exogenous supply. It may be tentatively concluded, for reasons previously given in respect to amino acids, that the nitrogenous substances derived from the intestinal digestion of microbial nucleic acids are waste products for the ruminant.

REFERENCES

Abdo, K. M., King, K. W. and Engel, R. W. (1964) Protein quality of rumen micro organisms. *J. Anim. Sci.* **23**, 734–6.

Abou Akkada, A. R. and Howard, B. H. (1961) The biochemistry of rumen protozoa. 4. Decomposition of pectic substances. *Biochem. J.* **78**, 512–17.

Abou Akkada, A. R. and Howard, B. H. (1962) The biochemistry of rumen protozoa. 5. The nitrogen metabolism of *Entodinium. Biochem, J.* **82**, 313–20.

Adams, E. P. and Heath, T. J. (1963) The phospholipids of ruminant bile. *Biochim. biophys. Acta* **70**, 688–90.

Allison, M. J., Bryant, M. P., Katz, I. and Keeney, M. (1962) Metabolic function of branched-chain volatile fatty acids, growth factors for ruminococci. II. Biosynthesis of higher branched-chain fatty acids and aldehydes. *J. Bact.* **83**, 1084–93.

Anderson, W. R., Fletcher, T. L., Pitts, C. L. and Harkins, H. N. (1962) Isolation and assay of ovine gastrin. *Nature, Lond.* **193**, 1286–7.

Annison, E. F. (1954a) Some observations on volatile fatty acids in the sheep's rumen. *Biochem. J.* **57**, 400–5.

Annison, E. F. (1954b) Studies on the volatile fatty acids of sheep blood with special reference to formic acid. *Biochem. J.* **58**, 670–80.

Annison, E. F. (1956) Nitrogen metabolism in the sheep. Protein digestion in the rumen. *Biochem. J.* **64**, 705–14.

Annison, E. F. (1964) Plasma free fatty acids. In: *Metabolism and Physiological Significance of Lipids*. Ed. by Dawson, R. M. C. and Rhodes, D. N., p. 289, Wiley: London.

ANNISON, E. F., BROWN, R. E., LENG, R. A., LINDSAY, D. B. and WEST, C. E. (1965) Quantitative evaluation of the role of glucose, acetate, propionate, butyrate, D–β–hydroxybutyrate and plasma free fatty acids in the oxidative metabolism of fed and starved sheep. *Proc. Aust. Physiol. Soc.*, Feb., 1965, p. 1.

ANNISON, E. F., HILL, K. J. and LEWIS, D. (1957) Studies on the portal blood of sheep. 2. Absorption of volatile fatty acids from the rumen of the sheep. *Biochem. J.* **66**, 592–9.

ANNISON, E. F. and WHITE, R. R. (1962a) Further studies on the entry rates of acetate and glucose in sheep, with special reference to endogenous production of acetate. *Biochem. J.* **84**, 546–52.

ANNISON, E. F. and WHITE, R. R. (1962b) Formate metabolism in sheep. *Biochem. J.* **84**, 552–7.

ARMSTRONG, D. G. (1965) Carbohydrate metabolism in ruminants and energy supply. In: *Physiology of Digestion in the Ruminant*. Ed. by Dougherty, R. W. and others, p. 272, Butterworths: Washington.

ARMSTRONG, D. G., BLAXTER, K. L. and GRAHAM, N. MCC. (1957) The heat increments of mixtures of steam-volatile fatty acids in fasting sheep. *Br. J. Nutr.* **11**, 392-408.

ASH, R. W. (1961a) Acid secretion by the abomasum and its relation to the flow of food material in the sheep. *J. Physiol., Lond.* **156**, 93-111.

ASH, R. W. (1961b) Stimuli influencing the secretion of acid by the abomasum of sheep. *J. Physiol., Lond.* **157**, 185–207.

ASH, R. W. (1962) Omaso-abomasal re-entrant cannulae for sheep. *J. Physiol., Lond.* **164**, 4P–5P.

ASH, R. W. and DOBSON, A. (1963) The effect of absorption on the acidity of rumen contents. *J. Physiol., Lond.* **169**, 39–61.

BADAWY, A. M., CAMPBELL, R. M., CUTHBERTSON, D. P. and MACKIE, W. S. (1958a) Further studies on the changing composition of the digesta along the alimentary tract of the sheep. 2. Volatile fatty acids and energy relative to lignin. *Br. J. Nutr.* **12**, 384–90.

BADAWY, A. M., CAMPBELL, R. M., CUTHBERTSON, D. P. and MACKIE, W. S. (1958b) Further studies on the changing composition of the digesta along the alimentary tract of the sheep. 3. Changes in the omasum. *Br. J. Nutr.* **12**, 391–403.

BAILEY, R. W. (1962) α–Galactosidase activity of rumen bacteria. *Nature, Lond.* **195**, 79–80.

BAILEY, C. B. and BALCH, C. C. (1961) Saliva secretion and its relation to feeding in cattle. 2. The composition and rate of secretion of mixed saliva in the cow during rest. *Br. J. Nutr.* **15**, 383–402.

BAKER, F. and HARRISS, S. T. (1947) Microbial digestion in the rumen (and caecum), with special reference to the decomposition of structural cellulose. *Nutr. Abstr. Rev.* **17**, 3–12.

BAKER, F., NASR, H., MORRICE, F. and BRUCE, J. (1950) Bacterial breakdown of structural starches and starch products in the digestive tract of ruminant and non-ruminant mammals. *J. Path. Bact.* **62**, 617–38.

BALCH, C. C. (1960) Rumen digestion and herbage utilization. *Proc. Eighth Int. Grassld Congr.*, Reading, p. 528.

BALCH, C. C., KELLY, A. and HEIM, G. (1951) Factors affecting the utilization of food by dairy cows. 4. The action of the reticulo-omasal orifice. *Br. J. Nutr.* **5**, 207–16.

BALCH, D. A. and ROWLAND, S. J. (1957) Volatile fatty acids and lactic acid in the rumen of dairy cows receiving a variety of diets. *Br. J. Nutr.* **11**, 288–98.

BALDWIN, R. L., WOOD, W. A. and EMERY, R. S. (1962) Conversion of lactate–C^{14} to propionate by rumen microflora. *J. Bact.* **83**, 907–13.

BALDWIN, R. L., WOOD, W. A. and EMERY, R. S. (1963) Conversion of glucose–C^{14} to propionate by the rumen microbiota. *J. Bact.* **85**, 1346–9.

BARCROFT, J., MCANALLY, R. A. and PHILLIPSON, A. T. (1944) Absorption of volatile acids from the alimentary tract of the sheep and other animals. *J. exp. Biol.* **20**, 120–9.

BARRY, J. M., BARTLEY, W., LINZELL, J. L. and ROBINSON, D. S. (1963) The uptake from the blood of triglyceride fatty acids of chylomicra and low-density lipoproteins by the mammary gland of the goat. *Biochem. J.* **89**, 6–11.

BATH, I. H. and HEAD, M. J. (1961) The *in vitro* production, by rumen microorganisms, of volatile fatty acids from cellulose and hemicellulose labelled with ^{14}C. *J. agric. Sci., Camb.* **56**, 131–6.

BECKER, E. R. (1929) Methods of rendering the rumen and reticulum of ruminants free from their normal infusorian fauna. *Proc. natn. Acad. Sci. U.S.A.* **15**, 435–8.

BELL, F. R. and LAWN, A. M. (1955) Localization of regions in medulla oblongata of sheep associated with rumination. *J. Physiol., Lond.* **128**, 577–92.

BENDER, A. E. (1961) Determination of the nutritive value of proteins by chemical analysis. *Publs natn. Res. Coun., Wash.*, No. 843, pp. 407–21.

BENSADOUN, A., PALADINES, O. L. and Reid, J. T. (1962) Effect of level of intake and physical form of the diet on plasma glucose concentration and volatile fatty acid absorption in ruminants. *J. Dairy Sci.* **45**, 1203–10.

BENZIE, D. and PHILLIPSON, A. T. (1957) *The Alimentary Tract of the Ruminant.* Oliver & Boyd: Edinburgh.

BERGMAN, H. D. and DUKES, H. H. (1926) An experimental study of the mechanism of regurgitation in rumination. *J. Am. vet. med. Ass.* **22**, 600–12.

BLACK, A. L., KLEIBER, M., SMITH, A. H. and STEWART, D. N. (1957) Acetate as a precursor of amino acids of casein in the intact dairy cow. *Biochim. biophys. Acta* **23**, 54–9.

BLACKBURN, T. H. (1965) Nitrogen metabolism in the rumen. In: *Physiology of Digestion in the Ruminant.* Ed. by Dougherty, R. W. and others, p. 322, Butterworths: Washington.

BLACKBURN, T. H. and HOBSON, P. N. (1960) Proteolysis in the sheep rumen by whole and fractionated rumen contents. *J. gen. Microbiol.* **22**, 272–81.

BLACKBURN, T. H. and HUNGATE, R. E. (1963) Succinic acid turnover and propionate production in the bovine rumen. *Appl. Microbiol.* **11**, 132–5.

BLAND, D. E. and LOGAN, A. F. (1965) The properties of syringyl, guaiacyl and *p*-hydroxyphenyl artificial lignins. *Biochem. J.* **95**, 515–20.

BLAXTER, K. L. (1961) Energy utilization in the ruminant. In: *Digestive Physiology and Nutrition of the Ruminant.* Ed. by Lewis, D., p. 183, Butterworths: London.

BLAXTER, K. L. (1962) *The Energy Metabolism of Ruminants.* Hutchinson: London.

BLAXTER, K. L. (1964) The nutritional significance of endogenous secretion of nitrogen in ruminants. In: *The Role of the Gastro-intestinal Tract in Protein Metabolism.* Ed. by Munro, H. N., p. 143, Blackwell: Oxford.

BLAXTER, K. L. and CLAPPERTON, J. L. (1965) Prediction of the amount of methane produced by ruminants. *Br. J. Nutr.* **19**, 511–22.

BLAXTER, K. L. and MARTIN, A. K. (1962) The utilization of protein as a source of energy in fattening sheep. *Br. J. Nutr.* **16**, 397–407.

BLAXTER, K. L. and MITCHELL, H. H. (1948) The factorization of the protein requirements of ruminants and of the protein values of feeds, with particular reference to the significance of the metabolic fecal nitrogen. *J. Anim. Sci.* **7**, 351–72.

BLOCK, R. J. and MITCHELL, H. H. (1946) The correlation of the amino-acid composition of proteins with their nutritive value. *Nutr. Abstr. Rev.* **16**, 249–78.

BOYNE, A. W., CAMPBELL, R. M., DAVIDSON, J. and CUTHBERTSON, D. P. (1956) Changes in composition of digesta along the alimentary tract of sheep. *Br. J. Nutr.* **10**, 325–33.

BRIDGE, P. S. and SPRATLING, F. R. (1966) The importance of vitamin A in the feeding of farm ruminants. In: *Recent Advances in Animal Nutrition.* Ed. by Abrams, J. T., p. 143, Churchill: London.

BROBERG, G. (1957) Measurements of the redox potential in rumen contents. *Nord. VetMed.* **9**, 918–50.

BROUWER, E. (1961) Mineral relationships of the ruminant. In: *Digestive Physiology and Nutrition of the Ruminant.* Ed. by Lewis, D., p. 154, Butterworths: London.

BRYANT, M. P. (1959) Bacterial species of the rumen. *Bact. Rev.* **23**, 125–53.

BURAU, R. and STOUT, P. R. (1965) Trans-aconitic acid in range grasses in early spring. *Science, N.Y.* **150**, 766–7.

CAMPBELL, C. D., MCLAREN, G. A., ANDERSON, G. C., WELCH, J. A., SMITH, G. S. and BROOKS, J. (1959) Sarcosine and creatine as nitrogen sources for the ruminant, and the influence of supplemental vitamin B_{12}. *J. Anim. Sci.* **18**, 780–9.

CAMPLING, R. C., FREER, M. and BALCH, C. C. (1962) Factors affecting the voluntary intake of food by cows. 3. The effect of urea on the voluntary intake of oat straw. *Br. J. Nutr.* **16**, 115–24.

CARROLL, E. J. and HUNGATE, R. E. (1955) Formate dissimilation and methane production in bovine rumen contents. *Archs Biochem. Biophys.* **56**, 525–36.

CHALMERS, M. I., CUTHBERTSON, D. P. and Synge, R. L. M. (1954) Ruminal ammonia formation in relation to the protein requirement of sheep. I. Duodenal administration and heat processing as factors influencing fate of casein supplements. *J. agric. Sci., Camb.* **44**, 254–62.

CHALMERS, M. I. and SYNGE, R. L. M. (1954) The digestion of protein and nitrogenous compounds in ruminants. *Adv. Protein Chem.* **9**, 93–120.

CLUNIES ROSS, I. (1934) Passage of fluids through the ruminant stomach. II. With observations on the effect of long starvation on anthelmintic efficiency. *Aust. vet. J.* **10**, 11-23.

COATS, D. A., DENTON, D. A., GODING, J. R. and WRIGHT, R. D. (1956) Secretion by the parotid gland of the sheep. *J. Physiol., Lond.* **131**, 13–31.

COOK, R. M., BROWN, R. E. and DAVIS, C. L. (1965) Protein metabolism in the rumen. I. Absorption of glycine and other amino acids. *J. Dairy Sci.* **48**, 475–83.

COOMBE, J. B., TRIBE, D. E. and MORRISON, J. W. C. (1960) Some experimental observations on the toxicity of urea to sheep. *Aust. J. agric. Res.* **11**, 247–56.

COOP, I. E. and BLAKLEY, R. L. (1949) Metabolism and toxicity of cyanides and cyanogenetic glucosides in sheep. I. Activity in the rumen. *N.Z. Jl Sci. Technol. A.* **30**, 277–91.

COOP, I. E. and BLAKLEY, R. L. (1950) The metabolism and toxicity of cyanides and cyanogenetic glucosides in sheep. III. The toxicity of cyanides and cyanogenetic glucosides. *N.Z. Jl Sci. Technol. A.* **31**, 44–58.

CUNNINGHAM, M. D., MARTZ, F. A. and MERILAN, C. P. (1964) Effect of drinking water temperature upon ruminant digestion, intraruminal temperature and water consumption of non-lactating dairy cows. *J. Dairy Sci.* **47**, 382–5.

DANIELLI, J. F., HITCHCOCK, M. W. S., MARSHALL, R. A. and PHILLIPSON, A. T. (1945) The mechanism of absorption from the rumen as exemplified by the behaviour of acetic, propionic and butyric acids. *J. exp. Biol.* **22**, 75–84.

DECKER, P., GÄRTNER, K., HÖRNICKE, H. and HILL, H. (1961) Continuous measurement of urea synthesis and urea reflux into the rumen in relation to urine discharge by means of ^{14}C-urea in goats. (Trans. title). *Pflügers Arch. ges. Physiol.* **274**, 289–94.

DENTON, D. A. (1956) The effect of Na^+ depletion on the $Na^+:K^+$ ratio of the parotid saliva of the sheep. *J. Physiol., Lond.* **131**, 516–25.

DENTON, D. A. (1957) The study of sheep with permanent unilateral parotid fistulae. *Q. Jl exp. Physiol.* **42**, 72–95.

DIXON, M. (1951) *Manometric Methods as applied to the Measurement of Cell Respiration and other Processes.* 3rd ed. University Press: Cambridge.

DOBSON, A. (1959) Active transport through the epithelium of the reticulo-rumen sac. *J. Physiol., Lond.* **146**, 235–51.

DOBSON, A. (1961) Absorption from the rumen. In: *Digestive Physiology and Nutrition of the Ruminant.* Ed. by Lewis, D., p. 68, Butterworths: London.

DOBSON, A. and BULLEN, J. J. (1964) A method for the control of E_h and pH during bacterial growth. *J. gen. Microbiol.* **35**, 169–74.

DOBSON, A. and PHILLIPSON, A. T. (1958) The absorption of chloride ions from the reticulo-rumen sac. *J. Physiol., Lond.* **140**, 94–104.

DODSON, M. E. (1959) Oxalate ingestion studies in the sheep. *Aust. vet. J.* **35**, 225–33.

DOUGHERTY, R. W. and HABEL, R. E. (1955) The cranial oesophageal sphincter; its action and its relation to eructation in sheep as determined by cinefluorography. *Cornell Vet.* **45**, 459–64.

DOUGHERTY, R. W., STEWART, W. E., NOLD, M. M., LINDAHL, I. L., MULLENAX, C. H. and LEEK, B. F. (1962) Pulmonary absorption of eructated gas in ruminants. *Am. J. vet. Res.* **23**, 205–12.

DOWNES, A. M. (1961) On the amino acids essential for the tissues of the sheep. *Aust. J. biol. Sci.* **14**, 254–9.

DUNCAN, W. R. H. and GARTON, G. A. (1962) The C_{18} fatty acids of ox plasma lipids. *J. Lipid Res.* **3**, 53–5.

DUNLOP, R. H. (1961) A study of factors related to the functional impairment resulting from loading the rumen of cattle with high carbohydrate feeds. Ph.D. Thesis, University of Minnesota.

DUNLOP, R. H. and HAMMOND, P. B. (1965) D-Lactic acidosis of ruminants. *Ann. N.Y. Acad. Sci.* **119**, 1109–32.

EGAN, A. R. (1965) Nutritional status and intake regulation in sheep. II. The influence of sustained duodenal infusions of casein or urea upon voluntary intake of low-protein roughages by sheep. *Aust. J. agric. Res.* **16**, 451–62.

EKMAN, J. and SPERBER, I. (1953) The distribution of concentrations of bicarbonate (including carbon dioxide) and chloride in the omasum of cows. *K. LantbrHögsk. Annlr* **19**, 227–31.

ELLINGER, G. M. and PHILLIPSON, A. T. (1964) The nitrogenous compounds passing to the duodenum of sheep. In: *The Role of the Gastro-intestinal Tract in Protein Metabolism.* Ed. by Munro, H. N., p. 137, Blackwell: Oxford.

ELLIS, W. C. and Pfander, W. H. (1965) Rumen microbial polynucleotide synthesis and its possible role in ruminant nitrogen utilization. *Nature, Lond.* **205**, 974–5.

ELSDON, S. R., VOLCANI, B. E., GILCHRIST, F. M. C. and LEWIS, D. (1956) Properties of a fatty acid forming organism from the rumen of sheep. *J. Bact.* **72**, 681–9.

EL-SHAZLY, K. (1952) Degradation of protein in the rumen of the sheep. 2. The action of rumen microorganisms on amino acids. *Biochem. J.* **51**, 647–53.

FITZGERALD, L. R. (1957) Cutaneous respiration in man. *Physiol. Rev.* **37**, 325–36.

FORD, J. E., PERRY, K. D. and BRIGGS, C. A. E. (1958) Nutrition of lactic acid bacteria isolated from the rumen. *J. gen. Microbiol.* **18**, 273–84.

FORSYTH, G. and HIRST, E. L. (1953) Protozoal polysaccharides. Structure of the polysaccharide produced by the holotrich ciliates present in sheep's rumen. *J. chem. Soc.* **1953,** 2132–5.

FREER, M. and CAMPLING, R. C. (1965) Factors affecting the voluntary intake of food by cows. 7. The behaviour and reticular motility of cows given diets of hay, dried grass, concentrates and ground, pelleted hay. *Br. J. Nutr.* **19**, 195–207.

GAILLARD, B. D. E. (1962) The relationship between the cell-wall constituents of roughages and the digestibility of the organic matter. *J. agric. Sci., Camb.* **59**, 369–73.

GÄRTNER, K., DECKER, P. and HILL, H. (1961) On the passage of urea and ammonia through the rumen wall of the goat. (Trans. title). *Pflügers Arch. ges. Physiol.* **274**, 281–8.

GARTON, G. A. (1964) Aspects of lipid metabolism in ruminants. In: *Metabolism and Physiological Significance of Lipids.* Ed. by Dawson, R. M. C. and Rhodes, D. N., p. 335, Wiley: London.

GARTON, G. A. (1965) Digestion and assimilation of lipids. In: *Physiology of Digestion in the Ruminant.* Ed. by Dougherty, R. W. and others, p. 390, Butterworths: Washington.

GARTON, G. A. and OXFORD, A. E. (1955) The nature of bacterial lipids in the rumen of hay-fed sheep. *J. Sci. Fd Agric.* **6**, 142–8.

GOODALL, E. D. and KAY, R. N. B. (1965) Digestion and absorption in the large intestine of the sheep. *J. Physiol., Lond.* **176**, 12–23.

GORDON, J. G. (1958) The act of rumination. *J. agric. Sci., Camb.* **50**, 34–42.

GRAY, F. V. (1948) The absorption of volatile fatty acids from the rumen. II. Influence of pH on absorption. *J. exp. Biol.* **25**, 135–44.

GRAY, F. V., PILGRIM, A. F. and WELLER, R. A. (1954) Functions of the omasum in the stomach of the sheep. *J. exp. Biol.* **31**, 49–55.

HABEL, R. E. (1965) Anatomical and histological nomenclature of the ruminant stomach. In: *Physiology of Digestion in the Ruminant.* Ed. by Dougherty, R. W., and others, p. 15. Butterworths: Washington.

HALLIWELL, G. (1957) Cellulolysis by rumen microorganisms. *J. gen. Microbiol.* **17**, 153–65.

HALLIWELL, G. (1965) Hydrolysis of fibrous cotton and reprecipitated cellulose by cellulolytic enzymes from soil microorganisms. *Biochem. J.* **95**, 270–81.

HALLIWELL, G. and BRYANT, M. P. (1963) The cellulolytic activity of pure strains of bacteria from the rumen of cattle. *J. gen. Microbiol.* **32**, 441–8.

HANSEN, A. and SCHMIDT-NIELSEN, K. (1957) On the stomach of the camel, with special reference to the structure of its mucous membrane. *Acta anat.* **31**, 353–75.

HARDWICK, D. C., LINZELL, J. L. and PRICE, S. M. (1961) The effect of glucose and acetate on milk secretion by the perfused goat udder. *Biochem. J.* **80**, 37–45.
HARRISON, F. A. (1962) Bile secretion in the sheep. *J. Physiol., Lond.* **162**, 212–24.
HEALD, P. J. (1951) The assessment of glucose-containing substances in rumen micro-organisms during a digestion cycle in sheep. *Br. J. Nutr.* **5**, 84–93.
HEATH, T. J. and MORRIS, B. (1962) The absorption of fat in sheep and lambs. *Q. Jl exp. Physiol.* **47**, 157–69.
HENSCHEL, M. J., HILL, W. B. and PORTER, J. W. G. (1961) The development of proteolytic enzymes in the abomasum of the young calf. *Proc. Nutr. Soc.* **20**, 40–1.
HILL, K. J. (1960) Abomasal secretion in the sheep. *J. Physiol., Lond.* **154**, 115–32.
HILL, K. J. (1961) Digestive secretions in the ruminant. In: *Digestive Physiology and Nutrition of the Ruminant*. Ed. by Lewis, D., p. 48. Butterworths: London.
HILL, K. J. (1965) Abomasal secretory function in the sheep. In: *Physiology of Digestion in the Ruminant*. Ed. by Dougherty, R. W. and others, p. 221. Butterworths: Washington.
HILL, K. J. and MANGAN, J. L. (1964) The formation and distribution of methylamine in the ruminant digestive tract. *Biochem. J.* **93**, 39–45.
HIRD, F. J. R. and WEIDEMANN, M. J. (1964) Ketone-body synthesis in relation to age of lambs. *Biochem. J.* **93**, 423–30.
HIRST, E. L., MACKENZIE, D. J. and WYLAM, CLARE B. (1959) Analytical studies on the carbohydrates of grasses and clovers. IX. Changes in carbohydrate composition during the growth of lucerne. *J. Sci. Fd Agric.* **10**, 19–26.
HOBSON, P. N. and MANN, S. O. (1961) The isolation of glycerol-fermenting and lipolytic bacteria from the rumen of the sheep. *J. gen. Microbiol.* **25**, 227–40.
HÖRNICKE, H. (1964) The passage of substances through the ruminant stomach wall. (Trans. title). *Z. Tierphysiol. Tierernähr. Futtermittel K.* **19**, 92–117.
HÖRNICKE, H., WILLIAMS, W. F., WALDO, D. R. and FLATT, W. P. (1965) Composition and absorption of rumen gases and their importance for the accuracy of respiration trials with tracheostomized ruminants. In: *Energy Metabolism:* Proceedings of the 3rd Symposium held at Troon, Scotland, May, 1964. Ed. by Blaxter, K. L., p. 165. Academic Press: New York.
HOGAN, J. P. (1961) The absorption of ammonia through the rumen of the sheep. *Aust. J. biol. Sci.* **14**, 448–60.
HOGAN, J. P. (1964) The digestion of food by the grazing sheep. II. The production of ammonia in the rumen. *Aust. J. agric. Res.* **15**, 397–407.
HOGAN, J. P. (1965) The digestion of food by the grazing sheep. IV. The effect of mastication on the release of soluble plant nitrogen. *Aust. J. agric. Res.* **16**, 855–62.
HOUPT, T. R. (1959) Utilization of blood urea in ruminants. *Am. J. Physiol.* **197**, 115–20.
HOWARD, B. H. (1963) Hydrolysis of naturally occurring galactosides by some rumen protozoa. *Biochem. J.* **89**, 90P–91P.
HUNGATE, R. E. (1942) The culture of *Eudiplodinium neglectum*, with experiments on the digestion of cellulose. *Biol. Bull. mar. biol. Lab., Woods Hole*, **83**, 303–19.
HYDÉN, S. (1961) The use of reference substances and the measurement of flow in the alimentary tract. In: *Digestive Physiology and Nutrition of the Ruminant*. Ed. by Lewis, D., p. 35. Butterworths: London.
HYDÉN, S. and SPERBER, I. (1965) Electron microscopy of the ruminant forestomach. In: *Physiology of Digestion in the Ruminant*. Ed. by Dougherty, R. W., and others, p. 51. Butterworths: Washington.
JAYASURIYA, G. C. N. and HUNGATE, R. E. (1959) Lactate conversions in the bovine rumen. *Archs Biochem. Biophys.* **82**, 274–87.
JOHNS, A. T. (1951) Isolation of a bacterium, producing propionic acid, from the rumen of sheep. *J. gen. Microbiol.* **5**, 317–25.
JOHNSON, B. C., HAMILTON, T. S., ROBINSON, W. B. and GAREY, J. C. (1944) On the mechanism of non-protein-nitrogen utilization by ruminants. *J. Anim. Sci.* **3**, 287–98.
JOHNSON, R. R. and MCCLURE, K. E. (1964) *In vitro* and *in vivo* comparisons on the utilization of urea, biuret and diammonium phosphate by sheep. *J. Anim. Sci.* **23**, 208–13.

JOYNER, A. E., Jnr., KESLER, E. M. and HOLTER, J. B. (1963) Absorption from the omasum and subsequent metabolism of butyrate and acetate. *J. Dairy Sci.* **46**, 1108–13.

JURTSHUK, P., Jnr., DOETSCH, R. N. and SHAW, J. C. (1958) Anaerobic purine dissimilation by washed suspensions of bovine rumen bacteria. *J. Dairy Sci.* **41**, 190–202.

KAY, R. N. B. (1958) Continuous and reflex secretion by the parotid gland in ruminants. *J. Physiol., Lond.* **144**, 463–75.

KAY, R. N. B. (1966) The influence of saliva on digestion in ruminants. *Wld Rev. Nutr. Diet.* **6**, 292–325.

KAY, R. N. B. and HOBSON, P. N. (1963) Reviews of the progress of dairy science. Section A, Physiology. Part 1. The physiology of the rumen. Part 2. Rumen microbiology. *J. Dairy Res.* **30**, 261–313.

KEATING, E. K., HALE, W. H. and HUBBERT, F. Jnr. (1964) *In vitro* degradation of vitamin A and carotene by rumen liquor. *J. Anim. Sci.* **23**, 111–17.

KELLER, P. J., COHEN, E. and NEURATH, H. (1958) The proteins of bovine pancreatic juice. *J. biol. Chem.* **233**, 344–9.

KON, S. K. and PORTER, J. W. G. (1954) The intestinal synthesis of vitamins in the ruminant. *Vitams Horm.* **12**, 53–68.

KREBS, H. A., HEMS, R. and GASCOYNE, T. (1963) Renal gluconeogenesis. IV. Gluconeogenesis from substrate combinations. *Acta biol. med. germ.* **11**, 607–15.

KRZYWANEK, Fr. W. (1929) The rumen temperature in sheep. (Trans. title). *Pflügers Arch. ges. Physiol.* **222**, 89–96.

LADD, J. N. and WALKER, D. J. (1959) The fermentation of lactate and acrylate by the rumen microorganism L C. *Biochem. J.* **71**, 364–73.

LENG, R. A. and ANNISON, E. F. (1964) The metabolism of D(–)–β–hydroxybutyrate in sheep. *Biochem. J.* **90**, 464–9.

LENG, R. A. and LEONARD, G. J. (1965) Measurement of the rates of production of acetic, propionic and butyric acids in the rumen of sheep. *Br. J. Nutr.* **19**, 469–84.

LEVENSON, S. M., CROWLEY, L. V., HOROWITZ, R. E. and MALM, O. J. (1959) The metabolism of carbon-labeled urea in the germ-free rat. *J. biol. Chem.* **234**, 2061–2.

LEWIS, D. (1951a) Metabolism of nitrate and nitrite in the sheep. I. The reduction of nitrate in the rumen of the sheep. *Biochem. J.* **48**, 175–80.

LEWIS, D. (1951b) Metabolism of nitrate and nitrite in the sheep. II. Hydrogen donators in nitrate reduction by rumen microorganisms *in vitro*. *Biochem. J.* **49**, 149–53.

LEWIS, D. (1955) Amino-acid metabolism in the rumen of the sheep. *Br. J. Nutr.* **9**, 215–30.

LEWIS, D. and ELSDEN, S. R. (1955) The fermentation of L-threonine, L-serine, L-cysteine and acrylic acid by a gram-negative coccus. *Biochem. J.* **60**, 683–92.

LEWIS, D., HILL, K. J. and ANNISON, E. F. (1957) Studies on portal blood of sheep. I. Absorption of ammonia from the rumen of the sheep. *Biochem. J.* **66**, 587–92.

LINZELL, J. L. (1960) Mammary-gland blood flow and oxygen, glucose and volatile fatty acid uptake in the conscious goat. *J. Physiol., Lond.* **153**, 492–509.

LOCKWOOD, L. B., YODER, D. E. and ZIENTY, M. (1965) Lactic acid. *Ann. N.Y. Acad. Sci.* **119**, 854–67.

MCANALLY, R. A. (1942) Digestion of straw by the ruminant. *Biochem. J.* **36**, 392–9.

MCANALLY, R. A. and PHILLIPSON, A. T. (1944) Digestion in the ruminant. *Biol. Rev.* **19**, 41–54.

MCARTHUR, J. M. and MILTIMORE, J. E. (1961) Rumen gas analysis by gas-solid chromatography. *Can. J. Anim. Sci.* **41**, 187–96.

MCDONALD, I. W. (1948) The absorption of ammonia from the rumen of the sheep. *Biochem. J.* **42**, 584–7.

MCDONALD, I. W. (1952) The role of ammonia in ruminal digestion of protein. *Biochem. J.* **51**, 86–90.

MCDONALD, I. W. (1954) The extent of conversion of food protein to microbial protein in the rumen of the sheep. *Biochem. J.* **56**, 120–5.

MCDONALD, I. W. (1958) The utilization of ammonia-nitrogen by the sheep. *Proc. Aust. Soc. Anim. Prod.* **2**, 46–51.

McElroy, L. W. and Goss, H. (1940) A quantitative study of vitamins in the rumen contents of sheep and cows fed vitamin-low diets. I. Riboflavin and vitamin K. II. Vitamin B_6 (pyridoxine). *J. Nutr.* **20**, 527–50.

McNaught, M. L., Owen, E. C., Henry, K. M. and Kon, S. K. (1954) The utilization of non-protein nitrogen in the bovine rumen. 8. The nutritive value of the proteins of preparations of dried rumen bacteria, rumen protozoa and brewer's yeast for rats. *Biochem. J.* **56**, 151–6.

McNaught, M. L., Smith, J. A. B. and Black, W. A. P. (1954). The utilization of carbohydrates of seaweed by rumen microflora *in vitro. J. Sci. Fd Agric.* **5**, 350–2.

Magee, D. F. (1961) An investigation into the external secretion of the pancreas in sheep. *J. Physiol., Lond.* **158**, 132–43.

Marston, H. R. (1952) Cobalt, copper and molybdenum in the nutrition of animals and plants. *Physiol. Rev.* **32**, 66–121.

Masson, M. (1950) Microscopic studies of the alimentary microorganisms of the sheep. *Br. J. Nutr.* 4, viii-ix.

Masson, M. J. and Phillipson, A. T. (1951) Absorption of acetate, propionate and butyrate from the rumen of sheep. *J. Physiol., Lond.* **113**, 189–206.

Mayfield, E. D., Smith, J. L. and Johnson, B. C. (1965) Metabolism of acetate by sheep liver homogenates. *J. Dairy Sci.* **48**, 93–8.

Maymone, B., Tiberio, M., Dattilo, M. and Giacomini, A. (1961) Oxalic acid metabolism in ruminants on prolonged diet of *Oxalis cernua* Thunb. *VIII Int. Congr. Anim. Husb.* **3**, 58–61.

Moir, R. J. (1965) The comparative physiology of ruminant-like animals. In: *Physiology of Digestion in the Ruminant.* Ed. by Dougherty, R. W., and others, p. 1. Butterworths: Washington.

Moomaw, C. R. and Hungate, R. E. (1963) Ethanol conversion in the bovine rumen. *J. Bact.* **85**, 721–2.

Mullenax, C. H., Buck, W. B., Keeler, R. F. and Binns, W. (1966) Stimulating eructation and vomition in normal and bloated ruminants with alkaloidal extracts from *Veratrum spp. Am. J. Vet. Res.* 27, 211–22.

Munch-Petersen, E. (1964) A new culture medium for rumen bacteria. *Zentbl. Bakt. ParasitKde* 1. Orig. **193**, 353–62.

Murray, M. G., Reid, R. S. and Sutherland, T. M. (1962) The rate of passage of digesta from the reticulo-rumen of the sheep. *J. Physiol., Lond.* **164**, 26P–27P.

O'Dell, G. D., King, W. A., Cook, W. C. and Moore, S. L. (1963) Effect of physical state of coastal Bermuda grass hay on passage through digestive tract of dairy heifers. *J. Dairy Sci.* **46**, 38–42.

Otterby, D. E., Ramsey, H. A. and Wise, G. H. (1964) 1. Lipolysis of milk fat by pregastric esterase in the abomasum of the calf. 2. Source of lipolytic enzymes in the abomasum of the calf. *J. Dairy Sci.* **47**, 993–6 and 997–9.

Oxford, A. E. (1964) A guide to rumen microbiology. *Bull. N.Z. Dep. Scient. ind. Res.*, No. 160.

Pearce, G. R. and Moir, R. J. (1964) Rumination in sheep. 1. The influence of rumination and grinding upon the passage and digestion of food. *Aust. J. agric. Res.* **15**, 635–44.

Pennington, R. J. (1952) The metabolism of short-chain fatty acids in the sheep. 1. Fatty acid utilization and ketone body production by rumen epithelium and other tissues. *Biochem. J.* **51**, 251–8.

Pennington, R. J. and Sutherland, T. M. (1956) The metabolism of short-chain fatty acids in the sheep. 4. The pathway of propionate metabolism in rumen epithelial tissue. *Biochem. J.* **63**, 618–28.

Phillipson, A. T. (1952) The passage of digesta from the abomasum of sheep. *J. Physiol., Lond.* **116**, 84–97.

Phillipson, A. T. (1955) Sodium transport and its role in ruminant digestion. *Vet. Rec.* **67**, 1048–53.

Phillipson, A. T. (1964) The digestion and absorption of nitrogenous compounds in the ruminant. In: *Mammalian Protein Metabolism.* Ed. by Munro, H. N. and Allison, J. B., Vol. 1, p. 71. Academic Press: New York.

PHILLIPSON, A. T. and ASH, R. W. (1965) Physiological mechanisms affecting the flow of digesta in ruminants. In: *Physiology of Digestion in the Ruminant*. Ed. by Dougherty, R. W., and others, p. 97. Butterworths: Washington.

PILGRIM, A. F. (1948) The production of methane and hydrogen by the sheep. *Aust. J. scient. Res.* **B1**, 130–8.

RAMSEY, H. A. and DAVIS, C. L. (1965) Metabolism of N–butyrate by the adult goat. *J. Dairy Sci.* **48**, 381–90.

RAND, A. G. and Ernstrom, C. A. (1964) Effect of pH and sodium chloride on activation of prorennin. *J. Dairy Sci.* **47**, 1181–7.

REED, F. M., MOIR, R. J. and UNDERWOOD, E. J. (1949) Rumenal flora studies in the sheep. I. The nutritive value of rumen bacterial protein. *Aust. J. scient. Res.* **B2**, 304–17.

REGOECZI, E., IRONS, L., KOJ, A. and MCFARLANE, A. S. (1965) Isotopic studies on urea metabolism in rabbits. *Biochem. J.* **95**, 521–32.

REID, C. S. W., LYTTLETON, J. W. and MANGAN, J. L. (1962) Bloat in cattle. XXIV. A method of measuring the effectiveness of chewing in the release of plant cell contents from ingested feed. *N.Z. Jl agric. Res.* **5**, 237–48.

REID, J. T. (1953) Urea as a protein replacement for ruminants: a review. *J. Dairy Sci.* **36**, 955–96.

REID, R. L. (1963) The nutritional physiology of the pregnant ewe. *J. Aust. Inst. agric. Sci.* **29**, 215–23.

REID, R. L., HOGAN, J. P. and BRIGGS, P. K. (1957) The effect of diet on individual volatile fatty acids in the rumen of sheep, with particular reference to the effect of low pH and adaptation on high starch diets. *Aust. J. agric. Res.* **8**, 691–710.

REIS, P. J. and SCHINCKEL, P. G. (1964) The growth and composition of wool. II. The effect of casein, gelatin and sulphur-containing amino acids given per abomasum. *Aust. J. biol. Sci.* **17**, 532–47.

REPP, W. W., HALE, W. H. and BURROUGHS, W. (1955) The value of several non-protein-nitrogen compounds as protein substitutes in lamb fattening rations. *J. Anim. Sci.* **14**, 901–8.

RIDGES, A. P. and SINGLETON, A. G. (1962) Some quantitative aspects of digestion in goats. *J. Physiol., Lond.* **161**, 1–9.

RIEK, R. F. (1954) The influence of sodium salts on the closure of the oesophageal groove in calves. *Aust. vet. J.* **30**, 29–37.

SABINE, J. R. and JOHNSON, B. C. (1964) Acetate metabolism in the ruminant. *J. biol. Chem.* **239**, 89–93.

SANDER, E. G., WARNER, R. G., HARRISON, H. N. and LOOSLI, J. K. (1959) The stimulatory effect of sodium butyrate and sodium propionate on the development of the rumen mucosa in the young calf. *J. Dairy Sci.* **42**, 1600–5.

SATTER, L. D., SUTTIE, J. W. and BAUMGARDT, B. R. (1964) Dietary induced changes in volatile fatty acid formation from α–cellulose–C^{14} and hemicellulose–C^{14}. *J. Dairy Sci.* **47**, 1365–70.

SCHAFER, M. L. and KING, K. W. (1965) Utilization of cellulose oligosaccharides by *Cellvibrio gilvus*. *J. Bact.* **89**, 113–16.

SCHMIDT-NIELSEN, B. and O'DELL, R. (1959) Effect of diet on distribution of urea and electrolytes in kidneys of sheep. *Am. J. Physiol.* **197**, 856–60.

SCHWARTZ, HELEN M., SCHOEMAN, C. A. and FÄRBER, M. S. (1964) Utilization of urea by sheep. I. Rates of breakdown of urea and carbohydrates *in vivo* and *in vitro*. *J. agric. Sci., Camb.* **63**, 289–96.

SCOTT, T. W., WARD, P. F. V. and DAWSON, R. M. C. (1964) The formation and metabolism of phenyl-substituted fatty acids in the ruminant. *Biochem. J.* **90**, 12–24.

SELLERS, A. F. (1965) Blood flow in the rumen vessels. In: *Physiology of Digestion in the Ruminant*. Ed. by Dougherty, R. W., and others, p. 171. Butterworths: Washington.

SHORLAND, F. B., WEENINK, R. O., JOHNS, A. T. and MCDONALD, I. R. C. (1957) The effect of sheep-rumen contents on unsaturated fatty acids. *Biochem. J.* **67**, 328–33.

SIMONNET, H., LE BARS, H. and MOLLÉ, J. (1957) The cycle of urea administered orally to ruminants. (Trans. title). *C.r. hebd. Séanc. Acad. Sci., Paris.* **244**, 943–5.

SIROTNAK, F. M., DOETSCH, R. N., ROBINSON, R. Q. and SHAW, J. C. (1954) Aspartate dissimilation reactions of rumen bacteria. *J. Dairy Sci.* **37**, 531–7.

SMART, W. W. G. Jnr., BELL, T. A., MOCHRIE, R. D. and STANLEY, N. W. (1964) Pectic enzymes of rumen fluid. *J. Dairy Sci.* **47**, 1220–3.

SMITH, A. D. M. and FOULKES, M. (1966) Cyanide excretion in the rat. *Nature, Lond.* **209**, 919–20.

SMITH, P. H. and HUNGATE, R. E. (1958) Isolation and characterization of *Methanobacterium ruminantium* n. sp. *J. Bact.* **75**, 713–18.

SMITH, R. M. and OSBORNE-WHITE, W. S. (1965) Metabolism of propionate by sheep liver. Oxidation of propionate by homogenates. *Biochem. J.* **95**, 411–22.

SOMERS, M. (1957) Saliva secretion and its functions in ruminants. *Aust. vet. J.* **33**, 297–301.

SOMERS, M. (1961) Factors influencing the secretion of nitrogen in sheep saliva. 4. The influence of injected urea on the quantitative recovery of urea in the parotid saliva and the urinary excretions of sheep. *Aust. J. exp. Biol. med. Sci.* **39**, 145–56.

STACY, B. D. and WARNER, A. C. I. (1966) Balances of water and sodium in the rumen during feeding: osmotic stimulation of sodium absorption in the sheep. *Q. Jl exp. Physiol.* **51**, 79–93.

STEVENS, C. E. and SELLERS, A. F. (1960) Pressure events in bovine oesophagus and reticulorumen associated with eructation, deglutition and regurgitation. *Am. J. Physiol.* **199**, 598–602.

STEVENS, C. E., SELLERS, A. F. and SPURRELL, F. A. (1960) Function of the bovine omasum in ingesta transfer. *Am. J. Physiol.* **198**, 449–55.

STEWARD, F. C. and POLLARD, J. K. (1957) Nitrogen metabolism in plants. Ten years in retrospect. *A. Rev. Pl. Physiol.* **8**, 65–114.

STEWARD, F. C. and THOMPSON, J. F. (1954). In: *The Proteins, Chemistry, Biological Activity and Methods.* Ed. by Neurath, H. and Bailey, K., Vol. IIa, p.513. Academic Press, New York.

SUTTON, J. D., MCGILLIARD, A. D. and JACOBSON, N. L. (1963) Functional development of rumen mucosa. 1. Absorptive ability. *J. Dairy Sci.* **46**, 426–36.

SYM, E. A. (1938) Hydrolytic activity in the contents of the horse caecum and bovine rumen. *Acta Biol. exp., Vars.* **12**, 192–210.

TAMATE, H., MCGILLIARD, A. D., JACOBSON, N. L. and GETTY, R. (1962) Effect of various dietaries on the anatomical development of the stomach in the calf. *J. Dairy Sci.* **45**, 408–20.

TAYLOR, R. B. (1962) Pancreatic secretion in the sheep. *Res. vet. Sci.* **3**, 63–77.

TEMPLETON, JUDITH A. and DYER, I. A. (1964) Amylase and beta-glucanase activity of rumen fluid. *J. Anim. Sci.* **23**, 90–3.

TITCHEN, D. A. and REID, C. S. W. (1965) The reflex control of the motility of the ruminant stomach. In: *Physiology of Digestion in the Ruminant.* Ed. by Dougherty, R. W., and others, p. 68. Butterworths: Washington.

TULLOH, N. M., HUGHES, J. W. and NEWTH, R. P. (1965) Physical studies of the alimentary tract of grazing cattle. 1. Measurement of volume of water in the reticulo-rumen. *N.Z. J agric. Res.* **8**, 636–51.

TURNER, A. W. and HODGETTS, V. E. (1955a) Buffer systems in the rumen of the sheep. 1. pH and bicarbonate concentration in relationship to pCO_2. *Aust. J. agric. Res.* **6**, 115–24.

TURNER, A. W. and HODGETTS, V. E. (1955b) Buffer systems in the rumen of the sheep. 2. Buffering properties in relationship to composition. *Aust. J. agric. Res.* **6**, 125–44.

VALLENAS, P. A. (1965) Some physiological aspects of digestion in the Alpaca (*Lama pacos*). In: *Physiology of Digestion in the Ruminant.* Ed. by Dougherty, R. W., and others, p. 147. Butterworths: Washington.

VAN DEN HENDE, C., OYAERT, W. and BOUCKAERT, J. H. (1964) The metabolism of arginine, citrulline, ornithine, proline and delta-aminovaleric acid and phenylalanine by rumen bacteria. *Res. vet. Sci.* **5**, 491–6.

VAN SOEST, P. J. (1963) Ruminant fat metabolism with particular reference to factors affecting low milk fat and feed efficiency. A review. *J. Dairy Sci.* **46**, 204–16.

VAN SOEST, P. J. (1964) Symposium on nutrition and forage and pastures: new chemical procedures for evaluating forages. *J. Anim. Sci.* **23**, 838–45.

VAN SOEST, P. J. (1965) Symposium on factors influencing the voluntary intake of herbage by ruminants: voluntary intake in relation to chemical composition and digestibility. *J. Anim. Sci.* **24**, 834–43.

VAVICH, M. G., DUTCHER, R. A., GUERRANT, N. B. and BECHDEL, S. I. (1945) Utilization and excretion of ingested ascorbic acid by the dairy cow. *J. Dairy Sci.* **28**, 759–70.

VISEK, W. J. (1962) Studies on urea hydrolysis in birds and mammals. *Am. J. vet. Res.* **23**, 569–74.

WAITE, R. and BOYD, J. (1953) The water-soluble carbohydrates of grasses. *J. Sci. Fd Agric.* **4**, 197–204; 257–61.

WALKER, D. J. (1958) The purification and properties of the L–threonine deaminase of the rumen microorganism LC1. *Biochem. J.* **69**, 524–30.

WALSER, MACK. and BODENLOS, L. J. (1959) Urea metabolism in man. *J. clin. Invest.* **38**, 1617–26.

WARD, J. K., RICHARDSON, D. and TSIEN, W. S. (1961) Volatile fatty acid concentrations and proportions in the gastrointestinal tract of full-fed beef heifers. *J. Anim. Sci.* **20**, 830–2.

WARD, P. F. V., SCOTT, T. W. and DAWSON, R. M. C. (1964) The hydrogenation of unsaturated fatty acids in the ovine digestive tract. *Biochem. J.* **92**, 60–8.

WARNER, A. C. I. (1956) Proteolysis by rumen microorganisms. *J. gen. Microbiol.* **14**, 749–62.

WARNER, A. C. I. (1964) The breakdown of asparagine, glutamine and other amides by microorganisms from the sheep's rumen. *Aust. J. biol. Sci.* **17**, 170–82.

WARNER, A. C. I. (1966) Diurnal changes in the concentrations of microorganisms in the rumens of sheep fed limited diets once daily. *J. gen. Microbiol.* **45**, 213–15.

WARNER, A. C. I. and STACY, B. D. (1965) Solutes in the rumen of the sheep. *Q. Jl exp. Physiol.* **50**, 169–84.

WARNER, R. G. and FLATT, W. P. (1965) Anatomical development of the ruminant stomach. In: *Physiology of Digestion in the Ruminant*. Ed. by Dougherty, R. W., and others, p. 24. Butterworths: Washington.

WARNER, R. G., FLATT, W. P. and LOOSLI, J. K. (1956) Dietary factors influencing development of the ruminant stomach. *J. agric. Fd Chem.* **4**, 788–92.

WASHBURN, L. E. and BRODY, S. (1937) Growth and development, with special reference to domestic animals. 42. Methane, hydrogen and carbon dioxide production in the digestive tract of ruminants in relation to the respiratory exchange. *Res. Bull. Mo. agric. Exp. Stn.* No. 263.

WATSON, R. H. (1933) The threshold for the renal excretion of inorganic phosphates in the sheep. *Aust. J. exp. Biol. med. Sci.* **11**, 197–207.

WATSON, R. H. (1941) Studies on deglutition in sheep: a resumé of observations on the course taken by liquids through the stomach of sheep at various ages from birth to four years. *Aust. vet. J.* **17**, 52–8.

WEENINK, R. O. (1961) Acetone-soluble lipids of grasses and other forage plants. 1. Galactolipids of red clover (*Trifolium pratense*) leaves. *J. Sci. Fd Agric.* **12**, 34–8.

WEISKE, H., SCHRODT, M. and DANGEL, ST. V. (1879) On the importance of asparagine in animal nutrition. (Trans. title). *Z. Biol.* **15**, 261–96.

WELLER, R. A. (1957) The amino acid composition of hydrolysates of microbial preparations from the rumen of sheep. *Aust. J. biol. Sci.* **10**, 384–9.

WELLER, R. A., GRAY, F. V. and PILGRIM, A. F. (1958) Conversion of plant nitrogen to microbial nitrogen in the rumen of the sheep. *Br. J. Nutr.* **12**, 421–9.

WESTON, R. H. and HOGAN, J. P. (1967) The transfer of N from the blood to the rumen in sheep. *Aust. J. biol. Sci.* **20**, 967–73.

WILDE, P. F. and DAWSON, R. M. C. (1966) The biohydrogenation of α–linolenic acid and oleic acid by rumen microorganisms. *Biochem. J.* **98**, 469–75.

WILLIAMS, P. P., DAVIS, R. E., DOETSCH, R. N. and GUTIERREZ, J. (1961) Physiological studies of the rumen protozoan *Ophryoscolex caudatus* Eberlein. *Appl. Microbiol.* **9**, 405–9.

WILLIAMS, V. J. (1965) Microbial metabolism in the forestomachs and the large intestine of sheep. *Aust. J. agric. Res.* **16**, 77–91.

WILLIAMS, V. J. and MACKENZIE, D. D. S. (1965) The absorption of lactic acid from the reticulo-rumen of the sheep. *Aust. J. biol. Sci.* **18**, 917–34.

WILLIAMS, W. F., HOERNICKE, H., WALDO, D. R., FLATT, W. P. and ALLISON, M. J. (1963) Ruminal carbonate as a precursor of eructed methane and carbon dioxide. *J. Dairy Sci.* **46**, 992–3.

WILSON, A. D. (1963) The effect of diet on the secretion of parotid saliva by sheep. III. Observations of the secretion of saliva by grazing sheep. *Aust. J. agric. Res.* **14**, 808–14.

WISE, G. H., MILLER, P. G. and ANDERSON, G. W. (1940) Changes observed in milk 'sham-fed' to dairy calves. *J. Dairy Sci.* **23**, 997–1011.

WRIGHT, D. E. (1960) Hydrogenation of chloroplast lipids by rumen bacteria. *Nature, Lond.* **185**, 546–7.

WRIGHT, D. E. (1960) Pectic enzymes in rumen protozoa. *Archs Biochem. Biophys.* **86**, 251–4.

WRIGHT, E. (1955) Site of phosphorus absorption in the sheep. *Nature, Lond.* **176**, 351–2.

WRIGHT, M. J. and DAVISON, K. L. (1964) Nitrate accumulation in crops and nitrate poisoning in animals. *Adv. Agron.* **16**, 197–247.

CHAPTER 5

The Physiology, Biochemistry and Microbiology of Digestion and Metabolism of Nutrients in Pigs

P. M. RIIS and P. E. JAKOBSEN

Department of Animal Physiology, Rolighedsvej 25, Copenhagen, Denmark

THE pig is considered an omnivorous animal, its feed is both of vegetable and animal origin. Figure 1 shows schematically the alimentary canal and related organs in the pig. It is seen that the pig has a large caecum and colon indicating that the pig is more related to the herbivore than to the carnivore.

1. UPTAKE AND MASTICATION OF FEED

The pig takes up the feed with its tongue. Longer particles like grass or beets are, however, torn off with the molar teeth. Normally most of the feed, which is made up of ground concentrates, is mixed with milk or water in the trough. The pig dips its snout to the bottom of the trough and sucks in this mixed feed. Since, however, the angles of the mouth are situated far back, part of the mouth will usually be above the surface of the trough. The pig therefore sucks air in together with the mixed feed. This suction of air into the

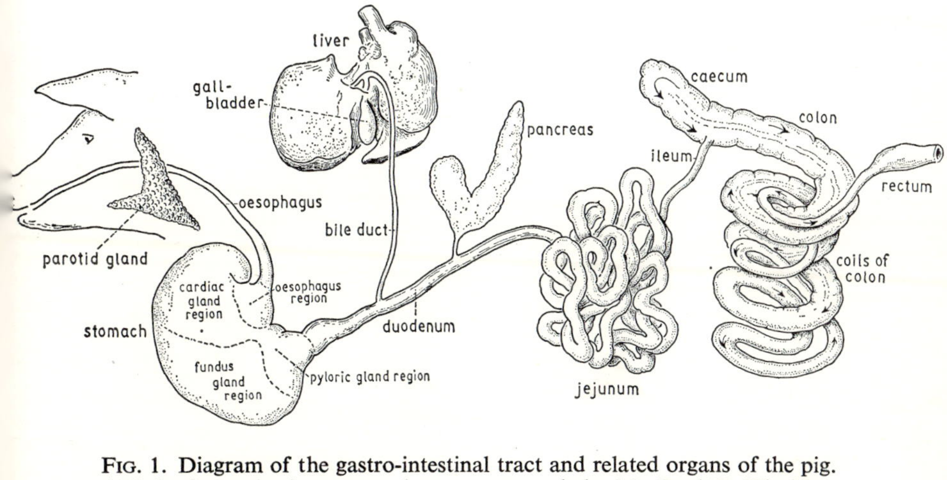

FIG. 1. Diagram of the gastro-intestinal tract and related organs of the pig. (All the figures in the present chapter were made by Mr. Poul H. Winther.)

mouth and rapid chewing movements with open lips cause the characteristic smacking and slurping sounds. It depends upon the type of feed how long the pig spends chewing.

During mastication the feed is mixed with saliva. The secretion of saliva is an intermittent process commencing after several chewing movements. The stimulus comes from receptors in the mouth. The exitation of these receptors by the feed starts an impulse in afferent nerve fibres leading to a centre in the medulla oblongata, from which efferent secretory fibres lead to the salivary glands. It has been found that about half of the total amount of saliva comes from the smaller salivary glands in the buccal and labial mucosae.[1,2] The average daily secretion of saliva amounts to about 15 l. However, as seen in Table 1, the amount of saliva depends very much upon the type of feed.[2]

TABLE 1

The Influence of Feed upon the Amount of Saliva in Pigs (after Behrens[2])

Feed	ml saliva during the first 5 min.
Maize	54·7
Wheat-bran	33·4
Potatoes	26·8
Mixed pig-feed	10·5
Beet-slices	5·2
Milk	0

Salivation in pigs can only become to a small extent a conditioned reflex. Only a little saliva is secreted by the smell or sight of good feed.

The parotid glands produce a serous, alkaline fluid, which normally contains small amounts of the enzyme amylase. Amylase or diastase catalyses the hydrolysis of starch. However, the amylase activity of pig saliva is many times less than that found in human saliva.

The secretion from the other salivary glands contains no enzyme. It is a water-clear, slimy, light alkaline product. The slimy consistency is due to a protein, mucin.[3] The saliva facilitates the transfer of feed through the oesophagus and this is the main function of saliva. The amylase in saliva initiates the digestion of carbohydrates and this digestion is continued for some time in the stomach until the acid in the gastric juice has inactivated the enzyme.

2. DIGESTION IN THE STOMACH

Figure 1 shows that the wall of the pig's stomach may be divided into four zones corresponding to four different types of mucosae.

Around the oesophageal opening the stomach is covered by a cutaneous mucosa. For comparison it may be mentioned that in all herbivorous animals the cutaneous mucosa of the oesophagus continues through the cardia and covers a smaller or greater part of the stomach wall. This has developed to the greatest extent in the ruminants in which the cutaneous zone comprises the three first parts of the stomach that is the rumen, the reticulum and the omasum. In the carnivore and in man all of the stomach mucosa carries a cylindrical epithelium containing glands.

Secretion in the Stomach

There are three different types of gastric glands namely cardiac glands, fundic glands and pyloric glands. It is unique for the pig that about half of the stomach is covered by a mucosa containing only the small cardia glands. These glands produce an alkaline mucous secretion which contains no enzymes. The secretion from the cardia glands does not contribute very much to the total amount of gastric juice. The main function of this secretion is to protect the gastric mucosa against the hydrochloric acid secreted from the fundic glands.

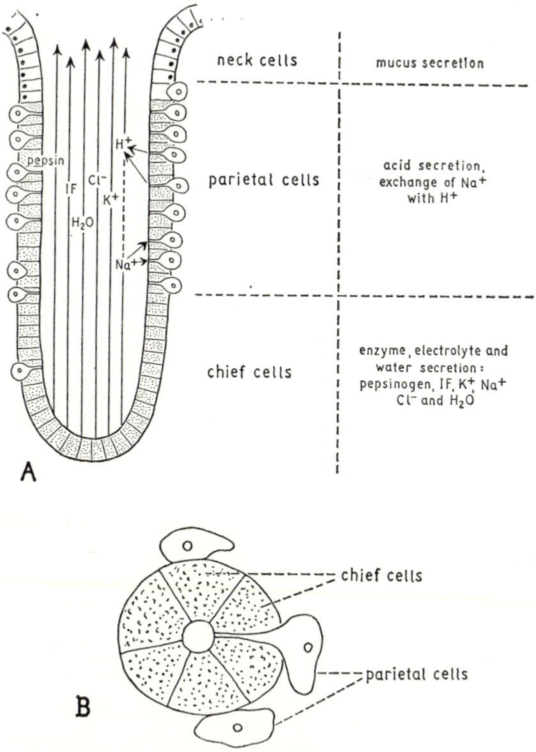

FIG. 2. A. Diagram showing the secretion from a fundic gland. B. Schematic drawing of cross-section of a fundic gland.

The main contributors to gastric juice are the fundic glands only a small part comes from the pyloric glands. The latter glands occupy a zone along the small curvature of the stomach and around the pylorus. The fundic region comprises about one-third of the stomach and is placed along the caudal part of the large curvature.

The fundic glands are long narrow tubules, which penetrate the whole thickness of the mucosa to the muscular layers of the stomach wall. Figure 2 shows schematically a fundic gland with three types of cells: (1) body chief or zymogenic cells, (2) parietal cells and (3) mucous neck cells. A fourth type, argentaffine cells may be present (4).

Many years ago Heidenhain devised a method for isolating a portion of the stomach. Later Pavlow showed how this could be done with retained innervation of the pouch. When such pouches are opened to the surface it is possible to collect pure gastric juice through the fistula uncontaminated by saliva or fluid intake.[5]

In studies especially with dogs it has been found that the secretion of gastric juice is intermittent. In the fasting animal with an empty stomach practically no secretion takes place. The secretion is initiated as soon as the animal starts chewing and swallowing feed and is continued as long as it is eating. This phase of gastric secretion is called the nervous phase. It requires an intact innervation of the stomach; the denervated gastric pouch (Heidenhain pouch) fails to respond on feed uptake. It was shown by feeding dogs prepared with both oesophageal fistulas and Pavlow pouches that the stimulus initiating and maintaining gastric secretion during feed uptake came from gustatory receptors in the mouth.[5] Feeding animals with oesophageal fistulas is called sham feeding because the feed does not reach the stomach, but it is passed out through the fistula.

The nervous phase of secretion continues only a short time after eating. However, both the innervated and denervated pouch secrete gastric juice as long as there is anything in the stomach. Since the denervated pouch also responds to food in the stomach it is evident that the stimulus is not mediated through the nervous system. It was found by Edkins[6] that the mediator responsible for this phase of gastric secretion, the chemical phase, was a hormone produced by the mucosa of the pyloric antrum. The pyloric antrum is the region of the stomach just before the pylorus. The hormone was called gastrin.

Various compounds are capable of releasing gastrin from the pyloric antrum. Partially digested products like peptones and dextrin are especially effective. For experimental and therapeutical purposes it is of interest to know that gastrin may be released by introduction into the stomach of alcohol and histamine.[7] Even mechanical distension of the pylorus may cause gastrin release.[8]

When the contents of the stomach are emptied into the small intestine the duodenal mucosa produces another hormone—secretin—which also stimulates the secretion of the fundic glands. All those compounds which release gastrin from the pyloric mucosa will also release secretin from the duodenal mucosa.[7]

Trautmann and Kirchhof [9] have studied gastric secretion in pigs prepared with stomach pouches. They found a continuous secretion also in the fasting pig. However, this was proved to be due to reflux of intestinal contents into the stomach. When this reflux was prevented and the stomach was completely empty no secretion took place. Under normal conditions the stomach does not empty completely from one feed to the next and gastric secretion is therefore normally continuous in the pig.

Pure gastric juice is an acid fluid containing various electrolytes and some organic components. In pigs the pH of gastric juice is about 1–2.[1] Figure 2 shows how the acid, which is hydrochloric acid, is secreted from the parietal cells. It is believed that the secretion involves exchange of sodium ions with hydrogen ions, however, the mechanism is not completely clear.[10]

In Fig. 2 the chief cells are presented as cylindrical cells filled with granules. These secretory granules contain pepsinogen, that is the precursor of the proteolytic enzyme, pepsin. As early as 1881 it had been shown by Langley[11] that the chief cells were the source of pepsinogen. In the acid gastric juice pepsinogen is rapidly converted into pepsin. Pepsin and the so-called 'intrinsic factor' (IF), which is necessary for the absorption of vitamin B_{12}, are the most important organic components of gastric juice.

Linderstrøm-Lang *et al.*[12] have investigated the distribution of enzymes in the gastric mucosa of the pig. In agreement with findings from other animals[13] they found the pepsinogen concentrated in the chief cells of the fundic and pyloric glands. In the chief cells of the pyloric glands they could also demonstrate a peptidase. A lipase was present in all cells. However, the lipase is bound to the cells and is not secreted with gastric juice. In gastric juice Trautmann and Kirchhof[9] found that lipase was normally present, but apparently this lipase originated from refluxed intestinal contents.

There is normally no carbohydrase activity in pure gastric juice and the amylase which is swallowed with saliva is inactivated by the hydrochloric acid. In a preparation of stomach mucosa from pigs Dahlquist[14] found a weak amylase activity, but he thought it was due to contamination of the preparation with saliva. However, in an extensive study on the composition of stomach juice from fasting pigs Heyenga[15] obtained in one case a sample which he thought originated from the cardiac glands. This sample was light alkaline and had a weak but significant amylase activity.

It is believed that the chief cells in addition to the pepsinogen and the 'intrinsic factor' also secrete the main part of the water and the electrolytes of the gastric juices excepting the hydrogen ions.

Biochemical Processes in the Stomach

Under normal conditions the pig's stomach is never empty. Hence, when it eats there is still some material from the last meal left in the stomach and recently swallowed feed will settle as a layer upon the contents already present in the stomach. By giving pigs feed with different colours successively, it has been found that half an hour after feeding the layers are very distinct in the stomach and even as long as 16 hours after feeding it is possible to distinguish a certain stratification of colours.[16] As a matter of fact an intimate mixing of gastric contents first takes place in the pyloric region where the peristaltic waves running through the stomach wall are strongest.

Because of this slow mixing of gastric contents it takes a long time before the feed comes into contact with gastric juice. The breakdown of starch by salivary amylase, which is initiated during chewing, will therefore continue for some time in the stomach. The pH-optimum for amylase is about 7, however, it will continue to act until the pH has fallen to 4·5.[17] This first phase of gastric secretion is called the amylolytic phase. Bacterial carbohydrases and carbohydrases of the feed will also show activity during the amylolytic phase of gastric digestion. Through the action of these enzymes starch and other polysaccharides are broken down to soluble carbohydrates such as erythrodextrin and to some extent to maltose. Cellulose is not digested in the stomach.

As stomach juice penetrates the swallowed feed the pepsin starts to break down the feed proteins. There will then be a phase with both amylolytic and proteolytic activity, the amylo-proteolytic phase of gastric digestion. Through the action of pepsin and HCl the feed proteins are broken down to peptides. Pepsin is an endopeptidase, it attacks peptide bonds in the interior of the peptide chain. The specificity of pepsin favours the hydrolysis of peptide linkages to aromatic amino acids. The pH-optimum of pepsin is about 2.[17]

When pH falls below 4 the amylase activity is completely inhibited and the only enzyme acting is pepsin. This last phase of gastric digestion is therefore called the proteolytic phase. It is evident from the above mentioned stratification of gastric contents that the various phases of digestion in the stomach go on concomitantly. In layers with recent swallowed feed there will be an almost pure amylolytic digestion. In other layers both amylolytic and proteolytic activity may be found and in layers close to the wall of the fundic and pyloric regions there will be a relatively pure proteolytic digestion.

As already mentioned, no lipases are secreted from the stomach glands. Gastric juice may show weak lipolytic activity due to lipases reflexed from the small intestine into the stomach. These lipases may liberate some free fatty acids from ingested lipids. However, the main change of feed lipids in the stomach is a result of gastric motility which by churning and kneading turn the lipids into a coarse emulsion.[18]

Gastric Motility and Gastric Emptying

Apparently there are no particular observations on gastric motility in pigs. However, it is believed that the movements of the stomach are similar in all mammals.[19] It is characteristic for the stomach that it adapts its capacity to the volume of the content. There is very little change in the intraluminal pressure when the stomach is filled. However, contents in the stomach will stimulate the motility of the organ.

In the empty stomach peristaltic contractions may be seen travelling through the wall at a couple of hours interval. When the stomach is filled with feed the peristaltic movement runs continuously with 2–3 waves each minute. These waves originate in the cardiac region and progress towards the pylorus.[19] Since the pig stomach is never empty under normal husbandry conditions it may be concluded that the pig's stomach never shows hunger contractions but always the above peristaltic type of motility. The peristaltic waves carry some of the material closest to the wall towards the pylorus, and new material is brought into contact with the stomach wall.

As already mentioned, only in the pyloric region are the contractions strong enough to cause an intimate mixing of the contents. In other regions of the stomach the upper layers are left almost unaffected.

Apparently each peristaltic wave causes an evacuation of stomach contents into the duodenum.[20] The material that leaves the stomach is called chyme.

In fluoroscopic studies with dogs[21] passage of chyme through the pylorus was observed as the peristaltic wave approached the pylorus but before any rise in the basal luminal pressure could be observed. The lumen pressure starts to rise when the wave reaches the antrum.[20] The pressure rise is due to contraction of the pyloric sphincter which causes an increased resistance to flow of chyme. During the early part of this pressure rise the chyme continues to flow. However, the sphincter muscle contracts more rapidly and more strongly than the rest of the pyloric region and soon closes the pylorus opening thus putting an end to the evacuation.[20]

During the period of chyme evacuation the duodenal motility is inhibited. However, the duodenal wall resumes its tone and rhythmic activity approximately at the same time as the sphincter muscles contract. Regurgitation of chyme from duodenum to the stomach is prevented by the sphincter and that is evidently the most important function of this structure. The sphincter plays no important part in the regulation of the rate of gastric emptying.

There are many investigations showing that gastric emptying is in some way regulated and that the rate of emptying depends upon the chemical composition of the feed or, rather, the chyme leaving the stomach.

Already in the last century it was found that addition of olive oil to a starch meal would delay gastric evacuation in human subjects[22] and as early as 1900 it was found that fats inhibited gastric motility. When confined alone to the stomach fat had no effect, but placed in the duodenum or upper jejunum it showed a marked inhibitory effect on the activity even of the denervated gastric pouch.[20,22] The effect on the denervated pouch showed that the inhibitory stimulus was carried by a hormone. The nature of this hormone has since been elucidated and it is called enterogastrone.[22] Under normal physiological conditions nervous reflexes may probably play some part in the inhibitory mechanism.

Although fat is the most potent inhibitor of gastric motility, other compounds may also release enterogastrone from the duodenal mucosa. Listed in order of decreasing inhibitory potency the following substances are active: fats, fatty acids, proteoses, peptones, amino acids, sugars and other products of starch digestion and hydrogen ions.[20]

3. DIGESTION IN THE SMALL INTESTINE

The gross anatomy of the small intestine is shown by the diagram in Fig. 1 which also shows the two large glands, the liver and the pancreas, connected to the first part of the intestine. By peristaltic movement of the small intestine the content is mixed and transferred through three parts: the duodenum, the jejunum and the ileum. In the small intestine the digestion is continued through the action of enzymes from intestinal glands and from pancreatic secretion. The enzyme action is supported by bile secretion from the liver.

The Formation and Secretion of Bile

There is a continuous formation of bile in the liver. Part of this bile is stored in the gall bladder and part of it flows directly through the duct to the small intestine. By appropriate stimulus the gall bladder contracts and delivers its content to the intestinal lumen.

Pig bile is a brown-yellow, viscid fluid containing various salts and organic substances. In liver bile the percentage of dry matter is about 3 whereas bladder bile has about 16 per cent dry matter. The higher dry matter content in the latter is due to absorption of water from the gall bladder. The major organic components of bile are the bile pigments and the bile salts, but significant amounts of lipids, mainly phospholipids and some cholesterol, are also present. The reaction is usually weak alkaline with a pH about 7–8. However, in the gall bladder the pH may be below 7.[7]

The bile salts are sodium or potassium salts of glycocholic or taurocholic acid which are formed by peptide linkages between cholic acid and glycine or taurine. Cholic acid is derived from cholesterol and is one of the main products of cholesterol metabolism.[17] Lecithin is the predominating phospholipid in bile but up to six different phospholipids have been demonstrated in bile.[23,24]

Bile salts play, together with the lipids of bile especially lecithin and lysolecithin, an important role in intestinal digestion. The secretion of bile pigments, bilirubin and biliverdin, is on the other hand an excretion of degradation products. They are degradation products of heme from destroyed erythrocytes. Also other compounds which like bile pigments are useless to the organism may be secreted with the bile. These substances are excreted with faeces whereas up to 90 per cent of the bile salts may be reabsorbed and returned to the liver through the portal circulation.[10]

Hilby [25] studied bile secretion in pigs and found that there was a continuous flow of bile through the duct to the intestine. The amount of bile delivered to the intestine decreased during fasting but increased again 4–14 hours after feeding. His studies indicated that bile formation and bile secretion in pigs are similar to that known from studies with other animals. [7,10]

Various compounds are capable of exciting bile formation in the liver. Most important is that the reabsorbed bile salts stimulate bile formation. The reabsorbed bile salts are apparently responsible for the maintenance of bile formation in the postabsorptive state.[10] Stimulation by absorbed fats and to a lesser extent amino acids will increase bile formation during digestive periods. As already mentioned instillation of acid material in the duodenum or the evacuation of chyme from the stomach causes release into the blood stream of a hormone, secretin.[22] This hormone has a marked stimulatory effect on bile formation.[26]

The application of dilute acids, fats or a meal to the duodenum will also cause contraction of the gall bladder and the delivery of stored bile to the intestine.[22] The bladder contraction is due to stimulation by a hormone which is also released from the duodenal mucosa.[27] This hormone is called cholecystokinin.[28,29]

Pancreatic Secretion

Pancreas is both an endocrine and exocrine gland. The endocrine secretions, insulin and glucagon are not considered here, since only the exocrine secretion is of interest in regard to digestion. The pig secretes from 7 to 15 l pancreatic juice per day. The juice is a clear, distinctly alkaline liquid which contains several proteolytic, lipolytic and amylolytic enzymes. The alkalinity is due to bicarbonate which is of importance to the neutralization of acid from the stomach.

There are three proteolytic enzymes in pancreatic juice, namely trypsin, chymotrypsin and a carboxypeptidase. Trypsin and chymotrypsin are secreted in an inactive form as zymogens called trypsinogen and chymotrypsinogen respectively. Trypsinogen is rapidly converted into the active enzyme by an enterokinase secreted by the intestinal glands. The conversion of trypsinogen to trypsin is autocatalytic and small amounts of trypsin will catalyse the conversion. Trypsin will also catalyse the activation of chymotrypsinogen. Both trypsin and chymotrypsin are endopeptidases and hydrolyse peptide linkages in the interior of the peptide chain. The carboxypeptidase attacks peptides from the carboxyl end of the chain. Like the other two proteolytic enzymes carboxypeptidase is secreted as a zymogen, a procarboxypeptidase.[17]

The lipolytic activity of pancreatic juice is due to a specific enzyme which removes only fatty acid residues linked to primary hydroxyl groups (α-groups) of glycerol.[30] The action of this enzyme thus causes the formation of β-monoglycerides and free fatty acids from triglycerides. This remarkable specificity of pancreatic lipase is used in studies on fatty acid distribution between α- and β-positions in triglycerides.[31] The rate of lipase action is increased by emulsifying agents such as bile salts, lecithin and lysolecithin.

In homogenates from pig pancreas Dahlquist[14] found a strong amylase activity. The effect of this enzyme resembles that of saliva amylase. Only a weak maltase activity was found and it was concluded that the amylase was the only carbohydrase of importance in pancreatic juice.

Pancreatic secretion is stimulated by the sensation of feed in the mouth, but the amount of juice secreted in response to this nervous stimulus is small.[19] During digestion the secretion is maintained by humoral stimulation. Probably the most studied effect of the earlier mentioned secretin is its effect on the pancreas. It has been shown by many investigators[22] that the pancreas responds to secretin with a production of a large volume of fluid. However, the enzyme concentration decreases with the increase in volume and at the height of the secretin effect the enzyme concentration is minimal.[32] It was believed that secretin controlled only the fluid and electrolyte output from the pancreas whereas the enzyme secretion was regulated by autonomic nerves in the vagus.[33] However, Harper and Voss[34] found that the increase in enzyme output from pancreas, when food or water entered the small intestine, was independent upon extrinsic innervation of the intestine. Following this observation it was demonstrated that a preparation from duodenal mucosa contained a substance which was capable of increasing enzyme concentration in pancreatic secretion. The substance was called pancreozymin.[35]

Secretion from the Small Intestine

Simple tubular glands, the so-called crypts of Lieberkühn are distributed throughout the entire length of the intestinal mucosa. The duodenal mucosa contains in addition branched coiled tubular glands called Brunner glands.[4]

The duodenum appears to be the only part of the intestine that gives an abundant secretion. In all species examined it has been found to be a viscous, sticky fluid with an alkaline reaction. The pH of duodenal juice from pigs was showed to range from 8·4 to 8·9.[36] The alkali and mucin or mucinlike substances in the secretion protect the intestinal mucosa from injury by the acid chyme coming from the stomach. Florey and Harding[36] found no appreciable enzyme activity in the secretion from the Brunner glands. However, it is possible that the crypts of Lieberkühn secrete several enzymes. Dahlquist[14] found a strong trehalase, lactase and cellobiase activity in duodenal preparations from pigs.

A study by Loscheidt[37] on the secretion and the effect of jejunal juice from pigs indicates that it resembles the secretion from the duodenum. From isolated loops of the jejunum in pigs Unger[38] obtained a serous to seromucous secretion with pH ranging from 7·4 to 8·7. It contained several enzymes, the presence of protease, lipase, maltase, saccharase, lactase, peroxidase and catalase were demonstrated. Apparently the activity of maltase and isomaltase as well as of saccharase is higher in jejunal secretion than in duodenal secretion whereas the opposite is true in the case of lactase and cellobiase.[14]

The secretion from intestinal glands appears to be continuous but increases after a meal. The increase follows a low flat curve with a maximum 1-2 hours after feeding and minimum about 8 hours later.[38] Florey and Harding[39,40] presented evidence indicating that the hormone, secretin, not only controlled pancreatic and bile secretions but the intestinal secretion as well. However, another hormone found by Nasset[41] called enterocrinin evidently contributes significantly to the control of intestinal glands.[41,42]

Biochemical Processes in the Intestine

When the acid chyme from the stomach is mixed with the alkaline secretion from the liver, the pancreas and the intestinal gland, the acid is partly neutralized and the pH of intestinal content rises slowly as the content is

Table 2

The pH of the Digestive Tract Liquor of the Pig[43]

Part of digestive tract	Stomach	Cranial ilium	Caudal ilium	Caecum	Colon* I	Colon* II
pH range	4·2—5·1	6·2—7·1	6·8—8·0	6·5—7·2	6·4—7·3	6·4—8·0

* The colon was divided at the flexure of the spiral.

passed down through the tract. Alexander[43] has measured pH in liquor from the digestive tract of horses and pigs immediately after the animals were slaughtered. His observations from 4 pigs are cited in Table 2. The pH of the duodenal contents was not measured, but Kolb[19] gives a value of 6·6 for the pH of the content in duodenum of swine.

As mentioned earlier, the feed lipids in the stomach are turned into a coarse emulsion. This emulsion is stabilized by detergents in the duodenum. The most important detergents are the bile salts and probably lysolecithin. At the oil-water interface of the emulsion droplets pancreatic lipase causes a rapid hydrolysis of the α-ester bond of triglycerides.[44] The products of this hydrolysis are free fatty acids, 2-monoglycerides, 1,2-diglycerides and slowly free glycerol. The liberated free fatty acids and monoglycerides form complexes or micelles with conjugated bile salts. These micelles are readily absorbed and represent probably the main pathway of fat absorption.[45] It may be seen from Fig. 5 that most of the fat is absorbed into the intestinal epithelial cells as a mixture of free fatty acids, monoglycerides and some diglycerides. Intact triglycerides may probably be absorbed to a very small extent (see also Chapter 10).

Phospholipids of feed as well as bile phospholipids are attacked by another pancreatic enzyme, phospholipase A or phosphatide acyl hydrolase.[46] In contrast to pancreatic lipase the phospholipase A attacks primarily the β- (or 2-) ester bond.[47] The hydrolysis by phospholipase A thus results in formation of lysolecithin. Lysolecithin is a very powerful detergent and may play a role in intraluminal micellar formation thus facilitating fat absorption.

The digestion of starch initiated by salivary amylase is continued by the action of pancreatic amylase. The product of the amylase action is the disaccharide maltose. Maltose is hydrolysed to glucose by intestinal maltase. The maltase hydrolyses only 1,4-glucosides. Ingested carbohydrates with different glucosidic linkages are attacked by the other carbohydrases present.[14] By the action of these enzymes the polysaccharides are broken down to glucose and other monosaccharides which are readily absorbed. However, it should probably be mentioned here that raw potato starch and cellulose are not digested by intestinal enzymes in the pig. Part of these latter substances are digested by bacterial fermentation in the large intestine.

The pepsin activity in the stomach results in formation of some free amino acids. However, the main products are large peptides. By the action of trypsin, chymotrypsin and carboxypeptidase these peptides are broken down to smaller units and the activity of the many peptidases secreted by intestinal glands completes the hydrolysis to amino acids.

4. PROCESSES IN CAECUM AND COLON

It is seen in Fig. 1 that the caecum and colon are relatively large organs. The caecum and the proximal part of colon are sacculated approximating that in typical herbivores. The secretion from the crypts of Lieberkühn in the mucosa of the large intestine is not very copious, it has a high mucin content but contains no significant amounts of digestive enzymes. In carnivores the caecum and colon are small simple tubes and the main process taking place there is an absorption of water. In herbivores and in the pig the digestion is continued by bacterial enzymes present in the large intestine. The bacterial fermentation in the alimentary tract leads to formation of substances which may be of beneficial effect to the host animal.[48]

Digestion of Cellulose and Production of Volatile Fatty Acids

It has been known for a long time that the pig can digest crude fibre to some extent and that this digestion is entirely dependent upon bacterial fermentation in the caecum and colon.[49] Until recently, however, little attention has been paid to the products of this fermentation and especially to the nutritional importance of these products in the pig.

Already 1906 Sheunert[50] demonstrated digestion of cellulose in the colon of pigs. Since then many investigations have shown that there is a wide variation in the digestibility of crude fibre in swine. The digestion coefficient for crude fibre of normal swine ration varies from 10 to 90 per cent.[49,50] Mangold[49] thought the variations were probably due to changes in intestinal flora. Apparently the form in which the cellulose is offered, especially the encrusting substances, play an important role.[48] Thus, pure cellulose or cellulose of straw after alkali or sulphite treatment is digested by more than 80 per cent in pigs.[52]

The amount of cellulose in the ration also affects its digestibility, furthermore, the cellulose content of the fodder may affect the digestibility of the other organic components.[51,52]

In extensive digestibility experiments with pigs Nordfeldt[53] found that the digestibility of crude fibre in straw meal decreased when the total ration contained more than 15 per cent straw meal. With 13 per cent straw meal the digestion coefficient for crude fibre was 19 per cent. Results from digestibility trials with oats-fed pigs[51,54] are in accordance with Nordfeldt's observations. Axelsson[55,56] thought that there was an optimum level for crude fibre content in swine rations and that this level was 6–7 per cent. However, the differences in weight gain and feed efficiency on various levels of crude fibre content were not great. In a recent publication on digestibility experiments in pigs Madsen[51] arrived at the conclusion that the normal amount of crude fibre in swine rations plays no significant role in the digestion and utilization of other nutrients in the ration.

The products of bacterial fermentation of cellulose are volatile fatty acids with acetic acid as the predominating acid, but significant amounts of propionic and butyric acids are also formed.[48]

Elsden *et al.*[57] have compared the concentration of volatile fatty acids (VFA) in the alimentary canal of various animals. They found that although the concentration of VFA in colon and caecum of pigs was lower than that found in the horse and the ox it was nevertheless relatively much higher in the pig than in the rabbit, guinea pig and other small herbivores. The average composition of mixed volatile fatty acids from the caecum of pigs was as follows: 62 per cent acetic, 28 per cent propionic and 10 per cent butyric acid.[57] A similar distribution has been found by other workers.[58,59]

Friend *et al.* have compared the concentrations of organic acids in the large intestine with that in other sections of the alimentary canal of the pig. Some of their results are shown in Table 3.

TABLE 3

The Concentration of the Organic Acids—Lactic and Volatile Fatty Acids in the Alimentary Canal of Pigs[60]

Section of tract	Stomach	Small intestine	Caecum	Colon
m-equiv. org. acids/100 ml fluid	1·6—0·6	3·0—2·5	14·6—15·0	22·5—17·7

In the stomach lactic acid accounted for 95–98 per cent of the total amount of the organic acids, in the colon and caecum 40–50 per cent of the total acids were acetic acid. In the baby pig[61] it was found that the concentration of organic acids was relatively high in the stomach and low in caecum and colon. As the pig aged, the concentration tended to decrease in the upper sections and increase in the lower sections of the tract.

The results of the investigations mentioned above leave no doubt that microbial digestion of cellulose with formation of volatile fatty acids takes place to a significant extent in the caecum and colon of the pig. It has also been found that removal of the caecum and the spiral colon causes an increase in the output of crude fibre and water in the faeces.[62] It may be noted also that large amounts of bicarbonate are secreted from the caudal ileum. The bicarbonate is apparently needed for neutralization of the produced acids.[43]

The Nutritional Significance of Microbial Fermentation in the Large Intestine

The amount and composition of intestinal microorganisms vary considerably with the diet. Willingale and Briggs[63] found from 10^8 to 10^9 microorganisms per gram caecal contents in normal pigs. The predominating species were lactobacilli and streptococci. Coli-bacteria were found in significant amounts and many other types were present in small amounts. It has been found that when these bacteria are incubated with carbohydrates *in vitro* they produce volatile fatty acids.[64]

The production of volatile fatty acids by microbial fermentation of cellulose in the large intestine of the pig was discussed in the previous section. In a review Cuthbertson and Phillipson[48] stated: "If the products of bacterial decomposition of cellulose in the large intestine of omnivores can be utilized as productively by pigs as by ruminants then fodder cellulose should have a value for fattening pigs not much inferior to that of oats."

By measurements of VFA-concentration in the blood draining the alimentary canal Barcroft *et al.*[65] obtained evidence for absorption of volatile fatty acids from the caecum and colon in pigs. Forbes and Hamilton[66] studied the metabolizable energy in several highly digestible sources of cellulose in restricted rations for pig and concluded that the organic acids produced were readily utilized by the animal. However, in agreement with many other investigations,[67] Cunningham *et al.*[68] found by addition of purified cellulose to a basal ration fed *ad libitum* that there was no net gain in energy from the consumption of cellulose. The apparent discrepancy between these findings may be explained by the fact that the level of feeding significantly affects the digestibility and utilization of cellulose.[69]

The many investigations [67] showing no effect of cellulose do not exclude the possibility that the cellulose or crude fibre of normal swine rations do contribute a little to the net energy of the ration. It should also be emphasized that the digestion of starch from some feeds is dependant upon the microbial fermentation in caecum and colon.[1] Apparently there are no investigations showing anything about the amount of volatile fatty acids which are normally absorbed from the intestinal gut and metabolized by the pig.

By his studies on the activity of the isolated intestinal flora from pigs Michel [64] found that it exerted a considerable catabolic activity. The catabolic activity was directed towards amino acids which the flora could deaminate and decarboxylate. However, it was mainly amides and non-essential amino acids that were catabolized. Furthermore, *in vivo*, the microbial catabolism of amino acids occurs in the large intestine while the main part of the digestible proteins are absorbed after hydrolysis in the small intestine. Hence, the catabolism of nitrogen compounds in the large intestine does not cause loss of absorbable amino acids. The deamination results in formation of ammonia which is readily absorbed. Since ammonia is a toxic substance it has to be converted into urea and excreted again. This process results in loss of energy. The extent to which it normally occurs in pigs is unknown.

An interesting question in regard to the microbiological activity in the intestine is the possible formation of dietary essentials, especially the B-vitamins.

By balance trials with pigs it has been found that an extensive synthesis of vitamin B_{12} took place in the intestinal tract.[70] Grzeszczak-Swietlikowski[71] obtained information indicating not only an extensive synthesis of cyanocobalamine and related compounds in the large intestine of the pig, but also significant absorption and utilization of the synthesized vitamins.

It is evident from many studies that the bacteria which are normally found in the intestine of pigs are capable of synthesizing most of the other vitamins belonging to the B-complex.[72,73] It may be presumed that the synthesis of B-vitamins by the intestinal flora in the pig plays a role in the normal supply of these essentials.

It is evident that the addition of antibiotics to the ration may have considerable influence upon the microbial activity in the gastro-intestinal tract. The beneficial effect which is sometimes observed by the addition[74] may be due especially to a depression of pathological microorganisms. The advantage of a general and continuous addition of antibiotics to swine rations is, for many reasons, doubtful.

5. RATE OF PASSAGE THROUGH GASTRO-INTESTINAL TRACT AND DEFAECATION

By peristaltic movement of the muscular layers in the wall of caecum and colon the intestinal content is slowly passed into the rectum. Roentgen-ray studies[75] have revealed that the large intestine of the pig, besides exhibiting peristaltic movements, shows segmental tonic contractions. This motor activity presumably causes an intimate mixing of the contents in the large intestine. No antiperistaltic movements were observed by the roentgen-ray studies.[75]

There are apparently no studies showing the time required for the passage of food residues through the large intestine. However, since the lumen of caecum and colon is wide compared to the lumen of the small intestine (see Fig. 1) it may be assumed that the rate of passage through the large intestine is relatively slow leaving good time for microbial fermentation.

By means of test meals coloured with fuchsin Lenkeit[76] has studied the time required for passage of food residues through the entire gastro-intestinal tract of pigs and other animals. When the test meal was fed in the morning the excretion of coloured substances commenced 11–13 hours after feeding and 56–63 per cent of the administered dye was excreted in the second 12-hour period. When the test meal was fed in the evening the first coloured substances appeared in the faeces 13–17 hours after administration and the main part was first excreted during the third 12-hour period. Three to four days were required for the excretion of the total amount of the dye.

The rate of passage through the gastro-intestinal tract evidently varies with the composition of the ration.[77] Since fats delay gastric emptying[19] rations rich in lipids presumably pass more slowly through the tract than do lipid-poor rations.

When the contents in the ampulla of the rectum have accumulated to a certain extent there arises the desire to defaecate and the contents are expelled through the anus.

Defaecation is a complex reflex act involving voluntary as well as involuntary muscles. The involvement of voluntary muscles means that defaecation may be aided or inhibited at will. When the sphincter muscles of the anus are relaxed the emptying of the rectum is accomplished by tonic contraction of muscles in its own wall and contraction of the abdominal muscles. It has been found that in man the contraction of the abdominal muscles with the glottis closed may raise the intra-abdominal pressure to as much as 200 mm Hg.[78]

The waste matter that leaves the rectum through the anus is called the faeces. It is composed of water, indigestible food residues, remains of digestive secretions, desquamated epithelial cells and numerous bacteria. The colour is mainly due to oxidized bile pigments and the characteristic odour is due to the content of indole and skatole.The water content varies normally between 55 and 75 per cent.[19] However, it need hardly be emphasized that the composition and amount of the faeces vary considerably with dietary conditions.

6. TRANSPORT OF NUTRIENTS FROM THE INTESTINE

After digestion in the intestine the nutrients pass through the intestinal epithelial cells into the blood or lymph draining the intestine. The passage through the epithelial cells may take place by passive diffusion. However, in many cases the transport is an active process requiring energy. In the case of active transport the nutrients are absorbed against a concentration gradient that is from lower to higher concentrations.

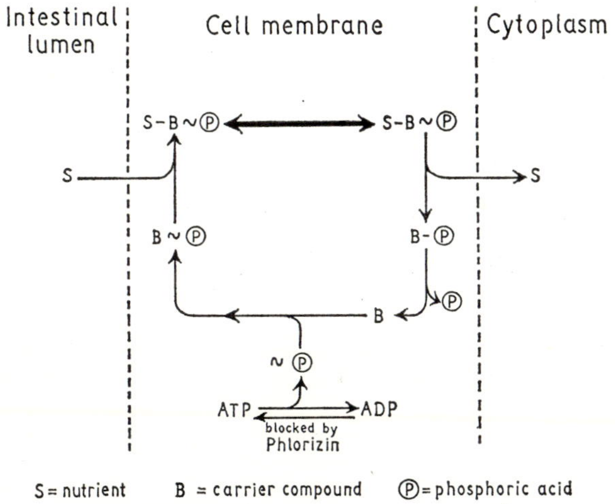

Fig. 3. General scheme for the active transport of nutrients.

In the newborn pig the intestinal epithelium is permeable to intact proteins, and antibodies ingested with colostrum may thus pass into the blood stream. The blood of the newborn piglet contains almost no γ-globulins and an early supply of antibodies in the colostrum is therefore of paramount importance for the health of the pig. However, it is just as important that within a few days, the intestinal epithelium becomes impassable for proteins and larger peptides. Absorption of proteins or other antigenic compounds by animals capable of antibody formation may cause allergy and anaphylactic reactions.

By digestion in the stomach and intestine the proteins are normally completely hydrolysed to amino acids. The transport of amino acids is an active process. Figure 3 shows a general scheme for the active transport of nutrients. The nutrient is bound to a carrier compound which carries it through the cell membrane into the cytoplasm of the epithelial cell. The energy required for the formation of the linkage between the carrier and the nutrient is derived by hydrolysis of ATP. It has been estimated that transport by absorption of 1 mole glycine requires 22 kcal. Reformation of ATP from ADP requires oxidative processes and administration of compounds such as phlorhizin which inhibits oxidative phosphorylation, interferes with absorption.

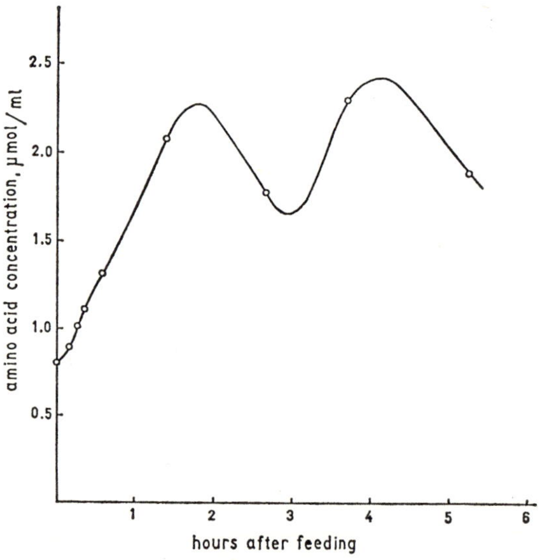

Fig. 4. The concentration of amino acids in the portal blood of a pig after ingestion of casein.

Three different transport systems (carrier compounds) for amino acids have been demonstrated, one for neutral, one for alkaline and one for acid amino acids. The transport systems consist apparently of proteins.[80] The stereospecificity of the transport systems favours the absorption of L-amino acids. However, D-amino acids are absorbed to some extent and the presence of D-amino acids may interfere with the absorption of the biologically active

L-form.[81] The activity of the transport systems is dependent upon pH[82] and is increased by pyridoxin and its derivatives.[83,84,85]

From the intestinal epithelial cells the amino acids are, *via* the portal circulation, transferred to the liver where they are metabolized or delivered to the systemic circulation. In experiments with a pig having a permanent catheter in the portal vein the concentration of amino acids was studied after ingestion of various proteins.[86] Figure 4 shows the results from a trial with casein. It is seen that the concentration increased shortly after ingestion of the protein. The concentration reached a maximum about 1·8 hours after feeding. It then decreased a little, but about 4½ hours after feeding there was a second maximum. The results are in agreement with findings from similar experiments with dogs.[87]

The first peak in Figure 4 was due to a rise in the concentration especially of neutral amino acids. Apparently it represents amino acids liberated by the action of pepsin and absorbed in the first part of the small intestine.[87,86] The second rise probably represents amino acids liberated by the action of pancreatic enzymes.

The blood velocity was not measured in the experiment and it is impossible to estimate the amount of amino acids transported with the portal blood during the periods after feeding. It may be noted that the blood flow through the portal system varies considerably with changes in absorptive states.[78]

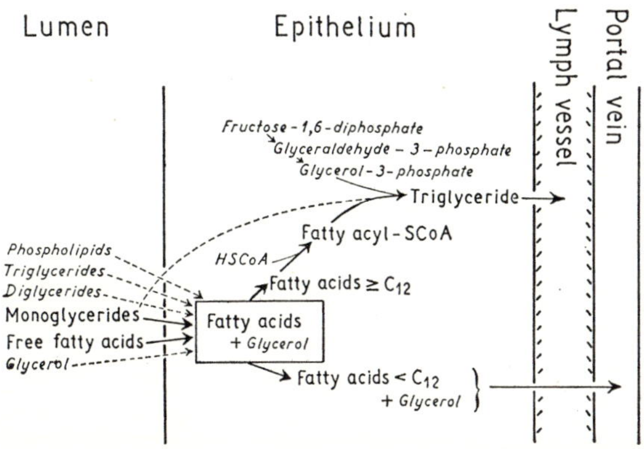

Fig. 5. Processes involved in the absorption of lipids and their transport through the intestinal epithelium.

The carbohydrates are absorbed as monosaccharides. Most of the hexoses are actively taken up by a carrier compound, whereas the pentoses are absorbed by passive diffusion. The pentoses are therefore absorbed much more slowly than the hexoses, especially glucose and galactose which show the greatest rate of absorption. By isomerases of the intestinal epithelium, galactose and fructose are converted into glucose and only the latter hexose appears in the portal blood.

Figure 5 shows the transport and metabolism of lipids in the epithelial cell. Most of the lipids are absorbed as free fatty acids and monoglycerides. Phlorrhizin and monoiodoacetic acid interfere with fat absorption indicating that it is an active process.

During transport lipases of the epithelial cell complete the hydrolysis and a pool of free fatty acids is formed.[88,89] According to the distribution theory developed by Frazer and his co-workers during the period 1938–50[90] part of the absorbed fat is transferred to the portal blood and part of it to the lymph.

A number of investigations with isotope labelled fatty acids have shown that these after ingestion are refound in the lymph as triglycerides irrespectively whether they are fed as free fatty acids or as glycerides.[91,92,93] As shown in Figure 5 it is mainly fatty acids having a chain length with twelve or more carbon atoms that follow this route. Lower fatty acids are passed into the portal blood as free fatty acids.

The finding of ingested fatty acids as lymph triglycerides shows that extensive synthesis of triglycerides takes place in the intestinal epithelium. Most of the glycerol for this synthesis is formed from fructose-1,6-diphosphate[94] and only a small part originates from ingested glycerides.[95,96,97,98]

The lymph lipids are delivered to the systemic circulation as small droplets called chylomicrons. Fatty acids transported with the portal blood are taken up by the liver.

7. INTERMEDIATE METABOLISM OF NUTRIENTS

In the previous section it was discussed how the nutrients after absorption were either carried with the portal blood to the liver or with the lymph to the systemic circulation. In the liver the nutrients are metabolized and to a certain extent stored. The liver contains thus a store of readily available nutrients and it delivers its store to the blood by which it is transported to the site of utilization. The storing capacity of the liver is limited and the main storage of energy takes place as fat deposition in adipose tissue.

Metabolism of Proteins and Amino Acids

Numerous experiments with tracer compounds have clearly shown that the proteins of the living organism are in a state of dynamic equilibrium. There is a continuous break-down and reformation of proteins, a catabolism and anabolism. In full-grown animals the catabolism and anabolism are in equilibrium and therefore, the N–balance = N in feed — N excreted is zero. In growing animals the anabolism normally exceeds the catabolism and the N-balance has a positive value.

The biological catabolism of proteins is not completely reversible and the dynamic state therefore causes a continuous loss of protein. The secretion of enzymes to the digestive tract represents also a loss of proteins. Hence a N-balance of zero does not mean that no protein is needed; there is a maintenance requirement for protein.

The amino acids required for protein synthesis in the cells are taken from the pool of free amino acids in the tissue fluid. This pool is supported by the amino acids coming from the liver with the blood. It has been found that blood plasma from 80 kg pigs contains about 7–9 mg amino-nitrogen corresponding to about 45–60 mg free amino acids per 100 ml.[99] When the rate of protein synthesis is high, the plasma concentration of amino acids is also high. Table 4 shows the concentration of amino acids in the plasma of four groups of pigs with increasing age. The plasma samples were taken in the morning before feeding and the Kjeldahl factor 6·25 was used for the calculation of the results from determination of α-amino-N.[100]

TABLE 4
Plasma Level of Free Amino Acids in Pigs[100]

No. of pigs	Weight of pigs (kg)	mg amino acid per 100 ml
11	< 20	80
9	21–40	46
12	41–90	38
8	> 90	39

The free amino acids of blood contain a relatively high proportion of essential amino acids. It is evident that a frequent supply of these acids which the cells are unable to synthesize is necessary. The concentration of free amino acids may be 5–10 times higher in the tissues than in the blood.[101]

It has been known for a long time that the synthesis of proteins is an energy-requiring process and recent years experiments have elucidated some of the complicated processes in protein synthesis. It is characteristic that in the biological synthesis of a certain protein the sequence of amino acids is constant.[17] The synthesis of proteins takes place in the cytoplasm of the cell on the surface of the ribosomes where the sequence of amino acids is coded by the bases of ribonucleic acid.[102]

The maximum rate of protein synthesis is a function of the physiological state of the animal. A deficient supply of essential amino acids interferes with the synthesis of protein and decreases its rate. An excess intake of the amount required for maximum rate causes an increased catabolism of amino acids.

The first step in the catabolism of amino acids is a deamination with formation of ammonia and a keto acid. The ammonia is transferred by the blood to the liver where urea is formed. Figure 6 shows the processes involved in urea formation. It is seen that formation of 1 molecule of urea requires 4 moles of

ATP. Urea formation thus represents a loss of energy which appears as heat. The main cause for the high specific dynamic action of proteins is apparently the energy loss through urea synthesis.

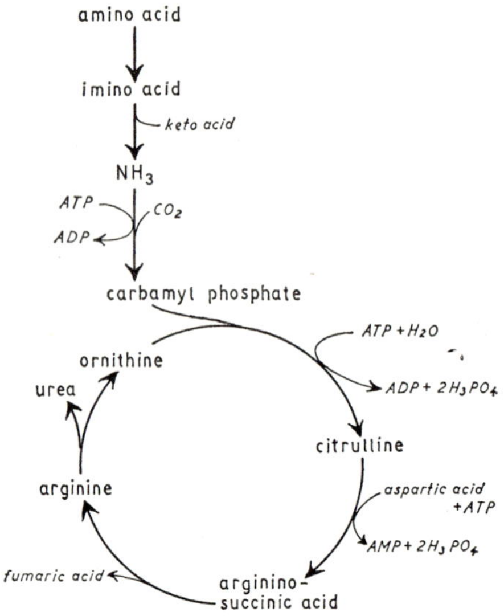

Fig. 6. The Krebs–Henseleit cycle and urea formation.

The nitrogen-free residue of the deaminated amino acids is oxidized to CO_2 and H_2O. Part of their combustion energy is stored in high-energy phosphate bonds. The energy stores in these bonds may be directly utilized for muscle contraction or other energy-consuming processes. The amount of energy in the high-energy bonds may therefore be called utilizable energy.

TABLE 5
Utilizable Energy of Amino Acids

Amino acid	Utilizable energy	
	kcal per molecule acid	kcal per g C converted to CO_2
Isoleucine	396	4·6
Leucine	360	4·2
Valine	432	6·0
Threonine	360	6·2
Lysine	396	4·6

The utilizable energy of nutrients may be calculated as described in the following section on carbohydrate metabolism. Table 5 shows the utilizable energy of some amino acids. It may be seen that each gram of amino-acid carbon oxidized to CO_2 gives about 5 utilizable kilocalories.

Metabolism of Carbohydrates

Part of the absorbed glucose is stored as glycogen in the liver. Liver glycogen represents a store of energy which is readily mobilized in case of emergency. When the energy requirement rises suddenly as for example by the onset of muscular work hydrolysis of liver glycogen prevents a rapid fall in blood glucose.

The concentration of glucose in mammalian blood varies normally within very narrow limits. Swine blood contains about 70–80 mg glucose per 100 ml.[99] It must be emphasized that the values normally reported for blood levels of glucose apply to whole blood. Glucose is evenly distributed in the entire extracellular space and it would probably be more relevant to use plasma values for glucose rather than whole blood values. It is noted that in the case of other nutrients plasma values are normally presented. The concentration of glucose in plasma is about 25 per cent higher than that in blood. Plasma from normal pigs contains 100–120 mg glucose per 100 ml.[103]

During absorption of glucose the blood level tends to increase. This causes an increased output of the hormone, insulin, from the islets cells of pancreas. Insulin stimulates the utilization of glucose. Hence, it increases the rate of glycogen formation as well as fat synthesis from glucose. In case of excess uptake of glucose fat deposition therefore is accelerated.

When energy requirement in the muscles rises rapidly the output of adrenaline from the adrenal medulla and of glucagon from the α-cells in the pancreas is stimulated. Glucagon accelerates hydrolysis of liver glycogen and adrenaline stimulates the hydrolysis of both liver and muscular glycogen as well as it increases oxidation of glucose in the tissues. Both the hormones thus increase the blood level of glucose whereas insulin tends to depress it.

Muscle tissues may cover their energy requirement by combustion of fatty acids. However, nerve tissues are dependant upon glucose metabolism exclusively. When the glycogen depot in the liver is exhausted the required glucose is formed mainly from amino acids. This process, the gluconeogenesis is stimulated by hormones from the adrenal cortex. Also growth hormone from pituitary gland takes part in the regulation of blood glucose and carbohydrate metabolism. Its action is antagonistic to insulin.

If blood glucose for some reason increases above a certain level which in non-ruminant mammalia is about 130 mg per 100 ml, glucose is excreted with the urine. The critical glucose concentration is known as the renal threshold. A constant glycosuria indicates disturbance of the endocrine regulation and is called diabetes mellitus.[5,7]

The utilization of glucose requires its transport into the intracellular space and subsequent formation of glucose-6-phosphate. The biochemical processes of glucose metabolism and metabolism of other carbohydrates are shown in Fig. 7.

The pyruvic acid formed by anaerobic glycolysis may undergo decarboxylation with the formation of acetyl-CoA or carboxylation with formation of oxaloacetic acid. Acetyl-CoA enters the Krebs' or citric acid cycle and is oxidized to CO_2 and H_2O. Oxaloacetic acid is a key substance not only in the entrance of the citric acid cycle but also in gluconeogenesis. The formation of phosphoenolpyruvate from pyruvic acid involves pathways over oxaloacetic acid.[104,105]

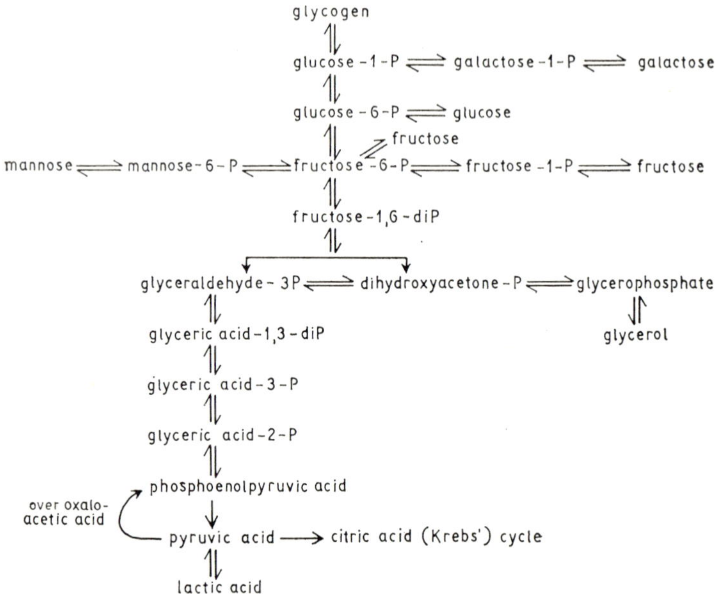

Fig. 7. Biochemical pathways of carbohydrate metabolism.

Figure 8 shows the oxidation of acetyl-CoA through the citric acid cycle. The hydrogen atoms removed from nutrients are accepted by coenzymes which thus become reduced. From the reduced coenzymes electrons are transferred through the cytocrome systems to oxygen and water is formed.

It is seen from Figs. 7 and 8 that 1 mole of glucose gives six high energy bonds and twelve equivalents of reduced coenzyme. *In vitro* experiments have shown that one equivalent of reduced coenzyme may give 3 moles of ATP. A total of 42 high energy bonds are thus formed by the complete oxidation of 1 molecule of glucose. However, the formation of glucose-6-P from glucose and of fructose-1,6-diP from fructose-1-P uses 2 moles of ATP. Hence, the net gain by combustion of 1 mole or 180 g glucose is 40 moles of ATP. Each mole ATP represents 10 kcal utilizable energy. Oxidation of 1 g glucose gives thus 2·2 utilizable kilocalories which is 58 per cent of the metabolizable energy.

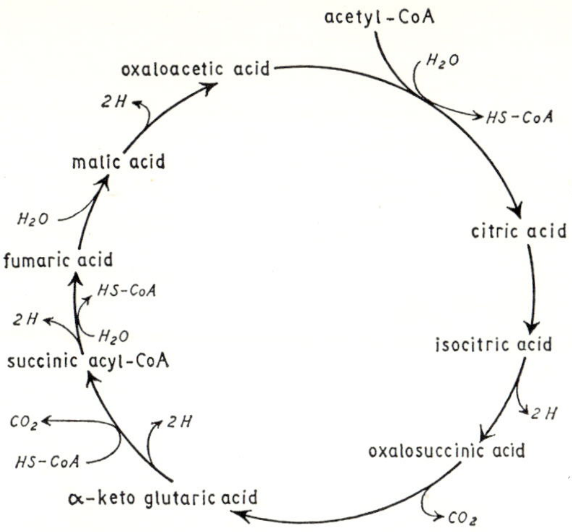

Fig. 8. The citric acid or Krebs' cycle and oxidation of acetyl-CoA.

Metabolism of Lipids

Like other substances in the animal the lipids are in a dynamic state of constant catabolism and anabolism. Hence, the fats absorbed and delivered to the systemic circulation as chylomicrons enter the pool in which the processes of this dynamic state take place (see also Chapter 10).

The lipids serve two major functions in the animal organism. Firstly they are essential elements in the structure of the living cell and secondly they represent the main store of potential energy.[107] Triglycerides of adipose tissue are typically energy deposits. Phospholipids of the cell membrane or of mitochondrial enzyme systems are typically structural lipids.[106,107] However, in many cases there are no sharp distinctions between depot fat and structural lipids. All the lipids in the organism represent more or less fatty acids in metabolism.

The fatty acids are transported from the site of deposition or absorption to the sites of metabolism as plasma lipids. Blood plasma from pigs contains on an average about 300 mg total lipids per 100 ml. About half of the total lipids is phospholipids, about 30 per cent is triglycerides and 20 per cent is cholesterol.[108] Besides the major fractions, triglycerides, phospholipids and cholesterol, total lipids in blood contains a small amount of non-esterified fatty acids (NEFA). In normal pigs weighing 80 kg the level of NEFA in the blood was found to be 0·3 to 0·7 micro-equivalents (μ-equiv.) per ml.[99] NEFA have a very rapid turnover and apparently they represent the main form in which fatty acids are transferred from fat depots to site of oxidation.[109] In the liver NEFA are esterified with formation of triglycerides, phospholipids and cholesterol esters.[103,110]

When lipid oxidation increases, the lipolytic activity in adipose tissue is stimulated causing an increased output of NEFA to the blood. In pigs fasted for 4 days Cunningham and Friend[99] found that the plasma level of NEFA increased from about 0·5 to 1·5 μ-equiv. per ml.

Several hormones affect lipid metabolism by stimulating the lipolytic activity in adipose tissue.[99,110] Also the synthesis of lipids appears to be under a homeostatic regulation and affected by the hormone balance. However, Masoro[111] found that no hormones had been demonstrated to exert a direct effect on lipogenic processes. On the other hand the adipokinetic activity of some pituitary substances and of catecholamines seems to be a direct effect upon the liberation of NEFA from adipose tissue.[99,110]

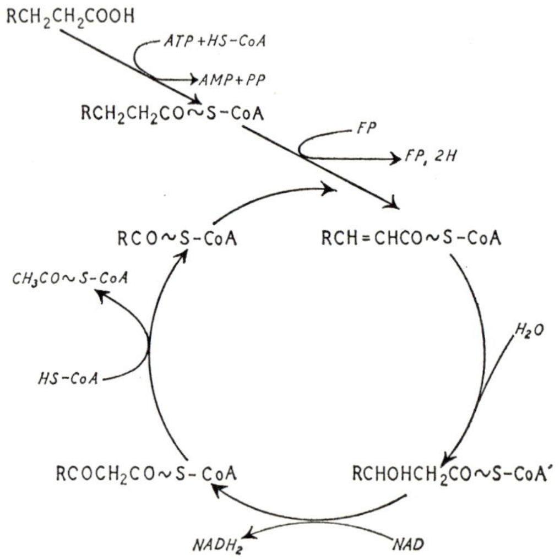

Fig. 9. The β-oxidation of fatty acids—the fatty acid oxidation spiral.

NEFA is taken up and metabolized by almost all tissues except nerve tissue. The oxidation of fatty acids takes place in the mitochondria by so-called β-oxidation. Before entering metabolic processes the fatty acids are converted to their acyl-CoA esters.[112,113,114] The processes of β-oxidation—the fatty acid spiral—are shown in Fig. 9. By each turn of the spiral the fatty acid chain becomes two carbon atoms shorter and a molecule of acetyl-CoA is formed.

The oxidation of 1 mole stearic acid gives 9 moles of acetyl-CoA and 16 equivalents of reduced coenzyme. Oxidation of 9 moles acetyl-CoA through the citric acid cycle (see Fig. 8) causes formation of 36 equivalents of reduced coenzyme and 18 moles CO_2. A total of 52 equivalents of reduced coenzyme are thus formed by the complete oxidation of 1 mole stearic acid. 52 equivalents of reduced coenzyme give 156 moles of ATP or 1560 kcal utilizable energy. 1560 kcal correspond to 58 per cent of the metabolizable energy in 1 mole stearic acid. It may be calculated that each gram of fatty acid carbon converted to CO_2 gives 7·2 utilizable kilocalories.

Only part of the depot fat and structural lipids originate from dietary lipids. About a century ago it was proved by balance trials that pigs can form fat from carbohydrates.[115,116] Later experiments have shown that in non-ruminant mammals carbohydrates are the main precursors in fatty acid synthesis.[117]

The immediate precursor in fatty acid synthesis is the acetyl-CoA formed by the breakdown of glucose or other substances. It was for some time believed that the synthesis of fatty acids took place by a direct reversal of the β-oxidation spiral. However, in recent years studies on fatty acid synthesis have shown that it takes place in the cytoplasm and follows a pathway as shown in Fig. 10.[118,119,120] $\left.\begin{matrix}HS\\HS\end{matrix}\right\}E$ in Fig. 10 symbolizes a multiple enzyme complex on which fatty acid synthesis takes place.

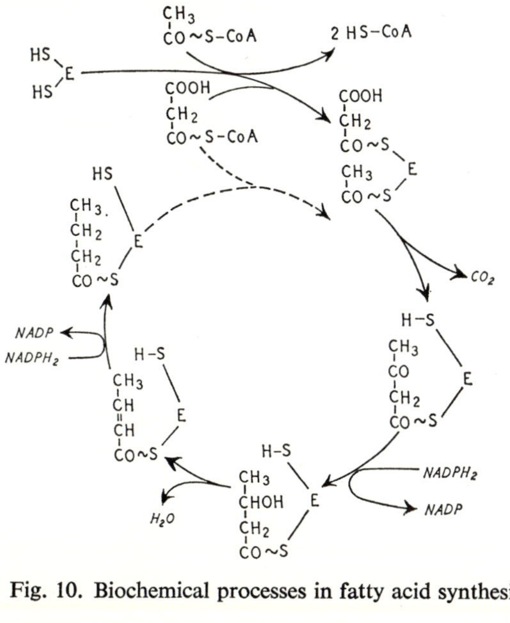

Fig. 10. Biochemical processes in fatty acid synthesis.

For the *de novo* synthesis of 1 mole stearic acid 8 moles of malonyl-CoA and 1 mole acetyl-CoA are used. The malonyl-CoA is formed by carboxylation of acetyl-CoA.[118,119]

It is noted that the product of the enzyme system is not an acyl-CoA but the free fatty acid. Consequently an accumulation of free fatty acid, i.e. a rise in the concentration of NEFA, will interfere with fatty acid synthesis.

Unsaturated fatty acids with one double bond are formed by dehydrogenation of the corresponding saturated fatty acid or rather its acyl-CoA.[120] Animal lipids contain in addition several fatty acids with two or more double bonds. One of the double bonds in these polyethenoic acids is placed in the 5–6 position to the methyl group and the animal organism is unable to dehydrogenate a fatty acid in this position. Since the polyethenoic acids are essential compounds of animal lipids, fatty acids having a double bond in the mentioned position must be considered as dietary essentials and are sometimes called vitagens.[117] The most important of these polyethenoic acids are linoleic, linolenic and arachidonic acid (see Chapter 10).

The synthesis of fatty acids is sometimes called lipogenesis. However, lipogenesis that is formation of lipids must include the esterification of the fatty acids. The formation of triglycerides and phospholipids proceeds stepwise with a diglyceride as common intermediate.[121,122] The diglyceride is formed from 2 molecules of acyl-CoA and glycerophosphate.

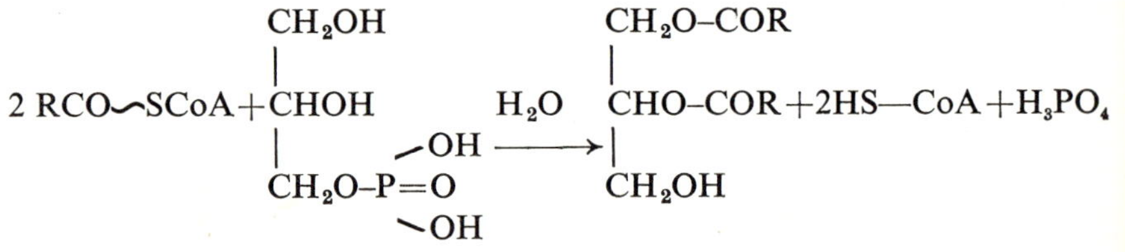

acyl-CoA glycerophosphate α,β-diglyceride

The α,β-diglyceride may be converted to triglyceride by addition of another acyl-CoA:

α,β-diglyceride $\frown$ RCO—SCoA ⟶ triglyceride + HS–CoA

Diglyceride with an unsaturated acid in the β-position may be used in formation of phosphatides. The formation of lecithin requires the addition of cytidine-diphosphate-choline:

α, β-diglyceride + cytidine — diphosphate — choline ⟶ lecithin + cytidine — monophosphate

Lecithin and other phospholipids contain relatively much of arachidonic acid. The fact that the phospholipids play an important role in intermediate metabolism[107] emphasizes the importance of a sufficient supply of essential fatty acids with the diet.

8. THE INFLUENCE OF FEED COMPOSITION UPON METABOLISM

In most cases of animal production it is of interest to obtain the maximum rate of protein synthesis. It is evident from the discussion of protein metabolism that this requires not only a supply of a certain amount of protein, estimated as N multiplied by 6·25, but also a certain quality of the protein.

The quality is expressed as the biological value (BV) which is defined as that percentage of the total amount of absorbed nitrogen which is used for synthesis of nitrogen containing compounds in the organism. According to the definition, the estimation of BV requires balance trials and determination of retained N as well as excreted endogenous nitrogen.[123] However, it is impossible in this way to determine the BV of all possible mixtures of proteins.[124]

Since the BV of a protein depends upon its content of essential amino acids Oser[125] suggested the use of an essential amino-acid index, EAAI, as an expression for BV. By comparing the results from several balance trials with the essential amino-acid content of the feed used in the trial Mitchell[126] found good correlation between the estimated EAAI and BV.

The main objection against EAAI is that it does not take the availability of the amino acids into account. The availability and the rate of amino-acid liberation in the digestive tract is an important factor for the value of a protein ration[86,124] (see also Chapter 8).

A method developed by Carpenter[127] for determination of the amount of free ε-amino groups in lysine of proteins was found to indicate the availability of amino acids.[128,129] Excellent correlations were found between the real BV and a combination of EAAI and 'available lysine'. The combination was called 'standard protein' and calculated as follows:[130]

$$\text{g standard protein} = \text{g N} \times 6{\cdot}25 \times \text{EAAI} \times \frac{\text{'available lysine'}}{\text{total lysine}}$$

It was found that Carpenter's method for determination of available lysine did not give reliable results with vegetable proteins. However, a modified procedure developed by Nielsen and Weidner [131] appears to be applicable to vegetable as well as animal proteins.[131,132,133]

An easy chemical method for determination of the biological value of proteins would make it possible to prepare rations always having a BV close to 100 and thereby secure a maximum rate of protein synthesis with a minimum loss of energy due to excretion of excess nitrogen.

In addition to protein synthesis, the production and deposition of fat probably are the most important processes of metabolism in the pig. The only factor of importance for the rate of fat deposition is the intake of net calories. An increased proportion of fat in the ration does depress lipogenesis,[111] but has not necessarily an effect on rate of fat deposition. The absorbed fatty acid may be directly transferred to the adipose tissue and deposited.

The old experiments of Munk[134] and Henriques and Hansen[135] showed the direct deposition of ingested fat. Ingested fatty acid may thus effect the composition of the lipids in adipose tissue. Table 6 shows the fatty acid composition of backfat from three groups of pigs fed a milk-barley ration. To the ration for two of the groups were added coconut oil and soybean oil respectively in amounts from 80 to 300 g daily. The experiment was carried out through the growth period from 60 to 90 kg live weight.[136] Especially in the case of soybean oil the composition of the backfat was markedly altered. It is seen that the concentration of linoleic acid increased from 4 per cent in the control group to 38·6 per cent in the soybean oil group.

TABLE 6
The Influence of Dietary Fat upon Composition of Back Fat[136]
(*values in % of total fatty acids*)

	Lauric acid	Myristic acid	Palmitic acid	Stearic acid	Oleic acid	Linoleic acid	Linolenic acid
Control group	—	2·2	30·3	17·6	41·5	8·4	trace
Coconut oil group	4·6	15·1	31·6	12·1	32·5	4·6	trace
Soybean oil group	—	2·0	16·2	11·3	29·3	38·6	2·8

By similar experiments with lactating sows it was found that some of the absorbed fatty acids were carried to the mammary gland and secreted with milk lipids. A change in the composition of milk lipids caused a change in the composition of the lipids in the suckling piglets. In one case it was found that the proportion of linoleic and linolenic acid in the carcass lipids from the baby pigs increased from 4 per cent to 24 per cent.[137]

The possibility of increasing the amount of essential fatty acids in animal products may be of interest. However, a change in the composition especially enrichment of essential fatty acids means also a change in keeping quality. Hence, in studies on the influence of feeding upon the composition of animal products possible alterations in the quality must be taken into account.

REFERENCES

1. MANGOLD, E. (1950) *Die Verdauung bei den Nutztieren.* Akademie-Verlag, Berlin.
2. BEHRENS, R. (1929/30) *Zur Sekretion der Parotis des Schweines.* Med. Vet. Diss., Hannover.
3. HENCKE, H. (1930/31) *Zur Sekretion der Glandula mandibularis und Glandula sublingualis monostomatica des Schweines.* Med. Vet. Diss., Hannover.
4. MAXIMOW, A. A. and BLOOM, W. (1957) *A Textbook of Histology.* W. B. Saunders, London.
5. BARD, P. (1956) *Medical Physiology*, 10th ed. C. V. Mosby, St. Louis.
6. EDKINS, J. S. (1906) *J. Physiol.* 34, 133.
7. SCHEUNERT, A. und TRAUTMANN, A. (1965) *Lehrbuch der Veterinär-Physiologie.* Paul Parey, Berlin.
8. WOODWARD, E. R. and DRAGSTEDT, R. R. (1960) *Physiol. Rev.* **40**, 490.

9. TRAUTMANN, A. und KIRCHHOF, H. (1937) *Archiv. wiss. pract. Tierheilkunde* **72**, 361.
10. HIGHTOWER, N. C. (1962) *Ann. Rev. Physiol.* **24**, 109.
11. LANGLEY, J. N. (1881) *J. Physiol.* **3**, 246 and 269.
12. LINDERSTROM-LANG, K., HOLTER, H. and OHLSEN, S. A. (1934) *Zeitschr. Physiol. Chem.* **227**, 1.
13. HIRSCHOWITZ, B. I. (1957) *Physiol. Rev.* **37**, 475.
14. DAHLQUIST, A. (1961) *Biochem. J.* **78**, 282.
15. HEYENGA, H. (1938) *Untersuchungen über Eigenschaften, Zusammensetzung und Wirkung des Magensaftes nüchternder Schweine.* Med. Vet. Diss., Hannover.
16. SCHEUNERT, A. und KIOK, F. (1921) *Zeitschr. Physiol. Chem.* **193**, 16.
17. FRUTON, J. S. and SIMMONDS, S. (1958) *General Biochemistry* 2nd ed. John Wiley, New York.
18. SENIOR, J. R. (1964) *J. Lipid Res.* **5**, 495.
19. KOLB, E. (1962) *Lehrbuch der Physiologie der Haustiere.* Veb. Gustav Fischer Verlag, Jena.
20. THOMAS, J. E. (1957) *Physiol. Rev.* **37**, 453.
21. WERLE, J. M., BRODY, D. A., LIGON, W. E., READ, M. R. and QUIGLEY, J. P. (1941) *Am. J. Physiol.* **131**, 606.
22. GREENGARD, H. (1948) *Hormones of the gastrointestinal tract.* p. 201 in *The Hormones* I.
23. BLOMSTRAND, R. (1960) *Acta Chem. Scand.* **14**, 1006.
24. NAKAYAMA, F. and BLOMSTRAND, R. (1960) *Acta Chem. Scand.* **14**, 1211.
25. HILBY, F. (1934) *Untersuchungen über die Gallensekretion des Schweines.* Med. Vet. Diss., Hannover.
26. TANTURI, C. A., IVY, A. C. and GREENGARD, H. (1937) *Am. J. Physiol.* **120**, 336.
27. IVY, A. C. and OLDBERG, E. (1928) *J. Am. Med. Assoc.* **90**, 445.
28. IVY, A. C., KLOSTER, G., LUETH, H. C. and DREWYER, G. E. (1930) *J. Am. Physiol.* **91**, 336.
29. IVY, A. C. (1943) *Gastroenterology* **3**, **433**.
30. MATTSON, F. H., BENEDICT, J. H., MARTIN, J. B. and BECK, L. W. (1952) *J. Nutrit.* **48**, 335.
31. LANDS, E. M. W. (1965) *Ann. Rev. Biochem.* **34**, 313.
32. GREENGARD, H., DUTTAN, D. F. and IVY, A. C. (1945) *Federation Proc.* **4**, 27.
33. MELLANBY, I. (1926) *J. Physiol.* **61**, 61.
34. HARPER, A. A. and VOSS, C. C. N. (1941) *J. Physiol.* **99**, 415.
35. HARPER, A. A. and RAPER, H. S. (1943) *J. Physiol.* **102**, 115.
36. FLOREY, H. W. and HARDING, H. E. (1934) *J. Path. and Bact.* **39**, 255.
37. LOSCHEIDT, W. (1938) *Untersuchungen über Sekretion und Wirkung des Ileumsaftes von Schwein.* Med. Vet. Diss., Hannover.
38. UNGER, H. (1938) *Untersuchungen über Sekretion, Zusammensetzung und Wirkung des Dünndarmsaftes (Jejunalsaftes) von Schwein.* Med. Vet. Diss,. Hannover.
39. FLOREY, H. W. and HARDING, H. E. (1935) *Proc. Royal Soc.* B **117**, 336.
40. FLOREY, H. W. and HARDING, H. E. (1935) *Quart. J. Exper. Physiol.* **25**, 329.
41. NASSET, E. S. (1938) *Am. J. Physiol.* **121**, 481.
42. FINK, R. M. (1943) *Am. J. Physiol.* **139**, 633.
43. ALEXANDER, F. (1962) *Res. Vet. Sci.* **3**, 78.
44. DESNUELLE, P. (1961) *Advan. Enzymol.* **23**, 129.
45. BORGSTROM, B. (1962) *Gastroenterology* **43**, 216.
46. MAGEE, W. L., GALLAI-HATCHARD, J., SANDERS, H. and THOMPSON, R. H. S. (1962) *Biochem. J.* **83**, 17.
47. DE HAAS, G. H., HENSKERK, G. H. T., VAN DEENEN, L. L. M., BAKER, R. W. R., GALLAI-HATCHARD, J., MAGEE, W. L. and THOMPSON, R. H. S. (1963) p. 244 in *Biochemical Problems of Lipids.* Ed. Frazer, A. C. Elsvier, Amsterdam, London and New York.
48. CUTHBERTSON, D. P. and PHILLIPSON, A. T. (1953) *Microbiology of digestion in Biochem. and Physiol. of Nutrition.* Ed. Bouren and Kiddle. Vol. II, p. 128. Academic Press, New York.
49. MANGOLD, E. (1934) *Nutrition Abs. and Rev.* **3**, 647.

50. SCHEUNERT, A. (1906) *Zeitschr. Physiol. Chemie* **48**, 9.
51. MADSEN, A. (1963) *Fordøjelighedsforsøg med svin* (Digestibility experiments with pigs). 337. Beretning fra Forsøgslaboratoriet, København.
52. WOODMAN, H. E. and EVANS, R. E. (1947) *J. Agric. Science* **37**, 202.
53. NORDFELDT, S. (1946) *Smältbarhetsförsök med svin*. Meddelande nr. 23. Lantbrukshögskolan, Norrtälje.
54a. THOMKE, S. (1960) *Annals Royal Agric. Coll. Sweden* **26**, 269.
&
54b. THOMKE, S. (1961) *Smältbarhetsförsök med havre til får, svin och höns*. Meddelande nr. 73. Kungl. Lantbrukshögskolan och Statens Lantbruksförsök, Uppsala.
55. AXELSSON, J. (1955) *Archiv für Tierernährung* **5**, 1.
56. AXELSSON, J. and ERIKSSON, S. (1953) *J. Animal Sci.* **12**, 881.
57. ELSDEN, S. R., HITCHCOCK, M. W. S., MARSHALL, R. A. and PHILLIPSON, A. T. (1946) *J. Exp. Biol.* **22**, 191.
58. THORBEK, G. (1960) *Årbog, Landøkonomisk Forsøgslaboratorium* p. 178 and (1961) p. 317.
59. FRIEND, D. W., CUNNINGHAM, H. M. and NICHOLSON, J. W. G. (1962) *Canad. J. Animal Sci.* **42**, 55.
60. FRIEND, D. W., CUNNINGHAM, H. M. and NICHOLSON, J. W. G. (1963) *Canad. J. Animal Sci.* **43**, 156.
61. FRIEND, D. W., CUNNINGHAM, H. M. and NICHOLSON, J. W. G. (1963) *Canad. J. Animal Sci.* **43**, 174.
62. TRAUTMAN, A. und KIRCHHOF, H. (1940) *Deutsche Tierärztl. Wch. schr.* **48**, 650.
63. WILLINGALE, J. M. and BRIGGS, C. A. E. (1955) *J. Appl. Bacteriol.* **18**, 284.
64. MICHEL, M. C. (1961) *Annales Biol. Animal. Biochem. Biophys.* **1**, 29.
65. BARCROFT, J., MCANALLY, R. A. and PHILLIPSON, A. T. (1944) *J. Exp. Biol.* **20**, 120.
66. FORBES, R. M. and HAMILTON, T. S. (1952) *J. Animal Sci.* **11**, 480.
67. BREIREM, K., HUSBY, M., PRESTHEGGE, K. and HOMB, T. (1958) *Zeitschr. Tierphysiol. Tierernährung u. Futtermittelkunde* **13**, 129.
68. CUNNINGHAM, H. M., FRIEND, D. W. and NICHOLSON, J. W. G. (1961) *Canad. J. Animal Sci.* **41**, 120.
69. CUNNINGHAM, H. M., FRIEND, D. W. and NICHOLSON, J. W. G. (1962) *Canad. J. Animal Sci.* **42**, 167.
70. RICHARDSON, D., CATRON, D. V., UNDERKOFLER, L. A., MADDOCK, H. M. and FRIEDLAND, W. G. (1951) *J. Nutrit.* **44**, 371.
71. GRZESZCZAK-SWIETLIKOWSKI, U. (1959) *International Symposium on B-vitamins*. Poznan IV, p. 221.
72. NAJJAR, V. A. and BARRETT, R. (1945) p. 23 in *Vitamins and Hormones* III.
73. EICHEL, H. und SCHICKETANZ, W. (1962) *Arc. Tierernährung* **12**, 37.
74. HASSLER, G. (1961-2) *Jahrb. Arbeitsgemeinschaft Fütterungberatung* **4**, 171.
75. TRAUTMAN, A. und ASHER, TH. (1941) *Archiv wiss. u. prakt. Tierheilkunde* **76**, 119.
76. LENKEIT, W. (1932) *Berliner Tierärztl. Wchschr.* **48**, 17.
77. MARCER, J. H., POND, W. G., LOOSLI, J. K. and LOWREY, R. S. (1962) *J. Animal Sci.* **21**, 49.
78. DUKES, H. H. (1955) *The Physiology of Domestic Animals*. 7th ed. George Banta Comp., Menasha, Wisconsin.
79. HEINZ, E. and WALSH, P. M. (1958) *J. Biol. Chem.* **233**, 1488.
80. HEINZ, E. (1961) *Coll. Gesellsch. Physiol. Chem.* **12**, 175 .
81. PAINE, C. M. and HEINZ, E. (1960) *J. Biol. Chem.* **235**, 1080.
82. CHRISTENSEN, H. N., RIGGS, T. R., FISCHER, H. and PALATINE, I. M. (1952) *J. Biol. Chem.* **198**, 1.
83. QUASTEL, J. H. and FRIEDHANDLER, L. (1955) *Arch. Biochem.* **56**, 424.
84. JACOBS, F. A. and HILLMAN, R. S. L. (1958) *J. Biol. Chem.* **232**, 445.
85. JACOBS, F. A., COEN, L. J. and HILLMAN, R. S. L. (1960) *J. Biol. Chem.* **235**, 1372
86. NIELSEN, J. J. (1965) *Bestemmelse af aminosyrers fordøjelighed* (thesis). Copenhagen.
87. DENTON, A. E. and ELVEHJEM, C. A. (1954) *J. Biol. Chem.* **206**, 455.
88. ASWORTH, C. T. and JOHNSTON, J. M. (1963) *J. Lipid Research* **4**, 454.

89. DAVENPORT, H. W. (1961) *Physiology of the Digestion Tract*. Year Book Med. Publish. Chicago.
90. FRAZER, A. C. (1946) *Physiol. Rev.* **26**, 103.
91. BERGSTROM, S., BORGSTROM, B., CARLSTAR, A. and ROTTENBERG, M. (1950) *Acta Chem. Scand.* **4**, 1142.
92. BORGSTROM, B. (1951) *Acta Chem. Scand.* **5**, 643.
93. REISER, R. and BRYSON, M. J. (1951) *J. Biol. Chem.* **187**, 87.
94. BUELL, G. C. and REISER, R. (1959) *J. Biol. Chem.* **234**, 217.
95. BORGSTROM, B. (1955) *J. Biol. Chem.* **214**, 671.
96. MATTSON, F. H. and VOLPENHEIN, R. A. (1962) *J. Biol. Chem.* **237**, 53.
97. MATTSON, F. H. and VOLPENHEIN, R. A. (1964) *J. Biol. Chem.* **239**, 2774.
98. SKIPSKI, V. P. and MOOREHOUSE, M. G. (1965) *Arch. Biochem. Biophys.* **107**, 525.
99. CUNNINGHAM, H. M. and FRIEND, D. W. (1965) *J. Animal Sci.* **24**, 41.
100. RIIS, P. M. (1957) *Beretning fra Sterilitetsinstituttet* p. 17. Copenhagen.
101. VAN SLYKE, D. D. and MEYER, G. M. (1913) *J. Biol. Chem.* **16**, 213.
102. WATSON, J. D. (1963) *Science*, **140**, 17.
103. RIIS, P. M. Unpublished observations.
104. KREBS, H. A. (1954) *Bull. John Hopkins Hosp.* **95**, 19.
105. UTTER, M. F. and KEECH, D. B. (1960) *J. Biol. Chem.* **235**, PC 17.
106. FRAZER, A. C. (1953) *Lipid metabolism* p. 212. In *Biochem. Physiol. Nutrition: I.* Ed. Bourne, G. H. and Kidder, G. W. Academic Press, New York.
107. ROSSITER, R. J. and STRICKLAND, K. P. (1960) *The metabolism and function of phosphatides*, p. 69. In *Lipide Metabolism.* Ed. Bloch, K. John Wiley, New York.
108. DEUEL, H. J. (1955) *The Lipids* II. Interscience, New York.
109. FREDERICKSON, D. S. and GORDON, R. S. (1958) *Physiol. Rev.* **38**, 584.
110. HIRSCH, R. L., RUDMAN, D., IRELAND, R. and SKRALY, R. K. (1963) *J. Lipid Research* **4**, 289.
111. MASORO, E. J. (1962) *J. Lipid Research* **3**, 149.
112a. LYNEN, F. (1952-53) *Harvey Lectures* **48**, 210 &
112b. LYNEN, F. (1955) *Ann. Rev. Biochem.* **24**, 653.
113. GREEN, D. E. and GIBSON, V. M. (1960) Chapter 7, p. 301 in *Metabolic Pathways I.* Ed. Greenberg, D. M. Academic Press, New York.
114. MAHLER, H. R., WAKIL, S. J. and BOCK, R. M. (1953) *J. Biol. Chem.* **204**, 453.
115. LAWES, J. B. and GILBERT, J. H. (1866) *Phil. Mag.* **32**, 439.
116. MEISSL, E. and STROHMER, F. (1883) *Sitzungsberichte Wien Akad. Wissenschaft. Matem.-Nat.* **88**, 205.
117. DEUEL, H. J. (1957) *The Lipids: III.* Interscience, New York.
118. WAKIL, S. J. (1962) *Ann. Rev. Biochem.* **31**, 369.
119. WAKIL, S. J. (1958) *Am. Chem. Soc.* **80**, 6465.
120. VAGELOS, P. R. (1964) *Ann. Rev. Biochem.* **33**, 139.
121. KENNEDY, E. P. (1957) *Ann. Rev. Biochem.* **26**, 119.
122. POPJAK, G. (1958) *Ann. Rev. Biochem.* **27**, 533.
123. MITCHELL, H. H., HAMILTON, T. S., BEADLES, J. R. and SIMPSON, F. (1945) *J. Nutrit.* **29**, 13.
124. MASON, V. C. (1963) *Amino Acids in Nutrition* (thesis). Copenhagen.
125. OSER, B. L. (1951) *J. Am. Diet Assoc.* **27**, 396.
126. MITCHELL, H. H. (1954) *Wiss. Abh. Deutsche Akad. Landwirt. Wiss.* V/2, 279.
127. CARPENTER, K. J., MORGAN, C. B., LEA, C. H. and PARR, L. J. (1962) *Brit. J. Nutrit.* **16**, 451.
128. MADSEN, A., MASON, V. C. and WEIDNER, K. (in print) *Acts Agriculturæ Scand.*
129. DANGOUMAU, A. and DEBRUYNE, H. (1961) *Les Industries l'Alimentation Animale.* Paris No. 124, 39.
130. JAKOBSEN, P. E. (1965) *Zeitschr. Tierphysiol. Tierernährung u. Futtermittelkunde* **20**, 168.
131. NIELSEN, J. J. and WEIDNER, K. (1966) *Determination of available lysine by column chromatography* (in preparation).
132. EGGUM, B. O. (1963) *Årbog, Landøkonomisk Forsøgslaboratorium* p. 400. Copenhagen

133. EGGUM, B. O. (1965) *Proteinbewertung. Das Verhältnis zwischen der Aminosäurezusammensetzung und der biologischen Wertigkeit.* Tagungsbericht, Giessen.
134. MUNK, J. (1884) *Kirchows Archiv* **95**, 407.
135. HENRIQUES, V. and HANSEN, C. (1899) *44. Beretning fra Forsøgslaboratoriet.* Copenhagen.
136. CHRISTENSEN, K. D. (1963) *Acta Agriculturæ Scand.* **13**, 249.
137. CHRISTENSEN, K. D. (1965) *Årbog, Landøkonomisk Forsøgslaboratorium* p. 327. Copenhagen.

CHAPTER 6

Digestion, Absorption and Metabolism in Poultry

W. BOLTON

Agricultural Research Council Poultry Research Centre, Edinburgh

INTRODUCTION

DIGESTION comprises those physical and chemical actions that the food undergoes between the times of its entry into the mouth and its absorption from the gut. In the bird they include swallowing; storage in the crop; grinding in the gizzard; degradation by enzymes, whether present in the food or secreted by the bird; and the action of microorganisms.

Absorption covers the processes by which the digested food passes through the wall of the alimentary canal.

Metabolism is the process whereby the individual produces heat, body cells or their products, or eggs from the absorbed nutrients.

Relatively little basic research has been carried out on these processes in the bird, although certain facets, e.g. calcium metabolism, have been studied. Of the work that has been carried out, the greater part has used the domestic fowl as the test animal. Most of our knowledge of nutrition in the fowl has come from experiments designed to determine requirements for nutrients, and from feeding trials where the yield was measured either as liveweight gain or as rate of egg production; in only a few experiments of this type has the yield been assessed in terms of protein, fat and carbohydrate.

For these reasons the material presented in this chapter serves more to demonstrate the areas in which basic research is urgently needed than to give a clear picture of the processes of digestion, absorption and metabolism in the fowl.

ANATOMY OF THE DIGESTIVE TRACT

The digestive tract of the bird consists of a tube running from the beak to the cloaca, to which are associated certain diverticula or organs: crop, pancreas, liver and two caeca. The tube is modified along its length into various regions; starting from the beak these are: mouth, pharynx, oesophagus, proventriculus, gizzard, duodenum, small intestine and large intestine (see Fig. 1).

The beak of the bird is horny and devoid of teeth. The opposing edges of the beaks of fowls and turkeys are smooth and the beaks are pointed. In water fowl the beaks are spatulate, the opposing edges are inclined at an angle of 45° and are grooved; this arrangement permits the bird to squeeze water out of the mouth whilst retaining food. The lower section is joined loosely to the skull, thereby permitting very wide distention of the mouth.

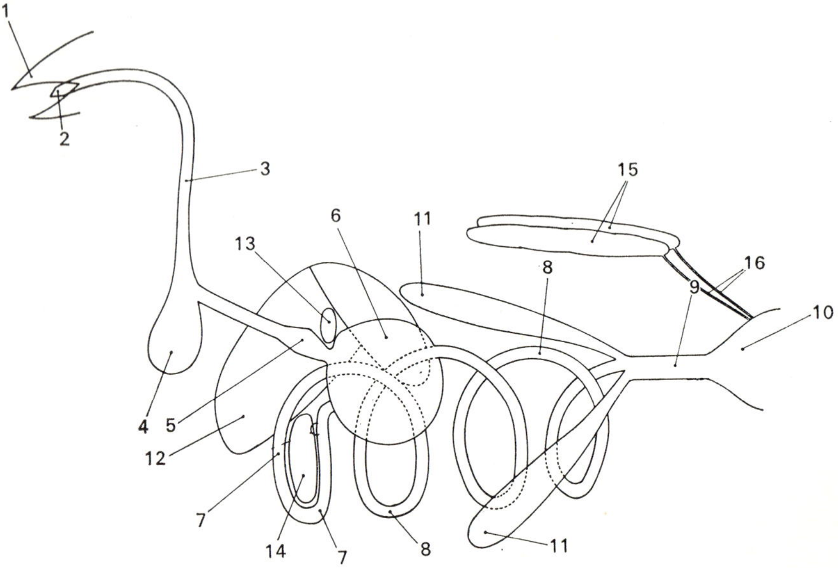

FIG. 1. Digestive tract of the fowl. 1. Beak; 2. Tongue; 3. Oesophagus; 4. Crop; 5. Proventriculus; 6. Gizzard; 7. Duodenum; 8. Small intestine; 9. Large intestine; 10. Cloaca; 11. Caeca; 12. Liver; 13. Gall Bladder; 14. Pancreas; 15. Kidneys; 16. Ureters.

The salivary glands are diffuse structures found (a) at the bottom of the oral cavity, (b) at the angle of the mouth, (c) in the tongue, (d) in the roof of the mouth and (e) in the pharyngeal canal (Schauder, 1923; Calhoun, 1954). Calhoun (1954) found only mucus-secreting cells in the glands, although Vermeulen (1929) had reported that serous cells were present in the glands of some graminiferous birds.

There is no area in the mouth comparable with the soft palate of mammals and the dividing line between the mouth and the pharynx is usually taken as the last row of papillae on the hard palate. The pharynx extends from the mouth to the oesophagus and its mucous membrane is similar in structure to that of the mouth (Biester and Schwartz, 1959).

The oesophagus is a relatively wide and distensible tube running from the pharynx to the proventriculus. The region is well-supplied with mucus-secreting cells. In fowls and turkeys a ventral diverticulum, the crop, is found at the junction of the cervical and thoracic portions. The crop is mostly glandless. Waterfowl have a long spindle-shaped dilatation in the cervical region.

The proventriculus is a small organ with very thick walls. The mucous membrane lining the organ is folded longitudinally. It is lined with cells showing features of the parietal and chief cells of mammals and also of the oxyntic cells of the amphibians; Selander (1963) studied these cells using the electron microscope and concluded that they secreted pepsin and hydrochloric acid, thus confirming the observations of Klug (1893), Plenk (1932), Yamamoto (1937) and Luppa (1962).

In the junction between the proventriculus and the gizzard the deeper, proventricular folds end abruptly, while the superficial cells gradually assume the characteristics of those of the gizzard.

The gizzard is a flattened, spheroidal organ made up from two thick muscles, opposing one another, joined by two thin muscles. The thick muscles contain long, tubular glands; the lumina of these glands contain a substance with the staining properties of keratohyaline (Calhoun, 1954). This material, with cellular debris, forms the corrugated layer of horny material lining the gizzard. The gizzard normally contains pebbles. These enable the organ, under muscular contraction, to function like a ball-mill and grind the food.

The duodenum is a U-shaped loop enclosing the pancreas. At the junction of the duodenum with the small intestine the secretions of the pancreas and the liver are added to the ingesta. The duodenum is lined with a thick, soft and highly vascular mucous membrane which has a velvety appearance.

The jejunum and ileum cannot be distinguished and are usually referred to as the small intestine. The lining is similar to that of the duodenum, but the villi are shorter. About half-way down the small intestine is found Meckel's diverticulum, the vestige of the yolk-sac.

Two caeca are found at the junction of the small and large intestines. They are long, blind sacs which normally contain a dark-brown, viscous fluid. The villi are well-developed near the mouths of the sacs, but are shorter and broader towards the fundus.

The large intestine is very short and runs in an almost straight line from the caecal orifices to the cloaca. The lining is similar to that of the small intestine.

The cloaca is a tubular cavity opening to the exterior, and common to the digestive, urinary and reproductive systems. The lining of the cavity is similar to that of the small and large intestines.

THE PASSAGE OF FOOD THROUGH THE TRACT

Food seized by the beak is swallowed whole. It is not possible for the bird to provide the muscular aid that mammals use in swallowing; instead, the bolus of food is propelled backwards by raising the head and giving it a quick, forward thrust. When a bird has been deprived of food long enough to empty the crop and gizzard, the first beakfuls of food pass directly to the gizzard; thereafter food not required to keep the gizzard full remains in the crop. During the process of filling the crop, the organ increases in size from a semicircular disc to a large, oblate spheroid. After the meal is finished, the food gradually passes from the crop into the gizzard. There does not appear to be any mixing of food in the crop when foods of the same consistency are eaten (Browne, 1922). When pellets are fed, the food in the anterior part of the crop and in the posterior oesophagus becomes much more moist than that in the posterior part of the crop (Bolton, unpublished).

The gizzard contracts rhythmically and grinds the wet food to a paste which passes to the duodenum; the grinding is aided by the provision of insoluble grit.

Food in the duodenum is propelled posteriorly by peristaltic contractions of the gut until it reaches the junction of the small and large intestines and caeca. At this point some of the liquid of the ingesta passes into the caeca, the remainder passes into the large intestine and is excreted via the cloaca as faeces. Urine from the kidneys is also excreted via the cloaca; most of the liquid in the urine is absorbed by the faeces and the unabsorbed residues appear as a white cap.

THE RATE OF PASSAGE OF FOOD THROUGH THE GUT

The time of transit of food through the gut has been estimated by two procedures: in one the birds were deprived of food long enough to empty the crop; food mixed with a marker was given and the interval between feeding and the appearance of the marker in the droppings was noted; in the other the food was marked with barium sulphate and the passage of the food through the gut was followed by X-rays.

Steinmetzer (1924) adopted the X-ray technique and concluded that food reached the crop in a few seconds. It took 30 sec to reach the gizzard, 15 min to the duodenum and 2 hr before the first marked droppings were voided. Halnan (1949) reported work which he and Barclay had carried out in 1929. They drilled holes in maple peas, filled the holes with a mixture of barium sulphate and gelatine and measured the transit time by X-rays. They found that food reached the gizzard in 15 to 30 sec and the duodenum in 10 min, thus confirming Steinmetzer's (1924) observations.

Kaupp and Ivey (1923) adopted the marker system and concluded that lampblack was the best of those they tried. They reported average transit times for laying birds of 3 hr 46 min, for growers 3 hr 52 min, for non-layers 8 hr and for broody hens 11 hr 44 min. Monson, Dietrich and Elvehjem

(1950) related the time of transit to the carbohydrate source in the diet. They found that lactose-based diets passed through the gut most rapidly, those based on dextrin were slowest and those on glucose were intermediate. Tuckey, March and Biely (1958) found that the inclusion of up to 12 per cent of fat did not affect transit time, but noted that if birds were excited the passage of food through the gut was slowed down.

The length of time that food can remain in the crop has been studied by several groups of workers. Brown (1904) fed 50 g maize to birds and found that it was 24½ hr before the crop was again empty: 30 g took about 12 hr. The time for 30 g was confirmed by Schwartz and Teller (1924). Card, Keith and Mitchell (1926) compared whole and ground maize and found that the crop took longer to empty when the ground maize was fed. Heuser (1945) disagreed with Card, Keith and Mitchell (1926); in his experiments maize meal stayed in the crop the shortest period of time, followed by cracked maize; whole maize stayed longest. The times were: maize meal 21 hr, cracked maize 24 hr and whole maize 28 hr.

THE HYDROGEN ION CONCENTRATIONS IN THE TRACT

The food eaten by poultry is slightly acid (see Table 1). If the crop was

TABLE 1
pH Values of Poultry Foods

Food	pH
Barley	5·80
Maize	6·10
Wheat	6·45
Bran	6·45
Fish meal	6·35
Grass	5·76
Yeast	5·80
Breeders' pellets	5·55
Layers' mash	5·92

empty before the bird began feeding, the pH of crop contents taken immediately after a meal was about 0·2 pH units higher, (i.e. more alkaline) than that of the food (Bolton, unpublished). Under the action of microorganisms, acid is produced in the crop and the pH gradually falls, reaching a final value of about 4 (Bolton, 1965).

The pH of the contents of the proventriculus is in the region 3·1 to 6·1, with a mean value of about 4·4 (Mussehl, Blish and Ackerson, 1933; Olson and Mann, 1935; Mayhew, 1935; Heller and Penquite, 1936; Farner, 1942; Buckner, Insko and Henry, 1944). However, the birds examined by these authors all had food in the digestive tract, and the pH of the contents of the proventriculus was only slightly lower than that of the crop contents.

The gizzard is the most acid portion of the gut and pH values as low as 2·6 have been reported (Farner, 1942); the highest is 5·0 (Olson and Mann, 1935). Examination of the contents of the gizzards of ten birds which had been offered food *ad libitum* gave a mean pH value of 3·6, with a range of 3·1 to 4·3 (Bolton, unpublished).

When the ingesta enter the duodenum they become more alkaline, pH range 5·6 to 6·3. Comparison of the duodenal contents of the ten birds referred to above showed a remarkable consistency of pH; the range was 6·20 to 6·40, with a mean value of 6·34.

The anterior half of the small intestine is normally slightly acid, with a pH of about 6: in the posterior half the reaction becomes almost neutral or slightly alkaline (pH 6·9 to 8·4). The gut contents become acid again (pH 6·3 to 6·9) in the large intestine.

In the caeca the pH is dependent partially upon the length of time the material has been present, for it is known (Hill, Annison and Noakes, 1964) that acetic, propionic and lactic acids are formed there by microbial action. The pH values reported in the literature range from 5·5 to 8·2.

SECRETIONS OF THE BIRD INTO THE GUT

The cells lining the mouth and anterior oesophagus secrete mucus, and the amylase ptyalin is secreted by the salivary glands in the mouth. The amylase was detected by Shaw (1913), and its presence has been confirmed by Vermeulen (1929) and Leasure and Link (1940). Jung and Pierre (1933) did not detect hydrolysis of starch to sugar in the crop, but Bolton (1965) found that about half the starch could be converted to sugar in this organ.

Browne (1922) claimed that the secretion from the few glands near the oesophageal inlet to the crop was profuse. Plimmer and Rosedale (1922) and Hamilton and Mitchell (1924) reported that an amylase is produced by cells in the same region. The presence of proteolytic and amylolytic enzymes in the crop has been both reported and denied (see Groebbels, 1932). The difference of opinion may be due, at least in part, to the method used for obtaining the sample: when birds are killed by cervical dislocation the agonal spasms may cause ingesta to be forced up and down the digestive tract and this can lead to spurious findings. Death by an overdose of anaesthetic avoids this complication.

Shaw (1912–13) reported that secretions from the proventriculus hydrolysed protein, but he did not detect either amylase or lipase. Collip (1922) states that production of acid by the proventriculus was noted by Colin in 1871 and secretion of pepsin by Wilczewski in 1870 and by Paira in 1900.

Cheney (1938) and Friedman (1939) found that gastric secretion was intermittent. Starvation decreased the flow of gastric juice, whereas feeding stimulated it (Collip, 1922). The rate of secretion can be increased by administration of drugs such as histamine (Keeton, Koch and Luckhardt, 1920; Cheney, 1938; and Friedman, 1939) and alcohol (Friedman, 1939). The total acidity and free acid in the gastric juice of fed birds were 0·22 per cent and 0·07 per cent respectively according to Farner (1943).

The secretions of the pancreas and liver are added at the junction of the duodenum and small intestine; as the pH in this region is about 6·3, it has been assumed that any proteolytic digestion in the duodenum is caused by pepsin acting at a pH well above its optimum, which is 1·8. However, the absence from the duodenum of proteolytic enzymes other than pepsin has not been proved.

Shaw (1912–13) and Plimmer and Rosedale (1922) demonstrated the presence of proteolytic, lipolytic and amylolytic enzymes in the pancreas. Langendorff (1879) cannulated the pancreatic ducts to obtain samples of pancreatic juice, and found that birds deprived of pancreatic secretions died in 6 to 12 days, thus demonstrating the essential nature of the pancreas for the maintenance of life.

The bile is acid, with a pH of about 6 (Mayhew, 1935; Heller and Penquite, 1936; Farner, 1942; and Buckner, Insko and Henry, 1944). The amylase content of bile from the gall bladder is higher than that of bile from the liver (Farner, 1942). Bile also contains mucin and bile salts.

The cells of the duodenum, small intestine and large intestine secrete mucus. Those of the duodenum and small intestine also secrete enzymes, but one cannot be certain of their identity because it is difficult to obtain samples of intestinal juice free from pancreatic juice and bile. Such evidence as has accumulated indicates the presence of the enzymes that are present in mammalian intestines, viz. erepsin, enterokinase, lipase and amylases. It appears that the cells of the large intestine and caeca do not secrete enzymes.

DIGESTION

Food seized by the beak is moistened with saliva and swallowed. If the bird has been deprived of food long enough to empty the crop, the first few mouthfuls of food pass directly to the gizzard via the oesophagus and proventriculus. Thereafter the gizzard is kept supplied with food and the excess is stored in the crop.

Food in the crop is acted upon by the enzymes present in the food, by the amylase secreted by the salivary glands and by the microflora of the crop. Starch is hydrolysed to sugar (Browne, 1922; Bolton, 1965) and glucose can be absorbed from the crop (Soedarmo, Kare and Wasserman, 1961). The microflora convert some of the sugar to acetic acid, lactic acid and alcohol; under exceptional conditions the concentration of lactic acid may be abnormally low and in this event the crop contents are rich in propionic and butyric acids. Fermentation in the crop is therefore similar in many respects to that which takes place in a silo during ensilage. Acetic, propionic and butyric acids (Bolton and Dewar, 1965) and lactic acid (Bolton and Dewar, 1962) are absorbed by the bird and presumably used as energy sources. Alcohol has been shown to stimulate the secretion of gastric juice (Friedman, 1939). The amount of nitrogenous secretion added to the ingesta at this stage of digestion is negligible (Bolton, unpublished).

The capacity of the proventriculus is small, and food does not remain in it for long; it appears to be concerned solely with the addition of gastric juice to the ingesta. The gastric juice contains pepsin and hydrochloric acid, and peptic digestion of the proteins proceeds in the gizzard.

The gizzard undergoes rhythmic contractions and these, aided by the insoluble grit normally present, grind the food to a smooth paste which passes into the duodenum.

In the duodenum the moisture content of the ingesta increases from the 60 per cent of the gizzard to about 88 per cent (Bolton, unpublished). There is also a tremendous outpouring of nitrogenous material into the lumen of the duodenum and this raises the total nitrogen content to about nine times the amount ingested (Bolton, 1961a). Examination of the contents of the duodenum has shown that about 30 per cent of the total nitrogen is present as mucin (Bolton, 1964). Much of the remainder may represent cellular debris due to erosion of the lining of the gut. The occurrence of a good deal of erosion is suggested by the fact that when a highly digestible diet is fed to birds, the cells lining the duodenum are long and have pointed apices, but if the food is fibrous and of low digestibility the cells are short and have blunted apices (Hallsworth and Coates, 1962).

Protein hydrolysis and absorption are rapid in the anterior portion of the small intestine and by the time the ingesta reach Meckel's diverticulum an amount of nitrogen equal to that added in the duodenum has been absorbed (Bolton, 1961a). A similar phenomenon, although smaller in scale, occurs with fat and, on a scale which can only be detected when diets extremely low in starch are fed, with carbohydrates (Bolton, unpublished).

The mechanism of protein hydrolysis is not clear. The pH in the duodenum and the anterior part of the small intestine is about 6, which is well away from the optima for pepsin and trypsin.

The posterior part of the small intestine is normally neutral or alkaline and it can be assumed that trypsin will there be the active protease.

At the junction of the small and large intestine and the caeca, some of the ingesta pass into the caeca, which fill and empty at intervals. Hill *et al.* (1964) have shown that acetic and propionic acids are formed in the caeca and in this they resemble the rumen. Comparison of digestibility coefficients of intact and caecectomized birds has shown that some of the more insoluble carbohydrate fractions of the food undergo breakdown in the caeca (Radeff, 1928; Henning, 1929).

The large intestine is short and ingesta do not stay there for any length of time; it is generally regarded as not playing a role in digestion.

THE CONTROL OF FOOD INTAKE

Thirst and hunger are the prime initiators of movement in birds, and a relationship exists between them. Thus birds which find the drinking water unacceptable to them due to the presence in it of certain drugs also reduce their food intake (Bolton, unpublished).

Before food is eaten it must be recognized by the bird as food. Birds reared on mash often fail to recognize pellets as food, even when placed in the same trough, and will starve for periods in excess of two days rather than sample pellets (Bolton, unpublished).

Preferences for material recognized as food exist in the fowl, but these are not due to taste, for the fowl has very few, if any, taste buds in the tongue. Fowls readily consume wet mash, they appear to prefer light colours to dark, and are attracted by shiny surfaces. Pellets are eaten readily, but foods which are dusty, or which absorb water readily are accepted with apparent reluctance. When birds are fed mixed grains they do not exhibit a constant behaviour pattern, some birds eating one of the grains first, others preferring a different grain.

Several suggestions have been made to explain why an animal stops eating. Lepkovsky, Feldman and Sharon (1962) mention that in mammals two types of sensory stimuli are present, one from the mouth and the other from the stomach, and that these are integrated by the brain. The sensations from the mouth are transmitted via receptors in the teeth, chewing muscles and tongue muscles, those in the stomach are activated by the stretching of the walls and travel via the vagus nerve. However, such a mechanism cannot be present in birds, because they do not chew food, and the vagus nerve does not have branches in the crop.

Feldman *et al.* (1957) produced aphagic White Leghorns by making lesions in the lateral hypothalamus. The birds lost the flighty nervous temperament characteristic of White Leghorns but retained the 'escape' characteristic, for they tried to avoid capture. When caught they did not struggle to get free. Some of the birds could be maintained by force-feeding; in others this failed and when they were examined *post mortem* it was noted that the food had not gone past the proventriculus. Hyperphagic birds were produced in 1961 by Lepkovsky and Feldman by making lesions in another part of the hypothalamus, thus establishing that the fowl has a central nervous control over food intake.

That this is one control is now beyond dispute, but other mechanisms may play important roles under certain circumstances. The available blood glucose level (van Itallie, Beaudoin and Mayer, 1953; Mayer, 1955) and the relationship between temperature control and the rate of food utilization (Brobeck, 1948; Passmore and Ritchie, 1957) have been proposed as factors in the short-term control of appetite; the state of the fat reserves characteristic of the animal has been suggested for long-term control (Kennedy, 1952–3). Control over food intake is not absolute even at best and it is common knowledge that even light-bodied hens lay down fat at the end of the laying year. Light-bodied birds have a greater degree of control than heavy-bodied birds; indeed, with the very heavy broiler breeding stock it is necessary to limit the amount of food offered if the birds are to be prevented from becoming excessively fat.

When high-producing layers are changed from a medium- to a high-energy diet there is a period of over-consumption before adjustment of intake occurs (Hill, 1958b). This period may last several weeks.

Appetite can be affected by the administration of drugs. Hill, Carew and van Tienhoven (1958) found that the implantation of diethylstilboestrol was followed by a 20 per cent increase in food intake; the digestibility and yield of metabolizable energy were not affected. Since the increased fat production was directly related to food intake, and the slope of the line relating metabolizable energy intake to tissue energy gain was the same as for normal chicks, the efficiency of energetic utilization was not altered.

Appetite is impaired when one amino acid is deficient. Birds fed these diets do not increase in body fat, but tend to have a lower body fat content (Combs, 1965).

WATER

Water has a high specific heat and is continuously circulated throughout the body via the blood stream. Therefore heat produced in one part of the body, e.g. by muscular activity, is rapidly carried over the whole body and excessive local rises in temperature are avoided. The heat produced is lost to the body through evaporation in the lungs and air sacs: thanks to its high latent heat of evaporation, the amount of water that has to be evaporated to remove unit amount of heat is small. During periods of muscular activity and during heat stress the rate of breathing, and hence the rate of evaporation of water, is increased; in the fowl the rate of respiration may rise to four times normal.

Nearly all physical and chemical changes in the body occur in aqueous media. The absorption of nutrients from the gut, their transport to the sites of anabolism and the removal of waste products are typical examples of physical changes; chemical changes include hydrolysis of starch to sugars and proteins to amino acids, and the oxidation of carbohydrates to carbon dioxide and water.

During the chemical changes through which the bird obtains energy from proteins, fats and carbohydrates, oxygen from the air combines with hydrogen in the nutrients to form metabolic water. It has been estimated that the oxidation of 100 g protein results in the formation of 42 g metabolic water, 100 g fat produces 100 g water, and 100 g carbohydrate yields 60 g water (Crampton and Lloyd, 1959).

Several factors have been found to influence the bodily need for water; high levels in the food of minerals, especially common salt, and protein increase the demand, as does the inclusion of large amounts of indigestible organic matter or of mucilage; or an increase in the environmental temperature. Water intake is affected also by the rate of egg production; an egg contains about 70 per cent water. Tyler (1958) has shown that thirst is an individual characteristic with a wide range of variation: of the birds he examined, one drank 674 ml water per day while another was satisfied with 208 ml.

Ross (1960) found that when broiler chicks aged from 0 to 6 weeks were restricted in access to water to three periods of 30 min per day, they achieved final liveweights lower than those of chicks reared with unlimited access to water. A case where ducklings 3 to 4 weeks old were deprived of water for several hours due to a leaking water trough was reported by Ranby (1960); when water was offered to them they drank excessively; some died, and others rolled on their backs and paddled the air vigorously with their feet.

ENERGY

The fowl is a homeothermic animal and to maintain homeostasis it must produce heat which is eventually lost to the atmosphere. Heat production is accomplished by the oxidation of fat, protein and carbohydrate digested from the diet. Foods are not all digested equally well by poultry, nor are their compositions the same; attempts have therefore been made to develop methods for assessing the amount of energy in a food which is of value to the bird.

Early workers in this field adopted the division of carbohydrates into crude fibre and nitrogen-free extractives, which was originally proposed by Henneburg and Stohmann (1860) for assessing the values of coarse fodders for fattening cattle. Halnan (1934) produced a table of poultry starch equivalents. Fraps, in a series of bulletins (1928, 1931, 1939, 1940, 1941, 1942, 1944, 1946), determined the net energy for growth and/or fattening of fowls and termed these values 'Productive Energy Contents'. Both workers were able to calculate energy contents from the results of chemical analysis for single foods, but not for compounded diets.

Deighton and Hutchinson (1940) showed that the net energy contents of oats and maize were about 80 per cent of their metabolizable energy contents. Later workers (Davidson, McDonald and Williams, 1957; Hill and Anderson, 1958) repeated Fraps' work and found, as he had reported, that the error of a single estimation was large. Metabolizable energy determinations, on the other hand, were highly reproducible, both within and between laboratories. Hence, although on theoretical grounds Productive Energy is the better measure, practical considerations have led to the general adoption of Metabolizable Energy. Tables of the metabolizable energy contents of poultry foods have been published by Hill *et al.* (1960); Hill and Renner (1960); Potter and Matterson (1960); Stutz and Matterson (1962); Sibbald, Slinger and Ashton (1962); Sibbald and Slinger (1962a); Cullen, Rasmussen and Wilder (1962); Potter *et al.* (1962) and Sibbald and Slinger (1962b).

Three groups of workers adopted a different approach. Metabolizable energy determinations require the use of birds as well as laboratories, so they attempted to correlate the metabolizable energy contents of foods, as estimated from experiments on birds, with their chemical compositions as determined in the laboratory. All groups reached substantially the same conclusion: that metabolizable energy estimates of an accuracy sufficient for dietary formulation can be calculated from knowledge of the protein, oil, starch and sugar contents. Carpenter and Clegg (1956) calculated a regression of biologically determined metabolizable energy contents on chemical composition and put forward formula (1).

$$\text{(1) ME (kcal/kg)} = 38{\cdot}0 \times \text{crude protein per cent} + 85{\cdot}5 \times \text{oil per cent} + 41{\cdot}8 \times \text{starch per cent} + 38{\cdot}0 \times \text{sugar per cent} + 53$$

For a series of 16 foods, ranging in metabolizable energy content from 1600 to 3400 kcal/kg, the standard deviation of the difference between the determined and calculated metabolizable energy values was 122 kcal/kg. Bolton (1962) put forward a formula equivalent to (2). This has a standard deviation of only 46 kcal/kg for the foods examined by Carpenter and Clegg.

$$\text{(2) ME (kcal/kg)} = 35{\cdot}3 \times \text{crude protein per cent} + 79{\cdot}5 \times \text{oil per cent} + 40{\cdot}6 \times \text{starch per cent} + 36{\cdot}9 \times \text{sugar per cent} + 199$$

Sibbald *et al.* (1963) from an examination of 103 foods suggested formula (3), but this has a standard deviation of 193 kcal/kg.

$$\text{(3) ME (kcal/kg)} = 35{\cdot}2 \times \text{crude protein per cent} + 78{\cdot}5 \times \text{oil per cent} + 41{\cdot}0 \times \text{starch per cent} + 35{\cdot}5 \text{ sugar per cent}$$

The total weight of food eaten per day by a bird tends to rise as the metabolizable energy content of its food is reduced. However, at some point, which is not clearly defined, the daily intake reaches such a high bulk that it becomes limited by the capacity of the gut. For chicks the ME level corresponding with this limiting bulk is below 1580 kcal/kg, because Hill and Dansky (1954) obtained good growth at this level; for layers it is below 1900 kcal ME/kg if mash is fed and below 1450 kcal ME/kg when pellets are fed (estimated from the results of Lillie, Haynes and Bird, 1951).

Because the total food intake per day tends to be regulated by the metabolizable energy content, and because the requirements of the other nutrients are properly expressed as absolute amounts per day, it follows that the percentage of these other nutrients in the diet is of little importance; what is important is their relationship to the metabolizable energy content. Estimation of the amount of metabolizable energy needed per day by the bird is therefore of great importance. Mitchell and Haynes (1927) measured the basal heat production of 28 hens and found that this had an average value of 55 kcal per kg body weight at 21°C; for 19 cocks the average was 56 kcal per kg body weight. They also found that the net energy was equal to 85 per cent of the metabolizable energy, hence the metabolizable energy requirements per day were 65 kcal per kg body weight for hens and 66 kcal for cocks. Eriksson (1954) maintained non-laying Rhode Island Red hens at 15°C or 20°C and found the metabolizable energy requirement per 2 kg bird was 180 kcal at the lower temperature and 160 kcal at the upper. The energy needed for maintenance has been estimated indirectly by means of partial correlations by Brody, Funk and Kempster (1938), Byerly (1941) and Bolton (1959). The values they obtained agreed well with each other, but since the birds in each case had been housed for long periods in uninsulated, unheated houses the average results, set out in Table 2, must be regarded as maxima.

TABLE 2
Energy Requirements for Maintenance

Liveweight (kg)	Metabolizable energy (kcal/day)
1·5	185
2·0	250
2·5	310
3·0	370

Hill (1962) found that the requirement for dietary energy varied throughout the year. If the amount needed per day in spring and autumn is given the arbitrary value of 100, that needed in winter is 105 and in summer 95.

The estimation of the energy requirement for maintenance is largely an academic exercise; birds are fed for production, and the product is meat and/or eggs. The Agricultural Research Council (1963) sets out a curve for liveweight versus cumulative metabolizable energy intake. From this it can be deduced that a broiler at 1·5 kg liveweight has used 7·3 kcal ME/g bodyweight, and at 2·0 kg it has consumed 7·8 kcal ME/g. The same bulletin sets out the requirement per 57 g egg as 122$\pm$ 10 kcal metabolizable energy.

FATS

The fats present in poultry diets fulfil two functions: they are sources of essential fatty acids and of energy. Turpeinen (1938) reported that arachidonic acid was the most effective of the essential fatty acids and that linoleic and linolenic acids served as precursors: such acids occur in adequate amounts in normal poultry foods.

Fowls can tolerate high levels of fat; Cruikshank (1934) successfully fed 28 per cent mutton fat and Combs *et al.* (1958) reared broiler chicks on a diet containing 50 per cent fat. However, fats suffer from a disadvantage: unsaturated fats oxidise rapidly once they are incorporated in a diet. That this should be so can be understood when it is realised that compounding it in a diet increases the surface area of fat exposed to the air by between 2400 and 24,000 times. This disadvantage can be reduced by adding an antioxidant to the fat before compounding.

The value of a fat as an energy source will depend largely on the efficiency with which it is digested and absorbed. An increase in the level of dietary fat is accompanied by a decrease in its digestibility (Duckworth, Naftalin and Dalgarno, 1950). These workers reported that when fat formed 12 per cent of the diet, 95 per cent of it was digested when it was liquid linseed oil but only 80 per cent when it was solid mutton fat. Bolton and Murty (1960) found that the inclusion of 10 per cent fat in the diets of young chicks was accompanied by a reduction in the digestibility of the carbohydrates; however, there was no reduction with adults. Digestibility of fat has been increased by the addition of ox bile (Fedde, Waibel and Burger, 1959) and of a non-ionic detergent (Bolton, 1961b), but not by a cationic or an anionic detergent (Bolton, 1961b). Renner and Hill (1961) compared the digestibilities of intact fats and of the fatty acid mixtures obtained from these by hydrolysis and found that the fatty acid mixtures were less well absorbed. They also found that the greater the proportion of saturated fatty acids in the mixture, the poorer was the absorption. The reasons why lard was well absorbed by young chicks, even though the main fatty acid was the saturated palmitic acid, were investigated in detail (Renner and Hill, 1961). It was found that feeding the palmitic acid as glyceryl monopalmitin, methyl palmitate, mixed sodium and potassium salts or in conjunction with glycerol increased the amount absorbed. Most of the palmitic acid in lard is in the 2-position (Mattson and

Lutton, 1959), so a further experiment was carried out in which la compared with a synthetic triglyceride prepared by hydrolysing lar re-esterifying the fatty acids with glycerol; in the synthetic compound palmitic acid is evenly distributed over the 1-, 2- and 3-positions. T experiment showed that the synthetic triglyceride was less well digested tha the lard, and therefore that the location of the palmitic acid in the 2-position was an important factor in its absorption by the bird.

The addition of supplemental fat to the diets of young chicks resulted in stimulation of liveweight gain (Rand, Scott and Kummerow, 1958). To investigate this effect, paired feeding techniques were used with diets calculated to be isocaloric (Carew, 1961). The birds fed the fat-supplemented diet showed the greater rate of liveweight gain, but the gain was entirely due to fat deposition; protein gain was unaffected.

PROTEIN

During the processes of digestion the dietary proteins are broken down into amino acids and amino sugars before being absorbed. However, nitrogenous material in the gut comes additionally from numerous other sources. Breakdown of proteins is accomplished by enzyme action; the enzymes are secreted by the bird and are nitrogenous compounds. The passage of food through the gut is aided by the secretion of mucin, a nitrogenous compound. When food is highly digestible the cells lining the lumen of the duodenum are long and have pointed ends, but if the diet is fibrous the cells are short and have blunt ends (Hallsworth and Coates, 1962); so erosion of the lumen must be considered normal and cellular debris containing nitrogen is thereby added to the ingesta. Bacteria thrive in the gut, especially in the crop and caeca; bacteria contain 20 per cent crude protein in their dry matter. However, all the nitrogenous compounds added by the bird to the ingesta must ultimately come from dietary nitrogenous compounds.

Some of the amino acids and amino sugars are used in the formation of the bird; those which are not so used are deaminated and used as energy sources. Those used to form the bird may reappear as discarded feathers, eggs or secretions into the gut. The last of these is the endogenous N secretion. The nitrogen in those deaminated is mainly voided as nitrogenous compounds in the urine. Some bacteria will be digested, some will be voided in the faeces.

The undigested residue of the endogenous N secretion is the metabolic faecal nitrogen and this, together with the undigested dietary N and bacterial N, forms the faecal N. The faecal N, therefore, consists of N residues originating from outside the bird, the undigested dietary N, and N residues from inside the bird, the metabolic faecal N. The relative magnitudes of the two excretions will depend on the amounts eaten or secreted and on the efficiencies with which they are digested. This information is shown diagrammatically in Fig. 2.

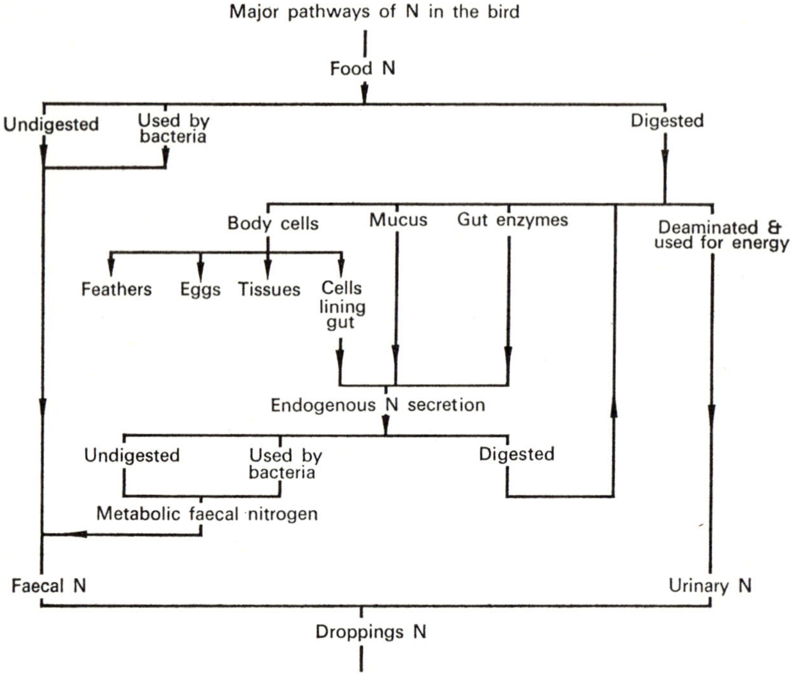

FIG. 2. Major pathways of N during digestion.

If the endogenous N secretion is small in relation to the dietary N, the undigested food N will probably form most of the faeces N; if the endogenous N secretion is large in relation to the dietary N, the faecal N may well approximate closely to the metabolic faecal N. This aspect was examined by Bolton (1961a), who found that, when normal diets were fed, the endogenous N in the duodenum was equal to about 8½ times the dietary N; but when diets high in indigestible organic matter were fed, the endogenous N in this region of the gut was about 50 times the dietary N. About 30 per cent of the N in the duodenum is present as mucin N (Bolton, 1964).

During digestion, proteins are broken down into amino acids. A series of Figs. 2 could therefore be drawn, one for each amino acid, and these would be greatly simplified diagrams, for some amino acids, the dispensable amino acids, can be manufactured by the bird from suitable N sources; others, the indispensable amino acids, cannot be, and must be supplied in the food. The position is further complicated because some amino acids can be used by the bird for purposes other than protein anabolism; thus methionine can also be used as a precursor of cystine and it can serve as a methyl donor; phenylalanine can be used as a precursor of tyrosine; and glycine can serve as a chelating agent. Diagrams depicting even the main pathways followed by amino acids during digestion would therefore be complex.

The secretion of nitrogenous compounds into the gut, and their rapid digestion and absorption, creates a metabolic pool of nitrogenous compounds which can serve several purposes: the amino-acid composition in the small intestine is kept almost constant, thereby allowing the absorptive processes to function optimally; the nitrogenous compounds will neutralize the acidity of the ingesta when they pass from the gizzard to the duodenum (Bolton, 1964); the nitrogenous compounds in circulation form a metabolic pool on which the bird can draw, therefore it can exist for short periods of time on diets which are deficient in one or more amino acids, e.g. during the digestion of the grain moiety when a mash plus grain feeding regime is adopted.

Digestibility can be measured in two ways: the apparent digestibility and the true digestibility. The apparent digestibility is determined by measuring the amount of the nutrient in unit weight of food, and subtracting from this the amount of the nutrient in the faeces voided after the digestion of this unit weight of food. For nitrogen the apparent digestibility would be given by the formula:

$$\text{Apparent digestibility} = \frac{(\text{Dietary N} - \text{Faecal N})}{\text{Dietary N}}$$

It has been demonstrated above that not all of the faecal N can be regarded as being undigested food N, some is metabolic faecal N. In the determination of true digestibility the amount of faecal N is corrected for the amount of metabolic faecal N in the faeces:

$$\text{True digestibility} = \frac{(\text{Dietary N} - (\text{Faecal N} - \text{Metabolic faecal N}))}{\text{Dietary N}}$$

Digestibility trials with poultry are complicated by the circumstance that faeces and urine are voided through the same orifice at the same time. Faecal N has been measured in either of two ways: by cutting the large intestine near the cloaca and bringing the cut end to the exterior through the body wall, thereby permitting the separate collection of faeces and urine; or by a chemical separation. Two types of chemical separation have been adopted; direct, in which the (urinary) uric acid is first converted into a soluble compound and the protein is precipitated; and indirect, in which the faecal N is estimated as the difference between the total N and the urinary N; since it is impracticable to determine the amounts of all the nitrogenous compounds in urine, it is customary to determine the uric acid N, and sometimes the ammoniacal N, and multiply by factors. A full discussion of the various procedures adopted is given by Ekman, Emanuelson and Fransson (1949). All the procedures are subject to criticism. A bird with the large intestine through to the body wall is not normal, even though it may lay eggs; however, it is the best test animal available and the chemical procedures that have been developed use the operated bird as control. Direct chemical methods assume that the faecal N is predominantly protein material, and indirect methods assume that the ratio of uric acid N to urinary N is constant. It is doubtful if either assumption is valid.

None of the procedures adopted has included a correction for the amount of dietary N used by the microflora of the gut and voided as bacterial N, nor has the magnitude of the error involved in the neglect of this term been investigated.

These criticisms apply equally to determinations of apparent and true digestibility. An additional criticism can be levelled at estimates of the latter. Metabolic faecal N has been estimated by measuring the amount of N in the faeces from a colostomized bird when the bird was fed a diet devoid of nitrogenous compounds. The N-free diet consisted only of starch, fat and minerals and was devoid of indigestible organic matter. The erosion of the cells lining the gut was therefore minimal, and so the estimate of metabolic faecal N was minimal.

Studies of apparent digestibility have shown that the treatment a food receives before it is offered to the animal can have a profound effect on its nutrient worth. Thus the heating of soya-beans to destroy the "trypsin-inhibitor" increases the amounts of amino acids in the meal, or in a diet containing the meal, which the bird can absorb (Evans, McGinnis and St. John, 1947; Bouthilet *et al.*, 1950; Hill and Renner, 1958). Overheating fish and meat meals results in the side-chains of lysine, and possibly of methionine and arginine, combining with carbohydrate to form linkages which resist the processes of digestion. Both phenomena have been investigated. It has been found that raw soya-beans do not inhibit the secretion of trypsin; indeed, the converse is true and trypsin secretion is increased (Lyman and Lepkovsky, 1957; Lyman, 1957). Prolonged feeding with raw soya-beans results in an enlargement of the pancreas (Chernick, Lepkovsky and Chaikoff, 1948). The bad effects of raw soya-beans can be overcome by the addition of denatured, and therefore inactive, trypsin to the diet (Borchers and Ackerson, 1951). Moreover, Fisher, Johnson and Ferdo (1957) have reported that increasing the amount of raw soya-bean in the diet reduced the growth inhibition found when heated soya-bean meal was used. The raw soya-bean meal may therefore exert its effect by increasing the secretion of trypsin and thereby reducing the size of the metabolic pool of amino acids below the level necessary for normal digestion.

When proteins are heated in the presence of carbohydrates the aldehyde group of a sugar condenses with the ε-amino group of lysine to form a complex which is not attacked by the digestive processes. The formation of the condensate has been demonstrated by dye-binding, or by the Van Slyke test

(Eldred and Rodney, 1946; Lea and Hannan, 1950). It is possible to measure the specific drop in lysine amino groups by reaction with fluorodinitrobenzene to give a dinitrophenyl compound: and this reaction is the basis of an estimation for available lysine developed by Carpenter (1960). The results from protein-rich foods of animal origin with this chemical test have shown a close correlation with 'Gross Protein Values' and with 'Net Protein Utilization Values' determined with chicks by Boyne, Carpenter and Woodham (1961). However, destruction of the dinitrophenyl-lysine compound occurs when excess carbohydrates are present, and the 'available lysine' values have not shown such a close correlation with the results of tests on birds fed protein-rich foods of vegetable origin (Carpenter, personal communication).

Due to the increasing emphasis on amino-acid content instead of on crude protein content, attempts have been made to measure the amounts of each amino acid present in the food, and the proportions of these that the bird can obtain from the food. Methods for determining the total amounts of amino acids in foods have been based on acid or alkali hydrolysis followed by determination of the amino acids in the hydrolysate by microbiological (Schiaffino, 1959) or chromatographic methods (Moore and Stein, 1951). Methods have been used to measure the amino-acid contents of foods which are available to the bird based on enzymic hydrolysis followed by microbiological assay and on microbiological assays using a proteolytic organism (Fernell and Rosen, 1956; Ford, 1960, 1962, 1964). Both approaches have given results which showed close correlation with those from feeding trials. Uwaegbute and Lewis (1963) measured the availability of a number of amino acids in one trial by the following procedure. A group, A, of chicks is reared on a diet containing 20 per cent protein and having the amino-acid levels recommended by the National Research Council of U.S.A. (1960) and the growth rate is measured. A second group, B, is reared on a diet containing 16 per cent of protein plus free amino acids to bring the levels up to those in the diet of group A; the growth rate should be the same. Further groups are fed diets containing the 16 per cent protein plus the test protein plus graded amounts of free amino acids, and the growth rates are measured. In this way a group, C, is found which has the same growth as groups A and B. Knowing the total amino-acid composition of the test protein and the amounts of free amino acids that were added to the diet of group C, the availability of the amino acids in the test protein can be calculated. The procedure has been further refined by measuring N retention as well as growth. (See also Chapter 8).

Other groups of workers have set out to determine the amounts of each indispensable amino acid necessary for optimal growth or for egg production. They have estimated the amino-acid requirements in two ways. First, the amount of the amino acid in a natural diet known to be deficient in it, but adequate in respect of all other nutrients, is determined; graded amounts of the amino acid are added and the minimal amount giving maximal yield, usually measured as liveweight increase, is taken as the requirement. Second, a purified diet is formulated, containing adequate amounts of all nutrients except the amino acid under test; graded levels of this amino acid are added and the lowest level of addition giving maximal yield is taken as the requirement. In the first method the inclusion of natural proteins in the diet has raised the question of the availability of the amino acids in these proteins; when free amino acid diets are fed, it has been found that rats can absorb them so efficiently that they reach the sites of anabolism in amounts greater than can be dealt with efficiently (Geiger, el-Rewi and Geiger, 1952; Maddy and Swift, 1952), this may also be true of fowls (Klose, Stokstad and Almquist, 1938). In the assessment of amino-acid requirements the Agricultural Research Council of Great Britain (1963) state that they placed equal weight on the results from each type of trial. Both approaches are subject to one criticism: the level at which maximum yield occurs is the level at which some other nutrient becomes the limiting factor. This is a real problem, for it has been found (cf. Hill, 1958) that increasing the level of the second limiting amino acid results in a decreased rate of liveweight gain.

The percentage requirements for two of the amino acids, lysine (Schwartz, Taylor and Fisher, 1958) and methionine (Baldini and Rosenberg, 1955) have been shown to be increased when the dietary level of metabolizable energy is increased. The percentages of the other amino acids needed by the bird when diets of differing energy are fed have not been investigated. It is apparent that the amount of any amino acid needed by the bird in one day will be equal to the sum of that (a) which it excretes, plus (b) that stored in the body as protein, plus (c) that used for purposes other than protein anabolism. If two diets of different metabolizable energy content are offered, the bird will tend to consume less total weight of the diet with the higher content. Since the amounts of amino acid required are the same, the percentages in the higher metabolizable energy diet will have to be increased to satisfy the demand. The Agricultural Research Council of Great Britain (1963) recommends that the percentages of amino acids be varied in direct proportion to the metabolizable energy content.

Requirements are stated for the L form except for methionine and phenylalanine, for which the D and L forms are of equal value (Fell, Wilkinson and Watts, 1959). These workers reported that the D forms of leucine, threonine and tryptophan are of no value on their own, but can be equal to up to 40 per cent of the L form when both are present in the diet; that D-histidine is without effect on growth, and that the addition of D-lysine antagonizes the action of the L form and depresses growth. For isoleucine and valine the evidence is contradictory; D-isoleucine has been reported as being similar to leucine, threonine and tryptophan (Fell *et al.*, 1959), and to be of no value (Leveille and Fisher, 1960), whereas D-valine has been reported to have no value (Grau and Peterson, 1946), to be capable of replacing part of the L form (Fell *et al.*, 1959) and to be equal to the L form (Leveille and Fisher, 1960).

If the amino acids absorbed from the gut and used for protein anabolism represent the greatest proportion of the total absorbed, it would be expected that requirements should be closely related to the amino-acid content of the yield, whether the yield was liveweight gain or eggs. Williams *et al.* (1954) showed that the carcass amino-acid composition was comparable with the dietary amino-acid requirements as determined by feeding trials. Scott (1959) confirmed this, and also showed that as the pullet matures and begins to lay eggs, the dietary indispensable amino-acid requirements shift to levels which correspond to those in eggs (Scott, 1962).

When all dietary contents of amino acids are deficient, the bird tends to eat more food in an attempt to compensate, and thereby increases its intake of metabolizable energy above that needed for growth. The excess energy is laid down as fat. If only one amino acid is deficient the retardation in growth rate is similar to that when all are deficient, but the body fat content does not increase, and may even show a tendency to decrease (Combs, 1965).

MINERALS

Calcium

The maintenance requirement of adult hens is 0·1 g/day (Tyler and Willcox, 1942). A laying hen also needs calcium for the production of egg-shells. Each egg-shell contains between 2·0 and 2·2 g calcium. Burmester, Scott and Card (1939) removed eggs from the shell-gland at known intervals of time after their arrival there, and, knowing the calcium content of the shell and of normal egg-shells from the same bird, were able to calculate the percentage shell deposited. They showed that shell deposition was slow during the first 3 hr; thereafter the rate increased and from the fourth to the twentieth hours was constant. Comar and Driggers (1949) and Driggers and Comar (1949) used radioactive ^{45}Ca as a tracer and showed that between 60 and 75 per cent of the calcium deposited as egg-shell came directly from the food; the remainder must come from the skeleton. Common (1938) demonstrated that the pullet stored sufficient calcium for about 6 egg-shells in its skeleton during the pre-laying period of ten to fourteen days.

All the calcium deposited as egg-shells must ultimately come from the food, and the efficiency with which birds retain calcium has been investigated. Common (1943) found that when the daily intake of calcium was less than 0·9 g/day, about 70 per cent was retained; between 0·9 and 3·5 g/day retention was 50 per cent. Hurwitz and Griminger (1960) observed 60 per cent retention when the diet contained 2·7 per cent calcium. Grimminger (1961) reported 56 per cent retention when the diet contained 2·27 per cent calcium, but only 38 per cent at a dietary level of 3·54 per cent.

Hurwitz and Grimminger (1961) also showed that the level of serum alkaline phosphatase increased as pullets began to lay, and was higher when the diet was low in calcium; they also found that the size of the parathyroid gland paralleled the level of serum alkaline phosphatase. Serum alkaline phosphatase levels decreased as the calcium intake increased and was minimal when the dietary calcium equalled the calcium requirements.

Common (1941) demonstrated the presence of carbonic anhydrase in higher concentration in the shell-gland than in other parts of the oviduct; and Benesch, Barron and Mawson (1944) and Gutowska and Mitchell (1945) found greater carbonic anhydrase activity in the shell-glands of good layers than of poor layers; these latter observations were not confirmed by Mueller (1961) who suggested that the results may have been due to the presence of carbonic anhydrase inhibitors in the diets.

Tyler (1946) found that calcium becomes soluble in the gizzard, is absorbed in the small intestine and that there is some excretion into the large intestine. Kienholz *et al.* (1961) fed a semi-purified breeders' diet containing 4·0 per cent calcium and 10 ppm zinc and obtained symptoms of zinc deficiency in the embryos. When the level of zinc was increased to 18 ppm the embryos were normal. Berg, Bearse and Merrill (1962) reported that an increase in the percentage calcium in the diet from 2·25 per cent to 3·25 per cent was accompanied by a decrease in hatchability and abnormal feather development of the embryos suggestive of zinc deficiency. However, Kienholz, Sunde and Hoekstra (1964) could not detect differences in the zinc contents of eggs due to the inclusion of 2·25 per cent, 4·0 per cent or 6·0 per cent in the diet, although the hatchability was impaired at the high levels of calcium.

Davis (1959) reviewed the literature on the effects of diets rich in calcium on the availabilities of manganese, iron, copper, iodine and magnesium. He concluded that high levels of calcium could have an adverse effect on the absorption of these minerals from the diet. Sell (1965) reported that increasing the calcium level of a chick diet from 0·6 per cent to 2·0 per cent, and the phosphorus level from 0·4 per cent to 1·4 per cent reduced the absorption of iron from 65 per cent to 47 per cent. The utilization of calcium has been found to be influenced by several factors: vitamin D, ascorbic acid, fat, zinc, protein, and calcium lactate. It is proposed to review only recent work on these aspects.

Coates and Holdsworth (1961) and Martin and Patrick (1962) showed that the presence of vitamin D in the diet increased twofold the absorption of calcium and its deposition in the bones of young chicks. Using laying hens as the experimental animal, Hertelendy and Taylor (1960) found that birds fed diets deficient in vitamin D produced eggs the shells of which became progressively thinner; then the rate of egg production decreased until it became about one egg per month.

Thornton and Moreng (1959) obtained an increase in egg-shell thickness during warm ($84 \pm 4°F$) weather when 22 mg ascorbic acid were added to each kg food; they also noted that the birds receiving the supplement had a higher daily food intake. Attempts to repeat this work have been made by Helbacka (1961), Harms and Waldroup (1961), Sullivan and Gekle (1962) Hunt and Aitken (1962) and Perek and Kendler (1962), but without success.

Kent, Reid and Worden (1960) reported that the inclusion of calcium lactate in the diets of laying hens improved egg production. Blount (1960) and Bartlett (1962) failed to confirm this finding. Martin and Patrick (1962) found that additional methionine in the diet resulted in a greater transfer of calcium to the skeleton.

Some of the chemicals used as coccidiostats have been found to have an adverse effect on egg-shell production. Benesch, Barron and Mawson (1944) dosed laying hens with 0·1 to 0·2 g sulphanilamide per kg body weight and found that the birds laid eggs with thin, pitted egg-shells; sulphapyridine did not have these adverse effects. Scott, Jungherr and Matterson (1944) studied the effects of sulphanilamide in detail. They found that 0·03 per cent sulphanilamide in the diet was without adverse effects, but 0·25 per cent resulted in a cessation of laying. Between these levels the decrease in shell thickness was proportional to the dose of sulphanilamide; the level of blood calcium was not affected, hence they concluded that the adverse effects occurred during secretion. Mann and Keilin (1940) found that sulphanilamide inhibited carbonic anhydrase. Gutowska and Pozzani (1945) tested the ability of a number of compounds to inhibit carbonic anhydrase. They found that sodium thiocyanate inhibited the enzyme, but manganese sulphate increased its activity. Bernard and Genest (1945) rationalized the situation when they showed that unsubstituted sulphonamides inhibited carbonic anhydrase, but substituted ones did not.

The discovery that calcium salts antagonize the action of the antibiotics (Price and Zolli, 1959; Price, Zolli and Hare, 1959; Stokstad *et al.*, 1959), combined with the widespread use of antibiotics as yield-promoters in broiler farming, has led to a re-examination of the calcium requirements of young chicks. The studies have been extended to cover possible relationships of calcium content to the metabolizable energy content of the food, and to the effects of supplemental fat in the diet, since insoluble calcium soaps may be formed in the gut and resist the digestive processes. Edwards *et al.* (1960) reported that there was a relationship between calcium percentage and the metabolizable energy content of the diet except in diets containing high levels of natural fats. The requirement was 0·9 per cent calcium for 2·5 Mcal metabolizable energy per kg diet, and 1·0 per cent for 3·2 Mcal.

Phosphorus

About 80 per cent of the phosphorus in the body is found in the skeleton, the remainder in the soft tissues. There is continual competition between the skeleton and the soft tissues for phosphorus and when the supply is insufficient to meet the demand, the soft tissues take precedence.

All unrefined foods contain phosphorus, but in cereals especially much of it is present as phytin, the calcium/magnesium salt of phytic acid. Common (1940) showed that phytin was incompletely hydrolysed in the intestine of the fowl and that considerable amounts of dietary phytin could be recovered from the droppings even when calcium supplements were not included in the diet; the addition of tricalcium phosphate or calcium carbonate increased the amount of dietary phytin recovered from the droppings. O'Dell and Savage (1960) suggested that phytin was the factor present in soya beans which renders zinc non-available to the bird, and Savage *et al.* (1964) found that the excretion of faecal zinc was greatly increased by supplemental dietary phytin. Early workers assumed that phosphorus present as phytin was almost non-available to the bird, but recent work by Ashton, Evans and Williams (1960) and by Temperton and Cassidy (1964a,b,c) has indicated that this assumption is invalid, and that about 40 per cent of the phytin phosphorus is utilized by the bird. A paper by Singsen and Mitchell (1945) sheds some light on the problem. These workers showed that the ability to utilize phytin phosphorus was influenced by the source of vitamin D; irradiated animal sterols were one and one-third more effective than cod-liver oil for promoting calcification; and they suggested that irradiated animal sterols contained some positive factor, or cod-liver oil some negative one, associated with the utilization of phytin. In their experiments Ashton *et al.* (1960) and Temperton and Cassidy (1964a,b,c,) used fish liver oil as the source of vitamin D.

Calcium phosphate is known to adsorb manganese in the gut (Schaible and Bandemer, 1942) and excessive amounts of phosphorus in the diet should be avoided. The phosphorus requirement for chicks has been estimated by many groups of workers (the Agricultural Research Council, 1963, lists ten references) and the requirements found range from 0·2 per cent to 0·6 per cent. During egg production there is a continual cycle of deposition of calcium and phosphorus in the skeleton and removal from it. Bone salt consists of tricalcium phosphate in physical association with carbonate (Brasseur, Dallemagne and Melon, 1946) and Tyler (1948), by an examination of the existing data, showed that the ratio of calcium to phosphorus was greater for this 'labile' skeleton than for normal fowl bone. Hurwitz and Grimminger (1962) estimated the phosphorus requirement of the laying hen using a balance technique and found that this was as low as 0·3 g/day when the phosphorus was supplied as inorganic phosphate.

For elements other than calcium and phosphorus there are two requirements for the laying bird: the first, which is normally the lower, is that amount which must be present in the diet to permit the hen to lay at her full genetical potential; the second is the amount which must be present before the egg contains the amount necessary for development of a chick. Ideally the newly-hatched chick should contain in its body that amount necessary for optimal well-being, but this aspect has been grossly neglected.

Magnesium

Magnesium is present chiefly in the skeleton, but it also occurs in the muscles where it is believed to be essential for muscular contraction (Kruse, Orent and McCollum, 1932). The digestion and absorption of magnesium have been little studied, although it has been estimated that only about 55 per cent of the mineral in isolated soya-bean protein is available to the fowl (McWard and Scott, 1961). It is also known that high levels of calcium or phosphorus increase the requirement of the chick for magnesium (Nugara and Edwards, 1961).

Studies with diets low in magnesium have been carried out on chicks and ducklings. Bird (1949) and Scott, Moeller and Hinners (1956) found that the growth-rate and feathering of chicks were adversely affected; muscle tone was decreased and there was incoordination of movements and convulsions, suggesting a disorder of the cerebellum. Van Reen and Pearson (1953) fed a magnesium-low diet to day-old ducklings and killed them after 14 days. They found that the activities of the alkaline phosphatase and the diphosphonucleotidase of the brain tissue were below the normal values.

Iron and Copper

Iron is present in red blood cells and copper is essential for haematopoeisis (Elvehjem, 1935); the two elements will therefore be considered together.

There are interactions of iron, copper and manganese. Hoogendoorn (1940) observed that supplemental manganese and iron increased egg production when birds were housed indoors, but not outdoors; that copper improved hatchability both indoors and outdoors; and that iron oxide and copper sulphate were antagonistic to the effect of manganese sulphate in increasing resistance to disease.

The egg contains about 1 mg iron. During incubation most of this iron is transferred from the egg to the embryo which, on hatching, will contain up to 0·75 mg (Ramsay, 1951). Before the hen comes into lay the level of plasma iron rises, and that of haemoglobin falls, but the decrease in haemoglobin would not release sufficient iron for egg production (Ramsay and Campbell,

1954). These authors suggested that the efficiency of absorption of iron is increased as the hen comes into lay, and reported that laying hens could absorb up to 20 per cent of the dietary iron.

All the iron in the egg is found in the yolk (Lesne, Zizine and Briskas, 1938) but copper is found in yolk and albumen (Elvehjem and Hart, 1929).

Beck (1961) reported that the liver was the main store of copper; Ramsay (1952) examined the livers of newly hatched chicks but found no appreciable store of iron.

A deficiency of iron (Davies, Norris and Kratzer, 1962) or of copper (Gallacher, 1957) adversely affected feather pigmentation and Nickerson (1946) has isolated a red, iron-containing pigment from feathers.

Molybdenum

The fowl is uricotelic and excretes nitrogen mainly as uric acid. Molybdenum was found in the enzyme xanthine oxidase, which converts hypoxanthine to xanthine to uric acid (De Renzo *et al.*, 1953). Higgins, Rickert and Westerfeld (1956) used tungsten as a molybdenum antagonist and showed that molybdenum was essential for optimal growth of chicks and for maintaining xanthine oxidase activity. When a tungsten-supplemented diet was fed to hens the xanthine oxidase activities of the liver, kidney and intestine were reduced; egg production and hatchability were not affected, but the chicks from the eggs grew more slowly than those from eggs of hens fed the unsupplemented diet.

Reid *et al.* (1956) concluded that the molybdenum in soya-bean protein was largely unavailable to fowls.

When the molybdenum in the diet exceeds 300 mg/kg food the growth rates of chicks and poults are depressed (Kratzer, 1952); above about 2000 mg/kg anaemia may be observed (Davies *et al.*, 1957). These adverse effects can be partially counteracted by supplementing the diet with inorganic sulphate (Miller, Pearson and Denton, 1958).

Manganese

Manganese is necessary for satisfactory egg production. Caskey and Norris (1938) found that hatchability of the egg and the ash content of the egg shell were low when the diet was deficient in manganese. Christiansen, Halpin and Hart (1939) found that levels of manganese adequate for satisfactory egg shell production in summer were unsatisfactory in winter; Couch, James and Sherwood (1947) concluded that there was a supplementary relationship between vitamin D and manganese. Bolton (1955) showed that manganese was higher in the liver of the laying hen than in the non-layer.

Gallup and Norris (1939b) examined the incubated egg. They found that the amount of manganese per g tissue in the embryo increased up to 5 days, and then remained constant until the chick hatched. If the breeders' diet is deficient in manganese, examination of those chicks failing to hatch shows that the incidence of micromelia and chondrodystrophy are high, whilst amongst the newly hatched chicks there will be birds with head-retraction (Caskey and Norris, 1939; Golding, Schaible and Davidson, 1940).

During the early weeks of post-embryonic life the feeding of a manganese-deficient diet will cause the birds to develop perosis (see Tyler, 1949). However, manganese deficiency is not the only cause of this disorder; Jukes (1940a,b) showed that choline was involved; McElroy and Jukes (1940) implicated biotin and Briggs and Lillie (1946) reported that thyroxine was a complicating factor. Perosis is a disorder in which the calcification of the bones, especially the long bones of the legs, is affected. Wiese *et al.* (1938) and Wiese *et al.* (1939) found that the blood and bone phosphatase activities of birds with manganese perosis were lower than those of non-perotic chicks. Wilgus and Patton (1939a) confirmed that manganese activated phosphatase, but thought that its action was related to bone cartilage. Hegsted, Mills, Elvehjem and Hart (1941) found that phosphatase levels were normal in choline perosis of chicks. Leach and Muenster (1962) examined the composition of bone cartilage; they found a severe reduction in the amount of mucopolysaccharide in the epiphyseal cartilage of the chick and also in the organic matrix of the other portions of the bone when manganese was deficient; the mucopolysaccharides concerned were mainly in the galactosamine fraction.

Wilgus and Patton (1939b), Gutowska, Parrot and Slesinski (1941) and Brown and McCracken (1965) found that about 30 per cent of dietary manganese is absorbed from the gut.

Poultry are resistant to manganese toxicity; hens can tolerate 1000 mg Mn/kg without ill-effects (Gallup and Norris, 1939a). However, 4800 mg Mn/kg food caused toxic symptoms in chicks (Heller and Penquite, 1937).

POTASSIUM

Potassium occurs mainly in the cells, soft tissues and muscles, and is also concerned in the regulation of the pH of the blood. When the diet is deficient the growth-rate is retarded, weakness and loss of use of the legs is noticed and there is excessive excretion of urates; death can be accompanied, or preceded, by tetanic tremors; examination post mortem shows that the ureters are conjested with urates. More potassium is needed for optimal bone calcification when the phosphorus content of the diet is marginal than when it is adequate (Gillis, 1948). Leach and Norris (1958) showed that the requirement of the chick for potassium increased proportionally with the metabolizable energy content of the diet, and that an inter-relationship also existed between the potassium and protein contents.

Burns, Cravens and Phillips (1953) studied the toxicity of excessive levels of potassium and reported that retardation of the growth-rate occurred when the potassium intake greatly exceeded that of sodium. However, it should be noted that these workers added the potassium and sodium as bicarbonates, and bicarbonate has been shown to be toxic for chicks (Delaplane, 1934) and for poults (Scrivener, 1946).

Zinc

Zinc is a constituent of several enzymes, e.g. carbonic anhydrase, carboxypeptidase, glutamic dehydrogenase, alcohol dehydrogenase and alkaline phosphatase, and acts as an activator of some enzyme systems (Vallee, 1959). It was shown (O'Dell and Savage, 1957; Morrison and Sarrett, 1958) to be an essential nutrient for poultry by the use of purified diets and a zinc-free environment; when the diet was deficient the growth-rate was retarded, feathering was poor, and there was dermatitis of the feet; in severe deficiency the long bones were shortened and thickened, due apparently to abnormal development of the epiphyseal cartilage. Day and Hill (1959) and Sullivan (1961) noted that turkey poults were more sensitive to zinc deficiency than chicks, showing especially disorders of the hock joint.

The hen's egg contains about 1 mg zinc, mainly in combination with the vitellin of the yolk; the albumen and shell contain negligible amounts (Tupper, Watts and Wormall, 1954). The laying hen can transfer about 30 per cent of the dietary zinc to the egg (Dewar, 1962). When the diet of the breeding hen is deficient in zinc, embryonic abnormalities occur; these usually feature faulty development of trunk or limbs, shortened limbs being common; if the embryo survives long enough to develop a head it may suffer from microphthalmia (Blamberg *et al.*, 1960).

Zeigler *et al.* (1961) observed that the zinc in diets based on casein/gelatin mixtures was more readily available than in those based on soya-bean protein. Lease *et al.* (1960) found that zinc in sesame meal and autoclaved peas was largely unavailable to the bird. It has been suggested that the compound which renders the zinc in soya-bean protein unavailable to the bird is phytic acid (O'Dell and Savage, 1960), and in support of this view it has been shown that the excretion of zinc is increased when phytic acid is added to the diets of hens (Savage *et al.*, 1964).

Poultry can tolerate high levels of zinc. When zinc is given as the oxide, carbonate or sulphate, levels up 1000 mg Zn/kg food had no adverse effects on chicks aged 0 to 5 weeks; at 1500 mg/kg the growth-rate was adversely affected, but mortality did not become severe until the dietary level reached 3000 mg/kg (Roberson and Schaible, 1960).

Cobalt

Cobalt is an essential constituent of vitamin B_{12}, which is referred to in Chapter 11. This section will deal with cobalt only as an inorganic element.

When cobaltous chloride in increasing doses was given to turkey poults the resting metabolism was reduced by up to 16·5 per cent; the respiratory quotient was increased from 0·71 to 0·87, suggesting stimulation of protein metabolism (Nedelnuic *et al.*, 1959). Lee and Wolterink (1955) reported that cobalt fed as cobaltous sulphate is readily absorbed from the gut; if it is injected intravenously, at least 15 per cent is voided in the faeces, indicating that it can be excreted into the gut.

Burns and Salmon (1956) found that chicks fed a diet adequate in all known vitamins except B_{12} but containing added cobalt grew faster than those fed the diet without supplemental cobalt. Davis, Briggs and Sloan (1953) found that supplemental dietary cobalt did not affect the growth rates, efficiency of food utilization or mortality of chicks fed a diet containing adequate vitamin B_{12}, but addition of 10 to 100 mg cobaltous sulphate per kg diet increased the growth rate when vitamin B_{12} was deficient; they suggested that the cobalt may exert its effect via the synthesis of vitamin B_{12} by the intestinal flora.

Masuhara and Migicorvsky (1963) reported that when supplemental vitamin D was added to the diet of young chicks the percentage of orally administered cobalt recoverable from the blood, liver and bones was increased.

Iodine

Iodine is used by the thyroid gland in the formation of thyroxine. For this purpose supplemental iodide and iodate are of equal value to the bird (Brumbaugh, Mehring and Titus, 1959). Keenan, Nikolaiczuk and Oliver (1953) found that 39 per cent of the iodine ingested was excreted in the faeces within 24 hr, and 17 per cent was stored in the thyroid gland. Nozaki and Ochima (1960), who administered radioactive iodine to a laying hen, reported that more was stored in the ovary than in the thyroid. Rogler *et al.* (1961) found an inverse relationship between the iodine content of the hen's diet and the size of the thyroid gland of the embryo; the range of content in the diet was 35 to 125 mg/kg.

Klein (1953) added iodized cod liver oil to the diets of 2- and 3-year-old hens and found that the rate of egg production was increased. Supplemental dietary iodinated casein has been reported to depress the basal metabolic rate (Mellen and Hill, 1953) and to increase oxygen consumption (Strite and

Yacowitz, 1956). Thornton and Deeb (1961) found that supplemental dietary iodinated casein increased the level of blood ascorbic acid; they suggested that the basal metabolic rate is associated with the ability to synthesize ascorbic acid.

The blood-protein-bound iodine (PBI) of the fowl is much lower than that of other domestic animals; Mellen and Hardy (1956) reported 1·2 μg PBI per 100 ml blood, and Bumgardner and Shaffner (1957) reported 1·06 μg; the blood of Holstein cows contained 4·1 μg.

Sodium and Chlorine

Sodium occurs mainly in the extracellular fluids, chlorine occurs in both fluids and cells. Fowls are more sensitive than mammals to salt poisoning (Krakower and Goettsch, 1945), probably because the renal glomerulus is morphologically different and the effective area of filtration is smaller. Blaxland (1946) showed that resistance to salt poisoning increased with age, and Field and Evans (1946) found that salt was more toxic when given in the drinking water than in the food.

Barlow, Slinger and Manery (1947) measured the salt and water concentrations of the muscle and skin from day-old to maturity. They found a rapid decrease in the concentrations during the first month of life; the water and salt concentrations of the skin rose again after the second laying year.

Waxler (1941) determined the blood chloride of chickens infected with coccidiosis, and reported that it rose about six to seven days after infection; similar rises were found to occur during artificial haemorrhage.

VITAMINS

Most of the work on vitamins in poultry nutrition has been concerned with the estimation of vitamin requirements; comparatively little has dealt with vitamin metabolism.

Tables of requirements have been published by the National Research Council of U.S.A. (1960) and the Agricultural Research Council of Great Britain (1963). Descriptions of the symptoms exhibited by poultry when the diets are deficient have appeared in many publications. It is not proposed to repeat this information in this section, except insofar as it has a bearing on the understanding of metabolic functions.

Vitamin A

Vitamin A exists in two forms: vitamin A_1, which is found mainly in salt-water fish: and vitamin A_2, found mainly in fresh-water fish; for poultry the latter has about 40 per cent of the value of the former (Ewing, 1963). There are four compounds (the pro-vitamins A) which the bird can convert into vitamin A: β-carotene, α-carotene, γ-carotene and cryptoxanthin; β-carotene has twice the potency of any of the other three. Naturally occurring vitamin A is a mixture of *cis* and *trans* forms; synthetic vitamin A is all in the *trans* form. (Of the six isomers of vitamin A, if the *trans* form is taken to have a potency of 100, the *cis* forms have potencies ranging from 75 to 15, cf. Ewing, 1963.) Ewing (1963) also reports that investigations have shown that fish oils contain some extraneous material which reacts with the Carr–Price reagent to give a blue colour: in practical tests equal weights of vitamin A mixed in a mash as gelatin-coated synthetic A and fish-liver oil had biopotencies of 100 and 60 respectively when freshly mixed, and 90 and 48 respectively after 6 weeks' storage.

Thompson, Coates and Kon (1950) showed that carotene was converted into vitamin A by the intestinal wall; when β-carotene was fed to vitamin A-depleted chicks, vitamin A was found in the gut wall 0·5 hr after feeding, the shortest time studied, but did not appear in the liver for 3·5 hr. The efficiency of conversion has been studied by measuring the requirements for optimal growth, and the amount of vitamin A stored in the liver. Gurcey, Boucher and Callenbach (1950) found that crystalline vitamin A acetate was four times as efficient for supporting growth as the corresponding levels of crystalline carotene; for liver storage 30,000 i.u. as carotene per lb food were equivalent to 4000 i.u. as crystalline vitamin A acetate. Ely (1959) found that vitamin A had a biopotency 2·6 times as great as that of carotene. Marusich and Bauernfeind (1963) fed β-carotene and vitamin A palmitate as gelatin beadlets, since Olsen *et al.* (1959) had shown that these were the most stable, and found that, at levels of inclusion that were low but sufficient to support optimal growth, the two were equal in biopotency; but at high levels the vitamin A was between 2·5 and 4 times more effective than carotene, as measured by liver storage.

A natural source of carotene, such as dried grass, is normally included in poultry diets when yellow-yolked eggs or yellow-fleshed carcasses are desired. Such natural sources contain carotenoid pigments other than carotene. Gillam and Heilbron (1935) examined eggs from birds fed high levels of yellow maize and found 0·015 mg carotene, 0·19 mg cryptoxanthin and 1·79 mg xanthophyll per 100 g yolk. When laying birds are allowed free access to pasture olive-yolked eggs may be produced. Davies (1951) showed that the pigment present was a chlorophyll derivative.

Fell and Mellanby (1952) used tissue culture to study the effect of hypervitaminosis A on bone formation: after 2 days in fowl plasma containing 1550 to 3247 i.u. vitamin A per 100 ml, growth of embryonic bone began to decline so that the shaft of the bone became shorter and more slender than in the controls; the explants also changed in consistency, becoming soft and rubbery. The same technique was adopted to study keratinization of the epidermis (Fell and Mellanby, 1953): ectoderm from the trunk and limbs of chick embryos developed normal squamous keratinizing epithelium when grown in unsupplemented fowl plasma; but when the medium contained excess vitamin A the explants resembled normal nasal mucosa. From the results of these and other experiments, Fell, Mellanby and Pele (1956) suggested that vitamin A may cause the production in the cartilage cells of an enzyme that produces soluble sulphated mucopolysaccharide instead of chondroitin sulphate.

Vitamin D

Vitamin D, standardized by rat assay, gave inconsistent results when tested for its ability to promote calcification in chicks, and thus led to the recognition of two forms of vitamin D: ergocalciferol (vitamin D_2) and cholecalciferol (vitamin D_3) (Waddell, 1934). Ewing (1963) lists seventeen compounds which are antirachitically active. Vitamin D_3 is about thirty-five times more active for chicks than vitamin D_2 (Remp and Marshall, 1938; McChesney, 1943). McChesney (1943) also found that 49 per cent of vitamin D_2 was recoverable from the faeces, but only 36 per cent of vitamin D_3; in the carcass he found 12 per cent of the dose of vitamin D_2 but 27 per cent of the vitamin D_3.

Vitamin D_3 is formed by the action of ultraviolet light on 7-dehydrocholesterol, and birds exposed to ultraviolet light or sunlight have a greatly reduced requirement (Koch and Koch, 1941).

Vitamin D, calcium and phosphorus are involved in bone formation. A deficiency of any one of these, or an excess of calcium, can cause rickets (Wilgus and Thorp, 1940). About 99 per cent of the calcium and 80 per cent of the phosphorus is found in the skeleton, there is continuous competition between the skeleton and the soft tissues for phosphorus, and vitamin D is thought to be necessary for the transfer of phosphorus from the soft tissues to the skeleton. Rosenberg (1945) described the following changes in metabolism during vitamin D deficiency: phosphorus excretion is increased, and this can be detected when other symptoms are absent: phosphorus metabolism becomes more intense, with an increase in serum phosphatase but a decrease in serum phosphorus; osteoblast formation ceases; the cartilage continues to grow but the osteoid material remains uncalcified. In acute rickets there is removal of phosphorus from the bones.

Vitamin E

Three diseases of chicks have been associated with a deficiency of vitamin E: encephalomalacia, exudative diathesis and muscular dystrophy. (For a fuller account see Chapter 31.)

Machlin and Gordon (1960) found that diets high in linoleic and arachidonic acids caused the concentration of these acids to increase in the tissues. In the absence of vitamin E or of certain antioxidants from the diet, autoxidation of the fatty acids took place in the tissues, cell structures became disrupted and encephalomalacia developed. Kokatnur *et al.* (1960) and Nishida *et al.* (1960) found that dietary supplements of 12-oxo-*cis*-9-octadecenoic acid increased the incidence of encephalomalacia, and that intravenous injections of the hydroperoxides of methyl linoleate or linoleic acid caused cerebellar disorders in five hours; chicks adequately supplied with vitamin E remained normal. Encephalomalacia, then, is prevented by the antioxidant activity of vitamin E.

Scott *et al.* (1955) and Nesheim and Scott (1958) found that exudative diathesis could be prevented by vitamin E or selenium, but that other antioxidants did not prevent it.

Nesheim, Calvert and Scott (1960) found that muscular dystrophy could be produced only when the diet was deficient in vitamin E, methionine and cystine but contained an adequate amount of arginine; selenium gave partial protection.

Bieri *et al.* (1960) raised the question whether or not vitamin E was required by chicks; they reared chicks from hatching to 24 weeks old on diets that were almost devoid of vitamin E, without any of the deficiency signs developing.

The role of vitamin E in hatchability has not been fully investigated. Jensen *et al.* (1953) found that supplemental vitamin E in a turkey breeders' diet greatly increased hatchability. High embryonic mortality was reported in eggs from turkey hens fed the unsupplemented diet, with a peak of mortality from 24 to 28 days of incubation (Jensen, Carver and McGinnis, 1955). Ferguson, Atkinson and Couch (1954) and Ferguson *et al.* (1955) noted a cloudiness of the central portion of the lens of the eye, some poults being blind, when the breeders' diet was unsupplemented with the vitamin.

Mecchi, Pool and Klose (1953) found that supplemental vitamin E improved the stability of the fat of birds, and Ames (1956) suggested that peroxides and polymers were formed in the fat when the vitamin was deficient.

Goldhaber, Zacharias and Kinsey (1950) showed that vitamin E enhanced the absorption of xanthophylls and Ames (1956) suggested that the vitamin improved the utilization of the carotenoids and vitamin A.

Dinning, Sime and Day (1955, 1956) found that vitamin E had an essential role in the metabolism of nucleic acids and Goldstein and Scott (1956) concluded that the vitamin was necessary for the synthesis of blood plasma protein of the chick. Michaelis and Wollman (1950) suggested that it acted catalytically as part of an enzyme system which transfers energy from cytochrome *b* to cytochrome *c* in the respiration of body cells.

Vitamin K

Dam, Schønheyder and Tage-Hansen (1936) and Schønheyder (1936) showed that the increased blood-clotting time of chicks fed certain diets was due to a reduced level of prothrombin. The deficient factor was vitamin K and the disease associated with it was haemorrhagic disease. Two forms of the vitamin occur naturally: vitamin K_1 or 2-methyl-3-phytyl-1:4-naphthoquinone and vitamin K_2 or 2-methyl-3-difarnesyl-1:4-naphthoquinone; the latter can be synthesized by the microflora of the gut. The side-chain is not necessary for biological activity and most experiments on the vitamin have been carried out using the synthetic vitamin K_3, menaphthone, 2-methyl-1:4-naphthoquinone, or its water-soluble form, menaphthone sodium bisulphite.

Work by Dam *et al.* (1938) showed that the deoxycholic acid of the bile was concerned in the absorption and transport of the vitamin; a vitamin K-deoxycholic acid preparation was effective when injected subcutaneously into deficient chicks, whereas an emulsion of the vitamin was effective when injected intramuscularly, but not subcutaneously.

Caskey (1953) reported that two types of haemorrhagic disease exist, only one of which is cured by dietary supplementation with vitamin K. Nelson and Norris (1959, 1960, 1961a, 1961b) carried out extensive studies on the vitamin K requirements of chicks. They found that menaphthone sodium bisulphite was about equal to vitamin K_1 in maintaining normal blood clotting times when the birds were fed normal diets; when the diets contained sulphaquinoxaline the synthetic form was inferior to the natural: there was an interrelationship with genetic make-up, environment and nutrition upon the vitamin K requirement, especially in the presence of sulphaquinoxaline: under some conditions the action of the drug could not be overcome even by increasing the menaphthone sodium bisulphite content of the diet to 50 times the requirement in unsupplemented food.

Menaphthone sodium bisulphite is not stored in the tissues, so an attack of coccidiosis can operate adversely in three ways when the chick is dependent upon a dietary supply of this form: the chick eats less, therefore the daily intake is reduced; the intestinal wall is injured, so absorption is impaired; and treatment with sulphaquinoxaline causes an increased requirement. Vitamin K_1, on the other hand, is retained in the body for a considerable period of time (Dam, Prange and Sondergaard, 1954; Taylor *et al.*, 1956). The richest natural sources of the vitamin are soya beans and dried green crops; solvent-extraction of soya beans in the manufacture of soya-bean meal removes most of the vitamin if not all and feeding broilers on depigmenting diets containing minimum amounts of dried grass renders them more liable to attacks of the disease.

Biotin

Biotin can be replaced by oxybiotin, but between 3 and 5 times as much are needed for the same biological response (McCoy, Felton and Hofmann, 1946; Moore, Luckey, Elvehjem and Hart, 1946).

A reduction in the utilization of pyruvic and succinic acids by duckling heart slices *in vitro* was noted by Olsen *et al.* (1948); the normal rate could be restored by injecting biotin into ducklings, but not by the addition of the vitamin to the medium; they concluded that the vitamin is required for the synthesis of a respiratory enzyme.

Coates *et al.* (1951a) found that the symptoms of biotin deficiency in chicks could be reduced by supplementing the diet with penicillin.

Biotin is found in the yolk and albumen of the egg (Couch *et al.*, 1948); that in the albumen exists in combination with avidin and is not available until the avidin, a protein, has been denatured and thereby rendered biologically inactive.

When the carbohydrate of the diet is supplied as dextrin the intestinal microflora of the adult bird can synthesize biotin and this is available to the bird; if the carbohydrate is sucrose or lactose the synthesis is not promoted (Couch *et al.*, 1948).

Choline

The role of choline as an antiperotic factor has been mentioned under manganese. Ewing (1963) reports that betaine and methionine can replace choline for this purpose if natural foods are in the diet, but not in purified diets.

In addition to being an antiperotic agent, choline serves as a methyl donor and as a lipotropic agent. As a methyl donor it can be replaced by a number of compounds: arsenocholine, mono-, di-, and triethylcholine, ethanolanine, betaine, betaine aldehyde and methionine (cf. Hogan, 1949–50); choline esters are inactive (Welch and Welch, 1938). Choline derivatives which contain alkyl groups other than the methyl group (Welch, 1940) and arsenocholine (Welch and Welch, 1938) can replace the lipotropic function of choline. Choline is present in the phospholipids of cells, and serves as a building unit in the synthesis of lecithin and sphingomyelin (Rosenberg, 1945).

Lucas, Norris and Heuser (1946) and Ringrose and Davis (1946) found that laying hens could synthesize substantial amounts of choline.

Melass, Pearson and Sherwood (1946) reported that raising the dietary level of choline to 1 per cent caused a reduction in the growth rates of chicks; 4 per cent further retarded the growth rate, but other evidence of toxicity was not noted.

Folic Acid

March and Biely (1955) could not find evidence that folic acid aided fat absorption or had a lipotropic effect, although the folic acid requirement was increased with diets high in both protein and fat (March and Biely, 1956). Klain, Hill and Branion (1957) fed diets high and low in folic acid and lysine to Broad Breasted Bronze turkeys, and found that both nutrients were involved in feather pigmentation. Spivey-Fox and Ludwig (1951) determined the amount of formiminoglutamic acid (a metabolite of histidine) in the droppings of birds and found that it increased when the diet was deficient in folic acid.

Waibel, Sunde and Cravens (1952) reported that supplemental penicillin G at 5 and 200 mg/kg diet increased the folic acid content of hens' eggs by 30 and 38 per cent respectively. Evans *et al.* (1952) determined the folic acid content of eggs after various periods of storage and found that 16 per cent of the vitamin was lost in 6 months.

Cropper and Scott (1964) state that as much as 85 per cent of the folic acid in natural poultry foods exists in the form of conjugates; however, all of the folic acid, free and conjugated, in dried brewers' yeast was available.

Nicotinic Acid

Krehl, Elvehjem and Strong (1944) found that the chick was unable to utilize the precursor of nicotinic acid in cereals. The work of Briggs (1945), Briggs, Groschke and Lillie (1946) and of Groschke, Anderson and Briggs (1948) showed that nicotinic acid in the diet of the chick could be replaced by tryptophan. Schweigert, German and Garber (1948) found that the chick embryo could synthesize nicotinic acid from tryptophan.

Coates *et al.* (1951a) reported that supplementation of the chick's diet with penicillin accentuated a deficiency of nicotinic acid; Biely and March (1952) found that chlortetracycline had a sparing effect.

Pantothenic Acid

Snell and Quarles (1941) concluded that the pantothenic acid content of chick embryos at different ages was constant; this was confirmed by Pearson, Melass and Sherwood (1945). Snell, Pennington and Williams (1940) showed that tissues of chicks fed diets adequate in pantothenic acid contained between two and five times the concentrations of tissues from chicks fed deficient diets. Taylor, Thacker and Pennington (1941) reported that the percentage of haemoglobin in the blood of embryos was reduced when the breeders' diet was deficient in pantothenic acid.

Pyridoxine

Schlenk and Snell (1945) suggested that pyridoxine was part of a co-enzyme involved in transamination; Green, Leloir and Nocito (1945) considered that pyridoxal phosphate was the co-enzyme.

Ames, Sarma and Elvehjem (1947) worked with rats and reported that the enzyme activity was only 40 per cent of normal when the diet was deficient; Brin and Olsen (1951), working with ducks, showed that the aspartic acid-glutamic acid transaminase activity was reduced by 30 per cent during a deficiency. Goswani and Robblee (1958) found that the deficiency in the chick reduced the activity of this enzyme system to between 40 and 53 per cent of normal.

Fuller and Hill (1964) reported that pyridoxine deficiency impaired the utilization of D-methionine; in adequate diets D- and L-methionine were of equal value to the chick.

Daghir and Balloun (1961) studied the cholesterol contents of tissues of chicks fed pyridoxine-deficient and adequate diets; they found that serum cholesterol and the cholesterol content of the heart were higher in birds fed deficient diets; although the aortas were heavier in the deficient birds, the cholesterol content was the same as that of adequately fed birds. They suggest that their results indicate a direct action of pyridoxine on cholesterol metabolism.

Riboflavin

The riboflavin requirement for chick growth has been assessed as the minimal level promoting optimal growth, the minimal level preventing curled-toe paralysis and that giving maximal storage in the liver, kidney, heart, leg muscles and breast muscles. Bolton (1944) used natural sources, and in 1947 repeated the work using crystalline riboflavin added to a diet made up of natural foods low in the vitamin. The estimated requirements were the same in both experiments, thus confirming the view that the naturally occurring forms of riboflavin (riboflavin, flavinmononucleotide and flavin-adenine-dinucleotide) were utilized equally well by the chick. The experiments also confirmed the results of Hodson (1940), who showed that the riboflavin contents of the livers of chicks increased with increase in the riboflavin contents of the diets.

Kleiber and Jukes (1942) found that nitrogen retention by chicks was reduced when the diet was deficient in riboflavin.

Hutt (1951) investigated causes of defective feathering in newly hatched chicks when the breeding hens were fed diets considered adequate in riboflavin. He found that black-feathered birds were affected, but not those with barred plumage, when Rhode Island Red males were mated with Barred

Plymouth Rock females; that the defective feathering was typical of riboflavin deficiency; and that the cause was genetic in origin.

Snell and Quarles (1941) and Clandinin (1946) reported that there was no change in the riboflavin content of the egg during incubation.

Bolton (1950, 1951, 1952, 1953a) studied the transfer of riboflavin from the bird to the egg. He found that most of the riboflavin in the fowls' blood, and in the egg, was present as the free vitamin. In the liver, kidney and heart most of the vitamin was present as flavin-adenine-dinucleotide; in the magnum about half was present as free vitamin. The amount of free riboflavin in the magnum showed a relationship to the ovulatory cycle; just prior to ovulation the content was maximal, as the egg passed through the magnum the organ lost an amount of free riboflavin approximately equal to that in the egg, and the magnum content rose again during shell secretion.

Thiamine

In any book on vitamins the experimental production of polyneuritis in hens by feeding them polished rice is mentioned. The work was carried out by Eijkman (1897) and laid the foundations of the whole concept of vitamins.

When the diet is deficient, excessive amounts of lactic and pyruvic acid accumulate in the tissues, and Peters (1936) showed that thiamine *in vitro* could restore the lost oxidative power of avitaminous tissues respiring in solutions containing these acids. Lohmann and Schuster (1937) proved that a pyrophosphate ester of thiamine is the co-enzyme needed by the enzyme carboxylase, which is concerned with the breakdown of pyruvic acid in the intermediary metabolism of carbohydrates. It is now well recognized that the biological demand for thiamine is related to the carbohydrate content of the diet, and not its content of metabolizable energy.

All the thiamine in the egg is originally deposited in the yolk (Westenbrink and Van Leer, 1940–1; Scrimshaw *et al.*, 1944) and the former authors concluded that the hen can transfer more thiamine to the egg at the start of the laying year than at the end; they also stated that nearly all the thiamine disappears when a fertile egg is incubated, although Scrimshaw *et al.* (1944) were unable to confirm this.

Vitamin B_{12}

Bosshardt *et al.* (1946a) and Bosshardt *et al.* (1946b) published results which, in the light of later knowledge, showed that vitamin B_{12} enhanced the utilization of protein. Henry and Kon (1951) found that the Biological Value of casein was increased from 82 to 89 by supplementing the diet with vitamin B_{12}; digestibility was unaffected.

Bosshardt, Paul and Barnes (1950) reported that fat had a sparing action on the requirement for vitamin B_{12} when birds were fed *ad libitum*. Frölich (1954) found that under-heated soya bean meal in the diet increased the biological demand for the vitamin.

Vitamin B_{12} is required for the synthesis of labile methyl groups from glycine (Johnson, Firth and Mistry, 1955); but is not involved in direct transmethylations in that it is not necessary for the reduction of homocystine to homocysteine (Mistry *et al.*, 1955).

When chicks and microorganisms were used as the test organisms in biological assays of the vitamin, the results did not always agree; the results of Coates *et al.* (1951b) showed that microbiological assay of intestinal contents indicated a higher level of vitamin B_{12} than did chick assay. Chromatography of intestinal contents revealed a number of substances closely allied chemically to vitamin B_{12} which were biologically active for the microorganism, but not for the chick (Ford, Kon and Porter, 1951).

Moore and Doran (1961) showed that a derangement of lipid components in chick embryos occurred when the breeders' diet was deficient in vitamin B_{12}, and thereby indicated the importance of the vitamin in lipid metabolism.

EGG PRODUCTION

The testes of the male bird are intra-abdominal organs, and consist of two bean-shaped bodies lying ventral to the anterior lobes of the kidneys. The size of the testes depends upon the age of the bird, and upon whether or not the bird is in breeding condition. The convex surfaces are directed towards the sternum, and on the slightly concave surface of each there is a small, flattened projection which corresponds to the epididymis of mammals. A wavy duct, the deferent duct, runs from each epididymis to the same small papilla in the cloaca. The accessory organs found in mammals are absent.

Two ovaries and oviducts can be seen in the female embryo, but only one, the left, develops and becomes functional in the adult. The ovary lies ventral to the anterior half of the kidneys and is firmly attached to the dorsal wall of the lumbar region. The anterior portion of the oviduct, the infundibulum, is a delicate, funnel-shaped structure lying ventrally to, and enclosing, the ovary. Immediately succeeding the infundibulum is the albumen-secreting region of the oviduct, or magnum, which forms about half the total length. The isthmus joins the magnum with the shell-gland, and the shell gland is connected to the cloaca by the vagina.

If a laying hen is killed, opened and the reproductive system is examined, five follicular stages will be found in the ovary: primary follicles, growing follicles, mature follicles, discharged follicles and atretic follicles. At hatching the number of ova in the ovary is astronomical and may run into millions (Bradley and Grahame, 1960). At the onset of lay some of these primary follicles begin to grow. Up to about 5mm in diameter growth is slow, but about 8 to 10 days before the first egg is laid the yolks begin to increase in size rapidly and successively and the final size of about 32 mm diameter is reached in about 7 days. At this stage the mature follicle ruptures along the stigma and the yolk is released, leaving a discharged follicle. If for some reason the follicle does not rupture the yolk material is reabsorbed and the follicle becomes atretic.

The released yolk falls into the infundibulum which gently propels it towards the magnum region. If the egg is fertilized, the sperm penetrates the germ-cell in the neck region of the funnel-shaped infundibulum. As the yolk passes down the magnum region it receives the layers of albumen, then passes through the isthmus, where the shell membranes are laid down, into the shell-gland. During the passage through the isthmus, and the early stages of shell formation, water containing various salts in solution passes through the shell membranes into the egg. Warren and Conrad (1942) showed that the deposition of the ground colour of egg-shells, when present, began slowly and was deposited at an ever-increasing rate, about 75 per cent being laid down in the last 5 hr; superficial pigments were laid down shortly before oviposition.

In the interval between the completion of shell-formation and oviposition the shell-gland coats the egg with a thin, transparent film, the cuticle (Richardson, 1935). After the completion of cuticle deposition the egg is excreted *via* the vagina.

The times spent by the yolk in the various regions of the oviduct are: infundibulum—18 min, magnum—3 hr, isthmus—$1\frac{1}{4}$ hr, shell-gland—21 hr, and vagina and cloaca—a few seconds. When the diet is deficient in protein, egg size is decreased. A decrease in egg weight was also noted when riboflavin or nicotinic acid (Gleaves *et al.*, 1961) and linoleic acid (Shutze and Jensen, 1963) were deficient in the diets. Birds in a warm environment lose less heat than those in a cold environment, and tend to eat less food. To maintain the daily intake of protein, or other nutrients at a constant level, it is necessary to vary the percentage in the diet with ambient temperature. When the diet was kept constant, Hutchinson (1953) found that short periods of 99°F reduced egg weight more than continuous temperatures of 90°F. Meuller (1961) reported that temperatures of 90°F reduced egg weight to 85 per cent of that from the same birds at 55°F; and Campos, Wilcox and Shaffner (1960) found a slow rise in temperature (5°F/day) reduced egg weights more than a fast rise (4°F/hr) to 100°C. Before concluding that the effects are due to environmental temperature only, it would be necessary to repeat the experiments keeping the daily intakes of amino acids and other nutrients constant.

The inclusion in the diet of low levels of tranquillizers has been reported to increase egg size, but high levels reduced it (Eoff *et al.*, 1961). Attempts to influence egg size by the inclusion of dietary supplements of thyroid hormone (Savage *et al.*, 1952; Aloufa, 1953; Miller, Sanford and Clegg, 1962) or stilboestrol (Talmadge, 1957) have either had no apparent effect, or have reduced egg weight. Antibiotics were without effect (Peterson and Lampman, 1952; Miller, 1958; Heywang and Varich, 1959), nicarbazin reduced egg weight (Polin, Ott and Zeissig, 1958) while the inclusion of gossypol caused egg size to diminish rapidly and then egg production ceased (Narain, Lyman and Couch, 1957).

As the bird passes from the non-laying to the laying condition the comb and wattles enlarge and redden. Examination of the cloaca during this period reveals that it enlarges and changes from almost dry to moist. Palpation of the pelvic bones is an aid to telling the condition; in the non-layer the apices of the bones are about one-finger's-breadth apart, in the layer three fingers can pass between them.

Internally the changes are distinct. The pelvic bones can be seen to have become wider apart in the layer, which permits oviposition. The ovary and oviduct enlarge many hundred-fold and yolks in various stages of development can be seen in the ovary. Blood plasma in the non-layer is an almost transparent liquid, in the layer it is opalescent. The liver enlarges and becomes yellower and more friable, and spicules of calcium salts can be seen in the marrow cavities of the laying hen, or in the later stages of the process of coming into lay. Except for the enlargement and reddening of the comb and wattles, these changes take place during the 8 to 10 days before the first egg is laid.

Concurrently with them, changes take place in the chemical composition of some parts of the body as reserves are established to meet demands of egg production. The opalescence of the blood plasma in the laying bird has been shown to be due to an increase in lipid content (Warner and Edmond, 1917). The blood lipids comprise a number of fractions: cholesterol, neutral fat, phospholipids and fatty acids. With the exception of cholesterol, these are much higher in the blood of the laying hen than the non-layer: thus the neutral fat increased from 47 mg to 922 mg per 100 ml whole blood; total fatty acids from 329 mg to 1564 mg; phospholipids from 282 mg to 642 mg (Lorenz, Entenman and Chaikoff, 1938). A plasma lipophosphoprotein complex which can be precipitated by diluting the plasma with nine times its volume of water was found by McIndoe (1957) to appear about 8 to 9 days before the first egg was laid. Heald and McLachlan (1963) isolated phosvitin from the plasma of the laying hen and demonstrated (Heald and McLachlan, 1964, 1965) that it was synthesized by the liver. Using cocks as controls, Husbands and Brown (1965) reported that more triglyceride was present in the plasma of layers, but less cholesterol ester.

The increase in size and friability of the liver has been shown to be due to increase in liver fat and liver protein (Common, Rutledge and Bolton, 1947) and partly due to an increase in the amount of water.

Sturkie and Newman (1951) found that the plasma protein levels were unchanged, although ovalbumin is present in the plasma of the layer and not in the non-layer (Kaminshi and Durieux, 1956; Williams, 1962), as was the level of phytic acid (Oshima, Taylor and Williams, 1964), and of blood sugar (Bell, 1957). Although the plasma alkaline phosphase levels of blood plasma were found to be the same in layers and non-layers by Common (1936), Bell (1960) reported that there was an increase during the prelaying period.

Hertelendy and Taylor (1964) could not detect changes in the citric acid contents of liver, kidney, heart, brain, pancreas or lung due to laying but found that the isthmus and shell gland of the laying hen were extremely rich in citric acid (400 mg/100 g fresh tissue) and in calcium (100 mg/100 g fresh tissue); they reported that the plasma citric acid level fell during the onset of lay.

Serum riboflavin levels increase with the onset of lay (Common and Bolton, 1946) as does that of manganese and the amount of manganese in the liver (Bolton, 1955). The total calcium includes a portion which is filtrable and a non-filtrable portion. The filtrable portion is mainly inorganic calcium. Attempts have been made to partition the non-filtrable portion, but the methods have been criticized (McDonald and Riddle, 1945). When the bird comes into lay the blood plasma calcium level increases from around 10 mg Ca/100 ml plasma to about 25 mg Ca/100 ml plasma, and the increase is mainly in the non-filtrable portion with little or no change in the filtrable portion (Heller, Paul and Thomson, 1934; Greenberg *et al.*, 1936; McDonald and Riddle, 1945).

The egg consists of four main parts; starting from the inside these are: yolk, albumen, shell membranes and shell.

The yolk is spherical, normally yellow in colour, and on it may be seen the germinal disc; in an infertile egg this is known as the blastodisc, and in a fertile egg as the blastoderm. In a 57 g egg, there are about 18·2 g yolk, and this contains 8·75 g water, 3·04 g protein and 6·01 g lipid (Bolton, 1961c). The compositions of the protein and lipid fractions are set out in greater detail in Tables 3 and 4 (from Bolton, 1961c).

Egg yolk has been shown to contain some substance which increases the clotting time of blood (Carr and Dewar, 1964).

Four distinct parts can be seen in the albumen: an inner thin albumen surrounding the yolk, the thick albumen, the outer thin albumen which lies between the thick albumen and the shell, and the chalazae, which run from the yolk to the poles of the egg and hold the yolk in position. Albumen makes up about 59 per cent of the egg, so in the 57 g egg there are about 33·6 g. Nineteen proteins have been identified in egg albumen (Lush, 1961).

TABLE 3
Protein in the Hen's Egg

		%
(a) Albumen	Ovalbumin	63·8
	Conalbumin	13·7
	Ovomucoid	10·1
	Lysozyme	3·5
	Ovoglobulin	7·5
	Ovomucin	1·4
	Avidin	0·1
(b) Yolk	Livetin, α and β	22
	Livetin, γ	22
	Vitellenin	25
	Vitellin	24
	Phosvitin	7

TABLE 4
Lipids in the Hen's Egg (*g*)

Neutral lipids	3·71
Phospholipids	2·01
Sterols (cholesterol)	0·29
Lecithin	1·47
Lysolecithin	0·12
Sphingomyelin	0·05
Phosphatidylethanolamine	0·30
Lysophosphatidylethanolamine	0·04
Inositol phospholipid	0·01
Plasmologen	0·02
Amino-acid phospholipid	0·004

The egg shell and shell membranes adhere very tightly together. The shell membranes can be examined by taking them from an egg which, for some biological reason, does not possess a shell: the absence of shell must first be confirmed by microscopic examination of a section of the membrane. Once the shell has been laid down attempts to remove the membrane nearly always result in some organic matter of the shell being removed with it.

The membranes contain 20 per cent water, 70 per cent organic matter, all of which is protein, and 10 per cent inorganic matter (Romanoff and Romanoff, 1949). The protein is a mixture of keratin and mucin (Bolton, 1961c). The egg shell contains two types of protein, pore protein and matrix protein, both of which are collagen, and is covered on the outside by the mucin of the cuticle. Egg shells plus membranes contain about 36 per cent calcium, 55 per cent carbonate and 0·6 per cent magnesium, with about 0·1 per cent each of phosphorus and citric acid.

The relative proportions of yolk, albumen and shell vary with the species, the breed of bird, the age of the bird, and the individual. When pullets first come into lay the egg-producing mechanisms are frequently not fully in tune, and double-yolked eggs are found. Normal, single-yolked eggs at this stage are made up of yolk 27 per cent, albumen 62 per cent and shell 11 per cent. As the bird settles down into a regular egg-laying condition, the percentage of yolk increases to 32 and the albumen decreases to 57, while the shell stays at 11 (Bolton, 1959). An examination of the eggs of pullets, yearling and aged hens showed that the major changes with increasing age were a decrease in the outer thin albumen, which was missing from eggs from 5-year-old hens (Greenwood and Bolton, 1956), and an increase in the porosity of the shell (Bolton, 1957). This increase in porosity was measured by determining the loss of weight of the egg in either a dry atmosphere or an egg-incubator; the eggs from the aged hens lost about three times as much weight as those from pullets.

The hen lays an egg in an attempt to reproduce. Once the egg is laid the embryo is wholly dependent upon the egg substance for its development into a chick. The egg from a hen kept for breeding should contain all the vitamins and all the trace minerals necessary for successful attainment of the objective. To obtain these levels the breeder offers the bird a diet which is well supplied with these nutrients. Such a diet is necessarily more expensive than the diet necessary for the production of eggs destined for human consumption; in Tables 5 and 6 (Bolton, 1961c) the vitamin and trace mineral contents of hens eggs are set out as ranges, the actual levels will depend upon the contents of the poultry foods that are the most economical to feed.

TABLE 5
Vitamin Contents of Hen's Eggs, minus shells
(amount per egg)

	Yolk and albumen, range
Vitamin A (i.u.)	0·7—400
Vitamin D (i.u.)	5—850
Vitamin E (μg)	288—1350
Vitamin K (μg)	42
Biotin (μg)	8·3—13·7
Choline (mg)	230—290
Folic acid (μg)	4·5—4·6
Nicotinic acid (μg)	27—38
Pantothenic acid (μg)	445—835
Pyridoxine (μg)	105—150
Riboflavin (μg)	83—330
Thiamin (μg)	35—328
Vitamin B_{12} (mμg)	125—700

HORMONES AND EGG PRODUCTION

Many of the changes which occur during the onset of lay, except the growth of the ovary, can be induced by injecting gonadal hormones into immature females. Injection of androgen stimulates the comb, which then becomes similar in appearance and histology to that of the male (Bolton,

TABLE 6
Mineral Contents of Hen's Eggs—minus shells
(amount per egg)

	Yolk and albumen, range
Calcium (g)	0·34
Phosphorus (g)	0·13
Magnesium (g)	0·05
Copper (mg)	0·05—0·033
Iron (mg)	1·0
Manganese (mg)	0·005—0·025
Potassium (g)	0·08
Sodium (g)	0·07
Zinc (mg)	0·7—1·0

1953b). Administration of oestrogen is followed by increases in the plasma levels of riboflavin and vitamin A, and in liver fat and protein (Common, Rutledge and Bolton, 1947; Common, Bolton and Rutledge, 1948). Ranney and Chaikoff (1951) injected oestrogen into intact and hepatectomized fowls and showed that the presence of the liver was necessary for the induction of lipaemia, and that it was the site formation of phospholipids. Stamler *et al.* (1950) implanted oestrogen into 6-week-old cockerels and reported that the cholesterol contents of the plasma, liver, kidney, adrenals, gut and lungs were increased, although Fleischman and Fried (1945) had concluded that oestrogen does not increase the cholesterol content.

Common, Rutledge and Hale (1948) reported that the plasma calcium level was increased by injections of oestrogen, but that oestrogen plus androgen were necessary before calcium retention was increased. Bolton (1953b) found that it was necessary to administer progesterone in addition to androgen and oestrogen before the comb and magnum became similar to those of the laying hen.

The production of gonadal hormones in the bird is controlled by the secretions of the anterior lobe of the pituitary. These secretions are similar in function to those of mammals in most respects, and only the main differences will be mentioned. Prolactin injections cause hens to go broody but the full expression of broodiness is only seen in those breeds of fowls with a natural tendency to broodiness (Riddle, Bates and Lahr, 1935). Growth hormone has not been demonstrated in avian pituitaries (Sturkie, 1954). Nalbandov, Meyer and McShan (1951) found that the combs of male chickens which atrophied after hypophysectomy could be made to grow continuously by administering fowl pituitary substances, but for only 10–12 days by the administration of mammalian luteinizing hormone. On the basis of this result they postulated the presence of an Interstitial-Cell-Stimulating Hormone in the avian, but not in the mammalian, pituitary.

Stimulation of the pituitary starts a chain reaction; the pituitary hormones stimulate the other endocrine glands and the production of eggs is increased.

The pituitary can be stimulated by light. Platt (1955) demonstrated that restricting light to 8 hr per day from 3 to 6 months of age resulted in an increased rate of lay later in life. King (1958) described a programme of light restriction to 6 hr for the first 16–18 weeks of life, followed by weekly increments of light and demonstrated that this prolonged the period of high rate of lay. It is now common practice in poultry farming to rear birds intended as future egg producers to about 18 weeks of age in windowless houses with a lighted period of 6–8 hr per day. Thereafter the length of the daily lighted period is increased by 20 min per week to a maximum 18 hr per day.

NON-NUTRIENT ADDITIVES

Two non-nutritional types of substance may be added to poultry foods: chemicals to reduce the loss due to attacks by vermin; and chemicals to increase the yield.

Insecticides, etc.

When foods are stored, they become subject to attacks by rats, mice, insects and fungi. Suitably designed containers prevent the ingress of rats and mice; to minimize insect attacks it is normal practice to use insecticides, and most insecticides used are fumigants. Only three of the fumigants used (ethylene dibromide, ethylene dichloride and tetrachloromethane) have been thoroughly tested for their subsequent effects on poultry, and ethylene dibromide was found to have an adverse effect on egg size (Bondi, Olomucki and Calderon, 1955; Bierer and Vickers, 1959; Caylor and Laurent, 1960; Fuller and Morris, 1962). The results of Fuller and Morris (1962) showed that when ethylene dibromide was ingested there was no effect for about 2 weeks and then egg size decreased; the rate of decrease increased with increasing rate of ingestion of ethylene dibromide. When the ethylene-dibromide-treated food was replaced by untreated food, egg size continued to decrease for about 2 weeks, and then increased. Carter (1963) plotted the relative decrease of egg size against rate of intake of ethylene dibromide and obtained a straight line graph. Extrapolation showed that it was directed to a point which did not differ significantly from the origin; the concept of some threshold rate of ingestion below which the ethylene dibromide would not have any adverse effect on egg size did not, therefore, appear to hold good. Ethylene dibromide is known to remain strongly sorbed on to grain, even after prolonged periods of airing, and the available evidence is a strong contra-indication of its use. Stadelman *et al.* (1965) reported that when 0·1 to 0·15 mg lindane, dieldrin, heptachlor or DDT were added per kg food of laying hens for 14 days the insecticides could be detected in the tissues;

increasing the level of addition to 10 to 15 mg per kg for 5 days gave residues which persisted in the tissues and eggs for more than 17 weeks, but less than 26.

When groundnuts are improperly stored after harvesting they may become attacked by the fungus *Aspergillus flavus*, and aflatoxin accumulates in the product. When such meals were incorporated in poultry diets they were found to cause the deaths of turkey poults, ducklings and young chicks (Blount, 1961; Sargent, Sheridan and O'Kelly, 1961). Forgacs and Carll (1955) and Forgacs *et al.* (1958) found that the haemorrhagic syndrome of fowls could be produced by feeding certain moulds. Schumaier, *et al.* (1961) extended this work and examined a number of moulds for toxin production. The moulds were grown on wheat, extracted with diethyl ether and the lipid-soluble material fed to chicks. They showed that the syndrome could be produced by feeding mycotoxins produced by *Aspergillus flavus oryzae* and *A. chevalier*.

Yield Promoters

When a chemical is added to the food and the yield of a group of birds fed this food is significantly greater than the yield of a similar group of birds fed the unsupplemented food the chemical is said to be a yield-promoter. Most experiments on possible yield-promoters have been carried out with young birds, and the yield-promoting property of many substances has been manifested only by the increased growth rate. Substances which have been shown to promote growth increase are: antibiotics, surface-active agents, organic arsenical compounds and the unknown growth factors.

Antibiotics. Antibiotics are complex substances produced by microorganisms which can inhibit the growth of some other microorganisms. They are distinct chemical entities and act upon bacteria in different ways; it is believed that their growth-promoting property is due to their ability to influence bacterial multiplication. Three possible mechanisms have been suggested: (i) suppression of organisms which produce sub-clinical infections; (ii) stimulation of the growth of organisms which synthesize essential nutrients; and (iii) reduction in the total numbers of bacteria which compete with the host for food in general, or for one or more specific nutrients.

Forbes and Park (1959) showed that penicillin supplementation of the diet does not improve the growth rates of germ-free chicks, and Forbes, Supplee and Combs (1958) found that neither penicillin nor oleandomycin supplements increased the growth-rate of germ-free turkeys.

Lev, Briggs and Coates (1957) examined the caecal flora of chicks reared with or without penicillin in the diet during the first 48 hr after hatching. They used birds in two environments: in one the chicks receiving penicillin had negligible numbers of *Clostridium welchii* in the caeca, although these organisms were plentiful in the caeca of control birds; in the other *Cl. welchii* were present in both groups in about equal numbers, but those from the birds fed penicillin were not able to produce lecithinase, the α-toxin normally produced by *Cl. welchii*.

Coates, Davies and Kon (1955) reported that the walls of the small intestine of birds receiving penicillin were thinner than those of controls.

Lactobacilli form the major part of the microflora in the crop (Lev and Briggs, 1956; De Santiago y Redel, 1960). Eyssen, De Prins and De Somer (1962) compared the effectiveness of various antibiotics against lactobacilli with their growth-promoting properties for chicks. They found that virginiamycin was superior to chlortetracycline and oxytetracycline for both purposes, and suggested that the antibiotics exert their major effect in the anterior part of the alimentary canal.

Surface-active agents. Brewers in the U.S.A. began using supplements of a surface-active agent to improve the extraction of the wort and Ely (1951) examined the substances to see if they had any deleterious effects, using chick growth as the criterion. He found that supplementation of diets with the surface-active agent increased the growth-rates of chicks; the observation was confirmed by Ely (1952), Ney and Newell (1954) and Lillie, Sizemore and Denton (1958).

Surface-active agents are of three types: non-ionic, anionic and cationic. Non-ionic detergents are usually polyoxyethylene derivatives of fatty acids or of fatty acid esters of polyhydric alcohols or ethers; anionic detergents are sodium or potassium salts of secondary alkyl sulphates; and cationic detergents are quaternary ammonium compounds. Bacteria do not develop resistance to surface-active agents, though they do to antibiotics, and surface-active agents are not used in human medicine.

It is to be expected that the emulsifying power of a detergent would aid the initial stages of fat digestion and studies on their mode of action were carried out using diets containing 10 per cent supplemental fat (Bolton, 1961b). It was found that the inclusion of a non-ionic detergent aided the digestion of fat, while leaving the digestibilities of protein and carbohydrate unaffected; the digestibility of fat was impaired slightly by the inclusion of an anionic detergent, and seriously by a cationic detergent; the cationic detergent had an adverse effect on carbohydrate digestibility. Bangham and Dawson (1959) found that the addition to the diet of a detergent which produced a positive charge at the surface (cationic) impeded digestion, while an anionic detergent aided digestion and concluded that the active end of the enzyme lipase carries a positive charge. The non-ionic detergent would aid emulsification of the fat, the anionic detergent would be a competitor for the positively charged enzyme, while the cationic detergent would tend to combine with the fatty acid ions; in the last two cases the result would be a reduction in the digestibility of the fat.

Arsenicals. The most commonly used arsenical compounds used as yield promoters in poultry diets are 3-nitro-4-hydroxyphenylarsonic acid, *p*-aminophenylarsonic acid (arsanilic acid), 4-nitrophenylarsonic acid and arsenosobenzene.

Coates, Davies and Kon (1955) found that the addition of 20 mg arsanilic acid per kg food produced a significant reduction in the weight of the guts of chicks compared to those receiving the unsupplemented diet; the decrease was similar to that produced by a penicillin supplement. Morrison, Hunsaker and Aitken (1954) also noted a similarity between arsenicals and antibiotics; they found that 3-nitro-4-hydroxyphenylarsonic acid produced a growth response in houses which had been used for poultry for some years, but not in new houses. Combs, Romoser and Bishop (1954) and West (1956) reported that supplementary arsanilic acid and antibiotics had an additive effect in stimulating chick growth. Frost and Spruth (1956) compared the bacteriostatic effects of sodium arsanilate, 3-nitro-4-hydroxyphenylarsonic acid, chlortetracycline and procaine penicillin: they found the minimal inhibitory concentrations, in mg/ml, for *Eschericha coli*, *Streptococcus faecalis*, *Proteus vulgaris* and *Salmonella enteritidus* for the first three were 10,000; 7000 and between 3 and 100 respectively; those for *Clostridium perfringens* for all four substances were 1000, 100, 0·01 to 0·02 and 0·08 respectively. Arsenicals are, therefore, much poorer bacteriostatic substances than the antibiotics.

The normal level of inclusion in diets for chicks or poults as yield promoters is between 50 and 100 mg/kg food; Abbott, Bird and Cravens (1954) used levels as high as 500 mg/kg without depressing the growth-rate.

Dietary arsenic accumulates in the tissues. When birds are fed diets containing the normal yield-promoting level the carcasses contain sufficient arsenic to exceed the maximal permitted level in the U.K., although they do not exceed that of U.S.A.

REFERENCES

Abbott, O. J., Bird, H. R. and Cravens, W. W. (1954) *Poult. Sci.* **33**, 1245.

Agricultural Research Council (1963) *The Nutrient Requirements of Farm Livestock: No. 1. Poultry*. Agricultural Research Council, London.

Ames, S. R. (1956) *Poult. Sci.* **35**, 145.

Ames, S. R., Sarma, P. S. and Elvehjem, C. A. (1947) *J. Biol. Chem.* **167**, 135.

Ashton, W. M., Evans, C. and Williams, P. C. (1960) *J. Sci. Fd. Agric.* **11**, 722.

Baldini, J. T. and Rosenberg, H. R. (1955) *Poult. Sci.* **34**, 1301.

Bangham, A. D. and Dawson, R. M. (1959) *New Scient.* **5**, 1149.

Bartlett, B. E. (1962) *J. Agric. Vict. Dep. Agric.* **60**, 42.

Barlow, J. S., Slinger, J. S. and Manery, S. F. (1947) *J. Geront.* **2**, 110.

Beck, A. B. (1961) *Austr. J. Agric. Res.* **12**, 743.

Bell, D. J. (1957) *Biochem. J.* **66**, 137.

Bell, D. J. (1960) *Biochem. J.* **75**, 224.

Benesch, R., Barron, N. S. and Mawson, C. A. (1944) *Nature, Lond.* **153**, 138.

Berg, L. R., Bearse, G. E. and Merrill, L. H. (1962) *Poult. Sci.* **41**, 1328.

Bernard, R. and Genest, P. (1945) *Science, N.Y.* **101**, 617.

BIELY, J. and MARCH, B. (1952) *Science, N.Y.* **114**, 330.
BIERER, B. W. and VICKERS, C. L. (1959) *J. Am. Vet. Med. Ass.* **34**, 452.
BIESTER, H. E. and SCHWARTZ, L. H. (1959) *Diseases of Poultry*. 4th ed. Iowa State Press, Ames, Iowa.
BIERI, J. G., BRIGGS, G. M., POLLARD, C. J. and SPIVEY FOX, M. R. (1960) *J. Nutr.* **70**, 47.
BIRD, F. H. (1949) *J. Nutr.* **39**, 13.
BLAMBERG, D. L., BLACKWOOD, U. B., SUPPLEE, W. C. and COMBS, G. F. (1960) *Proc. Soc. Exp. Biol. Med.* **104**, 217.
BLAXLAND, J. D. (1946) *Vet. J.* **102**, 157.
BLOUNT, W. P. (1960) *Vet. Rec.* **72**, 90.
BLOUNT, W. P. (1961) *J. Br. Turkey Fed.* **9**, 52.
BOLTON, W. (1944) *J. Agric. Sci. Camb.* **34**, 198.
BOLTON, W. (1947) *J. Agric. Sci. Camb.* **37**, 316.
BOLTON, W. (1950) *Nature, Lond.* **166**, 912.
BOLTON, W. (1951) *Proc. IX World's Poult. Cong. Paris* **2**, 27.
BOLTON, W. (1952) *Br. J. Nutr.* **6**, 171.
BOLTON, W. (1953a) *J. Agric. Sci. Camb.* **43**, 120.
BOLTON, W. (1953b) *J. Endocr.* **9**, 440.
BOLTON, W. (1955) *Br. J. Nutr.* **9**, 170.
BOLTON, W. (1957) *J. Agric. Sci. Camb.* **49**, 384.
BOLTON, W. (1959) *J. Agric. Sci. Camb.* **52**, 364.
BOLTON, W. (1961a) *Proc. Nutr. Soc.* **20**, xxvi.
BOLTON, W. (1961b) *J. Agric. Sci. Camb.* **57**, 83.
BOLTON, W. (1961c) *Biochemists' Handbook*. Ed. C. Long. Spon, London.
BOLTON, W. (1962) *Proc. XII. World's Poult. Congr. Sydney* **2**, 38.
BOLTON, W. (1964) *The Role of the Gastrointestinal Tract in Protein Metabolism*. Ed. H. N. Munro. Blackwell, Oxford.
BOLTON, W. (1965) *Br. Poult. Sci.* **6**, 97.
BOLTON, W. and DEWAR, W. A. (1962) *Proc. XII World's Poult. Congr. Sydney* **1**, 117.
BOLTON, W. and DEWAR, W. A. (1965) *Br. Poult. Sci.* **6**, 103.
BOLTON, W. and MURTY, V. N. (1960) *J. Agric. Sci. Camb.* **55**, 203.
BONDI, A., OLOMUCKI, E. and CALDERON, M. (1955) *J. Sci. Fd. Agric.* **6**, 600.
BORCHERS, R. and ACKERSON, C. W. (1951) *Proc. Soc. Exp. Biol. Med.* **78**, 81.
BOSSHARDT, D. K., PAUL, W. J. and BARNES, R. H. (1950) *J. Nutr.* **40**, 595.
BOSSHARDT, D. K., AYRES, M. M., YDSE, L. C. and BARNES, R. H. (1946a) *J. Nutr.* **32**, 93.
BOSSHARDT, D. K., PAUL, W. J., O'DOHERTY, K. and BARNES, R. H. (1946b) *J. Nutr.* **32**, 641.
BOUTHILET, R. J., HUNTER, W. L., LUHMAN, C. A., AMBROSE, D. and LEPKOVSKY, S. (1950) *Poult. Sci.* **29**, 837.
BOYNE, W. A., CARPENTER, K. J. and WOODHAM, A. A. (1961) *J. Sci. Fd. Agric.* **12**, 832.
BRADLEY, O. C. and GRAHAME, T. (1960) *The Structure of the Fowl.* Oliver and Boyd, Edinburgh.
BRASSEUR, H., DALLEMAGNE, H. J. and MELON, J. (1946) *Nature, Lond.* **157**, 453.
BRIGGS, G. M. (1945) *J. Biol. Chem.* **161**, 749.
BRIGGS, G. M., GROSCHKE, A. C. and LILLIE, R. J. (1946) *J. Nutr.* **32**, 659.
BRIGGS, G. M. and LILLIE, R. J. (1946) *Proc. Soc. Exp. Biol. Med.* **61**, 430.
BRIGGS, G. M., LUCKEY, T. D., TEPLY, L. J., ELVEHJEM, C. A. and HART, E. B. (1943) *J. Biol. Chem.* **148**, 517.
BRIN, M. and OLSEN, B. E. (1951) *J. Biol. Chem.* **166**, 81.
BROBECK, J. R. (1948) *Yale J. Biol. Med.* **20**, 545.
BRODY, S., FUNK, E. M. and KEMPSTER, H. L. (1938) *Res. Bull. Mo. Agric. Exp. Sta.* No. 278.
BROWN, E. W. (1904) *U.S.D.A. Bur. An. Ind. Bull.* 56.
BROWN, W. O. and MCCRACKEN, K. J. (1965) *J. Agric. Sci. Camb.* **64**, 305.
BROWNE, T. G. (1922) *J. Comp. Path. Therap.* **35**, 12.
BRUMBAUGH, J. H., MEHRING, A. L. and TITUS, H. W. (1959) *Poult. Sci.* **38**, 326.
BUCKNER, G. D., INSKO, W. M. and HENRY, A. H. (1944) *Poult. Sci.* **33**, 457.
BUMGARDNER, H. L. and SHAFFNER, C. S. (1957) *Poult. Sci.* **36**, 207.

BURMESTER, B. R., SCOTT, H. M. and CARD, L. E. (1939) *Proc. VII World's Poult. Congr. Cleveland, Ohio*, 99.
BURNS, C. H., CRAVENS, W. W. and PHILLIPS, P. H. (1953) *J. Nutr.* **50**, 317.
BURNS, M. H. and SALMON, W. D. (1956) *J. Agric. Food Chem.* **4**, 257.
BYERLY, T. C. (1941) *Bull. Md. Agric. Exp. Sta.* No. A1.
CALHOUN, M. L. (1954) *The Microscopic Anatomy of the Digestive System of the Chicken.* Iowa State College Press, Ames, Iowa.
CAMPOS, A. C., WILCOX, F. H. and SHAFFNER, C. S. (1960) *Poult. Sci.* **39**, 119.
CARD, L. E., KEITH, M. H. and MITCHELL, H. H. (1926) *Illinois Stat. Rep.* 80.
CAREW, L. B. (1961) Ph.D. Thesis, Cornell, cited by Hill, F. W. (1962).
CARPENTER, K. J. (1960) *Biochem. J.* **77**, 604.
CARPENTER, K. J. and CLEGG, K. M. (1956) *J. Sci. Fd. Agric.* **7**, 45.
CARR, J. G. and DEWAR, W. A. (1964) *Br. Poult. Sci.* **5**, 161.
CARTER, T. C. (1963) Personal communication.
CASKEY, C. D. (1953) *Proc. Nutr. Counc. Amer. Feed. Man. Assoc.* 8, cited by Scott, M. L. (1962). *Poult. Comment* **19**, (4).
CASKEY, C. D. and NORRIS, L. C. (1938) *Poult. Sci.* **17**, 433.
CASKEY, C. D. and NORRIS, L. C. (1939) *Proc. Soc. Exp. Biol. Med.* **40**, 590.
CAYLOR, J. F. and LAURENT, C. K. (1960) *Poult. Sci.* **39**, 216.
CHENEY, G. (1938) *Am. J. Dig. Dis.* **5**, 104.
CHERNICK, S., LEPKOVSKY, S. and CHAIKOFF, I. L. (1948) *Am. J. Physiol.* **155**, 33.
CHRISTIANSEN, J. B., HALPIN, J. G. and HART, E. B. (1939) *Science, N.Y.* **90**, 356.
CLANDININ, D. R. (1946) *Poult. Sci.* **25**, 223.
COATES, M. E., DAVIES, M. K. and KON, S. K. (1955) *Br. J. Nutr.* **9**, 110.
COATES, M. E., DICKINSON, C. D., HARRISON, G. F. and KON, S. K. (1951a) *Biochem. J.* **49**, 68.
COATES, M. E., FORD, J. E., HARRISON, G. F., KON, S. K., PORTER, J. W. G., CUTHBERTSON, W. J. F. and PEGLER, H. F. (1951) *Biochem. J.* **49**, lxvii.
COATES, M. E. and HOLDSWORTH, E. S. (1961) *Br. J. Nutr.* **15**, 131.
COLLIP, J. B. (1922) *Am. J. Physiol.* **59**, 435.
COMAR, C. L. and DRIGGERS, J. C. (1949) *Science, N.Y.* **109**, 282.
COMBS, G. F. (1965) *Proc. Md. Nutr. Conf.*
COMBS, G. F., ROMOSER, G. L. and BISHOP, R. W. (1954) *J. Nutr.* **53**, 511.
COMBS, G., SUPPLEE, W. C., QUILLAN, E. C., BLAMBERG, D. L., DONALDSON, W. E., ROMOSER, G. L. and HELBACKA, N. V. (1958) *Poult. Sci.* **37**, 1195.
COMMON, R. H. (1936) *J. Agric. Sci. Camb.* **26**, 472.
COMMON, R. H. (1938) *J. Agric. Sci. Camb.* **28**, 347.
COMMON, R. H. (1940) *J. Agric. Sci. Camb.* **30**, 113.
COMMON, R. H. (1941) *J. Agric. Sci. Camb.* **31**, 412.
COMMON, R. H. (1943) *J. Agric. Sci. Camb.* **33**, 213.
COMMON, R. H. and BOLTON, W. (1946) *Nature, Lond.* **158**, 95.
COMMON, R. H., BOLTON, W. and RUTLEDGE, W. A. (1948) *J. Endocr.* **5**, 263.
COMMON, R. H., RUTLEDGE, W. A. and BOLTON, W. (1947) *J. Endocr.* **5**, 121.
COMMON, R. H., RUTLEDGE, W. A. and HALE, R. (1948) *J. Agric. Sci. Camb.* **38**, 64.
COUCH, J. R., CRAVENS, W. W., ELVEHJEM, C. A. and HALPIN, J. G. (1948) *J. Nutr.* **35**, 37.
COUCH, J. R., JAMES, L. E. and SHERWOOD, R. M. (1947) *Poult. Sci.* **26**, 30.
CRAMPTON, E. W. and LLOYD, L. E. (1959) *Fundamentals of Nutrition.* Freeman, London.
CROPPER, W. J. and SCOTT, M. L. (1964) *Poult. Sci.* **43**, 1311.
CRUIKSHANK, E. M. (1934) *Biochem. J.* **28**, 965.
CULLEN, M. P., RASMUSSEN, O. G. and WILDER, O. H. M. (1962) *Poult. Sci.* **41**, 360.
DAGHIR, N. J. and BALLOUN, S. L. (1961) *Poult. Sci.* **40**, 1392.
DAM, H. GLAVIND, J., LEWIS, L. and TAGE-HANSEN, E. (1938) *Scand. Arch. Phys.* **79**, 121.
DAM, H., PRANGE, I. and SONDERGAARD, E. (1954) *Acta Pharmac. Tox.* **10**, 58.
DAM, H., SCHØNHEYDER, F. and TAGE-HANSEN, E. (1936) *Biochem. J.* **30**, 1075.
DAVIDSON, J., MCDONALD, I. and WILLIAMS, R. B. (1957) *J. Sci. Fd. Agric.* **8**, 173.
DAVIES, A. W. (1951) *Proc. IX World's Poult. Congr., Paris*, **4**, 20.
DAVIES, P. N., NORRIS. L. C. and KRATZER, E. H. (1962) *J. Nutr.* **78**, 445.

DAVIES, R. E., KURNICK, A. A., READ, B. L. and COUCH, J. R. (1957) *Poult. Sci.* **36**, 1111.
DAVIS, G. K. (1959) *Fed. Proc.* **18**, 1119.
DAVIS, R. L., BRIGGS, G. M. and SLOAN, H. J. (1953) *Proc. Soc. Exp. Biol. Med.* **82**, 175.
DAY, E. J. and HILL, J. E. (1959) *Poult. Sci.* **38**, 646.
DEIGHTON, T. and HUTCHINSON, J. C. D. (1940) *J. Agric. Sci. Camb.* **30**, 463.
DELAPLANE, G. F. (1934) *Vet. Alumni Quart. Ohio Univ.* **21**, 149.
DE RENZO, E. C., KALEITA, E., HEYTLER, P., OLESON, J. J., HUTCHINGS, B. L. and WILLIAMS, J. H. (1953) *J. Am. Chem. Soc.* **75**, 753.
DE SANTIAGO Y REDEL, E. (1960) *Arch. Zootecnia.* **9**, 210.
DEWAR, W. A. (1962) *Proc. Nutr. Soc.* **21**, xix.
DINNING, J. S., SIME, J. T. and DAY, P.L. (1955) *J. Biol. Chem.* **217**, 205.
DINNING, J. S., SIME, J. T. and DAY, P. L. (1956) *J. Biol. Chem.* **222**, 215.
DRIGGERS, J. C. and COMAR, C. L. (1949) *Poult. Sci.* **28**, 420.
DUCKWORTH, J., NAFTALIN, J. M. and DALGARNO, A. C. (1950) *J. Agric. Sci. Camb.* **40**, 39.
EDWARDS, H. M., DUNAHOO, W. S., CARMEN, J. L. and FULLER, H. L. (1960) *Poult. Sci.* **39**, 1389.
EIJKMAN, C. (1897) *Virchows Arch. path. Anat. Physiol.* **148**, 523.
EKMAN, P., EMANUELSON, H. and FRANSSON, A. (1949) *K. LantbrHogsk. Annlr.* **16**, 749.
ELDRED, N. R. and RODNEY, G. (1946) *J. Biol. Chem.* **162**, 261.
ELVEHJEM, C. A. (1935) *Phys. Revs.* **15**, 471.
ELVEHJEM, C. A. and HART, E. B. (1929) *J. Biol. Chem.* **84**, 131.
ELY, C. M. (1951) *Science, N.Y.* **114**, 523.
ELY, C. M. (1952) *Feedstuffs, Lond.* **24**, 24, 26, 28, 46–50.
ELY, C. M. (1959) *Poult Sci.* **38**, 1316.
EOFF, H. J., DAVIES, R. E., FERGUSON, T. M. and COUCH, J. R. (1961) *Poult. Sci.* **40**, 1315.
ERIKSSON, S. (1954) *K. LantbrHogsk. Annlr.* **21**, 385.
EVANS, R. E., DAVIDSON, J. A., BAUER, D. and BUTTS, H. A. (1952) *Poult. Sci.* **31**, 915.
EVANS, R. J., MCGINNIS, J. and ST. JOHN, J. L. (1947) *J. Nutr.* **33**, 661.
EWING, W. R. (1963) *Poultry Nutrition.* Ewing. Pasadena, Calif.
EYSSEN, H., DE PRINS, C. and DE SOMER, P. (1962) *Poult. Sci.* **41**, 227.
FARNER, D. S. (1942) *Poult. Sci.* **21**, 445.
FARNER, D. S. (1943) *Poult. Sci.* **22**, 79.
FEDDE, M. R., WAIBEL, P. E. and BURGER, R. E. (1959) *Poult. Sci.* **38**, 1203.
FELDMAN, S. E., LARRSON, S., DIMICK, M. K. and LEPKOVSKY, S. (1957) *Am. J. Physiol.* **191**, 259.
FELL, H. B. and MELLANBY, E. (1952) *J. Physiol. Lond.* **116**, 320.
FELL, H. B. and MELLANBY, E. (1953) *J. Physiol. Lond.* **119**, 470.
FELL, H. B., MELLANBY, E. and PELE, S. R. (1956) *J. Physiol. Lond.* **134**, 178.
FELL, R. V., WILKINSON, W. S. and WATTS, A. B. (1959) *Poult. Sci.* **38**, 1203.
FERGUSON, T. M., ATKINSON, R. L. and COUCH, J. R. (1954) *Proc. Soc. Exp. Biol. Med.* **86**, 868.
FERGUSON, T. M., VAUGHT, H. P., CROSS, M. S., HANDER, O. J. and COUCH, J. R. (1955) *Poult. Sci.* **34**, 1193.
FERNELL, W. R. and ROSEN, G. D. (1956) *Br. J. Nutr.* **10**, 143.
FIELD, H. I. and EVANS, E. T. R. (1946) *Vet. Rec.* **58**, 253.
FISHER, H., JOHNSON, D. and FERDO, S. (1957) *J. Nutr.* **61**, 611.
FLEISCHMAN, W. and FRIED, I. A. (1945) *Endocrinology*, **36**, 406.
FORBES, M. and PARK, J. T. (1959) *J. Nutr.* **67**, 69.
FORBES, M., SUPPLEE, W. C. and COMBS, G. F. (1958) *Proc. Soc. Exp. Biol. Med.* **99**, 110.
FORD, J. E. (1960) *Br. J. Nutr.* **14**, 485.
FORD, J. E. (1962) *Br. J. Nutr.* **16**, 409.
FORD, J. E. (1964) *Br. J. Nutr.* **18**, 449.
FORD, J. E., KON, S. K. and PORTER, J. W. G. (1951) *Biochem. J.* **50**, ix.
FORGACS, J. and CARLL, W. T. (1955) *Vet. Med.* **50**, 172.
FORGACS, J., KOCH, H., CARLL, W. T. and WHITE-STEVENS, R. H. (1958) *Am. J. Vet. Res.* **19**, 744.
FRAPS, G. S. (1928) *Bull. Tex. Agric. Exp. Stn.* No. 372.

FRAPS, G. S. (1931) *Bull. Tex. Agric. Exp. Stn.* No. 437.
FRAPS, G. S. (1940) *Bull. Tex. Agric. Exp. Stn.* No. 589.
FRAPS, G. S. (1944) *Bull. Tex. Agric. Exp. Stn.* No. 665.
FRAPS, G. S. (1946) *Bull. Tex. Agric. Exp. Stn.* No. 678.
FRAPS, G. S. and CARLYLE, E. C. (1939) *Bull. Tex. Agric. Exp. Stn.* No. 571.
FRAPS, G. S. and CARLYLE, E. C. (1941) *Bull. Tex. Agric. Exp. Stn.* No. 600.
FRAPS, G. S. and CARLYLE, E. C. (1942) *Bull. Tex. Agric. Exp. Stn.* No. 625.
FRIEDMAN, M. H. F. (1939) *J. Cell. Comp. Physiol.* **13**, 219.
FROLICH, A. (1954) *Nature, Lond.* **173**, 132.
FROST, D. V. and SPRUTH, H. C. (1956) *Symposium on Medicated Feeds*, p. 136. Medical Encyclopedia Inc. New York.
FULLER, H. L. and HILL, R. L. (1964) *Poult. Sci.* **43**, 1320.
FULLER, H. L. and MORRIS, G. K. (1962) *Poult. Sci.* **41**, 645.
GALLACHER, C. H. (1957) *Austr. Vet. J.* **33**, 311.
GALLUP, W. D. and NORRIS, L. C. (1939a) *Poult. Sci.* **18**, 76.
GALLUP, W. D. and NORRIS, L. C. (1939b) *Poult. Sci.* **18**, 99.
GEIGER, E., EL REWI, I. and GEIGER, L. E. (1952) *Fed. Proc.* **11**, 444.
GILLAM, A. E. and HEILBRON, I. M. (1935) *Biochem. J.* **29**, 1064.
GILLIS, M. B. (1948) *J. Nutr.* **36**, 351.
GLEAVES, E. W., TONKINSON, L. V., MILLS, T. J., DUNKELGOD, K. E., ROBERTS, C. A., THAYER, R. H., SIRNY, R. J. and MORRISON, R. D. (1961) *Poult. Sci.* **40**, 1098.
GOLDHABER, P., ZACHARIAS, L. and KINSEY, V. E. (1950) *J. Nutr.* **42**, 453.
GOLDING, W. V., SCHAIBLE, P. J. and DAVIDSON, J. A. (1940) *Poult. Sci.* **19**, 263.
GOLDSTEIN, J. and SCOTT, M. L. (1956) *J. Nutr.* **60**, 349.
GOSWANI, M. N. D. and ROBBLEE, A. R. (1958) *Poult. Sci.* **37**, 96.
GRAU, C. R. and PETERSON, D. W. (1946) *J. Nutr.* **32**, 181.
GREEN, D. E., LELOIR, L. R. and NOCITO, V. (1945) *J. Biol. Chem.* **161**, 559.
GREENBERG, D. M., LARSON, C. E., PEARSON, P. B. and BURMESTER, B. R. (1936) *Poult. Sci.* **15**, 483.
GREENWOOD, A. W. and BOLTON, W. (1956) *J. Agric. Sci. Camb.* **47**, 40.
GRIMMINGER, P. (1961) Poultry and Livestock Comment (E. I. Dupont de Nemours & Co., Industrial and Biochemicals Dept., Wilmington, Delaware), **18**, No. 3, p.1.
GROEBBELS, F. (1932) *Der Vogel.* Erster Band: *Altmangswelt und Nahrungswelt* Verlag con Gebruder Borntraeger, Berlin.
GROSCHKE, A. C., ANDERSON, J. O. and BRIGGS, G. M. (1948) *Proc. Soc. Exp. Biol. Med.* **68**, 564.
GURCEY, R., BOUCHER, R. V. and CALLENBACH, E. W. (1950) *J. Nutr.* **41**, 565.
GUTOWSKA, M. S. and MITCHELL, C. A. (1945) *Poult. Sci.* **24**, 159.
GUTOWSKA, M. S. and POZZANI, U. C. (1945) *Fed. Proc.* **4**, 91.
GUTOWSKA, M. S., PARROT, E. M. and SLESINSKI, F. A. (1941) *Poult. Sci.* **20**, 379.
HALLSWORTH, E. G. and COATES, J. I. (1962) *J. Agric. Sci. Camb.* **58**, 153.
HALNAN, E. T. (1934) *Bull. Minist. Agric. Fish.* No. 7. H.M.S.O., London.
HALNAN, E. T. (1949) *Br. J. Nutr.* **3**, 245.
HAMILTON, T. S. and MITCHELL, H. H. (1924) *J. Agric. Res.* **27**, 605.
HARMS, R. H. and WALDROUP, P. W. (1961) *Poult. Sci.* **40**, 1345.
HEALD, P. J. and MCLACHLAN, P.M. (1963) *Biochem. J.* **87**, 571.
HEALD, P. J. and MCLACHLAN, P. M. (1964) *Proc. VIth Int. Congr. Biochem.* New York.
HEALD, P. J. and MCLACHLAN, P. M. (1965) *Biochem. J.* **94**, 32.
HEGSTED, D. M., MILLS, R. C., ELVEHJEM, C. A. and HART, E. B. (1941) *J. Biol. Chem.* **138**, 459.
HELBACKA, N. V. (1961) *Proc. Md. Nutr. Conf.* p. 46.
HELLER, V. G. and PENQUITE, R. (1936) *Poult. Sci.* **15**, 397.
HELLER, V. G. and PENQUITE, R. (1937) *Poult. Sci.* **16**, 243.
HELLER, V. G., PAUL, H. and THOMSON, R. B. (1934) *J. Biol. Chem.* **106**, 357.
HENNEBERG, W. and STOHMANN, F. (1860) *Beitrage zur Begrundung einer rationellen Futterung der Weiderkauer.* Vol. 1.
HENNING, H. J. (1929) *Landwn. VerStnen.* **108**, 253.

HENRY, K. M. and KON, S. K. (1951) *Biochem. J.* **48**, xi.
HERTELENDY, F. and TAYLOR, T. G. (1960) *Biochem. Biophys. Acta* **44**, 200.
HERTELENDY, F. and TAYLOR, T. G. (1964) *Comp. Biochem. Physiol.* **11**, 173.
HEUSER, G. F. (1945) *Poult. Sci.* **24**, 20.
HEYWANG, B. W. and VARICH, M. G. (1959) *Poult. Sci.* **38**, 999.
HIGGINS, E. S., RICKERT, D. A. and WESTERFELD, W. W. (1956) *J. Nutr.* **59**, 539.
HILL, F. W. (1958a) *Poult. Comment.* **15**, No. 2.
HILL, F. W. (1958b) *Feedstuffs, Lond.* **30**, Nov. 8.
HILL, F. W. (1962) *Nutrition of Pigs and Poultry.* Ed. J. T. Morgan and D. Lewis. Butterworths, London.
HILL, F. W. and ANDERSON, D. L. (1958) *J. Nutr.* **64**, 587.
HILL, F. W., ANDERSON, D. L., RENNER, R. and CAREW, L. B. (1960) *Poult. Sci.* **39**, 573.
HILL, F. W., CAREW, L. B. and VAN TIENHOVEN, A. (1958) *Am. J. Physiol.* **195**, 654.
HILL, F. W. and DANSKY, L. M. (1954) *Poult. Sci.* **33**, 112.
HILL, F. W. and RENNER, R. (1958) *Poult. Sci.* **37**, 1212.
HILL, F. W. and RENNER, R. (1960) *Poult. Sci.* **39**, 579.
HILL, K. J., ANNISON, E. F. and NOAKES, D. E. (1964) *Proc. Conf. U.K. Branch World's Poult. Sci. Assoc.*
HODSON, A. Z. (1940) *J. Nutr.* **20**, 377.
HOGAN, A. G. (1949–50) *Nutr. Abstr. Rev.* **19**, 751.
HOOGENDOORN, J. (1940) *Landbouwk Tijdschr's-Grav.* **52**, 330. (*Nutr. Abstr. Rev.* 1946–7, **16**, 732.)
HUNT, J. R. and AITKEN, J. R. (1962) *Poult. Sci.* **41**, 219.
HURWITZ, S. and GRIMMINGER, P. (1960) *J. Agric. Sci. Camb.* **54**, 373.
HURWITZ, S. and GRIMMINGER, P. (1961) *J. Nutr.* **73**, 177.
HURWITZ, S. and GRIMMINGER, P. (1962) *J. Sci. Fd. Agric.* **13**, 185.
HUSBANDS, D. H. R. and BROWN, W. O. (1965) *Comp. Biochem. Physiol.* **14**, 445.
HUTCHINSON, J. C. D. (1953) *Poult. Sci.* **32**, 692.
HUTT, F. B. (1951) *Genetics, Princeton* **36**, 213.
JENSEN, L. S., CARVER, J. S. and MCGINNIS, J. (1955) *Poult. Sci.* **34**, 1203.
JENSEN, L. S., HEUSER, G. F., SCOTT, M. L. and NORRIS, L.C. (1953) *Poult. Sci.* **32**, 907.
JOHNSON, B. C., FIRTH, J. and MISTRY, S. P. (1955) *Arch. Biochem. Biophys.* **54**, 467.
JUKES, T. H. (1940a) *J. Biol. Chem.* **134**, 789.
JUKES, T. H. (1940b) J. *Nutr.* **20**, 445.
JUNG, L. and PIERRE, M. (1933) *C. R. Seanc. Soc. Biol.* **113**, 115.
KAMINSHI, M. and DURIEUX, J. (1956) *Exp. Cell Res.* **10**, 590.
KAUPP, B. F. and IVEY, J. E. (1923) *J. Agric. Res.* **23**, 721.
KEENAN, D., NIKOLAICZUK, N. and OLIVER, W. F. (1953) *Poult. Sci.* **32**, 908.
KEETON, R. W., KOCH, F. E. and LUCKHARDT, A. B. (1920) *Am. J. Physiol.* **51**, 454.
KENNEDY, G. C. (1952–3) *Proc. Roy. Soc.* (B) **140**, 578.
KENT, S. E., REID, T. F. and WORDEN, A. N. (1960) *J. Agri. Sci. Camb.* **55**, 137.
KIENHOLZ, E. W., SUNDE, M. L. and HOEKSTRA, W. G. (1964) *Poult. Sci.* **43**, 667.
KIENHOLZ, E. W., TURK, D. E., SUNDE, M. L. and HOEKSTRA, W. G. (1961) *J. Nutr.* **75**, 211.
KING, D. (1958) *Pacif. Poultryman* pp. 15 and 64 (Feb.).
KLAIN, G. J., HILL, D. C. and BRANION, H. D. (1957) *Poult. Sci.* **36**, 405.
KLEIBER, M. and JUKES, T. H. (1942) *Proc. Soc. Exp. Biol. Med.* **49**, 34.
KLEIN, W. (1933) *Arch. f. geflügelk.* **7**, 65.
KLOSE, A. A., STOKSTAD, E. L. R. and ALMQUIST, H. J. (1938) *J. Biol. Chem.* **123**, 691.
KLUG, F. (1893) *Ung. Arch. Med.* **1**, 35.
KOCH, E. M. and KOCH, F. C. (1941) *Poult. Sci.* **20**, 33.
KOKATNUR, M. G., OKUI, S., KUMMEROW, F. A. and SCOTT, H. M. (1960) *Proc. Soc. Exp. Biol. Med.* **104**, 170.
KRAKOWER, C. A. and GOETTSCH, M. (1945) *Arch. Pathol.* **40**, 209.
KRATZER, F. H. (1952) *Proc. Soc. Exp. Biol. Med.* **80**, 483.
KREHL, W. A., ELVEHJEM, C. A. and STRONG, F. M. (1944) *J. Biol. Chem.* **156**, 13.
KRUSE, H. D., ORENT, E. R. and MCCOLLUM, E. N. (1932) *J. Biol. Chem.* **96**, 519.
LANGENDORFF, O. (1879) *Arch. Anat. Physiol.* **18**, 1.

LEA, C. H. and HANNAN, R. S. (1950) *Biochim. Biophys. Acta* **5**, 433.
LEACH, R. M. and MUENSTER, A-M. (1962) *J. Nutr.* **78** 51.
LEACH, R. M. and NORRIS, L. C. (1958) *Poult. Sci.* **37**, 1220.
LEASE, J. G., BARNETT, B. D., LEASE, E. J. and TURK, D. E. (1960) *J. Nutr.* **72**, 66.
LEASURE, E. E. and LINK, R. P. (1940) *Poult. Sci.* **19**, 131.
LEE, C. C. and WOLTERINK, L. F. (1955) *Poult. Sci.* **34**, 764.
LESNE, E., ZIZINE, P. and BRISKAS, S. (1938) *C. R. Seanc. Soc. Biol.* **128**, 957. (*Nutr. Abstr. Rev.* 1938–9, **8**, 701.)
LEPKOVSKY, S. and FELDMAN, S. E. (1961) Personal communication.
LEPKOVSKY, S., FELDMAN, S. E. and SHARON, I. (1962) *Proc. Nutr. Soc.* **21**, 65.
LEV, M. and BRIGGS, C. A. E. (1956) *J. Appl. Bact.* **19**, 224.
LEV, M., BRIGGS, C. A. E. and COATES, M. E. (1957) *Br. J. Nutr.* **11**, 364.
LEVEILLE, G. A. and FISHER, H. (1960) *J. Nutr.* **70**, 135.
LILLIE, R. J., HAYNES, S. K. and BIRD, H. R. (1951) *Poult. Sci.* **30**, 922.
LILLIE, R. J., SIZEMORE, J. R. and DENTON, C. A. (1958) *Poult. Sci.* **37**, 288.
LOHMANN, K. and SCHUSTER, P. (1937) *Biochem. Z.* **294**, 188.
LORENZ, F. W., ENTENMAN, C. and CHAIKOFF, I. L. (1938) *J. Biol. Chem.* **122**, 619.
LUCAS, H. L., NORRIS, L. C. and HEUSER, G. F. (1946) *Poult. Sci.* **25**, 373.
LUPPA, H. (1962) *Acta histochem.* **13**, 233.
LUSH, I. E. (1961) *Nature, Lond.* **189**, 981.
LYMAN, R. L. (1957) *J. Nutr.* **62**, 285.
LYMAN, R. L. and LEPKOVSKY, S. (1957) *J. Nutr.* **62**, 269.
McCHESNEY, E. W. (1943) *Proc. Soc. Exp. Biol. Med.* **52**, 147.
McCOY, R. H., FELTON, J. R. and HOFMANN, K. (1946) *Arch. Biochem.* **9**, 141.
McDONALD, M. R. and RIDDLE, O. (1945) *J. Biol. Chem.* **159**, 445.
McELROY, L. W. and JUKES, T. H. (1940) *Proc. Soc. Exp. Biol. Med.* **45**, 296.
MACHLIN, L. J. and GORDON, R. S. (1960) *Proc. Soc. Exp. Biol. Med.* **103**, 659.
McINDOE, W. M. (1957) *Biochem. J.* **67**, 19p.
McWARD, G. W. and SCOTT, H. M. (1961) *Poult. Sci.* **40**, 1177.
MADDY, K. H. and SWIFT, R. W. (1952) *J. Nutr.* **47**, 243.
MANN, T. and KEILIN, D. (1940) *Nature, Lond.* **146**, 164.
MARCH, B. and BIELY, J. (1955) *Poult. Sci.* **34**, 39.
MARCH, B. and BIELY, J. (1956) *Poult. Sci.* **35**, 550.
MARTIN, W. G. and PATRICK, H. (1962) *Poult. Sci.* **41**, 213.
MARUSICH, W. and BAUERNFEIND, J. C. (1963) *Poult. Sci.* **42**, 949.
MASUHARA, T. and MIGICOVSKY, B. B. (1963) *J. Nutr.* **80**, 332.
MATTSON, F. H. and LUTTON, E. S. (1959) *J. Biol. Chem.* **233**, 868.
MAYER, J. (1955) *Ann. N.Y. Acad. Sci.* **63**, 15.
MAYHEW, R. L. (1935) *J. Am. Vet. Med. Ass.* **86**, 148.
MECCHI, E. P., POOL, M. F. and KLOSE, A. A. (1953) *Poult. Sci.* **32**, 915.
MELASS, V. H., PEARSON, P. B. and SHERWOOD, R. M. (1946) *Proc. Soc. Exp. Biol. Med.* **62**, 174.
MELLEN, W. J. and HARDY, L. B. (1956) *Poult. Sci.* **35**, 1159.
MELLEN, W. J. and HILL, F. W. (1953) *Poult. Sci.* **32**, 994.
MEULLER, F. (1961) *Poult. Sci.* **40**, 1434.
MICHAELIS, L. and WOLLMAN, S. H. (1950) *Biochim. Biophys. Acta* **4**, 156.
MILLER, B. F. (1958) *Poult. Sci.* **37**, 1299.
MILLER, E. C., PEARSON, P. B. and DENTON, C. A. (1958) *Fed. Proc.* **17**, 485.
MILLER, B. F., SANFORD, P. E. and CLEGG, R. W. (1962) *Poult. Sci.* **41**, 989.
MISTRY, S. P., VADOPALAITE, J., CHANG, J., FIRTH, J. and JOHNSON, B. C. (1955) *J. Biol. Chem.* **212**, 713.
MITCHELL, H. H. and HAYNES, W. T. (1927) *J. Agric. Res.* **34**, 549.
MONSON, W. J., DIETRICH, L. S. and ELVEHJEM, C. A. (1950) *Proc. Soc. Exp. Biol. Med.* **75**, 256.
MOORE, J. H. and DORAN, B. M. (1961) *Biochim. Biophys. Acta* **49**, 617.
MOORE, P. R., LUCKEY, T. D., ELVEHJEM, C. A. and HART, E. B. (1946) *Proc. Soc. Exp. Biol. Med.* **61**, 185.

MOORE, S. and STEIN, W. H. (1951) *J. Biol. Chem.* **192**, 669.
MORRISON, A. B., HUNSAKER, W. G. and AITKEN, J. R. (1954) *Poult. Sci.* **33**, 491.
MORRISON, A. B. and SARRETT, H. P. (1958) *J. Nutr.* **65**, 267.
MUELLER, W. J. (1962) *Poult. Sci.* **41**, 1792.
MUSSEHL, F. E., BLISH, N. J. and ACKERSON, C. W. (1933) *Poult. Sci.* **12**, 120.
NALBANDOV, A., MEYER, R. K. and MCSHAN, W. G. (1951) *Anat. Rec.* **110**, 475.
NARAIN, R., LYMAN, C. M. and COUCH, J. R. (1957) *Poult. Sci.* **36**, 1351.
NATIONAL RESEARCH COUNCIL OF U.S.A. (1960) *Nutrient Requirements of Poultry.* National Research Council, Washington.
NEDELNUIC, V., POPESCU, F., ELORESCU, S. and TACU, A. (1959) *Lucr. stiint. Inst. Cercet. zootech.* **17**, 473; (1961) *Nutr. Abs. Rev.* **31**, 336.
NELSON, T. S. and NORRIS, L. C. (1959) *Poult. Sci.* **38**, 1094.
NELSON, T. S. and NORRIS, L. C. (1960) *J. Nutr.* **72**, 137.
NELSON, T. S. and NORRIS, L. C. (1961a) *Poult. Sci.* **40**, 392.
NELSON, T. S. and NORRIS, L. C. (1961b) *J. Nutr.* **73**, 135.
NESHEIM, M. C., CALVERT, C. C. and SCOTT, M. L. (1960) *Proc. Soc. Exp. Biol. Med.* **104**, 783.
NESHEIM, M. C. and SCOTT, M. L. (1958) *J. Nutr.* **65**, 601.
NEY, L. F. and NEWELL, G. W. (1954) *Poult. Sci.* **33**, 297.
NICKERSON, M. (1946) *Physiol. Zoöl.* **19**, 66.
NISHIDA, T., TSUCHIYAMA, H., INOUE, M. and KUMMEROW, F. A. (1960) *Proc. Soc. Exp. Biol. Med.* **105**, 308.
NOZAKI, H. and OCHIMA, M. (1960) *Bull. Nat. Inst. Agric. Sci. Japan* No. 19, p. 59.
NUGARA, D. and EDWARDS, H. M. (1961) *Poult. Sci.* **40**, 1438.
O'DELL, B. L. and SAVAGE, J. E. (1957) *Poult. Sci.* **36**, 489.
O'DELL, B. L. and SAVAGE, J. E. (1960) *Proc. Soc. Exp. Biol. Med.* **103**, 30.
OLOUFA, M. M. (1953) *Poult. Sci.* **32**, 391.
OLSEN, E. M., HARVEY, J. D., HILL, D. C. and BRANION, H. D. (1959) *Poult. Sci.* **38**, 929.
OLSEN, R. E., MILLER, O. N., TOPPER, Y. J. and STARE, F. J. (1948) *J. Biol. Chem.* **175**, 503.
OLSON, C. and MANN, F. C. (1935) *J. Am. Vet. Med. Assoc.* **87**, 151.
OSHIMA, M., TAYLOR, T. G. and WILLIAMS, A. (1964) *Biochem. J.* **92**, 42.
PASSMORE, R. and RITCHIE, F. J. (1957) *Br. J. Nutr.* **11**, 79.
PEARSON, P. B., MELASS, V. H. and SHERWOOD, R. M. (1945) *Arch. Biochem.* **7**, 353.
PEREK, M. and KENDLER, J. (1962) *Poult. Sci.* **41**, 677.
PETERS, R. A. (1936) *Br. Med. J.* **2**, 903.
PETERSEN, C. F. and LAMPMAN, C. E. (1952) *Poult. Sci.* **31**, 1067.
PLATT, C. S. (1955) *Poult. Sci.* **34**, 1045.
PLENK, H. (1932) *Der Magen:* in v. Möllendorf's *Handbuch der Mikroskopischen Anatomie des Menschen* **5**, i. Springer, Berlin.
PLIMMER, R. H. A. and ROSEDALE, J. L. (1922) *Biochem. J.* **16**, 23.
POLIN, D., OTT, W. H. and ZEISSIG, A. (1958) *Poult. Sci.* **37**, 898.
POTTER, L. M. and MATTERSON, L. D. (1960) *Poult. Sci.* **39**, 781.
POTTER, L. M., PUDELKIEWICZ, W. J., WEBSTER, L. and MATTERSON, L. D. (1962) *Poult. Sci.* **41**, 1745.
PRICE, K. E. and ZOLLI, Z. (1959) *Avian Dis.* **3**, 135.
PRICE, K. E., ZOLLI, Z. and HARE, J. H. (1959) *Poult. Sci.* **38**, 235.
RADEFF, T. (1928) *Arch. Geflügelk.* **2**, 312.
RAMSAY, W. N. M. (1951) *Biochem. J.* **49**, 494.
RAMSAY, W. N. M. (1952) *Biochem. J.* **51**, 289.
RAMSAY, W. N. M. and CAMPBELL, E. A. (1954) *Biochem. J.* **58**, 313.
RANBY, P. D. (1960) *Qd. Agric. J.* **86**, 383.
RAND, N. T., SCOTT, H. M. and KUMMEROW, F. A. (1958) *Poult. Sci.* **37**, 1075.
RANNEY, R. E. and CHAIKOFF, I. L. (1951) *Am. J. Physiol.* **165**, 600.
REID, B. L., KURNICK, A. A., SVACHA, R. L. and COUCH, J. R. (1956) *Proc. Soc. Exp. Biol. Med.* **93**, 245.
REMP, D. G. and MARSHALL, I. H. (1938) *J. Nutr.* **15**, 525.
RENNER, R. and HILL, F. W. (1961) *J. Nutr.* **74**, 254.

RICHARDSON, K. C. (1935) *Trans. Roy. Soc. (London)* **225**, (B), 149.
RIDDLE, O., BATES, R. W. and LAHR, E. L. (1935) *Am. J. Physiol.* **111**, 352.
RINGROSE, R. C. and DAVIS, H. A. (1946) *Poult. Sci.* **25**, 646.
ROBERSON, R. H. and SCHAIBLE, P. J. (1960) *Poult. Sci.* **39**, 893.
ROGLER, J. G., PARKER, H. E., ANDREWS, F. N. and CARRICK, C. W. (1961) *Poult. Sci.* **40**, 1554.
ROMANOFF, A. L. and ROMANOFF, A. J. (1949) *The Avian Egg*, Chapman & Hall, London.
ROSENBERG, H. R. (1945) *Chemistry and Physiology of the Vitamins*. Interscience, New York.
ROSS, E. (1960) *Poult. Sci.* **39**, 999.
SARGENT, K., SHERIDAN, A. and O'KELLY, J. (1961) *Nature, Lond.* **192**, 1096.
SAVAGE, J. E., TURNER, C. W., KEMPSTER, H. L. and HOGAN, A. G. (1952) *Poult. Sci.* **31**, 22.
SAVAGE, J. E., YOKE, J. M., PICKETT, E. E. and O'DELL, B. L. (1964) *Poult. Sci.* **43**, 420.
SCHAIBLE, P. J. and BANDEMER, S. L. (1942) *Poult. Sci.* **21**, 8.
SCHAUDER, W. (1923) *Anatomieder Hausvögel.* Martin, P. *Lehrbuch der Anatomie der Haustiere*, **4**, 339. Schickhardt and Ebner, Stuttgart.
SCHIAFFINO, S. S. (1959) *J. Ass. Agric. Chem.* **42**, 525.
SCHLENK, F. and SNELL, E. E. (1945) *J. Biol. Chem.* **157**, 425.
SCHØNHEYDER, F. (1936) *Biochem. J.* **30**, 890.
SCHUMAIER, G., PANDA, B., DEVOLT, H. M., LAFFER, N. C. and CREEK, R. D. (1961 *Poult. Sci.* **40**, 1132.
SCHWARTZ, H. G., TAYLOR, M. W. and FISHER, H. (1958) *J. Nutr.* **65**, 25.
SCHWARTZ, C. and TELLER, H. (1924) *Fermentforschung*, **7**, 269.
SCHWEIGERT, B. S., GERMAN, H. L. and GARBER, M. H. (1948) *J. Biol. Chem.* **174**, 383.
SCOTT, H. M. JUNGHERR, E. and MATTERSON, L. D. (1944) *Poult. Sci.* **23**, 446.
SCOTT, H. M., MOELLER, M. W. and HINNERS, S. W. (1956) *Poult. Sci.* **35**, 1169.
SCOTT, M. L. (1959) *J. Am. Diet. Ass.* **35**, 247.
SCOTT, M. L. (1962) *Proc. Cornell Nutr. Conf.* p. 108.
SCOTT, M. L., HILL, F. W., NORRIS, L. C., DOBSON, D. C. and NELSON, T. S. (1955) *J. Nutr.* **56**, 387.
SCRIMSHAW, N. S., THOMAS, W. P., MCKIBBEN, J. W., SULLIVAN, C. R., and WELLS, K. C. (1944) *J. Nutr.* **28**, 235.
SCRIVENER, L. H. (1946) *J. Am. Vet. Med. Ass.* **108**, 27.
SELANDER, U. (1963) *Acta anat.* **55**, 299.
SELL, J. L. (1965) *Poult. Sci.* **44**, 550.
SHAW, T. P. (1912–13) *Am. J. Physiol.* **31**, 439.
SHUTZE, J. V. and JENSEN, L. S. (1963) *Poult. Sci.* **42**, 921.
SIBBALD, I. R., CZARNOCKI, J., SLINGER, S. J. and ASHTON, G. C. (1963) *Poult. Sci.* **42**, 486.
SIBBALD, I. R. and SLINGER, S. J. (1962a) *Poult. Sci.* **41**, 1282.
SIBBALD, I. R. and SLINGER, S. J. (1962b) *Poult. Sci.* **41**, 1612.
SIBBALD, I. R., SLINGER, S. J. and ASHTON, G. C. (1962) *Poult. Sci.* **41**, 46.
SNELL, E. E., PENNINGTON, D. and WILLIAMS, R. J. (1940) *J. Biol. Chem.* **133**, 559.
SNELL, E. E. and QUARLES, E. (1941) *J. Nutr.* **22**, 485.
SPIVEY-FOX, M. R. and LUDWIG, W. J. (1951) *Poult. Sci.* **40**, 1403.
SOEDARMO, D., KARE, M. R. and WASSERMAN, R. H. (1961) *Poult. Sci.* **40**, 123.
STADELMAN, W. J., LISKA, B. J., LANGLOIS, B. E., MOSTERT, G. C. and STEMP, A. R. (1965) *Poult. Sci.* **44**, 435.
STAMLER, J., BOLENE, C., DUDLEY, M. and LEVINSON, E. (1950) *Endocrinology*, **46**, 375.
STEINMETZER, K. (1924) *Pflügers Arch. ges. Physiol.* **206**, 500.
STEVENSON, E. L. (1961) *Proc. Distill. Feed Res. Counc.* **16**, 71.
STOKSTAD, E. L. R., HUHTANEN, C. N., WILLIAMS, W. L. and JUKES, T. H. (1959) *Poult. Sci.* **38**, 1251.
STRITE, G. H. and YACOWITZ, H. (1956) *Poult. Sci.* **35**, 142.
STURKIE, P. D. (1954) *Avian Physiology*. Comstock, Ithaca, New York.
STURKIE, P. D. and NEWMAN, H. J. (1951) *Poult. Sci.* **30**, 240.
STUTZ, M. W. and MATTERSON, L. D. (1962) *Poult. Sci.* **41**, 1617.

SULLIVAN, T. W. (1961) *Poult. Sci.* **40**, 334.
SULLIVAN, T. W. and GEKLE, M. H. (1962) *Poult. Sci.* **41**, 1016.
TALMADGE, D. W. (1957) *Poult. Sci.* **36**, 1162.
TAYLOR, A., THACKER, J. and PENNINGTON, D. (1941) *Science, N.Y.* **94**, 542.
TAYLOR, J. D., MILLAR, G. J., JAQUES, L. B. and SPINKS, J. W. T. (1956) *Canad. J. Biochem. Physiol.* **34**, 1143.
TEEKEL, R. A. and WATTS, A. B. (1959) *Poult. Sci.* **38**, 791.
TEMPERTON, H. and CASSIDY, J. (1964a) *Br. Poult. Sci.* **5**, 75.
TEMPERTON, H. and CASSIDY, J. (1964b) *Br. Poult. Sci.* **5**, 81.
TEMPERTON, H. and CASSIDY, J. (1964c) *Br. Poult. Sci.* **5**, 87.
THOMPSON, S. Y., COATES, M. E. and KON, S. K. (1950) *Biochem. J.* **46**, xxx.
THORNTON, P. A. and DEEB, S. A. (1961) *Poult. Sci.* **40**, 1063.
THORNTON, P. A. and MORENG, R. E. (1959) *Poult. Sci.* **38**, 594.
TUCKEY, R., MARCH, B. E. and BIELY, J. (1958) *Poult. Sci.* **37**, 786.
TUPPER, R., WATTS, R. W. E. and WORMALL, A. (1954) *Biochem, J.* **57**, 247.
TURPEINEN, O. (1938) *J. Nutr.* **15**, 351.
TYLER, C. (1946) *J. Agric. Sci. Camb.* **36**, 275.
TYLER, C. (1948) *J. Agric. Sci. Camb.* **38**, 37.
TYLER, C. (1949–50) *Nutr. Abstr. Rev.* **19**, 263.
TYLER, C. (1958) *J. Agric. Sci.* **51**, 237.
TYLER, C. and WILLCOX, J. S. (1942) *J. Agric. Sci. Camb.* **32**, 62.
UWAEGBUTE, H. O. and LEWIS, D. (1963) *Proc. Int. Congr. Nutr. VI. Edinburgh*, p. 499.
VAN ITALLIE, T. B., BEAUDOIN, R. and MAYER, J. (1953) *J. Clin. Nutr.* **1**, 208.
VAN REEN, R. and PEARSON, P. B. (1953) *J. Nutr.* **51**, 191.
VALLEE, B. L. (1959) *Physiol. Rev.* **39**, 443.
VERMEULEN, H. A. (1929) *Handbuch der Geflügelkrankheiten und der Geflügelzucht*. Ed. T. can Heelsbergen. Enka, Stuttgart.
WADDELL, J. (1934) *J. Biol. Chem.* **105**, 711.
WAIBEL, P. E., SUNDE, M. L. and CRAVENS, W. W. (1952) *Poult. Sci.* **31**, 621.
WARNER, D. E. and EDMOND, H. D. (1917) *J. Biol. Chem.* **31**, 281.
WARREN, D. C. and CONRAD, R. M. (1942) *Poult. Sci.* **21**, 515.
WAXLER, S. H. (1941) *Am. J. Physiol.* **134**, 19.
WELCH, A. D. (1940) *J. Biol. Chem.* **137**, 173.
WELCH, A. D. and WELCH, M. S. (1938) *Proc. Soc. Exp. Biol. Med.* **39**, 7.
WEST, J. W. (1956) *Poult. Sci.* **35**, 835
WESTENBRINK, H. G. K. and VAN LEER, I. S. (1940–1) *Arch. néerl. Physiol.* **25**, 544.
WIESE, A. C., JOHNSON, B. C., ELVEHJEM, C. A. and HART, E. B. (1938) *Science, N.Y.* **88**, 383.
WIESE, A. C., JOHNSON, B. C., ELVEHJEM, C. A., HART, E. B. and HALPIN, J. G. (1939) *J. Biol. Chem.* **127**, 411.
WILGUS, H. S. and PATTON, A. R. (1939a) *J. Nutr.* **18**, 35.
WILGUS, H. S. and PATTON, A. R. (1939b) *Science, N.Y.* **89**, 393.
WILGUS, H. S. and THORP, F. (Jr.) (1940) *Colo. Exp. Stn. Ext. Serv. Circ.* 1602 (cited by Ewing, 1963)
WILLIAMS, H. H., CURTIN, L. V., ABRAHAM, J., LOOSLI, J. K. and MAYNARD, L. A. (1954) *J. Biol. Chem.* **208**, 277.
WILLIAMS, J. (1962) *Biochem. J.* **83**, 346.
YAMAMOTO, O. (1937) *Jap. J. Gastroent* (cited by Selander, 1963).
ZEIGLER, T. R., LEACH, R. M., NORRIS, L. C. and SCOTT, M. L. (1961) *Poult. Sci.* **40**, 1584.

CHAPTER 7

The Microflora of the Alimentary Tract of the Bird and its Significance in Nutrition

D. J. JAYNE-WILLIAMS and M. E. COATES

National Institute for Research in Dairying, Shinfield, Reading, England

ALTHOUGH interest in the 'normal' gut flora of the fowl dates back to the end of the last century, and despite the impetus given to such studies by the observation of growth stimulation by dietary antibiotics and other chemotherapeutic substances, information on the subject is rather scanty. It is not the present intention to review all the past work on the gut flora of the fowl; those who wish to do so may find the publications of Shapiro and Sarles (1949), Ochi, Mitsuoka and Sega (1964) and Smith (1965) useful starting places. Table 1 briefly summarizes much of the data on the numbers and

TABLE 1

Summary of Published Data on the Numbers and Types of Microorganisms Detected in Various Sections of the Gastro-intestinal Tract of Poultry

Organisms	Organisms/g in contents of				
	Crop	Duodenum	Ileum	Caecum	Rectum
STREPTOCOCCI	$0–10^9$	$10^2–10^6$	$10^3–10^7$	$10^5–10^{10}$	$10^5–10^7$
Strep. faecalis					
Strep. liquefaciens					
Strep. zymogenes					
Strep. faecium					
LACTOBACILLI	$10^7–10^{10}$	$10^4–10^8$	$10^5–10^9$	$10^8–10^{10}$	$10^7–10^{10}$
L. lactis					
L. acidophilus					
L. salivarius					
COLIFORMS	$10^5–10^7$	$10^2–10^6$	$10^3–10^6$	$10^7–10^{10}$	$10^5–10^7$
E. coli					
A. aerogenes					
CLOSTRIDIA	not determined	$0–10^5$	$0–10^7$	$0–10^9$	$0–10^8$
Cl. welchii					
BACTEROIDES	0	0	0	$0–10^9$	$0–10^7$

Other organisms: *Cl. sporogenes*, *Cl. paraputrificum*, *Cl. tertium*, *Eubacterium* sp., aerobic sporeformers, coryneforms, bifids, anaerobic cocci, micrococci, moulds and yeasts.

types of bacteria occurring in different sites of the gastro-intestinal tract of poultry and also serves to illustrate the considerable variations in counts reported, some of the possible reasons for which will be discussed below.

TECHNICAL PROBLEMS IN THE STUDY OF THE MICROFLORA OF THE GASTRO-INTESTINAL TRACT

It is generally accepted by microbiologists that formidable and sometimes insurmountable difficulties are encountered in the study of mixed populations of bacteria such as occur in soil, sewage, water or silage. Studies on mixed populations existing in association with macroorganisms, as, for example, in the gut of animals and on plant roots, are further complicated by interactions between host and microflora. The following sections seek to draw attention to some of the problems of technique and the ways in which results may be misinterpreted. Unfortunately, with our present knowledge, it is not possible to recommend solutions to many of the problems illustrated.

The methods available to the microbiologist for the enumeration of bacteria are known to be subject to a variety of errors. With colony counts there are sampling, dilution and pipetting errors, counting errors, errors due to the clumping of bacteria, to the use of media and oxygen conditions during incubation not favourable to all the organisms present, and to *in vitro* antagonism between different organisms. By way of illustration of the last-mentioned point, it has recently been shown (Jayne-Williams and Powell, unpublished data) that when *Streptococcus zymogenes*, a common component of the gut flora of chicks, develops on plates used for the enumeration of *Clostridium welchii*, counts of the latter organism are very drastically reduced: studies with pure cultures have confirmed the effect. Figure 1 illustrates the effect and also shows that other common streptococcal components of the gut have no influence on the recovery of the anaerobe. The application of ordinary counting techniques using ordinary media frequently result in sub-dominant groups and fastidious organisms being overlooked. The former may be enumerated if suitable selective media and methods are available, but selective agents are not without possible injurious effect on the organisms being selected. Thus Medrek and Barnes (1958) reported that a selective medium developed by Barnes (1956) for the enumeration and differentiation of streptococci from the pig proved in practice to be unsuitable for many strains of *Strep. bovis*. Similarly Barnes and Goldberg (1962) in discussing a selective medium they devised for the isolation of obligate Gram-negative anaerobes ('bacteroides') from poultry stated that the medium "... probably isolated only a few of the types present in the caeca of chickens, as several standard Gram-negative anaerobic strains tested failed to grow . . .". The dramatic effect of applying special techniques to the study of gut organisms is well illustrated by the gassing technique employed by Hungate (1950) for the

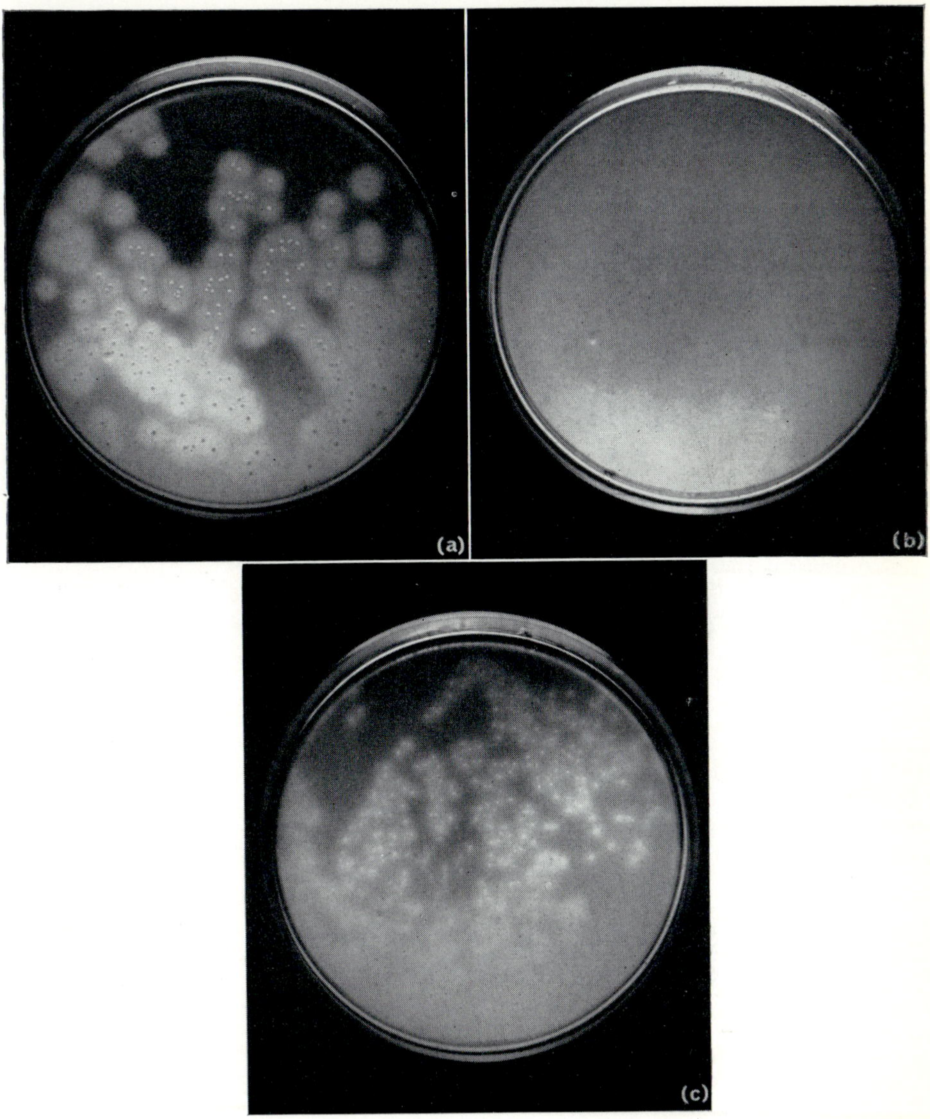
(a)
(b)
(c)

isolation, enumeration and study of the obligately anaerobic bacteria occurring in the rumen. Before the introduction of this technique—which is designed to protect the organisms from the harmful effects of contact with oxygen—the presence of these fastidious yet metabolically very active organisms in large numbers in the rumen was unsuspected; previously most attention had been paid to the organisms (such as lactobacilli, streptococci and coliforms) which both occur in high numbers and are relatively easy to cultivate.

In view of the inherent errors of counting methods, some of which have been outlined above, it follows that observed differences in counts apparently resulting from changes in the treatment of experimental animals (for example, the feeding of antibiotics) must be substantial to be significant. The broader or less defined the group of organisms being counted the less value will be the results obtained. Thus data referring to changes in counts of 'total aerobes' and 'total anaerobes' are virtually meaningless since many of the organisms occurring in the gut are able to grow both aerobically and anaerobically. However, there is a limit, imposed by time and the complexity of the techniques involved, to the extent to which changes in the numbers of individual species or strains of bacteria can be studied. For example, ideally one should study the changes in numbers of individual species or strains of streptococci, but because of the difficulties of doing this, most bacteriologists must be content with reporting the changes of 'streptococci' and overlook the possibility that a change in treatment may occasion no obvious change in the numbers as a whole but may result in one species or strain being replaced by another.

FACTORS AFFECTING THE ESTABLISHMENT OF THE GUT FLORA IN THE BIRD

Mention has already been made of the added complications that arise when mixed microbial populations existing in association with living animals are studied. In the bird, certain peculiarities in structure of the alimentary tract must be taken into account in any consideration of the establishment of a microflora and its possible nutritional significance. For instance, the crop, which can be regarded as an outgrowth of the oesophagus, forms a sac in which food may be retained or not, according to the state of fasting of the bird (Halnan, 1949). Thus it offers a site for possible microbial activity before the food undergoes the major processes of digestion. The small and large intestines are both short relative to those of mammals, and the bile duct enters distally to the duodenum. At the junction of the small and large intestines two caeca originate and in them, according to Barnes and Shrimpton (1957), the greatest number of microorganisms occur. The caeca are a site of major fermentative activity, as evidenced by gas production

(Beattie and Shrimpton, 1958) but the importance of this activity to the bird is doubtful, since caecectomy appears to have no detrimental effect (Beattie and Shrimpton, 1958; Sunde *et al.*, 1950).

In the fowl, obvious factors which may have some influence on the bacterial population are the conditions of rearing, age and diet of the birds. In chicks from healthy hens the gut is sterile at the time of hatching, but rapidly becomes contaminated with organisms as the bird begins to peck at particulate matter. It has been shown (Jayne-Williams and Powell, unpublished data) that with so-called day-old chicks, the gut flora before feeding varies enormously, both quantitatively and qualitatively, with the hygienic conditions of the incubator and brooder and with the time the chick spends in this environment before being moved to different quarters, and before being given food. The influence of hygiene in the rearing quarters has been clearly demonstrated by Coates *et al.* (1952), Bird, Lillie and Sizemore (1952), Hill, Branion and Slinger (1952), Lillie, Sizemore and Bird (1953), Morrison, Hunsaker and Aitken (1954), Elam *et al.* (1954) and Stokstad and Jukes (1958–9) with particular reference to antibiotic-relieved growth depression.

Microorganisms may be introduced into the alimentary tract by way of the diet, water, faecal particles and airborne organisms. These sources are very variable both with respect to the numbers and types of microorganisms they contribute. Once ingested, the organisms are presented with a range of environments varying from beneficial to inimicable. Conditions in the crop will generally be favourable for the proliferation of facultative anaerobes; at times sufficient oxygen may be present to inhibit the strict anaerobes or at other times there may not be enough to allow strict aerobes to grow. The extent of multiplication in the crop will be governed by the nature of the diet, its consistency and moisture content and the time it remains in this site. The crop flora of chicks is said to consist predominantly of lactobacilli (Eyssen, De Prins and De Somer, 1962; Bolton, 1965) which dissimulate sugars to form short chain acids, mainly lactic acid. The sugars may occur in the diet or be the result of hydrolysis of starch by enzymes secreted by glands in the mouth (Shaw, 1912–13, quoted by Bolton, 1965). Both sugars and acids are capable of being absorbed in the crop (Soedarmo, Kare and Wasserman, 1961; Bolton and Dewar, 1962, 1965) and acting as energy sources. It is possible that other products of bacterial metabolism (for example, vitamins) may also be produced there. Thus, as pointed out by Bolton (1965), the crop "... functions in a manner which bears certain points of resemblance to the rumen ...".

Bacterial proliferation in the crop may in some instances increase the chances of the organisms reaching the lower parts of the tract; conversely the reverse may hold, as it would probably do if the germination of resistant spores into sensitive vegetative forms were to be encouraged. The extent to which the adverse effect of the low pHs encountered in the proventriculus and gizzard—with cereal diets the pH can be as low as 2—reduce the bacterial load of the food mass will be governed by the nature and particle size of the diets (for instance, inside large particles bacteria may be afforded protection from the acidity), the types and numbers of bacteria involved and the time at which the food mass remains at the low pH. Surviving bacteria reaching the duodenum find the pH more favourable (pH 6), but then encounter secretions of bile, enzymes, etc., which may be deleterious: there may also be local sites of high osmotic concentration. Notwithstanding the possibility that environmental conditions in the proximal regions may be suitable for proliferation of bacteria, the absence of stasis in the healthy gut results in there being insufficient time for much multiplication to occur (Dixon, 1960; Meynell and Subbaiah, 1963, working with mammals). As the movement of intestinal contents slows down in the distal gut, surviving organisms will have the opportunity of proliferation. If digestion and absorption have been efficient the contents by this time will have a much reduced nutritive value but will be augmented by desquamated mucosal cells, secretions, etc. The nature of the pabulum, coupled with the anaerobic conditions which prevail and the bacterial interactions, will determine which of the surviving microorganisms will develop. Some organisms can undoubtedly maintain a 'foothold' in the lower gut without the need for continuous re-inoculation by ingested bacteria; from work with rats, it appears that others must be replenished (Gustafsson and Fitzgerald, 1960; Gustafsson, Fitzgerald and McDaniel, 1964). As the food mass slowly moves down the tract, and with some of it spending quite a time in the caeca, modification in the relative proportions of the different microbial types is likely to occur; this would account for the presence of large numbers of dead microorganisms in voided faeces. Thus the results of examining the flora of the lower gut will be very dependent on the site from which samples are taken. The probable influence of variations in intestinal motility obviously follows. Thus even if bacteria having potential harmful or beneficial effects on the host are detected in significant numbers in a given site, they may be passing through the site so rapidly that they do not have the opportunity of exerting their potential effects.

EFFECTS OF THE MICROFLORA ON CHARACTERISTICS OF THE GASTRO-INTESTINAL TRACT OF THE BIRD

Morphological Characteristics

In the chicken, a marked influence of the microflora of the gastro-intestinal tract on the utilization and metabolism of nutrients can be inferred from the results of many different investigations in poultry nutrition. No precise role has yet been assigned to any particular components of the flora, but it is clear from comparisons between germ-free and conventional chickens that in the presence of a microflora morphological characters of the gut differ from those seen in its absence. Since some of the differences affect the absorptive membrane, the possible consequences to digestion and uptake of nutrients must be considered.

On gross examination a thicker intestinal wall is apparent in conventional chicks. As the data in Table 2 show, the weight of the small intestine in

TABLE 2
Mean Weight and Length of Small Intestine in Groups of Germ-free and Conventional Chicks at 4 Weeks of Age

Type of diet	No. of chicks	Type of chick	Body wt. (g)	Intestinal wt. (g/g body wt.)	Intestinal length (cm/g body wt.)
Natural	17	Germ-free	265	0·033	0·32
	17	Conventional	255	0·043	0·35
Purified	27	Germ-free	327	0·027	0·24
	27	Conventional	350	0·032	0·25

(From Coates and Jayne-Williams 1966),

conventional chicks is greater than that in corresponding germ-free birds and, since the length is much the same in both groups, the greater weight must reflect an increased thickness. In the absence of microbial challenge the defence mechanisms in the germ-free birds remain dormant, and part of the increased weight of the gut in conventional birds can be accounted for by an increase in lymphoid tissue. The ileo-caecal lymph nodes, for instance, have been reported by Gordon, Wagner and Wostmann (1957–8) to be about half the weight of those in conventional chicks. In germ-free birds marked reduction in numbers of scattered reticuloendothelial cells in the ileal mucosa and submucosa was observed by Gordon and Bruckner-Kardoss (1958–9), but in birds monocontaminated at hatching with *Strep. faecalis* or *Cl. welchii* the number of free reticuloendothelial cells approached that of conventional birds. In further studies on the distribution of tissue components in the wall of the ileum, Gordon and Bruckner-Kardoss (1961) found little difference in the percentage of muscular or mucosal tissue between germ-free, conventional and monocontaminated chickens. The presence of a conventional flora, however, significantly augmented the amount of lamina propria, particularly the portion extending into the centres of the villi. Once again, monocontamination with *Strep. faecalis* or *Cl. welchii* brought about a trend towards the conventional picture. Other effects of the gut flora, so far only investigated in mammals, are an increased blood supply to the alimentary tract and more rapid renewal of the mucosal cells (Abrams, Bauer and Sprinz, 1963). Thus the intestinal wall of the conventional animal or bird has been regarded by Stokstad (1954) and others as being in a mild inflammatory condition, which could possibly influence its absorptive capacity. Enlargement of the lamina propria in the 'core' of the villus, for instance, must result in greater villus size and hence an overall increase in total mucosal surface; conversely the smaller, germ-free villus presents a greater proportion of mucosal to muscular and connective tissue. In the rat and mouse Heneghan (1963) has shown that xylose is more efficiently taken up from the germ-free intestine. There is no corresponding evidence with birds, but the finding of higher stores of vitamin A in the livers of germ-free chicks (Coates and Jayne-Williams, 1966) might indicate better absorption of this vitamin in the absence of a flora; there is also evidence of increased Ca absorption in the germ-free chick (Edwards and Boyd, 1963). The superior growth generally obtained in germ-free chicks (Forbes and Park, 1959; Coates *et al.*, 1963a) may in itself be a reflection of optimal absorption or utilization of nutrients from a gut unimpaired by local reaction to a burden of microorganisms.

Physiological Characteristics

As yet, little is known of the extent to which the presence of a microflora influences the physiological characteristics of the chicken gut. Some studies have been made of 5-hydroxytryptamine (5HT), which is generally accepted to be concerned in intestinal motility and vascular permeability. Phillips *et al.* (1961) found little difference in content of 5HT in the duodenum of germ-free and conventional chicks, but the amount in the ileal wall was significantly lower in the conventional birds. The results of Beaver and Wostmann (1962) were not entirely in agreement; they found only a small depressive effect of the gut flora on 5HT in the lower small intestine of chickens, although highly significant depressions occurred in similar tissue from rats and mice. The higher concentration of 5HT in the intestinal tract of germ-free chicks was accompanied by higher levels of monoamine oxidases (Lachapelle, 1963); neither was affected by monocontamination with *Cl. welchii.* The author suggests that utilization by intestinal bacteria of dietary tryptophan, from which 5HT is derived, might account for the lower concentration of 5HT in the tract of conventional chicks; the lower concentration of monoamine oxidase is explained in terms of an inhibition of the enzyme by bacterial amines absorbed from the intestinal tract.

Of the digestive enzymes, pancreatic proteases, amylase and lipase have been investigated by Lepkovsky *et al.* (1964), who found the enzyme activity of intestinal contents similar in germ-free and conventional chicks. In conventional caecal contents proteases and amylase were somewhat less but lipase concentration was higher, possibly due to some contribution of bacterial lipase. Germ-free caecal contents contained practically no protein nitrogen, a finding that might be expected from the increased level of proteases or, more particularly, from the absence of bacterial protein. It appears, therefore, that the presence of a microflora does not noticeably affect the pancreatic enzymes in the small intestine; there is no indication that the changes in composition and enzyme concentration of the caecal contents are nutritionally important to the bird.

RELATIONSHIP BETWEEN THE GUT MICROFLORA AND CHICK GROWTH

Growth-promoting Effect of Antibiotics

Despite the failure of some workers consistently to observe the phenomenon (e.g. Scott and Glista, 1950; Patrick, 1951), it is now generally recognized that dietary supplements of antibiotics and other antibacterial agents can improve the growth rates of chicks reared under practical conditions. The economic significance of this effect is such that a vast array of papers has been published on the influence of different materials on the growth rates and breeding performance of a variety of domestic birds kept under different systems of management. Too many papers have been published to permit a full survey at the present time; interested readers are referred to the reviews by Branion, Anderson and Hill (1952), Braude, Kon and Porter (1953), Jukes (1955), Taylor (1957) and François (1962).

Although the possibility of a systemic effect of antibiotics cannot be entirely ruled out, it is now generally agreed that the major part of their action on growth depends on their antimicrobial properties. This belief is founded on the failure of antibiotics to improve growth of chicks or poults in a germ-free environment (Forbes and Park, 1959; Forbes, Supplee and Combs, 1958; Coates, *et al.*, 1963a; Eyssen and De Somer, 1965). It is reasonable to suppose that the site of action of dietary antibiotics is in the gastro-intestinal tract and it has been suggested that the effect may be due to a reduction in the numbers of bacteria in the gut able to compete with the host for essential nutrients (see, for example, March and Biely, 1952), or that less fastidious organisms capable of elaborating growth factors which become available to the host are encouraged (see Romoser, Short and Combs, 1952; Hauser *et al.*, 1956). Excluding the possible contribution of the crop, the site for the greatest development of bacteria is the lower gut, a site which is posterior to the main sites for absorption of nutrients by the host; thus it may be supposed that, although the bacteria in the lower bowel may have the potential for denying the host of essential nutrients or of supplying growth factors, they do not have the opportunity of doing so.

The most widely accepted hypothesis to explain the effect of antibiotics is that growth-depressing organisms are eliminated or their metabolism altered. These organisms may be harmful by virtue of their ability to elaborate toxic materials or to invade the host tissue. Although different organisms have been implicated the position is at present far from clear-cut. For instance, Eyssen and De Somer (1963) have studied the action of dietary antibiotics in chicks given purified diets. With sucrose as the main source of carbohydrate they reported a 20–30 per cent growth depression in the absence of a dietary antibiotic. The depression was accompanied by a thickened intestinal wall, enteritis, and a greatly increased faecal excretion of lipid and carbohydrate. This 'malabsorption syndrome' was most severe at about the eighth day of life. When starch was substituted for sucrose neither the growth-depression nor the malabsorption occurred to any great extent. In subsequent studies no growth-depression or malabsorption syndrome was observed in germ-free chicks given the sucrose diet, nor could they be provoked by contamination with many of the common intestinal bacteria. However, in chicks monocontaminated with *Strep. faecalis* an effect was observed which could be exacerbated by simultaneous feeding of a bacteria-free filtrate of droppings from conventional chicks (Eyssen and De Somer, 1967). The authors favour the suggestion of an enterovirus, that could cause transitory damage to the mucosal cells, being present in the filtrate. Similar lines of investigation have been followed by Huhtanen and Pensack (1964) who confirmed the growth depression and malabsorption syndrome in conventional chicks given a sucrose diet and later isolated a strain of *Strep. faecalis* that caused a significant growth depression in germ-free birds (Pensack, 1964).

On commercial diets anaerobic sporeformers (the genus *Clostridium*) have been implicated in the growth depression counteracted by antibiotics. In early investigations of this problem Elam *et al.* (1954) fed heat-treated faeces (heated to destroy all but the spores of bacteria) from growth-depressed birds to chicks housed in an environment previously shown not to give rise to growth depression: in these circumstances the growth of the chicks was depressed and the relative counts of anaerobic sporeformers in the gut increased. Because of the disappearance of *Cl. welchii* from the gut of chicks fed penicillin, and as the organism could be demonstrated in the gut of depressed birds on the first day after feeding but only on the second day in non-depressed birds, Lev, Briggs and Coates (1957) concluded that this anaerobic sporeformer—a lecithinase-secreting organism pathogenic to man and animals in certain circumstances—was the cause of growth depression. Forbes, Park and Lev (1959) subsequently showed that germ-free chicks monocontaminated with a pure culture of *Cl. welchii* exhibited growth depression when compared with penicillin-fed or germ-free birds; however, it has not been possible to repeat these observations in this laboratory (unpublished data). In addition, our occasional failure to demonstrate *Cl. welchii* in the gut of depressed birds and the demonstration of this organism in the gut of non-depressed birds (unpublished), suggests that the mechanisms of this growth depression are in need of further elucidation.

By feeding uric acid to chicks Bare, Wiseman and Abbott (1964) have shown that a growth depression can be provoked that is alleviated by dietary antibiotic. They further showed that the effect of the antibiotic was to reduce the numbers of certain bacteria and encourage the growth of organisms (coliforms) able to dissimilate uric acid, in this instance the growth-depressing agent. However, in practice, birds not fed extra uric acid have very low (and probably non-toxic) levels of uric acid in the intestine so that it is likely that this growth-depressing mechanism does not operate without the intervention of the experimenter. Comparison of the growth rates of germ-free birds with those of birds monocontaminated with cultures of the uric acid-splitting coliforms would provide useful data.

Some workers have tacitly assumed the growth-depression of chicks relieved by antibiotics to be due in all circumstances to a single organism or a collection of organisms acting in association. The indications from the work quoted above, that the cause may be streptococci (with or without virus) on one occasion and anaerobic sporeformers on another, would thus be unacceptable. It is also noteworthy that Eyssen and De Somer (1965) failed to provoke growth depression in germ-free chicks given purified diets by monocontamination with *Cl. welchii.* However, it is probable that different diets will give rise to different gut floras, any component(s) of which may cause depression. Thus relative to the work quoted above it is likely that a diet of high sucrose content would tend to elect saccharolytic organisms (e.g. streptococci) or provide the organisms with a substrate from which they could more readily elaborate a toxic material; similarly a diet rich in protein, especially protein not readily digested by the host, would tend to elect or provide a substrate for quite different organisms (e.g. proteolytic anaerobes). The nature of the growth-depression may therefore be a function of the conditions of management and diet of the birds, and it is unlikely that a single explanation will be found to cover all circumstances in which dietary antibiotics are of benefit.

In considering possible mechanisms by which antibiotics improve the growth of young birds the similarities between the characteristics of antibiotic-treated and germ-free chicks cannot be ignored. As previously stated (see p. 248) germ-free birds grow significantly better than their conventional controls. The magnitude of growth improvement is greater than can be attained by feeding antibiotics in a conventional environment (Coates, *et al.*, 1963a). In birds given dietary antibiotics morphological characteristics of the alimentary tract tend towards those of germ-free birds (Gordon, Wagner and Wostmann, 1957–8). The intestinal wall is thinner, the ileo-caecal lymph nodes are smaller, scattered reticuloendothelial elements less in number and the lamina propria reduced, although in no instance does the effect of the antibiotic treatment equal that of the germ-free state. This seeming parallel between the condition of the intestinal absorptive membrane and chick growth is compatible with the 'malabsorption syndrome' described by Eyssen and De Somer (1963) and could reasonably account for much of the growth-promoting effect of dietary antibiotics. It may, however, be purely fortuitous and in the light of present knowledge no firm conclusions can be drawn as to the possible connection between gut structure and rate of growth in the chick.

Unidentified Growth Factors

For many years the existence has been claimed of hitherto unidentified growth factors for chicks in substances such as fish solubles, fermentation residues, dried whey and alfalfa. Failure of many attempts to isolate the active constituents has been partly accounted for by a lack of consistency in chick tests of the crude materials and concentrates prepared from them. In the course of three years' work with fish solubles, Barnett and Bird (1956) reported a variation in chick growth response from −10 per cent to +46 per cent. They noted that the larger responses occurred when the unsupplemented chicks grew poorly, whereas small responses were obtained when the control birds were growing well. A similar trend was observed by Harrison and Coates (1964). In the experiments of Barnett and Bird (1956) the loss of response to fish solubles coincided with failure of antibiotics to promote growth. The response was restored by feeding chick droppings, which brought about a depression in growth that was reversed by a dietary supplement of fish solubles. These findings led the authors to postulate that at least part of the growth-promoting action of fish solubles was brought about by a modification of the gut microflora. The suggestion was supported by Harrison and Coates (1964), who found a similarity in the pattern of response of chicks to fish solubles and penicillin, and in later (unpublished) experiments obtained no improvement in the growth of germ-free chicks given dietary supplements of fish solubles. In work with malt distillers solubles as a source of unidentified growth factors, Wakelam and Jaffé (1958) also noticed a similarity between the magnitude and occasion of responses to the solubles and to penicillin. In their experiments and those of Harrison and Coates (1964) the effects of the growth factors and of the antibiotic were supplementary, indicating that if the growth-promoting action were mediated through the gut microflora the responsible organism(s) must have been different from those implicated in the antibiotic response. This inference is not supported by Barnett and Bird (1956) who were able to reduce or eliminate the growth response to fish solubles by supplements of chlortetracycline.

In spite of some discrepancies in the results reported above, they lend support to the hypothesis that the intestinal microflora may be concerned in the response of chicks to the so-called unidentified growth factors. The idea is compatible with the reports of variations in response between laboratories, or at different times in the same laboratory, which could readily be explained in terms of changes in the microbial population of the gut or the environment.

Growth-depressing Effect of Raw Soyabean Meal

The unsatisfactory nature of raw soyabean and some other leguminous meals as a protein source for chicks and poults is well recognized. Such meals are known to contain toxic factors, including trypsin inhibitors, that are destroyed by heat treatment. Birds given raw soyabean meal in the diet grow very poorly and develop hypertrophy of the pancreas, but after heat treatment the meal supports normal growth and pancreas development. For many years attempts have been made to explain the detrimental effect on growth in terms of inhibition of trypsin and consequent impairment of protein digestion. Recently, reports that dietary antibiotics partially overcome the growth depression suggest that the gut microflora may be associated with the harmful effects of the meal. By supplementing diets containing raw soyabean meal with high levels (0·1 per cent) of a variety of antibiotics Braham, Bird and Baumann (1959) improved the growth of chicks by 31–51 per cent over that of birds given the unsupplemented diets; addition of antibiotics to similar diets containing the heated meal resulted in only very small improvements in growth. Similar results were obtained in turkeys (Linerode, Waibel and Pomeroy, 1961). Confirmation that microorganisms must play an important role in the growth-depressing effect of raw soyabean meal was obtained by Miller and Coates (1966) who showed that in a germ-free environment growth of chicks on diets containing raw soyabean meal was only slightly less than on similar diets with heated meal; in a conventional environment chicks given the raw meal grew significantly less well than their germ-free counterparts. Enlargement of the pancreas occurred to the same extent in germ-free and conventional birds. Thus the hypertrophic effect of raw soyabean on the pancreas, and part of its growth-depressing effect, seem to be independent of microbial action, but the growth depression is greatly exacerbated in the presence of a microflora. The inhibition of trypsin by some component of the raw meal must inevitably reduce protein digestion. The presence of undigested protein in the lower gut (cf. p. 245) might provide conditions suitable for certain organisms, such as proteolytic anaerobes, possibly implicated in the growth depression. Impaired fat absorption in young chicks given raw soyabean meal has been reported (Nesheim, Garlich and Hopkins, 1962) and it seems possible that a 'malabsorption syndrome', analogous to that described on sucrose diets by Eyssen and De Somer (1963), might be associated with some component(s) of the gut flora being elected or supplied with a substrate from which toxic materials can be elaborated.

METABOLIC ROLE OF MICROORGANISMS IN THE ALIMENTARY TRACT OF THE BIRD

In view of the number and variety of microorganisms in the gastro-intestinal tract it is not unreasonable to postulate their participation in almost any of the processes of metabolism of nutrients. The role of intestinal micro-organisms in a number of metabolic pathways has been studied in the rat, in which the flora has been shown to play an important part in the absorption and metabolism of lipids (Evrard *et al.*, 1964), the degradation of bile acids (Gustafsson and Norman, 1962) and cholesterol (Danielsson and Gustafsson, 1959), the reduction of bilirubin (Gustafsson and Lanke, 1960), hydrolysis of urea (Levenson *et al.*, 1959), formate oxidation (Baggiolini *et al.*, 1964), synthesis of vitamin K (Gustafsson *et al.*, 1962) and some of the vitamin B complex (Daft *et al.*, 1963) and degradation of choline (Prentiss *et al.*, 1961). It is feasible that intestinal bacteria are involved in the same processes in the chicken, but in view of the difference in gut structure it would be unwise to draw too close a parallel between mammals and birds. Metabolic processes that have been directly studied in birds are discussed in the following sections.

Vitamin Synthesis in the Gastro-intestinal Tract

The vitamin-synthesizing ability of microorganisms in the alimentary tract is well demonstrated in the ruminant, for which a dietary source of the vitamins B is made unnecessary by the activities of the ruminal bacteria. In single-stomached animals and birds there is evidence that vitamins are synthesized in the gastro-intestinal tract, but in most instances the full requirements of the host are not satisfied without coprophagy. The chick may need a dietary supplement of vitamins for one of the following reasons. First, insufficient amounts are synthesized by the organisms in the gut; second, the synthesized vitamins are bound, possibly to bacterial protein, and unavailable to the host without prior digestion; or third, the synthesis takes place at a site distal to the absorptive area of the small intestine. Microbial fermentation in the gut is most likely to occur in places where there is relative stasis, and where environmental conditions are most favourable. The crop is therefore a possible site of microbial activity, although there is no direct evidence of vitamin synthesis there. The caeca offer very favourable conditions for microbial activity and Couch *et al.* (1950) have shown that the levels of biotin, pantothenic acid, riboflavine, nicotinic acid and folic acid in caecal contents of hens and turkeys given a practical diet were much higher than those in the adjacent intestinal contents. However, as the caecal contents of the chicken pass through a very short colon before they are voided, there is little opportunity for absorption of any microbial products they may contain, except by direct uptake from the caeca.

The synthesis of vitamin B_{12} and other cobalamines by the gut microflora has been demonstrated by comparison of caecal contents and droppings from germ-free and conventional chicks (Coates, *et al.*, 1963b). The birds were housed on wire floors so that coprophagy was reduced to a minimum, and given a purified diet containing crystalline cyanocobalamin as the only source of vitamin B_{12}. No vitamin B_{12}-like factors were detected in caecal contents or droppings from germ-free birds, although high concentrations were present in corresponding samples from conventional birds; this finding is of little practical significance since most of the naturally-occurring analogues of vitamin B_{12} are without activity for the chick (Coates, *et al.*, 1956). Of more importance is the observation that the levels of true vitamin B_{12} in samples from conventional chicks were 100 times as high as in those from the germ-free birds. In later (unpublished) work with diets devoid of vitamin B_{12}, appreciable amounts of the vitamin were found in the caecal contents of conventional but not of germ-free chicks. However, the conventional birds did not grow better than their germ-free counterparts, nor was their mortality less or liver content of vitamin B_{12} higher. It appears, therefore, that the chick derives little benefit from the microbially-synthesized vitamin B_{12}. Similar unpublished investigations with riboflavine and pantothenic acid have led to the same conclusion regarding these vitamins. In contrast to these findings, indications that bacterially-synthesized folic acid is available to the chick were obtained by Miller and Luckey (1963), who compared germ-free chicks given a diet low in folic acid with similar birds monocontaminated at 7 days of age with a strain of *Escherichia coli* isolated from conventional chicken brooders. At the end of a further 2 weeks, the body weights, haemoglobin and haematocrit values and tissue contents of folic acid were significantly greater in the monocontaminated birds than in the germ-free controls. It appears, therefore, that folic acid was synthesized by the *E. coli* contaminant and, since coprophagy was prevented and vitamin-production by bacteria growing in the drinking water was shown to be negligible, direct absorption of the vitamin from the alimentary tract must be assumed.

The demonstration that conventional chicks are unable to utilize caecal vitamin B_{12} confirms and extends earlier work of Jackson *et al.* (1955), who failed to detect the uptake of isotopically-labelled vitamin B_{12} injected directly into occluded caeca in hens. However, when ^{35}S-methionine or sodium ^{35}S-sulphate were placed in an occluded caecum ^{35}S was recovered almost quantitatively in the blood, indicating that at least some molecules may pass through the caecal barrier. It is possible that folic acid may also be absorbed by this route.

There is little doubt that birds on deep litter or free range can derive a significant contribution to their vitamin requirement by ingestion of their own excreta (Kennard, Bethke and Chamberlin, 1948; Kennard and Chamberlin, 1948; Jacobs *et al.*, 1954). It is well known to investigators in the field of vitamin B_{12} that a deficiency of the vitamin cannot readily be demonstrated in birds having access to their own droppings. Faecal synthesis of some other vitamins has long been recognized. Almquist and Stokstad (1936) detected antihaemorrhagic activity in droppings from chicks given a diet devoid of vitamin K, and showed that the activity increased when the droppings were left to stand for 24 hr. Similarly the amount of riboflavine in hen droppings was shown by Lamoreux and Schumacher (1940) to increase during a 24-hr incubation period, and an organism was isolated that, on incubation with sterile droppings, increased their riboflavine content (Schumacher and Heuser, 1941). However, the vitamin-synthesizing activity in the excreta cannot be taken as a direct reflection of a similar procedure in the gut contents. After voiding, faeces will be subject to activity of aerobic organisms, whereas strictly anaerobic conditions prevail in the alimentary tract. Furthermore, the droppings are likely to be contaminated by organisms not normally found in the gut. The change from anaerobic to aerobic conditions and this subsequent contamination is likely to result in the production of a whole range of microbial metabolites that would not be formed in the gut: for these to exert an effect they would first need to be ingested.

Volatile Fatty Acid Synthesis

Analysis of the gut contents of laying hens by Annison, Hill and Kenworthy (1968) showed much higher concentrations of volatile fatty acids (VFA) in the caeca than in other parts of the digestive tract. Acetic and propionic acids were predominant, although small quantities of other acids were present. All were detected in portal blood, indicating that they had been absorbed from the caeca or possibly from other parts of the alimentary tract. In peripheral blood only acetic and propionic acids were found; their concentration was much the same in normal and caecectomized birds, suggesting that the VFA produced in the caeca were not the source of those circulating in the blood. This suggestion was confirmed by studies with germ-free chickens, the caeca of which contained only negligible amounts of VFA whereas the concentration in peripheral blood was similar to that found in conventional controls. Although these results provide an instance where direct uptake of nutrients, namely, VFA, from the caecum, appears to have taken place, nevertheless they indicate that the VFA of peripheral blood were not of caecal origin but were endogenously derived.

Cholesterol Metabolism

Studies *in vivo* in rats and man, and *in vitro* with intestinal microorganisms, have led to the conclusion that degradation of cholesterol and bile acids in the gut is brought about by microbial action (see, for example, Gustafsson *et al.*, 1957; Eneroth, Hellström and Ryhage, 1964). There are indications that the same may be true in the chicken, although the nature of the major dietary components appears to be important. In experiments with chicks, diets containing a high proportion of sucrose and 3 per cent added cholesterol had a hypercholesterolaemic effect compared with similar diets containing glucose; the effect was insignificant in the absence of the added cholesterol (Grant and Fahrenbach, 1959; Kritchevsky *et al.*, 1958). Since addition of antibiotics to the diets practically eliminated the difference by increasing the blood level of cholesterol in birds given the glucose diet, some influence of the gut microflora was suspected. Experiments with germ-free birds confirmed the suspicion, for the lower blood cholesterol levels observed in conventional birds given the glucose diet were increased in germ-free conditions to the same order as those supported by the sucrose diet (Kritchevsky *et al.*, 1959). These findings indicate a possible influence of dietary carbohydrate on organism(s) connected with cholesterol metabolism, but since the diet contained added cholesterol it is uncertain whether absorption of exogenous or metabolism of endogenous cholesterol, or both, were concerned. In comparing the effects of different carbohydrates in diets not supplemented with added cholesterol, Coates, Harrison and Moore (1965) found a small but consistent hypercholesterolaemia in germ-free birds given a purified diet containing starch. It was, however, compensated by a corresponding decrease in liver levels of cholesterol, so that the presence or absence of a microflora appeared to affect the partition between the cholesterol circulating in the blood and that stored in the liver. Birds given sucrose instead of starch had higher plasma cholesterol values in a germ-free environment but, in contrast to the findings in starch-fed birds, liver levels were also increased. Thus there appears to be some interrelationship between dietary carbohydrate, the gut microflora and metabolism of endogenous cholesterol in the chick, although the overall effect on the bird is small. In passing it may be pointed out that a marked depression in weight was noted in the conventional chicks given sucrose, in confirmation of the work of Eyssen and De Somer (1963) discussed on p. 249.

Coates, Harrison and Moore (1965) also compared the effects of high dietary levels of corn oil or butterfat on cholesterol levels in germ-free and conventional chicks. Very little difference was observed between any of the treatments, but when 1 per cent cholesterol was added to the diets a highly significant hypercholesterolaemia occurred in birds given butterfat. However, since it was seen in both germ-free and conventional environments it was apparently independent of the gut flora.

The influence of dietary protein on cholesterol metabolism in chicks was briefly studied by Kritchevsky *et al.* (1959), who observed higher serum cholesterol levels in chicks given casein than in birds that received soya protein in a purified diet. In a germ-free environment the levels of both groups rose to about the same extent.

Thus there are some indications that the presence of a microflora may lower the cholesterol content of blood and tissues in the chicken, possibly by degrading cholesterol and bile acids, and hence facilitating their elimination from the body. Nevertheless it is apparent that the influence of diet may considerably override that of the gut microflora.

Our colleagues Drs. J. E. Ford, R. Fuller, L. W. Powell and Mr. G. F. Harrison have collaborated in much of the unpublished work mentioned in this chapter. We are grateful to them for permission to quote the results here.

REFERENCES

ABRAMS, G. D., BAUER, H. and SPRINZ, H. (1963) *Lab. Invest.* **12**, 355.
ALMQUIST, H. J. and STOKSTAD, E. L. R. (1936) *J. Nutr.* **12**, 329.
ANNISON, E. S., HILL, K. J. and KENWORTHY, R. (1968) *Br. J. Nutr.* **22**, 207.
BAGGIOLINI, M., AEBI, H., SACQUET, E. and CHARLIER, H. (1964) *Helv. Physiol. Pharmac. Acta* **22**, 53.
BARE, L. N., WISEMAN, R. F. and ABBOTT, O. J. (1964) *J. Bact.* **87**, 329.
BARNES, E. M. (1956) *J. Appl. Bact.* **19**, 193.
BARNES, E. M. and GOLDBERG, H. S. (1962) *J. Appl. Bact.* **25**, 94.
BARNES, E. M. and SHRIMPTON, D. H. (1957) *J. Appl. Bact.* **20**, 273.
BARNETT, B. D. and BIRD, H. R. (1956) *Poult. Sci.* **35**, 705.
BEATTIE, J. and SHRIMPTON, D. H. (1958) *Q. J. Exp. Physiol.* **43**, 399.
BEAVER, M. H. and WOSTMANN, B. S. (1962) *Br. J. Pharmac. Chemother.* **19**, 385.
BIRD, H. R., LILLIE, R. J. and SIZEMORE, J. R. (1952) *Poult. Sci.* **31**, 907.
BOLTON, W. (1965) *Br. Poult. Sci.* **6**, 97.
BOLTON, W. and DEWAR, W. A. (1962) *Proc. World's Poult. Congr.* XII. Sydney, p. 117.
BOLTON, W. and DEWAR, W. A. (1965) *Br. Poult. Sci.* **6**, 103.
BRAHAM, J. E., BIRD, H. R. and BAUMANN, C. A. (1959) *J. Nutr.* **67**, 149.
BRANION, H. D., ANDERSON, G. W. and HILL, D. C. (1952) *Poult. Sci.* **32**, 335.
BRAUDE, R., KON, S. K. and PORTER, J. W. G. (1953) *Nutr. Abstr. Rev.* **23**, 473.
COATES, M. E., DAVIES, M. K., DAWSON, R., HARRISON, G. F., HOLDSWORTH, E. S., KON, S. K. and PORTER, J. W. G. (1956) *Biochem. J.* **64**, 682.
COATES, M. E., DICKINSON, C. D., HARRISON, G. F., KON, S. K., PORTER, J. W. G., CUMMINS, S. H. and CUTHBERTSON, W. F. J. (1952) *J. Sci. Fd. Agric.* **1**, 43.
COATES, M. E., FULLER, R., HARRISON, G. F., LEV, M. and SUFFOLK, S. F. (1963a) *Br. J. Nutr.* **17**, 141.
COATES, M. E., GREGORY, M. E., PORTER, J. W. G. and WILLIAMS, A. P. (1963b) *Proc. Nutr. Soc.* **22**, xxvii.
COATES, M. E., HARRISON, G. F. and MOORE, J. H. (1965) *Ernährungsforschung* **10**, 251.
COATES, M. E. and JAYNE-WILLIAMS, D. J. (1966) in *Physiology of the Domestic Fowl*, Eds. HORTON-SMITH, C. and AMOROSO, E. C., Edinburgh and London: Oliver & Boyd Ltd.
COUCH, J. R., GERMAN, H. L., KNIGHT, D. R., SPARKS, P. and PEARSON, P. B. (1950) *Poult. Sci.* **29**, 52.
DANIELSSON, H. and GUSTAFSSON, B. (1959) *Arch. Biochem. Biophys.* **83**, 482.
DAFT, F. S., MCDANIEL, E. G., HERMAN, L. G., ROMINE, M. K. and HEGNER, J. R. (1963) *Fed. Proc.* **22**, 129.
DIXON, J. M. S. (1960) *J. Path. Bact.* **79**, 131.
EDWARDS, H. M. and BOYD, F. M. (1963) *Poult. Sci.* **42**, 1030.
ELAM, J. F., JACOBS, R. L., FOWLER, J. and COUCH, J. R. (1954) *Proc. Soc. Exp. Biol. Med.* **85**, 645.
ENEROTH, P., HELLSTROM, K. and RYHAGE, R. (1964) *J. Lipid Res.* **5**, 245.

Evrard, E., Hoet, P. P., Eyssen, H., Charlier, H. and Sacquet, E. (1964) *Br. J. Exp. Path.* **45**, 409.
Eyssen, H. and De Somer, P. (1963) *J. Exp. Med.* **117**, 127.
Eyssen, H. and De Somer, P. (1965) *Ernährungsforschung* **10**, 264.
Eyssen, H. and De Somer, P. (1967) *Poult. Sci.* **46**, 323.
Eyssen, H., De Prins, V. and De Somer, P. (1962) *Poult. Sci.* **41**, 227.
Forbes, M. and Park, J. T. (1959) *J. Nutr.* **67**, 69.
Forbes, M., Park, J. T. and Lev, M. (1959) *Ann. N.Y. Acad. Sci.* **78**, 321.
Forbes, M., Supplee, W. C. and Combs, G. F. (1958) *Proc. Soc. Exp. Biol. Med.* **99**, 110.
Francois, A. C. (1962) *World Rev. Nutr. Diet.* **3**, 25.
Gordon, H. A. and Bruckner-Kardoss, E. (1958–9) *Antibiotics A.* 1012.
Gordon, H. A. and Bruckner-Kardoss, E. (1961) *Acta Anat.* **44**, 210.
Gordon, H. A., Wagner, M. and Wostmann, B. S. (1957–8) *Antibiotics A.* 248.
Grant, W. C. and Fahrenbach, M. J. (1959) *Proc. Soc. Exp. Biol. Med.* **100**, 250.
Gustafsson, B. E., Bergstrom, S., Lindstedt, S. and Norman, A. (1957) *Proc. Soc. Exp. Biol. Med.* **94**, 467.
Gustafsson, B. E., Daft, F. S., McDaniel, E. G., Smith, J. C. and Fitzgerald, R. (1962) *J. Nutr.* **78**, 461.
Gustafsson, B. E. and Fitzgerald, R. J. (1960) *Proc. Soc. Exp. Biol. Med.* **104**, 319.
Gustafsson, B. E., Fitzgerald, R. J. and McDaniel, E. G. (1964) *J. Nutr.* **84**, 155.
Gustafsson, B. E. and Lanke, L. S. (1960) *J. Exp. Med.* **112**, 975.
Gustafsson, B. E. and Norman, A. (1962) *Proc. Soc. Exp. Biol. Med.* **110**, 387.
Halnan, E. T. (1949) *Br. J. Nutr.* **3**, 245.
Harrison, G. F. and Coates, M. E. (1964) *Br. J. Nutr.* **18**, 461.
Hauser, M. M., Anderson, G. W., Pepper, W. F. and Slinger, S. J. (1956) *Poult. Sci.* **35**, 27.
Heneghan, J. B. (1963) *Am. J. Physiol.* **205**, 417.
Hill, D. C., Branion, H. D. and Slinger, S. J. (1952) *Poult. Sci.* **31**, 920.
Huhtanen, C. N. and Pensack, J. M. (1964) *Poult. Sci.* **43**, 1331.
Hungate, R. E. (1950) *Bact. Rev.* **14**, 1.
Jackson, J. T., Mangan, G. F., Machlin, L. J. and Denton, C. A. (1955) *Proc. Soc. Exp. Biol. Med.* **89**, 225.
Jacobs, R. L., Elam, J. F., Fowler, J. and Couch, J. R. (1954) *J. Nutr.* **54**, 417.
Jukes, T. H. (1955) *Proc. 1st Int. Conf. Antib. in Agric.* 162. Publication 397. Nat. Acad. Sci.–Nat. Res. Council, Washington, D.C.
Kennard, D. C., Bethke, R. M. and Chamberlin, V. D. (1948) *Poult. Sci.* **27**, 477.
Kennard, D. C. and Chamberlin, V. D. (1948) *Poult. Sci.* **27**, 240.
Kritchevsky, D., Grant, W. C., Fahrenbach, M. J., Riccardi, B. A. and McCandless, R. F. J. (1958) *Arch. Biochem. Biophys.* **75**, 142.
Kritchevsky, D., Kolman, R. R., Guttmacher, R. M. and Forbes, M. (1959) *Arch. Biochem. Biophys.* **85**, 444.
Lachapelle, R. C. (1963) *Bact. Proc.* 68.
Lamoreux, W. F. and Schumacher, A. E. (1940) *Poult. Sci.* **19**, 418.
Lepkovsky, S., Wagner, M., Furuta, F., Ozone, K. and Koike, T. (1964) *Poult. Sci.* **43**, 722.
Lev, M., Briggs, C. A. E. and Coates, M. E. (1957) *Br. J. Nutr.* **11**, 364.
Levenson, S. M., Crowley, L. V., Horowitz, R. E., and Malm, O. J. (1959) *J. Biol. Chem.* **234**, 2061.
Lillie, R. J., Sizemore, J. R. and Bird, H. R. (1953) *Poult. Sci.* **32**, 466.
Linerode, P. A., Waibel, P. E. and Pomeroy, B. S. (1961) *J. Nutr.* **75**, 427.
March, B. and Biely, J. (1952) *Poult. Sci.* **31**, 177.
Medrek, T. F. and Barnes, E. M. (1958) *J. Appl. Bact.* **21**, 79.
Meynell, G. G. and Subbaiah, T. V. (1963) *Brit. J. Exp. Path.* **44**, 197.
Miller, H. T. and Luckey, T. D. (1963) *J. Nutr.* **80**, 236.
Miller, W. S. and Coates, M. E. (1966) *Proc. Nutr. Soc.* **25**, iv.
Morrison, A. B., Hunsaker, W. G. and Aitken, J. R. (1954) *Poult. Sci.* **33**, 491.

NESHEIM, M. C., GARLICH, J. D., and HOPKINS, D. T. (1962) *J. Nutr.* **78**, 89.
OCHI, Y., MITSUOKA, T. and SEGA, T. (1964) *Zbl. Bakt.* (1 Orig.) **193**, 80.
PATRICK, H. (1951) *Poult. Sci.* **30**, 549.
PENSACK, J. M. (1964) Communication to IVth European Congress of the Feedstuffs Industry, Berlin.
PHILLIPS, A. W., NEWCOMB, H. R., SMITH, J. E. and LACHAPELLE, R. (1961) *Nature, Lond.* **192**, 380.
PRENTISS, P. G., ROSEN, H., BROWN, N., HOROWITZ, R. E., MALM, O. J. and LEVENSON, S. M. (1961) *Arch. Biochem. Biophys.* **94**, 424.
ROMOSER, G. L., SHORT, M. S. and COMBS, G. F. (1952) *Poult. Sci.* **31**, 932.
SCHUMACHER, A. E. and HEUSER, G. F. (1941) *Poult. Sci.* **20**, 272.
SCOTT, H. M. and GLISTA, W. A. (1950) *Poult. Sci.* **29**, 921.
SHAPIRO, S. K. and SARLES, W. B. (1949) *J. Bact.* **58**, 531.
SHAW, T. P. (1912–13) *Am. J. Physiol.* **31**, 439.
SMITH, H. W. (1965) *J. Path. Bact.* **89**, 95.
SOEDARMO, D., KARE, M. R. and WASSERMAN, R. H. (1961) *Poult. Sci.* **40**, 123.
STOKSTAD, E. L. R. (1954) *Physiol. Rev.* **34**, 25.
STOKSTAD, E. L. R. and JUKES, T. H. (1958–9) *Antibiotics A.* 998.
SUNDE, M. L., CRAVENS, W. W., ELVEHJEM, C. A. and HALPIN, J. G. (1950) *Poult. Sci.* **29**, 10.
TAYLOR, J. H. (1957) *Vet. Rec.* **69**, 278.
WAKELAM, J. A. and JAFFE, W. P. (1958) *Br. J. Nutr.* **12**, 147.

CHAPTER 8

Protein Quality for Non-Ruminants

ANTHONY A. WOODHAM

Rowett Research Institute, Bucksburn, Aberdeen

THE primary function of dietary protein is to supply amino acids in a form suitable for the building and maintenance of body protein and the ability of the protein of a particular feeding stuff to fulfil this function is a measure of its quality. There has been a tendency in the past to confuse the terms 'nutritive value' and 'protein quality'. The former is much the less precise expression embracing as it does digestibility, energy, vitamins, growth factors and bacterial contamination as well as the influence of other components in the diet. Nevertheless as these factors may influence the release of amino acids from the protein or their absorption by the animal, or may mask the true protein quality, some consideration must be given to them in protein quality studies.

For well over half a century it has been appreciated that protein feeding stuffs differ in quality and early investigations of the problem involved comparisons of such materials in practical feeding trials. Unfortunately owing to ignorance of mineral, vitamin and energy requirements as well as to the lack of a recognized standard testing technique the early results are difficult to interpret. In spite of this it was soon recognized that while many amino acids could be synthesized by animals from other nitrogen-containing constituents of the feed, some were either incapable of being produced in this way or were formed at too slow a rate to meet the animals requirements adequately. For optimum performance it was necessary then that appropriate protein feeds containing the essential amino acids be included in the diet (Willcock and Hopkins, 1906; Osborne and Mendel, 1912). From consideration of the digestive systems of ruminants and non-ruminants it will be apparent that the developed microbiological activities in the former are likely to make them much less exacting in their requirements for individual amino acids. The rumen microorganisms are capable of producing changes in the nature of the ingested protein and in addition the microbial protein is itself digested along with the food. Hence the mature ruminant can provide for itself a number of the amino acids which must be present in the diet fed to the young ruminant and non-ruminant. This discrimination makes the study of protein quality of particular importance for non-ruminant livestock.

Factors influencing protein quality. The difficulties involved in assessing the quality of a protein feed are due to the variety of factors which affect the ultimate value of the protein to the animal. Ideally the feeding stuff should be able to provide sufficient of each of the essential amino acids when incorporated at a practical level in the diet to meet the requirements of the animal. The content of amino acids in the whole diet should be balanced to ensure that a large excess of any one is avoided, as over-provision of a particular amino acid can be as harmful as a shortage. In consequence the quality of any particular protein must be considered in relation to the total diet in which it is to be fed. The necessary amino acids must be in a form that is available to the animal and as availability may be affected by the treatment received by the feeding stuff prior to incorporation in the diet, processing is one of the principal factors governing protein quality. Since even an ideal mixture of amino acids is of no value unless it is capable of being incorporated into the animal body in adequate amounts the palatability of the diet and its digestibility are of importance. Some natural proteins inhibit the action of digestive enzymes and the content of such inhibitors as well as of haemagglutinins and toxic products of bacterial and fungal contamination is important. Finally it must be remembered that the amino acid requirement for optimum growth or production is dependent upon the species, age and physiological state of the animal. It is unlikely then that any simple measurement will permit the complete assessment of the quality of a given feeding stuff for all classes of non-ruminants. The various influencing factors will now be considered in more detail.

The balance of amino acids. A protein source which provided in available form all of the essential amino acids, approximately ten in number, needed by the non-ruminant animal would be ideally balanced only if they were present in the exact proportions required. Such a perfect protein would be difficult to devise as it would pre-suppose a knowledge of the amino acid composition of the various animal tissues and the rate of turn-over of each amino acid. The latter changes with change in metabolic activity and age, and the protein source would need constant revision to keep pace. The use of pure amino acids to maintain the exact balance is not an ideal solution. Apart from economic considerations which in most cases rule out this expedient in any case, synthetic diets may produce deleterious results which as yet are not fully understood, and furthermore the absorption of free amino acids may differ from that of the naturally provided acid. A reasonable compromise may be achieved by the judicious use of mixtures of natural protein foods, taking into account any information available on the requirements of the particular animal concerned and the amino acid composition of the feeding stuffs. In practice, with diets in general use in a particular area it will be known which of the essential amino acids are most likely to be in short supply and appropriate measures can be taken to remedy the deficiency. In the U.S.A. much of the success of the broiler chicken industry has been due to the recognition that soyabean and maize are ideally complementary feeding stuffs for the growing bird. Maize provides adequate methionine but insufficient lysine for optimum growth while soyabean, not a good source of methionine, is comparatively rich in lysine. All cereals which may be grown in the U.K. are deficient in both sulphur-containing amino acids and lysine but fish meal added in relatively small amount can make good the shortage. It must be emphasized that none of these mixtures constitutes the ideal protein foodstuff. Some amino acids are present in unnecessarily large amounts but so long as the excess does not have a deleterious effect on the performance of the animals it may be economically preferable to a more correctly balanced but more exotic protein mixture. Providing that the amino acid requirements are satisfied, a considerable amount of an unbalanced protein can be tolerated (Harper, 1959). Commercially available feeding stuffs are limited in number so that the position is necessarily simplified, and it would appear that providing the levels of methionine, cystine, lysine and in certain cases arginine and tryptophan, are controlled, the levels of the remaining essential amino acids will be satisfactory in mixed diets. Changing economic conditions will certainly make it necessary to keep the situation under review and it seems highly likely that as the problem of amino acid imbalance becomes better understood, methods of analysis improve and unusual sources of cheap protein become more readily available, diets for non-ruminants will tend more and more towards ideal balance.

The availability of amino acids. Total amino acid analysis is now becoming a routine operation though it suffers still from some drawbacks which will be more fully considered in the section on laboratory tests of protein quality. Chief among the criticisms levelled at it is the fact that while nutritive value can be shown to be directly related to part of the amino acid content it is not directly related to the total levels of certain important amino acids. Biologically or chemically estimated 'available' lysine may be only a fraction of the total level of lysine present in the foodstuff. Owing to harsh treatment received during processing, for example, this important amino acid may be bound within the protein in such a way that although it is released by acid hydrolysis and can therefore be chemically determined, it is not capable of being utilized by the animal. Lysine is a dibasic amino acid having a free amino group at the end of a straight chain of carbon atoms. This reactive amino group is vulnerable and has been shown to react readily with carbohydrates on comparatively mild heat treatment—the 'browning' reaction. The lysine–carbohydrate complex is not broken down during digestion, so the valuable lysine is lost (Eldred and Rodney, 1946; Carpenter and March, 1961). The chemical determination of available lysine value (ALV) which will be described in the section on laboratory tests, thus gives a truer picture of the protein quality of a feeding stuff than does the total lysine content. Estimations of total methionine and cystine are also misleading as it has been shown that severe heat treatment reduces their availability (Donoso *et al.* 1962). There is evidence that other essential amino acids may be partially unavailable, too, but it is generally difficult to visualize the reason. There is no doubt that a fuller understanding of the causes and the precise nature of the unavailability of individual amino acids will contribute markedly towards the optimum utilization of feeding stuffs.

Palatability and digestibility. Even an ideally balanced protein is of no value to the animal if its constituents are not absorbed by the tissues. The intake of food must not be restricted by flavour or physical form. Too much starchy material can cause gumming of the beaks of young birds and there is evidence that rancid fat in the diet can cause chicks to eat less. Diets containing fish meal are more acceptable than otherwise identical diets in which the fish meal is replaced by groundnut. Such differences in food intake are very noticeable in protein evaluation experiments. The physical form of the diet can be very important. An unacceptable diet in powdered form will often be eaten if it is pelleted and this is particularly true in the case of low energy diets for poultry. Chicks fed a pelleted ration may consume 15 per cent more feed and gain 25 per cent more weight with a 10 per cent higher efficiency of food utilization than chicks fed the same ration as a meal (Hussar and Robblee, 1962). Pelleting also leads to economy of energy expenditure by the feeding bird. Jensen *et al.* (1962) showed that both chicks and turkey poults fed mash occupied 14 and 19 per cent respectively of their 12-hr day eating while those fed pellets occupied 5 and 2 per cent respectively. These are important considerations in these days of a highly competitive broiler industry when care taken to produce a well-balanced diet may be vitiated by feeding it in an unsuitable form. Again a protein feed may be a potential source of considerable amounts of needed amino acids and yet be so indigestible that they pass through the body without being released. Examples of such materials are cereal straws, seed husks and feathers.

THE REASONS FOR DIFFERENCES IN PROTEIN QUALITY

Because of their energy content cereals form the bulk of all rations for pigs and poultry in this country although they may provide considerably less than half of the total protein. The protein content may differ widely from sample to sample, maize and rice being in general lower than the other cereals but all can be made to vary by choice of sowing time, fertilizer treatment and genetic selection. Maize, for example, has yielded strains containing respectively as high as 20 per cent and as low as 8 per cent of protein. Mere increase in quantity of protein does not, however, necessarily indicate an improvement in protein quality and this dictum is especially true of cereals, which contain amongst their proteins the nutritionally inferior zein (maize), gliadin (wheat and rye) and hordein (barley). These are similar in containing low levels of lysine and sulphur amino acids and any increase in cereal protein level, by whatever method it is achieved, is at the expense of overall quality because in practice the addition involves an increase in the proportion of these deficient proteins. For example, in a high protein maize, as distinct from an average sample, the zein content is increased from about one-quarter of the total protein to over one-third.

With oilseed meals and legumes quality may well be influenced by variety, growth conditions and stage of maturity (Daniellson, 1952), but data are scarce. In the case of the oilseeds this is at least partly due to the fact that for so long these by-products of the oil expressing industry were considered to be of minor importance. Probably the chief reason for variable quality in plant protein concentrates is directly concerned with the processing of the seeds.

The quality of feeding stuffs manufactured from animal by-products is more likely to depend on differences in raw materials than is that associated with concentrates of plant origin. Varying proportions of offal in fish meals and of bone, skin and tendon in meat meals result in products which might be expected to differ considerably in protein composition. Poor quality in meat meals has been associated with a high collagen content indicated by hydroxyproline measurements.

Duration and conditions of storage can have a deleterious effect on protein quality though it would seem that commercial turnover is so rapid that trouble should not normally arise from this source. Dried milk powder is, however, particularly susceptible to storage deterioration and due allowance should be made for this.

The presence of toxic or growth depressing constituents within the foodstuff may affect the protein quality. Some, such as enzyme inhibitors, may be controlled to some extent by appropriate processing. Alternatively genetic selection of seed may be practised in an attempt to reduce the influence of undesirable characteristics. The production of glandless, and therefore gossypol-free, cottonseed is a case in point.

THE EFFECTS OF PROCESSING ON PROTEIN QUALITY

Heat Treatment of Soyabean Meals

The subjection of oilseed meals to heat during processing is unavoidable. If the oil is removed from the seed by simple crushing and pressing considerable heat is generated within the press cake, while solvent extraction of the crushed seeds necessarily involves heating even if only to remove solvent from the meal at the end of the process. In fact for the sake of efficiency the solvent is normally used at the boiling point, though this can be comparatively low. For example, hexane is very commonly used at a temperature of around 50°C. Not only, however, is the application of heat unavoidable, but it is also generally desirable. Osborne and Mendel commented on the superiority of heated soyabean meal as early as 1917 and the fact has been accepted and remarked upon in numerous reports since 1930 on soyabean meal quality for chicks, poults, rats and pigs. Nevertheless as recently as 1961 it was found necessary to compare meal prepared with just sufficient heat to remove residual solvent, with meal which had been more vigorously heated (Bornstein, Ben-Adam and Lipstein, 1961). The mild treatment was at that time commonly used in Israel and the work reported confirmed the superiority of meal which had been moistened and then toasted for 30 min. The same workers showed that addition of skimmed milk powder or fish meal did not compensate for the deleterious effect of the underheated soyabean meal. The early work of Hayward, Steenbock and Bohstedt (1936) and Wilgus, Norris and Heuser (1936) demonstrated convincingly the necessity for heating, the fact that dry heating was ineffective, and that excessive heating lowered nutritive value markedly. The beneficial effect gained by heat treatment is a resultant of the destruction of several heat labile growth-impairing factors on the one hand and of damage to the protein on the other. The growth depressing effect of raw soyabean meal has been attributed to an inhibition of intestinal proteolysis, to haemagglutinin activity, and also to an inadequate balance of amino acids in the diet (Fisher and Johnston, 1958).

Recent studies on the reason for the superiority of heated soyabean meal over the raw meal include investigations of the effect of heating on the availability of essential amino acids, the addition of enzymes to the unheated meal and the effect of unheated soyabean on amino acid requirements. Borchers (1958) and Hayward (1959) found that when fed at the 40 per cent level to rats, unheated and heated soyabean meal gave similar growth and this prompted the hypothesis that at the more usual 25 per cent level of feeding certain of the amino acids were not available in adequate amounts or alternatively the requirement for them was increased in the raw meal. Fisher and Johnson (1958) reported that a mixture of fourteen amino acids was effective in overcoming the growth-depressing properties of the raw meal in chick rations. Booth and co-workers (1960) have reported that four amino acids, methionine, threonine, valine and tyrosine, counteract poor growth in rats and suggested that the low nutritive value of the unheated soyabean meal may be due to excretion of amino acids in the faeces as a result of excessive endogenous secretions of protein into the gut. Observations of pancreatic hypertrophy and increased pancreatic enzymes in the intestines of chicks and rats following injection of unheated soyabean meal support this (Lepkovsky, Bingham and Pencharz, 1959).

However, Saxena, Jensen and McGinnis (1962a) fed amino acids varying from four to fourteen in number along with raw soyabean meal and failed to overcome completely the growth depression, pancreatic hypertrophy and low feed utilization in chicks. They concluded that growth depression was not a function of reduced food consumption. The same workers (1962b) later found that isolated autoclaved soyabean protein added to raw meal did not improve growth rate either, while several fractions of raw meal did not depress chick growth or cause pancreatic hypertrophy. They also found (1963) that chicks fed unheated soyabean meal excreted more undigested protein in the faeces than birds fed heated meal. Saxena and co-workers (1963) found that raw meal supplemented with all known nutrients was as effective as heated meal when fed to laying birds, as regards egg size and yield, feed efficiency and body weight, but there was a tendency to large blood spots within the yolk. Bornstein and Lipstein (1963) however, have reported a significant decrease in egg production and live weight gain when underheated soyabean meal was fed to laying pullets, and attributed this to a lowered feed intake caused either by reduced palatability or to a poorer amino acid balance.

Borchers (1961) has demonstrated that supplementation with methionine, threonine and valine will restore the poorer growth rate exhibited by rats fed 25 per cent raw soyabean meal to the level exhibited by rats fed properly heated meal. Ascarelli and Gestetner (1962) concluded that the availability of methionine was improved by heating.

Evans, Bandemer and Bauer (1962) found that more cystine and methionine were liberated from raw soyabean protein by enzymic digestion than from heated protein, except when the raw protein contained trypsin inhibitors. Partial acid hydrolysis released less cystine from autoclaved than from raw protein, but similar amounts of methionine were released.

The same authors had concluded (1961) from *in vitro* digestions with trypsin and erepsin that the autoclaving of soyabean protein leads to the formation of lysine-aspartic acid and lysine-glutamic acid linkages which are resistant to mild hydrolysis. Kwong, Barnes and Fiala (1962) determined methionine and nitrogen in the faeces of the rat fed raw, or heated soyabean meal or isolated protein, and found that both were poorly absorbed when the raw flakes were fed although the percentage absorption did not decrease when the level of unheated flakes was increased. The unheated soyabean flakes did not cause a decrease in the absorption of added methionine, and the authors concluded that faecal losses of endogenous nitrogen and the limiting amino acid methionine do not account for the low nutritive value of unheated soyabean products. More recently Kwong and Barnes (1963) have shown that unheated soyabean blocks the utilization of cystine by rats, but does not impair the availability of methionine, and the blockage appears to be caused by trypsin inhibitors.

Birk and Gertler (1961) were able to prepare a material having no haemagglutinating properties and with 50 per cent of original trypsin inhibiting activity by extraction of raw soyabean meal with hydrochloric acid at pH 4·2, but the extracted material was only slightly better than raw meal for chick growth. Using metabolizable energy as an indicator of the extent of digestion or absorption of the meal Brambila, Nesheim and Hill (1961) showed that trypsin supplementation had no effect on the growth depression or pancreatic hypertrophy caused by raw soyabean meal in chicks. Nesheim, Brambila and Hopkins (1961) demonstrated that in the presence of heated soyabean oil meal, soyabean oil was over 94 per cent absorbed whilst in the presence of raw soyabean meal the absorption of oil was only 49 per cent at 2 weeks of age, rising to 73 per cent at 3 weeks of age. They also showed that trypsin inhibitor concentrates depressed fat absorption even at levels which produced only slight growth depression.

Barnes, Kwong and Fiala (1961) concluded that in the absence of coprophagy some factor other than the trypsin inhibitor is the major contributor to growth inhibition caused by unheated soyabean meal in rats. They found a direct relationship between the 'Inhibitory Index'—the per cent decrease in maximal growth rate of rats fed high levels of unheated soyabean—and trypsin inhibitor activity, but not urease activity. There was no correlation when coprophagy was prevented. Borchers (1963) found that after incubation with papain, or in water alone, raw soyabean meal exhibited increased growth-inhibiting properties and concluded that this inhibitor exists in bound form and can be released by enzymes or microorganisms associated with the raw meal. He points out however that *in vitro* conditions may lead to the liberation or formation of factors not released or formed *in vivo*.

Heat Treatment of Cottonseed Meals

When it was first noted that heating caused an improvement in the quality of cottonseed meal, it was suggested that heat led to the inactivation of the toxic constituent gossypol by liberating it from the pigment glands and then causing it to react with the seed protein, the resulting 'bound' gossypol being non-toxic. However, it was not possible to obtain a satisfactory correlation between free gossypol and nutritive value. Further work suggested that the effect of heat is not only to deactivate the gossypol but also to damage the protein. After the discovery that nutritive value was related to bound gossypol content (Lyman, Chang and Couch, 1953) a method was devised for its removal without heat or other drastic treatment (Baliga and Lyman, 1957). The same workers formed a cottonseed protein–gossypol complex without heat and showed that the nutritive value of the protein for rats and chicks was reduced to less than half its original value. They also observed that the only type of free amino group in cottonseed globulin is the lysine ξ-amino group and that on reaction with gossypol this disappears (Lyman and Baliga, 1958). Conkerton *et al.* (1957) showed that on autoclaving cottonseed meal under conditions which would damage the protein and reduce its nutritive value for poultry the content of lysine was greatly reduced. Martinez and Frampton (1958) showed that cottonseed containing 4·1–4·2 g lysine per 16 g N yielded meal containing only 3·4–3·9 g/16 g N after processing, and concluded that the quantity of lysine destroyed was dependent upon the severity of the processing and not upon the type of processing used. In a study of seventeen cottonseed meals evidence was found that solvent extraction gave meals of superior quality to expeller meals (Duckworth, Woodham and McDonald, 1961) though Fisher *et al.* (1962) found them to be equivalent. They considered that poor protein digestibility was a factor affecting the nutritive value of cottonseed meal. Others found no nutritional differences between commercial meals although noting that normal processing variations caused changes detectable by liberation of amino groups in tryptic digests or by available lysine estimation (Ascarelli and Gestetner, 1962).

In the absence of added gossypol the solubility of the nitrogen in the meal but not the rate of live weight gain was reduced when cottonseed and soyabean meal which had been autoclaved for 2 hr at 17 lb pressure were fed to young pigs in diets based on maize (Clawson *et al.*, 1961). The addition of free gossypol at the 0·030 and 0·036 per cent levels respectively, depressed growth and caused mortality.

Increasing the autoclaving time of meals containing 0·026 per cent added free gossypol to 3 hr reduced growth significantly and this was restored to that obtained on an unautoclaved meal by the addition of 0·2 per cent L-lysine. Increasing the free gossypol level to 0·36 per cent reduced growth rate and caused the death of four out of six pigs. Free gossypol fed with soyabean meal resulted in superior performance and lower mortality than gossypol fed with cottonseed or peanut meal at all levels of free gossypol, and efficiency of feed utilization was not affected by the presence of gossypol in any of the mixtures (Kornegay *et al.*, 1961).

Martinez, Frampton and Cabell (1961) investigated meal made from glandless (and therefore gossypol-free) cottonseed. Raffinose and/or gossypol were added to this material and after heating in an autoclave it was noted that the amount of gossypol bound to the meal increased, but the increase was less in the presence of raffinose. Raffinose alone reduced the available lysine content and nutritive value on heating, and 10 per cent of the sugar was equivalent to 1 per cent of gossypol in reducing the level of available lysine in the protein.

Heat Treatment of Other Oilseed Meals

The quality of other oilseeds may be affected by heating in a similar manner to soyabean meal and cottonseed meal but the amount of work carried out on these recently is relatively small. The presence of enzymic inhibitors in groundnut meal is well known and the nutritional importance of available lysine has been studied recently. Bensabat *et al.* (1958) determined the available lysine value (ALV) in groundnut meal at various stages of processing and different cooking temperatures. They found that ALV was higher after cooking 1 hr at 111–112°C than after 2 hr at 120–121°C, and also that it decreased progressively at each stage of the process. For example, samples from the cooker after 2 hr had a value of 2·9 g/16 g N; from the conditioner 2·7, cake from the screw press 2·3, and finally meal from the hammer mill 1·9. ALV has been used to distinguish between groundnut meals exposed to different severities of processing (Mauron and Bujard, 1960).

Davidson and Boyne (1962) showed that a groundnut meal of high protein quality for chicks as judged by Gross Protein Value was not improved in quality for laying hens by the addition of DL-methionine while one of low quality for chicks was significantly improved. The poor groundnut could be made equivalent to fishmeal as a protein supplement by the addition of both lysine and methionine.

The quality of processed rapeseed meal has been studied by Clandinin and Tajcnar (1961) who found that the lower the cooking and conditioning temperatures, the higher was the content of fat and lysine. They concluded that if expeller processing is used only sufficient heat should be employed during the cooking and conditioning stages to permit reduction of the oil content to about 6 per cent if lowered protein quality is to be avoided. Fisher and co-workers (1962) concluded that supplementation of rapeseed with lysine could provide a useful protein source for chicks and the same authors showed that safflower meal was improved in terms of its growth promoting ability by supplementation with methionine and lysine. Butterworth and Fox (1963) have demonstrated that heat treatment of coconut meal lowers NPU, ALV and the solubility of nitrogen in 6N-HCl. The destruction of ricin and inactivation of the allergen of castor meal by heating for 1 hr at 15 psi has been accomplished by Jenkins (1963), but there is some doubt as to the effect of this treatment on the nutritive value. Bandemer and Evans (1963) have studied the destruction of amino acids during the heating of a variety of seeds for 6 hr at 15 psi and found that while cereals underwent little change in lysine content, most materials lost small amounts of the basic amino acids and tyrosine (psi = lb./square inch).

Heat Treatment of Meat and Fish By-products

Variation in the nature of the raw materials is likely to be a more important factor in the case of animal by-products than in the case of oilseeds. Varying proportion of muscle and offal in fish meals and differing contents of bone, skin and tendon in meat meals result in products which would be expected to differ considerably in nutritive value. Like oilseeds, however, the animal proteins are heated during processing and this can have a serious effect in particular upon the availability of lysine, the amino acid of particular importance in the feeding of non-ruminants. Rand *et al.* (1960) showed that composition, excessive heat, fat removal and duration of storage all had significant effects on the quality of fish meals, and in particular meals made from scales, skins or bones were poorer in quality than those made from whole fish or muscle. Carpenter, Ellinger and Shrimpton (1954) reported that meals of poor quality could be produced by drastic processing but that short periods at high temperature were not necessarily damaging, and Grau *et al.* (1960) claimed that low temperature drying did, in general, produce meals of better quality than high temperature drying. Njaa (1961) concluded that the slight destruction of amino acids in most cases was of little importance in ordinary processing, but that the possibility of appreciable destruction of sulphur amino acids could not be excluded.

Available lysine value is now known to be a reliable guide to heat damage in fishmeals (Laksesvela, 1958; Boyne, Carpenter and Woodham, 1961; Ascarelli and Gestetner, 1962). Olley and Watson (1961) found that neither azeotropic drying nor solvent washing reduced ALV and that formaldehyde treatment of fish guts yielded material with low ALV compared with that of untreated whole herring. When lysine availability is reduced by poor processing other amino acids are also affected (Ousterhout, Grau and Lundholm, 1959). Olley and Watson (1961) found little or no destruction during the later stages of drying when the moisture was low, confirming the findings of Carpenter *et al.* (1962) that greatest damage occurs if heating is carried out in the presence of moisture. However, Miller *et al.* (1961) found no difference in the growth, food or protein efficiency, or carcase composition of rats fed diets supplemented with either steam-cooked or fried sardines. Morrison (1963) found that steaming fish flour for 30 min or supplementation with either methionine or histidine increased weight gains in weanling rats. Summers and Fisher (1962) observed that properly processed menhaden meal was equivalent in quality to egg albumin, or to soyabean and sesame meals when supplemented with the appropriate limiting amino acids. The same authors found that poor quality in four meat meals was associated with a high collagen content as indicated by hydroxyproline estimations. Eastoe and Long (1960) noted that cooked bones contained about 83 per cent collagen, meat protein 58 per cent, and commercial meat and meat/bone meals 50–65 per cent. They suggested that hydroxyproline analysis would be a useful method for determining quality in meat products. Ascarelli and Gestetner (1962), on the other hand, found no correlation between hydroxyproline and nutritive value as measured by Net Protein Utilization (NPU) in fish meals. Woodham (unpublished results) similarly found no correlation between hydroxyproline content and Gross Protein Value (GPV) for a series of meat and meat/bone meals fed as supplements to cereals for chicks. The value of pork protein heated with water for 24 hr at 110°C was lower on biological assay than losses of lysine (total or available) or sulphur amino acids would warrant, and balance studies indicated that the fall in lysine utilization was largely accounted for by reduced digestibility, while falls in nitrogen and sulphur were due chiefly to reduced retention (Donoso *et al.*, 1962). The authors suggested possible reasons for the effects noted, including the Maillard reaction, cross-linkage reactions and disintegrating reactions. On dry autoclaving of air-dried meat protein, digestibility decreased with increasing carbohydrate content, but when reconstituted with water such decreases were largely eliminated and the proteins showed greater susceptibility to proteolysis (Schroeder, Iacobellis and Smith, 1961). Heller *et al.* (1961) using rats found that the amount of available lysine in pork, beef, and lamb muscle ranged from 74–92 per cent of the total in either raw samples or after standard cooking procedures. Severe heat treatment, however, involving autoclaving for 16 hr substantially reduced the amount of available lysine in the products.

Enzymic processing of fish meal as well as high-pressure acid processing has been used to yield 'liquid fish' which gave values intermediate between herring meals and condensed herring solubles when fed to chicks as a source of supplementary protein (March *et al.*, 1961).

Mould Contamination

An aspect of oilseed processing which has come into prominence recently concerns the development of toxins in the products due to mould infestation as a result of faulty handling of the kernels after harvesting, and unsatisfactory storage conditions. When oilseeds were first used as animal feedstuffs warnings were given that mould growth could cause damage resulting in meals which were not merely unpalatable but also injurious, and groundnut meal was one of the materials particularly referred to in this connection (see review, Spensley, 1963). Since then a number of outbreaks of livestock poisoning has occurred in which groundnut meal has been suspected as the cause but these were never thoroughly investigated and the possible association of mould growth with toxicity was forgotten.

In 1960, however, the cause of an outbreak of disease leading to high mortality amongst turkeys and ducks in the United Kingdom was eventually traced to certain batches of groundnut meal imported from Brazil (Blount, 1961). The series of investigations which led ultimately to the isolation and identification of the toxin produced by the mould *Aspergillus flavus* and known now as aflatoxin, is an outstanding example of the rapid progress which can be achieved when scientists of different disciplines pursue aspects of a common topic. It was soon shown that groundnut meal from other areas than Brazil were affected (Sargeant, Allcroft and Carnaghan, 1961) and that in addition to turkeys and ducklings, piglets and calves were highly susceptible (Loosmore and Harding, 1961; Loosmore and Markson, 1961).

Chickens are little affected by the toxin, for Barnes and Woodham (1963) have reported that in a series of groundnut meals aflatoxin levels bore no relationship to quality as measured by a protein quality test based on chick growth. There was no mortality under the conditions used.

Previous to the isolation of aflatoxin, *Aspergillus flavus* had been isolated from mouldy maize which had been suspected of killing hogs (Burnside *et al.*, 1957) and these workers had succeeded in growing the mould on autoclaved maize, yielding a toxic product which killed pigs and mice. Kulik (1954) obtained a fraction from peas infected with *Aspergillus flavus* which killed cats and rabbits.

Toxic symptoms similar to those produced by *Aspergillus flavus* contamination of groundnut meal have been observed in poults fed soyabean meal which had been stored under conditions which encouraged the growth of moulds (Richardson *et al.*, 1962). It was not established whether the effect was related to aflatoxin toxicity though the predominant fungi isolated from the soyabean meal included *Penicillium* sp. and *Aspergillus glaucus*. Supplementation of the mouldy soyabean meal with lysine produced normal growth. These cases are exceptional and the principal crop to be attacked is undoubtedly groundnut.

Important to the question of controlling and preventing infection is a knowledge of the requirements of *Aspergillus flavus*. A common mould, it can be isolated from tropical soils and is characterized by its somewhat higher requirement for moisture than the majority of moulds. At around 30°C it will grow at 80–85 per cent relative humidity and above, corresponding to a moisture content in the kernels of 9 per cent or more (Austwick and Ayerst, 1963). Thus infestation can only occur prior to the drying of the kernels after harvesting when the moisture content falls below the critical level, or subsequently if the kernels are stored under damp conditions. While in the ground the growing nuts have a moisture content of 25 per cent at least and must possess a protective mechanism against attack. If left in the ground after maturing, however, infestation can occur. Harvesting during rain and unsatisfactory drying conditions of the harvested crop will predispose towards infection. Toxicity normally only affects a proportion of the kernels so that factors which affect only a proportion of the whole crop merit consideration. Such a factor is damage to the shells and there is evidence that kernels from broken shells are more toxic than those from intact shells. Thus shelling before drying is hazardous particularly if the nuts are deliberately wetted before shelling. In this connection it is interesting that Duerden and Cutler (1957), observing that the traditional mortar and pestle type of shelling used in Nigeria resulted in a build-up of free fatty acids in the nuts, drew attention to the necessity for shelling before storage due to transport problems in the area. Davis (1961) noted the appearance of seed storage fungi on groundnuts having more than 10 per cent moisture, and associated this with an unusually high tocopherol and low total nitrogen content. Changes in stored peanuts caused by fungi have also been reported by Ward and Diener (1961).

A survey of drying methods carried out recently led to the conclusion that nuts should be harvested at maturity without breaking of the shell, cured and dried in the shell by methods which allow a rapid loss of moisture to 7 per cent without shell-breakage, and the unshelled or shelled nuts, meal or cake transported by methods which do not allow the re-absorption of moisture (Blatchford and Hall, 1963; Spensley, 1963). It does not seem likely that deactivation of toxic groundnut meal will be economically possible, so the primary need is to prevent the initial development of the contamination.

Toxicity and Processing

The preferred aliphatic amine for removing gossypol from cottonseed meal is *n*-octylamine but the subsequent removal of the amine presents a problem. Toxicity tests for poultry have been carried out by Clark and DuBose (1960) who found no toxicity unless the level of the amine was such that the chicks would not voluntarily accept the ration. This is of interest in the light of the findings of Greenberg *et al.* (1959) on the extraction of soyabean meal and cottonseed meal with several amines. It has been known for many years that extraction with trichloroethylene yields a meal which may be toxic for cattle, sheep, chickens and guinea pigs. Commercial trichloroethylene contains an inhibitor to prevent the corrosive action of degradation products and this is sometimes an organic amine. These workers found that extraction with amines produced a meal which was far more toxic for the chick than meals produced by trichloroethylene extraction, the amines producing the worst meals being *n*-butylamine, diethylamine and triethylamine.

The toxicity of soyabean meal extracted with trichloroethylene is not due to residual solvent and recent developments have led to the conclusion that the toxic factor is associated with the protein of the meal. McKinney and co-workers (1957) found that S-(dichlorovinyl)-L-cysteine produced the same syndrome in calves as trichloroethylene-extracted soyabean meal. Seto and co-workers (1958) developed a bio-assay technique which allows the study of the toxic factor, and confirmed that it is an organic compound associated with the protein fraction of the meal, from which it can be liberated by pancreatic digestion. It is stable at pH 1·5 but unstable at pH 12. Rehfeld and co-workers (1958) have further pointed out that trichloroethylene extraction of meat scrap endows it with toxic properties for cattle.

Work on the quality of sesame meal (Carter, Cirino and Allen, 1961) has demonstrated that the problem of oxalate content can be overcome by decortication which completely removes the portion of the seed containing the oxalate. Separation of endosperm and spermoderm particles from solvent-extracted undecorticated meals was effective in reducing oxalate content although some oxalate-containing particles of the spermoderm epiderm did pass through the 60 mesh screen.

Studies of toxic effects resulting from solvent extraction during fish flour preparation have been made by Morrison, Sabry and Middleton (1962). Rat growth studies in which the test flours were added as supplements at 5, 10 and 20 per cent levels indicated toxic effects revealed by histological examination of the livers of the rats fed at the 20 per cent level with flour which had been extracted with chloroform and isopropyl alcohol. The quality of fish flour produced by isopropyl alcohol extraction was superior to that produced by ethylene dichloride though the latter was not toxic.

THE BIOLOGICAL EVALUATION OF PROTEIN FEEDING-STUFFS

The concept of biological value of proteins was introduced originally by Thomas in 1909 and the technique was later refined by Mitchell (1923). It was based on measurements of nitrogen gained or lost by the animal and obtained by analysis of food, faeces and urine. The method has been considered by most workers to be reliable in practice and sound in theory, but it is laborious to carry out and very time-consuming. Simpler methods have generally been preferred. Differences in nutritive value of proteins are reflected in differing growth rates and growth rate is a comparatively easy measurement. Another approach is based on the effect of dietary protein on the composition of blood and tissues. No single biological test has as yet been accepted by all, but some are more widely used than others.

All biological tests can be divided into two classes—those which evaluate the test protein in a diet in which it provides all the protein, and those in which the test protein is mixed with other protein feedstuffs. While the second method approaches more closely to conditions found in practice, the first avoids the difficult question of the effect of different basal protein mixtures.

METHODS IN WHICH THE TEST MATERIAL IS FED AS THE SOLE PROTEIN SOURCE

1. *The balance-sheet method.* Mitchell's refinement of the Biological Value concept used growing rats as the experimental animal, enabling the measurement of requirements for both maintenance and growth. Nitrogen balance,

was determined from the equation $B = I\ (U + F)$ where B is the balance, I = nitrogen intake and U and F are the nitrogen excretions in urine and faeces respectively. Alternatively the retained nitrogen may be measured by carcass analysis. Differences found between 'balance-sheet' and carcass analysis results are possibly attributable to the cumulative effect of small errors but despite considerable investigation this point remains obscure (Duncan, 1966). Wallace (1959) has discussed the difficulties involved in reconciling nitrogen retention with body composition and Costa (1960) demonstrated that over long periods of time animals have apparently retained large quantities of nitrogen without changing in body weight. Frost reviewed the information available on the method in 1959 concluding that 'the biological value method presents fully as many imponderables and problems and is no more precise than the growth or repletion methods of protein assay'.

2. *Net protein utilization method.* A comparatively simple and easy method for the determination of NPU was described by Miller and Bender in 1955. Involving the feeding of 10 per cent protein diets to groups of rats for 10-day periods and determining the gain in body nitrogen over that of rats fed a low protein diet, results corresponding to biological value multiplied by digestibility were obtained. The method was further simplified by the discovery that moisture determinations on the carcass could be substituted for the nitrogen determinations because of the constancy of the N/H_2O ratio. NPU is defined as:

$$\frac{\text{body N of test group} - \text{body N of low-protein group} + \text{N consumed by low-protein group}}{\text{N consumed by test group}}$$

The NPU technique has been applied to chicks by Summers and Fisher (1962). Some workers have experienced considerable variability in NPU determination particularly between laboratories (Ford, 1960).

3. *Protein efficiency ratio* (*PER*). This is probably the most widely used of all methods for evaluating proteins. PER is simply the grammes gain in weight per gramme protein consumed by the test animal. The originators, Osborne, Mendel and Ferry reported in 1919 determinations of PER with rats at several levels of test protein and the maximum value obtained was taken as the best estimate of the value of the protein. Barnes and Bosshardt (1946) agreed that the maximum value was the best but Block and Mitchell (1946) disagreed. In practice most workers have tested protein at a single level, usually around 10 per cent. Assay periods have varied from 1 to 12 weeks, the majority being from 2 to 4 weeks.

With any biological procedure the interpretation of results obtained in different laboratories is always difficult. Variations in practical feeding conditions in different parts of the world almost invariably necessitate modifications of the techniques in order to conform with accepted ideas in the experimenter's own area. The performance of the test animals may differ between say, the U.S. and U.K., due to differences in strains and environmental conditions. Comparisons are difficult to make between results obtained in two laboratories even when a determined attempt is made to work under similar conditions. Further, no experimenter is ever entirely satisfied that the methods used by others cannot be improved.

An attempt to prescribe an 'official' protein quality test of the PER type has been made in the U.S. (Derse, 1962). Rats are fed diets containing approximately 10 per cent of protein all of which is provided by the test material. The experimental conditions are rigorously specified and a trial of the method in a number of laboratories showed that variability from one to another was reasonably small. Recently the carbohydrate source has been studied and it has now been recommended that either sucrose or starch be used, both giving good results (Derse, 1965). Although some workers have used the standard procedure it is by no means universal.

Methods in which the Protein is Fed as a Supplement to a Cereal Basal Diet

The gross protein value (GPV). This method was devised by Heiman, Carver and Cook in 1939 and has been modified for use with British diets by Duckworth, Woodham and McDonald (1961). The test material contributes 3 per cent of protein in a 11 per cent total protein ration, the remaining 8 per cent being provided almost entirely by cereals. Chick growth is measured over a 14-day period, a correction is applied for growth due to the cereals by subtracting the weight gains of birds fed the 8 per cent cereal portion of the diet only, and the result is expressed as a percentage of the value obtained when casein is used as the test material.

The principal advantage of the method is the fact that practical conditions are simulated to some extent and the result reflects the combinative value of test protein and cereal, a more meaningful assessment than any in which the test protein is fed alone. The disadvantage is the fact that the combination of cereals used must be arbitrary and it is conceivable that GPV could vary considerably with changes in the composition of the cereal mixture. The overall protein level of 11 per cent is designed to emphasize differences between concentrates, and the method has been criticized on the grounds that such differences may not be paralleled under practical feeding conditions at over 20 per cent total protein. There is evidence that this is not a serious objection (Woodham, unpublished results).

2. *Protein efficiency ratio determinations on mixtures of protein concentrates and cereals.* PER measurements have been made successfully on mixtures of cereals with test proteins, using both chicks and rats. Methods describing the applications using both animals are referred to by Bunyan and Woodham (1964). In the chick test the total dietary protein level was 18·6 per cent and the test protein provided two-thirds of the total. Quite satisfactory discriminations were obtained between test proteins under the near-practical conditions, thus removing one of the weaknesses referred to in connection with the GPV method. (Woodham, 1968.)

Biological Assays of Amino Acid Availability

Biological methods using chicks for assessing the availability of lysine (Carpenter *et al.*, 1963) and for methionine (Miller *et al.*, 1965) have been reported. Both involve a 10-day experimental feeding period.

The Effect of Protein Quality on Plasma or Liver Protein

Depletion of plasma or liver proteins by starvation, bleeding, etc., can be remedied by feeding protein and the rate of re-generation has been used as a measure of protein quality (Whipple, 1948; Campbell and Kosterlitz, 1948). Free amino acid concentrations have been measured in chick blood plasma and sesame seed meals (Smith and Scott, 1965).

A Standardized Biological Protein Quality Test

Informal discussions were held by a large group of workers from a number of countries who were all interested in biological evaluation of proteins at the 6th International Congress of Nutrition in Edinburgh in 1963. This emphasized the variety of tests favoured by different workers and the difficulty of obtaining agreement on a standard test. While some workers favoured placing the emphasis on assays of availability of particular amino acids, others felt that an overall test of protein quality would be of more general application, and there was disagreement, too, on the relative merits of tests of proteins as sole protein sources and as supplements to cereals. Nevertheless, in subsequent contacts within the group it has become apparent that there is a real desire for such a standardization and providing that a majority of workers are prepared to devote some time to collaborative work there seems a possibility that at least some uniformity in the conditions for the various tests may be achieved.

This survey of biological protein quality tests is not comprehensive, but it embodies the main evaluation procedures in use at the present time.

LABORATORY TESTS OF PROTEIN QUALITY

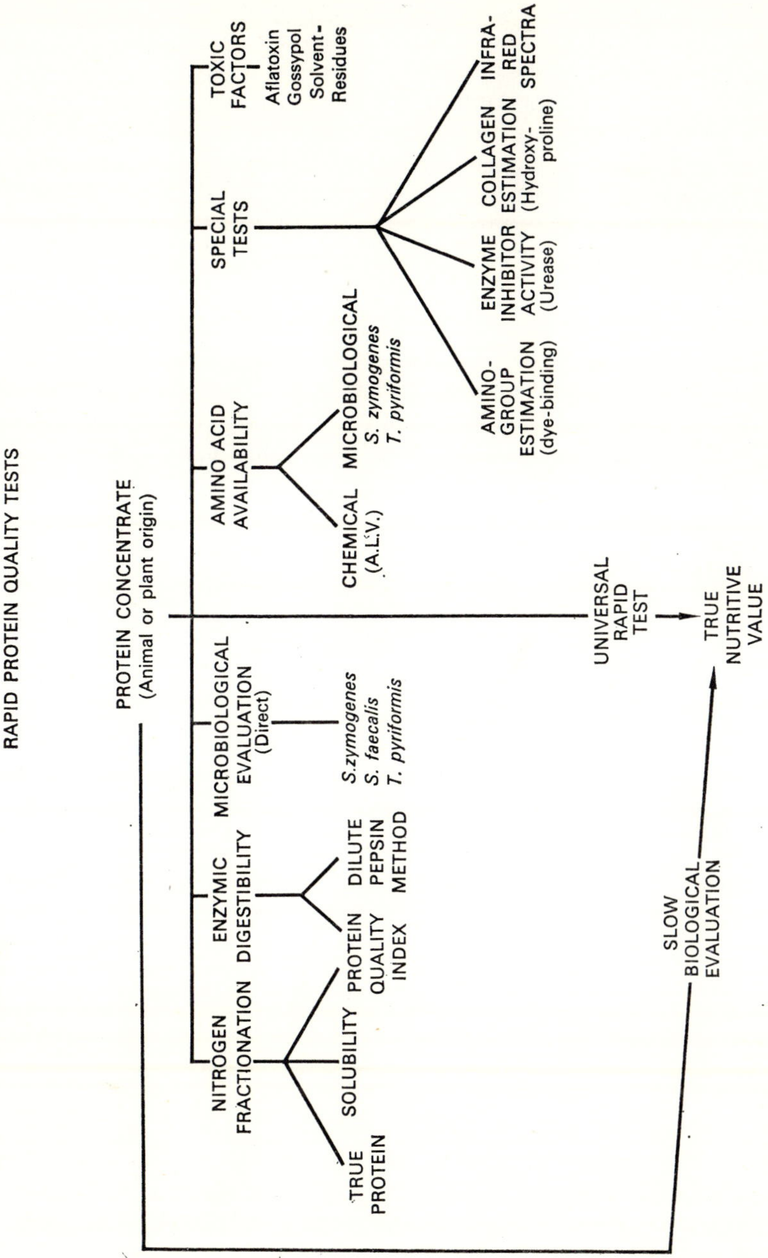

The impractability of evaluating individual feeding stuffs biologically coupled with the realization of the importance of protein in the feed of non-ruminants led to the adoption of a simple nitrogen analysis as the basis for costing and compounding farm feeds. Overall level of protein is of course an important indication of the potential value of the material, but unfortunately the nitrogen analysis does not take into account varying proportions of non-protein nitrogen which may be of little or no nutritive value; furthermore it will not discriminate between good and poor quality protein. Two meat meals may have an identical 'crude protein' (N × 6·25) content but one may be considerably inferior to the other in actual feeding value.

The need for a more informative test has been appreciated for many years but the difficulties involved are considerable. The test must be rapid and capable of routine application and furthermore it must give comparable grading with an acceptable biological criterion. Every biological evaluation method can be criticized on some grounds and it appears that some compromise will be necessary on this point.

A single feeding stuff is unlikely ever to be called upon to provide all the protein in a commercial ration for pigs or poultry and consequently the nature of the accompanying materials must be taken into consideration when assessing the nutritive value. In addition the value of the food will depend upon the function which it is called upon to perform. A protein supplement suitable for inclusion in rations for a laying hen may not necessarily be adequate for a fast growing broiler chicken or for a weanling pig. Thus a predictive test should indicate the value of the protein under a given set of circumstances and the more diverse these are the more generally useful the test will be. Because so many factors influence the quality of protein the discovery of a single and comparatively simple test might seem to be out of the question. However, it is possible that for a given class of protein feeding stuff one or two factors may be of over-riding importance. It has been repeatedly shown for example that carefully controlled heating is needed to produce optimum quality soyabean meal and it might be that a test which correctly indicates the extent of heating to which the meal has been submitted will provide all the information necessary for the commercial grading of soyabean samples intended for the supplementation of cereal based diets for broiler chicks.

Digestibility, Nitrogen Solubility and 'True' Protein

Not unnaturally the first attempts to improve upon a simple nitrogen analysis were based on measurements of digestibility, nitrogen solubility and so-called 'true' protein. Each of these attempts to exclude from consideration fractions of the total nitrogen which are likely to be of lesser importance physiologically. In 1935 Almquist, Stokstad and Halbrook proposed the Protein Quality Index (PQI) as a means of grading protein concentrate feeding stuffs. This is a factor obtained from four chemically determined values: the copper precipitable, hot water soluble, phosphotungstic acid precipitable and pepsin-indigestible fractions of the total nitrogen: and an attempt was made mathematically to apportion the contribution which each was considered to make to the total nutritive value. In this and in a later publication (Almquist, 1941) good correlations were reported between PQI and chick growth rate for various animal by-products. In 1945 Evans and St. John reported significant correlations between the PQI of peas, soyabean and cottonseed meals and a measurement of their supplementary value in cereal based rations for chicks—the gross protein value (GPV). In a collaborative investigation into the problem of rapid predictive tests which commenced in 1955 and was organized by the Agricultural Research Council, the PQI was one of the tests chosen for special study. In the first Progress Report (Boyne, Carpenter and Woodham, 1961) it was shown that there were good correlations between PQI and GPV for a range of 13 whale meals and 17 fish meals but not for 19 meat and meat/bone meals. The correlation coefficient for the fish meal range was considerably lowered when three non-typical samples were removed, indicating that the method is limited in use not only to certain classes of protein concentrate but to selected samples within a particular type (Barnes and Woodham, 1963).

The estimation of 'true' protein and pepsin-soluble nitrogen in the ranges of meat, fish, and whale meals already referred to yielded no significant correlations with either GPV or with net protein utilization (NPU) determined by the rat method of Miller and Bender. A nitrogen fractionation procedure devised by Lund and Sandstrom in 1943 was applied to soyabean meals by Evans and St. John (1945) who found a correlation between GPV and nitrogen solubility in 0·2 per cent KOH. For a more extensive range of soyabean meals as well as with ranges of sunflower seed and groundnut meals others found no such correlation (Boyne *et al.*, 1961). The same workers found, however, that for cottonseed meals all the Lund-Sandstrom fractions correlated with GPV to some extent. Olcott and Fontaine (1942) had noted a correlation between rat growth and nitrogen solubility in 3 per cent NaCl solution, but a similar correlation was not found for chicks (Lyman, Chang and Couch, 1953) though the same workers had noted a satisfactory correlation between chick growth and nitrogen solubility in 0·02 N sodium hydroxide. This was confirmed during the A.R.C. collaborative trial for a range of 17 cottonseed meals (Boyne *et al.*, 1961; Barnes and Woodham, 1963). A nitrogen solubility of 75 per cent indicates a cottonseed meal of the highest quality for non-ruminant feeding and a figure of 70 per cent has been widely accepted in the U.S. as the critical one below which cottonseed meals should be diverted to ruminant feeding.

Considerable work has been carried out on *in vitro* digestibility measurements (Gehrt, Caldwell and Elmslie, 1955; Bondi and Birk, 1955). However, proteins are not always completely split by these methods and the results are difficult to relate to *in vivo* digestibilities. Failure to obtain correlation between *in vitro* pepsin digestibilities and animal feeding experiments for fish meals has led recently to an attempt to establish an improved and standardized pepsin digestibility test at an international level. It was noted that highly satisfactory results were obtained by workers in Germany and Austria using extremely dilute pepsin solutions—one-hundredth or one-thousandth of the quantity specified in the method recommended by the Association of Official Agricultural Chemists (1960). Good correlations were obtained between results of the modified test and chemically determined available lysine values, but there are indications which suggest that the sensitivity of the test is such that it discriminates between meals which are indistinguishable in animal tests (Olley, 1964). Lovern (1965) has pointed out that the low pepsin strength method measures rate of digestion rather than extent of digestion and it has been repeatedly emphasized that pepsin solubilization alone is insufficient as a guide to quality—solubility of the protein in water and acid must also be taken into account. The amounts of essential amino acids released by pepsin followed by pancreatin digestion of protein feeding stuffs which included oilseed meals and leaf protein concentrates have been shown to correlate extremely well with literature figures for biological values determined with rats (Akeson and Stahmann, 1965).

Amino Acid Composition

Almost certainly the major factor determining the nutritive value of a protein concentrate is its ability to supply essential amino acids in the correct proportions for metabolic processes. It is not, however, correct to assume that total amino acid analysis is in itself a sufficient predictive test of quality. In the first place inherent errors in the technique may yield misleading results. Total amino acid analysis by microbiological assay methods are too tedious for routine use and in any case give somewhat variable values. This method has now been almost completely superseded by ion-exchange chromatographic techniques. These are speedy and give reproducible values but suffer from two principal disadvantages. The conditions for the preliminary acid hydrolysis which is necessary to release the amino acids from the protein prior to separation on the column, have not been satisfactorily settled, and the times, temperatures and acid strengths used by individual workers constitute a compromise. Hydrolysis must be continued until all the constituent amino acids have been released. Some are liberated quickly but may be partially destroyed during the further prolonged heating required to liberate others. Hydrolysis in sealed tubes may be a partial answer and Krampitz (1960) has found hydrolysis under mild conditions with a stannous chloride catalyst gives good recoveries of amino acids. Secondly, the extent to which an animo acid is released by acid hydrolysis is not necessarily matched by the digestive processes of the animal. In other words, the total level of an amino acid indicated by the chromatographic method may give a false indication of the nutritive value of the protein because a proportion is never released in the digestive process and is therefore unavailable for metabolic use.

Amino acid analysis has been used as the basis of chemical scoring methods for predicting nutritive value. Block and Mitchell in 1946 based the original 'Chemical Score' on the amount of the essential amino acid in greatest deficit in a protein compared to the level present in whole egg protein. This is an index of the value of the protein for growth only, because it assumes that the absence of an essential amino acid renders the protein completely unavailable even for tissue maintenance and experimental observation does not entirely support this. Gelatin and zein each lack one essential amino acid but are not completely valueless for tissue maintenance, and Oser (1959) has recognized this in his Essential Amino Acid Index. The EAAI is the geometric mean of the ratios of each of ten essential amino acid levels in the protein to the corresponding levels in egg protein. McLaughlan *et al.* (1959) proposed a simplified chemical score based on analysis for lysine, methionine and cystine only, recognizing that these three essential amino acids account for the vast majority of the deficiencies in protein feeding stuffs. Such chemical scoring methods could be useful for screening purposes but as Friedman concluded in 1958, other factors such as digestibility and anti-tryptic substances prevent the use of total amino acid analysis alone as a protein quality predictor. In addition there is a factor closely linked with the total amino acid content which is of even more importance and this is the amino acid availability.

Amino Acid Availability

Amino acid availability may be measured with rats or chicks as already described, or, in the laboratory it may be measured by the use of micro-organisms. In addition there is a specific chemical method for the measurement of the availability of lysine. So far chemical methods have not been developed for other essential amino acids.

The chemical estimation of available lysine (ALV) is now firmly established as a routine chemical procedure used widely in commercial laboratories everywhere and its importance cannot be over-estimated. The method relies upon the reaction of 1-fluoro-2 : 4-dinitrobenzene (FDNB) with the unbound ε-amino groups of lysine and gives a measure of the proportion of the total lysine which is available for use in the metabolic processes of the animal. This application to animal protein concentrates by Carpenter and Ellinger in 1955 of the reaction used by Sanger (1945) in his fundamental studies on the structure of insulin, subsequently modified by Carpenter (1960) provides a rapid test, the results of which have been shown to give extremely good correlations with chick growth tests (Carpenter *et al.*, 1957; Boyne *et al.*, 1961). The method has been used recently to evaluate animal protein concentrates in the absence of biological checks in order to study the possibility of allocating a mean figure applicable to the majority of samples of a given type (Pritchard, McLarnon and McGillivray, 1964). Wide ranges were found and average figures could be applied to only about 50 per cent of the samples in each class. This is in agreement with the findings of Duckworth, Woodham and McDonald (1961). Bunyan and Woodham (1964) reported the use of the ALV as a screening test for the selection of fish meals of differing nutritive value for an experiment designed to investigate the validity of applying to pigs conclusions drawn from chick experiments, and Laksesvela (1958) has found good correlations between ALV and chick growth estimations for a series of experimentally heated herring meals. Olley and Watson (1961) used available lysine content of a number of fish meals in a study of the effects of preservation, putrefaction and drying methods. Parallel digestibility studies revealed that high digestibility is not always to be associated with a high available lysine content.

The work mentioned so far concerns only animal protein concentrates. Certain difficulties arise when materials high in carbohydrate are examined and this has resulted in a comparative dearth of reports on the available lysine content of plant protein foods. Prolonged heating in the presence of carbohydrate, inevitable during the hydrolysis stage of the FDNB test, leads to disappearance of the N^{ε}-dinitrophenyllysine. Losses may be of the order of 25–30 per cent in the case of groundnut and soyabean meals. Allowance can be made for these losses after carrying out recovery checks using pure N^{ε}-dinitrophenyllysine heated under identical conditions, but the large corrections required do not inspire confidence in the result. It is probably justfiable to use the method to distinguish between plant proteins which differ fairly widely in nutritive value, but absolute figures should be quoted with appropriate reservations. The use of the method for plant protein evaluation has been reported by Carpenter and March (1961)—groundnut biscuit; Butterworth and Fox (1963)—coconut meal; Jones, Livingston and Cadenhead (1965)—soyabean meal; Anantharaman and Carpenter (1965); Woodham and Dawson (1966)—groundnut flour. Erbersdorfer and Zucker (1964) have claimed that their modified method allows the estimation of ALV in mixed feeds high in carbohydrate providing that the lysine level is greater than 1 per cent. Modifications introduced specifically to make the method more suitable for use on plant proteins have been suggested by Rao, Carter and Frampton (1963) who separated the N^{ε}-dinitrophenyllysine from other components of the hydrolysis mixture on an ion-exchange column. This method has been criticised by El-Nockrashy (1965) who found unidentified material was responsible for some of the absorption attributed to N^{ε}-dinitrophenyllysine. Mauron and Bujard (1963) guanidinated with *O*-methylisourea prior to hydrolysis in order to convert the lysine with free epsilon amino groups into homoarginine which is stable to acid hydrolysis. Good results were obtained by the guanidination procedure for milk powders and the results for groundnut and soyabean meals appeared promising. Paper chromatography has been used to separate the hydrolysis products (Baliga, Bayliss and Lyman, 1959). With cottonseed meals this was found to give results which correlated satisfactorily with protein quality evaluations using rat protein repletion tests. Two approaches have now been suggested by Lyman and Thomas (1965). One involves the suppression of the destructive action of carbohydrates by dilute hydrolysis conditions with prior addition of thiodiglycollic acid. This procedure is said to be involved and unsuitable for general use, and a second approach involving treatment of samples with amylase followed by solvent extraction to remove DNFB-sugar complexes before hydrolysis is thought more promising.

Microorganisms have been used to measure the availability of some amino acids. The availability of lysine and of methionine has been measured using the organism *Tetrahymena pyriformis* W. A simplified method has been described by Stott, Smith and Rosen (1963) and modifications have been reported by the same authors in Bunyan and Woodham (1963) where results are reported for three fish meals.

The organism *Streptococcus zymogenes* has been shown to possess an absolute requirement for methionine, leucine, isoleucine, arginine, histidine, tryptophan and valine, and Ford (1962) has described a method for measuring the availability of these amino acids. Lysine, threonine and phenylalanine are not indispensable for the organism and consequently it cannot be used for estimating their availability. Ford made the point that much more complete information is needed from animal tests. From the evidence presented there is a suggestion that the availability to the rat of the amino acids measured in whale and fish meals must parallel closely the availability to *Streptococcus zymogenes*. It may be noted also that fair correlations were found between GPV (chick) and available methionine (*Streptococcus zymogenes*) and between GPV and available tryptophan ($r = 0{\cdot}66$ and $0{\cdot}63$ respectively). Waterworth (1964) using Ford's method with certain modifications found good correlations between the chick GPV reported by Boyne *et al.* (1961) and the availabilities of arginine, histidine, leucine, isoleucine, methionine, valine and tryptophan for the same range of whale meals; with available histidine ($r = 0{\cdot}88$) and several others (r about $0{\cdot}6$) for the same range of fish meals; and with available methionine ($r = 0{\cdot}94$) and available histidine and leucine (r about 0·7) for the same range of meat meals. In the whale meal series and to some extent in the fishmeals the availability of each of the amino acids correlated with the availability of each of the others leading to the suggestion that a portion of the entire protein may be unavailable. Ford (1964) found that enzymic predigestion and fine grinding of test samples increased the values obtained for fish and whale meals, and comparing the results with those of Bunyan and Price (1960) he noted that the microbiological assay figures for available methionine and tryptophan and the chemically determined ALV were all closely correlated with digestibility and biological value. Recently in a series of meat, fish and whale meals which included some of those used in the ARC collaborative work, Miller *et al.* (1965) have found good agreement between available methionine determined by *Streptococcus zymogenes* and available methionine determined by a chick assay ($r = 0{\cdot}93$). These authors noted that the microbiological results were affected by the concentration and activity of the papain used for predigestion of the samples. Ford (1965) has submitted enzymically digested fish and whale meals to Sephadex-gel filtration and has noted that the digests of the poorer quality materials contained relatively more material of large molecular size which proved a poor source of microbiologically and chemically determined available amino acids. This approach seems to merit further attention.

MICROBIOLOGICAL METHODS FOR MEASURING OVERALL PROTEIN QUALITY

Microorganisms may be used to measure the quality of entire protein as distinct from the availability of individual amino acids. Three organisms have received most attention. Halevy and Grossowicz in 1953 used *Streptococcus faecalis* in order to compare the nutritive value of casein with that of egg albumin, gelatin, gluten and zein after *in vitro* digestion with a crude preparation of pancreatic proteinases. In 1960 Ford reported the use of *Streptococcus zymogenes* finding that values obtained for a variety of food proteins correlated closely with those obtained in biological tests. The ciliate protozoon *Tetrahymena pyriformis* was used by Rosen and Fernell in 1956 and this method has the advantage that no preliminary hydrolysis of the sample is needed.

Bunyan and Price (1960) obtained a good correlation between the growth of *Streptococcus faecalis* and Net Protein Utilization for whale meals. Boyne *et al.* (1961) confirmed this, finding good correlations also with GPV. They noted too that *Streptococcus zymogenes* gave good correlations with NPU and GPV for whale meals but with NPU only for meat meals ($r = 0{\cdot}73$). *Tetrahymena pyriformis* yielded a correlation with GPV for meat and whale meals. Waterworth (1964) found good correlations between relative nutritive values (RNV) determined with *Streptococcus zymogenes* and both the NPU and GPV ranges reported by Boyne *et al.* (1961) for the same samples of fish, whale and meat meals. Stott, Smith and Rosen (1963) described a simplification of the *Tetrahymena pyriformis* assay applied to 93 protein foodstuffs.

Dye Absorption Methods

In 1927 Chapman, Greenberg, and Schmidt titrated proteins such as casein and gelatin with six different dyes and established a quantitative relationship between the extent of dye binding and the free basic groups of arginine, lysine and histidine. Rawlins and Schmidt (1929) found a stoichometric relationship between protein level and absorption of basic dyes. Fraenkel-Conrat and Cooper (1944) devised methods for determining the total acidic and basic groups of proteins based upon their ability to combine with dyes in buffered acidic or alcoholic solution. Orange G was found especially satisfactory and in 1954 Udy used this to study wheat protein fractions. Frölich (1954) successfully used cresol red to detect under- and over-heating in soyabean meals and his results were confirmed by Olomucki and Bornstein (1960) and by Ascarelli and Gestetner (1962). Correlations between GPV and Orange G binding were observed for eight whale meals and eight fish meals, the correlation coefficients being 0·72 and 0·87 respectively, but no correlations were noted in a range of 11 groundnut meals (Boyne *et al.*, 1961). Bunyan and Price (1960) found good correlations between the extent of Orange G binding and NPU measured with rats by the method of Miller and Bender for ranges of meat and whale meals. Choppe and Kratzer (1963) have used Orange G for the evaluation of a range of 20 meat meals finding significant correlations ($r = 0{\cdot}68$) with the results of a chick growth test when the dye binding values were combined with estimates of nitrogen solubility in hot water.

MISCELLANEOUS TESTS

Meat by-products are notoriously variable in nutritive value and also in the nature of the raw materials used in their preparation. Large proportions of connective tissue are undesirable. Recent work has suggested that the availability of lysine is in indirect proportion to the amount of connective tissue present in cuts of raw meat (Dvorak and Vognarova, 1965). Hydroxyproline is a characteristic constituent of collagen, and Eastoe and Long (1960) suggested that it might be used as a measure of quality in meat products. Summers and Fisher (1962) attributed poor quality in four meat meals to the high collagen content as indicated by hydroxyproline analysis. Ascarelli and Gestetner (1962) on the other hand found no correlation between hydroxyproline content and rat NPU for fish meals, and Woodham (unpublished results) similarly found no correlation between hydroxyproline content and chick GPV for a series of 31 meat and meat/bone meals. Grant (1964) has published a method for estimating hydroxyproline by the Technicon Auto Analyser.

Measurements of antitryptic or urease activity were found useful by Ascarelli and Gestetner (1962) for distinguishing between underheated and properly heated soyabean meals but Boyne *et al.* (1961) failed to predict nutritive value in a series of soyabean meals with the urease test when it was used in conjunction with solubility measurements which might be expected to indicate overheating. Medium urease values did not apparently indicate the meals of highest nutritive value as had been suggested by earlier workers (Croston, Smith and Cowan, 1955).

Miller and Carpenter (1964) determined total sulphur in a series of seven meat, fish and whale meals but found no correlation with rat NPU probably because cystine and methionine together accounted for less than two thirds of the total sulphur in five of the samples. Similar results were reported for meat and whale meals by Boyne *et al.* (1961).

An interesting development reported recently (Johnston and Watts, 1965) is the prediction of quality in glandless cottonseed meals by means of their infrared spectra. Changes in the spectrum were noted corresponding to the differing nutritive values of the samples.

The Choice of Test

A variety of tests have been mentioned all of which have been found useful in the hands of some workers for predicting nutritive value in one or more of the classes of protein concentrate commonly incorporated in pig and poultry diets. None so far is acceptable to all workers for all protein materials and it may well be that no simple test will ever emerge as universally acceptable. The nearest approach to such a test is probably the chemical estimation of available lysine, accepted widely now as a good indicator of quality for animal by-products, and there is little doubt that in time a modification acceptable to all will extend the test to plant proteins. Provision of lysine in available form is undoubtedly a desirable characteristic in a protein concentrate, but obviously the test will not be useful in predicting the value of a supplement to a particular basal diet which is, for example, low in the sulphur amino acids. Undoubtedly a useful addition to the nutritionist's armoury would be a similar test for available methionine and cystine. Microbiological evaluations have exposed the possibility of non-availability in other essential amino acids and have underlined the fact that total amino-acid analysis can be very misleading. One of the most significant of recent developments has been the emergence of microbiological procedures as very promising tools in this type of investigation. Recent collaborative work amongst microbiologists has shown that attention to detail can result in considerable reductions in the extent of inter-laboratory variability, but much work remains to be done especially with regard to finding the extent to which results for the availability of particular amino acids for microorganisms may be applied to large animals.

Though the ideal is a single rapid test such as the highly successful solubility measurement which appears to be quite adequate for the assessment of cottonseed meals, it seems that, in general, combinations of tests are more likely to provide us with useful information. The successful linking of Orange G binding and hot water solubility has been mentioned and another useful combination of tests is embodied in the Protein Quality Index.

Tests such as the urease and aflatoxin tests are specific for particular types of concentrate and their inclusion here emphasises the fact that any protein quality evaluation method must be related to the material under investigation. They are further arguments against the likelihood of the eventual emergence of a single simple test for all materials. The search for new tests should not, however, be allowed to slacken and the infrared spectra study may well be capable of extension to other plant proteins.

Doubts concerning the validity of particular tests frequently rise when small ranges of test samples are used. In work of this type the number of samples should be as large as possible and even when highly significant correlations with biological tests are noted for ranges of say twenty samples, conclusions should still be drawn only with considerable reservations. One or two samples deviating inexplicably from a general pattern may not affect the correlation coefficient very much but they do tend to destroy confidence in the test and prejudice its adoption by others. On the other hand perfect correlations cannot be expected because of the inherent variability of the biological criteria which must be used.

FUTURE PROSPECTS

It is clear from consideration of the various aspects of protein quality which have been mentioned that, while the position at present is still unsatisfactory, great progress has been made during the last half-century. The extent of the problem is now realised and an increasing number of workers are coming into the field, so that we can safely forecast that improvements will continue. Economic considerations alone, resulting from the huge increase in the world's population and necessitating the optimum use of all protein materials for human consumption will inevitably result in increasing stringency in the use of proteins in animal feeding. Some have gone so far as to claim that mankind will eventually be unable to afford the luxury of feeding plant proteins to animals on the grounds that such use is wasteful. Certainly it will be only permissible if carried out with extreme efficiency and careful planning. This can only come about from accurate knowledge of the implications of protein quality.

Exciting developments are already in the offing. The production of protein by controlled fermentation of hydrocarbon oils is already an economic proposition and feeding trials using these materials are being carried out at the present time. There is little evidence as yet to indicate the potential nutritive value of these products but there seems no reason to suppose that they will not be suitable for feeding non-ruminant livestock. However, other industrial uses for the products now seem possible without going the length of fermentation and microbiological production of protein. The use of isolated leaf protein concentrate is being investigated in Africa and Jamaica. This is an obvious solution to the problem of protein shortage in developing countries which have abundant vegetation and although in the first instance effort will be aimed at incorporation of the concentrate directly into human diets, lack of acceptability as well as increasing production may eventually lead to use in animal feeding. Cost of production is a difficulty at the moment but improved technology may overcome this. We know that leaf protein concentrate can be of high protein quality and it has been shown that good results can be obtained when the material is mixed with groundnut meal and doubtless with other oilseeds in pig nutrition (Duckworth and Woodham, 1961; Duckworth, Hepburn and Woodham, 1961).

The processing of protein concentrates is continually being improved. Until fairly recently the only developments in oilseed processing were concerned with producing higher yields of oil, but now the value of the residue after oil expression is such that manufacturers have realized the necessity of adjusting processing conditions so that the residual meals are of best possible quality for animal feeding. It is reasonable to suppose that such improvements will continue. Recently fish meal of very poor quality was produced in South American countries where this industry was new, but with increasing experience the quality has improved. Similar considerations apply to the question of toxins in feeding meals. Choice of solvent for extraction, control of harvesting and storage conditions for seeds to reduce the danger of fungal contamination and the use of proper heating or other treatment to destroy harmful enzyme inhibitors may all be expected to result in better quality products in the future.

With regard to feeding stuffs from plants it may be envisaged that attention will be directed in the future towards improvement in strains from the point of view of protein quality. Genetic selection of oilseeds and cereals has hitherto been carried out with the object of increasing yields or producing disease resistance. Now that livestock feeding has become an increasingly important use for these materials attention is being given to this aspect. The development of a strain of groundnut meal which is low in trypsin inhibitor activity and containing protein which possesses a more balanced amino acid composition than those available at present could be of tremendous significance. Work along these lines is at present being carried out as part of the British contribution to the International Biological Programme and should assist in the better utilization of natural protein resources.

REFERENCES

AKESON, W. R. and STAHMANN, M. A. (1965) *J. Ag. Fd. Chem.* **13**, 145.
ALMQUIST, H. J. (1941) *J. Nutr.* **21**, 347.
ALMQUIST, H. J., STOKSTAD, E. L. R. and HALBROOK, E. R. (1935) *J. Nutr.* **10**, 193.
ANANTHARAMAN, K. and CARPENTER, K. J. (1965) *Proc. Nutr. Soc.* **24**, xxxii.
ASCARELLI, I. and GESTETNER, B. (1962) *J. Sci. Fd. Agric.* **13**, 401.
AUSTWICK, P. K. C. and AYERST, G. (1963) *Chem. and Ind.*, No. 2, 55.
BALIGA, B. P., BAYLISS, M. E. and LYMAN, C. M. (1959) *Arch. Biochem. Biophys.* **84**, 1.
BALIGA, B. P. and LYMAN, C. M. (1957) *J. Amer. Oil Chem. Soc.* **34**, 21.
BANDEMER, S. L. and EVANS, R. J. (1963) *J. Ag. Fd. Chem.* **11**, 134.
BARNES, M. McC. and WOODHAM, A. A. (1963) *J. Sci. Fd. Agric.* **14**, 109.
BARNES, R. H. and BOSSHARDT, D. K. (1946) *Ann. N.Y. Acad. Sci.* **47**, 273.
BARNES, R. H., KWONG, E. and FIALA, G. (1961) *Fed. Proc.* **20**, 369.
BENSABAT, L., FRAMPTON, V. L., ALLEN, L. E. and HILL, R. A. (1958) *J. Agric. Fd. Chem.* **6**, 778.
BIRK, Y. and GERTLER, A. (1961) *J. Nutrit.* **75**, 379.
BLATCHFORD, S. M. and HALL, D. W. (1963) *Tropical Sci.* **5**, 6.
BLOCK, R. J. and MITCHELL, H. H. (1946) *Nutr. Abstr. and Rev.* **16**, 249.
BLOUNT, W. P. (1961) *Turkeys*, **9**, 52.
BONDI, A. and BIRK, Y. (1955) *J. Sci. Fd. Agric.* **6**, 543.
BOOTH, A. N., ROBBINS, D. J., RIBELIN, W. E. and DE EDS, F. (1960) *Proc. Soc. Exp. Biol. Med.* **104**, 681.
BORCHERS, R. (1958) *J. Nutrit.* **66**, 229.
BORCHERS, R. (1961) *J. Nutrit.* **75**, 330.
BORCHERS, R. (1963) *Proc. Soc. Exp. Biol. Med.* **112**, 84.
BORNSTEIN, S., BEN-ADAM, Z. and LIPSTEIN, B. (1961) *J. Sci. Fd. Agric.* **12**, 80.
BORNSTEIN, S. and LIPSTEIN, B. (1963) *J. Sci. Fd. Agric.* **14**, 719.
BOYNE, A. W., CARPENTER, K. J. and WOODHAM, A. A. (1961) *J. Sci. Fd. Agric.* **12**, 832.
BRAMBILA, S., NESHEIM, M. C. and HILL, F. W. (1961) *J. Nutrit.* **75**, 13.
BUNYAN, J. and WOODHAM, A. A. (1964) *Brit. J. Nutr.* **18**, 537.
BUNYAN, J. and PRICE, S. A. (1960) *J. Sci. Fd. Agric.* **11**, 25.
BURNSIDE, J. E., SIPPEL, W. L., FORGACS, J., CARLL, W. T., ATWOOD, M. B. and DOLL, E. R. (1957) *Amer. J. Vet. Res.* **18**, 817.
BUTTERWORTH, M. H. and FOX, H. C. (1963) *Brit. J. Nutr.* **17**, 445.
CAMPBELL, R. M. and KOSTERLITZ, H. W. (1948) *J. Physiol.* **107**, 383.
CARPENTER, K. J. (1960) *Biochem. J.* **77**, 604.
CARPENTER, K. J. and ELLINGER, G. M. (1955) *Biochem. J.* **61**, xi.
CARPENTER, K. J., ELLINGER, G. M., MUNRO, M. I. and ROLFE, E. J. (1957) *Brit. J. Nutr.* **11**, 162.
CARPENTER, K. J., ELLINGER, G. M. and SHRIMPTON, D. H. (1954) *Proc. Nutr. Soc.* **13**, xx.
CARPENTER, K. J. and MARCH, B. E. (1961) *Brit. J. Nutr.* **15**, 403.
CARPENTER, K. J., MARCH, B. E., MILNER, C. K. and CAMPBELL, R. C. (1963) *Brit. J. Nutr.* **17**, 309.
CARPENTER, K. J., MORGAN, C. B., LEA, C. H. and PARR, L. J. (1962) *Brit. J. Nutr.* **16**, 451.
CARTER, F. L., CIRINO, V. O. and ALLEN, L. E. (1961) *J. Am. Oil Chem. Soc.* **38**, 148.
CHAPMAN, L. M., GREENBERG, D. M. and SCHMIDT, C. L. A. (1927) *J. Biol. Chem.* **72**, 707.
CHOPPE, W. and KRATZER, F. H. (1963) *Poult. Sci.* **42**, 642.
CLANDININ, D. R. and TAJCNAR, E. W. (1961) *Poult. Sci.* **40**, 291.

CLARK, S. P. and DUBOSE, R. T. (1960) *J. Ag. Fd. Chem.* **8**, 147.
CLAWSON, A. J., SMITH, F. H., OSBORNE, J. C. and BARRICK, E. R. (1961) *J. Anim. Sci.* **20**, 547.
CONKERTON, E. J., MARTINEZ, W. H., MANN, G. E. and FRAMPTON, V. L. (1957) *J. Ag. Fd. Chem.* **5**, 460.
COSTA, G. (1960) *Nature*, **188**, 549.
CROSTON, C. B., SMITH, A. K. and COWAN, J. C. (1955) *J. Amer. Oil Chem. Soc.* **32**, 279.
DANIELLSON, C. E. (1952) *Acta Chem. Scand.* **6**, 149.
DAVIDSON, J. and BOYNE, A. W. (1962) *Brit. J. Nutr.* **16**, 541.
DAVIS, N. D. (1961) *J. Amer. Oil Chem. Soc.* **38**, 516.
DERSE, P. H. (1962) *J. Ass. Off. Agric. Chem.* **45**, 418.
DERSE, P. H. (1965) *J. Ass. Off. Agric. Chem.* **48**, 847.
DONOSO, G., LEWIS, O. A. M., MILLER, D. S. and PAYNE, P.R. (1962) *J. Sci. Fd. Agric.* **13**, 192.
DUCKWORTH, J., HEPBURN, W. R. and WOODHAM, A. A. (1961) *J. Sci. Fd. Agric.* **12**, 16.
DUCKWORTH, J. and WOODHAM, A. A. (1961) *J. Sci. Fd. Agric.* **12**, 5.
DUCKWORTH, J., WOODHAM, A. A. and MCDONALD, I. (1961) *J. Sci. Fd. Agric.* **12**, 407.
DUERDEN, J. C. and CUTLER, J. R. (1957) *J. Sci. Fd. Agric.* **8**, 600.
DUNCAN, D. L. (1966) *Recent Advances in Animal Nutrition*, p.51, Ed. Abrams. Churchill, London.
DVORAK, Z. and VOGNAROVA, I. (1965) *J. Sci. Fd. Agric.* **16**, 305.
EASTOE, J. E. and LONG, J. E. (1960) *J. Sci. Fd. Agric.* **11**, 87.
ELDRED, N. R. and RODNEY, G. (1946) *J. Biol. Chem.* **162**, 261.
EL-NOCKRASHY, A. S. (1965) *Die Stärke*, **17**, 89.
ERBERSDORFER, VON H. and ZUCKER, H. (1964) *Z. Tierphysiol. tierernähr futtermittelk.* **19**, 244.
EVANS, R. J., BANDEMER, S. L. and BAUER, D. H. (1961) *J. Fd. Sci.* **26**, 663.
EVANS, R. J., BANDEMER, S. L. and BAUER, D. H. (1962) *J. Agric. Fd. Chem.* **10**, 416.
EVANS, R. J. and ST. JOHN, J. L. (1945) *J. Nutr.* **30**, 209.
FISHER, H. and JOHNSON, D. (1958) *Arch. Biochem. Biophys.* **77**, 124.
FISHER, H., SUMMERS, J. D., WESSELS, J. P. H. and SHAPIRO, R. (1962) *J. Sci. Fd. Agric.* **13**, 658.
FORD, J. E. (1960) *Brit. J. Nutr.* **14**, 485.
FORD, J. E. (1962) *Brit. J. Nutr.* **16**, 409.
FORD, J. E. (1964) *Brit. J. Nutr.* **18**, 449.
FORD, J. E. (1965) *Brit. J. Nutr.* **19**, 277.
FRAENKEL-CONRAT, H. and COOPER, M. (1944) *J. Biol. Chem.* **154**, 239.
FRIEDMAN, L. (1958) *J. Ass. Off. Agric. Chem.* **41**, 188.
FRÖLICH, A. (1954) *Nature*, **174**, 879.
FROST, D. V. (1959) *Protein and Amino Acid Nutrition* p. 225. Ed. A. A. Albanese. Academic Press, N.Y.
GEHRT, A. J., CALDWELL, M. J. and ELMSLIE, W. D. (1955) *J. Agric. Fd. Chem.* **3**, 159.
GRANT, R. A. (1964) *J. Clin. Path.* **17**, 685.
GRAU, C. R., KARRICK, N. L., LUNDHOLM, B. D. and BARNES, R. H. (1960) *Com. Fish. Rev.* **22**, 12.
GREENBERG, J., TAYLOR, D. J., BOND, H. W. and SHERMAN, J. F. (1959) *J. Agric. Fd. Chem.* **7**, 573.
HALEVY, S. and GROSSOWICZ, N. (1953) *Proc. Soc. Exp. Biol. Med.* **82**, 567.
HARPER, A. E. (1959) *J. Nutrit.* **68**, 405.
HAYWARD, J. W. (1959) *Feedstuffs* **31**, 18.
HAYWARD, J. W., STEENBOCK, H. and BOHSTEDT, G. (1936) *J. Nutrit.* **11**, 219.
HEIMAN, V., CARVER, J. S. and COOK, J. W. (1939) *Poult. Sci.* **18**, 464.
HELLER, B. S., CHUTKOW, M. R., LUSHBOUGH, C. H., SIEDLER, A. J. and SCHWEIGERT, B. S. (1961) *J. Nutrit.* **73**, 113.
HUSSAR, N. and ROBBLEE, A. R. (1962) *Poult. Sci.* **41**, 1489.
JENKINS, F. P. (1963) *J. Sci. Fd. Agric.* **14**, 773.

JENSEN, L. S., MERRILL, L. H., REDDY, C. V. and McGINNIS, J. (1962) *Poult. Sci.* **41**, 1414.
JOHNSTON, C. and WATTS, A. B. (1965) *Poult. Sci.* **43**, 957.
JONES, A. S., LIVINGSTON, R. M. and CADENHEAD, A. (1965) *Anim. Prod.* **7**, 286.
KORNEGAY, E. T., CLAWSON, A. J., SMITH, F. H. and BARRICK, E. R. (1961) *J. Anim. Sci.* **20**, 597.
KRAMPITZ, G. (1960) *Z. Tierphysiol. tierernähr futtermittelk*, **15**, 40.
KULIK, Y. I. (1954) *Rev. Med. Vet. Mycol.* **3**, 782.
KWONG, E. and BARNES, R. H. (1963) *J. Nutrit.* **81, 392**.
KWONG, E., BARNES, R. H. and FIALA, G. (1962) *J. Nutrit.* **77**, 312.
LAKSESVELA, B. (1958) *J. Agric. Sci.* **51**, 164.
LEPKOVSKY, S., BINGHAM, E. and PENCHARZ, R. (1959) *Poult. Sci.* **38**, 1289.
LOOSMORE, R. M. and HARDING, J. D. J. (1961) *Vet. Record* **73**, 1362.
LOOSMORE, R. M. and MARKSON, L. M. (1961) *Vet Record* **73**, 813.
LOVERN, J. A. (1965) *J. Ass. Off. Agric. Chem.* **48**, 60.
LUND, A. P. and SANDSTROM, W. M. (1943) *J. Agric. Res.* **66**, 349.
LYMAN, C. M. and BALIGA, B. R. (1958) *Processed Plant Protein Foodstuffs* Chap. 17, p. 517. Ed. Altschul.
LYMAN, C. M., CHANG, W. Y. and COUCH, J. R. (1953) *J. Nutrit.* **49**, 679.
LYMAN, C. M. and THOMAS, M. C. (1965) *J. Ass. Off. Agric. Chem.* **48**, 858.
MARCH, B. E., BIELY, J., McBRIDE, J. R., IDLER, D. R. and McLEOD, R. A. (1961) *J. Fish Res. Board Canada*, **18**, 113.
MARTINEZ, W. H. and FRAMPTON, V. L. (1958) *J. Agric. Fd. Chem.* **6**, 312.
MARTINEZ, W. H., FRAMPTON, V. L. and CABELL, C. A. (1961) *J. Agric. Fd. Chem.* **9**, 64.
MAURON, J. and BUJARD, E. (1960) *Ann. Nutrit. Aliment*, **14**, 291.
MAURON, J. and BUJARD, E. (1963) *Proc. 6th Int. Congr. Nut., Edinburgh*, p. 489, Livingston, Edinburgh.
McKINNEY, L. L., WEAKLEY, F. B., ELDRIDGE, A. C., CAMPBELL, R. E., COWAN, J. C., PICKEN, J. C. and BIESTER, H. E. (1957) *J. Amer. Chem. Soc.* **79**, 3932.
McLAUGHLAN, J. M., ROGERS, C. G., CHAPMAN, D. G. and CAMPBELL, J. A. (1959) *Can. J. Biochem. Physiol.* **37**, 1293.
MILLER, E. L. and CARPENTER, K. J. (1964) *J. Sci. Fd. Agric.* **15**, 810.
MILLER, E. L., CARPENTER, K. J., MORGAN, C. B. and BOYNE, A. W. (1965) *Brit. J. Nutr.* **19**, 249.
MILLER, D. S. and BENDER, A. E. (1955) *Brit. J. Nutr.* **9**, 382.
MILLER, S. A., GOLDBLITH, S. A., GOSWAMI, M. N. D., DYMSZA, H. A. and PROCTOR, B. E. (1961) *J. Nutrit.* **74**, 70.
MITCHELL, H. H. (1923) *J. Biol. Chem.* **58**, 873.
MORRISON, A. B. (1963) *Fed. Proc.* **22**, 551.
MORRISON, A. B., SABRY, Z. I. and MIDDLETON, E. J. (1962) *J. Nutrit.* **77**, 97.
NESHEIM, M. C., BRAMBILA, S. and HOPKINS, D. T. (1961) *Fed. Proc.* **20**, 368.
NJAA, L. R. (1961) *J. Sci. Fd.Agric.* **12**, 757.
Official Methods of Analysis of the Association of Official Agricultural Chemists. 9th ed. (1960) p. 286. Ed. W. Hurwitz. A.O.A.C. Washington, D.C.
OLCOTT, H. S. and FONTAINE, T. D. (1942) *Industr. Eng. Chem.* (*Anal.*), **34**, 714.
OLLEY, J. (1964). Unpublished results. Torry Research Station. Interim Record No. 13. Available on request from the Torry Research Station, Aberdeen.
OLLEY, J. and WATSON, H. (1961) *J. Sci. Fd. Agric.* **12**, 316.
OLOMUCKI, E. and BORNSTEIN, S. (1960) *J. Ass. Off. Agric. Chem.* **43**, 440.
OSBORNE, T. B. and MENDEL, L. B. (1912) *J. Biol. Chem.* **12**, 473.
OSBORNE, T. B. and MENDEL, L. B. (1917) *J. Biol. Chem.* **32**, 369.
OSBORNE, T. B., MENDEL, L. B. and FERRY, E. L. (1919) *J. Biol. Chem.* **37**, 223.
OSER, B. L. (1959). In *Protein and Amino Acid Nutrition* p. 281. Ed. A. A. Albanese. Academic Press, N.Y.
OUSTERHOUT, L. E., GRAU, C. R. and LUNDHOLM, B. D. (1959) *J. Nutr.* **69**, 65.
PRITCHARD, H., McLARNON, J. and McGILLIVRAY, R. (1964) *J. Sci. Fd. Agric.* **15**, 690.
RAND, N. T., COLLINS, V. K., VARNER, D. S. and MOSSER, J. D. (1960) *Poult. Sci.* **39**, 45,
RAO, S. R., CARTER, F. L. and FRAMPTON, V. L. (1963) *Anal. Chem.* **35**, 1927.

RAWLINS, L. M. C. and SCHMIDT, C. L. A. (1929) *J. Biol. Chem.* **82**, 709.

REHFELD, C. E., PERMAN, V., SAULTER, J. H. and SCHULTZE, M. O. (1958) *J. Agric. Fd. Chem.* **6**, 227.

RICHARDSON, L. R., WILKES, S., GODWIN, J. and PIERCE, K. R. (1962) *J. Nutrit.* **78**, 301.

ROSEN, G. D. and FERNELL, W. R. (1956) *Brit. J. Nutr.* **10**, 156.

SANGER, F. (1945) *Biochem. J.* **39**, 507.

SARGEANT, K., ALLCROFT, R. and CARNAGHAN, R. B. A. (1961) *Vet. Record* **73**, 865.

SAXENA, H. C., JENSEN, L. S. and MCGINNIS, J. (1962a) *J. Nutrit.* **77**, 259.

SAXENA, H. C., JENSEN, L. S. and MCGINNIS, J. (1962b) *Fed. Proc.* **21**, 387.

SAXENA, H. C., JENSEN, L. S. and MCGINNIS, J. (1963) *Poult. Sci.* **42**, 788.

SAXENA, H. C., JENSEN, L. S., SPENCER, J. V. and MCGINNIS, J. (1963) *Poult. Sci.* **42**, 788.

SCHROEDER, L. J., IACOBELLIS, M. and SMITH, A. H. (1961) *J. Nutr.* **73**, 143.

SETO, T. A., SCHULTZE, M. O., PERMAN, V., BATES, F. W. and SAUTER, J. H. (1958) *J. Agric. Fd. Chem.* **6**, 49.

SMITH, R. E. and SCOTT, H. M. (1965) *J. Nutrit.* **86**, 37–45.

SPENSLEY, P. C. (1963) *Endeavour*, **22**, 75.

STOTT, J. A., SMITH, H. and ROSEN, G. D. (1963) *Brit. J. Nutr.* **17**, 227.

SUMMERS, J. D. and FISHER, H. (1962) *J. Sci. Fd. Agric.* **13**, 496.

THOMAS, K. (1909) *Arch. Anat. Physiol. Lpz., Physiol. Abstr.* 219.

WALLACE, W. M. (1959) *Fed. Proc.* **18**, 1125.

WARD, H. S. and DIENER, U. L. (1961) *Phytopath.* **51**, 244.

WATERWORTH, D. G. (1964) *Brit. J. Nutr.* **18**, 503.

WHIPPLE, G. H. (1948) *Haemoglobin, Plasma Protein and Cell Protein.* C. C. Thomas, Springfield Ill.

WILGUS, H. S., NORRIS, L. C. and HEUSER, G. F. (1936) *Ind. Eng. Chem.* **28**, 586.

WILLCOCK, E. G. and HOPKINS, F. G. (1906) *J. Physiol.* **35**, 58.

WOODHAM, A. A. (1968) *Brit. Poult. Sci.* **9**, 53.

WOODHAM, A. A. and DAWSON, R. (1966) *Proc. Nutr. Soc.*, **25**, viii.

CHAPTER 9

Morphology of the Absorptive Surfaces of the Alimentary Tract

B. F. FELL

Rowett Research Institute, Bucksburn, Aberdeen

THE alimentary canal is lined throughout with epithelial cells; in the normal animal they probably form an unbroken surface from mouth to anus, with the exception of defects that may occur in the tonsillar crypts. The epithelial surface is not smooth, but by numerous papillæ, folds and villi, by pits, crypts and glands it is expanded into a large area. Absorption of water and certain drugs, at least, probably can occur from any part of the surface, but in the main food becomes available for absorption only after it has been broken down into soluble and comparatively simple substances. This is the process of digestion; it is accomplished by a series of enzymic hydrolyses and depends to an important extent in the ruminant on the action of microorganisms in the reticulo-rumen and cæcum. Digestion in the ruminant is a continuous process that is possible only because the products of microbial fermentation are not allowed to accumulate. The rumen is not only a fermentation chamber; as an essential adaptation it is also an organ of absorption.

In some primitive animals the processes of digestion are generalized; the cells of the alimentary canal individually phagocytose and digest particles of food (Yonge, 1937). In the mammal, the alimentary canal shows considerable structural and physiological organization and can be divided into relatively specialized areas, yet the epithelium of the mammalian intestine retains the ability to ingest large molecules and even particulate matter.

The first part of the alimentary canal through which appreciable absorption occurs is the reticulo-rumen. This structure is analogous to the œsophageal area or fore stomach of the horse and pig stomach (McAnally and Phillipson, 1944) and it is formed by dilatation of the primitive gut.

The mucosa of the blind sacs of the rumen is extended into permanent papillæ and in the reticulum, into permanent honeycomb-like ridges. The edges of the ruminal pillars, the medial wall of the cardia and a large part of the wall of the middle of the dorsal sac are not papillated in the ox rumen (Grossman, 1949). The papillæ may be well developed at birth in the calf (Tamate, *et al.*, 1962), but in the neonatal lamb they are not distinct entities and there are no inter-papillary spaces. The papillæ have a strong core of connective tissue and are not independently retractile. They increase the mucosal surface area and presumably provide the friction necessary to effect the movement of ingesta.

Contrary to what one might expect in an organ with marked powers of absorption, the rumen is lined with stratified squamous epithelium; it does not contain goblet cells or glands and a suitably fluid medium of favourable pH and osmotic pressure results from the copious inflow of buffered saliva and ingested fluids, balanced by absorption of water and other substances through the ruminal wall (see Chapter 4).

The ruminal epithelium is similar to that of the epidermis and is continuously renewed by cell division in the basal layer. The cells of the basal layer are columnar and are close to the capillaries of the lamina propria. They are succeeded by layers corresponding to epidermal stratum spinosum, stratum granulosum and stratum corneum (Dobson *et al.*, 1956).

In the fœtus, the epithelium of all the fore stomachs is stratified and non-cornified, and the cells contain much glycogen (Gobetto, 1957, 1958). Glycogen is present in the epithelium of the rumen and reticulum in the newborn calf (Tamate *et al.*, 1964; Habel, 1963), but this has disappeared by 4 weeks of age. Habel suggests that it is a remainder of the fœtal condition and that it is related to age but not to diet. However, in older calves glycogen may occur below the basement membrane, in the lamina propria of the fore stomachs.

Wardrop (1961a) found that in sheep all three fore stomachs showed the same general histological changes during the transition from fœtal to post natal life. The epithelium changed from a stratified cuboidal type in the fœtal specimens to a keratinized, stratified squamous epithelium in the post natal specimens. The basal layer of the epithelium became folded to form papillæ and the honeycomb-like structure of the reticulum, and in the rumen and reticulum further folding of this layer took place to form papillary bodies. The latter greatly increase the area of the basal layer of epithelium and so serve to increase the extent of contact between epithelium and capillary bed. It is possible that their formation may increase the absorptive area to a greater extent than could be achieved by an increase in the outer surface of the epithelium. There is an increase in the amount of keratinization of the fore-stomach epithelium after birth, but to an extent that is probably influenced by diet.

In the fully differentiated epithelium, the outermost layer of keratinized cells is irregular and may be lost through friction and desquamation. The stratum corneum contains droplets of lipid that are probably triglycerides produced within the cell during the process of keratinization, and it is possible that significant amounts of lipids are continuously added to the ingesta by epithelial desquamation (Habel, 1959). In the stratum corneum there is an intercellular substance that is periodic acid–Schiff positive, and resistant to diastase digestion. It is probably a glycoprotein and resembles that described in similar epithelial tissues in other situations (Habel, 1963). The epithelial cells become swollen during the keratinization process and the outer strata may consist entirely of swollen cells. In other areas the stratum corneum consists of compact, squamous cells. It is possible that the histological appearance of the epithelium may be affected to some extent by the osmotic value of the rumen contents.

Tamate *et al.* (1964) made a histological study of the effects of diet on the development of the calf stomachs. They found that in the newborn calf receiving colostrum the epithelium of the fore stomachs was thick, with well-developed pegs extending into the submucosa. The stratum corneum was thin and the cells of the stratum granulosum contained few keratohyalin granules. At 4 weeks of age the epithelial pegs had become shorter and the stratum corneum was thin or absent. Little further change occurred in the epithelium of calves that were fed on milk only for 12 weeks. The effects of diets other than milk on the maturation of the ruminant fore stomachs are discussed later.

The epithelial cells contain numerous mitochondria, particularly in the basal layers, where they are often elongated. As is usual in epithelial cells, the longer mitochondria are orientated parallel to the axis of the cell body. Dobson *et al.* (1956) compared the disposition of the mitochondria with those in the renal tubules and suggested that they may be directly involved in the transport of sodium ions.

Mitochondria are the chief sites of oxidative phosphorylation by which energy is made available to the cell; cytochrome oxidase and many of the Krebs cycle enzymes are located there (Novikoff, 1961), and a great part of the mitochondrial membrane is probably composed of regularly spaced multienzyme systems. Recent work, reviewed by Lehninger (1962) shows that mitochondria isolated *in vitro* can actively cause transport and accumulation of certain electrolytes and water. These transport processes, which may be important in the movement of water and ions in intact cells, are dependent on respiratory energy and require both respiratory substrate and molecular oxygen. Examination of the ruminal epithelium by electron microscopy shows that there is a continuous basement membrane and that the cells are for the most part separated by spaces of 0·1 μ or more (Lindhé and Sperber, 1959; Hydén and Sperber, 1965). This arrangement would appear to allow free diffusion through the epithelium, a process that would be limited mainly by the permeability of the continuous basement membrane. Blood capillaries occur below the basement membrane, and in many cases the distance from a basal cell to the capillary is less than 0·5 μ. Pinocytotic vesicles are common in the capillary endothelium. Thus on morphological grounds one might expect the ruminal mucosa not only to be capable of active transport of water and electrolytes, but also to allow passive diffusion of substances in solution in the direction governed by the electrical and concentration gradients between blood and rumen contents.

Under the epithelium, and extending into the papillæ, there is a rich network of capillaries (Barcroft, McAnally and Phillipson, 1944a; Habel and Biberstein, 1952; Dobson *et al.*, 1956), through which absorbed substances pass *via* the portal system to the liver. The possibility of lymphatic absorption from the rumen was considered by Gray (1947), and by Elsden and Phillipson (1948). It was explored by Kiddle, Marshall and Phillipson (1951), but the volatile fatty acid content of thoracic duct lymph was found to be similar to that of arterial blood. Kiddle, Marshall and Phillipson (1951) concluded that as lymph flow is small compared to blood flow, lymph as a route of absorption of short-chain fatty acids can be ignored.

In the submucosal and muscular layers of the rumen wall there are nerve plexuses containing ganglia, and receptors that are sensitive to touch and to muscular tension (Iggo, 1955; Ash and Kay, 1959). Hill (1959) found a close correspondence between reticulo ruminal sensory areas and the disposition of free branched nerve endings in the epithelium. The endings were extracellular and usually terminated just beneath the outermost keratinized layers of epithelial cells. There may also be in the fore stomach receptors that are sensitive to the concentration of free fatty acids (Ash, 1959).

The receptors transmit impulses that modify salivary secretion, and possibly influence food intake by hypothalamic connections (Kay, 1963).

ABSORPTION FROM THE RUMEN

Digestion in the rumen requires the continuous removal of the end-products of bacterial fermentation and constant regulation of the osmotic pressure of the rumen liquor. Such regulation is achieved partly by the passage of water across the ruminal epithelium; there is evidence that water can pass in either direction and that it probably follows the osmotic gradient between the blood and rumen contents (Parthasarathy and Phillipson, 1953; Tsuda, 1964). The absorption of inorganic ions from the rumen has been reviewed by Dobson (1961). It appears that chloride will pass across the ruminal wall in either direction, generally behaving as a freely diffusible ion, and moving in the direction determined by a combination of the effects of electrical and concentration gradients. Chloride can be absorbed against its concentration gradient because of a potential that makes the plasma of the sheep about 30 mV positive to the rumen contents (Dobson and Phillipson, 1958). Sodium, however, does not behave in this way, and large quantities can pass from the rumen contents against both an electrical and a concentration gradient (Dobson, 1959). The movement of sodium against its electrochemical gradient requires energy, and Dobson (1959) concluded that sodium transport is linked to the metabolism of the epithelial cells. Inorganic phosphate passes in both directions across the ruminal wall and it appears that the high concentration of phosphate in the rumen is maintained by active transport through the ruminal epithelium (Scarisbrick and Ewer, 1951; Smith *et al.*, 1955; Smith *et al.*, 1956).

There is good evidence that steam volatile fatty acids, the main products of the bacterial fermentation of carbohydrate in the rumen, are rapidly absorbed through the rumen wall and appear in the blood draining towards the portal vein (Barcroft, McAnally and Phillipson, 1944b; Annison, Hill and Lewis, 1957). Absorption is more rapid when acid conditions prevail in the rumen, and the free acids are absorbed more rapidly than the corresponding anions (Danielli *et al.*, 1945-6; Gray, 1947; Masson and Phillipson, 1951).

The most important of the ruminal volatile fatty acids are acetic, propionic and butyric acids, and Annison and Lewis (1959) and Dobson (1961) concluded that their movement across the ruminal epithelium is probably by a simple diflusion process, governed only by the concentration gradient between the blood and the rumen contents. Danielli *et al.* (1945-6) suggested that material may move through the ruminal epithelium both by passage through water-filled pores (interstices between the cells of the epithelium) and by diffusion through the lipid membrane of the epithelial cells. It is probable that absorption from the rumen can be affected by the functional state of the mucosal capillary bed, and blood flow through the rumen wall is known to be influenced by the pH of the rumen contents and by adrenaline (Dobson and Phillipson, 1956). Such factors may be important in experiments on the isolated, perfused rumen, where some degree of anoxia is possible (McCarthy *et al.*, 1958; Lindsay, 1959).

Conflicting results have been obtained from experiments designed to investigate the relative rates of absorption of different volatile fatty acids from the rumen (Annison and Lewis, 1959; Barnett and Reid, 1961). The confusion was partially resolved by the important observations of Kiddle, Marshall and Phillipson (1951) that there is a constant difference between the butyric acid–acetic acid ratio in the rumen contents, and in the blood. There is always relatively more acetic acid, and less butyric acid, in the blood leaving the rumen than in the rumen contents. The most probable explanation appeared to be that part of the butyrate absorbed is directly metabolized by the wall of the rumen, and this hypothesis led to Pennington's (1952) work on the metabolism of volatile fatty acids by ruminal epithelium. Pennington and his co-workers found good evidence that ruminal epithelium of the sheep isolated under aerobic conditions *in vitro*, can metabolize relatively large amounts of fatty acids, particularly butyric acid, and they showed that the epithelium of the fore stomachs is an important source of ketone bodies (Pennington, 1952; Pennington and Pfander, 1957). Ketone bodies were not formed from propionic acid, but later work showed that propionic acid was metabolized to succinate in the presence of CO_2. Other short-chain fatty acids that occur in the rumen are metabolized also to some extent (Annison and Pennington,

1954). Pennington (1952) calculated that an appreciable proportion of ruminal butyric acid is removed by the epithelium before it can enter the circulation; he suggested that the ruminal epithelium may utilize the energy derived from fatty acid metabolism in connection with the active absorption of substances from the rumen.

There is evidence from the experiments of Sutton, McGilliard and Jacobson (1963) and Sutton, McGilliard, Richard and Jacobson (1963) with calves, and of Walker and Simmonds (1962) with lambs that the mucosa of the immature rumen has a low metabolic activity. Immaturity may be due to extreme youth or to lack of the stimuli that normally are associated with the ingestion of solid foods. Sutton and his colleagues found that there was a greater absorption of acetic acid, relative to an inert marker, from the rumen of calves fed for 16 weeks on milk, hay and grain than there was from the rumen of similar calves fed on milk. Changing the diet from milk to solids at 18 weeks of age quickly stimulated the absorptive powers of the rumen mucosa, showing that the diet and not age is the critical factor. At 16 weeks on the two diets there was a significantly higher *in vitro* conversion of volatile fatty acids to ketones by pieces of ruminal mucosa from the calves receiving solids than there was by mucosa from the milk-fed calves. The metabolic pattern of rumen mucosa from calves fed on milk, hay and grain resembled that of adult tissue, whereas the mucosa from milk-fed calves utilized substantially less volatile fatty acid. Walker and Simmonds (1962) found that utilization of butyrate by the rumen wall in the newborn lamb was lower than in the adult. Both the anatomical and the biochemical development of the rumen wall were retarded in lambs that received milk only before slaughter at 7 and 9 weeks of age.

In homogenates of ruminal epithelium of cattle and sheep, Onodera, Tsuda and Umezu (1964) found enzymes that probably play an important role in the metabolism of the rumen contents. These were succinic dehydrogenase, glutamic dehydrogenase, alkaline phosphatase and ketone body-forming enzymes. Succinic and glutamic dehydrogenase activities in the ruminal epithelium were generally higher than in other parts of the alimentary canal, and alkaline phosphatase activity was higher in the rumen epithelium than in the liver in both adult cattle and sheep. Ketone body-forming enzymes were low in the young calf, but activity was high in the rumen of the adult cow. Habel (1963) found by histochemical studies that the upper strata of the epithelium in the fore stomachs gave a strong alkaline phosphatase reaction, except in the newborn calf, where there was little activity.

It is well known that in the rumen there is an important cycling of nitrogen by the absorption of ammonia through the rumen wall. This reaches the liver via the portal vein and re-enters the rumen as urea in the saliva, and across the ruminal mucosa. The nitrogen is utilized for protein synthesis by the ruminal microorganisms. There is evidence, however, that the rumen mucosa can fix and utilize some of this ammonia by its glutamic dehydrogenase activity. Ruminal mucosa *in vitro* can synthesize L-glutamate by the reductive amination of alpha ketoglutarate (McLaren *et al.*, 1961).

Shoji, Miyazaki and Umezu (1964) suspected that a proportion of volatile fatty acids might be metabolized into other substances during absorption by the ruminal epithelium, but comparisons of arterial and ruminal vein blood failed to give evidence of esterified fatty acid or cholesterol synthesis in the goat.

The experiments of Armstrong, Blaxter and Graham (1957) indicate a variation in the rate of absorption of volatile fatty acids due possibly to the metabolic state of the ruminal epithelium. It was found that when acetic acid alone was infused into the rumen of starved sheep it accumulated there due to poor absorption, but that when a mixture of volatile fatty acids was infused, the rate of absorption was markedly increased. These results were possibly due to a low rate of epithelial metabolism in the starved animal, and to the stimulation of metabolism by butyric and propionic acid. Armstrong, Blaxter and Graham (1957) suggested that acid absorption is a process that necessitates an actively metabolizing rumen epithelium.

Although sugars are rapidly destroyed in the rumen by bacterial fermentation (Phillipson and McAnally, 1942), the ruminal epithelium is able to absorb glucose (Rankin, 1940); certain large molecules such as those of sodium ortho-iodo-hippurate also can be absorbed (Barcroft, McAnally and Phillipson, 1944a).

THE INFLUENCE OF DIET ON THE DEVELOPMENT OF THE RUMEN

The embryological development of the ruminant stomach has been reviewed recently by Warner and Flatt (1965). Although the differentiation of the stomach compartments is discernible in very small embryos, and the rumen is the dominant organ early in gestation, the abomasum is the heaviest organ at birth, when the fore stomachs are relatively small and non-functional (Sisson and Grossman, 1963; Dukes, 1947; Savage and McCay, 1942; Becker, Arnold and Marshall, 1951; Godfrey, 1961a). In the suckling animal, milk passes via the oesophageal groove directly into the abomasum. Paradoxically, the rumen papillae are fairly well developed in the neonatal calf (Tamate *et al.*, 1962), possibly due to stimulatory factors diffusing across the placenta, but they quickly regress if the animal is kept on a milk diet. The reticulo rumen of calves, lambs and kids fed on milk remains anatomically immature (Wiese *et al.*, 1947; Warner *et al.*, 1953; Warner, Flatt and Loosli, 1956; Tamate, 1956–7; Wardrop, 1960, 1961a, b; Walker and Simmonds, 1962), and the evidence indicates that the ruminal mucosa in a structurally immature state is characterized by low metabolic activity.

When roughage is fed, the fore stomachs rapidly increase in capacity and usually also in weight (Warner *et al.*, 1953, 1956), although Blaxter *et al.* (1952) found that the ruminal enlargement in calves fed on roughage was not accompanied by an increase in weight over that in milk-fed controls, and they concluded that the enlargement was due to stretching. Warner, Flatt and Loosli (1956), in a review of the problem, note that many workers have believed that a coarse and bulky diet is necessary for the proper development of the rumen. They record experimental results, however, that show that ruminal development and maturation of the ruminal epithelium are associated with the consumption of "dry food" and with the release of chemical factors by microbial action, and are not due to direct action of coarse materials. The same conclusion was reached independently by Brownlee (1956). Warner, Flatt and Loosli (1956) found that the rumen of milk-fed calves remained immature and the papillae were less than 2 mm in height after 16 weeks on the diet. Failure to develop rumen papillae after 11 months on liquid diet was reported by Wing and Ammerman (1960). Warner, Flatt and Loosli (1956) found that the addition of nylon bristles had no effect on papillary development. On the other hand, feeding grain (7 per cent fibre) caused rapid and marked papillary growth; the papillae approached 1 cm in height and the rumen wall was very similar in appearance to that of calves fed on hay (27 per cent fibre). In calves reared on milk, the fore-stomachs grew only in proportion to the body weight, but calves fed on grain showed a marked increase in the volume of the reticulo rumen and omasum per unit of body weight. The hay diet resulted in a greater total capacity of the reticulo rumen and omasum than that found in the other groups, but the increase in fat-free dry matter was not greater than that resulting from feeding grain.

In the normal calf fed on milk, grain and hay, or milk and grass, the greatest rate of rumen development expressed as a percentage of live weight is over the period 2–6 weeks of age (Kesler, Ronning and Knodt, 1951; Godfrey, 1961a). Godfrey found that at 17 weeks of age the reticulo rumen was still gaining in weight relative to the body weight of the calf. Grossman (1949) concluded that in the calf the fore stomachs approach adult proportions in 4–6 months and that they achieve definitive form at about $1\frac{1}{2}$ years of age.

Effective rumination begins at 2 weeks after birth and rumination time increases rapidly for the next 6 weeks in calves that are provided with roughage (Swanson and Harris, 1958). Martin *et al.* (1959) fed calves from 5 days to 12 weeks of age on casein, fat, glucose, starch and hay and found that appreciable volatile fatty acid production occurred when the animals were about 6 weeks old. It seems probable that in calves with free access to roughage the qualitative development of rumen function is complete by at least 2 months of age (Godfrey, 1961b).

In sheep, the rumen had the fastest rate of growth of the fore stomachs in lambs suckling ewes at pasture, and the fore stomachs reached approximately adult proportions and function at about 8 weeks of age (Wardrop and Coombe, 1960, 1961). The rumen of lambs on pasture contained appreciable amounts of herbage after the first 21 days. The observations of Wardrop and Coombe (1960) were confirmed by Church, Jessup and Bogart (1962), who noted progressive histological changes in the organization of the papillary tissue. Omar, Reagor and Kunkel (1964) found that in creep-fed sucking lambs the fore stomach reached adult proportions in 5–9 weeks, but that the mucosal component of the rumen wall continued to develop after 9 weeks. There was a steady increase in size, and decreasing density of papillae in the ruminal mucosa, and in general the linear dimensions of the papillae were proportioned to body weight, i.e. the papillae did not show the changes in relative growth shown by the whole reticulorumen. The proportions of propionate and butyrate found in the rumen fluids were positively correlated with papillary development; this correlation was partly due to a steady increase in the butyrate fraction with age.

Wardrop (1961b) noted that the histology of the rumen wall was normal in a milk-fed lamb but the papillary body formation, stratum granulosum and stratum corneum were less well developed than the corresponding structures in a lamb of the same age that had been fed on lucerne chaff. Similar differences were found in the histology of the reticulum, and the structure of the omasum in the milk-fed lamb resembled that found in a newborn animal.

Histological studies were made by Tamate *et al.* (1964) on the effect of various diets on the development of the calf stomach. There were few changes in calves fed for 12 weeks on milk only, but in calves fed on milk, hay and grain there was a rapid cornification of the epithelium of the fore stomachs. Similar changes resulted when the solids were replaced with volatile fatty acids.

It is possible that the physical form of the ration also may influence the appearance of the ruminal epithelium. Thompson, Kintner and Pfander (1958) compared the effects of a completely pelleted ration containing 50 per cent roughages with conventional forms of the same diet fed to lambs. The pelleted ration resulted in increased papillary size in the rumen, hypertrophy of the epithelium and thickening of the keratinized layer.

Care must be taken in such dietary studies to distinguish between a possibly beneficial stimulation of the ruminal mucosa and the pathological state of ruminal parakeratosis. This is discussed in the next section.

Recent work shows that thickening of the ruminal mucosa can to some extent be dissociated from the other anatomical manifestations of maturation. Brownlee (1956) noted that the development of rumen muscle in calves fed on milk was not accompanied by a corresponding papillary development. He examined the rumen of cattle kept on various diets and concluded, as did

Warner, Flatt and Loosli (1956), that the development of papillae depended on the nature of the food but was not stimulated by roughage *per se*; it was influenced by the energy value of the foods or by the rapidity of fermentation. Animals fed on kale and other foods relatively low in fibre had greater papillary development than that resulting from eating hay.

Flatt, Warner and Loosli (1958) found that the presence of plastic or cellulose sponges, inserted through rumen cannulae, did not stimulate development of rumen papillae; neither did the consumption by young calves of considerable amounts of wood shavings (Smith, 1961). However, the introduction of a large number of sponges, each small enough to be regurgitated, did cause considerable increase in the ruminal musculature and capacity (Tamate *et al.*, 1962). Harrison *et al.* (1960) reared calves on high-roughage and high-concentrate diets until there was a dense development of ruminal papillae, and then substituted a milk diet. Under these conditions, the rumen papillae regressed and virtually disappeared. Evidence of the independence of mucosal and muscular growth was obtained in that the mucosal regression was the more rapid, and in other calves that consumed milk and shavings, muscular development was extensive in the absence of mucosal development. The concept of chemical stimulation of the ruminal epithelium by the products of microbial fermentation was supported by the further studies of Flatt, Warner and Loosli (1958). These workers demonstrated that whilst sponges did not stimulate the development of rumen papillae in young calves, there was a marked development of the mucosa when solutions of the sodium salts of acetic, propionic, butyric and lactic acids were introduced through rumen cannulae. Similar results were obtained by Sander *et al.* (1959) and by Tamate *et al.* (1962). Sander *et al.* (1959) found that butyrate and propionate caused the greatest development of the ruminal tissues: there was very little papillary response when glucose solutions were used.

Tamate *et al.* (1962) studied organ development and papillary growth in the floor of the anterior dorsal blind sac of the rumen in calves fed on milk with various supplements. They concluded that there was a reciprocal relationship between papillary length and number of papillae/cm^2 and that in some instances, at least, this was due to the fusion of papillae during hypertrophy. The general impression obtained from the various experiments that involved periodic introduction of volatile fatty acid salts into the calf rumen is that the stimulation of papillary growth is appreciable only if the acids are given in sufficient concentration and frequency and over a period of several weeks. The minimum amount of volatile fatty acid required to stimulate papillary growth was estimated by Tamate and his colleagues to be about 8 molar equivalents per week.

Calves fed on a diet containing salts of volatile fatty acids can absorb appreciable quantities of volatile fatty acid at 3 weeks of age, which is earlier than normal (Martin *et al.*, 1959). Sutton, McGilliard and Jacobson (1961, 1963) found that calves fed on milk, alfalfa pellets and grain could absorb volatile fatty acid at an earlier age, and at a greater rate, than similar calves fed on milk. One may conclude, therefore, that the changes in the ruminal epithelium brought about by volatile fatty acids are associated with enhanced powers of volatile fatty acid absorption. However, the depressing effect of starvation on absorption, and other experimental results discussed by McGilliard, Jacobson and Sutton (1965) shows that the realization of the absorptive potential of the rumen wall requires the continuous presence of volatile fatty acids. McGilliard, Jacobson and Sutton (1965) suggested that volatile fatty acids stimulate the metabolism of the mucosa and that the enhanced metabolism leads independently both to the structural changes and to increased absorptive ability.

Possible mechanisms of the effect of volatile fatty acids on the ruminal epithelium are discussed by Sander *et al.* (1959). They suggest that the fatty acids have physiological or biochemical properties that affect the ruminal mucosa during absorption. They support this view by reference to the work of Pennington and his colleagues on the metabolism of volatile fatty acids by ruminal epithelium. However, glucose, which is also metabolized by sheep ruminal epithelium *in vitro* (Pennington and Sutherland, 1956), did not have an appreciable effect on the ruminal walls of calves when introduced into the rumen by Sander *et al.* (1959). Very little glucose would be present in the normal rumen (Elsden and Phillipson, 1948), but blood-borne glucose would be available to the epithelium.

The absorption of volatile fatty acids from the rumen appears to be accompanied by an increase in the blood flow in the rumen wall (Dobson and Phillipson, 1956). This is a possible means by which fatty acids can affect the trophic state of the epithelium. Furthermore, Phillipson (1959) found that the relative rates of increase in blood flow due to fatty acids was butyrate—propionate—acetate, which is the order of activity in which the acids can stimulate the ruminal mucosa in the normal intact animal.

The type of food eaten by ruminants, or possibly the intensity of microbial activity in the rumen, has an effect on the colour of the ruminal epithelium. It is a common experience to find that young ruminants fed on milk, or on certain synthetic diets, lack the pigmentation that is seen in the ruminal mucosa in animals consuming conventional diets.

In both lambs (Sinclair and Kunkel, 1959) and calves (Tamate *et al.*, 1962) the pigment, when present, appears to be restricted to the layer of cornified cells. Sinclair and Kunkel (1959) found that the growth and development of lambs as a whole can be correlated with ruminal development. They reported that there was also a significant correlation between gain in body weight and mucosal colour in the rumen and they suggested that the pigment may provide a means of evaluating the activity of the ruminal flora.

When pigment is present in the keratinized layer of the calf fore stomachs, the same tissue gives a positive reaction for iron. The omasum is usually similar to the reticulo rumen in intensity of pigmentation, but there is a sharp dividing line between the epithelium of these structures and the unpigmented œsophageal and abomasal epithelium (Brownlee and Elliot, 1961). Brownlee and Elliot found that calves reared on a milk and hay diet had a straw-coloured epithelium in the first three stomachs, and the epithelium gave a negative reaction for iron. Similar calves fed on a ration of concentrate and hay had a darkened epithelium that contained iron. Brownlee (1962) associated pigmentation of rumen epithelium with concentrate feeding and he suggested that it may indicate lowered rumen function.

RUMINAL LESIONS IN INTENSIVE FEEDING

Lesions may occur in the ruminal epithelium of cattle under some forms of intensive management, and particularly so after fattening on diets containing much starch. The cause of the lesions has not been fully elucidated, but it is very probable that the damage is due to persistent conditions of unphysiological acidity. The lesions are inflammatory and acute rumenitis, with epithelial necrosis, hyperæmia and hæmorrhage in the papillæ, may progress to papillary necrosis and ulceration. The condition can develop into, or be superimposed upon chronic rumenitis, with loss of pigment, loss of papillæ and scar formation. There is a marked infiltration of inflammatory cells in the affected areas, and in the acute condition the epithelium contains vesicles that are commonly filled with leucocytes. There is usually hyperkeratosis, with clumping of villi. Liver abscesses occur in association, and it is probable that pathogenic organisms enter through the eroded areas of ruminal mucosa and pass to the liver via the portal vessels. Under American conditions, at least, the liver abscesses are probably due to invasion by *Spherophorus necrophorus* (*Fusiformis necrophorus*) and colonies of the organism occur in the necrotic areas of ruminal epithelium (Jensen, Flint and Griner, 1954; Jensen, Deane, Cooper, Miller and Graham, 1954; Smith, 1963). The organism is part of the normal alimentary microbial flora and it presumably plays a secondary role in the ruminal lesions, probably in much the same way that fungi may invade ulcer craters in the abomasum of cattle (Jarrett, 1955). *F. necrophorus* has been recorded also as a secondary invader of the omasal mucosa (see discussion by Brownlee and Elliot, 1960).

The syndrome of rumenitis with liver abscesses occurs commonly in Britain in animals fed on diets containing a high proportion of barley and it closely resembles the disease described by American workers (Wieser, 1965). It is not certain, however, that *F. necrophorus* is the cause of the liver abscesses in this country. Jensen, Connell and Deem (1954) found significantly more rumenitis in cattle when rations were changed rapidly from roughage to concentrate than when the change was made over a longer period; the incidence also varied with the proportion of concentrate in the diet.

There have been several reports on the production of parakeratotic lesions in the rumen of lambs and cattle fed on concentrate-containing rations that have been ground and pelleted, and it appears that the particle size within the pellet may be of primary importance. The condition was described first by Jensen *et al.* (1958) in lambs fattened on a pelleted feed. Jensen *et al.* described ruminal parakeratosis as a non-contagious disease characterized by hardening, enlargement and clumping of mucosal papillæ. Microscopically there is an excessive accumulation of nucleated and keratinized epithelial squames on the papillæ. Affected lambs showed relatively poor weight gains. Garrett *et al.* (1961) found that the incidence and severity of rumen parakeratosis was greatest in steers receiving a finely ground diet in $\frac{1}{4}$ in. or $\frac{5}{8}$ in. pellets. Addition of antibiotic to the feed did not affect the incidence of the condition, but feeding long oat hay with the pellets reduced the severity of the lesions.

The relationship between physical form of diet and ruminal lesions is discussed by Beardsley (1964).

The rapid transition from diets high in roughage to diets composed mainly of cereals may result in acute toxic effects. This syndrome can follow the sudden introduction of *ad lib.* feeding of rations containing 85 per cent of barley (Macdonald, Bradley and McCrea, 1962), and is probably due to acidosis (Broberg, 1960).

Certain hazards are associated with the practice of early weaning. Phillipson and Cuthbertson (1956) could find little reason to advise delay in the treatment of young ruminants as adult for longer than 6 weeks with calves, or 3 weeks with lambs, from the point of view of feeding provided adequate milk is given. They recognized that there is some danger of excess lactic acid production due to the preponderance of lactobacilli in the rumen. However, Hall (1962) has drawn attention to the high incidence of impaction of the rumen and abomasum due to gastric dysfunction and indigestion of fibre, in early weaned calves.

HISTAMINE AND THE RUMINANT

The effects of high-starch diets, and overeating, introduce reference to histamine production and absorption in the rumen. Usually, amino acids including histidine, are deaminated by the ruminal microorganisms with the release of ammonia (for review see Lewis, 1955), but there is evidence that when glucose is added to the rumen there is a sharp increase in the concentration of amino acids and other basic components, including histamine (van der

Horst, 1961). Sanford (1963) reported an increase in the histamine content of the ruminal fluid in sheep that had been dosed with glucose. The increase appeared to depend on a fall in pH, and where high acidity persisted there was a marked increase in the histamine content. Dain, Neal and Dougherty (1955) reported the occurrence of histamine and tyramine in the rumen ingesta of experimentally overfed sheep. Histamine levels of more than 70 μg/ml of ingesta were obtained, and there was a direct correlation between the amount of histamine in the ingesta and the signs of toxicity. It is well established that the intravenous injection of histamine results in rumen paralysis, and predisposition to bloat is probable in this situation. The ingesta of normal sheep contains very little histamine (Dain, Neal and Dougherty, 1955).

Lactobacilli possessing active amino-acid decarboxylases were isolated by Rodwell (1953) from horse-stomach and sheep-rumen contents. Certain strains of these bacteria produced relatively large amounts of histamine when incubated anærobically in wheat-digest histidine broth. Rodwell found histamine present in abnormally high concentrations in the rumen of sheep after wheat feeding and in the stomach of horses that had gorged on wheat. However, not all the histamine that occurs in the rumen is of endogenous origin. Sjaastad and Stormorken (1963) found that grass silage contains relatively large amounts of histamine.

Shinozaki (1957) obtained evidence that histamine is not absorbed to a significant extent from the rumen of sheep. There were no clinical effects after introducing histamine through a rumen fistula, and histamine that was placed in a rumen pouch did not decline in concentration. On the other hand, Sjaastad and Stormorken (1963) found that there was a sharp peak in the concentration of histamine in the rumen liquor of sheep after feeding silage. The decline was not due to microbial metabolism and there was a large increase in the urinary excretion of acetylhistamine. These workers obtained evidence that moderate quantities of histamine can be metabolized by the rumen wall. This capacity would be exceeded when the rumen contained much histamine, when detoxification would probably occur in the liver.

It is interesting to speculate on the situation that might occur in intensively fed cattle that have erosions of the rumen and multiple liver abscesses.

The state of the ruminal epithelium is of particular interest in zinc deficiency, when the epithelium of the skin and œsophagus is markedly affected. In several species the epidermis shows the pathological changes of acanthosis and hyperkeratinization, or of parakeratosis (Follis, 1948; Ott *et al.*, 1964). It appears that the ruminal epithelium could be affected directly by the deficiency, and also that it could be influenced by changes in volatile fatty acid production in the rumen, such as those found by Ott *et al.* (1964) in zinc-deficient lambs. Parakeratotic changes in the ruminal epithelium of zinc-deficient calves have been reported by Miller and Miller (1960, 1962).

THE OMASUM

The omasum is situated between the reticulum and the abomasum and it appears to regulate the onward passage of ingesta by a valve-like action. The numerous experiments that have been conducted on the effect of diet on maturation of the ruminant fore stomach show in general that omasal development, like that of the reticulo rumen, is stimulated by feeding fermentable substances. The percentage of total stomach tissue contributed by the omasum changes very little with age in sheep, but increases in cattle for about a year (Warner and Flatt, 1965).

There is a wide variation in the number and arrangement of laminæ in the ox omasum (Becker, Marshall and Arnold, 1963), but there are usually about 100 laminæ of different sizes or orders that resemble the pages of a book. Layers of smooth muscle run up the centre of each lamina into the terminal thickening. The laminæ are prominent from 70 days of fœtal life in sheep, the major changes with age being an increase in length and the development of papillæ (Wardrop, 1961a). At birth, the laminæ in the lamb are closely packed and have smooth outlines. The stratum corneum consists of a thin layer of flattened cells. Little change occurs in lambs that are fed on milk only, but when there is access to solid food the laminæ increase in length, the papillæ become more prominent and the stratum corneum increases in depth (Wardrop, 1961b). The semi-liquid ingesta leaving the reticulo rumen must pass through the neck or between the folds of the organ. Filtration and trituration have been regarded as its main function, but the omasum is now looked upon as an effective site of absorption. However, McAnally and Phillipson (1944) noted that short-circuiting the organ by anastomosis of the reticulo rumen and abomasum did not seriously impair the health and digestion of the animal, and Brownlee and Elliot (1960) reported that animals can readily adjust themselves to loss of omasal function due to extensive adhesion or fenestration of the laminæ.

The laminæ, which form the major part of the organ, are covered in a relatively thin non-glandular stratified squamous epithelium (Favilli, 1937; Brownlee and Elliot, 1960). The mucous membrane is thin on the floor of the organ and in goats it may contain mucous glands (Habel and Biberstein, 1952). The surfaces of each fold are studded with macroscopic papillæ and these are largest near the reticulo-omasal orifice. Some of the papillæ on the surface of the larger laminæ have a horny cap, and under this the epithelium is thickened and shows a distinct stratum spinosum. The cores of the larger papillæ have metachromatic staining properties that may be associated with the presence of mucin and water absorption (Habel and Biberstein, 1952; Brownlee and Elliot, 1960). There is a rich blood supply to the organ, and the mucous membrane has a well-developed capillary bed. (Favilli, 1937; Barcroft, McAnally and Phillipson, 1944a).

Characteristically, the ingesta found in the omasum is high in dry matter, and fine particles of food occur between the laminæ. Favilli (1937) suggested that contractions of the muscular wall squeeze fluid out of the ingesta; the fluid is retained between the leaves by swelling of the free border of the leaves and by contraction of the omasal groove, and absorbed. Favilli concluded that the omasum has a very limited significance as a grinding organ.

This view is supported by the observations of Balch, Kelly and Heim (1951), who suggested that much of the ingesta passes through the neck of the omasum and that muscular contractions of the wall expell liquids that then pass, with particulate matter, between the lamellæ of the organ.

Gray, Pilgrim and Weller (1954) calculated that 33–64 per cent of water and 40–69 per cent of fatty acids were absorbed from the digesta that entered the omasum of sheep, and Badawy *et al.* (1958b) found that 48–55 per cent of water was absorbed in the organ. However, Phillipson and Cuthbertson (1956) did not rule out the possibility that fluid material from the reticulum, or fluid squeezed out of ingesta in the omasum, could enter the abomasum in advance of the more solid material. This view is supported by the experiments of Oyært and Bouckært (1961) who found by omasal cannulation that part of the liquid leaving the reticulum passes directly into the abomasum.

Most of the protozoa that enter the omasum are destroyed there, and there may be an extensive destruction of starch (Weller and Gray, 1954). Smiles and Dobson (1955) reported that during the passage of ingesta through the sheep stomachs, degenerative changes in rumen organisms begin in the omasum and become more marked in the abomasum. Probably the osmotic changes and mechanical effects in the omasum are sufficient to kill the organisms; the starch granules that are released from protozoa are then exposed to enzymes of microbial origin that are present in solution.

Absorption of steam volatile fatty acids occurs through the omasal epithelium, and significant amounts are present in the blood draining the omasum, so that the material entering the abomasum is practically free from these substances. Volatile fatty acids are not present in the blood draining the abomasum and small intestine of sheep (Barcroft, McAnally and Phillipson, 1944b).

The idea that one of the functions of the omasum is to absorb water is supported by the observation that there is a considerable concentration of inorganic phosphate, soluble calcium and magnesium in the omasal liquor of slaughtered sheep, when these are compared with the corresponding values in rumen liquor (Garton, 1951).

THE GLANDULAR STOMACH

In animals with simple stomachs the organ functions in part as a temporary food reservoir, and gastric function is intermittent; in the ruminant the reticulo rumen is normally never empty and there is a continuous supply of ingesta to the abomasum—the true glandular stomach.

The gastric mucosa is characterized by the presence of gastric glands; in the horse and pig there is an œsophageal region or fore stomach that is analogous to the non-glandular fore stomach of the ruminant.

The fore stomach in the horse is relatively large and is lined with stratified squamous epithelium. It is probable that some microbial fermentation of ingesta and absorption of microbial products occurs in the cardia of the pig stomach (Barcroft, McAnally and Phillipson, 1944b) and in the horse stomach (Rodwell, 1953).

The non-glandular œsophageal region is much smaller in the pig than in the horse, and much of the proximal part of the stomach is occupied by a thin mucosa that contains much lymphoid tissue and many branched and coiled mucous glands.

Generally, the main part of the body of the stomach is occupied by a relatively thick mucosa containing simple, tubular fundic (peptic) glands that open into pits on the surface of the mucosa. The secretions of the fundic glands are mucus from cells in the neck of the gland, hydrochloric acid from the large parietal cells and pepsin from the zymogenic (chief, or peptic) cells in the body of the gland (Habel and Biberstein, 1952). The zymogenic cells are influenced by the secretions of the adrenal cortex and quickly show atrophic changes in the adrenalectomised animal (Baker and Bridgman, 1954).

The distal end of the stomach is occupied mainly by coiled and branched mucin-secreting pyloric glands.

The abomasum is characterized by the presence of numerous permanent longitudinal folds that begin abruptly near the omaso-abomasal orifice and become smaller towards the pyloric region. The surface of the pyloric region is smooth apart from a few small rugæ.

The mucosa covering the folds presents a large surface area and contains very many fundic glands.

Sommerville (1956) has described the distribution of parietal and zymogenic cells in the sheep abomasum. The numbers of zymogenic cells remained relatively constant from the start of the peptic zone to within about 5 cm of the pyloric region, where they declined abruptly. The parietal cells were less numerous than the zymogenic cells, and they decreased in number with distance from the anterior end of the abomasum. This distribution is similar to that described by Hill (1951) in the goat, except that Hill found the maximum number of cells close to the cardiac region, with a gradual fall towards the pylorus.

In both goats and sheep there is about the omaso-abomasal orifice, a narrow zone of mucous glands that may be regarded as forming the cardiac region (Hill, 1951; Sommerville, 1956). At the junction of fundic and pyloric regions there is an intermediate zone which consists of mixed glands. This region in sheep is devoid of zymogenic cells but not of parietal cells.

The mucosa of the abomasum contains considerable amounts of lymphoid tissue. In the deeper parts of the gastro-intestinal mucosa, in a peripheral position about the gastric glands and intestinal crypts, there are argentaffin cells. There is evidence that these cells are associated with the formation of 5-hydroxytryptamine (Campbell, 1959).

In the sheep abomasum the argentaffin cells occur most commonly in the middle of the fundic region of the mucosa, and are absent from the pyloric region (Sommerville, 1956). In the goat, however, argentaffin cells are most numerous in the cardiac zone and in the first few centimetres of the fundic region. In the remainder of the fundic region only a few scattered cells occur, but they are again numerous amongst the pyloric glands (Hill, 1951). The distribution of argentaffin cells in the pig stomach resembles that in the goat; in the horse stomach they are confined to the pyloric region (Jacobson 1939).

The surface of the mucosa, except in the stratified squamous zone of the horse and pig fore stomachs, is covered with a single layer of columnar epithelial cells that continue into the gastric pits and are joined by transitional stages with the cuboidal cells of the glandular epithelium. The surface epithelium commences abruptly at the border of the stratified squamous epithelium of the œsophageal mucosa or at the omasa-abomasal junction. The epithelium is renewed by mitotic divisions in the necks of the glands and in the region connecting the gastric pits with the glands. Some of the cells produced by these divisions are mucous neck cells. Others differentiate into typical surface cells in the pits and migrate along the basement membrane to replace worn-out or extruded elements of the surface epithelium. Mitoses occur also at the junction of pits and pyloric mucous glands. The subject of renewal of cell populations in the digestive and other systems has been reviewed by Leblond and Walker (1956). The epithelial cells contain numerous rod-shaped mitochondria and in some species, have a striated border. They secrete mucus.

Leucocytes occur between the epithelial cells, apparently in the course of migration into the lumen. Plasma cells and eosinophilic granular leucocytes are common in the tissues of the lamina propria.

The Schollenleukozyten of Weill (1919), although not regular constituents of the mucosa, commonly occur between the surface—and glandular—epithelial cells of the abomasum. Maximow (1924) was not able to reach a definite conclusion about the nature of these cells but recent workers have tended to group them with the plasma cells (Pearse, 1949; Kirkman, 1950; White, 1954). Such classification implies that they may be concerned in the immune response, such as local antibody production in the mucosa (Pierce, 1959). Sommerville (1956) found some indication of an indirect association between the presence of these cells and nematode infestations of sheep. Fell, Campbell and Boyne

(1964) did not attempt to distinguish between Schollenleukozyten and plasma cells containing Russell bodies.

The stomach is richly supplied with blood and lymph vessels, and there are many autonomic ganglia in the submucosa and between the layers of the tunica muscularis. The vascular bed has many submucosal arterio-venous connections and arterio-venous shunting of blood readily occurs (Barclay and Bentley, 1949).

Hill (1956) has presented an excellent description of the development of the gastric mucosa in sheep, in which he draws attention to the existence of a mechanism whereby protein digestion is delayed until after the absorption of colostrum. Hill found that zymogenic cells gradually increased in number in the fœtal lamb from 85 days to term, but that parietal cells were very sparse until after birth. The abomasal contents were neutral in reaction in the fœtus and at birth, but the parietal cells rapidly increased in number during the first 3 days of life, and many mitoses were seen in the middle and lower third of the gastric glands. At 36 hr *post partum* the pH was adequate for the activation of pepsin. In the rat, which can rapidly absorb antibodies from the alimentary canal for at least 18 days after birth (Halliday, 1955), the gastric mucosa was very poorly developed at birth and differentiation proceeded slowly until at 25 days the mucosa resembled that of the adult rat. A comparatively low pH was not attained until after the normal weaning time of 20–25 days. The special ability of the newborn to absorb antibodies is not solely due to the lack of enzymatic activity, however, as is shown by Halliday's (1955) data on selective absorption of macromolecules. Antibody transference occurs *in utero* in the guinea-pig and in this species the gastric glands were well defined in late fœtal life and were fully differentiated at birth. The stomach contents of very young guinea-pigs had a pH of 1·0–2·0.

These results, as Hill points out, show that the absorption of proteins from colostrum is ensured not only by the special permeability of the intestinal epithelium in the newborn, but also by the delayed development of proteolytic digestion. The presence of a trypsin inhibitor in colostrum appears to protect further the proteins by shielding them from intestinal proteolytic secretion (Laskowski and Laskowski, 1951).

Tamate *et al.* (1964) found that the development of the glands in the abomasum of the newborn calf was similar in some respects to that described by Hill (1956) in the lamb. At birth, the fundic glands were short and both zymogenic and parietal cells were small or non-functional in appearance. Parietal cells were few in number; the zymogenic cells were relatively numerous but were deficient in secretory granules. By the third day *post partum* the cells had acquired adult characteristics and the fundic glands had lengthened considerably. The proportion of parietal cells to total cells in the body of the glands had increased by the fourth week to about 30 per cent, and thereafter remained at much the same level up to 12 weeks of age.

Diet had a significant effect on the development of the abomasum in that in milk-fed calves, the combined lengths of gastric pit and glandular neck was considerably greater than the length of the body of the gland, whereas in calves of the same age reared on milk and solid foods, or on milk and volatile fatty acids, the glands increased markedly in total length and also the body of the gland lengthened at the expense of the neck.

Tamate *et al.* suggested that as in the ruminal mucosa, the development of the fundic glands is stimulated by volatile fatty acids.

The continuous nature of abomasal secretion has been emphasized by Hill (1961a, b, 1965) in recent reviews of ruminant physiology. There is a more or less continuous passage of digesta from the rumen to the abomasum, and as there is no cephalic phase of abomasal secretion in the sheep, there is no obvious relation of secretion to feeding (Hill, 1955, 1960). There is a reduction in the amount of abomasal secretion in the starved animal, after removal of the rumen contents or ligation of the omasum, indicating that the normal continuous secretion is not due to an innate function of the glands but is a response to the stimulus of ingesta entering the abomasum (Hill, 1955). Ash (1961) concluded that the inflow of food from the fore stomachs, acid secretion by the fundic glands and flow of material from the abomasum are normally integrated, and that the acidity of the abomasal contents is of major importance in the control of acid secretion.

Large amounts of volatile fatty acids are produced in the rumen by microbial action and Hill (1960) found that volatile fatty acid and in particular, acetic acid, is a potent stimulus to gastric secretion. The volatile fatty acid levels in the abomasal ingesta are usually very low, as the absorption of these acids takes place almost entirely in the reticulo rumen and omasum (Barcroft, McAnally and Phillipson, 1944b).

Fell, Campbell and Boyne (1964) reported that there is a considerable increase in gastro-intestinal weight and total nitrogen content, due to a general hypertrophy of the tissues of the stomach and intestinal wall, in sheep that are lactating at pasture. The weight of the abomasum and other alimentary organs gradually returns to normal after weaning. Fell, Campbell and Boyne (1964) suggested that the hypertrophy is a response to the increase in food consumption normally associated with lactation. During the regressive phase there is a distinct atrophy and even a degeneration of the stomach mucosa and associated with these changes there is a massive infiltration by plasma cells containing Russell bodies and by Schollenleukozyten.

There are analogies to this cellular reaction. Although plasma cells are known to be closely associated with the immune response, and are the probable source of circulating antibodies, they also occur prominently in degenerating tissues where it is doubtful or at least not proven that they produce specific antibodies, e.g., in the vicinity of neoplasms and in the ligated kidney. If they are stimulated to occur in such large numbers by the onset of extensive atrophic changes in the mucosa, then it is possible that the normal occurrence of these cells and Schollenleukozyten in the abomasum is due to patchy ageing or degeneration of the mucosa with subsequent regeneration. On the other hand, Sommerville (1956) has obtained some evidence of a relationship between these cells and the extent of gastro-intestinal parasitism. The well-known phenomenon of a spring rise in nematode egg output is now known to be closely associated with lactation in the host (Gibson, 1965). It is possible that the increased food intake, or the hypertrophied alimentary canal, leads to an increase in the worm burden and that at some stage this stimulates an immunological reaction by the cells of the gastro-intestinal mucosa.

The copious secretions of the abomasum fully replace the fluids absorbed in the omasum of both cattle and sheep (Garton, 1951; Hydén, 1961). Hogan (1964) estimated the rates of flow of ingesta by reference to the marker ^{51}Cr EDTA and found that there was a considerably greater flow from the abomasum than from the reticulum in sheep eating large quantities of lucerne hay or green feed.

Hill (1961a), in discussing the very large output of abomasal secretions in cattle and sheep, points out that such production is made possible by the large area of mucosa formed by the abomasal folds. In animals with simple stomachs there are quiescent periods between meals during which there is a re-synthesis of pepsinogen granules (Babkin, 1950), whereas in ruminants the output is uninterrupted. Hill suggested that there may be phasic activity in the abomasum so that small segments of mucosa may enter a temporary resting phase.

Pepsin and rennin are the main proteolytic enzymes secreted by the stomach. The main secretion by the calf abomasum is rennin, and the enzyme has been purified and crystallized from extracts of the mucosa (Berridge, 1945). There is a decline in rennin production, and an increase in pepsin production, when the diet of calves is changed from milk to solid food (Berridge *et al.*, 1943). Berridge *et al.* (1943) obtained evidence that this change is considerably delayed in animals kept on a milk diet, indicating that food is as important as age in determining the character of abomasal secretion.

The specificity of the action of rennin on casein means that other proteins are not coagulated in the stomach. The flow of material from the abomasum of the young ruminant normally is regulated by clot formation, and severe diarrhœa results when clotting is inhibited or when diets low in casein are given. Bacteria play only a secondary role in such dietary scouring (Blaxter and Wood, 1953).

Hill (196lb) points out that the location of the rennin-secreting cells in the mucosa is not known; possibly the zymogenic (peptic) cells contain the precurors of both rennin and pepsin either in separate subcellular systems or in the same granules.

The acid secretions of the stomach not only provide the optimum pH for the activation of pepsin but also destroy ingested bacteria, including ruminal organisms (Smiles and Dobson, 1955) and pathogenic bacteria whose portal of entry is the alimentary canal. Smith (1954) demonstrated that the nature of the diet can have a profound influence on the bactericidal action of the gastric juice in the chicken gizzard. Another important effect of gastric acidity is to increase the solubility of inorganic elements of the ingesta (Garton, 1951).

Impaction of the abomasum with sand or fibrous material may occur, and, as in so many other species, the mucous membrane of the ox stomach is subject to erosions and peptic ulceration (Pinsent, 1962).

THE SMALL INTESTINE

The small intestine is a long, tubular and muscular organ of digestion, absorption and excretion through which food is propelled by peristalsis. It begins at the pylorus and ends at the lesser curvature of the cæcum. It is lined with a continuous sheet of simple columnar epithelial cells that have marked powers of absorption.

Conventionally, the small intestine is divided into duodenum, jejunum and ileum. Anatomically, the duodenum is characterized by its relatively fixed position and by its short mesenteries. The remainder of the small intestine is relatively mobile and no distinct point exists at which to distinguish jejunum from ileum.

The duodenum is distinguished histologically by the presence of duodenal glands (Brunner's glands) in the submucosa. The anatomical and histological criteria do not agree, as is shown by Table 1.

TABLE 1
Length of the Duodenum in the Domestic Animals

Anatomical duodenum (Sisson and Grossman, 1938)		Histological duodenum (Habel and Biberstein, 1952)
Horse	1 m	5–6 m
Ox	1 m	4–5 m
Sheep	60–70 cm	60–70 cm
Pig	60 cm	3–5 m

The duodenal glands are branched, tubuloalveolar glands that have a voluminous mucoid secretion. Elias (1947) has described the marked species differences that occur in the finer anatomy of these glands amongst the domestic animals. One probable function of the duodenal glands is to protect the intestine against the effects of the gastric secretions and Florey *et al.* (1939) have shown in the pig that the duodenum is more resistant to the effects of undiluted gastric juice than is the rest of the small intestine. A fully functional duodenum can resist the effects of gastric juice even in the absence of bile and pancreatic juice. Occasionally, epithelial damage in the duodenum can result in replacement by mucin-containing columnar cells that resemble those of the gastric epithelium, and it is possible that these are derived from the ducts of the duodenal glands (Florey *et al.*, 1939).

The surface area of the intestine is greatly increased by the presence of villi, which are finger- or leaf-like projections of the mucosa. The villi show pulsating movements in the living animal by which lymph is pumped from the central lacteals into the efferent lymphatics. Villus contraction is due to the action of strands of smooth muscle that originate in the muscularis mucosæ and insert in the tissues of the lamina propria, and extension is due to the pull of transverse elastic fibres. Yoffey and Courtice (1956) have provided an extensive review of the anatomy and physiology of the lymphatic drainage of the alimentary canal.

The villi are covered with columnar epithelial cells (principal cells; chief cells) interspersed with goblet cells. The goblet cells contain mucus, but this is usually discharged by the time the escalating cells have reached the villus apex. Between the villi are the simple, tubular crypts of Lieberkühn. Small numbers of Paneth cells and argentaffin cells are present in the crypts. The Paneth cells probably secrete carboxypeptidases and the argentaffin cells are usually associated with 5-hydroxytryptamine. Villi and crypts occur throughout the small intestine; the villi become smaller towards the distal end of the intestine.

The epithelium of the intestinal surface and of the crypts is continuous and is composed of similar elements, but the cells of the crypts are relatively undifferentiated.

CELL RENEWAL AND MATURATION IN THE INTESTINAL EPITHELIUM

Numerous mitoses occur amongst the cells in the crypts of Lieberkühn, and there is a considerable amount of evidence that the cells so formed migrate from the crypts on to the villi and are eventually extruded from the villus tips into the intestinal lumen. Friedman (1945) found that irradiation of the rat duodenum inhibited cell division in the crypts, but not the maturation of goblet cells. The goblet cells ripened in the crypts, where they were arrested, causing the so-called mucous change. After 2 or 3 days, normal migration was resumed and there appeared strata of fully formed goblet cells at successive levels along the villi. The phenomenon of the high mitotic index in the crypts was investigated by Leblond and Stevens (1948) in the rat, and it was found that there is continuous mitotic activity throughout the day in the epithelial cells of the crypts and that there is a migration of cells to the tips of the villi. Leblond and Stevens easily found degenerating epithelial cells in the succus entericus after a period of fasting. Although the constant migration and shedding of cells causes interruption of the epithelium at the villus tip, the basement membrane remains continuous.

Calculations by Leblond and Stevens, based on their assessment that about 3 per cent of the cells are dividing at any one time and that the duration of mitosis is 1 hr 8 min, reveal that the time taken for the total renewal of the epithelium, or average life of the cells, is about 36 hr. By autoradiography, Leblond, Stevens and Bogoroch (1948) found that animals killed 2 hr after injection of ^{32}P had strong activity over the crypts, indicating uptake of the isotope into newly synthesized DNA; in animals killed 24 hr after the injection the activity was mainly over the cell nuclei of the villus epithelium. The rate of renewal of intestinal epithelium in the cat was studied by McMinn (1954). McMinn confirmed the conclusions of Leblond and his collaborators and calculated that the villus epithelium is renewed in 2·25 days in the duodenum and 2·75 days in the ileum. The mature cells were shed from "extrusion zones" at the tips of the villi. McMinn concluded that the crypt cells can be regarded as a self-replenishing reservoir that is constantly delivering cells to the villus group, in which cell divisions do not occur. McMinn found also that the rate of epithelialization of "excision ulcers" in the cat ileum was closely similar to the normal rate of epithelial growth.

Further evidence for the rapid renewal of the intestinal epithelium was provided by studies on the uptake of ^{35}S-methionine by growing cells (Leblond, Everett and Simmons, 1957). The rate of division and subsequent migration of cells can be followed also by observations based on the uptake of ^{3}H-thymidine into DNA. In some organs, more cells become labelled than eventually divide, indicating that non-dividing tissues of the adult may periodically renew their DNA, but it appears that all cells that are labelled in the intestinal epithelium do eventually divide (Pelc, 1964). Autoradiography of tissue removed 1 hr after the injection of ^{3}H-thymidine showed numerous radioactive nuclei in the crypts of Lieberkühn, and at 24 hr many were up the sides of the villi. At 48 hr after the injection, radioactive cells were at the tops of the villi and in the intestinal lumen (Hughes *et al.*, 1958; Messier and Leblond, 1960). Leblond and Walker (1956) point out that if the data obtained in the rat should apply to man, then the mass of cells released daily from the surface epithelium into the gastro-intestinal tract would weigh about half a pound. Nasset (1965) calculates that the amount of endogenous protein contributed by shed cells in man may be up to 90 g/day. To this one may add the well-known observation that herbivores have a relatively long alimentary canal.

It is commonly thought that the upward migration of cells onto the villi is due to the pressure resulting from mitosis, but Grad and Stevens (1949) found that cell loss from the villus tips continued in rats and mice suffering from radiation damage by which cell renewal from the crypts was suspended. It is possible that the normal cell extrusion is the result of muscular contractions, and that as cells are lost from the extrusion zones, new cells glide along the basement membrane to cover the gap (Leblond and Walker, 1956). Leblond and Stevens (1948) found that intestinal cell renewal continued in rats that were starved for 5 days, and Hooper (1956) noted that isolated loops of rat ileum rapidly filled with shed cells and mucus, indicating that irritation by food particles is not the prime cause of cell loss.The constant mitotic rate in the crypts, and the normally intact surface epithelium constitute further evidence against the idea that the normal intestinal cell loss is due to abrasion, and Hooper (1956) suggests that cell turnover in epithelia is probably an inherent property that can be modified by internal influences such as hormones, and by external influences such as diet.

The relatively undifferentiated cells that proliferate in the crypts are strongly basophilic. As the cells migrate up the sides of the villi they become columnar rather than cuboidal, less basophilic, and acquire a brush border and an increased content of alkaline phosphatase and other enzymes associated with digestion or absorption (Clark, 1959; Wilson, 1962; Padykula, 1962).

The intestinal epithelial cells have a high metabolic activity and energy is used both for synthetic activity and for the active transport of materials across the cell wall. The enzymes concerned with the active intracellular life of the epithelial cells have been reviewed by Spencer and Knox (1960), who concluded that the large energy demands of the intestine are supplied mainly from carbohydrate metabolism. Spencer and Knox found evidence also that the pattern of enzyme distribution in the mucosa appears to change along the length of the intestine.

Shnitka (1960), Wilson (1962) and Padykula (1962) have reviewed some aspects of the distribution and histochemistry of hydrolytic enzymes associated with digestion, and other enzymes of the gastro-intestinal tract.

THE FREE SURFACE OF THE EPITHELIAL CELL—MORPHOLOGY AND ENZYMIC ACTIVITIES

The morphology and histochemistry of the free surface of the cells of the intestinal epithelium is of considerable interest because it is through this surface of the cell that absorption must occur. At the free surface of the mature absorptive cell there is a narrow striated border (brush border). For many years the nature of this structure was uncertain. Baker (1942) reviewed the literature on this subject and described the free surface of the intestinal epithelium in amphibia, reptiles, birds and mammals. He was the first to publish photomicrographs rather than drawings of the striations. The difficulty was that to some people, working at the extreme limit of microscopical resolution, the striations resembled pores or canals through the border; to others, rodlike extensions of the free surface. Baker (1951) was able to demonstrate spindle-shaped particles of lipid in the free surface of the cells during the absorption of fat by the intestinal epithelium of the mouse.

It is now known, from observations with the electron microscope (Granger and Baker, 1950) that in the epithelial cell of the rat intestine, the striated border is composed of numerous, closely packed, small rod-like processes projecting into the intestinal lumen. It has been estimated that they increase the absorptive surface by a factor of 30 (Granger and Baker, 1950), 24 (Palay and Karlin, 1959a) or, in the mouse, 14 (Zetterqvist, 1956). By histochemical procedures it has been shown in several species that the striated border and the supra-nuclear zone of the intestinal epithelial cell are sites of intense alkaline phosphatase activity. The activity is weak in the crypts and becomes stronger as the cells mature and pass up the sides of the villi (Deane and Dempsey, 1945; Emmel, 1946; Ruyter, 1952; Johnson and Kugler, 1953). The striated border gives a distinct bilaminar reaction (Deane and Dempsey, 1945; Johnson and Kugler, 1953), and in many species there is a gradient of decreasing intensity in both the general reaction and in the bilaminar reaction, from the pylorus to the ileo-cæcal valve (Ruyter, 1952; Johnson and Kugler, 1953).

Possibly several different alkaline phosphatases are present, and the evidence that they may be concerned in the transport of phosphorylated sugars through the mucosa is reviewed by Knox, Auerbach and Lin (1956) and by Shnitka (1960). There is now evidence, however, that the striated border is the site of several intracellular hydrolytic enzymes (Miller and Crane, 1961; Holt and Miller, 1961; Hübscher, West and Brindley, 1965) and it is probable that the phosphatases of the striated border are also digestive enzymes and not solely concerned with absorption.

Johnson and Kugler point out that the presence of a gradient in alkaline phosphatase content in the small intestine of several species may explain the observation of Fisher and Parsons (1950) that in the rat, even when the gradient of mucosal surface area in the small intestine and the gradient of glucose utilization by the tissues of the intestinal mucosa are taken into account, there remains a gradient of translocation of glucose across the mucosa. Alkaline phosphatase may also be concerned with the absorption of fatty acids and amino acids (see review by Knox, Auerbach and Lin, 1956).

It is possible in many instances to correlate the development of alkaline phosphatase activity with functional capacity in the intestinal epithelium (Moog, 1962). In the chick, which can eat a varied diet at hatching, the maximum level is reached soon after emergence from the shell. The relatively independent neonatal guinea-pig is fully equipped with alkaline phosphatase, but in the newborn mouse there are two peaks in alkaline phosphatase activity, one at birth and the second at about weaning, 18 days later. Moog has shown that the pre-weaning accumulation of phosphatase in the mouse can be advanced in the duodenum by as much as 1 week by the administration of cortisone acetate and that it is almost completely prevented by adrenalectomy at 11 or 12 days of age. Adrenocorticotrophic hormone also can bring about a precocious rise in the alkaline phosphatase content of the epithelium in the duodenum. It is well known that some young mammals acquire passive immunity after birth by the absorption of milk globulins (see review by Brambell, Hemmings and Henderson, 1951). Calves obtain antibodies exclusively by absorption through the small intestine within the first 2 days after birth (Deutsch and Smith, 1957) and in pigs the transient period of gamma globulin absorption is about 24 hr (Payne and Marsh, 1962). The absorptive cells of the small intestine of such animals contain large inclusions resembling droplets of colostrum during this period (Smith, 1925; Comline, Roberts and Titchen, 1951; Comline, Pomeroy and Titchen, 1953; Hill and Hardy, 1956; Hill, 1956). Similar material also appears in the lacteals and thoracic duct but not in the portal blood. Using fluorescein-labelled gamma globulin, Payne and Marsh (1962) were able to follow the uptake of globulin and its subsequent transfer to lymphatics, in the young pig. The normal ability of the sucking animal to absorb antibodies is lost at 17 days in the mouse and 20 days in the rat (Halliday, 1956, 1959). In both mouse and rat this period of decreasing absorption coincides with an increase in alkaline phosphatase activity in the duodenum (Halliday, 1959). Halliday also found that when young rats were injected with cortisone acetate the reduction in antibody absorption occurred at least 9 days before normal, and this was associated with precocious development of alkaline phosphatase activity in the duodenum. Payne and Marsh (1962) also correlated the abrupt change in intestinal permeability in the suckled pig with the development of alkaline phosphatase activity in the intestinal epithelium. The intestine of the newborn pig remained permeable to proteins for an extended period if the pigs were starved, but starved pigs injected with cortisone acetate failed to absorb gamma globulin.

Moog (1962) points out that the rise in alkaline phosphatase activity stimulated in the mouse duodenum by cortisone is not due to the accelerated synthesis of an enzyme already present, but to the production of a new alkaline phosphatase that otherwise appears at about weaning as part of the normal differentiation process. It appears probable that the association between phosphatase increase and the suppression of antibody absorption is due to general maturation of the cells, and that this process may be influenced by the thyroid as well as by the pituitary and adrenal glands.

THE STRIATED BORDER AS A DIGESTIVE SURFACE

It has been known for a long time that some substances are absorbed from the intestine much more rapidly than would be expected from the hydrolytic activity of the intestinal juice (Cajori, 1933; Borgström *et al.*, 1957) and that the succus entericus is not rich in digestive enzymes when the pancreatic output is excluded. Hydrolytic properties are added with scrapings of the mucosa (Wright *et al.*, 1940) and in general, the digestive activity of the contents of the small intestine can be correlated with the amount of suspended matter. Such observations have led to the concept that digestion in the small intestine may be an intracellular process, and that much of the activity that is present in the succus entericus is due to the disintegration of shed epithelial cells.

Substantial evidence in support of this view has come recently from the isolation of intestinal cell brush borders by Miller and Crane. Preparations of brush border containing a small proportion of the total protein of the original cell homogenate were found to contain a high percentage of the total invertase, maltase, aminopeptidase and alkaline phosphatase activities of the intestinal mucosa (Miller and Crane, 1961; Holt and Miller, 1961). Miller and Crane believe that their evidence justifies the characterization of the brush border as a digestive surface. They showed that the liberation of monosaccharides by the digestive hydrolysis of disaccharides takes place within the brush border and they suggested that in the brush border there is an efficient coupling of digestive hydrolysis and active absorption. The predominant localization of peptidase activity in the brush border suggests that terminal digestion of proteins as well as carbohydrate occurs in this intracellular locus. The presence of enzymic activities in the brush border fraction of intestinal cell homogenates was confirmed by Hübscher, West and Brindley (1965), who worked with the guinea-pig intestine. Hübscher, West and Brindley (1965) found that invertase and alkaline ribonuclease were exclusively localized in the brush border. A number of other enzymes were true constituents of the brush border but also occurred in other subcellular structures. These authors summarize recent results on the intracellular distribution of hydrolytic enzymes of the intestinal mucosa.

PATHOLOGICAL AND AUTOLYTIC CELL SHEDDING IN THE INTESTINE

The high metabolic activity and turnover rate of the intestinal epithelium renders the tissue very susceptible to the effects of irradiation (Friedman, 1945), anti-metabolites (Zamcheck, 1960; Williams, 1963) and anoxia (Fell, Boyne and Cuthbertson, 1961). Traumatic and other forms of shock can result in damage to the intestinal epithelium, and in gastric erosions or ulceration, probably due to vascular stasis and local anoxia. Prolonged vascular engorgement or spasm can have a similar effect. Under such conditions the epithelial cells may be shed from the villi in large numbers or sheets of cells, apparently due to autolysis, and diffuse capillary hæmorrhage may follow.

The intestinal epithelium undergoes a very similar type of lysis and shedding as an early *post mortem* change that appears whilst the intestines are still showing peristaltic movements, and the shed cells cause, or contribute to, an increase in the nitrogen content of the intestinal digesta (Badawy *et al.*, 1957; Badawy *et al.*, 1958). Zetterqvist (1956) analysed by electron microscopy the early *post mortem* changes in the cells of the mouse jejunum as part of a study on methods of fixation. Although he did not mention epithelial cell shedding, he found that within 5 min of death the mitochondria become rounded, with loss of ground-substance and destruction of their inner membranes, whilst vacuoles appear in the cytoplasm. These changes are accentuated with the passage of time. Zetterqvist refers to the earlier literature on light, phase-contrast and electron microscopy to show that in various tissues the first signs of cell injury are usually manifested by changes in the mitochondria and that these appear within a short time after the supply of oxygen to the tissues has been interrupted. The cell shedding described by Badawy *et al.* (1958) occurs within about 10 min of the cessation of respiration, and is accelerated if the animal is exsanguinated (Badawy *et al.*, 1958; Campbell *et al.*, 1958; Fell, 1961). This phenomenon has led in the past to confusion in pathological investigations, the more so because it can be exaggerated by slow fixation even in freshly killed animals (Fell, 1961). Miller and Crane (1961) found that intestinal cell brush borders could be successfully separated and studied only in the presence of EDTA, which appeared to inhibit a metal ion-dependent autolytic process.

THE ADDITION OF SERUM PROTEINS TO THE DIGESTA

Nitrogen is added to the digesta by the mucosal secretion of mucoproteins and other substances, and by the secretions of the accessory digestive glands, as well as by the shedding of epithelial cells and leucocytes into the gut lumen. A comparatively recent discovery is that there is in addition a constant passage of serum proteins, particularly albumin, across the intestinal wall, and that the alimentary canal is probably the main site of albumin catabolism. Possibly half the total albumin degradation takes place in the alimentary canal, most of it in the small intestine. Jarnum (1963), Freeman (1964) and Wetterfors (1965) have reviewed the evidence for this phenomenon. Jarnum points out that the appropriate conditions of high blood flow and considerable capillary permeability are present in the alimentary canal, and there is a high concentration of protein in the intestinal lymph. Occasionally the transudation may reach pathological proportions and result in the hypoproteinæmia of protein-losing gastroenteropathy (Jarnum, 1963). The route by which the proteins enter the intestinal lumen is not known; diffuse passive leakage between the epithelial cells is one possibility. The rate of leakage is probably influenced by the functional state of the capillary bed and possibly by the extent of the surface area of the crypts and surface epithelium. An important nutritional consequence of the transudation, and of the constant addition of endogenous nitrogen from digestive secretions and shed epithelial cells, is that from the breakdown of the proteins there is derived a metabolic reserve or pool of amino acids and there is evidence that the amino-acid pattern of the intestinal contents is remarkably constant regardless of diet (Nasset, Schwartz and Weiss, 1955; Nasset, 1965). Substantial amounts of plasma albumin pass into isolated re-entrant loops of jejunum in sheep (Campbell *et al.*, 1961) and calves (Nielsen and Dich, 1965).

MATURATION OF INTESTINAL CELL FUNCTION AND EARLY WEANING

The practice of early weaning depends to an important extent on the ability of the young animal to digest food other than milk. Complications in the early weaning of pigs have led to a number of studies on the maturation of the enzymic digestive systems of the pig alimentary canal; this subject has been reviewed recently by Braude (1964). In the first few weeks of life pigs are deficient in proteolytic and amylolytic enzyme systems, due in part to low levels of pepsin and immaturity of the pancreas (Lewis *et al.*, 1957; see review by Braude, 1964). Several workers have suggested that in the very young pig there may be insufficient gastric acidity for pepsin activation, but the data of Walker (1959) and Hartman *et al.*, (1961) do not show a change in acid secretion associated with the improved digestion of maturity. Glucose can be absorbed at birth, and lactose is digested by young pigs, but sucrose and maltose are not digested during the first week of life at least, and the inclusion of such disaccharides in the diet results in severe diarrhœa.

Lactase and acid and alkaline phosphatases occur in the intestinal mucosa of the fœtal pig and increase in amount during gestation (Sprague *et al.*, 1963). Sprague *et al.* (1963) found that enzyme activity was maximal in the middle portion of the small intestine. Bailey, Kitts and Wood (1956) found considerable lactase activity in the newborn pig and through the first 2 weeks *post partum*. There was then a sharp decline in activity which reached minimum levels in 3-4 weeks, whereas sucrase and maltase activity increased from negligible levels at birth to a maximum after about 25 days. Hartman *et al.* (1961) found in the young pig that the proteinase activity of intestinal contents, and the maltase and sucrase activities of the intestinal tissues, had reached their maximum activity per unit weight at 7 weeks of age. Their data did not explain the better utilization of certain proteins by the older animal, and it seems possible that the mature pig may possess a more complete hydrolytic enzyme system that includes enzymes not yet investigated in the neonatal animal. There is considerable intestinal lactase activity in the newborn calf, and this declines with age, but in contrast to the pig, the calf intestine does not develop sucrase activity (Huber *et al.*, 1961). This finding fits in well with data on the poor digestion of sucrose by calves and with the known fermentative activity of the adult ruminal microflora.

MORPHOLOGICAL ADAPTATIONS OF THE ALIMENTARY CANAL

The amount and composition of the diet cause morphological adaptations in the intestinal wall in addition to functional changes in enzyme adaptation that may occur. The pancreas is influenced to an important extent both in its structure and enzymic secretions, particularly by the protein level of the diet (Davies, 1948; Thompson and Trowell, 1952; Adams, Fernand and Schnieden, 1958; Ramalingaswami, 1964; Snook, 1965).

In acute starvation in the mouse there is degeneration of the brush border, impaired differentiation of the intestinal epithelium and a reduction in the rate of intestinal cell renewal (Sun, 1927; Brown, Levine and Lipkin, 1963). Nissim (1961) has described degenerative and atrophic changes in the intestine in starvation and under-nutrition, and the data of Fell, Campbell and Boyne (1964) indicate that there may be some degree of atrophy in the alimentary canal of Scottish hill sheep on winter grazing. It is well established that low protein–low calorie diets lead eventually to digestive disturbances and to morphological changes in the gastro-intestinal tract. Platt, Heard and Stewart (1964) have described the histological appearance of the atrophied intestine in protein- and calorie-deficient animals. In the severely deficient pig showing a kwashiorkor-like syndrome there were striking changes in the jejunum, where the villi were shorter and wider than normal, the epithelial cells were reduced to cuboidal form and the mucosa was infiltrated with lymphocytes and plasma cells. Very similar changes are found in kwashiorkor (Stanfield, Hutt and Tunnicliffe, 1965) and in experimental kwashiorkor in the Rhesus monkey (Ramalingaswami, 1964). Platt, Heard and Stewart (1964) also refer to characteristic changes that occur in the oral and œsophageal epithelium and in the squamous epithelium of the pig fore-stomach in protein–calorie deficiency. The changes can be readily detected in smears from the buccal mucosa.

Animals fed on fibrous or high-residue diets generally show considerable hypertrophy of the stomach and large intestine with relatively little change in the small intestine (Addis, 1932; Wierda, 1950), but animals with hyperphagia induced by experimental lesions of the hypothalamus show generalized hypertrophic and hyperplastic changes in the alimentary canal (Brobeck, Tepperman and Long, 1942–3). Somewhat similar results are obtained in rats trained by intermittent starvation to show marked hyperphagia during the periods of free access to food (Brownlee and Moss, 1959; Fábry and Kujalová, 1960). After 26 weeks of such experimental feeding the intestines were longer, wider in section and 36 per cent heavier, with an increased total nitrogen content. Histochemical studies (Lojda and Fábry, 1959) revealed an increase in the alkaline phosphatase content of the jejunal epithelium and an increase in the area of transverse sections of the intestine. The stomach was considerably hypertrophied. Fábry and Kujalová point out that in simple chronic under-nutrition the digestive system is one of the organs that suffers considerable weight loss, indicating that there must be a special adaptive mechanism at work in the intermittently starved rats.

During lactation in the rat there is an increase in the weight and nitrogen content of the alimentary canal when compared with that of unmated controls (Poo, Lew and Addis, 1939). Boyne, Chalmers and Cuthbertson (1953) showed that this increase greatly exceeded that found in pregnancy. Fell, Smith and Campbell (1963) extended these observations by histological and chemical studies, finding considerable hypertrophy and hyperplasia of the intestinal mucosa, with a progressive increase in the total nitrogen content of the alimentary canal during lactation; on weaning there was a subsequent decline towards the values found in pregnancy. In further studies, Campbell and Fell (1964) were able to abolish the alimentary hypertrophy of lactating rats by controlled feeding and they suggested that the hypertrophy is due to the increase in food consumption normally associated with lactation. Campbell and Fell found that the apparent coefficient of digestibility of crude protein was the same in lactating rats feeding *ad lib.* (with marked alimentary hypertrophy), in lactating rats on restricted feeding and in non-lactating controls, and they suggested that the hypertrophy is an adaptive response that enables the animal to maintain a constant coefficient of digestibility at increased levels of food consumption. The abomasum and small intestine of Scottish hill sheep were found to increase rapidly and progressively in weight and total nitrogen content during lactation and to show histological changes very like those noted in the rat. There was a striking regression in these organs after the lambs were weaned (Fell, Campbell and Boyne, 1964).

There is a marked enlargement of the remaining small intestine in dogs after extensive resection of the organ (Flint, 1912). Flint suggested that the hypertrophic and hyperplastic changes were due to compensatory proliferation caused by decreased absorption and that there is a relationship between the mucosal surface area and the nutrition of the animal. In one instance a functional segment showed the hypertrophy whilst a short-circuited portion did not, indicating that the cause of the enlargement was probably not an hormonal stimulus. Similar proliferative changes were obtained by Loran, Althausen and Irvine (1956) in rats with "minimal" (10–15 per cent) resection of the small intestine. More recent studies by Loran and Althausen (1960) and Loran and Crocker (1963) have shown that in rats subjected 2 months previously to resection of the lower ileum amounting to 10 per cent of the small intestine, the villi along the entire length of the intestine were increased in height, the increase in any part being inversely related to the distance from the site of resection. In addition, the rate of epithelial cell migration up the villi was abnormally high, the life of the epithelial cells was shortened, the proportion of intestinal crypt cells incorporating ^{3}H-thymidine was considerably greater then normal, and uniformly high throughout the shortened intestine, and the time during which cells were retained in the crypts was markedly reduced. Loran and Crocker (1963) postulated that these persistent and generalized effects of partial resection are due to the action of an humoral factor which they tentatively called "intestinal epithelial growth hormone".

In germ-free animals the weight of the stomach and small intestine is reduced, and the wall is thinner than normal. This appears to be because the stroma, lymphoid tissue and defensive elements of the animal are relatively underdeveloped, and the intestinal epithelium is not appreciably altered. The food intake and conversion rates of germ-free animals is similar to that of conventional animals (see review by Luckey, 1963). A similar reduction in the weight of the small intestine occurs in pigs and chickens that are fed on antibiotics (Coates, Davies and Kon, 1955; Braude, *et al.*, 1955). The small intestine is of comparable length in germ-free animals, in animals fed on antibiotics and in conventional animals.

Many facets of intestinal digestion and absorption have been adequately and frequently reviewed in recent years. Some aspects are complex, as in the case of fat digestion where lipolytic enzymes, bile salts and the intestinal epithelium are inseparably involved. Recent concepts of fat digestion, absorption and biochemistry are to be found in the reviews of Senior (1964) and Garton (1967, also Chapter 10) and in the remarkable morphological studies of Palay and Karlin (1959b).

THE LARGE INTESTINE

There are no villi in the cæcum and colon, but in other respects the mucosa and epithelium are structurally similar to those of the small intestine. There are many goblet cells in the epithelium. Fell *et al.* (1963, 1964) found that during hypertrophy of the cæcum in the rat and sheep the epithelial cells increased in height, the mucosa developed microscopic folds and the crypts became tortuous; normally the crypts in the cæcum and colon are remarkably straight. In germ-free animals the cæcum is usually greatly enlarged, perhaps due to lack of tone in the muscular layers (Luckey, 1963).

The large intestine is in all the domestic animals an important site of water resorption: it is also an organ in which microbial fermentation of carbohydrate occurs. In ruminants the fermentation that takes place in the cæcum and colon is negligible compared with that in the rumen, but in non-ruminant herbivores there may be an important production of volatile fatty acids from fermentable material reaching the lower ileum and large intestine (Hungate *et al.*, 1959). Barcroft *et al.* (1944b) found volatile fatty acids in significant amounts in blood draining the cæcum of the sheep, pig, pony and rabbit and in the veins of the colon in the pig and pony. They regarded this as a clear indication that fermentation of carbohydrate occurs or recommences in the large intestine. The cæcum and colon are very large in the horse, where they constitute a substantial fermentation chamber for ingested cellulose. Alexander (1952) found considerable quantities of volatile fatty acids in the cæcum and ventral colon, with slightly smaller amounts in the dorsal colon. Relatively more propionic acid was produced in the ventral colon than in the dorsal colon. The amounts of volatile fatty acids in the large intestine were reduced by fasting, particularly in the cæcum and ventral colon, and increased again on re-feeding; some animals showed an increase within 3 hr of feeding. The cæcum is not indispensable, however, and Lloyd, Dale and Crampton (1957) found that removal of the cæcum from pigs at 6 weeks of age did not affect the growth of the pig and made but slight difference to the digestibility of several dietary components.

Volatile fatty acids are utilized by the wall of the cæcum in sheep (Pennington, 1952) and lambs (Walker and Simmonds, 1962) but with slight production of ketone bodies. Walker and Simmonds found that the cæcal wall of the fœtal and newborn lamb utilized butyrate at a higher rate than did adult sheep tissue.

Much water is removed from the digesta in the large intestine of the sheep and of the digestible dry matter in the ration, about 70 per cent is removed between the mouth and the duodenum and 19 per cent between the ileo-cæcal valve and the anus (Hogan, 1957).

Boyne *et al.* (1956) and Badawy *et al.* (1958a) found that resorption of water in the sheep large intestine resulted in a considerable rise in the dry matter content of the digesta over the values found in the small intestine, with a further concentration between the cæcum and colon. Characteristically, the highest values for volatile fatty acids were obtained in the main sites of microbial activity—the reticulo rumen, the cæcum and colon. Badawy *et al.* (1958a) described changes in the digesta relative to lignin. Using polyethylene glycol as a reference substance, Hydén (1961) found that considerable quantities of water, sodium, chloride, potassium and phosphate were absorbed in the cæcum and that absorption of water and potassium particularly, continued in the remainder of the intestine.

REFERENCES

ADAMS, C. W. M., FERNAND, V. S. V. and SCHNIEDEN, H (1958) *Br. J. Exp. Path.* **39**, 393.
ADDIS, T. (1932) *Am. J. Physiol.* **99**, 417.
ALEXANDER, F. (1952) *Quart. J. Exp. Physiol.* **37**, 205.
ANNISON, E. F., HILL, K. J., and LEWIS, D. (1957). *Biochem. J.* **66**, 592.
ANNISON, E. F. and LEWIS, D. (1959) *Metabolism in the Rumen.* Methuen & Co. Ltd., London, pp. 124-31.
ANNISON, E. F. and PENNINGTON, R. J. (1954) *Biochem. J.* **57**, 685.

ARMSTRONG, D. G., BLAXTER, K. L. and GRAHAM, N. McC. (1957) *Br. J. Nutr.* **11**, 392.
ASH, R. W. (1959) *J. Physiol. Lond.* **147**, 58.
ASH, R. W. (1961) *J. Physiol. Lond.* **156**, 93.
ASH, R. W. and KAY, R. N. B. (1959) *J. Physiol. Lond.* **149**, 43.
BABKIN, B. P. (1950) *Secretory Mechanism of the Digestive Glands.* 2nd Ed. Paul B. Hoeber, New York, pp. 19-41.
BADAWY, A. M., CAMPBELL, R. M., CUTHBERTSON, D. P. and FELL, B. F. (1957) *Nature, Lond.* **180**, 756.
BADAWY, A. M., CAMPBELL, R. M., CUTHBERTSON, D. P., FELL, B. F. and MACKIE, W. S. (1958) *Br. J. Nutr.* **12**, 367.
BADAWY, A. M., CAMPBELL, R. M., CUTHBERTSON, D. P., and MACKIE, W. S. (1958a) *Br. J. Nutr.* **12**, 384.
BADAWY, A. M., CAMPBELL, R. M., CUTHBERTSON, D. P. and MACKIE, W. S. (1958b) *Br. J. Nutr.* **12**, 391.
BAILEY, C. B., KITTS, W. D. and WOOD, A. J. (1956) *Can. J. Agric. Sci.* **36**, 51.
BAKER, J. R. (1942) *Quart. J. Microsc. Sci.* **84**, 73.
BAKER, J. R. (1951) *Quart. J. Microsc. Sci.* **92**, 79.
BAKER, B. L. and BRIDGMAN, R. M. (1954) *Am. J. Anat.* **94**, 363.
BALCH, C. C., KELLY, A. and HEIM, G. (1951) *Br. J. Nutr.* **5**, 207.
BARCLAY, A. E. and BENTLEY, F. H. (1949) *Gastroenterology* **12**, 177.
BARCROFT, J., MCANALLY, R. A. and PHILLIPSON, A. T. (1944a) *J. Exp. Biol.* **20**, 132.
BARCROFT, J., MCANALLY, R. A. and PHILLIPSON, A. T. (1944b) *J. Exp. Biol.* **20**, 120.
BARNETT, A. J. G. and REID, R. L. (1961) *Reactions in the Rumen.* Edward Arnold Ltd. London, pp. 70–75.
BEARDSLEY, D. W. (1964) *J. Anim. Sci.* **23**, 239.
BECKER, R. B., ARNOLD, P. T. D. and MARSHALL, S. P. (1951) *J. Dairy Sci.* **34**, 329.
BECKER, R. B., MARSHALL, S. P. and ARNOLD, P. T. D. (1963) *J. Dairy Sci.* **46**, 835.
BERRIDGE, N. J. (1945) *Biochem. J.* **39**, 179.
BERRIDGE, N. J., DAVIS, J. G., KON, P.M., KON, S. K. and SPRATLING, F. R. (1943) *J. Dairy Res.* **13**, 145.
BLAXTER, K. L., HUTCHESON, M. K., ROBERTSON, J. M. and WILSON, A. L. (1952) *Br. J. Nutr.* **6**, i.
BLAXTER, K. L. and WOOD, W. A. (1953) *Vet. Rec.* **65**, 889.
BORGSTRÖM, B., DAHLQVIST, A., LUNDH, G. and SJOVALL, J. (1957) *J. Clin. Invest.* **36**, 1521.
BOYNE, A. W., CAMPBELL, R. M., DAVIDSON, J. and CUTHBERTSON, D. P. (1956) *Br. J. Nutr.* **10**, 325.
BOYNE, A. W., CHALMERS, M. I. and CUTHBERTSON, D. P. (1953) *Hoppe—Seyler's Z. physiol. Chem.* **295**, 424.
BRAMBELL, W. F. R., HEMMINGS, W. A. and HENDERSON, M. (1951) *Antibodies and Embryos.* University of London Press, London.
BRAUDE, R. (1964) *Problems in Pig Rearing.* European Association for Animal Production. Commission on Pig Production.
BRAUDE, R., COATES, M. E., DAVIES, M. K., HARRISON, G. F. and MITCHELL, K. G. (1955) *Br. J. Nutr.* **9**, 363.
BROBECK, J. R., TEPPERMAN, J. and LONG, C. N. H. (1942-3) *Yale J. Biol, Med.* **15**, 831.
BROBERG, G. (1960) *Acute Overeating with Cereals in Ruminants.* Lovisa. Ab. Lovisa Nya Tryckeri.
BROWN, H. O., LEVINE, M. L. and LIPKIN, M. (1963) *Am. J. Physiol.* **205**, 868.
BROWNLEE, A. (1956) *Br. Vet. J.* **112**, 369.
BROWNLEE, A. (1962) *Vet. Rec.* **74**, 1516.
BROWNLEE, A. and ELLIOT, J. (1960) *Br. Vet. J.* **116**, 467.
BROWNLEE, A. and ELLIOT, J. (1961) *Vet. Rec.* **73**, 384.
BROWNLEE, A. and MOSS, W. (1959) *Br. Vet. J.* **115**, 225.
CAJORI, F. A. (1933) *Am. J. Physiol.* **104**, 659.
CAMPBELL, A. C. P. (1959) In: *Modern Trends in Pathology.* Ed. D. H. Collins, Butterworths, London, pp. 231-47.

Campbell, R. M., Cuthbertson, D. P., Fell, B. F. and Mackie, W. S. (1958) *Z. Tierernähr.* **13**, 311.
Campbell, R. M., Cuthbertson, D. P., Mackie, W., McFarlane, A. S., Phillipson, A. T. and Sudsaneh, S. (1961) *J. Physiol. Lond.* **158**, 113.
Campbell, R. M. and Fell, B. F. (1964) *J. Physiol. Lond.* **171**, 90.
Church, D. C., Jessup, G. L., Jr. and Bogart, R. (1962) *Am. J. Vet. Res.* **23**, 220.
Clark, S. L., Jr. (1959) *J. Biophys. Biochem. Cytol.* **5**, 41.
Coates, M. E., Davies, M. K. and Kon, S. K. (1955) *Br. J. Nutr.* **9**, 110.
Comline, R. S., Roberts, H. E. and Titchen, D. A. (1951) *Nature, Lond.* **168**, 84.
Comline, R. S., Pomeroy, R. W. and Titchen, D. A. (1953) *J. Physiol. Lond.* **122**, 6P.
Dain, J. A., Neal, A. L. and Dougherty, R. W. (1955) *J. Anim. Sci.* **14**, 930.
Danielli, J. F., Hitchcock, M. W. S., Marshall, R. A. and Phillipson, A. T. (1945-6) *J. Exp. Biol.* **22**, 75.
Davies, J. N. P. (1948) *Lancet* **i**, 317.
Deane, H. W. and Dempsey, E. W. (1945) *Anat. Rec.* **93**, 401.
Deutsch, H. F. and Smith, V. R. (1957) *Am. J. Physiol.* **191**, 271.
Dobson, A. (1959) *J. Physiol. Lond.* **146**, 235.
Dobson, A. (1961) In: *Digestive Physiology and Nutrition of the Ruminant.* Ed. Lewis, D. London, Butterworths, pp. 68-75.
Dobson, M. J., Brown, W. C. B., Dobson, A. and Phillipson, A. T. (1956) *Quart. J. Exp. Physiol.* **41**, 247.
Dobson, A. and Phillipson, A. T. (1956) *J. Physiol. Lond.* **133**, 76P.
Dobson, A. and Phillipson, A. T. (1958) *J. Physiol. Lond.* **140**, 94.
Dukes, H. H. (1947) *The Physiology of Domestic Animals.* 6th Ed. Comstock Publishing Co., Inc., Ithaca, pp. 307-9.
Elias, H. (1947) *Am. J. Vet. Res.* **8**, 311.
Elsden, S. R. and Phillipson, A. T. (1948) *A. Rev. Biochem.* **17**, 705.
Emmel, V. M. (1946) *Anat. Rec.* **95**, 159.
Fábry, P. and Kujalová, V. (1960) *Acta anat.* **43**, 264.
Favilli, N. (1937) *Dt. tierärztl. Wschr.* **45**, 592.
Fell, B. F. (1961) *J. Path. Bact.* **81**, 251.
Fell, B. F., Boyne, R. and Cuthbertson, D. P. (1961) *J. Path. Bact.* **82**, 445.
Fell, B. F., Campbell, R. M. and Boyne, R. (1964) *Res. Vet. Sci.* **5**, 175.
Fell, B. F., Smith, K. A. and Campbell, R. M. (1963) *J. Path. Bact.* **85**, 179.
Fisher, R. B. and Parsons, D. S. (1950) *J. Anat.* **84**, 272.
Flatt, W. P., Warner, R. G. and Loosli, J. K. (1958) *J. Dairy Sci.* **41**, 1593.
Flint, J. M. (1912) *Bull. Johns Hopkins Hosp.* **23**, 127.
Florey, H. W., Jennings, M. A., Jennings, D. A. and O'Connor, R. C. (1939) *J. Path. Bact.* **49**, 105.
Follis, R. H., Jr. (1948) *The Pathology of Nutritional Disease.* Blackwell, Oxford, pp. 59-63.
Freeman, T. (1964) In: *The Role of the Gastrointestinal Tract in Protein Metabolism.* CIOMS Symposium, Ed. H. N. Munro, Blackwell, Oxford.
Friedman, N. B. (1945) *J. Exp. Med.* **81**, 553.
Garrett, W. N., Meyer, J. H., Lofgreen, G. P. and Dobie, J. B. (1961) *J. Anim. Sci.* **20**, 833.
Garton, G. A. (1951) *J. Exp. Biol.* **28**, 358.
Garton, G. A. (1967) *Wld. Rev. Nutr. Diet.* **7**, 225-50.
Gibson, T. E. (1965) *Vet. Rec.* **77**, 1034.
Gobetto, A. (1957) *Atti Soc. ital. sci. vet.* **11**, 782.
Gobetto, A. (1958) *Aunali Fac. Med. vet. Univ. Pisa* **11**, 56.
Godfrey, N. W. (1961a) *J. Agric. Sci. Camb.* **57**, 173.
Godfrey, N. W. (1961b) *J. Agric. Sci. Camb.* **57**, 177.
Grad. B. and Stevens, C. E. (1949) *Cancer Res.* **9**, 620.
Granger, B. and Baker, R. F. (1950) *Anat. Rec.* **107**, 423.
Gray, F. V. (1947) *J. Exp. Biol.* **24**, 1.
Gray, F. V., Pilgrim, A. F. and Weller, R. A. (1954) *J. Exp. Biol.* **31**, 49.
Grossman, J. D. (1949) *J. Am. Vet. med. Ass.* **114-15**, 416.

HABEL, R. E. (1959) *Am. J. Vet. Res.* **20**, 437.
HABEL, R. E. (1963) *Am. J. Vet. Res.* **24**, 199.
HABEL, R. E. and BIBERSTEIN, E. L. (1952) *Fundamentals of the Histology of Domestic Animals*. Translation and revision of the work by A. Trautmann and J. Fiebiger. Baillière, Tindall & Cox, London.
HALL, S. A. (1962) *Vet. Rec.* **74**, 814.
HALLIDAY, R. (1955) *Proc. R. Soc. B.* **143**, 408.
HALLIDAY, R. (1956) *Proc. R. Soc. B.* **145**, 179.
HALLIDAY, R. (1959) *J. Endocr.* **18**, 56.
HARRISON, H. N., WARNER, R. G., SANDER, E. G. and LOOSLI, J. K. (1960) *J. Dairy Sci.* **43**, 1301.
HARTMAN, P. A., HAYS, V. W., BAKER, R. O., NEAGLE, L. H. and CATRON, D. V. (1961) *J. Anim. Sci.* **20**, 114.
HILL, K. J. (1951) *J. Anat.* **85**, 215.
HILL, K. J. (1955) *Quart. J. Exp. Physiol.* **40**, 32.
HILL, K. J. (1956) *Quart. J. Exp. Physiol.* **41**, 421.
HILL, K. J. (1959) *Quart. J. Exp. Physiol.* **44**, 222.
HILL, K. J. (1960) *J. Physiol. Lond.* **154**, 115.
HILL, K. J. (1961a) In: *Digestive Physiology and Nutrition of the Ruminant*. Ed. D. Lewis, Butterworths, London, pp. 48-58.
HILL, K. J. (1961b) *Vet. Rev.* **7**, 83.
HILL, K. J. (1965) In: *Physiology of Digestion in the Ruminant*. Papers presented at the Second International Symposium on the Physiology of Digestion in the Ruminant. Ed. R. W. Dougherty, Butterworths, Washington, pp 221-30.
HILL, K. J. and HARDY, W. S. (1956) *Nature, Lond.* **178**, 1353.
HOGAN, J. P. (1957) *J. Physiol. Lond.* **139**, 25P.
HOGAN, J. P. (1964) *Aust. J. Agric. Res.* **15**, 384.
HOLT, J. H. and MILLER, D. (1961) *J. Lab. Clin. Med.* **58**, 827.
HOOPER, C. E. S. (1956) *J. Histochem. Cytochem.* **4**, 531.
HORST VAN DER, C. J. G. (1961) *Nature, Lond.* **191**, 73.
HUBER, J. T., JACOBSON, N. L., ALLEN, R. S. and HARTMAN, P. A. (1961) *J. Dairy Sci.* **44**, 1494.
HUBSCHER, G., WEST, G. R. and BRINDLEY, D. N. (1965) *Biochem. J.* **97**, 629.
HUGHES, W. L., BOND, V. P., BRECHER, G., CRONKITE, E. P., PAINTER, R. B., QUASTLER, H. and SHERMAN, F. G. (1958). *Proc. Nat. Acad. Sci. U.S.A.* **44**, 476.
HUNGATE, R. E., PHILLIPS, G. D., MCGREGOR, A., HUNGATE, D. P. and BUECHNER, H. K. (1959) *Science, N.Y.* **130**, 1192.
HYDÉN, S. (1961) In: *Digestive Physiology and Nutrition of the Ruminant*. Ed. D. Lewis, Butterworths, London, pp. 35-47.
HYDÉN, S. and SPERBER, I. (1965) In: *Physiology of Digestion in the Ruminant*. Papers presented at the Second International Symposium on the Physiology of Digestion in the Ruminant. Ed. R. W. Dougherty, Butterworths, Washington, pp. 51-60.
IGGO, A. (1955) *J. Physiol. Lond.* **128**, 593.
JACOBSON, W. (1939) *J. Path. Bact.* **49**, 1.
JARNUM, S. (1963) *Protein—Losing Gastrœnteropathy*. Blackwell, Oxford.
JARRETT, W. F. H. (1955) *Vet. Rec.* **67**, 815.
JENSEN, R., CONNELL, W. E. and DEEM, A. W. (1954) *Am. J. Vet. Res.* **15**, 425.
JENSEN, J., DEANE, H. M., COOPER, L. J., MILLER, V. A. and GRAHAM, W. R. (1954) *Am. J. Vet. Res.* **15**, 202.
JENSEN, R., FLINT, J. C. and GRINER, L. A. (1954) *Am J. Vet. Res.* **15**, 5.
JENSEN, R., FLINT, J. C., UDALL, R. H., DEEM, A. W. and SEGER, C. L. (1958) *Am. J. Vet. Res.* **19**, 277.
JOHNSON, F. R. and KUGLER, J. H. (1953) *J. Anat.* **87**, 247.
KAY, R. N. B. (1963) *J. Dairy Res.* **30**, 261.
KESLER, E. M., RONNING, M. and KNODT, C. B. (1951) *J. Anim. Sci.* **10**, 969.
KIDDLE, P., MARSHALL, R. A. and PHILLIPSON, A. T. (1951) *J. Physiol. Lond.* **113**, 207.
KIRKMAN, H. (1950) *Am. J. Anat.* **86**, 91.

KNOX, W. E., AUERBACH, V. H. and LIN, E. C. C. (1956) *Physiol. Rev.* **36**, 164.
LASKOWSKI, M. Jr., and LASKOWSKI, M. (1951) *J. Biol. Chem.* **190**, 563.
LEBLOND, C. P., EVERETT, N. B. and SIMMONS, B. (1957) *Am. J. Anat.* **101**, 225.
LEBLOND, C. P. and STEVENS, C. E. (1948) *Anat. Rec.* **100**, 357.
LEBLOND, C. P., STEVENS, C. E. and BOGOROCH, R. (1948) *Science, N.Y.* **108**, 531.
LEBLOND, C. P. and WALKER, B. E. (1956) *Physiol. Rev.* **36**, 255.
LEHNINGER, A. L. (1962) *Physiol. Rev.* **42**, 467.
LEWIS, D. (1955) *Br. J. Nutr.* **9**, 215.
LEWIS, C. J., HARTMAN, P. A., LIU, C. H., BAKER, R. O. and CATRON, D. V. (1957) *J. Agric. Fd. Chem.* **5**, 687.
LINDHÉ, B. and SPERBER, I. (1959) *K. LantbrHögsk. Annlr.* **25**, 321.
LINDSAY, D. B. (1959) *Vet. Rev.* **5**, 103.
LLOYD, L. E., DALE, D. G. and CRAMPTON, E. W. (1957) *J. Anim. Sci.* **16**, 1039.
LOJDA, Z. and FÁBRY, P. (1959) *Acta histochem.* **8**, 289.
LORAN, M. R. and ALTHAUSEN, T. L. (1960) *J. Biophys. Biochem. Cytol.* **7**, 667.
LORAN, M. R., ALTHAUSEN, T. L. and IRVINE, E. (1956) *Gastroenterology* **31**, 717.
LORAN, M. R. and CROCKER, T. T. (1963) *J. Cell Biol.* **19**, 285.
LUCKEY, T. D. (1963) *Germfree Life and Gnotobiology*. Acad. Press, London, pp. 337-51.
MACDONALD, D., BRADLEY, R. and MCCREA, C.T, (1962) *Vet. Rec.* **74**, 1280.
MARTIN, W. G., RAMSEY, H. A., MATRONE, G. and WISE, G. H. (1959) *J. Dairy Sci.* **42**, 1377.
MASSON, M. J. and PHILLIPSON, A. T. (1951) *J. Physiol. Lond.* **113**, 189.
MAXIMOW, A. A. (1924) *Physiol. Rev.* **4**, 533.
MCANALLY, R. A. and PHILLIPSON, A. T. (1944) *Biol. Rev.* **19**, 41.
MCCARTHY, R. D., SHAW, J. C., MCCARTHY, J. L., LAKSHMANAN, S. and HOLTER, J. B. (1958) *Proc. Soc. Exp. Biol. Med.* **99**, 556.
MCGILLIARD, A. D., JACOBSON, N. L. and SUTTON, J. D. (1965) In. *Physiology of Digestion in the Ruminant.* Papers presented at the Second International Symposium on the Physiology of Digestion in the Ruminant. Ed. R. W. Dougherty, Butterworths, Washington, pp. 39-50.
MCLAREN, G. A., ANDERSON, G. C., MARTIN, W. E. and COOPER, W. K. (1961) *J. Anim. Sci.* **20**, 942.
MCMINN, R. M. H. (1954) *J. Anat.* **88**, 527.
MESSIER, B. and LEBLOND, C. P. (1960) *Am. J. Anat.* **106**, 247.
MILLER, D. and CRANE, R. K. (1961) *Biochim. Biophys. Acta.* **52**, 293.
MILLER, J. K. and MILLER, W. J. (1960) *J. Dairy Sci.* **43**, 1854.
MILLER, J. K. and MILLER, W. J. (1962) *J. Nutr.* **76**, 467.
MOOG, F. (1962) *Fedn. Proc. Fedn. Am. Socs. Exp. Biol.* **21**, 51.
NASSET, E. S. (1965) *Fedn. Proc. Fedn. Am. Socs. Exp. Biol.* **24**, 953.
NASSET, E. S., SCHWARTZ, P. and WEISS, H. V. (1955) *J. Nutr.* **56**, 83.
NIELSEN, K. and DICH, J. (1965) *Acta Vet. scand.* **6**, 249.
NISSIM, J. A. (1961) *Lancet ii*, 132.
NOVIKOFF, A. B. (1961) In: The Cell, Biochemistry, Physiology, Morphology. Ed. J. Brachet and A. E. Mirsky, Vol. II, Academic Press, New York, pp. 299-404.
OMAR, E. M., REAGOR, J. C. and KUNKEL, H. O. (1964) *J. Anim. Sci.* **23**, 729.
ONODERA, R., TSUDA, T. and UMEZU, M. (1964) *Tohoku, J. Agric. Res.* **15**, 99.
OTT, E. A., SMITH, W. H., STOB, M. and BEESON, W. M. (1964) *J. Nutr.* **82**, 41.
OYAERT, W. and BOUCKAERT, J. H. (1961) *Res. Vet. Sci.* **2**, 41.
PADYKULA, H. A. (1962) *Fedn. Proc. Fedn. Am. Socs. Exp. Biol*, **21**, 873.
PALAY, S. L. and KARLIN, L. J. (1959a) *J. Biophys. Biochem. Cytol.* **5**, 363.
PALAY, S. L. and KARLIN, L. J. (1959b) *J. Biophys. Biochem. Cytol.* **5**, 373.
PARTHASARATHY, D., GARTON, G. A. and PHILLIPSON, A. T. (1952) *Biochem. J.* **52**, xvi.
PARTHASARATHY, D. and PHILLIPSON, A. T. (1953) *J. Physiol. Lond.* **121**, 452.
PAYNE, L. C. and MARSH, C. L. (1962) *Fedn. Proc. Fedn. Am. Socs. Exp. Biol.* **21**, 909.
PEARSE, A. G. E. (1949) *J. Clin. Path.* **2**, 81.
PELC, S. R. (1964) *J. Cell Res.* **22**, 21.
PENNINGTON, R. J. (1952) *Biochem. J.* **51**, 251.

PENNINGTON, R. J. and PFANDER, W. H. (1957) *Biochem. J.* **65**, 109-11.
PENNINGTON, R. J. and SUTHERLAND, T. M. (1956) *Biochem. J.* **63**, 353.
PHILLIPSON, A. T. (1959) quoted by Sander *et al.* (1959).
PHILLIPSON, A. T. and CUTHBERTSON, D. P. (1956) *VIIth Int. Cong. Anim. Husb., Madrid.*
PHILLIPSON, A. T. and MCANALLY, R. A. (1942) *J. Exp. Biol.* **19**, 199.
PIERCE, A. E. (1959) *Vet. Rev.* **5**, 17.
PINSENT, P. J. N. (1962) *Vet. Rec.* **74**, 1282.
PLATT, B. S., HEARD, C. R. C. and STEWART, R. J. C. (1964) In: *The Role of the Gastrointestinal Tract in Protein Metabolism.* CIOMS Symposium. Ed. H. N. Munro, Blackwell, Oxford.
POO, L. J., LEW, W. and ADDIS, T. (1939) *J. Biol. Chem.* **128**, 69.
RANKIN, A. D. (1940) Thesis, Cornell. Cited in *Vet Bull.* **11**, 328 (1941).
RAMALINGASWAMI, V. (1964) *Nature, Lond.* **201**, 546.
RODWELL, A. W. (1953) *J. Gen. Microbiol.* **8**, 224.
RUYTER, J. H. C. (1952) *Acta Anat.* **16**, 209.
SANDER, E. G., WARNER, R. G., HARRISON, H. N. and LOOSLI, J. K. (1959) *J. Dairy Sci.* **42**, 1600.
SANFORD, J. (1963) *Nature, Lond.* **199**, 829.
SAVAGE, E. S. and MCCAY, C. M. (1942) *J. Dairy Sci.* **25**, 595.
SCARISBRICK, R. and EWER, T. K. (1951) *Biochem. J.* **49**, lxxix.
SENIOR, J. R. (1964) *J. Lipid Res.* **5**, 495.
SHINOZAKI, K. (1957) *Tohoku J. Agric. Res.* **8**, 149.
SHNITKA, T. K. (1960) *Fedn. Proc. Fedn. Am. Socs. Exp. Biol.* **19**, 897.
SHOJI, Y., MIYAZAKI, K. and UMEZU, M. (1964) *Tohoku J. Agric. Res.* **15**, 91.
SINCLAIR, J. H. and KUNKEL, H. O. (1959) *Proc. Soc. Exp. Biol. Med.* **102**, 57.
SISSON, S. and GROSSMAN, J. D. (1963) *The Anatomy of the Domestic Animals.* 3rd Ed. W. B. Saunders Co., Philadelphia, p. 456.
SJAASTAD, Ø. V. and STORMORKEN, H. (1963) *Nature, Lond.* **197**, 907.
SMILES, J. and DOBSON, M. J. (1955) *J. Roy. Microsc. Soc.* **75**, 244.
SMITH, L. DS. (1963) *Bull. Off. Int. Epizoot.* **59**, 1517.
SMITH, R. H. (1961) *J. Agric. Sci. Camb.* **56**, 105.
SMITH, T. (1925) *J. Exp. Med.* **41**, 81.
SMITH, H. W. (1954) *Brit. J. Exp. Path.* **35**, 447.
SMITH, A. H., KLEIBER, M., BLACK, A. L. and BAXTER, C. G. (1955) *J. Nutr.* **57**, 507.
SMITH, A. H., KLEIBER, M., BLACK, A. L. and LOFGREEN, G. P. (1956) *J. Nutr.* **58**, 95.
SNOOK, J. T. (1965) *Fedn. Proc. Fedn. Am. Socs. Exp. Biol.* **24**, 941.
SOMMERVILLE, R. I. (1956) *Aust. Vet. J.* **32**, 237.
SPENCER, R. P. and KNOX, W. E. (1960) *Fedn. Proc. Fedn. Am. Socs. Exp. Biol.* **19**, 886.
SPRAGUE, J. I., ULLREY, D. E., WADILL, D. G., MILLER, E. R., ZUTAUT, C. L. and HOEFFER, J. A. (1963) *J. Anim. Sci.* **22**, 121.
STANFIELD, J. P., HUTT, M. S. R. and TUNNICLIFFE, R. (1965) *Lancet* **i**, 519.
SUN, T. P. (1927) *Anat. Rec.* **34**, 341.
SUTTON, J. D., MCGILLIARD, A. D. and JACOBSON, N. L. (1961) *J. Dairy Sci.* **44**, 1184.
SUTTON, J. D., MCGILLIARD, A. D. and JACOBSON, N. L. (1963) *J. Dairy Sci.* **46**, 426.
SUTTON, J. D., MCGILLIARD, A. D., RICHARD, M. and JACOBSON, N. L. (1963) *J. Dairy Sci.* **46**, 530.
SWANSON, E. W. and HARRIS, J. D. (1958) *J. Dairy Sci.* **41**, 1768.
TAMATE, H. (1956-7) *Tohoku J. Agric. Res.* **7**, 209.
TAMATE, H., MCGILLIARD, A. D., JACOBSON, N. L. and GETTY, R. (1962) *J. Dairy Sci.* **45**, 408.
TAMATE, H., MCGILLIARD, A. D., JACOBSON, N. L. and GETTY, R. (1964) *Tohoku J. Agric. Res.* **14**, 171.
THOMPSON, G. B., KINTNER, L. D. and PFANDER, W. H. (1958) *J. Anim. Sci.* **17**, 1220.
THOMPSON, M. D. and TROWELL, H. C. (1952) *Lancet* **i**, 1031.
TSUDA, T. (1964) *Tohoku J. Agric. Res.* **15**, 83.
WARDROP, I. D. (1960) *J. Agric. Sci. Camb.* **55**, 127.
WARDROP, I. D. (1961a) *J. Agric. Sci. Camb.* **57**, 335.

Wardrop, I. D. (1961b) *J. Agric. Sci. Camb.* **57**, 343.
Wardrop, I. D. and Coombe, J. B. (1960) *J. Agric. Sci., Camb.* **54**, 140.
Wardrop, I. D. and Coombe, J. B. (1961) *Aust. J. Agric. Res.* **12**, 661.
Walker, D. M. (1959) *J. Agric. Sci. Camb.* **52**, 352.
Walker, D. M. and Simmonds, R. A. (1962) *J. Agric. Sci. Camb.* **59**, 375.
Warner, R. G., Bernholdt, H. F., Grippin, C. H. and Loosli, J. K. (1953) *J. Dairy Sci.* **36**, 599.
Warner, R. G. and Flatt, W. P. (1965) In. *Physiology of Digestion in the Ruminant.* Papers presented at the Second International Symposium on the Physiology of Digestion in the Ruminant. Ed. R. W. Dougherty, Butterworths, Washington, pp. 24-38.
Warner, R. G., Flatt, W. P. and Loosli, J. K. (1956) *J. Agric. Fd. Chem.* **4**, 788.
Weill, P. (1919) *Arch. Mikrosk. Anat. Entw. Mech.* **93**, 1.
Weller, R. A. and Gray, F. V. (1954) *J. Exp. Biol.* **31**, 40.
Wetterfors, J. (1965) *Acta med. scand. Suppl.* 430.
White, R. G. (1954) *Br. J. Exp. Path.* **35**, 365.
Wierda, J. L. (1950) *Anat. Rec.* **107**, 221.
Wiese, A. C., Johnson, B. C., Mitchell, H. H. and Nevens, W. B. (1947) *J. Nutr.* **33**, 263.
Wieser, M. F. (1965) Personal communication.
Williams, A. W. (1963) *J. Path. Bact.* **85**, 467.
Wilson, T. H. (1962) *Intestinal Absorption.* W. B. Saunders Co., Philadelphia, pp. 13-17.
Wing, J. M. and Ammerman, C. B. (1960) *J. Dairy Sci.* **43**, 1012.
Wright, R. D., Jennings, M. A., Florey, H. W. and Lium, R. (1940) *Quart. J. Exp. Physiol.* **30**, 73.
Yoffey, J. M. and Courtice, F. C. (1956) *Lymphatics, Lymph and Lymphoid Tissue.* 2nd Ed. Arnold, London.
Yonge, C. M. (1937) *Biol. Rev.* **12**, 87.
Zamcheck, N. (1960) *Fedn. Proc. Fedn. Am. Socs. Exp. Biol.* **19**, 855.
Zetterqvist, H. (1956) *The Ultrastructural Organisation of the Columnar Absorbing cells of the Mouse Jejunum.* Diss. Med. Karolinska Institutet, Stockholm.

CHAPTER 10

Lipid Metabolism of Farm Animals

G. A. GARTON
Rowett Research Institute, Bucksburn, Aberdeen.

I. INTRODUCTION

THE farm animals with which this chapter on lipid metabolism is concerned comprise three ruminant species (ox, sheep and goat) and two species of simple-stomached animals, the horse and the pig. In conformity with the broad terms of reference of this Encyclopaedia the emphasis will be on the more nutritional aspects of the subject and, where possible, topics will be considered from the comparative point of view with respect to ruminants on the one hand and non-ruminants on the other.

The term 'lipids' cannot be defined rigidly, but, as used here, it is taken to embrace substances containing chemically-bound fatty acids (usually in ester combination), together with free fatty acids and other naturally-occurring substances (such as sterols and hydrocarbons) which have similar solubility properties. The water-soluble, steam-volatile fatty acids (acetic, propionic and butyric acids) which, as end-products of fermentation in the alimentary tract, play such an important part in the overall biochemical economy of herbivorous animals are not generally considered as lipid. Many aspects of their biochemistry are dealt with elsewhere in this volume and so passing reference only will be made here to a few aspects of their metabolism. Fundamental processes of lipid metabolism (such as the synthesis and breakdown of long-chain fatty acids) which are apparently common to all animals will not be described in detail since accounts of these are readily available in modern textbooks of biochemistry (e.g. Cantarow and Schepartz, 1962) and, at a more advanced level, in books and reviews devoted specifically to lipids (e.g. Bloch, 1960; Dawson and Rhodes, 1964). Complete references to the original literature pertaining to the extensive subject matter of this chapter are not given; references are largely restricted to recent papers, reviews and books which give detailed bibliographies.

II. DIETARY LIPIDS

Leaves of such plants as grasses and legumes make up almost the entire diet of grazing ruminants and horses; fresh herbage also often provides a considerable part of the daily ration of the pregnant sow. The lipid content of leaves is about 5 to 10 per cent of the dry matter and for the most part is present in the chloroplasts. This lipid was, until quite recently, considered to consist mainly of triglycerides but it has now been found to comprise mono- and di-galactosyl derivatives of diglycerides, together with smaller amounts of phospholipids, free fatty acids and sterols; triglycerides are present in very small amounts only.

Monogalactosyl-diglyceride	Digalactosyl-diglyceride
CH_2O —— Galactose	CH_2O —— Galactose —— Galactose
|	|
CHO —— Fatty acid	CHO —— Fatty acid
|	|
CH_2O —— Fatty acid	CH_2O —— Fatty acid

Of the total weight of lipid, fatty acids comprise about 50 per cent and consist mainly of palmitic acid and the C_{18} unsaturated acids, oleic, linoleic and linolenic acids; in the galactoglycerides linolenic acid predominates. Examples of the composition of the total fatty acids of the lipids of clover-rich pasture (Shorland, Weenink and Johns, 1955) and mixed pasture grasses (Garton, 1960a) are given in Table 1.

TABLE 1
Fatty Acid Composition of the Lipids (mainly galactoglycerides) of Pasture
(Values as per cent by weight of total fatty acids)

Fatty acid	Source of lipid	
	Clover-rich pasture	Mixed grasses
Saturated:		
Myristic	—	1·1
Palmitic	8·9	15·9
Stearic	2·8	2·0
Others	3·9	0·5
Unsaturated:		
Palmitoleic	7·9	2·5
Oleic	9·5	3·4
Linoleic	8·1	13·2
Linolenic	58·9	61·3

In that unsaturated fatty acids can undergo structural changes in the digestive tract it should be noted that, as they occur in plant lipids, oleic acid (octadec-9-enoic acid), linoleic acid (octadeca-9,12-dienoic acid) and linolenic acid (octadeca-9,12,15-trienoic acid) possess double bonds of the *cis* configuration and that no positional isomers of these acids are present in normal fodders. Double bonds having the *trans* configuration were thought until recently not to occur in plant fatty acids; however, it has now been found that very small amounts of a C_{16} unsaturated acid with a *trans* bond (hexadec-3-enoic acid, an isomer of palmitoleic acid) are present in the phosphatidylglycerol fraction of the lipids of green leaves.

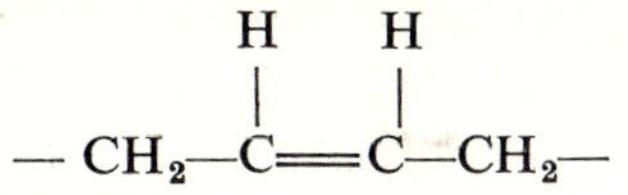

cis configuration

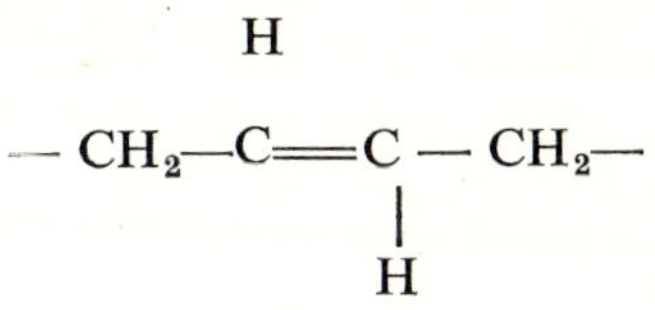

trans configuration

Pasture plants preserved as hay or silage for stall-feeding of ruminants and horses retain their lipid content and fatty acid composition providing they are properly prepared and stored.

The so-called 'concentrate' feeds, prepared largely from plant seeds, form an important part of the diet of stall-fed ruminants and horses and provide the main component of the rations of trough-fed (bacon) pigs. Such materials include maize meal, linseed meal, groundnut meal, soyabean meal, coconut meal, barley, oats, rye and wheat; of these barley is used mostly in pig feeds, oats mainly for horses and the oilseed meals in rations for dairy cattle and pigs.

The lipid content of these feeds varies widely (see Evans, 1960) and in the case of the oilseeds in particular, on the extent to which processing (e.g. solvent extraction, mechanical pressing) has been carried out. In contrast to leaves, the lipids of seeds consist mainly of triglycerides, though C_{18} unsaturated fatty acids again predominate; linoleic acid is often a major component, notably in the seed glycerides of graminaceous plants. Typical examples taken from Hilditch and Williams (1964) and Aylward and Showler (1962) of the composition of the total fatty acids of some of these concentrates are shown in Table 2.

Lipids are, of course, present in many other ingredients which may be used for the making of feeds for domestic animals, though usually their contribution to the total intake of lipid is relatively small, except perhaps in the case of some fish meals and meat meals.

Though, as yet, the addition of lipids as such to feedingstuffs is not a common commercial practice, it has been done on an experimental basis for a considerable time. For example, lipids such as tallow or seed oils have been included as a concentrated source of energy in rations for fattening cattle and attempts have been made to improve milk fat output by incorporating various triglycerides or long-chain fatty acids in the diet of dairy cattle; in other studies with cows triglycerides have been sprayed on pasture or given as a 'drench' in attempts to control bloat.

TABLE 2
Fatty Acid Composition of the Triglycerides of Some Vegetable 'Concentrate' Feeds
(Values as per cent by weight of total fatty acids)

Fatty acid	Source of fatty acids				
	Rye	Barley	Oats	Maize	Linseed
Saturated:					
Myristic	trace	trace	trace	0·4	trace
Palmitic	21·0	22·0	15·9	12·7	5·6
Stearic	trace	trace	trace	2·7	5·8
Unsaturated:					
Oleic	7·0	7·2	40·4	30·7	21·6
Linoleic	63·7	70·8	43·7	53·5	12·5
Linolenic	8·3	—	trace	—	54·5

This general account of dietary lipids would be incomplete without mention of the importance of the lipids (triglycerides) of milk in the nutrition of the unweaned young. The lipid content of the milk of the cow, ewe and goat is usually in the range 3 to 5 per cent, whilst that of mare milk is lower (1 to 2 per cent) and that of sow milk is somewhat higher (5 to 7 per cent).

III. DIGESTION AND ABSORPTION

A. NON-RUMINANTS

As far as is known very little, if any, published information is available concerning the processes of digestion and absorption in the horse and the pig. However, it is reasonable to suppose that the pattern of events is similar to that in other simple-stomached mammals in whose alimentary tract little or no changes take place in lipids until the digesta reach the small intestine. Here, in the jejunum, enzymes are present which can liberate fatty acids from ester combination and, in the presence of bile, lipid absorption takes place. For many years the role of enzymes (particularly pancreatic lipase) and of conjugated bile acids in the digestion and absorption of lipids has been a contentious topic and it is only recently that some measure of agreement has been reached regarding the details of these important processes. It is not intended here to recapitulate research in this field—suffice it to refer the reader to reviews by Schmidt-Nielsen (1946) and Senior (1964) in which the evidence leading to current views is discussed. It is now apparent that, when the diet does not contain abnormally large amounts of lipid, very little absorption of intact, unhydrolysed droplets (pinocytosis) occurs and that absorption takes place from a micellar solution (pH usually 6·3 to 6·5) containing fatty acids, monoglycerides and conjugated bile acids. This micellar solution consists of soluble, molecular aggregates of the bile acids together with the products of the hydrolytic activity of pancreatic lipase towards triglycerides, *viz.* free fatty acids and 2-monoglycerides, the enzyme acting specifically on the acids esterified at positions 1 and 3 of the triglyceride molecule.

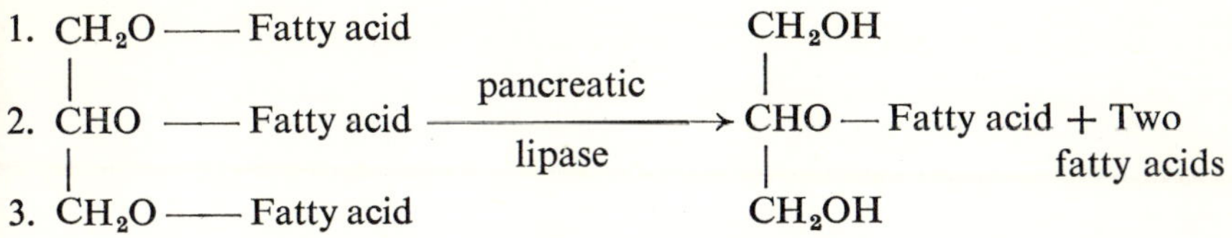

The production of a fine emulsion of triglyceride particles is promoted by the presence of bile salts and this facilitates enzyme action. It is probable that triglycerides in the diet of pigs and horses are digested and presented for absorption in this way and that galactoglycerides will also yield 2-monoglycerides and free fatty acids, the galactose residues being removed by glycosidases present in the intestinal lumen (cf. Conchie and MacDonald, 1959).

The fatty acids and monoglycerides are absorbed into the jejunal epithelial cells though the bile acids are not and they return, as it were, to the intestinal lumen there to effect the micellar solution of more of the products of digestion of glycerides; the bile acids which accompany the residues of the digesta into the ileum are absorbed from this portion of the intestine. Once inside the epithelial cells the fatty acids and monoglycerides recombine to form the triglycerides of the chylomicrons which are delivered to the venous circulation by way of the intestinal lymph and the thoracic duct.

The dietary lipids of the adult horse and pig do not normally contain short-chain fatty acids which, of themselves, are water-soluble. Such acids when fed, as such or in glyceride combination, to other simple-stomached animals are absorbed from the intestine directly into the portal blood and do not participate in the synthesis of chylomicron triglycerides. This is presumably the fate of the short-chain fatty acids liberated from milk fat by pancreatic lipase in the intestine of the unweaned foal and piglet.

B. Ruminants

Though investigations some time ago showed that, under the influence of rumen microorganisms, dietary carbohydrates and proteins can undergo extensive hydrolytic and fermentative changes in the rumen, it is only relatively recently that lipids, too, have been found to be profoundly affected by microorganisms in this part of the alimentary tract (for reviews see Garton, 1965; 1967). The changes include (i) hydrolytic release of esterified fatty acids,

(ii) hydrogenation of the double bonds of unsaturated fatty acids and (iii) fermentation of glycerol and galactose liberated from lipids. Thus virtually complete digestion of dietary lipids can be effected before the ingesta pass from the rumen through the abomasum (true stomach) to the small intestine —a situation in marked contrast to that in the simple-stomached animal.

It seems that fatty acids can be hydrolytically released in the rumen from any class of lipid; thus, following the original demonstration that triglycerides undergo hydrolysis, it has been shown that fatty acids can be liberated from galactoglycerides, phospholipids, sterol esters, methyl esters and from synthetic detergents such as Tween 80 (poly-oxyethylene sorbitan mono-oleate). In studies *in vitro* with triglycerides (e.g. linseed oil, maize oil) it was found that no lipolysis took place in rumen contents which had been boiled or in saliva (which provides much of the liquid of rumen contents); enzyme activity was confined mainly to the rumen bacteria (Garton, Lough and Vioque, 1961). Subsequently lipolytic bacteria were isolated from sheep rumen contents and studied in pure culture (Hobson and Mann, 1961). These organisms are strictly anaerobic, Gram-negative curved rods which can ferment glycerol, fructose and ribose but which cannot metabolize the long-chain fatty acids which they release.

Unlike pancreatic lipase, bacterial enzyme activity towards triglycerides does not result in the accumulation of 2-monoglycerides. Very little, if any, mono- or diglycerides can be detected at any stage during hydrolysis so it appears that if these are produced they have but a transitory existence. So far attempts to obtain cell-free lipolytic enzyme preparations from rumen bacteria have proved unsuccessful except in the case of lysolecithinase activity, making it seem likely that these enzymes are intimately associated with the cell walls of viable organisms.

Galactosidase activity has been demonstrated in mixed rumen organisms and cell-free preparations of rumen bacteria have been shown to remove the galactose residues from both mono- and di-galactosyl glycerol, though not from galactosyl glycerides, i.e. when esterified fatty acids are present in the molecule. It thus appears that, in the rumen, lipolysis of galactoglycerides must precede bacterial galactosidase action. This does not apply to rumen protozoa of which some species can effect the release of galactose from galactosyl glycerides as well as being able to liberate the sugar from galactosyl glycerol.

The free glycerol and galactose released from triglycerides, galactoglycerides or phospholipids can both be fermented by the rumen bacteria to yield short-chain fatty acids. Whilst galactose shares a common fate with dietary glucose and fructose in being metabolized to a mixture of acetic, propionic and butyric acids, glycerol yields mostly propionic acid which, incidentally, accounts for its antiketogenic action when used therapeutically in sheep and cattle because it acts as a precursor of liver glycogen.

Hydrogenation of unsaturated fatty acids in the rumen involves mainly the C_{18} acids, oleic acid, linoleic acid and linolenic acid since, as noted earlier (p.336), these acids predominate in dietary lipids. Complete hydrogenation of each acid yields stearic acid which is thus usually a major component of the total fatty acids (mostly free fatty acids) in the rumen contents when these are examined not long after the ingestion of a feed. It is likely that the acids are for the most part hydrogenated following their lipolytic release, though it has been demonstrated that hydrogenation can take place (somewhat less rapidly) while they are still present in ester combination.

Hydrogenation is not, however, always complete and can result in the formation of geometrical (*trans*) and positional isomers of the common unsaturated acids. The apparent migration of double bonds along the carbon chain of the molecule and the appearance of *trans* bonds (see p. 337) is consistent with a dynamic process of hydrogenation and dehydrogenation which shows a remarkable resemblance to purely chemical hydrogenation as carried out in the presence of a metal catalyst such as nickel. Some results *in vitro* with rumen organisms suggest that the formation of a *trans* mono-unsaturated acid (octadec-13-enoic acid or octadec-14-enoic acid) may be necessary before stearic acid can be produced.

Though the mechanism of the reaction is not yet fully understood, it appears that certain rumen bacteria are capable of effecting the hydrogenation of the double bonds of linoleic and linolenic acids, whilst other species may be involved in the hydrogenation of mono-unsaturated C_{18} acids (Polan, McNeill and Tove, 1964). Hydrogenation is apparently non-specific; experiments *in vitro* have shown that the reduction of unsaturated fatty acids can be partially inhibited in an atmosphere of carbon dioxide or in the presence of acrylate, crotonate or fumarate, indicating that these substances can successfully compete for available metabolic hydrogen. In this connection it is of interest to note that, in studies with sheep, Czerkawski and Blaxter (1965) have demonstrated that methane production (which in part results from the reduction of carbon dioxide by hydrogen or a hydrogen donor) can be effectively lowered when relatively large amounts of C_{18} unsaturated fatty acids are infused into the rumen. The reduction of unsaturated fatty acids by rumen bacteria *in vitro* is completely inhibited by cyanide or arsenite, indicating that iron and sulphydryl groups are important constituents of the bacterial hydrogenase systems.

The main changes in the principal dietary lipids which are effected by rumen microorganisms are summarized in Fig. 1.

In addition to lipids of dietary origin, the lipids of rumen contents include those present as structural components of the bacteria and protozoa. These lipids are not only of interest in relation to the metabolic activities of the organisms themselves, but also because of the contribution they make to the lipids in the digesta leaving the rumen. It has been estimated that at least

140 g of microbial lipid can be available daily for intestinal digestion in the cow.

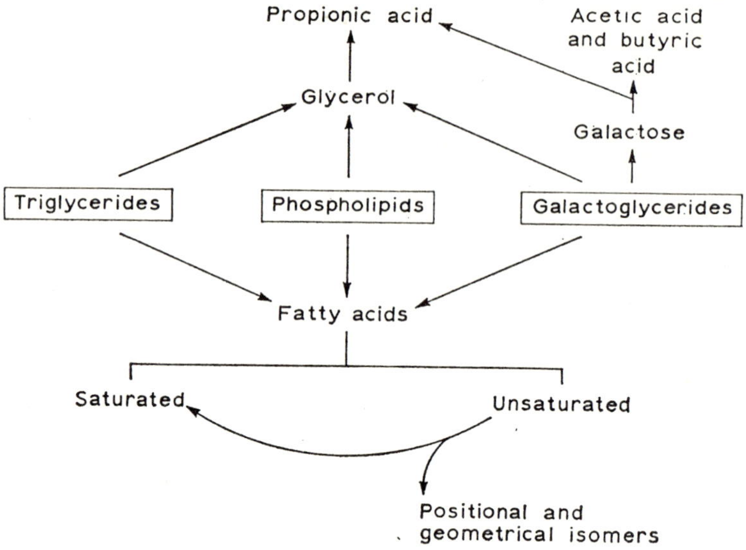

FIG. 1. Fate of dietary lipids in the rumen [from Garton (1964), by kind permission of the publishers, John Wiley, London].

Other than *n*-fatty acids with an even number of carbon atoms, rumen bacterial lipids contain significant amounts of *n*-acids having an odd number of carbon atoms and branched-chain acids with 13, 14, 15, 16 and 17 carbon atoms. These branched-chain acids comprise members of the *iso* and *anteiso* series.

$$CH_3-\underset{}{\overset{\overset{\displaystyle CH_3}{|}}{CH}}-(CH_2)_n-COOH$$

Iso acid

$$CH_3-CH_2-\overset{\overset{\displaystyle CH_3}{|}}{CH}-(CH_2)_n-COOH$$

Anteiso acid
(the substituent methyl group being attached to the ante-penultimate carbon atom)

Of the branched-chain acids those having 15 carbon atoms predominate; they have been shown by Keeney, Katz and Allison (1962) to comprise 44 per cent of the total fatty acids of the rumen bacterium *Ruminococcus flavifaciens* and about 20 per cent of the phospholipid fatty acids of bovine rumen mixed bacteria. Viviani and Lenaz (1963) found that C_{15} *anteiso* acid accounted for 8·4 per cent of the total fatty acids in the rumen contents of sheep fed on a lipid-free ration. Synthesis of the branched-chain fatty acids takes place, at least in part, as a result of chain-lengthening of the short-chain acids *iso*butyric acid, *iso*valeric acid and 2-methylbutyric acid which arise in the rumen by the bacterial deamination and decarboxylation of the amino-acids valine, leucine and *iso*leucine respectively; some rumen bacteria require these short-chain, branched acids as growth factors. Branched, long-chain fatty aldehydes are also synthesized by rumen bacteria, though in amounts much less than those of the corresponding fatty acids.

Bacteria in general do not apparently synthesize fatty acids with more than one double bond and, though some workers have reported their presence in preparations of rumen bacteria, this was probably occasioned by the absorption or adsorption of lipids of dietary origin.

Some species of rumen protozoa have been shown to contain up to 9 per cent of lipid (dry matter basis) and it has been demonstrated *in vitro* that long-chain fatty acids can be assimilated by these organisms. The occurrence in protozoal lipids of fatty acids and fatty aldehydes similar to those in rumen bacteria (Katz and Keeney, 1964) suggests that they may be derived, at least in part, from bacteria ingested by the protozoa. The formation *de novo* of polyunsaturated fatty acids in protozoa is a possibility suggested by the presence of C_{18} di-unsaturated acid in the total microbial lipids of sheep maintained on a lipid-free ration. Since dietary polyunsaturated fatty acids may be virtually 'lost' as a result of hydrogenation in the rumen, any such acids made available by the intestinal digestion of protozoal lipid could be of considerable importance to the host animal.

Though the water-soluble, short-chain fatty acids produced by the fermentative degradation of cellulose in the rumen are absorbed through the rumen wall, only an insignificant fraction of the free long-chain acids is assimilated in this way. These acids, together with feed particulate matter and associated microorganisms, pass more or less continuously from the rumen through the omasum and abomasum into the small intestine. In the abomasum most of the microorganisms disintegrate and this probably facilitates the subsequent digestion of their structural lipids.

As in non-ruminant animals, lipid absorption from the small intestine is greatly reduced in the absence of pancreatic juice and virtually abolished in the absence of bile (Heath and Morris, 1963). Though the micellar absorption of long-chain fatty acids is clearly dependent on the presence of conjugated bile acids, the need for pancreatic juice is perhaps not so readily apparent. Whereas in the non-ruminant animal pancreatic lipase effects the hydrolysis of triglycerides to yield 2-monoglycerides which promote the formation of mixed micelles, very little if any dietary glyceride usually survives complete hydrolysis in the rumen. It seems likely that the need for pancreatic juice in the ruminant may depend on its being a source of phospholipases which can progressively hydrolyse biliary lecithin (Leat, 1965; Lennox, Lough and Garton, 1965), of which some 10–15 g daily enter the intestine of the sheep (Adams and Heath, 1963); the lysolecithin formed as an intermediate in the hydrolysis of lecithin may well promote the formation of mixed micelles by acting as a detergent to render soluble the solid particles of long-chain fatty acids and by taking part itself in mixed micelle formation.

The fatty acids of dietary origin, together with those resulting from the hydrolysis of biliary lecithin and of microbial lipids, are for the most part converted in the intestinal mucosa to triglycerides which comprise 70 to 80 per cent of the lipids of chylomicrons. Most of the rest of the chylomicron lipid consists of phospholipids. Though some of the glycerol required for the synthesis of triglycerides and phospholipids may be derived from lysolecithin, it is probable that most is provided by endogenously-produced glycerophosphate. The principal fatty acids present in the triglycerides and phospholipids of the chylomicrons of sheep intestinal-duct lymph (Heath, Adams and Morris, 1964) and thoracic-duct lymph (Feliński *et al.*, 1964) are shown in Table 3. The proportion of these acids reflect those in the intestinal lumen and the relatively high content of stearic acid in both classes of lipid is noteworthy; in the triglycerides the stearic acid is found preferentially esterified to the 1- and 3-positions of the glycerol molecule (Garton and Duncan, 1965). It has been calculated that, in sheep fed on hay and vegetable concentrates, the amount of stearic acid in daily transport from the intestine can range from 5·6 to 7·4 g. The C_{18} unsaturated acids contain the geometrical and positional isomers of oleic, linoleic and linolenic acids which originate in the rumen and, though not shown in Table 3, acids of microbial origin (branched-chain acids and *n*-acids with an odd number of carbon atoms) are present in the lymph lipids in minor proportions.

TABLE 3

Fatty Acid Composition of the Triglycerides of Sheep Intestinal-duct Lymph and Thoracic-duct Lymph

(Values as per cent by weight of total fatty acids; minor and trace components not included)

Fatty acid	Source of fatty acids			
	Intestinal-duct lymph		Thoracic-duct lymph	
	Triglycerides	Phospholipids	Triglycerides	Phospholipids
Saturated:				
Palmitic	20·2	13·0	14·2	18·2
Stearic	42·5	35·6	45·1	23·8
Unsaturated:				
C_{18} mono-unsaturated	26·5	20·6	21·9	16·2
C_{18} di-unsaturated	4·1	18·5	4·8	20·2
C_{18} tri-unsaturated	1·6	1·3	4·5	8·4

Before the development of an active rumen the digestion of milk fat by the sucking animal apparently takes place in the small intestine under the influence of pancreatic lipase, though it has been shown that limited lipolysis occurs in the abomasum of the calf; the enzyme responsible (so-called 'pre-gastric' esterase) is secreted into the saliva and brings about the release of short-chain acids, particularly butyric acid, from milk fat (Otterby, Ramsey and Wise, 1964).

IV. COMPOSITION OF TISSUE AND MILK LIPIDS, WITH SPECIAL REFERENCE TO THE INFLUENCE OF DIET

Once the chylomicrons enter the blood stream from the thoracic duct they are taken up by many tissues and organs, in particular by the liver, by adipose tissue and by lactating mammary tissue. The chylomicron triglycerides undergo hydrolysis and the released fatty acids can be oxidized or can participate along with endogenously-synthesized fatty acids in the *de novo* formation of triglycerides, phospholipids and cholesterol esters (see Robinson, 1964; Tove, 1965).

A. Non-ruminants

(i) *Tissue lipids*

The depot fats of the horse are characterized by the presence of palmitic acid and the C_{18} unsaturated acids, oleic, linoleic and linolenic acids as major components, the relative proportions of which may vary according to the fatty acid composition of the dietary lipids. As the typical values in Table 4 show, the depot glycerides of the pasture-fed horse contain more linolenic acid than linoleic acid (Gupta and Hilditch, 1951), whereas the converse holds for horses which had been stall-fed on oats (Holmberg and Rosenqvist, 1949). These acids are identical with those found in plant lipids and have thus apparently undergone no structural modification during their digestion and absorption. As compared with the internal depot glycerides those from subcutaneous tissue, such as that between the shoulder blades (withers), show a notably high content of palmitoleic acid (Dahl, 1957). The significance of this is not clear, though it is of interest to note that another outer tissue, namely hoof, also contains lipids having a high content of the same acid.

The main components of the triglycerides derived from the mesenteric tissue of the pasture-fed horse were identified as palmito-oleo-linolenins (approx. 30 per cent) and glycerides with one oleic group and two other (usually different) unsaturated acyl groups. With respect to the intramolecular structure of horse glycerides Mattson and Lutton (1958) found that, in a sample the total acids of which comprised 39 per cent saturated acids and 61 per cent unsaturated acids (i.e. comparable to those shown in Table 4), the acids esterified at the 2-position consisted of 21 per cent of saturated acids

TABLE 4
Fatty Acid Composition of Horse Depot Glycerides
(Values as per cent by weight of total fatty acids)

Fatty acid	Source of glycerides		
	Mesenteric tissue of a pasture-fed horse	Perinephric tissue of a horse stall-fed on oats	Subcutaneous (wither) tissue of four horses
Saturated:			
Myristic	4·5	2·8	4·4
Palmitic	25·9	20·3	26·0
Stearic	4·7	7·2	2·1
Others	0·6	0·6	—
Unsaturated:			
Palmitoleic	6·8	3·7	14·2
Oleic	33·7	38·8	35·2
Linoleic	5·2	22·1	6·4
Linolenic	16·3	2·2	9·6
Others	2·3	2·1	2·1

and 79 per cent of unsaturated acids. In the liver glycerides of the pasture-fed horse linolenic acid is present to a somewhat greater extent than in the glycerides of extra-hepatic tissues, though this does not apply to the corresponding liver phospholipids in which linoleic acid is the predominant component (Shorland, Bruce and Jessop, 1952).

It is not surprising in view of the economic importance of the pig as a source of food for man that its tissue lipids have been very extensively studied. It was recognized many years ago that the nature of the dietary lipids could markedly influence the fatty acid composition of the tissue lipids of the pig and that a product undesirable to the consumer could readily result from the assimilation of too much of certain oils; soft fat in bacon develops when certain vegetable oils such as maize oil are fed and the inclusion in the feed of too much oil of marine origin, such as may be present in some fish meals, can result in unacceptable 'taint' of bacon and lard. Before discussing these effects in more detail it is perhaps worthwhile to direct attention to the fatty acid composition of the tissue glycerides which the pig elaborates *de novo* from acetate derived mainly from the catabolism of dietary carbohydrate. Given a diet which is virtually lipid-free the pig lays down considerable amounts of depot glycerides, particularly along the back. Representative recent analyses (Leat *et al.*, 1964) of the fatty acids of such glycerides are shown in Table 5; many other similar analyses are recorded and discussed by Dahl (1958a) and Hilditch and Williams (1964).

It is evident that 90 to 95 per cent of the total fatty acids produced endogenously by the pig consist of palmitic acid, stearic acid and oleic acid; no linolenic acid is present and the very small amounts of linoleic acid probably

TABLE 5

Fatty Acid Composition of Depot Glycerides of Pigs (mean of two animals) fed from Weaning on a Semi-synthetic Lipid-free Ration

(Values as per cent by weight of total fatty acids)

Fatty acid	Source of glycerides		
	Subcutaneous (back) tissue		Perinephric tissue
	Outer layer	Inner layer	
Saturated:			
Myristic	1·0	1·1	1·1
Palmitic	23·1	25·0	27·0
Stearic	12·5	14·5	18·0
Others	0·4	0·4	0·5
Unsaturated:			
Palmitoleic	4·0	3·7	3·6
Oleic	56·5	53·2	47·8
Linoleic	0·3	0·2	0·2
Others	1·7	1·4	1·4

represent the conserved residue of that which was ingested before the animals were weaned. Evidence suggesting that the relative proportions of oleic and stearic acids could be related to the tocopherol intake of the animals (Bratzler *et al.*, 1950) could not be confirmed in subsequent studies (Duncan *et al.*, 1959; Leat, 1961). It is noteworthy (Table 5) that the unsaturation of the tissue glycerides (largely a reflection of their content of oleic acid) is greater in the outer layer of subcutaneous tissue than in the inner layer which, in turn, is more unsaturated than the perinephric glycerides. This well-established 'gradient' can be correlated with the temperature of the tissue at site of glyceride deposition.

Though the feeding to pigs of diets containing glycerides with a high content of saturated fatty acids (e.g. beef tallow) occasions very little change in the composition of the depot glycerides compared with that found when little or no dietary lipid is given, the feeding of glycerides containing a relatively high content of unsaturated fatty acids has very pronounced effects. Presumably the lack of effect of saturated fatty acids is due to their being but poorly absorbed. The unsaturated acids appear in considerable amount in the depot glycerides, mainly at the expense of oleic and palmitic acids. The extent of incorporation of these acids of exogenous origin parallels the normal unsaturation gradient, being greatest in the outer (subcutaneous) tissue and least in the internal depots, though these are still very markedly affected. Many examples of the effects of feeding vegetable and marine oils are quoted by Hilditch and Williams (1964); that given in Table 6 (from Leat *et al.*, 1964) was selected because the basal lipid-free diet to which 10 per cent of maize oil was added was the same as that used in the experiments referred to in Table 5. In addition to the presence of linoleic acid as a major component of the depot glycerides, a number of minor components also reflected the influence of the dietary oil, notably the presence of linolenic acid and of arachidonic acid—the latter having been produced endogenously from linoleic acid.

TABLE 6

Fatty Acid Composition of Depot Glycerides of Pigs (mean of two animals) fed from Weaning on a Semi-synthetic Diet Containing 10 per cent of Maize Oil

(Values as per cent by weight of total fatty acids)

Fatty acid	Source of glycerides			
	Subcutaneous (back) tissue		Perinephric tissue	Maize oil as fed
	Outer layer	Inner layer		
Saturated:				
Myristic	1·1	1·0	1·1	0·1
Palmitic	19·0	18·7	18·1	12·0
Stearic	10·9	11·5	14·8	2·6
Others	0·4	0·4	0·2	0·2
Unsaturated:				
Palmitoleic	1·8	1·9	2·0	0·4
Oleic	36·7	37·5	40·2	28·9
Linoleic	27·0	26·1	24·0	54·2
Linolenic	0·4	0·5	0·4	1·1
Arachidonic	0·3	0·3	0·2	—
Others	2·2	2·3	2·2	—

The amounts of dietary unsaturated glycerides do not have to be as great as 10 per cent of the ration to exert an effect on depot glyceride composition; for example, Dahl (1958b) found that when barley and wheat bran, which respectively contained 1·7 and 3·5 per cent of glycerides, were the sole source of dietary lipid for bacon pigs, the back fat contained 7 per cent of linoleic acid and that when oats (containing 5·5 per cent of oil) replaced barley in the ration the level of linoleic acid in the back fat rose to 16 per cent. Such high proportions of linoleic acid in bacon fat and lard make them soft and readily susceptible to oxidative rancidity.

At the other extreme when excessively large amounts of dietary glycerides were given, as when a pig received 4 lb of whale oil daily, the depot glycerides were found to contain 40 to over 60 per cent of apparently unchanged whale oil glycerides (Garton, Hilditch and Meara, 1952).

The foregoing study did not, however, preclude the possibility that the glycerides had undergone hydrolysis in the intestine to be subsequently reassembled giving triglycerides virtually indistinguishable from those of the original whale oil. To investigate this possibility Garton and Duncan (1954) fed to pigs a diet containing similar large amounts of glycerides in the form of a mixture of equal parts of cod liver oil and lard. If intestinal hydrolysis of these glycerides took place followed by resynthesis in the intestinal mucosa these newly-formed mixed glycerides would not, of course, be representative of either cod liver oil or lard. However, considerable amounts of a fraction almost identical with cod liver oil were isolated from the depots and it was evident that the ingested cod liver oil had been absorbed unchanged, probably in particulate form. Nevertheless, it should be emphasized that such an absorption mechanism may well not obtain when the usual very much smaller amounts of triglycerides are present in the diet.

Much work of a purely chemical nature has been done in the past on the overall fatty acid composition of fractions of the total triglycerides of pig adipose tissue and, more recently, on the intramolecular disposition of the component acids (see Coleman, 1963; Hilditch and Williams, 1964). Unlike almost all other triglycerides of animal (or plant) origin palmitic acid is present predominantly in the 2-position, with oleic acid mainly in the 1- and 3-positions; monounsaturated-disaturated glycerides and diunsaturated-monounsaturated glycerides usually constitute 80 to 90 per cent of the total triglycerides. The reason any given fatty acid is esterified to a particular hydroxyl group of the glycerol moiety in any natural triglyceride is not yet understood, though there are indications (see Lands, 1965) that inroads into this problem may soon be made.

(ii) *Milk lipids*

The fatty acid composition of the milk lipids of the horse (Hilditch and Jasperson, 1944) and of the colostral and milk lipids of the pig (DeMan and Bowland, 1963) are shown in Table 7; though traces of other lipids are present, the milk lipids consist essentially of triglycerides.

The mare milk glycerides, which were obtained from a pasture-fed animal, show several interesting features. These include the presence, as in ruminant milk fats (Table 9), of a considerable proportion of short-chain fatty acids ranging from butyric to lauric acid and the occurrence, as in horse depot fat (Table 4), of linoleic and linolenic acids in amount approaching a quarter of the total acids. The presence of short-chain acids is apparently associated, as in ruminants, with the assimilation and metabolism in mammary tissue of acetate produced in the alimentary tract by the bacterial fermentation of dietary cellulose, whilst the high content of linoleic and linolenic acids reflects the occurrence of these acids in dietary lipids.

In contrast, sow colostral and milk glycerides do not contain significant proportions of short-chain acids and are characterized by their relatively high content of unsaturated acids, in particular of oleic and linoleic acids, though at the same time the proportion of palmitic acid is greater than in mare milk fat. Colostrum contains less palmitic acid and more linoleic acid than normal milk. Similar findings to those of DeMan and Bowland (1963) were made by

TABLE 7

Component Fatty Acids of Horse Milk Lipids and of Pig Colostral and Milk Lipids
(Values as per cent by weight of total fatty acids; unidentified and trace components omitted)

Fatty acid	Source of lipid		
	Mare milk	Sow colostrum	Sow milk
Saturated:			
Short chain	3·9*	—	—
Capric	5·5	—	0·2
Lauric	5·6	—	0·3
Myristic	7·0	1·4	3·3
Palmitic	16·1	22·5	30·3
Stearic	2·9	5·7	4·0
Unsaturated:			
Palmitoleic	7·5	5·0	9·9
Oleic	18·7	41·7	35·3
Linoleic	7·6	20·9	13·0
Linolenic	16·1	2·4	2·5
Others	8·8†	—	—

* Comprising 0·4 per cent butyric acid, 0·9 per cent hexanoic acid and 2·6 per cent octanoic acid.

† Including 0·9 per cent decenoic acid, 1·0 per cent dodecenoic acid and 1·8 per cent tetradecenoic acid.

Lindberg and Tollerz (1964) whose gas chromatographic analyses showed some twenty-two different fatty acids to be present in sow colostrum and milk; they also noted that arachidonic acid was present to a greater extent in colostral fatty acids than in milk fatty acids, the values being 1 to 2 per cent and about 0·5 per cent respectively of the total fatty acids. It seems likely from the few results available that fatty acids of dietary origin can influence sow milk fat in a similar way to that in which they affect depot fat; DeMan and Bowland (1963) found that the back fat and milk fat of a sow given a diet containing 15 per cent of tallow showed the same differences compared with the controls, namely a lower content of palmitic and linoleic acids and an enhanced content of oleic acid. Salmon-Legagneur (1962) found that supplementing the diet of lactating sows with 17 per cent of lard promoted an increased output of milk fat with a higher content of oleic and linoleic acids and a correspondingly lower content of saturated acids than in normal milk fat.

B. RUMINANTS

(i) *Tissue lipids*

The tissue lipids of adult ruminant animals are characterized by several features, notably a relatively high content of stearic acid in internal depot glycerides, the occurrence of isomeric forms of unsaturated fatty acids (particularly those of the C_{18} group) and the presence of small amounts of branched-chain fatty acids and of acids having an odd number of total carbon atoms. Furthermore, the component fatty acids of the tissue lipids, especially the triglycerides, do not normally reflect the fatty acid composition of normal dietary lipids and remain relatively refractory to changes in the amounts and fatty acid composition of such lipids. This is in marked contrast to the pattern of events in non-ruminant herbivores and is occasioned largely by events in the rumen which, as discussed earlier (p. 341), lead to the production from dietary lipids of structurally-modified unsaturated fatty acids and the formation of saturated fatty acids, especially of stearic acid.

Many analyses of the fatty acids of the adipose tissue glycerides of ox and sheep have been published (see Hilditch and Williams, 1964); representative values for the proportions of the principal fatty acids present in perinephric and subcutaneous tissue glycerides (Hilditch and Pedelty, 1941; Hartman and Shorland, 1961) are given in Table 8.

TABLE 8
Fatty Acid Composition of the Depot Glycerides of Ox and Sheep
(Trace components excluded; values as per cent by weight of total fatty acids)

Fatty acids	Source of glycerides			
	Ox adipose tissue		Sheep adipose tissue	
	Perinephric	Subcutaneous	Perinephric	Subcutaneous
Saturated:				
Myristic	3·3	2·6	2·3	3·1
Palmitic	27·0	24·6	26·2	28·3
Stearic	31·7	7·2	27·1	13·5
Unsaturated:*				
C_{14}	0·4	2·1	0·3	0·2
C_{16}	1·7	7·7	1·0	0·6
C_{18}	34·2	55·3	42·0	52·7

* Mostly mono-unsaturated.

Palmitic, stearic and C_{18} mono-unsaturated acids usually comprise about 90 per cent of the total acids and, whilst the proportion of palmitic acid is relatively constant, there is an apparent reciprocal relationship between the proportions of stearic acid and C_{18} mono-unsaturated acid. As in pig adipose tissue, the subcutaneous glycerides of ruminants are more unsaturated than those of the internal tissues, due mainly to an enhanced content of C_{18} mono-unsaturated acid. In the ox (as in the horse) C_{16} mono-unsaturated acid can also contribute to the unsaturation of external tissue glycerides (Table 8) and breast (brisket) glycerides have been found (Dahl, 1958a) to contain as much as 17·5 per cent of this acid, accompanied by a correspondingly low content of stearic acid (5·0 per cent). This suggests that the stearic acid content of the depot glycerides varies not just with the C_{18} mono-unsaturated acid, but rather with the total content of unsaturated acids. Whilst no increment in the proportion of C_{16} mono-unsaturated acid has been reported in sheep subcutaneous glycerides (usually obtained from the region of the back and flanks of the animal), it is noteworthy that recent (unpublished) studies in the writer's laboratory have shown that some 5 to 10 per cent of C_{16} mono-unsaturated acid is present in glycerides extracted from the ears and lower part of the legs of adult sheep. Neat's foot oil (prepared from ox and sheep hooves) has been known for some time to have a relatively high content (10 to 12 per cent) of C_{16} mono-unsaturated acid.

Infrared spectra of the unsaturated fatty acids of ox and sheep depot glycerides indicate that a considerable proportion (10 to 20 per cent) have double bonds of the *trans* configuration, some of which have been identified, e.g. octadec-*trans* 10-enoic acid and octadec-*trans* 11-enoic acid. Of the small amount (1 to 4 per cent) of unsaturated fatty acids having more than one double bond which are normally present in ruminant tissue glycerides, most is in the form of geometrical and positional isomers of linoleic acid. Other minor component acids, additional to those evidently resulting from the assimilation of acids derived from rumen bacterial lipids, include *inter alia* capric acid, lauric acid, arachidic acid and phytanic acid (3,7,11,15-tetramethylhexadecanoic acid), the latter probably arising from the metabolism of the phytol moiety of feed chlorophyll.

Though much less information is available on the fatty acid composition of goat lipids it is evident that, in general, they resemble those of the sheep (see Rajagopal and Achaya, 1964).

The major glyceride types in ruminant tissue lipids are usually oleopalmitostearins and palmitodioleins, though fully saturated glycerides are present also in variable amount depending on the proportion of stearic acid in the total fatty acids, e.g. palmitodistearins, whilst accounting for only about 5 per cent of the external tissue glycerides of the sheep, can comprise up to 15 per cent of perinephric glycerides. Studies on the intramolecular disposition of the fatty acids of the depot and liver glycerides of ox and sheep have shown that C_{18} mono-unsaturated acid is preferentially esterified to the secondary alcohol group (2-position) of the glycerol molecule, with the saturated acids (particularly stearic acid) being mainly present in the 1- and 3-positions.

Though the apparent failure of dietary unsaturated oils of vegetable origin (e.g. soyabean oil, maize oil) to affect the composition of depot glycerides of the adult ruminant to any marked extent is due to hydrogenation of the unsaturated C_{18} acids by rumen bacteria, somewhat increased amounts of stearic acid may appear in some tissue lipids, e.g. in the brisket glycerides of cattle given supplementary ground soyabeans (Tove and Mochrie, 1963). The fatty acids of dietary marine oils (e.g. menhaden oil, herring oil) are also not deposited in tissue lipids, a recent experiment being that of Dahl (1958a) who, over a period of 6 weeks before the animal was slaughtered, fed 5 kg of herring oil to a cow and subsequently found no trace in the depot glycerides of the highly unsaturated C_{20} and C_{22} fatty acids which characterize herring oil. If, in this study, C_{20} and C_{22} saturated acids were formed in the rumen, they, too, were not absorbed though in other investigations it has been shown that dietary saturated acids of shorter chain-length (e.g. lauric and myristic acids of coconut oil) can be incorporated into depot (and milk) lipids of the cow.

However, before the development of an active rumen the weanling ruminant assimilates dietary fatty acids like a non-ruminant and depot fat composition can be markedly influenced in the adult animal when unsaturated oils are administered through a duodenal cannula, thereby by-passing the rumen. Depot glycerides of an unusually high degree of unsaturation were produced in an experiment by Tove and Matrone (1962) in which sheep were fed a semi-synthetic ('purified') diet; changes in the rumen flora which resulted in an diminution in the extent to which dietary C_{18} fatty acids were hydrogenated were apparently responsible for this observation.

In addition to depot fat, the fatty acid composition of the lipids of various other tissues has been investigated. Extensive studies on plasma lipids of the ox (e.g. Lough and Garton, 1957; Duncan and Garton, 1963) and the sheep (Horgan and Masters, 1963; Garton and Duncan, 1964) have shown *inter alia* that the cholesterol esters—the major lipid present—are characterized by their high content of C_{18} di- and tri-unsaturated acids. Compared with the unsaturated fatty acids in the plasma triglycerides, the proportion of *trans* acids in the cholesterol esters is low and much of the C_{18} acids appear to consist of linoleic and linolenic acids. The acids apparently represent those of dietary origin which have escaped hydrogenation in the rumen (possibly within protozoa) and which subsequently become preferentially esterified to cholesterol in the intestinal mucosa or elsewhere.

A study correlating age and rumen development with the fatty acid composition of triglycerides, cholesterol esters, free fatty acids and phospholipids of plasma, liver and other tissues of the sheep was made by Masters (1964a, b). The steady increase in stearic acid levels and decrease in levels of C_{18} mono-unsaturated acid expected with rumen development were most marked in depot glycerides, though similar changes were noted in the neutral lipid fractions of heart, kidney and liver. In tissues such as heart muscle, skeletal muscle, kidney and liver the major lipid fraction consists of phospholipids [see Garton (1960b) and Masters (1964c)]. Changes in sheep tissue lipids which occur in response to prolonged protein depletion have been investigated by Horgan and Masters (1963); one of the main findings was a lowering of the C_{18} tri-unsaturated fatty acid content of liver lipids, accompanied by a corresponding increase in the proportion of C_{18} mono-unsaturated acid, possibly related to a disturbance of the rumen flora involved in hydrogenation of dietary unsaturated acids [cf. the previously-mentioned study of Tove and Matrone (1962)].

(ii) *Milk lipids*

The chemistry and biochemistry of the milk lipids of ruminants, and of the cow in particular, have attracted a great deal of attention for many years and the resulting literature is so extensive that only a brief outline can be included here. Recent reviews include those of Folley and McNaught (1961), Garton (1963) and Peeters and Lauryssens (1964). Not only do the milk glycerides of ruminants contain all the wide variety of fatty acids found in tissue lipids, but they also contain a relatively high content of short-chain fatty acids (especially butyric acid) not found in any other lipid. Examples of typical analyses (previously cited by Garton, 1963) showing the main component acids of cow, ewe and goat milk fats are shown in Table 9; because of the wide variation in the molecular weights of the acids present in these milk fats, the values are given as molecular percentages (rather than percentages by weight) in order to show the relative numbers of molecules of each constituent acid.

TABLE 9

Component Fatty Acids of Cow Colostral and Milk Lipids and of Sheep and Goat Milk Lipids

(Values as molecular percentages of total fatty acids; unidentified and trace components not included)

Fatty acid	Source of lipid			
	Cow colostrum	Cow milk	Sheep milk	Goat milk
Saturated:				
Butyric	2·6	10·5	7·5	7·5
Caproic	1·6	4·6	5·3	4·7
Caprylic	0·5	1·3	3·5	4·3
Capric	1·6	2·7	6·4	12·8
Lauric	3·2	2·6	4·5	6·6
Myristic	9·5	9·6	9·9	11·8
Palmitic	31·7	23·4	21·6	24·1
Stearic	11·8	9·7	10·3	4·7
Longer chain	0·6	0·6	0·8	0·4
Unsaturated:				
C_{10}	0·1	0·3	0·2	0·3
C_{12}	0·2	0·2	0·2	0·3
C_{14}	0·7	1·0	0·6	0·8
C_{16}	2·7	2·1	2·0	2·2
C_{18}	31·0	30·4	25·9	19·3
Longer chain	1·8	1·0	1·3	0·2

The fatty acid compositions of the milk fats of the three species of ruminants are fairly similar, though cow milk fat usually has relatively more butyric acid and less of the other short-chain acids than sheep or goat milk fat. Compared with normal bovine milk fat, the colostral glycerides contain relatively less of the shorter-chain acids and more palmitic acid. Goat milk fat shows a notably high content of capric acid accompanied by a correspondingly lower content of stearic acid and unsaturated C_{18} acids. The unsaturated acids in Table 9 are grouped according to chain length since they include, as in depot glycerides, several positional and geometrical isomers. Most of these are mono-unsaturated, though the C_{18} acids in particular contain small amounts of di-unsaturated acids (including linoleic acid) and traces of tri-unsaturated components. The unsaturated acids of chain length greater than C_{18} include arachidonic acid (all *cis*-eicosa-5,8,11,14-tetraenoic acid).

The unsaturated and saturated acids of bovine milk glycerides are apparently distributed at random among the available glycerol molecules. Monopalmitoglycerides predominate and the saturated acids of chain-length shorter than C_{16} are present mainly in glycerides which also contain two different long-chain fatty acids, e.g. palmitic or stearic acid, together with C_{18} mono-unsaturated acid. Studies on the intramolecular distribution of the acids indicate some concentration of lauric, myristic and palmitic acids in the 2-position of the glycerol molecules, with a corresponding tendency for stearic acid and C_{18} mono-unsaturated acid to be esterified preferentially in the 1- and 3-positions.

Dietary lipids, other than those of pasture or normal stall feeds, can influence the fatty acid composition of milk fat to some extent depending on the amounts administered and on the nature of their constituent fatty acids. Thus the ingestion by the lactating cow of certain acids, such as the lauric and myristic acids of coconut oil and palm kernel oil, can lead to the appearance in the milk fat of enhanced proportions of these acids and the feeding of rape oil (containing erucic acid) and of marine oil (containing polyunsaturated long-chain acids) results in the incorporation into the milk glycerides of small amounts of these otherwise abnormal acids. Marine oils also tend to lower milk fat output, possibly due to a deleterious effect on rumen bacteria responsible for acetate production. As with depot fats, the inclusion in the cow's diet of moderate amounts of oils rich in C_{18} unsaturated fatty acids does not significantly affect milk fat composition, though the intravenous infusion of such an oil (cottonseed oil) has been shown to bring about a marked increase in the linoleic acid content of milk fat. For further discussion of the effects of supplementary dietary oils on milk fat composition see Hilditch and Williams (1964), McDowall, Patchell and Reid (1957) and McDowall, Reid and Patchell (1957).

Of more practical importance is the effect on the output of milk fat of different 'production' rations fed to dairy cows. Since this topic is considered in some detail in Chapter 19, suffice it to note here that a certain minimum of feed lipid is apparently required and that this is probably of the order of

300 to 400 g/day (Dijkstra, Dammers and Frens, 1953). Of as much, if not perhaps more, importance is the carbohydrate (cellulose, starch) part of the ration which, in the rumen, yields the volatile fatty acids (particularly acetate) required for the synthesis of higher fatty acids in mammary tissue. Rations which for one reason or another fail to yield a sufficient supply of acetate (or β-hydroxybutyrate) result in a diminished output of milk fat and a reduction in the proportion of its component short-chain fatty acids [see reviews by Rook (1959) and Van Soest (1963)]. It is interesting to note that some years ago Smith and Dastur (1938) found that fasting caused a reduction of 24 per cent in the molecular proportion of C_4—C_{14} fatty acids in cow milk fat, while at the same time the C_{18} fatty acids (stearic acid plus mono-unsaturated acid) increased by a similar proportion to give glycerides resembling in composition those of depot fat.

Though it is well established (see Garton, 1963) that, under normal feeding conditions, about half the fatty acids of milk fat are derived from plasma glycerides and the remainder by *de novo* synthesis in the mammary tissue, it is not yet clear why the milk fats of herbivores (and in particular of ruminants) should contain a significantly greater proportion of short-chain acids than is present in the milk fats of other species. Their presence is evidently associated with the abundant supply of acetate to the mammary gland and the arresting of chain-lengthening, possibly as a result of the rapid esterification of the shorter-chain acids as they are elaborated.

V. ASPECTS OF THE PHYSIOLOGICAL SIGNIFICANCE OF LIPIDS

A. Essential Fatty Acids

It has been known for many years that animals (particularly growing animals) cannot thrive on a lipid-free diet and that, in the absence of certain dietary unsaturated fatty acids, growth is retarded or arrested, skin lesions develop and abnormalities of pregnancy and lactation can occur. The 'essential fatty acids' (EFA) are linoleic and arachidonic acids, but in that arachidonic acid can be produced *in vivo* from linoleic acid, only the latter is truly an essential dietary constituent. When linoleic acid is not present in sufficient amount in the diet the animal attempts, albeit in vain, to compensate for failure to produce arachidonic acid (eicosatetraenoic acid) by converting oleic acid into an unusual C_{20} tri-unsaturated acid (eicosatrienoic acid). The ratio of eicosatrienoic acid : eicosatetraenoic acid in tissues, e.g. in plasma or serum is now recognized as a useful index of the EFA status of an animal.

By determining this ratio in the serum, liver and heart tissue of pigs which were reared from 10 to 200 lb live weight on diets containing linoleic acid in amount such that it provided from 0·07 to 3·50 per cent of the calories, Leat (1962) concluded that about 1 per cent of the calories as linoleic acid was required to prevent the metabolic lesion from developing. As previously observed by other workers, Leat noted skin lesions in the EFA-deficient animals but pointed out that, due to the variable nature of these lesions, they were not a good criterion of EFA status. An interesting incidental observation was that no eicosatrienoic acid was present in the depot glycerides of the EFA-deficient pigs.

Cunningham and Loosli (1954) found that when 2-day-old lambs and kids were fed on a fat-free synthetic milk they became dull and listless and muscular inco-ordination developed before death supervened within a few weeks; control lambs given the same milk with the addition of 2 per cent of lard survived successfully, as did kids given the milk plus 0·25 per cent of lard or linoleic acid. The lard, though not a good source of linoleic acid, was evidently supplying sufficient to permit normal growth. Similar findings with respect to calves have been reported by Lambert *et al.* (1954). Recent studies by Leat (1964) indicate that colostrum is probably of considerable importance as a source of linoleic acid for the new-born ruminant; in the suckling lamb, for example, dramatic changes in the fatty acid composition of the plasma lipids were observed within a very short time after birth, the linoleic acid content rising from 0·7 per cent at birth to 13·7 per cent within 48 hr.

B. Fatty Acids as a Source of Energy

Much of the energy requirements of animals result from the oxidation, in the tricarboxylic acid (Krebs) cycle, of acetate derived from the metabolism of all the major classes of foodstuffs—carbohydrates, proteins and lipids. Dietary fatty acids provide a particularly concentrated source of such energy and, in this respect, milk fat is of special significance for the suckling animal. Depot glycerides, on the other hand, were considered until quite recently to represent a store of energy which the animal utilizes only under starvation conditions; however, in the past decade, their physiological significance as a normal energy supply for the non-fasting animal has been recognized (see Annison, 1964). For this purpose the glycerides are readily mobilized to yield free fatty acids (FFA) which are released into the blood plasma for transport, as albumin complexes, to tissues (notably muscle) where they are oxidized. The concept of 'calorie homeostasis' was introduced by Fredrickson and Gordon (1958) to account for the relative concentrations in plasma of FFA and other oxidizable subtrates, e.g. the inverse relationship noted in man and dog between plasma values for FFA and for glucose. Inverse relationships between FFA and artificially-raised plasma levels of glucose, of lactate and of β-hydroxybutyrate in the sheep have been observed [see

Lindsay (1961) and Annison (1964)]; administration of acetate was accompanied by depressed plasma FFA values and, under conditions when plasma acetate values were low, the values for FFA were correspondingly high. Similarly, in the plasma of lactating cows, Adler and Wertheimer (1962) noted a reciprocal relationship between glucose and FFA. It should, however, be emphasized that in absolute terms the concentration of FFA in plasma in any nutritional state is very low, though the turnover time is rapid.

In experiments in which fasting sheep received continuous infusions of ^{14}C-labelled palmitate into the blood stream, West and Annison (1964) noted that the oxidation of this fatty acid alone was responsible for about 5 per cent of the total production of carbon dioxide. Palmitate accounts for about 20 per cent of plasma FFA in the sheep so, assuming that the other components of the FFA (largely stearic acid and C_{18} mono-unsaturated acid) are similarly oxidized, the contribution of FFA to the total production of carbon dioxide would be about 25 per cent; cf. values of 21–31 per cent in the dog. The effects of glucose and insulin on plasma FFA levels are almost identical in sheep and dogs (Annison, 1964), indicating that the normal control of fatty acid deposition and mobilization (see Steinberg, 1963; Freinkel, 1964) is very similar in both ruminants and non-ruminants.

REFERENCES

Adams, E. P. and Heath, T. J. (1963) *Biochim. Biophys. Acta*, **70**, 688.

Adler, J. H. and Wertheimer, E. (1962) *Bull. Res. Council Israel*, **10E**, 97.

Annison, E. F. (1964) In *Metabolism and Physiological Significance of Lipids*, p. 289. Ed. by Dawson, R. M. C. and Rhodes, D. N., John Wiley, London, New York and Sydney.

Aylward, F. and Showler, A. J. (1962) *J. Sci. Food Agric.* **13**, 492.

Bloch, K. (1960) *Lipide Metabolism.* John Wiley, New York and London.

Bratzler, J. W., Loosli, J. K., Krukovsky, V. N. and Maynard, L. A. (1950) *J. Nutr.* **42**, 59.

Cantarow, A. and Shepartz, B. (1962) *Biochemistry*. 3rd ed. W. B. Saunders Co., Philadelphia and London.

Coleman, M. H. (1963) In *Advances in Lipid Research*, vol. 1, p. 1. Ed. by Paoletti, R. and Kritchevsky, D., Academic Press, New York and London.

Conchie, J. and MacDonald, D. C. (1959) *Nature, Lond.* **184**, 1233.

Cunningham, H. M. and Loosli, J. K. (1954) *J. Anim. Sci.* **13**, 265.

Czerkawski, J. W. and Blaxter, K. L. (1965) *Proc. Nutr. Soc.* **24**, xvii.

Dahl, O. (1957) *Acta Chem. Scand.* **11**, 1073.

Dahl, O. (1958a) *Svensk. kem. Tidskr.* **70**, 43.

Dahl, O. (1958b) *Acta Agric. Scand.* **8**, 106.

Dawson, R. M. C. and Rhodes, D. N. (Eds.) (1964) *Metabolism and Physiological Significance of Lipids.* John Wiley, London, New York and Sydney.

DeMan, J. M. and Bowland, J. P. (1963) *J. Dairy Res.* **30**, 339.

Dijkstra, N. D., Dammers, J. and Frens, A. M. (1953) *Verslag. Landbouwk. Onderzoek* (The Netherlands), No. 59.

Duncan, W. R. H. and Garton, G. A. (1963) *Biochem. J.* **89**, 414.

Duncan, W. R. H., Garton, G. A., McDonald, I. and Smith, W. (1959) *Brit. J. Nutr.* **14**, 371.

EVANS, R. E. (1960) *Rations for Livestock*. Min. Agric. Fisheries and Food, Bull. 48. H.M.S.O., London.
FELIŃSKI, L., GARTON, G. A., LOUGH, A. K. and PHILLIPSON, A. T. (1964) *Biochem. J.* **90,** 154.
FOLLEY, S. J. and MCNAUGHT, M. L. (1961) In *Milk: The Mammary Gland and its Secretion,* vol. 1, chapter 12. Ed. by Kon, S. K. and Cowie, A. T., Academic Press, New York and London.
FREDRICKSON, D. S. and GORDON, R. S. (1958) *Physiol. Rev.* **38,** 585.
FREINKEL, N. (1964) In *Metabolism and Physiological Significance of Lipids,* p. 455. Ed. by Dawson, R. M. C. and Rhodes, D. N., John Wiley, London, New York and Sydney.
GARTON, G. A. (1960a) *Nature, Lond.* **187,** 511.
GARTON, G. A. (1960b) *Nutr. Abstr. Rev.* **30,** 1.
GARTON, G. A. (1963) *J. Lipid Res.* **4,** 237.
GARTON, G. A. (1964) In *Metabolism and Physiological Significance of Lipids,* p. 335. Ed. by Dawson, R. M. C. and Rhodes, D. N., John Wiley, London, New York and Sydney.
GARTON, G. A. (1965) In *Physiology of Digestion in the Ruminant,* p. 390. Ed. by Dougherty, R. W., Butterworths, Washington.
GARTON, G. A. (1967) *World Rev. Nutr. Dietet.* **7,** 225.
GARTON, G. A. and DUNCAN, W. R. H. (1954) *Biochem. J.* **57,** 120.
GARTON, G. A. and DUNCAN, W. R. H. (1964) *Biochem. J.* **92,** 472.
GARTON, G. A. and DUNCAN, W. R. H. (1965) *Biochim. Biophys. Acta,* **98,** 436.
GARTON, G. A., HILDITCH, T. P. and MEARA, M. L. (1952) *Biochem. J.* **50,** 517.
GARTON, G. A., LOUGH, A. K. and VIOQUE, E. (1961) *J. Gen. Microbiol.* **25,** 215.
GUPTA, S. S. and HILDITCH, T. P. (1951) *Biochem. J.* **48,** 137.
HARTMAN, L. and SHORLAND, F. B. (1961) *New Zealand J. Sci.* **4,** 16.
HEATH, T. J. and MORRIS, B. (1963) *Brit. J. Nutr.* **17,** 465.
HEATH, T. J., ADAMS, E. P. and MORRIS, B. (1964) *Biochem. J.* **92,** 511.
HILDITCH, T. P. and JASPERSON, H. (1944) *Biochem. J.* **38,** 443.
HILDITCH, T. P. and PEDELTY, W. H. (1941) *Biochem. J.* **35,** 932.
HILDITCH, T. P. and WILLIAMS, P. N. (1964) *The Chemical Constitution of Natural Fats.* 4th ed. Chapman & Hall, London.
HOBSON, P. N. and MANN, S. O. (1961) *J. Gen. Microbiol.* **25,** 227.
HOLMBERG, J. and ROSENQVIST, U. (1949) *Svensk. kem. Tidskr.* **61,** 89.
HORGAN, D. J. and MASTERS, C. J. (1963) *Austral. J. Biol. Sci.* **16,** 905.
KATZ, I. and KEENEY, M. (1964) *Biochim. Biophys. Acta,* **84,** 128.
KEENEY, M., KATZ, I. and ALLISON, M. J. (1962) *J. Amer. Oil Chem. Soc.* **39,** 198.
LAMBERT, M. R., JACOBSON, N. L., ALLEN, R. S. and ZALETEL, J. H. (1954) *J. Nutr.* **52,** 259.
LANDS, W. E. M. (1965) In *Annual Review of Biochemistry,* p. 313. Ed. by Luck, J. M. and Boyer, P. D., Annual Reviews Inc., Palo Alto, Calif., U.S.A.
LEAT, W. M. F. (1961) *Brit. J. Nutr.* **15,** 259.
LEAT, W. M. F. (1962) *Brit. J. Nutr.* **16,** 559.
LEAT, W. M. F. (1964) *Biochem. J.* **93,** 22P.
LEAT, W. M. F. (1965) *Biochem. J.* **94,** 21P.
LEAT, W. M. F., CUTHBERTSON, A., HOWARD, A. N. and GRESHAM, G. A. (1964) *J. Agric. Sci.* **63,** 311.
LENNOX, A. M., LOUGH, A. K. and GARTON, G. A. (1965) *Biochem. J.* **96,** 27P.
LINDBERG, P. and TOLLERZ, G. (1964) *Acta vet. Scand.* **5,** 311.
LINDSAY, D. B. (1961) In *Digestive Physiology and Nutrition of the Ruminant,* p. 235. Ed. by Lewis, D., Butterworths, London.
LOUGH, A K. and GARTON, G. A. (1957) *Biochem. J.* **67,** 345.
MASTERS, C. J. (1964a) *Austral. J. Biol. Sci.* **17,** 183.
MASTERS, C. J. (1964b) *Austral. J. Biol. Sci.* **17,** 190.
MASTERS, C. J. (1964c) *Austral. J. Biol. Sci.* **17,** 200.
MATTSON, F. H. and LUTTON, E. S. (1958) *J. Biol. Chem.* **233,** 868.
MCDOWALL, F. H., PATCHELL, M. R. and REID, C. S. W. (1957) *New Zealand J. Sci. Technol.* (*A*) **38,** 1036.

McDowall, F. H., Reid, C. S. W. and Patchell, M. R. (1957) *New Zealand J. Sci. Technol. (A)* **38,** 1054.

Otterby, D. E., Ramsey, H. A. and Wise, G. H. (1964) *J. Dairy Sci.* **47,** 993.

Peeters, G. and Lauryssens, M. (1964) In *Metabolism and Physiological Significance of Lipids*, p. 351. Ed. by Dawson, R. M. C. and Rhodes, D. N., John Wiley, London, New York and Sydney.

Polan, C. E., McNeill, J. J. and Tove, S. B. (1964) *J. Bact.* **88,** 1056.

Rajagopal, N. S. and Achaya, K. T. (1964) *J. Sci. Food Agric.* **15,** 497.

Robinson, D. S. (1964) In *Metabolism and Physiological Significance of Lipids*, p. 275. Ed. by Dawson, R. M. C. and Rhodes, D. N., John Wiley, London, New York and Sydney.

Rook, J. A. F. (1959) *Proc. Nutr. Soc.* **18,** 117.

Salmon-Legagneur, E. (1962) *Ann. Zootech.* **11,** 295.

Schmidt-Nielsen, K. (1946) *Acta Physiol. Scand.* **12,** Suppl. 37, pp. 1–83.

Senior, J. R. (1964) *J. Lipid Res.* **5,** 495.

Shorland, F. B., Bruce, L. W. and Jessop, S. A. (1952) *Biochem. J.* **52,** 400.

Shorland, F. B., Weenink, R. O. and Johns, A. T. (1955) *Nature, Lond.* **175,** 1129.

Smith, J. A. B. and Dastur, N. N. (1938) *Biochem. J.* **32,** 1868.

Steinberg, D. (1963) In *The Control of Lipid Metabolism*, p. 111. Ed. by Grant, J. K. Academic Press, London.

Tove, S. B. (1965) In *Physiology of Digestion in the Ruminant*, p. 399. Ed. by Dougherty, R. W., Butterworths, Washington.

Tove, S. B. and Matrone, G. (1962) *J. Nutr.* **76,** 271.

Tove, S. B. and Mochrie, R. D. (1963) *J. Dairy Sci.* **46,** 686.

Van Soest, P. J. (1963) *J. Dairy Sci.* **46,** 204.

Viviani, R. and Lenaz, G. (1963) *Boll. Soc. ital. Biol. sper.* **39,** 1836.

West, C. E. and Annison, E. F. (1964) *Biochem. J.* **92,** 573.

CHAPTER 11

Mineral Nutrition and Metabolism

C. F. MILLS

Rowett Research Institute, Bucksburn, Aberdeen

THIS chapter will selectively review our recent knowledge of the function of minerals in the animal body, the nature of minerals in the diet and their subsequent absorption and transport in the body, the processes which govern mineral element homeostasis and finally the influence of diet composition on the physiological utilization of minerals. Limitations of space preclude exhaustive treatment of even these topics but where appropriate the reader will be referred to recent key reviews in other works.

THE DISTRIBUTION OF ELEMENTS IN THE ANIMAL BODY

Differences between different tissues in respect of mineral content are larger than between-species differences for the same tissue. Adult bone has a remarkably uniform composition between species; among the soft tissues muscle shows the smallest species differences (fish tissues, however, show notable differences from mammalian tissue with respect to their mineral content, being richer in calcium, sodium and chloride than muscle of mammalian species). A further species difference lies in the high concentrations of the trace element copper found in liver tissue of the ruminant animals, ox and sheep; further reference will be made to this point when the metabolism of the trace elements is being considered. A further noteworthy feature regarding the mineral content of ruminant tissues is the remarkably high content of sodium (*ca.* 1 mg/g fresh tissue) found in the muscle of the adult sheep. The significance of this situation is not known.

Widdowson and Dickerson (1964) have reviewed information on the contribution of individual tissues to the total mineral content of the body at different ages. Among the most interesting features is the relative constancy of the contribution of nervous tissue to the total body sodium, potassium and chloride during foetal development and the remarkable fall in this contribution during growth from birth to maturity. Although we are familiar with the concept of the skeleton as the major depot of calcium and phosphorus and to a lesser extent magnesium, the importance of this tissue as a reservoir of body sodium is frequently overlooked as is the fact that of the large proportion of total body sodium found in the skeleton of most species, about 40 to 50 per cent, is readily exchangeable with body fluids. Changes in bone composition resulting from changes in calcium and phosphorus nutrition will be considered in greater detail later. For a detailed description of the storage of other elements (including the trace elements and radioisotopes) the reader is referred to the review by Duckworth and Hill (1953). With the exception of muscle, there is a decline in the mineral contribution of all the soft tissues during growth from the full-term foetus to the adult. Dramatic changes in the trace metal content of individual organs take place in some species during early postnatal growth. There is, for example, a six- to tenfold decrease in the copper concentration of the human and porcine liver during growth to adulthood; in contrast, the decline in copper concentration in the bovine liver in this period is usually only twofold whereas in sheep a three- to fourfold *increase* in concentration normally takes place.

The influence of mineral deficiency and excess on the mineral composition of individual tissues will be considered later but it is appropriate to mention at this juncture that changes in mineral composition also result from a general restriction of the intake of food or water. Such changes generally reflect the changes in water balance and tissue wasting that result from inanition or dehydration. Tissues that are relatively inert retain a greater proportion of their mineral component with the result that the proportion of the total body content of a particular mineral in such a tissue may rise. Thus the contribution of the skeleton to the total calcium, phosphorus and sodium of the body rises as an animal is maintained on a low plane of nutrition. Changes of this nature may finally influence the gross mineral composition of the whole animal body and often render difficult the interpretation of data on mineral composition of individual tissues when the intake of protein and energy by the animal is insufficient to maintain the normal differential pattern of growth and development of tissues. The extent to which a tissue gains, retains or loses its mineral element component is determined by several processes, viz.: (i) the concentration of each element in the fluids bathing that tissue, (ii) the affinity for these elements of structural units of the tissue already in existence, (iii) the demand for these elements in functional components in process of synthesis and (iv) the release of elements from components subject to breakdown. The importance of these processes differs with different tissues and different elements and consequently it is not possible to present a simple description of the processes of mineral metabolism. As an example of these differences we may consider the elements sodium and copper. Sodium exists mainly in ionic form in the extracellular fluids of the body and, with the possible exception of some components of connective tissue and nervous tissue, the binding affinity of cell structures for this element is small. In contrast, the structural components of the cell have a very strong affinity for the element copper; acceptor proteins exist which have a binding affinity for this element several million times that for sodium or other alkali metals. Even the copper in solution within intracellular and extracellular fluid exists not as the free ion but as soluble organic complexes.

Although this concept of the avidity of organic constituents of animal tissues for mineral elements has been introduced in connection with its influence in the establishment of mineral element equilibrium between the tissues and their milieu, its consequences dominate the wider picture of mineral element metabolism as a whole and are responsible to a large degree for the many differences in the processes involved in the utilization of individual mineral elements by the animal. Parameters like ionic size, valency and oxidation/reduction potential influence the diffusion of elements between the body fluids and the cellular mass, determine the extent to which transport across membranes must be an active (i.e. energy consuming) or a passive process and, ultimately, determine the biological function of individual elements. However, the opportunities that exist for stable complex formation with binding groups on inert or metabolically active tissues broadly define at the outset the nature of the processes that must be involved if an element is to be successfully liberated from food materials during digestion and subsequently transported from the lumen of the digestive tract to its functional site. With this in mind consideration will now be given firstly to qualitative and quantitative aspects of mineral elements in feeding stuffs and secondly to the process involved in the liberation and absorption of the mineral component of the ration.

The Mineral Composition of Foods

For extensive data on the mineral composition of animal feeding stuffs the reader should consult U.S. National Research Council Publications Nos. 659, 449 and 585. (The most complete tabulation of data on the mineral content of human foods is to be found in U.K. Medical Research Council Special Report No. 297, *The Composition of Foods*, by McCance and Widdowson (1960).)

The following paragraphs summarize some of the salient features of data presented in these reports. Where possible comments will be made on the availability of minerals from different food materials. For convenience foodstuffs will be considered element by element. Unless otherwise stated all concentrations are expressed on a dry-matter basis.

Mineral Requirements

Space does not permit detailed discussion of the requirements of individual classes of farm livestock for each mineral element. An indication of the general range of the requirement under different circumstances will be given, however. It is recommended that the reader requiring more detailed information should consult the recent publications, *Nutrient Requirements of Farm Livestock* (Agricultural Research Council, 1963, 1965, 1967).

WATER METABOLISM

Water Absorption

The volume of water in the gastrointestinal lumen is remarkably constant for any one species of mammal. The percentage of the total body water found in the lumen decreases from *ca.* 20 per cent in the case of the guinea-pig, to *ca.* 1 per cent in man. The proportion of the total body water in the gastrointestinal tract is high in ruminant animals with voluminous fore stomachs (rumen, reticulum and omasum) and in animals with highly developed caeca (horse, rabbit).

Controversy exists on the extent to which the absorption of water from the digestive tract may be regarded as an osmotic process or, alternatively, an active, energy-consuming process. Although it has been clearly shown that the rate of water absorption is slower from hypertonic than from iso- or hypotonic solutions (e.g. Annegers, 1961; Parsons and Wingate, 1961) the work of others (e.g. Vaughan, 1960) supports a concept of active transport of water. It is quite probable that both passive and active transport may proceed—indeed only such a situation could clearly account for observations both on the influence of tonicity and solute ion flux on water transport and on the effects of energy sources such as glucose increasing water absorption unless energy utilization is blocked by inhibiting oxidative phosphorylation (Tidball, 1961). Recent work on interrelationships between water absorption and cation flux across the intestinal wall has been reviewed by Curran (1965).

Water Excretion

Changes in the output of water by the kidney in response to the salt and protein content of the diet are believed to be governed largely by the antidiuretic hormone, vasopressin, produced in the hypothalamic region of the mid brain. The release of vasopressin via the posterior pituitary into the bloodstream stimulated by a rise in the osmotic concentration of the blood causes an increased reabsorption of water by the kidney tubule, an increase in the solute concentration of the urine and a reduction in water loss from the blood. In the absence of vasopressin circulating in the bloodstream bathing the kidney, the concentration of solutes in the urine becomes lower than that of plasma. Molina, Farah and Kruse (1963) have shown that vasopressin decreases the content of protein bound –SH groups in the cytoplasm of kidney tubules and have suggested that this hormone acts by producing structural changes in the membrane of the kidney tubule. Vasopressin also increases the renal output of sodium and potassium in species other than man as will be described later.

The minimum daily output of water by the kidneys thus depends upon the osmotic load and the ability of the kidney tubule to achieve concentration of urine and reabsorption of water. Defective reabsorption as in tubular necrosis or diabetes mellitus (where the osmotic effect of the large amount of sugar present in the urine hinders reabsorption of water) may lead to excessive loss of water in the form of large volumes of urine of normal or near normal specific gravity. On the other hand, the disease diabetes insipidus in which there is a decreased output of vasopressin from the posterior pituitary leads to similar large increases in water output, but in this case in the form of urine with a low specific gravity.

The loss of water by evaporation from the skin and *via* respired air may be as great as that lost *via* the kidneys. In pigs in cool environments water lost by vaporization decreases from approximately 80 g/kg liveweight/day in young animals to 23 g/kg liveweight/day in older stock (Mitchell and Kelley, 1938). Obviously the excretion of water via these routes is greatly dependent upon environmental conditions and on metabolic activity. Newburgh and Johnston (1942) have reviewed work on the influence of environment on water loss by perspiration and expired air and Schmidt-Nielsen (1964) has written a detailed treatise on changes in water balance in several species in response to adverse climatic conditions.

The attempts of the animal to compensate for excessive loss of water by perspiration are well illustrated by the work of Winchester and Morris (1956) which showed that the water intake of cattle rises from about 3·1 l./kg dry matter ingested at temperatures between 10° and 40°F to 5·9 l./kg at 85°F. With closely clipped sheep Blaxter *et al.* (1959) found the intake rose from 1·4 l./kg dry matter ingested at 46° and 64°F to 3·0 l./kg at 82°F.

The Influence of Pregnancy and Lactation on Water Intake

The voluntary intake of water by sheep doubles as pregnancy advances and it has been suggested (A.R.C., 1965) that the water requirements of cattle may increase by 50 per cent during pregnancy and lactation.

Effects of Dehydration

The greatest control of the water content of the body is achieved, in part, by control of urinary water output. The minimum urinary loss is governed by the salt load to be excreted and the quantity of urea (arising from the nitrogen of dietary protein or from tissue protein catabolism). If water loss via other routes exceeds the regulatory capacity of the kidney, partial dehydration follows; losses of plasma water in excess of losses from other body sources result in an increase in plasma protein concentration and viscosity. Schmidt-Nielsen (1964) has suggested that the increase in blood viscosity and a decreased heart-stroke volume are major factors in decreasing the efficiency of heat loss from deep-seated tissues and cause the rapid rise in body temperature that occurs as dehydration proceeds. A similar failure of circulatory heat transfer through dehydration underlies the 'explosive' heat rise observed in animals maintained in a hot environment on a restricted water intake.

SODIUM METABOLISM

Sodium in Feeding Stuffs

Cereal grains are usually low in sodium content (0·01–0·07 per cent of the fresh weight); in contrast, most animal protein concentrates are rich in sodium (0·5–1 per cent of fresh weight) whereas protein concentrates derived from vegetable sources contain appreciably less (0·05–0·3 per cent). The sodium content of fresh grass varies appreciably (0·02–0·25 per cent of the fresh material) and is markedly depressed by increasing the potassium level of the soil. Fresh leguminous crops usually contain between 0·2 and 0·8 per cent sodium. In dairy cows at the peak of lactation the supply of sodium from grass may be insufficient to meet requirements if the content of this element is less than 0·04 per cent.

The greater part of the sodium in common feeding stuffs is freely soluble in water and it may be presumed that most of this element occurs in a freely ionizable form as its chloride, sulphate or phosphate or as ionizable salts of carboxylic acids in plant and animal tissues. There is no evidence which suggests that the availability of sodium in common feeding stuffs is less than 100 per cent although it may be presumed that a part of the sodium in feeding stuffs containing bone products (e.g. meat and bone meal) may not be so readily available until digestion of the bone matrix has taken place.

Absorption

Net absorption of sodium has been observed from the colon on a wide variety of sodium intakes (Sammons, 1961); rather less efficient absorption takes place in the small intestine. There is evidence that although the total flux into and out of the lumen of the duodenum is large there is little net uptake. In the ileum the magnitude of net uptake is inversely related to sodium intake whereas in the colon Field *et al.* (1955) found that between 78 and 90 per cent of the sodium present in the lumen was absorbed (net) when the sodium intake ranged between 73 and 1 m-equiv. per day. Ileal absorption of sodium is accompanied by a small net efflux of potassium into the lumen; the magnitude of this loss is greatly increased when the diet is low in sodium (Code *et al.*, 1960). A similar exchange of potassium for sodium takes place in the human colon.

Functions

Sodium is concerned with a very wide variety of physiological processes the most important of these being its contribution towards the regulation of body fluid volume by its influence on fluid osmolality and its role in nerve impulse transmission in which the development of an action potential in nerve is associated with a sudden outward flux of the Na^+ ion. The avidity of biological metal binding groups for sodium is low (as it is also for the other alkali metals) and consequently few complexes of sodium with cellular components are known. Among these are complexes with chondroitin sulphate, deoxyribonucleic acid and brain lipids. The proportion of the body pool of sodium that is present in skeletal tissues is firmly bound in the adult but not in the young animal.

Judging by the dramatic clinical effects of sodium deficiency other functions of this element exist which are as yet poorly characterized. Changes due to sodium depletion will be described later.

Sodium Excretion and Homeostasis

Under normal environmental circumstances about 90 to 95 per cent of excreted sodium is excreted *via* the urine. Only about 2 per cent is excreted *via* the faeces; the remainder is excreted in sweat and minor secretions. Sodium loss through sweat rapidly increases as the environmental temperature rises and may reach the point that clinical signs of sodium depletion appear (see later).

The output of sodium in the urine can be influenced by factors which change either the filtration rate of this ion in the kidney tubule or the efficiency of its reabsorption during passage down the tubule. Changes in tubular reabsorption appear to exercise the most important influence on sodium output. Abnormally high concentrations of other solutes in the glomerular filtrate also increase sodium output.

Undoubtedly the most effective control of sodium excretion is exercised by the steroid hormone, aldosterone, produced by the adrenal cortex. As little as 0·2 μg administered to an adrenalectomized rat decreases renal excretion of sodium by 50 per cent by, it is believed, increasing tubular reabsorption. The close relationship between sodium balance and aldosterone secretion is well illustrated by the hyperplasia of the adrenal gland and increased aldosterone secretion that result when the sodium intake is reduced. The mechanism whereby aldosterone is released is obscure but it is known that potassium loading increases aldosterone output and potassium depletion has the reverse effect.

Aldosterone and the structurally related deoxycorticosterone also influence sodium output in the faeces. Increased concentration of aldosterone or deoxycorticosterone decrease the faecal sodium/potassium ratio, most evidence suggesting that this effect results from increased reabsorption of sodium.

The posterior pituitary hormone, oxytocin, enhances the renal excretion of sodium and also facilitates the active transport of this ion through other membranes. Many other hormones have been shown to influence gross sodium output but in these cases a less direct mode of action appears to be involved.

Sodium Depletion

Diets low in sodium or excessive losses of this element in sweat or in urine (following kidney damage or in disease conditions associated with defective functioning of the adrenals leading to a low aldosterone output) may lead to sufficient depletion to lower the concentration of sodium in the extracellular fluid compartment with the result that water may be transferred from this compartment to tissue cells to restore osmotic equilibrium. Over hydration of these cells results. It will readily be seen that the magnitude of such osmotic changes will be a function of both sodium loss and water loss. Situations, like excessive perspiration, which lead to loss of both sodium and water in isotonic proportions do not have serious clinical repercussions until concentration of the blood plasma causes circulatory difficulties. In contrast the osmotic shifts which follow an intake of water without salt after exposure to heat or after violent exercise may readily lead to muscular cramp and shock. (Dauphinee (1962) gives a more detailed account of the relationships of sodium intake and excretion to osmotic changes and clinical effects in the body.)

A change of diet from one of normal to one with a low content of sodium is strikingly reflected by changes in the sodium/potassium balance of parotid saliva of the sheep. A rapid fall in sodium content is accompanied by a rapid increase in potassium content as a result of an increase in aldosterone output (Denton, 1956; Coghlan *et al.*, 1960). The influence of changes in diet on sodium conservation in the sheep has been clearly demonstrated by the work of Dobson (1965). For example, a change from hay to fresh frozen grass was accompanied by a large but transient increase in urinary sodium output; urinary sodium output fell when animals were changed from a diet of grass cubes to frozen grass. Beilhartz and Kay (1963) have shown that the appetite of the sodium-depleted sheep for sodium is governed by factors other than the concentration of this element in rumen fluid and blood plasma.

Thieler, Green and Du Toit (1927) demonstrated that a low sodium intake ($<$2 g/day) had no adverse effect on the growth of cattle but cases of abortion occurred in pregnant heifers.

The toxicity of sodium chloride to pigs and poultry when it is included in rations at levels in excess of 2 per cent is largely preventable by allowing free access to water (Todd, Lawson and Dow, 1964).

Sodium Requirements

The sodium requirement of the chick has been stated to be 0·11 per cent of the dry ration (McWard and Scott, 1961); that of growing pigs increases from 0·4 to 1·0 g/day as the liveweight increases from 5 kg to 90 kg (A.R.C., 1967). Recent studies of past experimental work have led to the suggestion that the sodium requirement of cattle probably ranges from *ca.* 7·7 g/day in the case of dry cows to 20 g/day for cows in full lactation while the requirement of sheep is probably about 0·9 g/day (A.R.C., 1965).

CHLORIDE METABOLISM

Chloride in Feeding Stuffs

The introductory remarks on sodium in feeding stuffs also apply to chloride. The element is distributed similarly to sodium among the common feeding stuffs and is believed to be equally available.

Absorption

Active absorption of the chloride ion has been repeatedly demonstrated in many mammalian species although what proportion of the total net uptake occurs by this process has not been determined. Diffusion of the ion down an electrochemical gradient from the lumen of the colon to the blood has been suggested by Curran and Schwartz (1959) as accounting for the greater part of the chloride absorption from the colon.

The main sites of chloride absorption appear to be the ileum and the proximal region of the large intestine (e.g. Annegers, 1961). The absorption of the chloride ion from isotonic saline in the ileum is accompanied by the secretion into the lumen of an approximately equivalent quantity of the bicarbonate ion with a consequent rise in pH.

The Distribution of Chloride in the Body

Cotlove and Hogben (1962) summarizing analytical data for the chloride content of a wide range of adult mammalian tissues showed that only 10 to 16 per cent of the total chloride is intracellular and of this between one-half and one-third is accounted for by chloride in the red blood cell. Other tissues with a relatively high intracellular chloride content are the skin and the viscera. The content of chloride in the extracellular fluid differs little between individual tissues; concentrations are mostly within the range of 100–110 m-equiv./l. found in blood plasma. Exceptions are the relatively lower concentrations of chloride in saliva and sweat and the higher concentration found in cerebrospinal fluid.

The secretion of chloride into the gastric juices as hydrochloric acid is a major route for the output of extracellular chloride. This net flux of chloride into the gastric lumen against adverse concentration and potential gradients is an active energy-consuming process. Chloride loss in the form of vomited gastric juices is a common cause of depletion of plasma chloride. The resulting ionic imbalance in turn leads to a compensatory increase in plasma bicarbonate and a resulting alkalosis. A sequel to this is often an increased urinary sodium output as the kidney simultaneously excretes the bicarbonate and sodium ions.

Chloride Excretion and Homeostasis

In most mammalian species approximately 90 to 95 per cent of the chloride output is excreted in the urine, 4 to 8 per cent in the faeces and 2 per cent via other routes. During renal excretion the pathways and regulation of chloride and sodium filtration and reabsorption are closely linked. Thus aldosterone secretion from the adrenal which largely governs sodium output also has a parallel influence on chloride output in the urine.

Functional Roles of the Chloride Ion

A major function of the chloride ion is that shared with the sodium ion in maintaining the tonicity of extracellular fluids. Its role in acid base balance will be considered later. The only enzyme known so far to be activated by the chloride ion is amylase—even here the action of chloride is not specific (Fischer and Stein, 1960).

Chloride Requirements

The chloride requirement of the growing chick is approximately 60 mg/kg of diet (A.R.C., 1963). The requirement of growing cattle gaining weight at *ca.* 0·5 kg/day rises from 1·8 to 10·9 g/day as the liveweight increases from 50 kg to 400 kg (A.R.C., 1965). The chloride requirement of growing sheep has been estimated to be within the range 0·35 to 1·6 g/day increasing with liveweight and rate of growth. The requirement of the pig cannot be stated with certainty.

POTASSIUM METABOLISM

Potassium in Foods

The pattern of distribution of this element among feeding stuffs is very different from that of sodium. Vegetable protein concentrates are rich in potassium (1–2 per cent on fresh weight basis) and the leaves and stems of the Graminae often contain between 1·5 and 3 per cent of this element. The potassium content of grass is markedly influenced by the level of potassium fertilization. The potassium/sodium ratio in crops fertilized with potassic fertilizers may rise to 50/1. Whole cereal grains and most vegetable crops contain between 0·1 and 0·5 per cent. The high content of potassium in many feeding stuffs and the fact that the greater part of the element in these materials appears to be in a freely soluble ionic form of high availability are responsible for the fact that potassium deficiency arising from an inadequate intake is unknown except on specially prepared synthetic diets.

Potassium in foods appears to be either associated with freely soluble materials or with materials from which its free ion can be liberated by mildly acid solutions.

Absorption and Distribution

What little evidence there is on the mechanism of potassium absorption from the digestive tract suggests that it is a passive process (Wilson, 1954). The direction of the net flux of potassium depends greatly on the sodium concentration of the contents of the lumen of the small intestine and colon. On diets of normal sodium content net uptake occurs in the small intestine with efflux in the colon, but it is apparent that as the sodium intake falls there is a large net efflux of potassium into the ileum. This situation gives rise to the apparent paradox of a low sodium diet leading to potassium depletion under circumstances which reduce the efficiency of potassium reabsorption—e.g. on diets which promote a rapid rate of passage of material through the distal small intestine.

Further metabolic relationships between sodium and potassium are apparent from the work of Dobson (1965) who showed that the rise in salivary potassium concentration in cows fed heavily fertilized grass was accompanied by a fall in sodium concentration.

Sheep may be divided into two major groups according to whether the red blood cell potassium concentration is *ca.* 95 m-equiv. potassium/l. (HK) or whether it is *ca.* 20 m-equiv./l. (LK). The distribution of these types is determined genetically and the LK type is dominant (Evans and King, 1955). The physiological significance of this finding is obscure.

Potassium is present in high concentrations within most tissue cells; the ratio of intra- to extracellular potassium is approximately 30:1. This ratio is considerably influenced, however, by the level of potassium intake and its rate of excretion and by the nature of tissue activity. For example, an appreciable uptake of potassium takes place while cells are synthesizing protein (a relationship is suspected between the adequacy of cellular potassium concentration and the accumulation of amino acids by the cell prior to protein synthesis). Conversely, catabolic activities and tissue trauma lead to the liberation of intracellular potassium to the plasma. Such increases in plasma potassium concentration are, however, transient because of the rapid excretion of the excess of plasma potassium in the urine.

Unlike sodium there is little potassium in bone; on the other hand, the concentration of potassium in muscle tissue is about four times greater than that of sodium.

Potassium Depletion

Excessive losses of potassium occur via the gastrointestinal tract in prolonged diarrhoea (e.g. Gamble, Wise and Hansen, 1948). Urinary losses may be excessive in situations which give rise to high outputs of aldosterone (e.g. as a consequence of adrenal tumours or after shock). The sodium-conserving effect of aldosterone exerted in the kidney tubule is accompanied by a net loss of potassium by excretion. Following surgical shock or tissue damage and their associated depletion of intra- and extracellular pools of potassium, the later period of tissue repair is associated with repletion of the intracellular component of tissue potassium. As Wilde (1962) has pointed out this process may be so rapid that hypokalaemia results with its attendant muscular weakness in the heart, intestinal and respiratory muscles. The rapid repletion of tissue protein reserves after a period of malnutrition can similarly lead to hypokalaemia through rapid uptake of this ion into developing cells.

An important route of loss of potassium in sheep is via the suint of wool—1 g of potassium may be lost every 4 days by this route (Farnworth, 1956).

Potassium Requirements

It has been suggested by Leach *et al.* (1959) that the potassium requirement of the chick is related to the metabolizable energy content of the diet rising from 0·2 to 0·34 per cent of the ration as the metabolizable energy rises from 2·8 to 4·25 Mcal/kg ration. No such relationship has yet emerged in studies with pigs where the requirement appears to be met by rations containing 0·15 per cent potassium (Hughes and Ittner, 1942). In the absence of indications that potassium deficiency arises under practical conditions no statements of potassium requirement of cattle have been made in a recent review (A.R.C., 1965), however, an indication of the requirement of calves for potassium is given by the work of Sykes and Alfredson (1940) who obtained abnormal electrocardiograms when the ration contained only 0·12 per cent potassium.

CALCIUM METABOLISM

CALCIUM IN FOOD AND IN THE DIGESTIVE TRACT

In contrast to most of the salts of the alkali metals which are present in foods in a freely soluble ionizable form, the salts of the alkali earth metals calcium and magnesium are less freely soluble and frequently form complex organic salts which differ greatly in the extent to which they can be degraded during passage down the digestive tract. In the case of calcium it is probably only with the salts of fatty acids, with the phosphates and with the phytates that poor solubility or complex formation has a marked influence on the extent to which calcium can be utilized from the diet. Other naturally organic occurring salts such as the oxalate and citrate appear sufficiently degraded during digestion to liberate calcium for absorption although work with rats suggests that in this species the oxalate salt is a poor source of calcium.

The efficiency with which calcium can be utilized from its fatty acid salts differs with the nature of the acid involved. In work with rats fed moderate calcium intakes the net utilization of the element from its stearic acid salt was only 25 per cent; from calcium oleate it was 90 per cent (Boyd, Crum and Lyman, 1932). Under practical conditions the adverse influence of lipid on calcium utilization is only apparent when high levels of fat ($>$ 10 per cent) are used in the diet. Hypocalcaemia and hypocalciuria in human subjects with steatorrhoea have been attributed to an interference with calcium utilization caused by an excess of fat in the intestinal lumen.

The importance of phytic acid and the phytates arising from cereals, grains, legume seeds and other vegetable components of the diet in depressing calcium utilization depends largely upon whether the animal in question secretes the hydrolytic enzyme phytase. In man, the pig and in poultry the intestinal phytase activity may not be sufficient to overcome completely the adverse effects of phytate particularly when this has been added to the diet as a soluble salt (e.g. as sodium phytate). In ruminants, the rat and the dog phytate hydrolysis is more extensive. The utilization of calcium from calcium phytate is improved by increasing the vitamin D intake—a situation which increases intestinal phytase activity according to Pileggi, De Luca and Steenbock (1955).

Though it has been suggested that the adverse effects of a wide calcium to phosphorus ratio in the diet on calcium and phosphorus utilization result from the formation of insoluble salts in the lumen of the digestive tract (Telfer, 1922–3) there is no evidence that wide variations in this ratio have much influence on faecal calcium loss. However, it is quite clear from the work of Coates and Holdsworth (1961) that when calcium is present solely as insoluble phosphate no absorption takes place so that the concept that the calcium : phosphorus ratio influences the solubility of intestinal calcium and thus its absorption cannot be readily discarded. No changes in endogenous faecal calcium output were found in sheep when the calcium : phosphorus ratio was varied between 0·8 and 6·0 (Lueker and Lofgreen, 1961). It has been suggested that for pigs optimal utilization of calcium is obtained on rations with calcium : phosphorus ratios between 2 :1 and 1 : 2 (Becker, 1955) provided vitamin D is not limiting. Further reference to the significance of calcium : phosphorus ratios will be made in the section dealing with phosphorus metabolism.

It has been shown in many species that the availability of calcium in milk is extremely high compared with other foods. Hansard, Lyke and Crowder (1961) have shown that the availability of calcium to the growing pig declines progressively from 99 per cent in the milk fed piglet to 27 per cent in mature animals fed a maize soya bean diet. Similar changes have been shown in calves (Hansard, Comer and Plumlee, 1954). Mellander (1950) has shown that calcium in the form of complexes with phosphopeptides derived from casein is very readily absorbed by young animals.

Absorption

The proximal small intestine seems to be the principal site of calcium absorption; evidence has been put forward which indicates that the uptake of calcium from the lumen of the gut is partly an active process inhibited by respiratory poisons and by tissue anoxia (e.g. Schachter, Dowdle and Schenker, 1960) and partly a process of passive diffusion (Harrison and Harrison, 1960).

In vitro and *in vivo* studies clearly show that vitamin D (or a metabolite of vitamin D) increases the net uptake of calcium particularly from diets high in phytate or with a wide calcium : phosphorus ratio. In studies with intestinal sacs it has been found that added vitamin D is not immediately effective in increasing calcium transport; a delay occurs, variously reported as being between 1 hour to 2–3 days with tissues from different species (e.g. see Coates and Holdsworth, 1961). The mechanism of action of vitamin D is not known. Harrison *et al.* (1960) have shown that it apparently increases the rate of both active and passive absorptive processes. This surprising finding may be resolved by the recent suggestions that vitamin D_3 may inhibit an energy-dependent translocation of calcium from the mucosa back to the intestinal lumen and may thus decrease the rate of calcium loss.

In work with chicks Webling and Holdsworth (1965) have closely studied the effects of bile in promoting calcium absorption. This effect, known for some time, was previously believed to be solely associated with the action of bile acids increasing vitamin D uptake. Webling *et al.* have, however, shown that the effect is also associated with the detergent-like action of the taurine-conjugated bile acids on sparingly soluble calcium salts in the lumen.

Unlike vitamin D_3, bile salts have no protective action against the development of rickets.

In discussing homeostatic control of calcium metabolism it is pertinent to consider briefly the behaviour of the structurally related alkaline-earth metal strontium. Intensive investigation of calcium/strontium relationships was stimulated by concern at the accumulation of radiostrontium from 'fall-out' by living tissues. Early work showed that, in contrast to calcium, there is only poor homeostatic control of strontium uptake and excretion (Barnes *et al.*, 1961). The overall ability of the animal body to discriminate against strontium and in favour of calcium increases with age; the observed discrimination ratio

$$\frac{(\text{Sr/Ca in body})}{(\text{Sr/Ca in diet})}$$

in man probably declines from 0·5 to 1 year of age to 0·25 at puberty (Bryant and Loutit, 1961). The greatest discrimination appears to take place during absorption of strontium and calcium from the digestive tract.

TISSUE CALCIUM: FORMS AND FUNCTIONS

Walser (1961) has shown that in man the plasma calcium is in the following forms: free ions, approximately 48 per cent of total; protein-bound, 46 per cent; mono-calcium phosphate, 1·6 per cent; calcium citrate, 1·7 per cent and unidentified complexes, 3·2 per cent. There exists a dynamic equilibrium between these forms; in turn the total calcium concentration of the blood is governed by dietary and hormonal agencies influencing absorption and the accumulation or release of calcium from skeletal stores.

Changes in the intracellular concentration of the free calcium ion in nerve and muscle tissue reflecting changes in the free ion concentration of plasma have profound physiological effects. A low calcium ion concentration in the nerve fibre produces hyperexcitability, spontaneous nerve fibre discharge and ultimately a failure of synaptic transmission. Low concentrations of calcium in nerve fibres are believed to increase the permeability of nerve cells to sodium and potassium ions (Frankenheuser and Hodgkin, 1957) and thus to influence the generation of nerve action currents. A further aspect of nerve fibre function influenced by calcium is the normal development of structural changes in nerve proteins following excitation. Luxoro, Rojas and Wittig (1963) have shown that this configurational change does not take place when the extracellular fluid is low in calcium.

That calcium is involved in the contractile process of muscle has been clearly established. It has been found that approximately 40 per cent of muscle calcium is present in a non-exchangeable form, but the precise role of the intracellular calcium ion pool in maintaining the contractile response to stimulation is not yet elucidated.

The dependence of these basic processes on the calcium ion concentration of tissue fluids rather than solely on the *total* calcium concentration explains why conditions leading to hypocalcaemia may in turn give rise to a variety of clinical disturbances (e.g. bradycardia, tetany) if there is a low level of the free ionic species, or, alternatively, to other less dramatic conditions if the level of the free ion has been maintained at the expense of 'bound' forms of calcium.

The calcium ion has been found to activate the proteolytic enzyme, trypsin *in vitro*. Whether trypsin activity declines when hypocalcaemia occurs *in vivo* is not known.

The role of calcium in skeletal structures will be discussed in sections dealing with homeostatic mechanisms and the effects of disordered calcium metabolism.

Homeostatic Mechanisms

Demands for calcium at different periods of the life cycle differ greatly. Demands are high during growth and maximal at the peak of lactation. Homeostasis of intra- and extracellular calcium is primarily achieved by deposition or, alternatively, resorption of bone salt (there being 24 to 28 per cent calcium in dry bone; Eastoe, 1961) under hormonal control of the parathyroid and thyroid glands. A low plasma calcium concentration stimulates the release of parathyroid hormone. Mobilization of bone calcium follows until the calcium level of the blood plasma bathing the parathyroids returns to normal. The possible mechanisms of action of parathyroid hormone have been recently reviewed by Aurbach and Potts (1964) (see also Chapter 13). Experimental evidence suggests one or other of the following processes may account for the hypercalcaemic effect of parathyroid hormone: (i) destruction of the bone matrix by stimulation of the enzyme collagenase may release bone mineral, (ii) an increase in the content of organic acids in interstitial fluids in bone may cause bone mineral to dissolve by a combined effect resulting from the development of acidity and a chelating action of these acids or (iii) an energy-dependent system sensitive to parathyroid hormone may stimulate active transfer of calcium from bone to interstitial fluid.

More recent work on hormonal regulation of tissue fluid calcium concentration has shown the presence of a hormone, thyrocalcitonin, in the thyroids of several species (Hirsch, Gauther and Minson, 1963; Care, Duncan and Webster, 1966). Thyrocalcitonin appears to act by inhibiting resorption of skeletal tissue (Milhaud, Perault and Mouktar, 1963). Perfusion of the thyroid of the pig with fluid with a low ionic concentration of calcium stimulates a rise in the concentration of calcium in the systemic plasma; the converse is also true. Intravenous injection of thryocalcitonin causes a rapid reduction in plasma calcium concentration. In respect of their action on skeletal calcium reserves parathyroid hormone and thyrocalcitonin thus act to achieve homeostasis but by quite different mechanisms. The influence of the sex hormones on calcium metabolism has been considered in Chapter 12. The influence of parathyroid hormone on excretion of calcium will be considered in the next section.

Moodie and Robertson (1962) have suggested that impairment of calcium absorption from the digestive tract as a consequence of bowel stasis may be an important contributory factor to naturally occurring hypocalcaemia in the lactating cow. Injections of hyoscine hydrobromide to induce intestinal stasis caused a marked fall in serum calcium concentration in old lactating cows; a smaller effect was observed in young lactating animals and no change occurred in dry cows suggesting that in these cases the rate of mobilization of calcium from the skeleton was adequate to compensate for a decrease in absorption from the digestive tract.

Faecal and Urinary Loss of Calcium

The urinary output of calcium normally reflects changes in the plasma calcium level although circumstances exist in which an abnormally high urinary output of calcium may occur in the absence of hypercalcaemia. In hypervitaminosis D an excessive output may arise even to the extent that calcification of renal tubules may result. Such a situation has been reported in rams being fed to a high plane of nutrition prior to exhibition and simultaneously receiving injections of vitamin D. In most species the output of calcium in a urine rich in the phosphate ion is dependent on the simultaneous excretion of the amino acids with which calcium can combine to form soluble complexes. When urinary citrate is low there is an increased tendency towards the formation of deposits of insoluble calcium salts in the kidney. In contrast, there are indications that under some circumstances, as yet poorly defined, renal deposits of calcium citrate can also form—until recently elevated urinary citrate levels have been regarded as beneficial in promoting calcium clearance from the renal tubule.

In contrast to the rat, cattle (Visek *et al*., 1953) and pigs (Moore and Tyler, 1955) do not appear to excrete calcium into the intestine in significant quantity. Faecal calcium in these species probably arises from incomplete digestion rather than from endogenous sources.

Skeletal Calcium and the Effects of Calcium Deficiency

Since these topics are so closely related to metabolic processes involving phosphorus they will be considered after discussion of aspects of phosphorus metabolism.

19-20

PHOSPHORUS METABOLISM

Phosphorus in Feeding Stuffs

Feeding stuffs of animal origin have the highest contents of phosphorus usually in a highly available form. Among common feeding stuffs fishmeal (1·5 to 3 per cent phosphorus in the fresh meal) and meat and bone meal (3 to 4 per cent phosphorus) are by far the richest sources. In the sources so far mentioned an appreciable proportion of the phosphorus is inorganic in nature arising from the calcium phosphates of skeletal tissue present in these meals. The remainder is in the form of phosphate complexes of proteins and lipids and phosphate esters of carbohydrates nearly all of which are readily degraded during digestion and utilized as readily as free phosphate (e.g. Kawin and Palmer, 1958).

The cereal grains usually contain between 0·3 and 0·4 per cent phosphorus; legume hays between 0·25 and 0·4 and grass hays between 0·15 and 0·25 per cent phosphorus. Vegetable protein concentrates usually range between 0·6 and 1·2 per cent phosphorus. Most cereal straws are extremely low in phosphorus (*ca.* 0·1 per cent).

A smaller proportion of the phosphorus in feeding stuffs of vegetable origin is present as inorganic phosphates than in materials of animal origin. Furthermore, much of the plant phosphorus may be present in the form of phytic acid (inositol hexaphosphoric acid), the phosphorus of which is of only limited availability depending upon the activity of intestinal phytase in the animal. The proportion of the total phosphorus in this form ranges from 80 per cent in the case of cereal brans to about 50 per cent in soya bean meal and 40 per cent in groundnut meal. Probably little of the phosphorus of grass and forage crops is in this form unless the plants are carrying mature seed heads. Estimates of the availability of phytic acid phosphorus to farm livestock are as follows: to poultry, about 40 per cent (A.R.C., 1963), to sheep, 30 to 60 per cent (Lofgreen, 1960; Tillman and Brethour, 1958) and to pigs, *ca.* 30 to 60 per cent (Chapman, 1957). In this context it should be stressed, however, that these average figures may oversimplify the question of phytic acid phosphorus availability. In rats, dogs and children the availability of phosphorus from this source is inversely related to the calcium level of the diet (Hoff-Jørgensen, 1946 a, b)—on a diet low in calcium almost 100 per cent utilization was recorded in one study (Pileggi, De Luca and Steenbock, 1955). The influence of phytic acid on trace metal availability will be considered later.

Wide differences in the availability of phosphorus from a variety of inorganic salts have been reported from studies with cattle, sheep, pigs and poultry. The generalization can be made that as with calcium, the more soluble the salt the higher the availability of the phosphorus. Thus the phosphorus of mono- or dicalcium phosphates is more available than that of tricalcium phosphate, bone meal or 'rock phosphate'. Again, as with calcium, in all the above species the availability of phosphorus from any one source decreases with increasing age of animal. For example, the availability of the phosphorus of dicalcium phosphate has been found to be 100 per cent for the weanling lamb and 48 per cent for the 2-year-old wether.

Phosphorus Absorption, Transport and Excretion

There is little knowledge of the mechanisms whereby phosphate is absorbed from the digestive tract although *in vitro* studies with rat and chick intestinal tissue suggest that the process may be energy-consuming and sensitive to oxygen tension. It is probable that the duodenum and upper jejunum are the principal sites at which phosphorus is absorbed judging by studies with the pig (Moore and Tyler, 1955). Considering the immediate fate of phosphorus after passage into the bloodstream there is controversy about the proportions of the quantity of this element present in plasma as free ions of orthophosphoric acid and that which is non-ionic in form. Estimates for the proportion of complexed phosphorus range from 12 per cent to approximately 50 per cent (see Irving, 1964). The onset of egg laying in poultry is accompanied by the appearance in the plasma of a phosphoprotein the function of which is probably to transport phosphorus required for egg production. That the synthesis of this fraction is under hormonal control is suggested by the work of Clegg *et al.* (1956) who found that it appeared in cockerel plasma after the administration of diethylstilboestrol.

The role of skeletal tissue in maintaining plasma phosphorus levels will be discussed later; apart from this regulatory mechanism a powerful control of plasma level is exerted by renal excretion under the influence of parathyroid hormone which influences the selective reabsorption of this element from the glomerular filtrate. In studies with a wide range of non-ruminant species it has been shown that parathyroidectomy results in an abrupt hypophosphaturia. Similarly, injection or infusion of parathyroid hormone decreases phosphate reabsorption from the glomerular filtrate during its subsequent passage through the kidney tubule.

Functions of Phosphorus

Between 20 and 25 per cent of the phosphorus in the body is present in the extraskeletal soft tissues in contrast to calcium where only *ca.* 1 per cent is found in tissues other than bone. Some of the phosphorus in soft tissues may be regarded as having a structural function, e.g. the phospholipids of the central nervous system, but the greater part is present as phosphate-containing metabolic intermediates in the synthesis and degradation of carbohydrates, proteins and the nucleic acids. In view of the vital nature of this role of phosphorus it is not surprising that in prolonged phosphorus depletion there is proportionally a much smaller change in the phosphorus content of soft tissues than in the skeleton (e.g. Steenbock and Bunkfeldt, 1951). It is probable that depletion of the phosphate pool available for the synthesis of the phosphate ester intermediates required for tissue growth and energy metabolism is responsible for the poor efficiency of food utilization observed in all classes of farm livestock suffering prolonged phosphorus deprivation.

Skeletal Calcium and Phosphorus

As mentioned in the preceding section the skeleton contains approximately 99 per cent of the total calcium of the body and *ca.* 75 per cent of the phosphorus. A variable proportion of these reserves are in equilibrium with plasma calcium and phosphorus and since the plasma calcium and phosphorus levels reflect both the intake of these elements in the diet and demands for lactation and growth it follows that skeletal accretion or resorption reflects these variables. The influence of changes in dietary composition and nutrient demands on bone metabolism has been extensively studied in many species of farm animal. Many of these studies have been aimed at either determining the extent to which skeletal reserves may become depleted without causing permanent skeletal damage or determining the levels of calcium and phosphorus intake needed to minimize or, if possible, entirely prevent loss of bone calcium and phosphorus during periods of peak demand for these elements.

In experiments designed to examine the effects of different levels of dietary calcium on skeletal changes in the ewe during pregnancy and lactation Benzie *et al.* (1955, 1956) fed rations providing 1·8, 4·7 and 7·6 g calcium/day and determined the radiographic density of selected bones during and at the end of the experiments and skeletal ash in slaughtered animals at intervals during and after the completion of lactation. They found that on low calcium diets approximately 18 per cent of the skeletal mineral had been lost by mid-lactation—a loss which was not completely made good even 2 months after the end of lactation when a lower calcium demand should have permitted skeletal repair. In contrast, only 6 per cent of skeletal mineral had been lost by mid-lactation in ewes fed 5 g calcium daily and the skeleton underwent rapid repair after milk production had ceased.

In a similar study (Benzie *et al.*, 1959) to determine the effects of differing intakes of phosphorus, more dramatic effects were found. A ration providing approximately 1·5 g phosphorus/day promoted a loss of 40 per cent of skeletal mineral by mid-lactation and less than half this loss had been made good 2 months after the end of lactation. Even a ration providing about 4·5 g phosphorus daily caused a loss of 18 per cent of the skeletal ash before mid-lactation but in this instance, without changing the phosphorus intake, skeletal repair was complete 2 months after lactation had ceased.

It has been repeatedly shown in lactating sheep that negative phosphorus balances frequently occur and it is doubtful whether it is economically desirable or even physiologically possible to prevent loss of skeletal mineral during the height of lactation. The above experiments clearly indicate the capacity of the animal for undertaking skeletal repair after lactation has ceased even with what might be considered to be only moderate intakes of calcium and phosphorus.

In later studies (McRoberts, Hill and Dalgarno, 1965 a, b, c) the effect of diets of similar ranges of calcium and phosphorus contents on mandibular and dental development of growing sheep was examined. Low calcium diets retarded the formation of both dentine and enamel of teeth whereas low phosphorus diets only slightly retarded the formation of dentine. Neither of these treatments significantly reduced the ash content of dentine or enamel. Calcium depletion for 12 months followed by repletion for a similar period caused differential growth of the mandible and skull and led to the protrusion of incisor teeth beyond the maxillary pad. Phosphorus deficiency leading to weak and poorly mineralized mandibles caused malocclusion of molar teeth. It is probable that these changes leading as they do to impaired mastication may have a bearing upon the marked loss of condition observed in phosphorus-deficient sheep maintained under hill grazing conditions.

The importance of rate of growth as determined by the intake of nutrients other than calcium and phosphorus upon the skeletal response to low intakes of these two elements has been clearly demonstrated by several workers. Much of this work has been reviewed by Benzie *et al.* (1960). The following are the authors' more important conclusions:

> (1) that there are combinations of energy, calcium, phosphorus and vitamin D intakes that support a rate of growth that falls short of realizing the animal's inherent potentialities but although the skeleton is undersized in relation to age there is little or no damage demonstrable by radiography or chemical analysis;
>
> (2) that there are other combinations of intakes (of these nutrients) that cause damage to the skeleton and these include simple deficiencies of calcium, phosphorus and vitamin D accompanied by sufficient energy and protein to sustain good rates of growth;
>
> (3) that the age at which the animal begins to eat an unbalanced diet, capable of causing bone disease, is of importance in determining the kind of damage that will predominate in the skeleton, younger animals being more liable to exhibit such gross clinical symptoms as frank rickets than older growing animals that are closer to the completion of skeletal development. . . .

Skeletal resorption in the high-yielding dairy cow under practical conditions usually is a consequence of an inadequate phosphorus intake and leads to joint abnormalities and lameness ('Bog Crook', 'Cruban' or 'milk lameness') (McTaggart, 1959).

Using radioisotopic techniques the size of the easily exchangeable calcium pool of the skeleton of the mature cow has been determined. Estimates range from 18 g calcium with a maximum exchange rate of 0·7 g/hr (Manston, 1964) to 41 g with an exchange rate of 1·4 g/hr (Garner, Jones and Sansom, 1960).

In studies with young pigs growing from an initial weight of 8 lb (3·6 kg) to a final weight of 25 lb (11·4 kg) Blair and Benzie (1964) investigated the influence of a wide range of calcium and phosphorus contents of a synthetic diet on skeletal development and bone density. Increasing the calcium and phosphorus contents of the diet from 0·4 and 0·6 per cent to 1·6 and 1·5 per cent, respectively, had no influence on the rate of weight gain of piglets. However, from studies of radiographic density, bone ash content and measurements of the rate of closure of epiphyses the authors concluded that for satisfactory skeletal development the available calcium content of the diet had to be as high as 1·3 per cent of the dry matter when the phosphorus content was 1·5 per cent.

It is perhaps questionable whether the attempt should be made in practice to obtain a maximum rate of accretion of skeletal mineral by incorporating such high levels of calcium and phosphorus since (i) performance as judged by other characteristics was not adversely influenced by lower intakes of calcium and phosphorus and (ii) the inclusion of high levels of calcium is known to affect adversely the utilization of some of the trace elements, notably zinc.

In poultry the first few weeks of egg production are associated with resorption of skeletal mineral (Morgan *et al.*, 1942); this draft is subsequently stopped probably by a reduction in the calcium content of the egg shell. This initial withdrawal of skeletal mineral was not stopped even by feeding high calcium diets.

The Calcium/Phosphorus Ratio

Many studies with non-ruminants have shown that increasing the calcium to phosphorus ratio of the diet may interfere with the absorption of phosphorus and, conversely, that a high phosphorus to calcium ratio may restrict calcium absorption. The magnitude of the effect of changes in this ratio (i) differs greatly from species to species, (ii) depends greatly upon the forms of calcium and phosphorus present in the diet (e.g. the utilization of phytate phosphorus is adversely affected by increasing the calcium/phosphorus ratio from 1:1 to 2 to 1 (Mitchell, 1947)) and (iii) depends upon the vitamin D status of the animal the effects being greater when the vitamin D intake is low.

In studies with growing pigs the effects of changes in the calcium/phosphorus ratio of from 0·2 to 4·0 have been examined. The general concensus of opinion is that changes in the ratio from 1:1 to 2:1 have little, if any, effect on calcium or phosphorus utilization and that ratios $>2:1$ may adversely affect phosphorus utilization particularly if the phosphorus content of the diet is below 0·4 per cent (e.g. Chapman *et al.*, 1962).

The above remarks also apply to the growing chick with the exception that in this case the optimum ratio has been more precisely defined as being between 1·5:1 and 2:1 when calculated on the basis of available phosphorus.

Growing ruminants appear to be less sensitive to changes in the calcium/phosphorus ratio. For example, Wise, Ordoveza and Barrick (1963) found that the calcium/phosphorus ratio of the diet of young ruminants had to be increased to 8:1 before an adverse effect on performance was apparent. The reason for this difference between ruminants and non-ruminants is not known. It has been suggested that the adverse effects of a high calcium/phosphorus ratio may be due to the precipitation of tricalcium phosphate at an alkaline pH in the digestive tract (Davis, 1963). If this is so there is some reason to suppose that such precipitation might be delayed in the ruminant until the digesta had passed the sites of calcium and phosphorus absorption in the duodenum and upper jejunum.

Despite this greater tolerance of wide calcium/phosphorus ratios by ruminants recognition of the facts that the ratio of these elements in the whole ruminant body is *ca.* 1·8:1 and in milk is approximately 1·3 or 1·4:1 makes it desirable to ensure that the available calcium/phosphorus ratio in all ruminant rations is maintained within the range of 1:1 or 2:1 particularly in rapidly growing or heavy-milking animals.

In conclusion it is noteworthy that the suggestion has been made that a wide calcium/phosphorus ratio markedly decreases conception rates in cattle—a situation which, it was claimed, reflected an impaired utilization of dietary manganese (Hignett, 1956). More recent work has, however, failed to confirm this suggestion (Littlejohn and Lewis, 1960).

Calcium and Phosphorus Requirements

For detailed information on calcium and phosphorus requirements at all stages of growth and during reproduction, lactation or egg production the reader is referred to the recent publications of the Agricultural Research Council on *Nutrient Requirements of Farm Livestock* (A.R.C., 1963, 1965, 1967). The following represents a severely abbreviated statement of requirements derived from these publications.

In cattle the dietary requirement for calcium is a function of rate of growth or stage of pregnancy or lactation. For a 100 kg animal requirement is nearly trebled by increasing the daily rate of gain from 0·33 to 1·00 kg/day the estimated needs for calcium being 11 and 27 g/day. Under the same circumstances the phosphorus requirement increases from 5·4 to 12·7 g/day respectively. The effects of lactation are even more striking, the calcium requirement of a 500 kg cow increasing from 32 g/day when producing 5 kg milk/day to 102 g when producing 30 kg. A more detailed description of nutrient requirements for lactation will be found in Chapter 13.

Calcium and phosphorus requirements of sheep are similarly influenced by rate of gain and physiological state. Demands for calcium range from 0·8 g/day in the young (5 kg) animal to 14 g/day in a 50 kg ewe in her first lactation. The corresponding figures for phosphorus are 0·5 and 9·3 g/day.

The calcium and phosphorus requirements of pigs decrease steadily from weaning to maturity requirements for both being approximately 1 per cent of the ration initially and 0·6 and 0·4 per cent respectively at 90 kg liveweight. The earlier discussion of uncertainties regarding optimum intakes to ensure rapid accretion of bone mineral early in life should be recalled in this context. The requirements of the lactating sow have been inadequately studied.

Growing chicks require between 0·9 and 1·5 per cent calcium and 0·2 to 0·4 per cent inorganic phosphorus. Requirement for calcium increases with metabolizable energy content of the diet and that for phosphorus decreases with increasing age. In formulating rations in which part of the phosphorus is supplied in plant material it has been suggested that a figure of 40 per cent should be assumed for the availability of phosphorus from this source (A.R.C., 1963). The same authors have suggested that the calcium requirement of the layer may be derived from the equation—

$$\text{Daily requirement (g)} = 0{\cdot}049 \times \% \text{ production} + 0{\cdot}1.$$

—and that the phosphorus requirement for the layer should be adequately met by 0·3 g inorganic phosphorus/day.

MAGNESIUM METABOLISM

Magnesium in Feeding Stuffs

Cereal grains usually contain less magnesium than most other common feeding stuffs (usually 0·15–0·2 per cent of the dry matter). Of this an appreciable proportion is found in the cereal bran (*ca.* 0·4 to 0·6 per cent magnesium). Most grasses contain 0·15–0·25 per cent and clovers contain rather more (0·4–0·5 per cent) magnesium. Vegetable protein concentrates usually contain 0·4–0·6 per cent. Milk and dried milk products have a low magnesium content; other animal protein concentrates based on offal have a very variable magnesium content depending upon the extent to which skeletal tissues, relatively rich in magnesium, have been incorporated with other animal tissues. If skeletal tissue is absent the magnesium content may be as low as 0·1 per cent.

Little is known about the form in which magnesium is present in feeding stuffs other than in green fodders which contain appreciable proportions of the magnesium-porphyrin, chlorophyll. The magnesium in chlorophyll is readily liberated, presumably as the free ion, by even the mildly acid conditions found in the rumen of cattle and sheep and certainly degradation is complete in the more acid environment of the abomasum (Davidson, 1954) or in the stomach of non-ruminants.

Magnesium Absorption

Studies of the forms in which magnesium occurs in the digestive tract of the sheep have shown that in all organs except the abomasum an appreciable proportion of this element is present in a non-ultrafilterable form (Storry, 1961a). During the passage of digesta from the acid milieu of the abomasum (pH 2–5) to the small intestine (pH 6–8) there was a marked fall in the proportion of the total magnesium which is ultrafilterable from 80 to 100 per cent in the abomasum to 15 to 60 per cent in the small intestine. The nature of the magnesium-binding components of the digesta was not elucidated (Storry, 1961b). This work suggests that if magnesium has to be in an ultrafilterable form before absorption, the most favourable conditions are to be found in the abomasum and duodenum.

Field (1961) found that in sheep the small intestine was an important site for magnesium absorption and in later work Care and van t'Klooster (1965) suggested that the main site was distal to the duodenum. In young calves, Smith (1962) detected absorption from both the small and large intestine. Although similarities exist in the behaviour of calcium and magnesium in the presence of materials in the digestive tract with which they can form non-ultrafilterable complexes it appears likely that they do not share a common absorptive mechanism since magnesium absorption is not enhanced by vitamin D either in the calf (Smith, 1962) or man (Wacker *et al.*, 1962).

Following observations that heavy applications of potassic fertilizers induced hypomagnesaemia in grazing cattle (e.g. Kemp and t'Hart, 1958) the suggestion has been frequently made that high concentrations of potassium limit the absorption of magnesium from the digestive tract. Recent work both with isolated ileal loops and with the intact animal does not support this suggestion (Care, 1965). Again, it has been suggested that the hypomagnesaemic response obtained on turning cattle out to grass in the spring might be related to high rumen ammonia concentrations precipitating magnesium as magnesium ammonium phosphate within the digesta or by, raising duodenal pH, decreasing the amount of ultrafilterable magnesium in the intestinal lumen (Head and Rook, 1957). More recent studies do not completely support these suggestions (e.g. Care, 1965).

As with calcium and phosphorus, there is evidence of a steady decline in the availability of magnesium with increasing age of both cattle and sheep (for review see A.R.C., 1965) but in this case it appears probable that the greater part of the change in availability is due to changes in the nature of the diet as the animal progresses from the suckling stage to the consumption of solid diets. For example, the availability to the calf of magnesium in milk is high (60 to 90 per cent) whereas that in grass is usually low (10 to 30 per cent). It is of interest that Suttle and Field (1966) recently found that raising the water intake of sheep markedly increases the apparent availability of magnesium from hay and cubed grass rations. Infusing water into the rumen significantly increased urinary magnesium output while decreasing faecal excretion of this element. These workers conclude that high water intakes associated with the consumption of lush spring herbage are probably not responsible for the development of hypomagnesaemia in animals grazing such pastures.

It appears possible from studies with man (McCance and Widdowson, 1942–3) that in non-ruminant species the availability of magnesium might be low in diets rich in phytate (see section on Calcium Metabolism).

Tissue Magnesium: Forms and Functions

A variable fraction (between 15 and 50 per cent) of the total magnesium of serum is bound to serum proteins. Between 50 and 75 per cent of the total body magnesium is found in the skeleton where the normal calcium/magnesium ratio is about 55:1. A large proportion of this skeletal magnesium is not readily exchangeable with magnesium in tissue fluids and, furthermore, the size of this exchangeable fraction decreases with increasing age of animal. From studies with growing rats Duckworth and Godden (1943) found that skeletal reserves are only poorly utilized when rapid bone growth is proceeding on diets high in calcium and phosphorus.

Magnesium plays a vital role as an activator of the many enzymes which transfer and split phosphorus groups. It is also involved in enzymic decarboxylations, in protein and peptide hydrolysis and in protein synthesis.

The magnitude of the effect of thyroxine in uncoupling oxidative phosphorylation is reduced when the magnesium content of the diet is increased judging from studies with rats (Vitale, Nakamura and Hegsted, 1957) thus suggesting that magnesium has an important regulatory role in energy-yielding reactions within animal cells.

Hyperirritability and convulsions are typical signs of magnesium deficiency in all species of farm animals indicating that magnesium has a role in the maintenance of normal nerve tissue function. A role in muscular function is suggested by the muscular weakness that develops in the deficient animal. Metastatic calcification occurs in magnesium deficiency in some species; it has been reported, for example, in milk-fed calves in the diaphragm, endocardium and blood vessels (Moore, Hallman and Scholl, 1938) but not in calves fed a magnesium-deficient milk substitute (Blaxter, Rook and MacDonald, 1954; Parr, 1957).

The relationship of magnesium to grass tetany in the ruminant will be considered in Chapter 35.

Magnesium Homeostasis

Thyro-parathyroidectomy in the sheep results in a reduction of the net absorption rate of magnesium from the ileum. Deoxycorticosterone administered to either normal or adrenalectomized sheep reduces the apparent availability of dietary magnesium (Care, 1965). Studies with hyperparathyroid human patients showing that this condition is associated with a negative magnesium balance suggest that magnesium metabolism may be subject to similar hormonal control in non-ruminant species.

A measure of homeostasis is achieved by accretion or resorption of magnesium from bone although the nature of the regulatory mechanism is obscure. Thus in young magnesium-deficient calves up to 60 per cent of the total bone magnesium may be mobilized to compensate for inadequate dietary intake. In contrast, mature cows suffering from hypomagnesaemia frequently show no depletion of bone magnesium suggesting that in the mature skeleton the release of this element from bone mineral is impaired. Saliva and gastric juice are rich in magnesium and Storry (1961a) has pointed out that small changes in rate of reabsorption of endogenous magnesium would greatly influence magnesium homeostasis. He demonstrated that in a 40-kg sheep with an extracellular pool of 16 m-equiv. of magnesium complete reabsorption of magnesium in the daily output digestive secretions containing 16 m-equiv. must be achieved to maintain magnesium balance. The studies referred to earlier on the effects of duodenal pH and water intake on magnesium absorption from the gut have already indicated how changes in nutrient intake may influence this process of reabsorption.

Little is known of the mechanisms controlling renal output of magnesium.

Magnesium Requirements

The magnesium requirement depends greatly upon the physiological state of the animal. Thus the requirement of a 500 kg animal has been calculated to lie between 8·2 and 9·5 g/day depending on the rate of growth whereas in a lactating cow of the same weight producing 20 kg milk/day the requirement is as high as 20 g/day (A.R.C., 1965). Similarly the magnesium requirement of the rapidly growing sheep is more than double the maintenance requirement (for a 30 kg animal gaining only 50 g/day the requirement is 435 mg/day; if the rate of growth increases to 400 g/day requirement becomes 1065 mg/day (A.R.C., 1965)).

In growing pigs the concensus of opinion from a range of trials to determine requirement appears to be that 0·04 per cent magnesium in the diet meets the needs of young stock (e.g. Mayo, Plumlee and Beeson, 1959). Requirements for pregnancy and lactation have not so far been determined according to the author's knowledge.

In chicks the requirement for magnesium has been stated to be 0·025 per cent; requirements for egg production have not been determined.

In conclusion it should be stressed that although in the rat, guinea-pig and other species it has been shown that changes in the calcium or phosphorus levels of the diet have a marked influence on magnesium requirement (e.g. O'Dell, 1960) there is a dearth of similar studies in farm livestock.

THE TRACE ELEMENTS

The group of elements now to be considered, by virtue of the fact that they are present in animal and plant tissues in concentrations lower than the elements just described, are known as the trace elements. The comments which follow will undoubtedly reflect our greater ignorance of their availability and function. This arises from the difficulties inherent in experimental work with these elements. Particular attention will accordingly be paid to the limitations of our knowledge especially since these limitations frequently prevent a clear understanding of how trace-element-deficiency disorders arise in practice.

IRON METABOLISM

Iron in Feeding Stuffs

As with most trace elements the content of iron in feeding stuffs is highly variable and is influenced greatly by the widespread possiblities for contamination in this case from ferrous metals during processing. The generalization can accordingly be made that the more extensive the processing of the feeding stuff the higher its iron content. Thus, for example, processed plant protein concentrates are rich in iron (often >800 ppm) whereas cereal grains, unmilled, are low in this element (usually <40 ppm). Fresh milk is an extremely poor source of iron usually containing only about 0·5 mg/l. but, again, the iron content rises as milk is processed and dried whole or skimmed milk has a very variable content. Colostrum contains appreciably higher levels of iron than milk produced later in lactation. The iron content of green forages is also variable depending largely upon the degree of soil contamination. Most forages fall within the range of 100–400 ppm iron in the dry matter but, as expected, higher values are frequently found in hays made from these crops. High concentrations (>1000 ppm) of iron are frequently found in 'natural' sources of calcium carbonate and phosphate used as mineral supplements due to coprecipitation of iron as its hydroxide, carbonate or phosphate during the formation of the mineral deposit.

Little is known about the forms of iron present in common feeding stuffs. In animal protein supplements a considerable proportion will undoubtedly be present in the form of the iron-porphyrin, myoglobin and depending on the amount of occluded blood, a proportion will be present in haemoglobin. In cereals and in other feeding stuffs rich in phytic acid it is reasonable to presume that an appreciable proportion of the iron present will be in the form of an iron–phytic acid complex.

Absorption and Availability of Iron

There is general agreement that under normal circumstances the most important site of iron absorption is in the duodenum although in iron-deficiency and protein-deficiency anaemia uptake by the mucosa of the jejunum and ileum has also been demonstrated (Wack and Wyatt, 1959).

It is beyond the scope of this review to give detailed consideration to the very large volume of past work on the mechanism of transport of iron across the intestinal mucosa and the response of this mechanism to iron demand and dietary iron supply. Originally it was postulated (Granick, 1946) that control of iron absorption was achieved by changes in the level of a specific protein, apoferritin, in mucosal cells which accepted iron from the intestinal lumen. More recent work suggests, however, that the formation of this complex may represent a mechanism for limiting excessive iron absorption but that iron in active transit through the mucosa is in the form of another complex which is as yet unidentified (Charlton *et al.*, 1965).

There have been few critical studies of factors influencing the availability of dietary iron to farm livestock. In studies with calves Hansard and Foote (1961) found the availability of iron in milk to be 26 per cent. A similar proportion (30 per cent) of the iron in milk was found to be utilized by pigs up to 60 days of age (Ullrey *et al.*, 1960b) whereas up to 60 per cent of an oral dose of 30 mg iron was absorbed during the first week of life (Braude *et al.*, 1962). In non-ruminant experimental animals iron absorption and availability have been shown to be depressed by high dietary levels of phosphate, by additions of soluble phytates (but not to any significant extent by the insoluble phytates present in cereal grains, legume seeds, etc.) and by high levels of calcium carbonate in the diet. A considerable volume of evidence indicates that ferrous iron is more readily absorbed than the ferric form. Much of this work is reviewed by Wiseman (1964). In all species the apparent availability of dietary iron is higher in the iron deficient than the normal animal. For example, in the sheep Georgi (1964) found that only about 1 to 3 per cent of a dose of ferrous sulphate administered orally or by infusion into the abomasum was absorbed by normal sheep whereas 16 per cent was absorbed by animals with depleted iron stores. In contrast, little iron (<3 per cent) was absorbed from haemoglobin administered by these routes whether the sheep were or were not depleted of iron.

Tissue Iron: Forms and Functions

The data of Granick (1958) indicate that 60–70 per cent of total body iron is accounted for as haemoglobin in the erythrocyte, 3–5 per cent in myoglobin of muscle, 7–15 per cent as the iron-storage protein ferritin, 0·2 per cent in respiratory enzymes containing the haem nucleus and 0·1 per cent as the β_1-globulin transferrin present in the plasma. Wide variations can exist from species to species in the proportion of body iron present as the insoluble iron-storage protein. Species differences also exist in the concentration of iron in the fat-free body; the pig being particularly notable for the low concentration in its body at birth.

As is now well established, the molecule haemoglobin consists of four ferroprotoporphyrin nuclei linked to the protein globin. Synthesis takes place in the bone marrow and there is good evidence that the utilization of iron for haemoglobin synthesis is in some way dependent upon the maintenance of adequate tissue reserves of copper judging from studies with pigs (Cartwright *et al.*, 1956). Suggestions have been made that copper is a component of an enzyme involved in the formation of the haemoglobin precursor, porphobilinogen. The practical implications of iron deficiency anaemia will be considered later.

Iron is also a constituent of other chromoproteins containing a haem nucleus. The cytochromes of the terminal respiratory chain contain iron, and iron deficiency in pigs and other species results in depletion of tissue cytochrome C and of catalase concentrations. Iron is also a component of many other enzymes, notably the flavoproteins and aconitase.

Reference has already been made to ferritin, the relatively readily mobilized iron-storage protein which is found in liver spleen, bone marrow and in the mucosa of the intestine. A structurally related iron protein, haemosiderin, of higher iron content arises in tissues when iron intake is excessive or when defects arise in iron mobilization within tissues.

Iron Homeostasis

Homeostasis is largely achieved through control of absorption and re-excretion via the mucosa of the intestine. Charlton *et al.* (1965) have suggested that, depending upon demand, ferritin deposits in mucosal cells may either yield their iron to the plasma to form the β_1-globulin transferrin or alternatively may be lost into the intestinal lumen when mucosal cells are exfoliated. For a review of conflicting evidence regarding mechanisms controlling intestinal absorption and excretion the reader is referred to the paper of Coons (1964). Loss of iron via the urine and bile accounts for only a small proportion of iron output and these routes are relatively unimportant in maintaining homeostasis.

In studies with sheep and pigs using radioactive iron Hansard and Hoskins (1963) found that plasma iron level, haemoglobin concentration and iron excretion were not affected by stage of gestation but during the last third of pregnancy the total iron content of the body and the absorbed tracer iron increased both in dam and foetus, particularly in the liver, spleen and kidney. Their results suggested an acceleration of iron absorption by intestine and placenta and a more extensive storage within the tissues.

Iron Deficiency and Iron Requirements

As a consequence of the low reserves of iron at birth and the low content of this element in milk the piglet is particularly dependent upon extraneous sources of supply to maintain haemoglobin production during early life. Until the need for iron supplements was demonstrated, piglet anaemia, often associated with respiratory distress and poor growth, was a common feature where litters were denied access to soil as in concrete rearing pens. Placental transfer of iron is usually inadequate to prevent the fall of haemoglobin level in the offspring even when the iron intake of the sow is high (e.g. Pond *et al.*, 1960). Effective control is achieved by either the oral administration of ferrous salts or intramuscular administration of iron complexes such as iron dextran (Linkenheimer *et al.*, 1960) into the young pig. Braude *et al.* (1962) have shown that a piglet must retain 21 mg iron/kg liveweight increase to maintain satisfactory tissue stores of iron. A concentration of 60 to 80 ppm iron in the dry matter of the ration is probably adequate to meet requirements during growth for the first 50 days of life. The iron requirements of older pigs are not known with certainty.

A mild iron-deficiency anaemia in areas where calves are fed large quantities of liquid milk and little solid food has been shown to exist by Blaxter, Sharman and MacDonald (1957). The microcytic normochromic anaemia developed in calves during the first 80 days of life both in calves on farms and in a laboratory experiment where the intake of iron was 2 to 4 mg/day. A recent report (A.R.C., 1965) suggests that the iron requirement of cattle and sheep should be met by concentrations of 30 ppm in animals receiving solid rations.

Iron deficiency is not commonly encountered in poultry under practical conditions although it may be readily produced experimentally in chicks fed diets containing less than 15 ppm iron (Matrone, 1960).

In conclusion, it should be stressed that although it has been shown experimentally that high intakes of phosphates, carbonates or low intakes of copper decrease the utilization of dietary iron the influence of these changes on iron requirement under practical conditions has yet to be determined. Recent studies (Suttle and Mills, 1966) indicate that iron utilization for haemoglobin synthesis is impaired in pigs fed high levels of copper (>250 ppm) as a growth stimulant.

COPPER METABOLISM

Copper in Feeding Stuffs

Most processed feeding stuffs have a variably high copper content arising from metallic contaminants particularly where the material is rich in protein and is processed in a moist state presenting favourable conditions for uptake of copper by the protein component. Most animal and vegetable protein concentrates contain between 10 and 30 ppm copper although concentrations exceeding 100 ppm are sometimes encountered. Cereal grains commonly contain 4–10 ppm the highest concentrations being found in the germ. Most grasses contain similar concentrations of copper while in legumes the concentration is usually rather higher ranging in most instances from 6 to 12 ppm. Rasheed and Seeley (1966) have established a positive relationship between the copper and protein contents of grasses and legumes.

There is little information on the nature of the copper compounds present in these feeding stuffs. In animal protein concentrates much of the copper will undoubtedly be directly associated with tissue proteins (see later). In grasses Mills (1954) has shown that the greater part of the soluble copper is present in the form of organic complexes which are stable in aqueous solution above pH 2·5 and the experiments of Deijs and Bosman (1961) suggest that at least a part of the copper in grass may be present as a complex formed between copper and phaeophytin, the magnesium-free residue of chlorophyll. It is not known whether copper in this form is available to the animal.

Copper Absorption and Homeostasis

In studies with pigs Bowland *et al.* (1961) found that copper absorption occurred in the small intestine. Absorption from the upper jejunum has been detected in the dog and man but no similar studies have been carried out with ruminants or poultry.

The addition of high levels of sulphide or carbonate to the diet of sheep markedly reduces copper absorption (Dick, 1954) presumably by forming the relatively insoluble copper sulphide or carbonate within the digestive tract or, in the case of carbonate, by increasing the pH of the contents of duodenum or ileum and so restricting the release of copper from the less stable organic complexes of copper present in the ingesta. The low availability of copper in the form of its sulphide has also been demonstrated in the pig (Bowland *et al.*, 1961).

Studies with both pigs and ruminants emphasize that under normal circumstances only between 2 and 10 per cent of ingested copper is absorbed—in contrast, injected copper is readily retained and only slowly excreted, the major pathway of excretion being via the bile in pigs (Bowland *et al.*, 1961), poultry (Beck, 1961) and probably in cattle (Comar, 1950). An indication of the relative importance of different pathways of copper excretion is given by the work of Mahoney *et al.* (1955) who, using dogs, found that excretion via the bile accounted for 7 to 10 per cent of an intravenously administered dose of radioactive copper; 0·5 per cent was excreted in the urine and 1·5 per cent entered the gastrointestinal tract via a route other than bile.

Tissue Copper: Forms and Functions

Much of the copper in blood plasma is associated with an α_2–globulin caeruloplasmin (Wintrobe, Cartwright and Gubler, 1953). This complex contains 8 atoms of copper and has a molecular weight of 151,000. Although it has the enzymic properties of an oxidase its precise function is unknown. The remainder of the plasma copper is loosely bound to albumin and it is this fraction which initially rises in concentration following copper administration, suggesting that it is directly associated with copper transport. Discrete copper proteins have also been isolated from cattle erythrocytes and liver but again the metabolic function of these proteins has not been defined. The element is widely distributed throughout all tissues and a small but obviously vital fraction is accounted for by the presence of copper in the respiratory enzyme, cytochrome oxidase. A dramatic decline in the activity of this enzyme is an early finding in copper deficiency in all species so far examined and is probably responsible for at least some of the clinical effects observed (see discussion of effects of copper deficiency). Copper is also a component of the enzyme uricase.

Certain organs, particularly the liver, kidney, heart and spleen, have high concentrations of copper relative to other tissues. The livers of normal cattle and sheep have variable but normally high concentrations (e.g. 50–1000 ppm in the dry matter) as compared with livers of pigs and poultry (*ca.* 10–30 ppm) —the reason for this difference is not known (for detailed discussion of species and age differences see Underwood, 1962).

The role of copper in the formation of the porphyrin nucleus of haemoglobin—interruption of which process leads to copper-deficiency anaemia in cattle and sheep (Cunningham, 1946; Bennetts and Beck, 1942), pigs (Lahey *et al.*, 1952) and poultry (Matrone, 1960)—and its role in maintaining the function of bone osteoblasts interruption of which leads to the formation of brittle, often deformed skeletal structures (Underwood, 1962), are as yet incompletely understood. Its roles in melanin pigment generation in hair and wool and in keratin formation in wool are also not completely resolved.

Copper-deficiency Diseases

Poor growth, anaemia and bone fragility are the commonest findings in copper-deficient pigs and poultry. Deficiency in these species is uncommon and usually arises when the copper content of the ration is low (e.g. <3 ppm).

In contrast, copper-deficiency disorders of ruminants are more common and can arise either from a low copper intake or from the presence in the diet of factors which interfere with copper utilization even where the copper intake is relatively high. The typical picture of copper deficiency in cattle is one of poor growth, anaemia, bone fragility and sometimes diarrhoea and myocardial fibrosis. A decrease in milk production often occurs and fertility is frequently poor (Allcroft and Parker, 1949; Alderman, 1963).

The effects of copper deficiency in sheep are quite different in that depressed growth rate and anaemia are rare. Instead, the most significant effects are during late foetal and early post-natal development in the lamb where copper deficiency causes degenerative changes in nervous tissue which, after birth, are manifested by an inco-ordination of the hind legs which may progress to complete immobilization, blindness and death. The pathology of this disease, known as 'swayback' or 'enzootic ataxia', has been described by Innes and Shearer (1940). The degenerative changes in the brain and spinal cord are associated with copper contents of less than 3 ppm and with a low activity of the copper enzyme, cytochrome oxidase (Mills and Williams, 1962; Howell and Davison, 1959). It is not yet clear whether the histologically abnormal distribution of myelin lipid in nervous tissue of these animals is a consequence of breakdown or a failure of synthesis. However, a close correlation between the clinical severity of the condition and the number of degenerating motor neurones in the 'red nucleus' of the brain stem showing a low cytochrome oxidase activity has been found by one group of workers (Fell, Mills and Boyne, 1965).

Our understanding of the aetiology of copper-deficiency disorders in ruminants is incomplete. As with non-ruminants, deficiency can arise simply from a low copper content of the diet but frequently it appears that deficiency occurs when the copper intake is not low. Dick (1954) has shown that elevated intakes of molybdenum and sulphate can induce a 'conditioned' copper deficiency; a situation which may result from a depression of copper solubility in the digestive tract (Mills, 1960). Many cases of copper deficiency undoubtedly arise in the field from this cause but, in others, molybdenum cannot be incriminated and the search for other substances which can interfere with copper utilization continues.

Copper Requirements

In young growing pigs Ullrey *et al.* (1960a) found that 6 ppm in the ration allowed normal growth and haemoglobin production whereas Carter, Miller and Brooks (1959) found that 3 ppm led to the development of skeletal abnormalities and a depression of the liver cytochrome content.

The copper requirement of the chick is approximately 3 ppm judging from the studies of Hill and Matrone (1958). The requirement of the laying hen is not known. The work of Davis, Norris and Kratzer (1962) suggests that rations rich in phytic acid may increase the apparent requirement of the chick for copper presumably as a consequence of the formation in the digestive tract of a non-available copper–phytic acid complex.

The copper requirements of ruminants are less well defined than those of pigs and poultry because of uncertainties about the nature of dietary components which can decrease availability.

Unpublished work from several research institutes suggests that when fed synthetic diets the copper requirement of the sheep at least may be as low as that of pigs or poultry. In practice, however, it is apparent that when fed conventional rations, ruminants have appreciably greater requirements. Australian work (Beck, 1941) supported by field observations from many other parts of the world suggests that in cattle and sheep at pasture a so-called 'uncomplicated' copper deficiency may be expected to arise in cattle or sheep if the copper content of pasture falls below about 4 ppm. In contrast, 'conditioned' deficiencies of both cattle and sheep are frequently encountered on pastures providing twice this concentration of copper.

COBALT METABOLISM

Cobalt in Feeding Stuffs

Cobalt in trace quantities is widely distributed in common feeding stuffs. In plant material the existence of stable cobalt–protein complexes has been detected, and in feeding stuffs derived from animal tissues cobalt is also known to be present in the form of cyanocobalamin (vitamin B_{12}) and related compounds, meat and fish meals being particularly rich sources of the latter.

It is beyond the scope of this chapter to consider the requirement of non-ruminant animals needing cobalt in the state of pre-formed vitamin B_{12}; the following discussion will deal only with ruminant species. The vitamin B_{12} requirements of pigs and poultry are considered in Chapters 29, 30 and 32 of this volume.

Absorption and Function of Cobalt in the Ruminant

Comar and Davis (1947 a, b) demonstrated that following the oral administration of physiological quantities of radioactive cobalt to cattle 80 per cent. of the dose was rapidly excreted via the faeces and only 0·5 per cent in the urine. After intravenous injection, however, 65 per cent of the activity appeared in the urine and 7 to 30 per cent in faeces and 5 to 15 per cent in the bile in different experiments.

It is now well known that to be physiologically effective in the ruminant, cobalt must be ingested. Phillipson and Mitchell (1952) showed that cobalt administered by injection or by infusion into the abomasum or duodenum was quite ineffective in preventing cobalt deficiency; a finding which stresses the probable importance of the accumulation of cobalt by rumen micro-organisms that had been detected earlier (Tosic and Mitchell, 1948). It has since become clear that these findings reflect firstly, the ability of rumen microorganisms to synthesize vitamin B_{12} (see review by Lester-Smith, 1962) and the ultimate dependence of the ruminant animal for cobalt in this form. Hoekstra, Pope and Phillips (1952) showed that the rumen contents of cobalt-deficient sheep contained only 0·09 μg vitamin B_{12} activity/g dry matter in contrast to 1·3 in cobalt-supplemented sheep. This difference was reflected by even more spectacular differences in the liver content of vitamin B_{12}. Supplemented sheep had liver concentrations of 0·9 μg/g fresh weight and deficient sheep 0·05 μg/g. There is good evidence, however, that much of the cobalt entering the rumen is converted into vitamin B_{12}-like compounds which are not physiologically utilized by the ruminant (Hine and Dawbarn, 1954) and even the absorption of true vitamin B_{12} appears to be poor, only about 3 per cent of that formed in the rumen being utilized (Marston, Allen and Smith, 1961).

The Function of Tissue Cobalt

As far as is yet known the only functional form of tissue cobalt is as a part of vitamin B_{12} and its derivatives. In addition to the well-established roles of this vitamin in transmethylation, erythropoesis and purine biosynthesis in several species, its role in tissue propionate metabolism is of particular importance to the ruminant which depends upon the efficient utilization of propionate produced in the rumen for a large part of its energy requirements. Smith and Monty (1959) found that in vitamin B_{12}-deficient rats there was a failure to convert propionate to succinate, the intermediate in the tricarboxylic acid cycle through which propionate is ultimately metabolized to yield energy. This defect was associated with a decline in the activity of the enzyme methylmalonyl coenzyme A-isomerase which mediates an intervening stage of this reaction and which was subsequently found to require vitamin B_{12} as a cofactor. Marston *et al.* (1961) later found that this defect occurred in cobalt-deficient sheep and further showed that there was a marked decline in the rate of utilization of intravenously injected propionate in these animals. Underwood (1962) has suggested that the high circulating blood propionate levels may be responsible for the failure of appetite that is such a dramatic feature of cobalt deficiency in the ruminant.

Cobalt Deficiency: Signs and Occurrence

Emaciation and anorexia in young animals are the typically non-specific signs of cobalt deficiency in ruminants. In later stages anaemia develops although this may not be a consistent finding in many borderline cases. Deficiency is not uncommon under field conditions and is usually found in areas where the soil parent material is derived from rocks low in cobalt status; typical examples are soils derived from Old Red Sandstone or Devonian rocks, from calcareous shell sands and from some limestones and acid granites. Cobalt deficiency is common in areas where the concentration of cobalt in pasture falls below 0·08 ppm in the dry matter. The vitamin B_{12} content of colostrum and milk of ewes is low under such conditions (O'Halleran and Skerman, 1961). Evidence is accumulating from field studies with cattle that fertility may be adversely affected (Alderman, 1963).

Cobalt Requirements of Ruminants

Marston (1952) suggested that the requirement of sheep for cobalt is probably met by rations containing 0·08 ppm in the dry matter. The requirement of young cattle is probably of the same order. Concentrations of 0·05 ppm in the diet are inadequate for both species.

ZINC METABOLISM

Zinc in Feeding Stuffs

Most cereal grains contain usually between 15 and 25 ppm zinc, most of this being associated with the bran and germ fractions which, when considered alone, may contain 60–80 ppm, or more if contamination has occurred during processing. Vegetable protein concentrates usually contain 40 to 80 ppm zinc. Roughages commonly contain 20 to 35 ppm while animal protein concentrates have highly variable contents much of which is extraneous zinc arising during processing—most animal tissues when uncontaminated contain not less than 30 ppm in the dry matter. Colostrum is usually high (>150 ppm in the dry matter) while milk not contaminated from galvanised utensils usually contains 30 to 50 ppm in the dry matter. Egg albumin is noteworthy for its very low zinc content—usually <3 ppm.

Little is known about the forms of zinc present in these feeding stuffs although its association with the protein fraction may be inferred. In cereal grains and legume seeds rich in phytic acid the presence of zinc-phytic acid complexes of poor availability to non-ruminants is probable (see section on zinc requirements and availability).

Zinc Absorption

Judging from studies with cattle using radioactive zinc this element is but poorly absorbed from the digestive tract only between 3 and 10 per cent of the administered dose being recovered in the tissues (Feaster *et al.*, 1954). Seventy per cent of the dose was recovered in the faeces and only 0·3 per cent in the urine. In contrast 20 per cent of the isotope was recovered in the faeces and 0·25 per cent in the urine following intravenous injection.

The mechanism whereby zinc is absorbed from the digestive tract is unknown, but it is of interest that both Kirchgessner (1962) and Hsu (1963) have found that low dietary levels of pyridoxin reduce zinc utilization although it is not clear whether this effect is simply due to a decreased efficiency of absorption.

Zinc in Tissues: Forms and Functions

In plasma, zinc is firmly bound to the globulin fraction (Wolff, 1956) and less firmly associated with the albumins. In all species the greater part of the zinc of the erythrocyte is in the form of the zinc enzyme, carbonic anhydrase. Other zinc enzymes are widely distributed in the animal body. Many of the dehydrogenases, the enzymes alkaline phosphatase and carboxypeptidase together with carbonic anhydrase in tissues other than the erythrocyte contain zinc as an essential component (for review see Vallee, 1962). It is highly probable that other important roles for zinc exist, which are not yet elucidated, for an appreciable volume of work with experimental animals suggests that this element may be intimately associated with the processes of cell division and growth.

Zinc is widely distributed in the animal body; among the soft tissues only the prostate and the choroid and the iris of the eye and possibly the pituitary accumulate the element in high concentrations (>50 ppm). One report suggests that corticotrophin from the pituitary of the pig has a strong affinity for zinc. Purified preparations contained up to 350 ppm zinc (Kocsis *et al.*, 1953). Relatively high concentrations are found in the skeleton, wool and fur. In the hen's egg the zinc is firmly bound to the phosphoprotein, vitellin.

Zinc Deficiency

In poultry, zinc deficiency is manifested by poor hatchability of eggs, malformed embryos, poor growth, defective feathering and retarded skeletal development together with the development of skin lesions associated with defective proliferation of keratin (see Supplee, Combs and Blamberg, 1958). In pigs, deficiency is similarly associated with poor growth both of soft and skeletal tissues and also by parakeratosis, alopecia, appetite failure and in males by testicular degeneration (Smith, Plumlee and Beeson, 1961). The sequence of events during the development of zinc deficiency in cattle and sheep is growth failure, excessive salivation, the development of parakeratosis of the skin around the eyes and on the neck and legs and the development of bony swellings around the hock joint. Hoof and horn growth is excessive and these structures are abnormally soft. These signs develop when the plasma zinc concentration falls below 0·4 μg/ml. Recent studies with ruminants show that the animal has little if any capacity for storing zinc in a physiologically available form. Plasma zinc concentrations fall below 0·4 μg/ml within 24 hr of feeding a zinc-deficient diet (Mills *et al.*, 1965; Mills *et al.*, 1966).

Zinc Requirements and Availability

Pigs and poultry maintained on synthetic diets containing less than about 20 ppm zinc rapidly develop signs of deficiency (Underwood, 1962; Smith *et al.*, 1961) but deficiency may arise at considerably higher dietary concentrations if components containing appreciable concentrations of phytic acid are present in the ration (Oberlease *et al.*, 1961). Such an increase in zinc requirement has been noted on rations containing soya bean meal, pea meal and sesame meal. The effects of phytic acid in reducing zinc availability are enhanced by high concentrations of calcium in the diet (see review by Forbes, 1960). It is, however, probable that calcium *per se* may have an adverse effect on zinc utilization. For example, Newland *et al.* (1958) found that the endogenous output of zinc in the faeces of pigs increased as the calcium content of the ration was increased above 1 per cent. Haaranen (1963) has suggested that the zinc requirement of cattle is similarly increased by high intakes of calcium but it is of interest that in the ruminant, phytic acid has no effect on zinc requirement (Dalgarno and Mills, unpublished observations).

Zinc deficiency has been reported in ruminants on rations providing between 18 and 120 ppm zinc (e.g. Legg and Sears, 1960; Grashuis, 1963) and yet in many areas where cattle and sheep maintain excellent health and productivity the pasture provides only 25–35 ppm. The nature of factors other than phytic acid and calcium which influence zinc requirement is under study.

Recent work has shown that the common British practice of feeding high levels of copper (250 ppm) to growing pigs as a growth-promoting stimulant may increase zinc requirements (Suttle and Mills, 1966).

MANGANESE METABOLISM

Manganese in Feeding Stuffs

Comments made earlier regarding the variability of trace element content in individual feeding stuffs arising from extraneous contamination also apply in the case of manganese. Maize grain has a particularly low manganese content (*ca.* 4 to 10 ppm); other cereal grains have contents usually within the range 9 to 50 ppm. This difference may underlie the greater need for manganese supplementation of poultry rations which contain high proportions of maize. This species difference is even more striking on comparing the manganese content of maize germ and wheat germ where the contents are approximately 15 and 120 ppm respectively. Most protein concentrates contain 10 to 40 ppm manganese values at the lower end of this range being found in animal by-products. Milk and milk products are particularly low in manganese content (usually 2 to 8 ppm in the dry matter). Fodder crops and hays usually fall within the range 20 to 100 ppm but in this instance it is stressed that the content depends greatly on soil conditions, low concentrations being generally found in crops growing on well-aerated organic soils of high pH.

Manganese Absorption and Excretion

Judging from work with laboratory animals and man, orally ingested manganese is poorly absorbed, over 90 per cent of the administered dose being recovered in faeces. Injected manganese is excreted via the bile and the other digestive tract secretions and urine appear to be relatively unimportant as pathways for manganese excretion (Wiseman, 1964). Murty (1957) has estimated the endogenous loss of manganese to be 6 mg/day in the sheep. In the pullet Hill (1965a) found that manganese in the droppings and gut contents accounted for about 92 per cent of the administered dose irrespective of whether the birds had previously been on diets of high or low manganese content.

Tissue Manganese: Forms and Functions

In all animal species so far examined, the skeleton, the liver and kidney have been found to have the highest concentrations of manganese; the level varying with the dietary manganese content but usually within the range 4 to 10 ppm. Muscle is low in manganese usually containing 0·8 to 1·2 ppm. Hill (1965b) has found that only if bone contains less than 1 ppm manganese are evidence of bone abnormalities and other signs of deficiency apparent.

The manganese content of milk depends on the stage of lactation, colostrum being particularly rich in this element (0·16 to 0·19 mg/l.). Later milk only contains about one-fifth of this concentration and reflects the manganese intake of the cow (Underwood, 1962).

We are largely ignorant of the forms in which manganese exists in tissues: indeed the work of Borg and Cotzias (1958) suggests that this element is highly mobile within the body, rapidly exchanging with administered (radioactive) manganese. The element has been found in preparations of the enzymes; arginase, prolidase and isocitric dehydrogenase and it acts as a cofactor for many other enzymes. Its precise functional role in osteogenesis is obscure.

The Occurrence and Signs of Manganese Deficiency

In practice deficiency usually arises through the inadvertent selection of dietary components low in manganese and particularly when the calcium and phosphorus content of the diet is high. In pigs and poultry manganese deficiency results in abnormal skeletal development with lameness and with short, thick leg bones showing a characteristic twist. Leach and Muenster (1962) have found that there is severe interruption of mucopolysaccharide formation in the epiphyseal cartilage of the leg bones of manganese–deficient chicks which may account for the observed deformities. The condition known as 'slipped tendon' in chicks in which the gastrocnemius tendon slips out of the intercondyloid groove is often a consequence of abnormal hock development; it should be noted, however, that deficiencies of choline, biotin, folic and nicotinic acids have also in the past been implicated in similar conditions.

Low intakes of manganese retard sexual maturity and produce infertility in pigs (Plumlee *et al.*, 1956). Lowered egg production and poor hatchability of eggs may arise from manganese deficiency in poultry even where the deficiency is not sufficiently severe to produce skeletal abnormalities (Schaible, Bandemeer and Davidson, 1938).

Although manganese deficiency has been produced experimentally in calves and cattle with similar effects on skeletal development and reproductive performance to those reported in pigs it remains doubtful whether this deficiency arises under commercial farming conditions. Bosch, Van der Grift and Hartmans (1965) have recently presented a very critical review of earlier work suggesting that manganese deficiency existed in calves and lactating cows in Holland. Their own evidence does not support this suggestion. In contrast, Munro (1957) working in Britain obtained an improvement in the reproductive performance of cows and heifers from manganese supplementation in an area where the pasture manganese content ranged from 11 to 26 ppm.

Manganese Requirements

Requirements for the growing chick have been suggested to be approximately 35 ppm (A.R.C., 1963). It is probable that the requirement rises as the dietary calcium and phosphorus level increases, particularly when these minerals are added in the form of bone meal; one study having indicated that 67 ppm is inadequate when the ration contains bone meal and has calcium and phosphorus contents of 3·2 per cent and 1·6 per cent respectively.

It is probable that 35–40 ppm manganese meets the requirements of growing and breeding pigs for this element.

IODINE METABOLISM

Iodine in Feeding Stuffs

The concentration of iodine in common feeding stuffs is highly variable. Protein sources of animal origin other than those derived from fish offal contain nutritionally insignificant quantities of this element. Protein concentrates of plant origin usually contain 100 to 300 μg/kg and common cereal grains 40 to 100 μg/kg. The iodine content of common forage crops depends greatly on the species and even upon the strain of plant. Underwood (1962) quotes results showing that two strains of white clover growing on the same soil type had mean levels of 40 and 1200 μg iodine/kg; the usual range of values found in roughages is 300 to 500 μg/kg.

Iodine Absorption and Excretion

It is probable that much of the iodine in common feeding stuffs is in the form of inorganic iodide although some undoubtedly will be present as iodine derivatives of the amino acids. Work with laboratory animals suggests that iodine in the form of iodide is rapidly and efficiently absorbed in the proximal third of the small intestine. Absorption of iodinated amino acids proceeds less rapidly. The kidney is the most important pathway of excretion; the iodine present in duodenal secretions is normally reabsorbed before the colon, and faecal losses are small.

Tissue Iodine: Forms and Functions

Whole blood iodine concentrations in farm livestock are extremely variable, a typical range being 3 to 20 μg/100 ml. The greater part of blood iodine is plasma protein-bound and, in turn, this component varies greatly in concentration with species, iodine level of the diet, stage of growth and pregnancy and the level of activity of the thyroid gland. Work on the influence of these parameters on plasma protein-bound iodine level has been reviewed by Underwood (1962).

The iodine of milk, most of which is dialysable in ruminant species, also reflects iodine intake. Kirchgessner (1959) has demonstrated an increase in concentration from 28 μg/l. in cows receiving 1·6 mg iodine/day to 267 μg/l. when the intake was 20 mg/day.

The greater part of the body iodine is concentrated in the thyroid gland (in most species *ca.* 80 per cent of the total) in the form of the iodinated colloid protein, thyroglobulin, containing mono-, di- and tri-iodotyrosines, thyroxine and trace quantities of other iodinated amino acid derivatives including tri-iodothyronine. It is unfortunately beyond the scope of this contribution to present detailed consideration of the physiological functions of the two most important thyroid hormones, thyroxine and tri-iodothyronine, other than to point out that they control cellular energy exchange and metabolic rate and tissue growth and development. All available evidence suggests that the nutritional role of this element is entirely related to its action as a component of the thyroid hormones.

Iodine Deficiency and Thyroid Activity

Deficiency in all farm animals is accompanied by thyroid hyperplasia ('goitre') and a histologically demonstrable decrease in the thyroglobulin concentration within the follicles of the thyroid gland.

A survey of published work on iodine requirements of poultry (Ewing, 1963) suggests that mild thyroid hyperplasia may exist without demonstrable effects on growth and reproductive performance. Laying hens, given a ration containing 0·01 to 0·02 ppm iodine, produced eggs of low hatchability and retarded embryonic development; 0·35 ppm was adequate to maintain good hatchability but even 0·75 ppm did not prevent thyroid hyperplasia in otherwise normal embryos (Rogler *et al.*, 1961).

Cases of foetal death, abortion or the production of weak lambs or calves associated with hypothyroidism have been reported in several instances where plants containing goitrogens have constituted an appreciable proportion of the ration during pregnancy (e.g. Allcroft, Scarnell and Hignett, 1954; Sinclair and Andrews, 1954). Goitre in such cases arises from either an inhibition of tri-iodothyronine production arising from the presence of 5-vinyl 2–thiooxazalidone or related compounds in plants such as the *Brassicae* or from the cyanogenetic glucosides present in some strains of clover and linseed which indirectly inhibit iodine accumulation by the thyroid gland. The nature of cultural and environmental factors which influence the goitrogenicity of these crops is as yet poorly defined. Goitrogens of the thiooxazalidone type readily pass from the bloodstream to the milk of the lactating animal and the milk in consequence possesses goitrogenic potency in some instances where large quantities of kale are fed.

Reproductive failure has been reported in both male and female cattle, sheep and pigs suffering from goitre (Devuyst *et al.*, 1961).

Iodine Requirements

Recent estimates of iodine requirements where the diet does not contain goitrogens are as follows: for the growing chick, 0·35 ppm (A.R.C., 1963), for pregnant and lactating ruminants, 0·8 ppm and for other ruminants, 0·12 ppm (A.R.C., 1965). The requirements of growing and breeding pigs are probably of the same order.

SELENIUM METABOLISM

Since nutritional relationships between selenium and α–tocopherol are being more extensively considered in Chapter 31 in relation to poultry the following discussion will be brief and will be restricted to ruminants and to pigs.

Selenium in Feeding Stuffs

Accurate estimates of the selenium content of common feeding stuffs are scarce because of analytical difficulties. Typical values for herbage plants are: 0·022 ppm *Agrostis* spp., 0·01 ppm *Lolium* spp. and *Dactylis* and 0·008 ppm for clovers.

Much higher concentrations are frequently found in plants growing on seleniferous soils. In this case pasture concentrations up to 100 ppm are by no means uncommon; levels may rise to 700 ppm in *Brassicae* and in accumulator plants such as *Astragalus bisulcatus* 0·2 per cent selenium has been found.

Nearly all the selenium in common feeding stuffs is in organic combination usually in the form of selenium analogues of the sulphur amino acids (see Shrift, 1961).

Selenium Absorption and Excretion

In ruminants there is circumstantial evidence that most forms of dietary selenium are reduced in the rumen and converted to insoluble forms before excretion mainly in the faeces (Cousins and Cairney, 1961). This situation contrasts with that found in non-ruminant species where urinary excretion and loss through the exhalation of dimethyl selenide are the most important excretory pathways.

Selenium Deficiency in Ruminants and Pigs

It is now well established that muscular dystrophy, in ruminants is preventable either by selenium or α–tocopherol administration (see review by Blaxter, 1963). Muscular dystrophy and liver necrosis in pigs resulting from the ingestion of toxic grain are also prevented by these supplements (Lannek *et al.*, 1960). In certain areas probably associated with soil types derived from rocks with a low selenium content the growth of lambs is improved by selenium administration even in the absence of specific signs of deficiency (Hartley and Grant, 1961; Blaxter, 1963).

The aetiology of many of these conditions is obscure in the absence of reliable information on the selenium content of the ration. Two postulates may be advanced to account for these disorders, viz.

(i) that livestock are in some circumstances fed rations of low α–tocopherol potency (as in certain areas of north-east Scotland) or containing tocopherol-binding factors which increase requirements for selenium; or

(ii) that the diet contains organic or inorganic antagonists which prevent efficient utilization of ingested selenium (in work with laboratory animals it has been convincingly shown that sulphate limits the uptake of selenate).

In the absence of further information on the selenium contents of diets in situations where selenium-responsive disorders exist and without a fuller understanding of the nature of selenium/α–tocopherol relationships it is not possible to state the quantitative requirements of any class of livestock for this element. In view of the great toxicity of selenium when present in the diet at concentrations in excess of *ca.* 2 ppm the risks inherent in its use with mineral supplements or fertilizers are too great to be acceptable in commercial practice. For review of the toxic hazards associated with over-ingestion of selenium see Rosenfeld and Beath (1964).

MINERAL TOXICITIES

Work on toxicosis arising from high dietary levels of copper, manganese, iron, zinc fluorine, cobalt, manganese, selenium and nitrates has been adequately reviewed by Comar and Bronner (1960–4), Underwood (1962) and Gallagher (1964).

REFERENCES

AGRICULTURAL RESEARCH COUNCIL (1963) *The Nutrient Requirements of Farm Livestock*, No. 1, *Poultry*, A.R.C.: London.

AGRICULTURAL RESEARCH COUNCIL (1965) *The Nutrient Requirements of Farm Livestock*, No. 2, *Ruminants*, A.R.C.: London.

AGRICULTURAL RESEARCH COUNCIL (1967) *The Nutrient Requirements of Farm Livestock*, No. 3, *Pigs*, A.R.C.: London.

ALDERMAN, G. (1963) *Vet. Rec.* **75**, 1015.

ALLCROFT, R. and PARKER, W. H. (1949) *Br. J. Nutr.* **3**, 205.

ALLCROFT, R., SCARNELL, J. and HIGNETT, S. L. (1954) *Vet. Rec.* **66**, 367.

ANNEGERS, J. H. (1961) *Am. J. Physiol.* **200**, 107.

AURBACH, G. D. and POTTS, J. T. (1964) *Proc. 2nd Intern. Congr. Endocrinology*, London, p. 1260.

BARNES, D. W. H., BISHOP, M., HARRISON, G. E. and SUTTON, A. (1961) *Intern. J. Radiation Biol*, **3**, 637.

BECK, A. B. (1941) *J. Dept. Agric. W. Aust.* **18**, 285.

BECK, A. B. (1961) *Aust. J. Agric. Res.* **17**, 743.

BECKER, N. (1955) *Fütter u. Fütterung*, p. 437.

BEILHARTZ, S. and KAY, R. N. B. (1963) *J. Physiol. Lond.* **165**, 468.
BENNETTS, H. W. and BECK, A. B. (1942) *Austral. Council Sci. Ind. Res. Bull.* **142.**
BENZIE, D., BOYNE, A. W., DALGARNO, A. C., DUCKWORTH, J., HILL, R. and WALKER, D. M. (1955) *J. Agric. Sci. Camb.* **46**, 425.
BENZIE, D., BOYNE, A. W., DALGARNO, A. C., DUCKWORTH, J., HILL, R. and WALKER, D. M. (1956) *J. Agric. Sci. Camb.* **48**, 175.
BENZIE, D., BOYNE, A. W., DALGARNO, A. C., DUCKWORTH, J., HILL, R. and WALKER, D. M. (1959) *J. Agric. Sci. Camb.* **52**, 1.
BENZIE, D., BOYNE, A. W., DALGARNO, A. C., DUCKWORTH, J., HILL, R. and WALKER, D. M. (1960) *J. Agric. Sci. Camb.* **54**, 202.
BLAIR, R. and BENZIE, D. (1964) *Br. J. Nutr.* **18**, 91.
BLAXTER, K. L. (1963) *Br. J. Nutr.* **17**, 105.
BLAXTER, K. L., GRAHAM, N. MCC., WAINMAN, F. W. and ARMSTRONG, D. G. (1959) *J. Agric. Sci. Camb.* **52**, 25.
BLAXTER, K. L., ROOK, J. A. F. and MACDONALD, A. M. (1954) *J. Comp. Path.* **64**, 157.
BLAXTER, K. L., SHARMAN, G. A. M. and MACDONALD, A. M. (1957) *Br. J. Nutr.* **11**, 234.
BORG, D. C. and COTZIAS, G. C. (1958) *J. Clin. Invest.* **37**, 1269.
BOSCH, S., VAN DER GRIFT, J. and HARTMANS, J. (1965) *Versl. Landbouwk. Onderzoek.* **666.**
BOWLAND, J. F., BRAUDE, R., CHAMBERLAIN, A. G., GLASCOCK, R. F. and MITCHELL, R. G. (1961) *Br. J. Nutr.* **15**, 59.
BOYD, O. F., CRUM, C. L. and LYMAN, J. F. (1932) *J. Biol. Chem.* **95**, 29.
BRAUDE, R., CHAMBERLAIN, A. G., KOTARBINSKA, M. and MITCHELL, R. G. (1962) *Br. J. Nutr.* **16**, 427.
BRYANT, F. J. and LOUTIT, J. F. (1961) *U.K.A.E.A. Res. Grp. Rept. R3718*, H.M.S.O.: London.
CARE, A. D. (1965) *Proc. Nutr. Soc.* **24**, 99.
CARE, A. D. and VAN T'KLOOSTER, A. T. (1965) *J. Physiol. (Lond.)* **177**, 174.
CARE, A. D., DUNCAN, T. and WEBSTER, D. (1966) *Endocrinology* (in press).
CARTER, J. H., MILLER, R. F. and BROOKS, C. C. (1959) *J. Anim. Sci.* **18**, 1502.
CARTWRIGHT, G. E., GUBLER, C. J., BUSH, J. A. and WINTROBE, M. M. (1956) *Blood*, **9**, 143.
CHAPMAN, H. L. (1957) *Iowa State Coll. J. Sci.* **31**, 375.
CHAPMAN, H. L., KASTELIC, J., ASHTON, G. C., HOMEYER, P. G., ROBERTS, C. Y., CATRON, D. V., HAYS, V. W. and SPEER, V. C. (1962) *J. Anim. Sci.* **21**, 112.
CHARLTON, R. W., JACOBS, P., TORRANCE, J. D. and BOTHWELL, T. H. (1965) *J. Clin. Invest.* **44**, 543.
CLEGG, R. E., ERICSON, A. T., HEIN, R. E., MCFARLAND, R. H. and LEONARD, G. W. (1956) *J. Biol. Chem.* **219**, 447.
COATES, M. E. and HOLDSWORTH, E. S. (1961) *Br. J. Nutr.* **15**, 131.
CODE, C. F., BASS, P., MCCLARY, G. B., NEWNUM, R. L. and ORVIS, A. L. (1960) *Am. J. Physiol.* **199**, 281.
COGHLAN, J. P., DENTON, D. A., GODING, J. R. and WRIGHT, R. D. (1960) *Postgrad. Med. J.* **36**, 76.
COMAR, C. L. (1950) in: *Copper Metabolism—a Symposium* (W. D. MCELROY and B. GLASS, Editors). Johns Hopkins: Baltimore, p. 191.
COMAR, C. L. and BRONNER, F. (1960–4) *Mineral Metabolism.* Academic Press: New York, **1A**, **1B**, **2A** and **2B**.
COMAR, C. L. and DAVIS, G. K. (1947a) *Arch. Biochem. Biophys.* **12**, 257.
COMAR, C. L. and DAVIS, G. K. (1947b) *J. Biol. Chem.* **170**, 379.
COONS, C. M. (1964) *Ann. Rev. Biochem.* **33**, 459.
COTLOVE, C. and HOGBEN, C. A. M. (1962) in: *Mineral Metabolism* (C. L. COMAR and F. BRONNER, Editors). Academic Press: New York, **2B**, 109.
COUSINS, F. B. and CAIRNEY, I. M. (1961) *Aust. J. agric. Res.* **12**, 927.
CUNNINGHAM, I. J. (1946) *N.Z. J. Sci. Tech.* **27A**, 372 and 381.
CURRAN, P. F. (1965) *Fedn. Proc. Fedn. Am. Socs. Exp. Biol.* **24**, 993.
CURRAN, P. F. and SCHWARTZ, G. F. (1959) *J. Gen. Physiol.* **43**, 555.
DAUPHINEE, J. A. (1962) in: *Clinical Nutrition* (N. JOLIFFE, Editor). Harper: New York, p. 398.

DAVIDSON, J. (1954) *J. Sci. Fd. Agric.* **5**, 86.

DAVIS, G. K. (1963) in: *The Transport of Calcium and Strontium across Biological Membranes* (R. H. WASSERMAN, Editor), Academic Press: New York, p. 129.

DAVIS, P. N., NORRIS, L. C. and KRATZER, F. H. (1962) *J. Nutr.* **77**, 217.

DEIJS, W. B. and BOSMAN, M. S. M. (1961) *Jaarb. Inst. Biol. Scheik. Ond. Landbouwg.* No. 159, p. 93.

DENTON, D. A. (1956) *J. Physiol. Lond.* **131**, 516.

DEVUYST, A., ARNOULD, R., VANBELLE, M., VERVACK, M. and MOREELS, A. (1961) *Agricultura* **9**, 407.

DICK, A. T. (1954) *Aust. J. Agric. Res.* **5**, 511.

DOBSON, A. (1965) in: *Physiology of Digestion in the Ruminant* (R. W. DOUGHERTY, Editor), London: Butterworth, p. 88.

DUCKWORTH, J. and GODDEN, W. (1943) *Biochem. J.* **37**, 595.

DUCKWORTH, J. and HILL, R. (1953) *Nutr. Abst. Rev.* **23**, 1.

EASTOE, J. E. (1961) in: *Biochemist's Handbook* (C. LONG, Editor), E. F. Spon: London, p. 715.

EVANS, J. V. and KING, J. W. B. (1955) *Nature, Lond.* **176**, 171.

EWING, R. (1963) *Poultry Nutrition*, Ewing: Pasadena.

FARNWORTH, A. J. (1956) *Aust. J. Appl. Sci.* **7**, 233.

FEASTER, J. P., HANSARD, S. L., MCCALL, J. T., SKIPPER, F. H. and DAVIS, G. K. (1954) *J. Anim. Sci.* **13**, 781.

FELL, B. F., MILLS, C. F. and BOYNE, R. (1965) *Res. vet. Sci.* **6**, 170.

FIELD, A. C. (1961) *Br. J. Nutr.* **15**, 349.

FIELD, H., SWELL, L., DAILEY, R. E., TROUT, E. C. and BOYD, R. S. (1955) *Circulation* **12**, 625.

FISCHER, E. H. and STEIN, E. A. (1960) in: *The Enzymes* (P. D. BOYER, H. LARDY and K. MYRBACK, Editors). Academic Press: New York, **4A**, 336.

FORBES, R. M. (1960) *Fedn. Proc. Fedn. Am. Socs. Exp. Biol.* **19**, 643.

FRANKENHEUSER, B. and HODGKIN, A. L. (1957) *J. Physiol. Lond.* **137**, 217.

GALLAGHER, C. H. (1964) *Nutritional Factors and Enzymological Disturbances in Animals.* Crosby Lockwood: London.

GAMBLE, A. H., WISE, H. F. and HANSEN, A. E. (1948) *Pediatrics*, **1**, 58.

GARNER, R. J., JONES, H. G. and SANSOM, B. F. (1960) *Biochem. J.* **76**, 572.

GEORGI, J. R. (1964) *Am. J. Vet. Res.* **25**, 952.

GRANICK, S. (1946) *Science*, **103**, 107.

GRANICK, S. (1958) in: *Trace Elements*, Proc. Conf. Wooster, Ohio (C. A. LAMB, O. G. BENTLEY and J. M. BEATTIE, Editors). Academic Press: New York.

GRASHUIS, J. (1963) *Landbouwk. Tijdschr.* **75**, 1127.

HAARANEN, S. (1963) *Feedstuffs*, March 1963, p. 17.

HANSARD, S. L., COMAR, C. L. and PLUMLEE, M. P. (1954) *J. Anim. Sci.* **13**, 25.

HANSARD, S. L. and FOOTE, L. E. (1961) *Fedn. Proc. Fedn. Am. Socs. Exp. Biol.* **20**, 296.

HANSARD, S. L. and HOSKINS, F. H. (1963) *Proc. 6th Int. Congr. Nutrition, Edinburgh* (C. F. MILLS and R. PASSMORE, Editors), p. 143.

HANSARD, S. L., LYKE, W. A. and CROWDER, H. M. (1961) *J. Anim Sci.* **20**, 292.

HARRISON, H. E. and HARRISON, H. C. (1960) *Am. J. Physiol.* **199**, 265.

HARTLEY, W. J. and GRANT, A. B. (1961) *Fedn. Proc. Fedn. Am. Socs. Exp. Biol.* **20**, 679.

HEAD, M. J. and ROOK, J. A. F. (1957) *Proc. Nutr. Soc.* **16**, 25.

HIGNETT, S. L. (1956) *Proc. Int. Cong. Animal Reproduction* (Cambridge).

HILL, C. H. and MATRONE, G. (1958) *Poult. Sci.* **37**, 1211.

HILL, R. (1965a) *Br. J. Nutr.* **19**, 171.

HILL, R. (1965b) *Br. J. Nutr.* **19**, 163.

HINE, D. C. and DAWBARN, M. C. (1954) *Aust. J. Exp. Biol. Med. Sci.* **32**, 641.

HIRSCH, P. F., GAUTHER, G. F. and MINSON, P. F. (1963) *Endocrinology* **73**, 244.

HOEKSTRA, W. G., POPE, A. L. and PHILLIPS, P. H. (1952) *J. Nutr.* **48**, 421.

HØFF-JORGENSEN, E. (1946a) *Biochem. J.* **40**, 189.

HØFF-JORGENSEN, E. (1946b) *Biochem. J.* **40**, 555.

HOWELL, J. MCC. and DAVISON, A. N. (1959) *Biochem. J.* **72**, 365.

HSU, J. M. (1963) *Fedn. Proc. Fedn. Am. Socs. Exp. Biol.* **22**, 378.
HUGHES, E. H. and ITTNER, N. R. (1942) *J. Agric. Res.* **64**, 189.
INNES, J. R. M. and SHEARER, G. D. (1940) *J. comp. Path.* **53**, 1.
IRVING, J. T. (1964) in: *Mineral Metabolism* (C. L. COMAR and F. BRONNER, Editors). Academic Press: New York, **2A** 249.
KAWIN, B. and PALMER, R. F. (1958) *Nature, Lond.* **181**, 127.
KEMP, A. and T'HART, M. L. (1958) *Netherl. J. Agric. Sci.* **5**, 4.
KIRCHGESSNER, M. (1959) *Z. Tierphysiol. Tierernähr. Futtermittelk.* **14**, 270.
KIRCHGESSNER, M. (1962) *Int. Z. VitaminForsch.* **32**, 1415.
KOCSIS, J. J., WALASZEK, E. J., GRAHAM, C. E. and GEILING, E. M. R. (1953) *Fedn. Proc. Fedn. Am. Socs. Exp. Biol.* **12**, 336.
LAHEY, M. E., GUBLER, C. J., CHASE, M. S., CARTWRIGHT, G. F. and WINTROBE, M. M. (1952) *Blood*, **7**, 1053.
LANNEK, N., LINDBERG, P., NILSSON, G., NORDSTRÖM, G., ORSTADIUS, K. and ABERG, B. (1960) *Zentbl. VetMed.* **7**, 403.
LEACH, R. M., DAM, R., ZEIGLER, T. R. and NORRIS, L. C. (1959) *J. Nutr.* **68**, 89.
LEACH, R. M. and MUENSTER, A. M. (1962) *J. Nutr.* **78**, 51.
LEGG, S. P. and SEARS, L. (1960) *Nature, Lond.* **186**, 1061.
LESTER-SMITH, E. (1962) in: *Mineral Metabolism* (C. L. COMAR and F. BRONNER, Editors). Academic Press: New York, **2B**, 349.
LINKENHEIMER, W. H., PATTERSON, E. L., MILSTREY, R. A., BROCKMAN, J. A. and JOHNSON, D. D. (1960) *J. Anim. Sci.* **19**, 763.
LITTLEJOHN, A. I. and LEWIS, G. (1960) *Vet. Rec.* **72**, 1137.
LOFGREEN, G. P. (1960) *J. Nutr.* **70**, 58.
LUEKER, C. E. and LOFGREEN, G. P. (1961) *J. Nutr.* **74**, 233.
LUXORO, M., ROJAS, E. and WITTIG, E. (1963) *J. Gen. Physiol.* **46**, 1109.
MCCANCE, R. A. and WIDDOWSON, E. M. (1942–3) *J. Physiol. Lond.* **101**, 44.
MCCANCE, R. A. and WIDDOWSON, E. M. (1960) *The Composition of Foods*, H.M.S.O.: London.
MCROBERTS, M. R., HILL, R. and DALGARNO, A. C. (1965 a, b, c) *J. Agric. Sci. Camb.* **65**, 1, 11, 15.
MCTAGGART, H. S. (1959) *Vet. Rec.* **71**, 709.
MCWARD, G. W. and SCOTT, H. M. (1961) *Poult. Sci.* **40**, 1026.
MAHONEY, J. P., BUSH, J. A., GUBLER, C. J., MORETZ, W. H., CARTWRIGHT, G. E. and WINTROBE, M. M. (1955) *J. Lab. Clin. Med.* **46**, 702.
MANER, J. H. and LOOSLI, J. K. (1960) *J. Anim. Sci.* **19**, 1286.
MANSTON, R. (1964) *Br. Vet. J.* **120**, 365.
MARSTON, H. R. (1952) *Physiol. Rev.* **32**, 66.
MARSTON, H. R., ALLEN, S. W. and SMITH, R. M. (1961) *Nature, Lond.* **190**, 1085.
MATRONE, G. (1960) *Fedn. Proc. Fedn. Am. Socs. Exp. Biol.* **19**, 659.
MAYO, R. H., PLUMLEE, M. P. and BEESON, W. M. (1959) *J. Anim. Sci.* **18**, 264.
MELLANDER, O. (1950) *Acta Soc. Med. Upsaliensis* **55**, 247.
MILHAUD, G., PERAULT, A. M. and MOUKTAR, M. S. (1963) *C.R. Nebd. Séance. Acad. Sci., Paris*, **261**, 813.
MILLS, C. F. (1954) *Biochem. J.* **57**, 603.
MILLS, C. F. (1960) *Proc. Nutr. Soc.* **19**, 162.
MILLS, C. F., DALGARNO, A. C., WILLIAMS, R. B. and QUARTERMAN, J. (1966) *Proc. Nutr. Soc.* **25**, xxi.
MILLS, C. F., QUARTERMAN, J., WILLIAMS, R. B. and DALGARNO, A. C. (1965) *Proc. Nutr. Soc.* **24**, xxi.
MILLS, C. F. and WILLIAMS, R. B. (1962) *Biochem. J.* **85**, 629.
MITCHELL, H. H. (1947) *J. Anim. Sci.* **6**, 365.
MITCHELL, H. H. and KELLEY, M. A. R. (1938) *J. Agric. Res.* **56**, 811.
MOLINA, G., FARAH, A. and KRUSE, R. (1963) *Am. J. Physiol.* **204**, 531.
MOODIE, E. W. and ROBERTSON, A. (1962) *Res. Vet. Sci.* **3**, 470.
MOORE, L. A., HALLMAN, E. T. and SCHOLL, L. B. (1938) *Arch. Path.* **26**, 820.
MOORE, J. H. and TYLER, C. (1955) *Br. J. Nutr.* **9**, 81.

MORGAN, C. L., MITCHELL, H. H., RINGROSE, R. C. and LEASE, E. J. (1942) *Poult. Sci.* **21**, 212.
MUNRO, J B. (1957) *Vet. Rec.* **69**, 125.
MURTY, V. N. (1957) *J. Sci. indust. Res.* **16C**, 120.
NEWBURGH, L. H. and JOHNSTON, M. W. (1942) *Physiol. Rev.* **22**, 1.
NEWLAND, H. W., ULLREY, D. E., HOEFFER, J. A. and LUECKE, R. W. (1958) *J. Anim. Sci.* **17**, 886.
OBERLEASE, D., MUHRER, M. E., O'DELL, B. L. and KINTNER, L. D. (1961) *J. Anim. Sci.* **20**, 945.
O'DELL, B. L. (1960) *Fedn. Proc. Fedn. Am. Socs. Exp. Biol.* **19**, 648.
O'HALLERAN, M. W. and SKERMAN, K. D. (1961) *Br. J. Nutr.* **15**, 99.
PARR, W. H. (1957) *Vet. Rec.* **69**, 71.
PARSONS, D. S. and WINGATE, D. L. (1961) *Biochim. Biophys. Acta* **46**, 170.
PHILLIPSON, A. T. and MITCHELL, R. L. (1952) *Br. J. Nutr.* **6**, 176.
PILEGGI, V. J., DE LUCA, H. F. and STEENBOCK, H. (1955) *Arch. Biochem. Biophys.* **58**, 194.
PLUMLEE, M. P., THRASHER, D. M., BEESON, W. M., ANDREWS, F. N. and PARKER, W. H. (1956) *J. Anim. Sci.* **15**, 352.
POND, W. G., LOWRY, R. S., MANER, J. H. and LOOSLI, J. K. (1960) *J. Anim. Sci.* **19**, 1286.
RASHEED, M. A. and SEELEY, R. C. (1966) *Nature, Lond.* (in press).
ROGLER, J. C., PARKER, H. E., ANDREWS, F. N. and CARRICK, C. W. (1961) *Poult. Sci.* **40**, 1554.
ROSENFELD, I. and BEATH, O. A. (1964) *Selenium-Geobotany, Biochemistry, Toxicity and Nutrition.* Academic Press: New York.
SAMMONS, H. G. (1961) *Biochem. J.* **80**, 30P.
SCHACHTER, D., DOWDLE, E. B. and SCHENKER, H. (1960) *Am. J. Physiol.* **198**, 263, 275.
SCHAIBLE, P. J., BANDEMER, S. L. and DAVIDSON, J. A. (1938) *Mich. Agric. Exp. Sta. Tech. Bull.* No. 159.
SCHMIDT-NIELSEN, K. (1964) *Desert Animals—Physiological Problems of Heat and Water.* Clarenden Press: Oxford.
SHRIFT, A. (1961) *Fedn. Proc. Fedn. Am. Socs. Exp. Biol.* **20**, 695.
SINCLAIR, D. P. and ANDREWS, E. D. (1954) *N.Z. Vet. J.* 1954, p. 72.
SMITH, R. H. (1962) *Biochem. J.* **83**, 151.
SMITH, R. M. and MONTY, K. J. (1959) *Biochem. Biophys. Res. Comm.* **1**, 105.
SMITH, W. H., PLUMLEE, M. P. and BEESON, W. M. (1961) *J. Anim. Sci.* **20**, 128.
STEENBOCK, H. and BUNKFELDT, R. (1951) *Arch. Biochem. Biophys.* **32**, 309.
STORRY, J. E. (1961a) *J. Agric. Sci. Camb.* **57**, 97; *Nature, Lond.* **190**, 1197.
STORRY, J. E. (1961b) *J. Agric. Sci. Camb.* **57**, 1.
SUPPLEE, W. C., COMBS, G. F. and BLAMBERG, D. L. (1958) *Poult. Sci.* **37**, 63.
SUTTLE, N. F. and FIELD, A. C. (1966) *Proc. Nutr. Soc.* **25**, xx.
SUTTLE, N. F. and MILLS, C. F. (1966) *Br. J. Nutr.* **20**, 135, 149.
SYKES, J. F. and ALFREDSON, B. V. (1940) *Proc. Soc. Exp. Biol. N.Y.* **43**, 575.
TELFER, S. V. (1922–3) *Quart. J. Med.* **16**, 45.
THIELER, A., GREEN, H. H. and DU TOIT, P. J. (1927) *J. Agric. Sci. Camb.* **17**, 291.
TIDBALL, C. S. (1961) *Am. J. Physiol.* **200**, 305.
TILLMAN, A. D. and BRETHOUR, J. R. (1958) *J. Anim. Sci.* **17**, 104.
TODD, J. R., LAWSON, G. H. K. and DOW, C. (1964) *J. Comp. Path.* **74**, 331.
TOSIC, J. and MITCHELL, R. L. (1948) *Nature, Lond.* **162**, 502.
ULLREY, D. E., MILLER, E. R., THOMSON, O. A., ZUTANT, C. L., SCHMIDT, D. A., RITCHIE, H. D., HOEFFER, J. A. and LUECKE, R. W. (1960a) *J. Anim. Sci.* **19**, 1298.
ULLREY, D. E., MILLER, E. R., THOMPSON, O. H., ACKERMANN, I. M., SCHMIDT, D. A., HOEFFER, J. A. and LUECKE, R. W. (1960b) *J. Nutr.* **70**, 187.
UNDERWOOD, E. J. (1962) *Trace Elements in Human and Animal Nutrition.* Academic Press: New York.
VALLEE, B. L. (1962) in: *Mineral Metabolism* (C. L. COMAR and F. BRONNER, Editors). Academic Press: New York, **2B**, 443.
VAUGHAN, B. E. (1960) *Am. J. Physiol.* **198**, 1235.
VISEK, W. J., MONROE, R. A., SWANSON, E. W. and COMAR, C. L. (1953) *J. Nutr.* **43**, 469.

VITALE, J. J., NAKAMURA, M. and HEGSTED, D. M. (1957) *J. Biol. Chem.* **228**, 573.
WACK, J. P. and WYATT, J. P. (1959) *Arch. Path.* **67**, 237.
WACKER, W. E. C., MOORE, F. D., ULLMER, D. D. and VALLEE, B. L. (1962) *J. Am. Med. Ass.* **180**, 161.
WALSER, M. (1961) *J. Clin. Invest.* **40**, 723.
WEBLING, D. D. and HOLDSWORTH, E. S. (1965) *Biochem. J.* **97**, 408.
WIDDOWSON, E. M. and DICKERSON, J. W. T. (1964) in: *Mineral Metabolism* (C. L. COMAR and F. BRONNER, Editors). Academic Press: New York, **2A**, 2.
WILDE, W. S. (1962) in: *Mineral Metabolism* (C. L. COMAR and F. BRONNER, Editors). Academic Press: New York, **2B**, 73.
WILSON, T. H. (1954) *Biochem. J.* **56**, 527.
WINCHESTER, C. F. and MORRIS, M. J. (1956) *J. Anim. Sci.* **15**, 722.
WINTROBE, M. M., CARTWRIGHT, G. E. and GUBLER, C. J. (1953) *J. Nutr.* **50**, 395.
WISE, M. B., ORDOVEZA, A. L. and BARRICK, E. R. (1963) *J. Nutr.* **79**, 79.
WISEMAN, G. (1964) *Absorption from the Intestine*. Academic Press: London.
WOLFF, H. (1956) *Klin. Wschr.* **34**, 409.

CHAPTER 12

Nutrition and Reproduction

G. E. LAMMING

University of Nottingham, School of Agriculture, Sutton Bonington, Loughborough, Leicestershire

INTRODUCTION

THE physical and economic productivity of farm species depends on the maintenance of an optimum level of reproductive activity. Since the nutritional status of the individual animal and of populations has a profound effect, both direct and indirect, on their reproductive processes, studies on the effects of nutrition on reproduction are of singular importance. This is particularly so with present trends of intensive animal production, which give rise to a situation where maximum reproductive performance is essential and yet the necessary population density increases competition for food, and may result in a decline in reproductive capacity.

Much literature has accumulated from observations on the various effects that nutrition has on man and other vertebrates. This chapter is concerned with the review of effects of inadequate nutrition on reproduction with particular reference to farm species. However, evidence gained from experiments in laboratory animals form such an integral part of knowledge on this topic that it cannot be divorced from our deliberations.

The term 'inadequate nutrition' must be defined at the outset since it can involve at least four distinct aspects. It commonly involves inanition, that is a shortage of food, or, less frequently, an excess intake of a balanced diet leading to obesity, both causing suboptimal reproductive performance. Alternatively, inadequate nutrition can involve an excess or lack of a particular nutrient which might normally be required in macro- or micro-quantities.

THE GENERAL EFFECTS OF NUTRITION ON REPRODUCTION

Poor nutrition affects all organ systems of the mammalian body. In the normal well-nourished animal the various organ systems function in an integrated fashion, although not necessarily at an optimum level for each system. For example, there is evidence that while a high intake of nutrients may provide an optimal rate of structural growth such a high food intake shortens reproductive life in laboratory animals and farm species. In the vast majority of cases, poor reproductive performance is due to a shortage of nutrients or lack of 'balance' within the diet.

In a state of optimal nutrition a partition of nutrients occurs which provides adequate quantities of substrate for those reactions controlling the various functions of the body. When nutrient intake declines below the optimum level, the different systems of the body are not all similarly affected and there are breed and sex differences in response. In general, as Hammond (1944) postulated, the reproductive system has a low priority for available nutrients under these conditions. However, the interval before signs of dysfunction appear varies with the circumstances, and reproductive performance is the first physiological function to decline, particularly in the female. For example, acute starvation over short intervals can result in anoestrus in mature rats before a large loss in liveweight occurs, while similar effects occur in chronic inanition in rats only after a body weight loss varying from 20 to 30 per cent (Krause, 1963). On the other hand, the mature male continues to reproduce satisfactorily throughout stages of weight loss in excess of this level, and libido and gametogenesis persist in the male to late stages of chronic inanition.

It has frequently been claimed that once pregnancy has been established the pregnant female will seriously deplete body reserves of energy and protein to maintain the pregnancy. The nutrient requirements for pregnancy are low and while the pregnant female may maintain liveweight, changes in body composition occur which might mask a loss in body energy and protein. In addition, the pregnant condition involves such an important change in the endocrine balance of the animal as well as providing an additional endocrine gland in the form of the placenta, that any differences in response might be explained by changes in the levels of circulating hormones, so the pregnant animal must be treated as a special case in relation to nutrition.

The effects of the plane of nutrition on reproduction may be defined into three categories, each with a variety of different effects. These may be defined as direct effects and indirect effects, the latter subdivided into major and minor influences.

The Direct Effects of Nutrition on the Reproductive System

These effects involve the primary and secondary sex organs, their individual functions and interrelationships, and also those endocrine glands (i.e. the anterior and posterior pituitary glands) which are directly concerned with the hormonal control of reproduction.

The Major Indirect Effects of Nutrition on Reproduction

These effects are concerned with the influence of nutrition on organ systems not primarily concerned with reproduction, but which, nevertheless, have functional relationships with the reproductive process, e.g. the nervous system and the endocrine system, other than the pituitary-gonad axis.

(i) The nervous system is involved in neurohumoral responses which control the various proccsses in reproduction. For example, there is considerable interest at the present time in hypothalamic control of hypophysial function, particularly in relation to the occurrence of oestrus and the timing of ovulation.

(ii) The endocrine system is intimately involved in reproduction and this includes endocrine glands other than the anterior pituitary-gonad axis. For example, nutrition influences the activity of the thyroid gland and the adrenal cortex, which could indirectly affect reproduction via changes in the level of hormone secretion.

Other Indirect Effects of Nutrition on Reproduction

Some indirect effects of inadequate nutrition influence organ systems more removed from the reproductive process, but these influences are nevertheless important; for example:

(i) Effects of nutrition on homeostatic regulation of appetite, leading to hypophagea and, therefore, a more direct influence on reproduction.

(ii) Effect of nutrition on muscular activity, e.g. producing dystocia at parturition, thus increasing the incidence of perinatal mortality.

(iii) Effect of the balance of nutrients or the level of intake on the incidence of metabolic disorders, e.g. obesity or starvation causing pregnancy toxaemia in sheep, thus leading to foetal death and abortion.

(iv) Effect of poor nutrition on subsequent lactation increasing post natal losses and, therefore, causing suboptimal reproductive performance.

(v) An effect of nutrition on homeostatic regulation of body temperature, e.g. loss of wool in sheep due to poor nutrition leading to an increased energy expenditure to maintain body temperature and therefore a more direct effect on reproduction. Supporting evidence is gained for this effect from observations that the maintenance requirements for sheep of similar weights varies, depending on whether they are housed indoors or whether they are maintained without shelter but with limited or unlimited exercise.

Obviously the environmental temperature must influence the energy expended for homeostasis, therefore animals of similar weight but under different conditions would require different proportions of their energy intake for those processes necessary for the survival of the individual. The environmental conditions would, therefore, influence the ability of a particular intake to support reproduction. This factor is particularly important in farm experiments in which studies are made of the influence of nutrient lack on reproduction. In the published reports of such work exact details of the environment and climatic conditions must be recorded.

Naturally the material discussed in this chapter will be concerned mainly with the first category (direct effects) with citation of pertinent literature from the second. We shall not deal with the third, except to draw the reader's attention to the importance of these indirect effects on the attainment of optimal reproductive performance.

The direct and some indirect effects of nutrition on reproduction can be illustrated in Fig. 1. An optimal state of nutrition involves the satisfaction of appetite and the absence of hunger. Psychological stresses in the well-nourished animal are known to influence the reproductive process, and hunger is a stress which farm animals may suffer, particularly where competition for available nutrients is increased by housing large numbers of animals intensively. Energy expenditure by animals in searching for food and water in times of shortage will lead to the diversion of nutrients away from the normal support of gonadal activity.

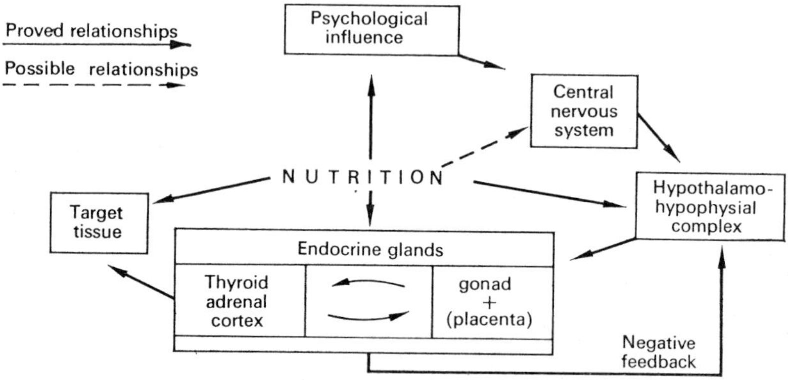

FIG. 1. Diagrammatic representation of some influences of nutrition on reproduction.

Many of the relationships shown between nutrition and reproduction have been determined by the use of rather crude methods of hormone assay and there is still much work to be completed before all the implications shown in Fig. 1 can be accepted without question, and especially those aspects involving the influence of nutrition on gonadotrophin secretion and release.

Although progress has been made in chemical and biological methods of assessing rates of secretion, distribution and catabolism of hormones of the thyroid, the adrenals and the gonads, present methods are not sufficiently accurate or sensitive to measure changes in the levels of circulating pituitary gonadotrophins in the intact animal. Since the effect of nutrient lack is generally to cause hyposecretion this increases the difficulties of accurate direct measurement of the effects of nutritional status on the rates of secretion of pituitary hormones, yet such information is essential for understanding the influence of nutrition on reproduction, or of clarifying or extending Fig. 1. However, new methods of assessing the levels of circulating hormones are now being developed which promise improvement in sensitivity, specificity and accuracy over present methods.

NUTRITION AND SEXUAL DEVELOPMENT IN THE GROWING ANIMAL

A high rate of growth in both laboratory and farm animals has frequently been associated with an early occurrence of puberty. Female rats from small litters (Engle, Crofts and Zeithanil, 1937) or from fast growing strains (Blunn, 1939) reached puberty earlier, while energy restriction delayed sexual maturity in rats (Conte Marotte, 1937; Scheer, Soule, Fields and Deuel, 1947), in guinea pigs (Bortolucci and Zamborini, 1934) and in rabbits (Adams, 1953; Campos Zamorano, 1955). The date of vaginal opening in rats was delayed by restricting the calorie or protein intake (Goettsch, 1949). Adams (1953) showed that rabbits reared on a high plane of nutrition responded to gonadotrophin injections at 12 weeks of age while litter sisters reared on a lower plane failed to do so, even though injected with larger quantities of hormone. A short period of increased food intake by low plane animals quickly resulted in an ovarian response similar to that of high plane controls.

The general effects observed in laboratory animals are also observed in farm species. Hammond, Jnr. (1944) has shown that nutrition is an important factor in determining whether or not a lamb experiences oestrus during the first winter and these observations were later confirmed by Hafez (1951, 1952) who postulated that a threshold of liveweight occurred below which ewe lambs would not mate during their first breeding season. Single lambs (with presumably a higher plane of nutrition) reached sexual maturity earlier than twins (Williams, 1954; Mourib, Ahmed and Hamada, 1956) and the weight gain made by ewe lambs mated at this time influences their current and subsequent breeding efficiency (Roberts, Davies and Williams, 1949; Davies, 1950, 1954; Esplin, Madsen and Phillips, 1940) and effects the incidence of barrenness and twinning (Coop and Clark, 1955).

Ewe lambs receiving a high plane of nutrition before puberty, have larger reproductive tracts in relation to body weight and a greater proportion of multiple ovulations (El Sheikh, Hulet, Pope and Casida, 1955; Foote, Pope,

Chapman and Casida, 1957), although a relatively higher intake of nutrients (flushing) does not affect ewe lambs as markedly as older sheep (Williams, 1954; Allen and Lamming, 1961b).

Similarly, with growing cattle, the plane of nutrition influences the onset of puberty, delays occurring following restriction of energy, protein (Reid, Trimberger, Asdell, Turk and Smith, 1951) or phosphorus (Palmer, Gullickson, Boyd, Fitch and Nelson, 1940) although as with ewe lambs (Allen and Lamming, 1961) and pigs (Burger, 1952) animals on restricted intake experienced first oestrus at lower liveweights (Sorensen, Bratton, Hansel and Hough, 1954). Similarly, Joubert (1954a,b) working in South Africa showed that under conditions of low-plane feeding, puberty was usually delayed but could be advanced nine months by supplementary feeding. Delayed occurrence of first heat and winter anoestrus in poorly-fed heifers has been widely noted (Quinlan, 1929; Allen, 1943; Joubert, *loc. cit* Durell, 1955).

In pigs severe restriction of energy intake delays sexual maturity and produces a lower ovulation rate (Haines, Warnick and Wallace, 1959; Goode, Warnick and Wallace, 1960) although the results obtained by different authors have been varied, and experimental conditions have not always been precisely defined or controlled.

Unrestricted intakes by growing gilts have been shown to cause a delay in puberty in some experiments (Christian and Nofziger, 1952; Self, Grummer and Casida, 1955) but not in others where full-fed gilts reached puberty earlier than controls (Robertson, Casida, Grummer and Chapman, 1951). *Ad libitum* intake of energy or of diets high in protein may delay puberty, but an early onset of heat with a concomitant high ovulation rate can be achieved by rearing pigs on a moderate plane of nutrition (i.e. 70 per cent of *ad libitum* intake), changing to a higher intake as sexual activity becomes apparent (Self *et al.*, 1955).

In all species studied the development of the reproductive tract is accelerated by a liberal nutrient allowance but there is little evidence that full recovery at least of potential reproductive activity cannot be achieved in undernourished animals, providing they subsequently receive an adequate plane of nutrition. McCance (1960), McCance and Widdowson (1962), Dickerson, Gresham and McCance (1964) and Widdowson, Mavor and McCance (1964) have used both pigs and rats to study the effects of prolonged undernutrition on the reproductive organs of the immature animal, and the response which can be obtained to rehabilitation. In severely undernourished male animals, the testes continued to enlarge, with the seminiferous tubules appearing normal, but the interstitial cells regressed and became unrecognizable. In the female the reproductive tract is enlarged and although the primordial follicles in the ovaries developed, ovulation did not occur and no corpora lutea were found. Refeeding produced an initial exaggeration in growth of the gonads but this gradually declined until the development of the reproductive tracts in these animals resembled those which had been fed normally. The results of a recent study by Van Demark and Mauger (1964) do not entirely agree with the above work in that their results indicate that as far as the bovine is concerned a lowered energy intake impaired the growth of the body, the endocrine glands, and the reproductive tract of the young animals. In their work, refeeding corrected the growth rate but did not result in complete recovery in the size and function of the testes. It should be emphasized that the work of McCance and his colleagues involved conditions more severe than those generally adopted and species differences in response to underfeeding and rehabilitation after severe undernutrition might well be expected.

PLANE OF NUTRITION AND REPRODUCTIVE PERFORMANCE IN THE MATURE ANIMAL

In the young animal puberty occurs normally while the animal is still making structural growth, provided that the dietary allowance is adequate. Where the diet fed to such animals after puberty is restricted, reproductive activity declines or ceases some time before somatic growth ceases. In natural environments there are seasonal depressions in nutrient intake which affect animals of all ages. However, the mature animal is more resistant than the immature animal to dietary restriction, especially if the onset is gradual so that acclimatization to a lower food intake may occur, and oestrous activity frequently continues until a major reduction in body weight has occurred. Thereafter ovulation may take place without the accompanying signs of oestrus. When comparing the results of different experiments where the weight losses at which anoestrus occurred were recorded, it is important to note the initial weight and condition of the experimental animals, for this is a most important criterion and the results obtained should be interpreted in the light of that information.

Dietary restriction of rats caused a body weight loss before anoestrus occurred of between 20 (Krause, 1963) and 50 per cent (Werner, 1939). Similar results have been recorded in mice and guinea pigs, the effects being more severe in younger animals (White, White, Mider, Kelley and Heston, 1944; Loeb, 1917, 1921). Similarly, reduction of protein intake below critical levels caused irregularity or failure of oestrous cycles (Guilbert and Goss, 1932; Arvy, Aschkenasy, Aschkenasy-Lelu and Gabe, 1946; Aschkenasy-Lelu and Aschkenasy, 1947.) All these effects are quickly reversed following realimentation of energy in laboratory animals (Scott and Scott, 1958; Ball, Barnes and Visscher, 1947; Macomber, 1933; Stephens and Allen, 1941; Krause, 1963) and in farm animals (Dickerson *et al.*, 1964; Durrell, 1951; Hafez, 1952b) although the weight gain before oestrous cycles reappeared varied between experiments. In general, oestrus occurred before the full weight had been regained. Similar results have been obtained following realimentation of protein-deficient animals (Guilbert and Goss, 1932).

Coop (1962) has studied the influence of liveweight and nutrition prior to mating on the fertility of sheep as measured by number of lambs at parturition. Coop's data were based on observations over fifteen years on a total of over eleven thousand sheep, and he showed that ewe and lamb mortality (3·8 and 12 per cent respectively) were independent of liveweight at mating, except at liveweights below 100 lb. However, he claimed that the incidence of twinning increased approximately linearly at 6 per cent per 10 lb increase in liveweight, a trend supported from slaughter data by Averill (1964) and from observations by Wallace (1958). However, Coop claimed that the poorer performance of young ewes compared to older animals could be explained almost entirely due to differences in liveweight at mating, although others are not prepared to agree at the present time. Most would agree that liveweight *per se* at mating can account for a considerable proportion of the variation of ovulation rate in sheep, and these observations are of importance in determining the economic value of 'flushing', the procedure of increasing the plane of nutrition prior to mating, a commercial practice used for many years, especially for sheep (Heape, 1899; Marshall, 1905; White and Roberts, 1922; Nichols, 1926; Clark, 1934; Underwood and Shier, 1941; Darroch, 1950; Foote *et al.*, 1957, 1959a, b). El Sheikh *et al.* (1955) showed that higher nutrient intakes increase ovulation rate and the number of follicles over 2 mm in diameter, a result confirmed by Allen and Lamming (1961a) who concluded that ovulation rate appeared to be influenced strongly by the level of nutrients available from the diet or from body reserves. When fat ewes were fed submaintenance diets there was no decline in ovulation rate while the animals had adequate body reserves. As the period of sub-maintenance feeding continued there was a fall in ovulation rate which could be restored by injections of gonadotrophins as Pregnant Mares' Serum. Feeding a submaintenance diet to mature sheep during the breeding season resulted in longer intervals between oestrus and eventually in ovulation without oestrus (Hafez, 1952a; Allen and Lamming, 1961a). More recently similar results on the effect of low plane feeding have been obtained with pigs (Heap, Lodge and Lamming, 1966).

NUTRITION AND REPRODUCTION IN THE FEMALE

The Anterior Pituitary Gland, the Gonads and their Functional Relationships

Various aspects of science may be more completely understood if the components are separated and examined independently. Indeed, progress in some branches of science may be measured by the detail in which each facet of a field is divided and assessed. In reproductive physiology such a division is meaningless if applied to the functions of the gonads, for inevitably the pituitary gland must also be considered and this applies particularly to the influence of nutrition on the activity of these endocrine glands (cf. Lamming, 1966, a review on nutrition and the endocrine system). Indeed, those instances where nutrition appears to influence the gonads independently of an effect on the anterior pituitary gland are so exceptional as to require re-examination.

As a general rule, nutritional influences on the activity of the gonads are mediated directly or indirectly via the pituitary gland and it is appropriate therefore to consider the pituitary-ovarian system as a functional inter-relationship which is readily affected by level of nutrition, rather than try to discuss the influence of nutrition separately on the gonads and pituitary gland.

Evidence about the effects of nutrition on the structure and secretory function of the anterior pituitary gland is conflicting but the subject was extensively and critically reviewed by Ershoff (1952). Little information has been added since, but later reviews have been prepared by Leatham (1958, 1961).

Inevitably, early observations on the influence of nutrition on hypophysial and gonadal function were based on circumstantial evidence gained from studies where the main interest was the influence of the plane of nutrition, or a lack or excess of specific nutrients on the processes of reproduction, lactation and growth. Usually lack of facilities or of suitable techniques limited the number of detailed observations of gonadal function which could be undertaken during such studies. In addition many of the observations on the influence of nutrition on secretory activity of the hypophysial-gonad axis were based on the results of a direct implantation of whole pituitary glands from experimental animals into test animals with subsequent measurement of the response of a target organ, and much of this evidence was obtained before the development of suitable statistical techniques, which can be applied to the more sensitive bio-assay procedures now available. Thus, many of the earlier observations and the theories derived from their examination must be discarded or treated with considerable reserve. Much of this earlier work was concerned with hypophysial function without adequate consideration of gonad activity in the same undernourished animal. Many of these studies (including those in our own laboratory) have been made on animals slaughtered by conventional methods without an adequate knowledge that the level of gonadotrophins in the pituitary glands represents the true state of affairs. Evidence derived from measurements of pituitary glands obtained after death is difficult to interpret and must be accompanied by evidence on the structure and activity of target organs of those animals subjected to nutritional experiments. Under these stricter conditions, many of the theories put forward by earlier workers must be discarded for lack of proof. What is really required are accurate determinations of circulating levels of hormones within the body of undernourished animals made serially under circumstances as near normal as possible.

In spite of the advances made in techniques for studying levels of hormones in glands obtained at autopsy, accurate routine methods of measuring the low values in the blood of intact animals are still in the process of development.

An exception to this general comment about the validity of previous theories must be made for the work of Mulinos and Pomerantz (1940) who noted the similarity between the effects of under-nutrition and of hypophysectomy and adopted the term 'pseudohypophysectomy' to cover the endocrinological syndrome of undernutrition. Mulinos and Pomerantz (1940, 1941) noted the general effects of nutrition on the structure and activity of the gonads and concluded that these effects were mediated through the pituitary gland; their work represents a landmark within this field. It is now known that inanition, protein deficiency, essential fatty acid deficiency, avitaminosis and lack of minerals all cause a lowered production or secretion of gonadotrophins which in turn impairs gonadal function. Thus when considering the influence of nutrition on the gonads it is necessary to determine whether the effect is mediated via the pituitary gland or whether the gonads are influenced directly. If gonadal function is impaired this can be caused either by a change in structure or by a change in the level of hormone secretions. More recently Mann (1960) has outlined some of the difficulties in this type of work in attempting to distinguish primary effects on the reproductive organs due to a lack of a specific nutrient, and secondary effects due to a variety of other causes including reduced appetite, inadequate intake, inanition and certain nervous or stress factors. He posed the question of whether or not a given syndrome resulting from deficiencies of a particular dietary constituent is the result basically of:

(i) A failure of hormone synthesis and/or release by an endocrine gland.
(ii) Loss of responsiveness on the part of the target organs, or
(iii) an increased or decreased rate of hormone metabolism and destruction in the animals' tissues.

The evidence available in the case of many syndromes of nutritional deficiency is incomplete in essential detail. Early investigators could only examine the effect of nutrition on the structure and number of gametes, libido, puberty or overall reproductive capacity. Due to a lack of suitable techniques they could not measure the influence of nutrition on gonadal hormone levels in circulating blood, or the sensitivity of tissues to injected hormones. Later investigators, although using techniques of a more sophisticated nature and with improved experimental designs, have not yet exploited fully the new assay techniques and other experimental methods which are now available. Collection of such evidence will not prove a simple task for, in general, inadequate nutrition leads to hyposecretion, thus magnifying the difficulties of accurate measurement.

(i) *The Effect of Inanition*

When the general effects of inanition on reproductive activity, and especially on the activity of the pituitary are being considered, the duration or severity of the nutritional stress, i.e. whether chronic or acute is important. Chronic inanition has been shown to result in increase in the size and activity of pituitary basophils with a decrease in activity of acidophils before general atrophy of the gland occurs, which in later stages is accompanied by some degeneration (D'Angelo, Gordon and Charipper, 1948). In chronic starvation there is anoestrus in rats when body weight has been lost (Werner, 1939; Rinaldini, 1949). On the other hand, acute starvation does not influence the relative weight of the pituitary in female rats (Meites and Reed, 1949). Adrenal hypertrophy after severe restriction of energy intake is thought to be due to increased production of adrenocorticotrophin by the pituitary at the expense of other 'less critical' hormones (D'Angelo, 1951). Thus there appears to be a variation in the degree of activity of the anterior pituitary gland dependent upon the duration and severity of the nutritional imbalance.

Many workers have considered it important to examine the effects of exogenous gonadotrophins given to undernourished animals in attempts to assess the sensitivity of target organs (Marrian and Parkes, 1929; Allen and Lamming, 1961a; Morris and Nalbandov, 1961; Greep, 1961). Most of them have claimed that the ovaries of intact underfed animals will respond normally to the injection of gonadotrophins, although Fielden and Brumby (1962) have claimed that in hypophysectomized mice there is a decrease in response of the ovaries to gonadotrophins, arising from mechanisms other than those normally involving the pituitary. McClure (1962) observed that the reproductive performance of mice was reduced by starvation for a short period early in gestation which caused death of the embryo. He concluded that this was due to depression of production or liberation of gonadotrophins and not due to effects on the sensitivity of target organs, since daily injection of human chorionic gonadotrophin or progesterone maintained the pregnancy. Huseby, Ball and Visscher (1945) found that mice kept on a calorie intake two-thirds that of control animals fed *ad lib.* showed pituitary, ovarian, uterine and mammary gland histology similar to those of hypophysectomized animals. The caloric restriction markedly reduced the incidence of mammary carcinoma in such mice. The explanation was put forward that caloric restriction leads to hypophysial insufficiency thus reducing the production of ovarian hormones.

More recent work has suggested that inadequate nutrition modifies the rate of release rather than the rate of synthesis of gonadotrophins. A tentative suggestion of this effect was made from work in sheep by Allen and Lamming (1961a). Srebnik and Nelson (1963), working on deficiency of protein, potassium or thiamine in female rats, found an increase in the level of gonadotrophins in the pituitary. Their findings were confirmed by Krause (1963) who reported an increase of pituitary follicle-stimulating hormone in underfed female rats. Both groups found concomitant failure of the oestrous cycle and postulated that the primary effect was to reduce the rate of release of gonadotrophins rather than of their synthesis (see also later reference to Srebnik, 1964). Casida (1963) drew attention to species differences in the response to plane of nutrition when measured in terms of ovulation rate, and stated that previous nutritional history also was important. In addition, sex and age effects are also apparent; for example, Srebnik and Nelson (1963) found in female protein-deficient rats there was a rise in pituitary gonadotrophins but in male protein-deficient rats Srebnik and Nelson (1962) found a decrease of gonadotrophins in the pituitary, more marked in young than in mature animals.

In spite of the difficulties of interpretation of results it is evident that partial inanition up to a level of 20-30 per cent liveweight loss leads to a cessation of cyclical activity in both laboratory and farm animals and at least in the initial stages this is accompanied by an increase in the level of pituitary gonadotrophins. Scheer *et al.* (1947) reported that low energy intake in rats and mice resulted in a poor development of the reproductive organs and the mammary gland together with irregular oestrous cycles and lowered fertility.

(*ii*) *The Effect of Obesity*

It has been known for some time that obese animals tend to be infertile. Various studies have been carried out using rations above the recognized feeding standards with a view to determining the optimum levels for growth, reproductive capacity and the normal development of libido. In pigs there appears to be a higher embryonic mortality associated with overfeeding (Christian and Nofziger, 1952). On the other hand, the provision of short periods of higher dietary allowance (flushing) is a practice which is widely used in the sheep industry, and it has been shown that it can increase lambing percentage by 10–20 per cent, but flushing of well-fed ewes has had no such effect (Nichols, 1926; Allen and Lamming, 1961a), and there is no evidence that flushed animals are more fertile than well-nourished animals.

(*iii*) *The Effect of Protein*

In considering the effects of protein deficiency on reproductive function, several factors must be borne in mind before any valid comparisons can be made of the results of different authors. In particular it is important when assessing the various reports to know the severity and duration of the deficiency, whether qualitative or quantitative in nature, and whether the nutritional imbalance occurred before or after puberty.

It has been known for some time that severe restriction of dietary protein quantitatively or qualitatively leads to a cessation of oestrus or to long and irregular cycles in the female (Guilbert and Goss, 1932). The effects are due to impaired secretion of gonadotrophins since the administration of exogenous gonadotrophin or pituitary implants alleviated the syndrome in both male and female. More recent work has been concerned with the influence of lack of protein at different stages of reproductive activity.

When rations low in protein were fed to young animals, puberty was delayed and occurred at a reduced body size (Reid, Loosli, Trimberger, Turk, Asdell and Smith, 1964). The use of similar diets did not adversely affect the gonads after puberty (Meacham, Cunha, Warnick, Hentges and Hargrove, 1963; Venkatachalam and Ramanatham, 1964) indicating a lower priority of the reproductive tract in comparison with somatic growth when competing for the limited supply of protein in the young animal. In the mature animal, with the high requirement for growth largely removed, this leaves adequate plasma amino acids for the limited requirements for gonadal activity. In mature animals, however, it has been shown that an intake of protein adequate for maintaining oestrous activity and gestation, is inadequate for maintaining lactation, but this seems a natural result if the relative requirements of protein for pregnancy and lactation are compared. Of course, similar degrees of protein deficiency in the lactating animal still attaining some somatic growth have a very severe effect. For example, a 10 per cent level of crude protein was found to be adequate for gestation in the rat but unsatisfactory for lactation (Nelson and Evans, 1953). A very low protein ration (5 per cent fed as casein) given to pregnant rats from the day of mating caused high embryonic mortality and poor development of the mammary gland (Ambegaokar and Chandran, 1959) and 8 per cent casein in the diet was insufficient to support pregnancy and lactation in rats (Shaw and Griffiths, 1963). Foetuses were absorbed by mid-gestation in animals fed on a protein-free diet (Nelson and Evans, 1953), but recently it has been shown by Kinzey and Srebnik (1963) that the administration of steroid hormones can counteract the effects of protein deficiency in the pregnant animal. These workers showed that in protein-deficient rats the placenta could successfully carry foetuses to term provided that oestrogen and progesterone were administered on days 5–9 of pregnancy. Pregnancy could also be maintained in hypophysectomized protein-deficient rats when supported with hormones between days 5–11. These facts would suggest that in the case of protein deficiency in mature rats there is a failure of pituitary gonadotrophin secretion, but the placenta may function normally in maintaining pregnancy after the 9th–11th day of gestation. This work is thus important in reinforcing the view that the major effect of protein lack is the production or release of gonadotrophins from the pituitary gland. In support of this theory, protein-deficiency has been shown to affect the histological appearance and the endocrine function of the pituitary, resulting in increase of follicle stimulating hormone in the gland (Srebnik, Nelson and Simpson, 1961). There was no evidence that protein-deficient rats could not respond as well as normally-fed rats to the injection of gonadotrophins (Srebnik, Nelson and Simpson, 1958). Injections of oestrone and progesterone have been shown to maintain pregnancy in protein-deficient rats (Nelson and Evans, 1954).

Srebnik (1964) found a sex difference in the response of rats to protein deficiency. The regression of the reproductive organs following protein deprivation in the female was very similar to that in hypophysectomized animals but in the male the reproductive organs were much less severely affected by absence of dietary protein than they were by removal of the pituitary gland. When hypophysectomy was superimposed on protein deficiency in the male, testis weight and function declined significantly. These findings support the hypothesis that the absence of dietary protein interferes more markedly with the gonadotrophin release from the pituitary gland in female rats than in male rats. In the female, such severe deficiencies of protein or caloric restriction lead to an increase in anterior pituitary FSH levels probably representing a continued synthesis of FSH at a low level unaccompanied by any significant release of the hormone.

Thus it appears that since protein is required from dietary sources for the growth and development of immature animals, a low protein intake will adversely affect the gonads after a relatively short period. Once maturity is reached, a low level of protein in the diet can be adequate for the maintenance of gonadal structure and function unless the deficiency is severe or prolonged. Under these circumstances regression of the gonads will occur and endocrine function will be impaired.

Some work has been completed on the use of high-protein diets. Richardson, Godwin, Wilkes and Cannon (1964) showed that diets containing higher levels of protein (18–25 per cent) were of equal merit in maintaining reproduction in rats but when the level was increased to 35 per cent the average number of young produced per female decreased.

(iv) *The Effects of Fats*

Since Evans and Burr (1927) first reported the adverse endocrine effects of fat deficiency on the gonads a considerable amount of work has been carried out on the effects of deficiency of essential fatty acids (EFA). Burr and Burr (1929, 1930) made an extensive study of the fat deficiency syndrome in both male and female rats and work since then has verified their results and supplied further details. Recently Aaes-Jørgensen (1961) and Panos (1963) have published excellent reviews on the subject.

EFA deficiency results in extensive changes in the structure and function of both primary and secondary sex organs, influencing ovulation, gestation and parturition. In EFA-deficient rats, ovulation, determined by vaginal smears, is irregular and may cease entirely. On the addition of supplementary EFA ovulation is resumed within a few days.

Evans, Lepkovsky and Murphy (1934) showed that EFA deficiency in the female has effects in addition to those concerning ovulation, for the gestation period was increased by one to three days in rats and smaller litters were produced, 20 per cent of the females failed to litter after implantation had occurred and 80 per cent of the neonates were stillborn in those which reached parturition. Christensen and Dam (1956) and Dam and Granados (1962) confirmed the deleterious effects of low fat or fat-free diets on fertility. Chronic deficiency of EFA rendered female mice totally incapable of producing young (Decker, Fillerup and Mead, 1950). The primary sex organs of both male and female animals have shown very clear changes in histological structure due to EFA deficiency and in some cases differences have been shown in the secondary sex organs as well. The ovaries of EFA-deficient rats differed histologically from those of normal animals. Panos and Finerty (1953) showed that although the ovarian weight, follicles and corpora lutea were normal, the oestrous cycle became irregular and large numbers of 'wheel' cells appeared in the interstitial tissue indicating a lack of gonadotrophic influence.

As with other aspects of poor nutrition in the female, there is considerable experimental evidence that the effects of EFA deficiency are mediated via the pituitary. Panos and Finerty (1954) showed that there was a significant decrease of the numbers of acidophilic cells within the pituitary gland of male rats fed on a fat-free diet. Using histochemical techniques it was also shown that there was an increase in both FSH and LH-secreting cells in the pituitary for a period of 9 weeks after rats are fed fat-free rations (Panos, Klein and Finerty, 1959), but then the FSH-producing cells decreased in number while there was a concurrent increase in the number of LH secretory cells. They claimed that injection of testosterone corrected the balance in FSH and LH production as shown by a marked increase in testicular weight while the histological structure of the pituitary gland returned towards normal.

Examination of the pituitary glands of female EFA-deficient rats showed changes in the basophilic and acidophilic cells which corresponded to the changes occurring in the male (Panos and Finerty, 1953). The irregularities in the oestrous cycle could be related to reduction in LH production, while the increase in basophilic cell number could be caused by reduced oestrogen production. This explanation is further supported by the results of Alfin-Slater and Bernick (1958) and Greenberg and Ershoff (1951) in a summary of their work also suggested that the deficiency causes impairment of pituitary gland function. Aaes-Jørgensen (1961) pointed out that while some general conclusions can be drawn from all these results, factors such as the EFA status of the animal prior to the experiment and the length of the experiment, must be taken into account when explaining the results of particular authors' observations, a view which few would dispute.

Fats in various proportions have been added to several iso-protein rations. With levels up to 15 per cent fat, larger numbers of viable young were produced, but above this level the parturition rate decreased with heavy losses in the first three days post-parturition. It was shown that the fat level *per se* was unimportant but that the energy-protein ratio exerted an effect. Parturition rate and survival were unaffected until 45 per cent fat was included (Richardson *et al.*, 1964).

Excess fatty acids as well as EFA-deficiencies are known to affect reproduction in poultry. Work by Schneider, Kurnick, Vavich and Kemmerer (1962) showed that cyclopropenoid fatty acids found in cotton seed oil retarded the growth of the ovaries and oviducts and also inhibited egg production.

(v) *The Effects of Mineral Deficiencies*

There is considerable evidence that deficiencies of minerals affect reproduction in the female, and certain minerals are important as enzyme cofactors in general metabolism. Deficiencies of these substances are manifest mainly in the foetal and early post-natal stages, since at this time there is an increase in metabolism and tissue synthesis which are essential for growth and development. This places a large demand on the synthetic systems, and thus the demand for essential minerals increases. It would appear from the comprehensive reviews of Hignett (1959), Moustgaard (1959) and Underwood (1962), to which the reader is referred, that in the sexually-mature animal, deficiency of an essential mineral resulting in specific endocrine failure, has not been demonstrated.

Of those minerals implicated in nutritionally-induced reproductive failures, calcium, phosphorus, manganese and copper have figured more prominently than others. However, no true pattern of effect emerges from the literature apart from their specific metabolic function. It was suggested by Orent-Keiles and McCollum (1931) that manganese did have a specific effect on the anterior pituitary gland affecting gonadotrophin production and/or release and it has been shown to be essential for the maintenance of ovarian activity in rats (Kemmerer, Elvehjem and Hart, 1931; Waddell, Steenbock and Hart, 1931). More recently a similar pathway has been claimed for calcium by Taylor, Morris and Hertelendy (1962) and Taylor (1963) working with calcium-deficient hens.

None of the specified minerals can be considered in isolation, as a complex of interrelationships appears to exist. An example of this complex relationship is shown by the work of Shils and McCollum (1943) who fed milk balanced in calcium, phosphorus and iron to young rats. When the calcium: phosphorus ratio was lowered, symptoms of manganese deficiency appeared. Other work has also shown that abnormal ratios of dietary calcium and phosphate have an influence on reproductive activity. Diets containing 0·22 per cent phosphate and four times as much calcium, delayed the onset of sexual maturity or caused complete cessation of oestrus (Guilbert and Hart, 1930). When the content of phosphate was increased to about 0·5 per cent without any change in calcium or when the latter was lowered to the same level as the phosphorus content, the function of the reproductive organs returned to normal. It was reported by Hignett and Hignett (1951, 1952, 1953) that in cattle sequential variation of the phosphorus intake caused gradation in ovarian activity, ranging from anoestrum to normal oestrous cycles, and that impaired fertility due to a high calcium: phosphorus ratio with a relatively low phosphorus content, may be counteracted by an adequate supply of vitamin D.

In the human female blood copper levels rise throughout pregnancy until at full term the level is 2–3 times that of the non-pregnant woman, returning to the normal level about three weeks post-partum (Nielson, 1944). It has also been shown that blood copper levels are under hormonal influence, the level in serum increasing significantly following administration of oestradiol or testosterone (Johnson, Kheim and Kountz, 1959). This work suggests that there may be a mechanism correlating blood copper levels with the activity of these steroid hormones.

(vi) *The Effects of Vitamin Deficiencies*

It is well established that deficiencies of the vitamins can affect reproduction in the female and it has been postulated that some of the effects observed are brought about via the pituitary gland (Nelson, Lyons and Evans, 1951; Wooten, Nelson, Simpson and Evans, 1955; Forker and Morgan, 1955). When effects on the pituitary gland have been examined more closely, however, findings have not been consistent. For example, in the case of vitamin E deficiency in rats, Mason (1944) suggested that the pituitary was not responsible for the observed reproductive disturbances while it has been shown in other experiments that the gonadotrophin contents of the pituitaries of vitamin E-deficient rats may be decreased (Rowlands and Singer, 1956) unchanged (Biddulph and Meyer, 1941) or increased (P'an, van Dyke, Kaunitz and Slanetz, 1949; Griesbach, Bell and Livingston, 1957) so that the situation remains confused.

Similarly, deficiencies of vitamin A, riboflavin and pyridoxine have been claimed to decrease the secretion of pituitary hormones in the rat (Serfaty and Olivereau, 1955) but such deficiencies may be accompanied by increased synthesis within the gland owing to the failure of feedback mechanisms resulting from secondary damage to target organs (Sutton, Krause and Hansard, 1940; Forker and Morgan, 1955).

Such variation in results may be due in part to the establishment of varying severities of deficiency in different experiments and to the different assay techniques employed.

Thus at the present time it is difficult to present a coherent picture with regard to vitamin deficiencies although the interrelationships of the vitamins, gonadal function and reproduction have received considerable attention from research workers. Earlier work on various aspects of the field has been reviewed by Blaxter and Brown (1952), Lutwak-Mann (1958) and Mann (1960). Developments in the field since the above reviews were written have not led to complete clarification of effects with regard to any of the vitamins, although some advances have been made.

Vitamin A deficiency has long been known to cause degenerative changes in the female reproductive tract. One interesting relationship which has come to light as a result of work on this subject is that between oestrogen and the manifestation of signs of vitamin A deficiency. Bo (1955, 1956) working with ovariectomized vitamin A-deficient female rats failed to show the metaplasia of the uterine epithelium normally observed in intact deficient females. When such ovariectomized, deficient rats were treated with oestrogen, the symptoms of metaplasia appeared and became progressively worse. This apparent synergistic effect of oestrogen in producing symptoms of vitamin A-deficiency has since been confirmed by Grangaud, Conquy and Nicol (1960) who found that ovariectomy in weanling rats fed on vitamin A-deficient diets postponed the appearance of the characteristic eye lesions and death.

Recently, Howell, Thompson and Pitt (1964) have studied the response of the pregnant female rat to a diet deficient in vitamin A alcohol but containing vitamin A acid. Deficiency produced necrosis of cells in the placental labyrinth and junctional zone. Initially, the foetus appeared unaffected but later underwent resorption.

The B-complex vitamins have received a certain amount of attention since Lutwak-Mann (1958) remarked upon the relative shortage of information on the relationship of these vitamins with gonadal function, but almost all of this work has been carried out on the male.

In the case of vitamin E, Blaxter and Brown (1952) have reviewed the evidence on the effects of the deficiency on the reproductive tract of laboratory and farm animals. Lutwak-Mann (1958) who extensively summarized more recent work on the subject, gives details of differences between the effects of different vitamins on gonad function and draws attention to those deficiencies which can be corrected by the use of exogenous gonadotrophins or by careful administration of the steroid sex hormones. Since 1958, little addition has been made to previous knowledge.

Nutrition and Intra-Uterine Development

Prenatal development has conventionally been divided into three stages, that of the fertilized ovum, the embryo and the foetus. The development of the ovum extends until the beginning of implantation of the free-living blastocyst. The embryo phase is the period during which implantation and differentiation of the embryo and placenta occur. Once placentation is completed the remaining period of intra-uterine development is the foetal phase. The chronological sequence of these events in domestic animals is not clearly defined because implantation appears to be a gradual process.

The subject of prenatal mortality has recently been reviewed by several authors, including Laing (1952); Casida (1953, 1956); Hanly (1961) and Boyd (1965). Giroud (1959) specifically reviewed the nutritional requirements for embryonic development and the repercussions of nutritional deficiencies.

The greatest embryonic loss in farm animals appears to take place at the time of implantation, in contrast to the situation in rabbits, rats and mice, where there appears to be two independent peaks of loss, one before and one after implantation. The cost and difficulties of completing adequate experiments may have precluded exact definition of the situation in domestic species, and the difference may be one of experimentation rather than fact. In the various species studied the process of implantation has a critical requirement for a correct balance of hormones and conditions suitable for blastocyst growth may not be suitable for implantation. Until placentation is completed the embryo is free living in the uterus and during this period the relationship between the embryo and endocrine status of the mother is of great importance. As embryonal and foetal development proceeds the influence of nutrition *per se* becomes more important, and under conditions where the supply of nutrients is limited, demands for nutrients for the development of the conceptus may lead to a series of abnormalities. These can include (a) embryonic or foetal death with abortion or resorption, (b) malformation of the embryo or foetus, or (c) a delay in the development or growth of the foetus. Much of the work concerning the effect of nutrition on embryonic and foetal development has only recorded the gross effects of nutritional deficiencies, without comcomitant investigation of the causes of the changes observed.

(i) *General Nutrition*

The effects of chronic inanition on the embryo and foetus are less marked than those of specific deficiencies because the maternal organism by virtue of the activity of the placenta buffers the embryo or foetus against deficiency, the extent depending upon the nutrient involved, the length and severity of the deficiency and the extent of maternal reserves. However, severe inanition rapidly interferes with conception and especially with implantation. Dyban,

Demkiv and Augustinovic (1963) noted in the rat that severe undernutrition during early pregnancy caused embryonic loss at the time of implantation. In the deprived rats the endometrium failed to develop normally and 15 days after mating both the endometrium and the myometrium were atrophic. If the rats were allowed a normal diet again from 3–9 days post-mating the level of embryonic mortality was decreased.

A dietary restriction in the rat to one-quarter of *ad lib.* intake led to some evidence of resorption, smaller litters, smaller mean foetal weights and a persistent failure of offspring to attain control weights during a year (Chow and Lee, 1964).

In mice, McLure (1961a) found that fasting for the 48 hour period from the end of the third to the fifth day after mating, resulted in a complete failure of pregnancy in all mice. Fasting for 48 hours at other times up to seven days after mating had less effect. Starvation over the critical 48 hour period resulted in death and degeneration, firstly of the deciduoma and then of the embryo (McLure, 1961b). Daily injections of HCG or progesterone protected pregnancy in fasted mice so that it lasted at least six days longer than in control fasted mice. Pregnancy could not, however, be maintained beyond the fourteenth or fifteenth day by this means. McLure suggested that the effect of fasting was mediated *via* the pituitary gland, being a failure of hypophysial gonadotrophic function and that inability to maintain pregnancy to term indicated a failure to establish the normal placenta-ovary relationship of the second half of pregnancy.

In addition to his work on the effects of fasting post-mating, McLure (1963) studied the effects of nutritional stress, in the form of a 24 hour fast, prior to mating. Females mated one and two days after the end of the fast had significantly lower littering rates than those mated at other times relative to the fasting period.

Bruce (1963) noted that fasting on the third day of pregnancy for 24 hours resulted in a block of 54 per cent pregnancies; the treatment appeared to inhibit pituitary luteotrophic activity and depress oestrogenic activity since the return of vaginal cornification only occurred when food was restored. Grouped females were able to withstand the stress better than singles, with the suggestion that grouping increased luteotrophic activity. Although these effects (i.e. starvation of *ad lib.* fed animals for short period following mating) are apparent in laboratory species the same treatment does not have similar effects in pigs (cf. Ray and McCarty (1965) cited later).

In the sheep Hafez (1952a, b.) found that feeding a sub-maintenance diet causing loss of up to 45 per cent of the initial weight resulted in irregular oestrus and interfered with conception. Under-feeding, resulting in a weight loss of 25 per cent during pregnancy, reduced the weight of the foetus and placental cotyledons at 90 days (Everitt, 1964). In contrast, a low level of feeding of pigs before and after mating produced a negligible difference of within-litter mortality by 28 days (Heap *et al.*, 1966), but there was a 25 per cent loss of total litters (King and Young, 1957). Hanson, Ferrin and Aunon (1953) found a lower farrowing rate with smaller piglets in sows fed on a low plane of nutrition. Reid *et al.* (1951) reported that 1·56 services were required per conception in cattle fed 65 per cent of the Morrison standard compared with 1·22 services in animals receiving the normal allowance.

Nutrition of the ewe has profound effects on growth and viability of the foetus, a lowered plane of nutrition being generally detrimental to the foetus in terms of growth, differentiation and viability (Wallace, 1946a, b; 1948a, b, c; Alexander, McCance and Watson, 1956; McClymont and Lambourne, 1958; Everitt, 1964; Taplin and Everitt, 1964). It has been claimed, however that the foetus is not consistently affected by feeding levels (Foote *et al.*, 1959a, b). Bellows, Pope, Chapman and Casida (1963) studied the effect of level of feed during gestation. Increased levels gave improved foetal survival in the Hampshire ewe mated to Columbia ram, but was detrimental in Columbia or Hampshire ewes mated to the Hampshire ram. Overall there was a depression of foetal survival as a result of an increased intake. They suggested that genetic effects and nutritional influences in their animals cancelled each other resulting in a non significant feed level versus breed-of-dam interaction. However, more work would be required with different breeds of sheep before this could be accepted as a breed effect directly influenced by nutrient intake.

While there is little doubt that severe undernutrition is detrimental to conception rate and that ovulation rates are higher where liberal dietary allowances are given there is also evidence that embryonic mortality may be increased by a high plane of nutrition. The situation has been extensively studied in pigs and several papers have shown that embryonic mortality is reduced by restricted feeding during the period before and after mating. Christian and Nofziger (1952) compared the performance of gilts fed *ad lib.* with animals fed 70 per cent of the *ad lib.* consumption from weaning to puberty. Ovulation rate was highest on *ad lib.* feeding but fertilization rate was decreased, and embryonic mortality increased, 7·4 live pigs per litter being born in the low-plane group compared with 4·7 live pigs per litter in the high-plane group. Starvation for 1–3 days after mating in pigs previously fed *ad lib.* did not significantly decrease ovulation rate (although that in restricted animals was consistently lower) but the average number of embryos was higher in restricted animals, implying that while ovulation rates are higher in animals fed *ad lib.* embryonic mortality is also higher (Ray and McCarty, 1965). Similar conclusions have been reported by Robertson *et al.* (1951); Self *et al.* (1953), Cybulko (1957); Hafez (1959) and Dimov (1963). Lodge (1966) has reviewed the available evidence in his chapter in this volume.

In sheep El-Sheikh *et al.* (1955); Foote *et al.* (1959a) and Hoxsey, Hoversland, Blackmore and Nan Horn (1960) found that a high plane of nutrition increased both ovulation rate and embryonic mortality. In cows Joubert (1954) noted that animals on a high plane of nutrition returned to service more quickly after parturition but required more services for conception. Pomeroy (1954) suggested that a high ovulation rate in pigs might lead to a high level of circulating progesterone which may increase embryonic loss due to a rapid passage of ova through the fallopian tube.

Squires, Dickinson and Mayer (1950) and Self *et al.* (1955) found that rate of gain in groups of pigs fed *ad lib.* or $\frac{2}{3}$ *ad lib.* up to mating was positively correlated with embryo survival. From these results they suggested that the fastest growing animals were less fat and thus had a higher rate of embryo survival.

(ii) *Energy Levels*

The effect of energy intake on reproduction in farm animals has been reviewed by Reid (1960) and results similar to those already considered have been obtained when diets containing high or restricted amounts of energy have been compared.

The effects of different energy levels fed to farm stock during pregnancy are not 'all or none' effects similar to those which accompany certain vitamin deficiencies, but are graded differences from high to low plane of energy intake. Ovulation rate and embryonic mortality were increased on the high-energy diets (Haines, Warnick and Wallace, 1955, 1958, 1959; Goode, Warnick and Wallace, 1960, 1965; Sorensen, Thomas and Gosset, 1961; Dean and Tribble, 1962). Wiltbank, Rowden, Ingalls, Gregory and Koch (1962) found that in beef cattle a low energy level fed post partum did not delay the involution of the uterus, but increased the interval from parturition to oestrus and the number of services required for conception. In later work Wiltbank, Rowden, Ingalls and Zimmerman (1964) studied the influence of energy intake in cows previously fed 50 per cent of the recommended allowance during pregnancy. They observed that conception rate was highest when a ration of 150 per cent of the NRC recommended level was fed to such animals immediately following parturition. This implied that the higher nutrient intake offset the effects of previous undernutrition. These authors reported that energy restriction seemed to exert its effect on oestrous behaviour and conception rather than on foetal development, although lighter calves were born following energy restriction (Wiltbank *et al.*, 1962).

Thus in spite of some references to the contrary, a high energy intake favours early puberty and a high ovulation rate but may increase embryonic mortality. Excessive or low energy intake is detrimental to conception rate, implantation and embryo survival. Nutritional influences on embryo and foetal development may be buffered by maternal reserves, but a restricted energy intake can influence the development of the placenta and foetus.

(iii) *Protein Levels*

Within limits, the pregnant animal may deplete body reserves of protein to offset a dietary shortage of protein or its constituent amino acids and thus maintain foetal growth at normal or near normal rates. The greater part of the increase in protein requirement of pregnancy occurs during the later stages. Most work has been completed in laboratory species, particularly the rat and Guilbert and Goss (1932) demonstrated that 5 per cent protein was minimal in this species and below this level almost all young were aborted. Normal intra-uterine development took place in rats on a negative nitrogen balance due to the type of protein fed (Champigny and Jacquot, 1961). These animals tended to retain more nitrogen during pregnancy, a phenomenon not found in the pig (Mitchell, Carroll, Hamilton and Hunt, 1931; Heap, 1966). Curtiss (1953) noted that low protein diets gave a high incidence of uncompleted pregnancies in the rat accompanied by weight loss during the first two weeks, and Nelson and Evans (1946, 1953) found that rats given purified diets supplemented with all known dietary essentials but containing casein as its protein source failed to maintain pregnancy. Symptoms were similar to those found in vitamin B_6-deficient and in hypophysectomized animals which suggested that failure was due to hormonal deficiencies. However, failure was not noted when protein deficiency commenced shortly after implantation. Kinzey and Srebnik (1963) found that pregnancy could be maintained in protein-deficient rats if ovarian hormones were supplied during the period before placental function was fully developed when it was presumed that protein deprivation was interfering with pituitary gonadotrophic function. Placental function appeared to be independent of protein deficiency; when protein deprivation was instituted after implantation pregnancy was maintained without treatment.

The effect of protein intake on intra-uterine growth in farm animals is not so clear. The situation in the sow was reviewed by Boaz (1962) who indicated that irregular breeding and a smaller number of pigs per litter was associated with low protein intakes, although Becze (1962) reported that low protein intakes fed to sows before mating had no effect on ovulation but decreased embryo survival. Reduction of protein after mating increased embryonic mortality only in those breeds with a 'genetically determined' high protein requirement and these losses were not reduced by progesterone treatment. Ripple, Rasmussen, Jensen, Norton and Becker (1964) claimed that different levels of protein fed during the latter half of pregnancy had no effect on numbers in the litter.

In sheep, the effect of restricted protein intakes during the second half of pregnancy had more serious effects on lamb viability in twins rather than singles (Thomson and Thomson, 1948). Lowered conception rates in cows fed 50 per cent of the U.S.A. NRC protein requirements were also noted (Witt, Warnick, Koger and Cunha, 1958).

Excess protein has been shown to cause histological changes of the ovary and genital tract (Tuchmann-Duplessis and Aschkenasy-Lelu, 1947) and lower the number of young born in rats (Richardson *et al.*, 1964).

(iv) *The Effects of Vitamins*

Numerous experiments on both laboratory species and farm stock have demonstrated the necessity for an adequate supply of both fat and water soluble vitamins for normal foetal development and viability of the offspring.

Vitamin A. In laboratory species there are many effects of vitamin A deficiency during pregnancy. In the rat diets deficient in vitamin A increased the length of gestation and caused difficult parturition (Mason, 1935). Foetal death and resorption also occurred, the foetal death apparently due to maternal deficiency, and very few young survived even if transferred to normal mothers. In cases where the deficiency did not cause foetal death, the offspring were generally malformed, most frequently with anophthalmos, harelip, malformed or misplaced kidneys, hernia of the diaphragm and cleft palate. A study of the effect of restoration of vitamin A during gestation led to the conclusion that vitamin A deficiency causes malformations during the period of active organ formation but not while the primordia are still undifferentiated (Warkany and Roth, 1948; Wilson, Roth and Warkany, 1953). A similar situation of foetal degeneration, resorption and congenital hydrocephalus also occurred in the rabbit (Lamming, Salisbury, Hays and Kendall, 1954; Millen Woollam and Lamming, 1954). Hydrocephalus occurred in up to 80 per cent of young rabbits born to mothers depleted of vitamin A and increased cerebrospinal fluid pressure was found in some young which were not hydrocephalic (Millen, Woollam and Lamming, 1953, 1954; Lamming, Woollam and Millen, 1954). It is an interesting feature of these malformations that in both the rat and rabbit the young are affected before the dam shows signs of avitaminosis. Lamming and Salisbury (1952) found that in rabbits with incipient vitamin A deficiency there was a reduction in the number of animals mating, although in those which mated ovulation rate was not reduced. A large number of eggs were found to be degenerating at 40 hours *post coitum*. Nalbandov (1952) produced similar degeneration by transplanting normal ova into unovulated does. It is suggested that many of the symptoms of abortion, resorption and egg degeneration in vitamin A-deficiency may result from a decreased supply of endogenous progesterone (Lamming *et al.*, 1954). Foetal malformation in the rat also occurs due to an excess intake of vitamin A (Giroud and Martinet, 1955).

Early work on the vitamin A requirements of farm animals has been reviewed by Hart (1941) and little work has been completed since then. Vitamin A deficiency in cattle caused foetal death, abortion, or the birth of non-viable calves, frequently with retention of the placenta. In cattle a suboptimal level of vitamin A for two or three generations resulted in abnormal reproductive behaviour and damage to the pituitary, adrenal and sex glands of later generations (Byers, Jones and Bone, 1956).

In sheep deprived of vitamin A, reproduction is not adversely affected by a low vitamin intake, unless the condition is prolonged. Under these circumstances lambs are stillborn or die shortly after birth. The deficiency did not become effective until after two successful pregnancies (Pierce, 1954) indicative that the animal is able to store considerable amounts of vitamin A.

In sows a pronounced vitamin A deficiency often leads to prenatal death or birth of pigs with a low viability, and congenital malformations similar to those found in laboratory species (Goodwin and Jennings, 1958; Braude, Kon, Mitchell and Thompson, 1951). The latter workers demonstrated that debilitation of the offspring did not occur until the sows were depleted by feeding a vitamin A-free diet for two years and had reared 3 normal litters. Under these circumstances the fourth litter was weak and had to be reared artificially and the piglets which survived subsequently produced weak litters.

Thus, while vitamin A is required for normal reproductive function and embryo development, the maternal stores commonly found in most farm animals usually prevent deficiency symptoms occurring.

Vitamin E. There appears to be species differences existing in the necessity of this vitamin for normal reproduction. The vitamin is necessary for normal embryonic development in the chicken (Ferguson and Couch, 1954; Ferguson, Atkinson and Couch, 1954) and the turkey (Jensen and MacGinnis, 1957). Deficiency caused degeneration of placental vessels in the rat and subsequent death of the embryo (Suomalainen, 1950) and the work of Cheng, Chang and Bairnson (1957) has demonstrated that embryonic malformations similar to those caused by vitamin A can occur due to absence of vitamin E from the diet. However, it is claimed that both goats (Wilson, Thomas and Cannon, 1935) and cattle (Gullickson, Palmer, Boyd and Olson, 1944) will reproduce normally when fed diets devoid of the vitamin.

Pantothenic acid. The importance of this vitamin to foetal development has been demonstrated by a number of workers. The signs noticed included implantation defects, resorption of the foetus and smaller litters with underdeveloped young. Abnormal embryos were found and the abnormalities included exencephaly and anophthalmia, ectocardia and various vascular alterations (Nelson and Evans, 1946; Hodgskiss, Ensminger, Coldby and Cunha, 1950; Lefebvres, 1951; Barboriak, Krehl, Cowgill and Whedon, 1957). In one tentative case it has been stated that in sows, a lowered pantothenic acid allowance seemed to be associated with a high gonadotrophin content and vice versa, but this is not well established (Ullrey, Becker, Terrill and Notzold, 1955).

Riboflavin. The influence of lack of vitamin B_2 (riboflavin) on embryo development is well marked. The consequences of this deficiency are a function of its duration and intensity. Long term deficiency leads mainly to resorption while a short one may allow normal development. An intermediate deficiency (i.e. feeding a deficient diet 20 days before mating in the rat and continued during gestation) appeared to have no effect on conception but gave rise to palate and limb abnormalities (Giroud, Levy and Lefebores-Boisselot, 1950; Gilman, Perry and Hill, 1952). In swine the deficiency leads to early parturition and the production of weak, deformed piglets (Ensminger, Bowland and Cunha, 1947; Miller, Ellis, Stevenson and Devey, 1953).

Many other vitamin deficiencies have effects on embryo development similar to those cited above, causing either resorption or having teratogenic effects. These have been extensively reviewed by Giroud (1959).

(v) *The Effects of Minerals*

Calcium and Phosphorus. Calcium and phosphorus deficiencies in the mother result in the birth of immature and stillborn foetuses in a number of species due to lack of development and calcification of the skeleton, e.g., sow (Hart and Steenbock, 1918) and rat (Bodansky and Duff, 1941a). On the other hand, Ellinger, Duckworth, Dalgarno and Quenouille (1952) found that severe calcium deficiency did not impair fertility or mothering instinct of the rat. The difference, however, may lie in the fact that these authors used a natural diet, while Bodansky and Duff used a synthetic diet.

There is also conflicting evidence in other species. In cows, Rushoff (1950) states that Jersey cows fed on a low calcium diet produced weak or dead calves, or aborted whilst Boda (1956), Boda and Cole (1956) showed that feeding low calcium diets as a preventive for milk fever appeared to have no effect on parturition or the subsequent health of calves.

Pigs fed calcium-deficient diets suffered periodically from loss of appetite and had difficulties in farrowing, from which few pigs survived (Evans, 1929).

Potassium and sodium deficiency. Stewart and Welt (1961) claimed that in rats, deprivation of potassium reduces maternal serum and muscle potassium but had no effect on foetal potassium, whilst on the other hand, sodium deficiency appears to cause resorption of the foetuses in the rat (Orent-Keiles, Robinson and McCollum, 1937).

Other trace elements. It has been claimed that the magnesium requirements are connected with the calcium content of the diet in the rat, an incorrect ratio causing abortion of the young, or causing them to be eaten alive.

Manganese appears to be indispensable to reproduction in rats, and Daniels and Everson (1935) showed that young animals die of congenital debility if the mother's diet is deficient in manganese, the young rats containing 65 per cent less manganese than control animals.

There are serious difficulties of investigation of trace-element deficiencies and one problem is the preparation of suitable deficient diets. As a result of these difficulties attempts to investigate the effects of micro nutrient deficiencies on reproduction in farm species have produced either negligible effects or inconsistent results (Grummer, Bentley, Phillips and Bohstedt, 1950; Bentley and Phillips, 1951). Thus, there are a number of other references to the effect of metallic elements on embryo development in farm stock but none are particularly conclusive. These have been reviewed by Marston (1952) and Hignett (1959) who suggested that complex interactions between ions take place to influence breeding performance.

NUTRITION AND REPRODUCTION IN THE MATURE MALE

The function of the male mammal in reproduction is twofold, to produce gametes and to deposit them at an appropriate time in the area where they proceed to participate in the events leading to fertilization. Failure of either function results in failure of the whole process of reproduction. In mammals the first process involves virtually a continuous proliferation of spermatozoa after puberty with a recognizable diurnal rhythm whilst the second involves the maintenance of libido which can be affected by physiological and psychological influences.

Whereas the female produces a small number of gametes in a rhythmic manner, in some species, the male produces vast numbers of gametes with seasonal variation. In mammalian species of agricultural importance the technique of artificial insemination has demonstrated that considerable economy of males can occur within a controlled population when judged by the criterion of successful reproduction. Thus, in the domestic species, infertility in the male is of less consequence than infertility in the female.

In the male as in the female, there appears to be an interaction of age and growth rate which influences the maturation of the reproductive tract. Thus liveweight is the major factor influencing the occurrence of puberty, this is well illustrated from experiments involving variations in the plane of nutrition (Jackson, 1932; Flipse, Snyder, Thacker and Almquist, 1953; Davies, Mann and Rowson, 1957; Dutt and Barnhart, 1959; Flipse and Almquist, 1961; Dickerson *et al.*, 1964; Van Demark and Mauger, 1964; Van Demark, Fritz and Mauger, 1964; Widdowson *et al.*, 1964). Work with monozygous twin bulls had indicated the extent of the 'carry-over' effects into maturity due to environmental conditions during the pre-pubertal phase (Davies *et al.*, 1957), and this is probably the most significant factor affecting the post-pubertal performance of the male. Thus the post-pubertal phase of male reproductive activity cannot be fully described without knowledge of the animals' pre-pubertal nutritional history. Much information has been obtained from the recent critical studies of Dickerson, Gresham and McCance (1964) and

Widdowson *et al.* (1964), and it is to be hoped that further work will clarify the interrelationships between the pre- and post-pubertal nutritional environment.

The current concept of the endocrine control of male reproductive activity can be summarized briefly as follows: the hypothalamus, acting under the control of higher nervous centres, regulates the production of Follicle Stimulating Hormone (FSH) and Interstitial Cell Stimulating Hormone (ICSH) giving rise to a balanced output of both. FSH controls the gametogenic activity of the germinal epithelium while ICSH acts on the interstitial cells and regulates their output of the male sex steroid, testosterone, which circulates in the bloodstream exerting an influence on the metabolic and the secretory activity of the accessory organs (seminal vesicles, prostate and bulbo urethral glands). By some unknown mechanism testosterone levels are metered by higher nervous centres, which in turn modify the secretory/release activity of the hypothalamo-pituitary complex. Thus a system of reciprocal control is thought to exist similar to that of the female. Impairment of male reproductive function may be caused, therefore, under the following circumstances:

(i) A primary effect of nutrition on the anterior pituitary leading to a low secretion and/or release of gonadotrophins.

(ii) Inactivation or lowered response of the gonads depressing spermatogenesis and testosterone synthesis.

(iii) Other effects.

When considering the influence of nutrition, failure of reproduction may be a result of a direct effect on the reproductive processes or an indirect effect due to a lowered metabolic activity.

The Influence of Nutrition on Pituitary Gonadotrophins

Although many studies on the interrelationship between nutrition and male reproduction have been carried out in farm animals, the most useful results have been obtained from work in laboratory animals. In 1940, it was shown that under conditions of chronic inanition there was an absolute failure of reproductive activity in mature male rats, with an endocrine pattern resembling that of hypophysectomy (Mulinos and Pomerantz, 1940). The studies of these workers are again classic in the field of nutrition and reproduction, their observations of the endocrine pattern of chronic inanition being substantiated by more recent studies in the female in which new assay techniques, developed since their work, have been used (Krause, 1963). In a further study Mulinos and Pomerantz (1941) used Human Chorionic Gonadotrophin (rich in ICSH activity) and testosterone therapy in 'nutritionally infertile' male rats. Their results are summarized as follows:

(a) Atrophy of the interstitial cells of the testis was shown by the appearance of a 'wheel-like' formation in the nuclei of these cells, considered to be a reliable index of an insufficiency of gonadotrophins. Injection of HCG resulted in the disappearance of the 'wheel-like' formation in the nuclei of the interstitial cells.

(b) Small doses of HCG or testosterone propionate caused a marked and prompt increase in weight of the accessory sex organs.

(c) Injection of HCG resulted in growth of the testes and a resumption of spermatogenesis, the latter occurring regardless of dose after approximately 40 days of treatment, indicating that one effect of chronic inanition is exerted on the primordial germ cells. HCG treatment would appear to exert its effect through the pituitary, since spermatogenesis did not occur in the atrophic testes of the hypophysectomized male rat following HCG injection.

(d) Such sterile animals when treated with HCG will both mate and successfully impregnate normal female rats despite the continued underfeeding.

From this excellent study Mulinos and Pomerantz concluded that in the chronically underfed male rat, atrophy of the genital organs and the consequent sterility is primarily due to an insufficiency of gonadotrophic hormones resulting from depression of pituitary gland activity. The development of techniques for the measurement of gonadotrophins in blood would provide the final piece of evidence to substantiate this work.

Since the work of Mulinos and Pomerantz, other workers have postulated a similar pathway for the effect of nutrition in mice (Goldsmith and Nigrelli, 1950) and bulls (Davies *et al.*, 1957). (See Fig. 2.) While Mulinos and Pomerantz were concerned mainly with the influence of plane of nutrition a similar explanation is appropriate for other nutrition syndromes, e.g. protein deficiency (Courrier and Raynaud, 1932; Lafon and Veillet, 1938; Rivero-Fontan, Paschkis, West and Cantarow, 1952; Leathem and Glasser, 1953; Scott, 1955), fat deficiency (Ferrando, Jaques, Malboux and Sollogub, 1955; Bieri, Briggs, Spivey-Fox, Pollard and Ortiz, 1956), in vitamin deficiency (Wolfe and Salter, 1931; Moore and Mark, 1936; Knoop, 1946; Elson and Koller, 1948; Lutwak-Mann and Mann, 1950; Kline and Dorfman, 1951; Maddock, Cohen and Wolbach, 1953) and in mineral nutrition (Follis, Day and McCollum, 1940; Boyer, Shaw and Phillips, 1942; Knoop, 1946).

Nutrition and Testicular Function

The validity of the above explanation concerning the primary effects of nutrition on the anterior pituitary gland depends on evidence that a response of the target organ can be obtained with exogenous gonadotrophin. In view of the levels of exogenous hormones usually injected in experiments such responses may mask a refractoriness of the target organ to physiological

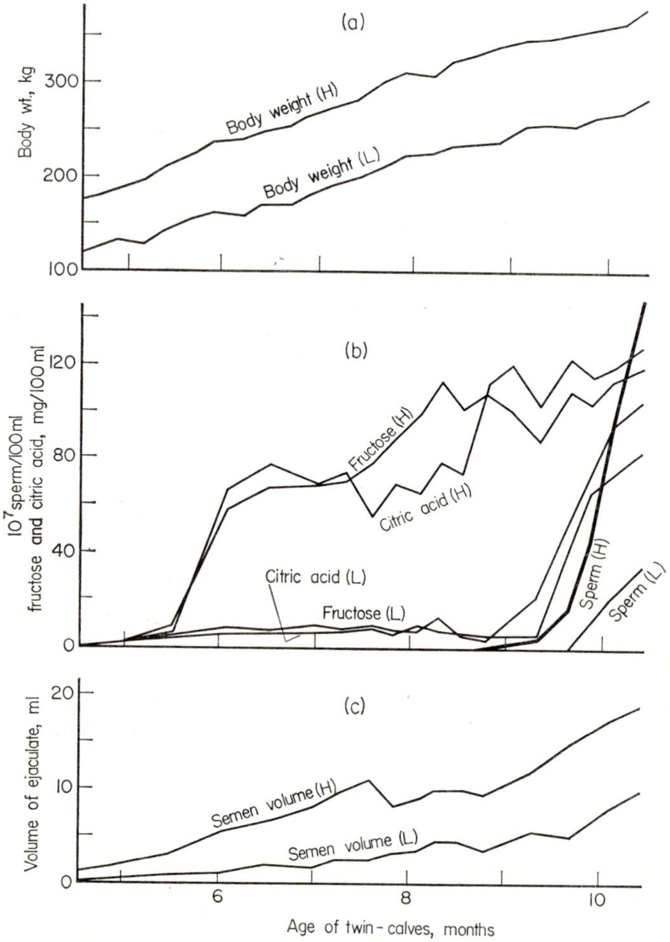

FIG. 2. *a* The effect of high (H) and and low (L) planes of nutrition on the body weight (kg); *b* the concentration of spermatozoa (10^7 sperm/100 ml.), fructose (mg/100 ml.) and citric acid (mg/100 ml.); and *c* the volume of electrically discharged semen, from the pair of identical (monozygous) bull-calves.

levels of hormones as a result of poor nutrition, a view expressed by Sir John Hammond (1940).

The work of Lutwak-Mann and Mann (1950) and Mann and Walton (1953) leads to the conclusion that spermatogenesis must function at a lower level of gonadotrophin secretion than the Leydig cells require to maintain a blood testosterone level necessary for adequate accessory gland activity.

As in the female there is a lack of data on the levels of gonadotrophins in the blood of normal male animals, and until this information is available conclusive evidence will be lacking whether or not target tissues of male animals are refractory to gonadotrophins as a result of nutrition, although some observations have been made on this point.

Some information on the effect of plane of nutrition on the sex glands of the castrated bovine to injected testosterone propionate has been obtained by the work of Mann, Rowson and Hay (1960) and described by Mann (1964). (See Figs. 2 and 3). This work provided evidence suggesting that underfeeding from 11 months to 16 months of age decreased the response of accessory organs of castrated male animals as shown by the levels of seminal citric acid and fructose.

Lowered activity of the germinal epithelium and/or cells of Leydig has been suggested or can be inferred from reports of the effect of protein deficiency (Holt, Albanese, Shettles, Kajdi and Wangerin, 1942; Horn, 1955) and vitamin deficiency (Evans and Burr, 1927; Marrian and Parkes, 1928; Evans, 1928; 1932; Bryan and Mason, 1940; Evans and Emerson, 1943; Erb, Andrews, Bullard and Hilton, 1944; Madsen, Eaton, Heemstra, Davis, Cabell and Knapp, 1946; Lindley, Brugman, Cunha and Warnick, 1949; Mayer and Truant, 1949; Goldin, Greenspan, Goldberg and Schoenberg, 1950; Traina, 1950; Mayer and Goddard, 1951; Ingelman-Sundberg, 1954; Barboriak, Krehl, Cowgill and Whedon, 1957; Lutwak-Mann, 1958; Fidanza, Bonomolo, De Cicco and Esposito, 1959; Cavazos, Jeffrey, Manning and Feagens, 1961; Howell, Thompson and Pitt, 1963; Pulludan, 1963; Arscott, Parker and Dickinson, 1965). On the basis of this extensive evidence, coupled with the knowledge of their function, it can be claimed that the vitamins as a group probably cause a failure of gonadal response to gonadotrophins.

Other Effects

In addition to the above explanation of reproductive failure in the male there has recently been an interest in the physiological effects of under nutrition on the liver. Interest in these effects has stemmed from work which shows a cessation of reproductive activity in animals suffering from liver failure due to cirrhosis or carcinoma (Morrione, 1944; Bennett, Baggenstoss and Butt, 1950; Meites, 1953). It is postulated that a progressive rise in blood testos-

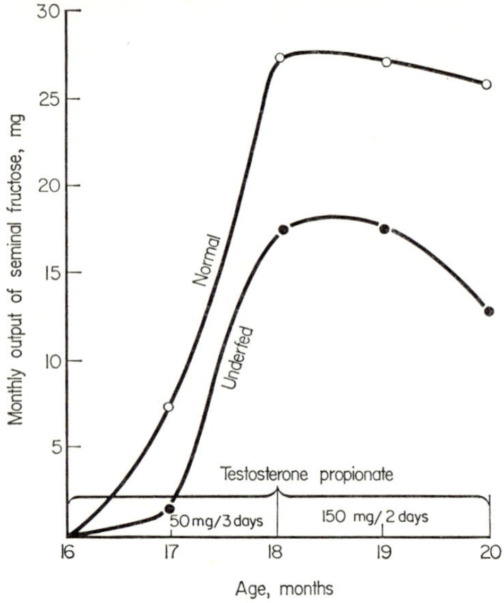

FIG. 3. Effect of reduced food intake on the output of seminal fructose (mg./month), studied in a pair of castrated, monozygous bull-calves. Testosterone propionate was injected in identical amounts into both the normally fed and underfed twin.

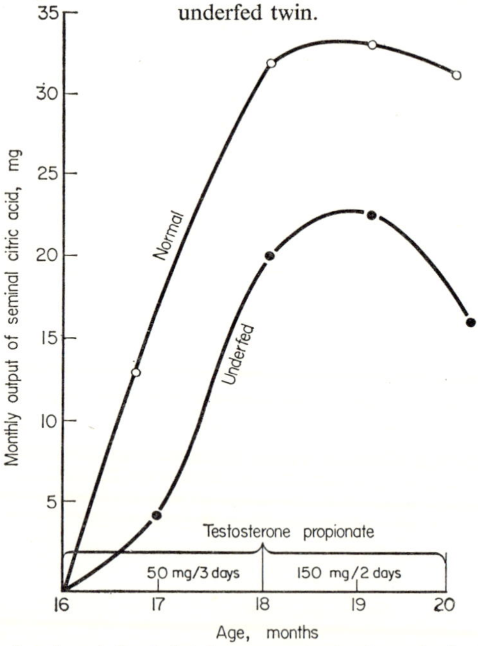

FIG. 4. Effect of reduced food intake on output of seminal citric acid (mg./month), studied in a pair of castrated, monozygous bull-calves. Testosterone propionate was injected in identical amounts into both the normally fed and underfed twin.

terone levels, due to impaired liver function, may cause an imbalance in the hormonal relationship of the pituitary-gonad axis. It is possible that liver function might be impaired during chronic malnutrition.

Work on the influence of nutrition on libido has recently been carried out in rats (Sachs, 1955). Sixteen male rats were tested for mating behaviour after 1–9 days of starvation. All males copulated after three days of food deprivation, the only significant effect being a delay preceding initial intromission and an increased frequency of mounts without intromission. After longer periods of starvation some males failed to achieve intromission or ejaculation; but even when starved for 8–9 days, eleven out of sixteen males ejaculated without marked changes in most measures of their sexual behaviour. On the other hand Bratton, Salisbury, Tanabe, Branton, Mercier and Loosli, (1948) reported that mature bulls fed diets low in carotene for long periods retained a high level of libido, even though they were unable to mate females due to muscular incoordination as a result of vitamin A deficiency. The incoordination also occurred before any marked impairment of semen production was evident. Work of this nature illustrates the remarkable resistance of the mature male in maintaining reproductive activity in spite of low levels of nutrition.

NUTRITION AND REPRODUCTION IN THE MALE WITH SPECIAL REFERENCE TO FARM ANIMALS

The interrelationships of nutrition and male reproduction have been studied in most farm animals but the majority of work has been carried out on cattle. It is important to stress that the most striking effects of low nutrient intake is pre-pubertal during the rapid growing phase, and that there is a considerable degree of 'carry over' into maturity after such low-plane nutritional regimes (Davies *et al.*, 1957).

Cattle—The U.S. National Research Council (NRC) recommendations for Total Digestible Nutrient (TDN) are more than adequate. Allowances greater than those suggested by NRC do not increase either reproductive function or semen quality (Branton, Bratton and Salisbury, 1947; Hiroe, Masaki, Tomizuka and Hanada, 1964). The effect of feeding less than NRC TDN allowances, depends on the degree of inanition and the stage of maturity. Seventy per cent (Hiroe *et al.*, 1964) and 60 per cent TDN (Van Demark, Fritz, Ewing and Mauger, 1960; Van Demark *et al.*, 1964) caused lowered semen output. In the above work Hiroe *et al.* (1964) claimed that the treatment produced a lower volume of spermatozoa, a lower sperm count per ejaculate and inferior chemical composition of the seminal plasma; their findings, however, are confused by conditions of low protein intake. The quality of roughage in the diet seems to have no detrimental effect on reproductive function or semen quality with the proviso that intake is not limited (Jones, Haag and Dougherty,

1942; Jones, Dougherty and Haag, 1945). When mature bulls were induced to lose 6·5 kg/week in liveweight over a period of 23 weeks, no differences against controls were observable in volume and density of semen, or motility and morphology of spermatozoa (Mann and Walton, 1953). Accessory gland function on the other hand was markedly affected, fructose and citric acid being reduced to 30 and 60 per cent of normal levels, respectively. On feeding, these values gradually returned to normal.

The work of Flipse and Almquist (1963) indicates that 10 per cent Digestible Crude Protein (DCP) is more than adequate for the maintenance of reproductive activity and semen quality. That the young bull can still reproduce even when fed diets low in protein has been well illustrated (Meacham *et al.*, 1963). The protein-deficient bulls (8·09–1·35 per cent DCP) had significantly smaller volumes of semen and a lower total number of sperm per ejaculate. Motility of spermatozoa varied but was not correlated with the diet, and motile sperm were present up to death. Libido was adversely affected, but this might have been due more to the induced weakness than to the dietary treatment. In a more specific study it was shown that the reduction of protein from 14 to 1·5 per cent significantly reduced the levels of fructose, citric acid and 5-nucleotidase in the semen.

Studies on the effect of vitamin A deficiency in male bovine reproduction are clear in the characterization of the deficiency, degeneration of the seminiferous tubules being noted with almost complete disappearance of spermatozoa from the lumen (Hart and Guilbert, 1937; Erb *et al.*, 1944; Bratton *et al.*, 1946; Madsen *et al.*, 1946; Erb, Andrews, Hauge and King, 1947). A loss of libido has also been noted (Hodgson, Hall, Sweetman, Wiseman and Converse, 1946) together with cystic pituitary glands (Sutton *et al.*, 1940; Hodgson *et al.*, 1946).

In the herbivora normal reproduction can be supported by diets deficient in vitamin E. Bulls, fed for long periods on heated natural rations, which failed to support reproduction in male rats, had normal testes and normal potency (Gullickson and Fitch, 1944; Gullickson, Palmer, Boyd, Nelson, Olsen, Calverly and Boyer, 1949; Gullickson, 1949).

Manganese and Iodine deficiencies have been reported to exert an effect in bulls. Ejaculates from bulls reared on a low manganese diet were small with a low sperm count, poor sperm motility and a short storage period. In contrast, bulls receiving the same ration plus a supplement of manganese sulphate had ejaculates of normal volume, high sperm count, excellent motility (Lardy, Phillips and Rupel, 1942). When a diet deficient in iodine was supplemented with potassium iodide (50 mg/day) bulls showed improved libido, increased semen output, increased percentage of living cells and a high motility rating (Knoop, 1946).

Sheep. Rams fed on a plane of nutrition 30 per cent less than that required to maintain constant bodyweight gave an ejaculate low in volume, but with a high sperm count after 16–20 weeks (Comstock and Brady, 1937), and when bodyweight was depressed so that fat was reduced to 12 per cent of the carcass by weight, the output of testosterone was reduced. Thus the effect of low plane of nutrition in rams follows the same pattern as in cattle.

The pattern of avitaminosis A in rams is essentially similar to that in cattle (Gunn, Sanders and Granger, 1942; Lindley *et al.*, 1949).

Pigs. Widely varying planes of nutrition may be tolerated by mature boars without detrimental affects on spermatozoa, but seminal plasma output is markedly reduced under conditions of low plane nutrition (Stevermer, Kovacs, Hoekstra and Self, 1961), thus following the pattern in cattle and sheep. Palludan (1963) showed that a similar reproductive pattern of avitaminosis A occurred in the pig, as in other species. In contrast with bulls, the boar exhibits normal spermatogenesis when fed a diet low in manganese (Plumlee, Thrasher, Beeson, Andrews and Parker, 1956).

Poultry. That inanition decreases the level of fertility in male chickens has been illustrated by Parker and McSpadden (1943). When an energy intake of 1584 kcal of ME/kg was fed for thirteen weeks the volume and fertilizing capacity of semen was reduced. At this level of energy intake, body weight loss was marked, and it is interesting to note that the males became completely infertile when their weight loss was 30 per cent (Parker and Arscott, 1964). Thirty per cent loss in bodyweight is the level generally quoted for nutritionally-induced anoestrum in female mammals.

The mature male chicken can maintain normal reproductive function on diets containing 6·9 per cent DCP (Arscott and Parker, 1963). Fertilizing capacity and sperm concentration of semen were adversely affected when male chickens were fed diets high in linoleic acid but without vitamin E or ethoxyquinone. The addition of vitamin E or ethoxyquinone to high linoleic acid diets overcame the adverse effects (Arscott *et al.*, 1965). This suggests that the role of vitamin E in poultry nutrition may be that of an antioxidant.

GENERAL SUMMARY AND CONCLUSIONS

As undernutrition develops the female is generally affected before the male. Mature males and to a less extent mature females show a remarkable resistance in terms of maintaining some reproductive activity during various stages of undernutrition and in chronic undernutrition animals may adapt themselves and maintain reproductive activity in spite of the lowered nutrient intake.

The question of anterior pituitary function occupies a dominant position in studies of the relationship of nutrition and reproduction and there is good evidence that inadequate nutrition can have adverse effects on pituitary function. Most signs of reproductive failure in both male and female can be traced to a change in the pattern of synthesis and/or release of pituitary gonadotrophins, although final proof, i.e. measurement of levels of circulating gonadotrophins is at present lacking. There is less evidence to indicate that the level of nutrition influences the response of target organs to either gonadotrophins or sex hormones, although circumstantial evidence indicates a positive effect in the case of vitamin deficiencies. Merely to inject exogenous hormones, and observe the response of the target tissue, is not satisfactory proof of normal function. The influence of nutrition on the metabolism of hormones remains a wide, but valid field for research.

While the majority of problems for research into the adverse effects of nutrition on reproduction are concerned with the female, ironically most progress has been made on studies in the male. This has been possible by the use of objective methods of assessing secondary sex-organ response to sex hormones, e.g. citric acid levels in accessory gland secretions in response to testosterone. Using such experimental techniques the influence of low nutrient intake in the growing male has been traced mainly to a change in pituitary gland activity, with consequent primary, secondary and tertiary effects. These methods offer the possibility of studying if there is a change in the response of target tissues to hormones as a result of nutrition and presents a wide field for further investigation. Unfortunately, in the female there is, as yet, no objective chemical test to measure oestrogen activity. This increases the difficulty of objectively evaluating the endocrine function of the gonads. Assessment of the gametogenic activity in the female is hampered by the discontinuous nature of the cycle. Secondary and tertiary responses cannot be studied as easily in the female as in the male.

The female shows particular susceptibility to poor nutrition during the period immediately preceding implantation. Once placental function is established the continuation of pregnancy is less influenced by poor nutrition, although the number and size of offspring may be affected. This basic difference may reside with those factors controlling hypophysial and placental gonadotrophin secretion. Hypophysial gonadotrophin activity is influenced by neuro-humoral factors, which might be selectively affected by poor nutrition, while placental gonadotrophin output proceeds independent of nervous control.

REFERENCES

AAES-JØRGENSEN, E. (1961) *Physiol. Rev.* **41**, 1–51.
ADAMS, C. E. (1953) *Mammalian Germ Cells*, a Ciba Foundation Symposium. Ed. Wolstenholme, Churchill, Ltd., London.
ALFIN-SLATER, R. B. and BERNICK, S. (1958) *Am. J. Clin. Nutr.* **6**, 613–24.
ALLEN, G. S. (1943) *Vet. Rec.* **55**, 168.
ALLEN, D. M. and LAMMING, G. E. (1961a) *J. Agric. Sci.* **56**, 67–9.
ALLEN, D. M. and LAMMING, G. E. (1961b) *J. Agric. Sci.* **57**, 87–95.

ALEXANDER, G., MCCANCE, I. and WATSON, R. H. (1956) *Proc. 3rd Int. Congr. Anim. Reprod., Cambridge, Engl.* Sect. 1, p. 5.

AMBEGAOKAR, S. D. and CHANDRAN, K. (1959) *Indian J. Med. Res.* **47**, 539.

ARSCOTT, G. H. and PARKER, J. E. (1963) *J. Nutr.* **80**, 311–4.

ARSCOTT, G. H., PARKER, J. E. and DICKINSON, E. M. (1965) *J. Nutr.* **87**, 63–8.

ARVY, L., ASCHKENASY, A., ASCHKENASY-LELU, P. and GABE, M. (1946) *C.R. Séanc. Soc. Biol. (Paris)* **140**, 730.

ASCHKENASY-LELU, P. and ASCHKENASY, A. (1947) *C.R. Séanc. Soc. Biol. (Paris)* **141**, 687–9.

AVERILL, R. L. W. (1964) *N.Z. J. Agric. Res.* **7**, 514–24.

BALL, Z. B., BARNES, R. H. and VISSCHER, M. B. (1947) *Am. J. Physiol.* **150**, 511–9.

BARBORIAK, T. T., KREHL, W. A., COWGILL, G. R. and WHEDON, A. D. (1957) *J. Nutr.* **63**, 591–9.

BECZE, J. (1962) *Allattenyésztés*, **11**, 93–96. *Nutr. Abstr. Rev.* **32**, 1396.

BELLOWS, R. A., POPE, A. L., CHAPMAN, A. B. and CASIDA, L. E. (1963) *J. Anim. Sci.* **22**, 101–8.

BENNETT, H. S., BAGGENSTOSS, A. H. and BUTT, H. R. (1950) *Am. J. Clin. Path.* **20**, 814–28.

BENTLEY, O. G. and PHILLIPS, P. H. (1951) *J. Dairy Sci.* **34**, 396–403.

BIDDULPH, C. and MEYER, R. K. (1941) *Am. J. Physiol.* **132**, 259.

BIERI, J. G., BRIGGS, G. M., SPIVEY FOX, M. R., POLLARD, C. J. and ORTIZ, L. O. (1956) *Proc. Soc. Exp. Biol. Med.* **93**, 237–40.

BLAXTER, K. L. and BROWN, F. (1952) *Nutr. Abstr. Rev.* **22**, 1–17.

BLUNN, C. T. (1939) *Anat. Rec.* **74**, 199.

BO, W. J. (1955) *Anat. Rec.* **124**, 241–9.

BO, W. J. (1956) *Anat. Rec.* **124**, 619–27.

BOAZ, T. G. (1962) *Vet. Rec.* **74**, 1482–93.

BODA, J. M. (1956) *J. Dairy Sci.* **39**, 66–72.

BODA, J. M. and COLE, H. H. (1956) *J. Dairy Sci.* **39**, 1027–54.

BODANSKY, M. and DUFF, M. B. (1941) *J. Nutr.* **21**, 235–42.

BORTOLUCCI, P. and ZAMBORINI, A. (1934) *Monit. Obstét. Ginéc.* 299–363.

BOYD, H. (1965) *Vet. Bull.* **35**, 251–66.

BOYER, P. D., SHAW, J. H. and PHILLIPS, P. H. (1942) *J. Biol. Chem.* **143**, 417–25.

BRANTON, C., BRATTON, R. W. and SALISBURY, G. W. (1947) *J. Dairy Sci.* **30**, 1003–13.

BRATTON, R. W., SALISBURY, G. W., TANABE, T., BRANTON, C., MERCIER, E. and LOOSLI, J. K. (1948) *J. Dairy Sci.* **31**, 779–91.

BRAUDE, R., KON, S. K., MITCHELL, K. G. and THOMPSON, S. Y. (1951) *Vet. Rec.* **63**, 671–3.

BRUCE, H. M. (1963) *J. Reprod. Fert.* **6**, 221–7.

BRYAN, W. L. and MASON, K. E. (1940) *Am. J. Physiol.* **131**, 263–7.

BURR, G. O. and BURR, M. M. (1929) *J. Biol. Chem.* **82**, 345–67.

BURR, G. O. and BURR, M. M. (1930) *J. Biol. Chem.* **86**, 587–621.

BURGER, J. F. (1952) *Onderstepoort J. Vet. Res.* Suppl. 2.

BYERS, J. H., JONES, I. R. and BONE, J. F. (1956) *J. Dairy Sci.* **39**, 1556–64.

CAMPOS-ZAMORANO, H. (1955) *Rev. Chilena. Educ. Fis.* **21**, 1074–9.

CASIDA, L. E. (1953) *Fertilisation Failure and Embryonic Death in Domestic Animals. Pregnancy Wastage.* Ed. E. T. Engle, Charles C. Thomas, Springfield.

CASIDA, L. E. (1956) 3rd Cong. Anim. Reprod. Plenary papers 19– .

CASIDA, L. E. (1963) *Proc. 6th Int. Congr. Nutr. Edinburgh*, p. 366.

CAVAZOS, L. F., JEFFREY, J. E., MANNING, J. P. and FEAGANS, W. M. (1961) *Anat. Rec.* **140**, 71–6.

CHAMPIGNY, O. and JAQUOT, R. (1961) *C.R. Hebd. Séanc. Acad. Sci. Paris.* **252**, 2765–7.

CHENG, D. W., CHANG, L. F. and BAIRNSON, T. A. (1957) *Anat. Rec.* **129**, 167–86.

CHOW, B. F. and LEE, C. J. (1964) *J. Nutr.* **82**, 10–18. *Nutr. Abstr. Rev.* **34**, 772.

CHRISTENSEN, F. and DAM, H. (1956) *Acta Physiol. Scand.* **36**, 82–6.

CHRISTIAN, R. E. and NOFZIGER, J. C. (1952) *J. Anim. Sci.* **11**, 789 (Abstr.).

CLARK, R. T. (1934) *Anat. Rec.* **60**, 125–59.

COMSTOCK, R. E. and BRADY, D. E. (1937) *Proc. Am. Soc. Anim. Prod.* 233–8.

CONTE-MAROTTA, R. (1937) *Boll. Soc. Ital. Biol. Sper.* **12**, 186–8.

COOP, I. E. and CLARK, V. R. (1955) *N.Z. J. Sci. Technol.* **37**, 214–28.

COOP, I. E. (1962) *J. Agric. Sci.* **58**, 179–86.

COURRIER, R. and RAYNAUD, R. (1932) *C.R. Séanc. Soc. Biol.* **109**, 881–3.
CURTISS, C. (1953) *Metabolism.* **2**, 344–53.
CYBULJKO, U. D. (1957) *Sv. inovodstvo.* **11**, No. 4, 40–2.
DAM, H. and GRANADOS, H. (1962) *Acta Pharmac. Tox.* **8**, 47–54.
D'ANGELO, S. A. (1951) *Endocrinology* **48**, 341–3.
D'ANGELO, S. A., GORDON, A. S., CHARIPPER, H. A. (1948) *Endocrinology* **42**, 399–411.
DANIELS, A. L. and EVERSON, J. G. (1935) *J. Nutr.* **9**, 191–203.
DAVIES, G. M. (1950) *J. Minist. Agric.* **57**, 379–83.
DAVIES, G. M. (1954) *Scott. J. Agric.* **34**, 149–53.
DAVIES, D. V., MANN, T. and ROWSON, L. E. A. (1957) *Proc. R. Soc. B.* **147**, 332.
DEAN, B. T. and TRIBBLE, L. F. (1961) *Bull. Mo. Agric. Exp. Stn.* No. 774, p. 44.
DECKER, A. B., FILLERUP, D. L. and MEAD, J. F. (1950) *J. Nutr.* **41**, 507–21.
DICKERSON, J. W. T., GRESHAM, G. A. and MCCANCE, R. A. (1964) *J. Endocr.* **29**, 111–8.
DIMOV, J. (1963) *Anim. Breed. Abstr.* **31**, 236.
DURRELL, W. B. (1951) *Can. J. Compt. Med.* **15**, 1–11.
DURRELL, W. B. (1955) *Can. J. Compt. Med.* **19**, 144–52.
DUTT, R. H. and BARNHART, C. E. (1959) *J. Anim. Sci.* **18**, 3–13.
DYBAN, A. P., DEMKIV, L. P. and AUGUSTINOVIC, M. S. (1963) *Dokl. Akad. Nauk. SSSR.* **149**, 145–6. *Nutr. Abstr. Rev.* **33**, 1074.
ELLINGER, G. M., DUCKWORTH, J., DALGARNO, A. C. and QUENOUILLE, M. H. (1952) *Br. J. Nutr.* **6**, 235–53.
EL-SHEIKH, A. S., HULET, C. V., POPE, A. L. and CASIDA, L. E. (1955) *J. Anim. Sci.* **14**, 919–29.
ELSON, L. A. and KOLLER, P. C. (1948) *Biochem. J.* **43**, 000.
ENGLE, E. T., CROFTS, R. C. and ZEITHANIL, L. C. E. (1937) *Proc. Soc. Exp. Biol. Med.* **37**, 437– .
ENSMINGER, M. E., BOWLAND, J. P. and CUNHA, R. J. (1947) *J. Anim. Sci.* **6**, 409–23.
ERB, R. E., ANDREWS, F. N., HAUGE, S. M. and KING, W. A. (1947) *J. Dairy Sci.* **30**, 687–702.
ERB, R. E., ANDREWS, F. N., BULLARD, J. F. and HILTON, J. H. (1944) *J. Dairy Sci.* **27**, 769–72.
ERSHOFF, B. H. (1952) *Vitams. Horm.* **10**, 79–140.
ESPLIN, A. C., MADSEN, M. A. and PHILLIPS, R. W. (1940) *Bull. Utah Agric. Exp. Stn.* No. 292, pp. 12.
EVANS, E. M. and EMERSON, G. A. (1943) *J. Nutr.* **26**, 555–67.
EVANS, H. M. and BURR, G. O. (1927) *Proc. Soc. Exp. Biol. Med.* **24**, 740.
EVANS, H. M. (1928) *J. Nutr.* **1**, 1–21.
EVANS, H. M. (1932) *Am. J. Physiol.* **99**, 477–86.
EVANS, H. M., LEPKOVSKY, S. and MURPHY, E. A. (1934) *J. Biol. Chem.* **106**, 431–50.
EVANS, R. E. (1929) *J. Agric. Sci.* **19**, 752–800.
EVERITT, G. C. (1964) *Nature (Lond.)* **201**, 1341–2.
FERGUSON, T. M., ATKINSON, R. L. and COUCH, J. R. (1954) *Proc. Soc. Exp. Biol. Med.* **86**, 868–71.
FERGUSON, T. M. and COUCH, J. R. (1954) *J. Nutr.* **54**, 361–70.
FERRANDO, R., JAQUES, F., MALBOUX, H. and SOLLOGUB, C. (1955) *C.R. Acad. Sci. (Paris)*, **241**, 107–9.
FIDANZA, A., BONOMOLO, A., DE CICCO, A. and ESPOSITO, L. (1959) *Quad. Nutr.* **19**, 364–72.
FIELDEN, E. D. and BRUMBY, P. J. (1962) *Aust. J. Biol. Sci.* **15**, 213–7.
FLIPSE, R. J., SNYDER, J. W., THACKER, D. L. and ALMQUIST, J. O. (1953) *Rep. Pa. Agric. Exp. Stn.* **104**.
FLIPSE, R. J. and ALMQUIST, J. O. (1961) *J. Dairy Sci.* **44**, 905–14.
FLIPSE, R. J. and ALMQUIST, J. O. (1963) *J. Dairy Sci.* **46**, 1416–9.
FOLLIS, R. H., DAY, H. G. and MCCOLLUM, E. V. (1940) *J. Nutr.* **20**, 181–95.
FOOTE, W. C., POPE, A. L., CHAPMAN, A. B. and CASIDA, L. E. (1957) *J. Anim. Sci.* **16**, 1100 (Abstr.).
FOOTE, W. C., POPE, A. L., CHAPMAN, A. B. and CASIDA, L. E. (1959a) *J. Anim. Sci.* **18**, 453–62.

FOOTE, W. C., POPE, A. L., CHAPMAN, A. B. and CASIDA, L. E. (1959b) *J. Anim. Sci.* **18**, 463–74.

FORKER, B. R. and MORGAN, A. F. (1955) *J. Biol. Chem.* **217**, 659–67.

GILMAN, J. P. W., PERRY, F. A. and HILL, D. C. (1952) *Can. J. Med. Sci.* **30**, 383–9.

GIROUD, A. (1959) *Wld. Rev. Nutr. Diet.* **1**, 233–63.

GIROUD, A., LEVY, G. and LEFEBORES-BOISSELOT, J. (1950) *Vitamin Forsch.* **22**,

GIROUD, A. and MARTINET, M. (1955) *C.R. Séanc. Soc. Biol.* **149**, 1088–90.

GOETTSCH, M. (1949) *Archs. Biochem.* **21**, 289–300.

GOLDIN, A., GREENSPAN, E. M., GOLDBERG, B. and SCHOENBERG, E. P. (1950) *Cancer*, **3**, 849–55.

GOLDSMITH, E. D. and NIGRELLI, R. F. (1950) *Trans. N.Y. Acad. Sci. II*, **12**, 236–7.

GOODE, L., WARNICK, A. C. and WALLACE, H. D. (1960) *J. Anim. Sci.* **19**, 643 (Abstr.).

GOODE, L., WARNICK, A. C. and WALLACE, H. D. (1965) *J. Anim. Sci.* **24**, 959.

GOODWIN, R. F. W. and JENNINGS, A. R. (1958) *J. Comp. Path. Ther.* **68**, 82–95.

GRANGAUD, R., CONQUY, T. and NICOL, M. (1960) *C.R. Séanc. Soc. Biol.* **154**, 587–90.

GREENBERG, S. M. and ERSHOFF, B. H. (1951) *Proc. Soc. Exp. Biol. Med.* **78**, 552–4.

GREEP, R. O. (1961) *Sex and Internal Secretions*, vol. 1, 3rd ed. Ed. W. C. Young. Ballière, Tindall & Cox Ltd., pp. 240–307.

GRIESBACH, W. E., BELL, M. E. and LIVINGSTON, M. (1957) *Endocrinology* **60**, 729–40.

GRUMMER, R. H., BENTLEY, O. G., PHILLIPS, P. H. and BOHSTEDT, G. (1950) *J. Anim. Sci.* **9**, 170–5.

GUILBERT, H. R. and HART, G. H. (1930) *Hilgardia*, **5**, 101 (Abstr.).

GUILBERT, H. R. and GOSS, H. (1932) *J. Nutr.* **5**, 251–65.

GULLICKSON, T. W. (1949) *J. N.Y. Acad. Sci.* **52**, 256.

GULLICKSON, T. W. and FITCH, J. B. (1944) *J. Dairy Sci.* **27**, 331–5.

GULLICKSON, T. W., PALMER, L. S., BOYD, W. L. and OLSEN, F. C. (1944) *J. Dairy Sci.* **27**, 634 (Abstr.).

GULLICKSON, T. W., PALMER, L. S., BOYD, W. L., NELSON, J. W., OLSEN, F. C., CALVERLY, C. E. and BOYER, P. D. (1949) *J. Dairy Sci.* **32**, 495–507.

GUNN, R. M. C., SANDERS, R. N. and GRANGER, W. (1942) *Bull. Coun. Scient. Ind. Res. Melb.* **148**, 140.

HAFEZ, E. S. E. (1951) *Nature* (*Lond.*) **168**, 1046–7.

HAFEZ, E. S. E. (1952a) *J. Agric. Sci.* **42**, 189–265.

HAFEZ, E. S. E. (1952b) *6th Int. Congr. Anim. Husb.* **3**, 35–7.

HAFEZ, E. S. E. (1959) *J. Am. Vet. med. Ass.* **135**, 606–14.

HAINES, C. E., WARNICK, A. C. and WALLACE, H. D. (1955) *J. Anim. Sci.* **14**, 124–6.

HAINES, C. E., WARNICK, A. C. and WALLACE, H. D. (1958) *J. Anim. Sci.* **17**, 879–85.

HAINES, C. E., WARNICK, A. C. and WALLACE, H. D. (1959) *J. Anim. Sci.* **18**, 347–54.

HAMMOND, J. (1940) *J. Minist. Agric.* **47**, 116–20.

HAMMOND, J. (1944) *Proc. Nutr. Soc.* **2**, 8–12.

HAMMOND, J., JNR. (1944) *J. Agric. Sci.* **34**, 97–105.

HANLY, S. (1961) *J. Reprod. Fert.* **2**, 182–94.

HANSON, L. E., FERRIN, E. F. and AUNAN, W. J. (1953) *J. Anim. Sci.* **12**, 919.

HART, E. B. and STEENBOCK, H. (1918) *J. Biol. Chem.* **33**, 313.

HART, G. H. and GUILBERT, H. R. (1937) *J. Am. Vet. med. Ass.* **91**, 193–200.

HART, G. H. (1941) *Nutr. Abstr. Rev.* **10**, 261–72.

HEAP, R. B., LODGE, G. A. and LAMMING, G. E. (1966) *J. Reprod. Fert.* In press.

HEAPE, W. (1899) *J. R. Agric. Soc.* **10**, 217–48.

HIGNETT, S. L. and HIGNETT, P. G. (1951) *Vet. Rec.* **63**, 603–9.

HIGNETT, S. L. and HIGNETT, P. G. (1952) *Vet. Rec.* **64**, 203–6.

HIGNETT, S. L. (1953) *Mem. Soc. Endocr.* **1**, 21–4.

HIGNETT, S. L. and HIGNETT, P. G. (1953) *Vet. Rec.* **65**, 21–5.

HIGNETT, S. L. (1959) *Vet. Rec.* **71**, 247–56.

HIROE, K., MASAKI, J., TOMIZUKA, T. and HANADA, A. (1964) *Bull. Natn. Inst. Anim. Indust. Japan*. Summaries of reports, No. 6, 1.

HODGSKISS, H. W., ENSMINGER, M. E., COLDBY, R. W. and CUNHA, T. J. (1950) *J. Anim. Sci.* **9**, 618–24.

Hodgson, R. E., Hall, S. R., Sweetman, H. J., Wiseman, H. C. and Converse, H. T. (1946) *J. Dairy Sci.* **29**, 669–87.

Holt, L. E., Jnr., Albanese, A. A., Shettles, L. B., Kajdi, C. and Wangerin, D. M. (1942) *Fedn. Proc.* **1**, 116–7.

Horn, E. H. (1955) *Endocrinology* **57**, 399–408.

Howell, J. McC., Thompson, J. N. and Pitt, G. A. J. (1963) *J. Reprod. Fert.* **5**, 159–66.

Howell, J. McC., Thompson, J. N. and Pitt, G. A. J. (1964) *J. Reprod. Fert.* **7**, 251–8.

Hoxsey, V., Hoversland, A. S., Blackmore, D. W. and Van Horn, J. L. (1960) *J. Anim. Sci.* **19**, 959 (Abstr.).

Huseby, R. H., Ball, Z. B. and Visscher, M. B. (1945) *Cancer Res.* **5**, 40.

Ingelman-Sundberg, A. (1954) *Acta Endocr.* **17**, 165–71.

Jackson, C. M. (1932) *Am. J. Anat.* **51**, 347–79.

Jensen, L. S. and MacGinnis, J. (1957) *Poult. Sci.* **36**, 1344–50.

Johnson, N. C., Kheim, T. and Kountz, W. B. (1959) *Proc. Soc. Exp. Biol. Med.* **102**, 98–9.

Jones, I. R., Haag, J. R. and Doucherty, R. W. (1942) *J. Dairy Sci.* **25**, 689–90.

Jones, I. R., Doucherty, R. W. and Haag, J. R. (1945) *J. Dairy Sci.* **28**, 311–20.

Joubert, D. M. (1954a) *J. Agric. Sci.* **44**, 5–64.

Joubert, D. M. (1954b) *J. Agric. Sci.* **45**, 164–72.

Kemmerer, C. A., Elvehjem, C. A. and Hart, E. B. (1931) *J. Biol. Chem.* **92**, 623.

King, J. W. B. and Young, G. B. (1957) *J. Agric. Sci.* **48**, 457–63.

Kinzey, W. G. and Srebnik, H. H. (1963) *Proc. Soc. Exp. Biol. Med.* **14**, 158–60.

Kline, I. T. and Dorfman, R. I. (1951) *Endocrinology*, **48**, 34–8.

Knoop, C. E. (1946) *J. Dairy Sci.* **29**, 555–6.

Krause, J. B. (1963) Ph.D. Thesis, Nottingham University.

Lafon, M. and Veillet, A. (1938) *C.R. Séanc. Soc. Biol.* **129**, 256–9.

Laing, J. A. (1952) *Rept. IInd Int. Congr. Anim. Reprod. Copenh.* **2**, 17–34.

Lamming, G. E. (1966) *Nutr. Abstr. Rev.* **36**, 1–13.

Lamming, G. E. and Salisbury, G. W. (1952) *6th Int. Congr. Anim. Husb.* **3**, 101–6.

Lamming, G. E., Salisbury, G. W., Hays, R. L. and Kendall, K. A. (1954) *J. Nutr.* **52**, 217–40.

Lamming, G. E., Woollam, D. H. M. and Millen, J. W. (1954) *Br. J. Nutr.* **8**, 363–9.

Lardy, H. A., Phillips, P. H. and Rupel, I. L. (1942) *J. Anim. Sci.* **1**, 79.

Leathem, J. H. (1958) *Recent Prog. Horm. Res.* **14**, 141–76.

Leathem, J. H. (1961) *Sex and Internal Secretions*, Ed. W. C. Young. vol. 1, 3rd edn. Ballière, Tindall & Cox Ltd., pp. 666–704.

Leathem, J. H. and Glasser, S. R. (1953) *Anat. Rec.* **115**, 341.

Lefebvres, J. (1951) *Annls. Méd.* **52**, 225.

Lindley, C. E., Brugman, H. H., Cunha, T. J. and Warnick, E. J. (1949) *J. Anim. Sci.* **8**, 591–602.

Lodge, G. A. (1966) Chapter (this book).

Loeb, L. (1917) *Biol. Bull.* **33**, 92.

Loeb, L. (1921) *J. Am. Med. Ass.* **77**, 1646.

Lutwak-Mann, C. and Mann, T. (1950) *Nature* (*Lond.*) **165**, 556–7.

Lutwak-Mann, C. (1958) *Vitams. Horm.* **16**, 35–75.

Macomber, D. (1933) *New. Engl. J. Med.* **209**, 1105, 1160, 1235, 1277.

Maddock, C. L., Cohen, J. and Wolbach, S. B. (1953) *A.M.A. Arch. Path.* **56**, 333–40.

Madsen, L. L., Eaton, O. N., Heemstra, L., Davis, R. E., Cabell, C. A. and Knapp, B. (1946) *J. Anim. Sci.* **5**, 391–2.

Mann, T. and Walton, A. (1953) *J. Agric. Sci.* **43**, 343–7.

Mann, T. (1960) *Proc. Nutr. Soc.* **19**, 15–8.

Mann, T. (1964) *The Biochemistry of Semen and of the Male Reproductive Tract.* Methuen & Co. Ltd., London.

Mann, T., Rowson, L. E. A. and Hay, M. F. (1960) *J. Endocr.* **21**, 361–72.

Marrian, G. F. and Parkes, A. S. (1928) *J. R. Microsc. Soc.* **48**, 257–69.

Marrian, G. F. and Parkes, A. S. (1929) *Proc. R. Soc. B.* **105**, 248–58.

Marston, H. R. (1952) *Physiol. Rev.* **32**, 66.

MARSHALL, F. H. A. (1905) *Proc. R. Soc. B.* 77, 58–62.
MASON, K. E. (1944) *Vitam. Horm.* **2**, 107.
MASON, K. E. (1935) *Am. J. Anat.* **57**, 303–49.
MAYER, J. and TRUANT, A. P. (1949) *Proc. Soc. Exp. Biol. Med.* **72**, 436–8.
MAYER, J. and GODDARD, J. W. (1951) *Proc. Soc. Exp. Biol. Med.* **76**, 149–51.
MCCANCE, R. A. (1960) *Br. J. Nutr.* **14**, 59–72.
MCCANCE, R. A. and WIDDOWSON, E. M. (1962) *Proc. R. Soc. B.* **156**, 326–37.
MCCLURE, T. J. (1961a) *J. Reprod. Fert.* **2**, 381–6.
MCCLURE, T. J. (1961b) *J. Comp. Path.* **71**, 16.
MCCLURE, T. J. (1962) *J. Reprod. Fert.* **4**, 241.
MCCLURE, T. J. (1963) *Nature* (*Lond.*) **199**, 504–5.
MCCLYMONT, G. L. and LAMBOURNE, L. J. (1958) *Proc. Aust. Soc. Anim. Prod.* **2**, 135–8.
MEACHAM, T. N., CUNHA, T. J., WARNICK, A. C., HENTGES, J. F., JNR. and HARGROVES, D. D. (1963) *J. Anim. Sci.* **22**, 115–20.
MEITES, J. and REID, J. O. (1949) *Proc. Soc. Exp. Biol. Med.* **70**, 513–6.
MEITES, J. (1953) *Iowa St. Coll. J. Sci.* **28**, 19–44.
MILLEN, J. W., WOOLLAM, D. H. M. and LAMMING, G. E. (1953) *Lancet*, **265**, 1234–6.
MILLEN, T. W., WOOLLAM, D. H. M. and LAMMING, G. E. (1954) *Lancet*, **267**, 679–83.
MILLER, C. O., ELLIS, N. R., STEVENSON, J. W. and DEVEY, R. (1953) *J. Nutr.* **51**, 163–70.
MITCHELL, H. H., CARROLL, W. E., HAMILTON, T. S. and HUNT, G. E. (1931) *Illinois Univ. Agric. Exp. Stn. Bull.* p. 375.
MOORE, R. A. and MARK, J. (1936) *J. Exp. Med.* **64**, 1–6.
MORRIONE, T. G. (1944) *Arch. Path.* **37**, 39–48.
MORRIS, T. R. and NALBANDOV, A. V. (1961) *Endocrinology* **68**, 687–97.
MOURIB, M. S., AHMED, I. A. and HAMADA, M. K. O. (1956) *Alexandria J. Agric. Res.* **4**, 85.
MOUSTGAARD, J. (1959) *Nutrition and Reproduction in Domestic Animals; Reproduction in Domestic Animals*. Ed. Cole, H. H. and Cupps, P. T., N.Y. pp. 170–224.
MULINOS, M. G. and POMERANTZ, L. (1940) *J. Nutr.* **19**, 493–504.
MULINOS, M. G. and POMERANTZ, L. (1941) *Endocrinology*, **29**, 267–75.
NEILSON, A. L. (1944) *Acta Med. Scand.* **118**, 87.
NELSON, M. M. and EVANS, H. M. (1946) *J. Nutr.* **31**, 497–507.
NELSON, M. M., LYONS, W. R. and EVANS, H. M. (1951) *Endocrinology*, **48**, 726–32.
NELSON, M. M. and EVANS, H. M. (1953) *J. Nutr.* **51**, 71–84.
NELSON, M. M. and EVANS, H. M. (1954) *Endocrinology*, **55**, 543–9.
NICHOLS, J. E. (1926) *J. Agric. Sci.* **16**, 365–75.
ORENT-KEILES, E. and MCCOLLUM, E. V. (1931) *J. Biol. Chem.* **92**, 651.
ORENT-KEILES, E., ROBINSON, A. and MCCOLLUM, E. V. (1937) *Am. J. Physiol*, **119**, 651–61.
PALLUDAN, B. (1963) *Acta Vet. Scand.* **4**, 136–55.
PALMER, L. S., GULLICKSON, T. W., BOYD, W. L., FITCH, C. P. and NELSON, J. N. (1940) *J. Dairy Sci.* **24**, 199–210.
P'AN, S. Y., VAN DYKE, H. B., KAUNTIZ, H. and SLANETZ, C. A. (1949) *Proc. Soc. Exp. Biol.* **72**, 523–8.
PANOS, T. C. and FINERTY, J. C. (1953) *J. Nutr.* **49**, 397–423.
PANOS, T. C. and FINERTY, J. C. (1954) *J. Nutr.* **54**, 315–29.
PANOS, T. C., KLEIN, G. F. and FINERTY, J. C. (1959) *J. Nutr.* **68**, 509–38.
PANOS, T. C. (1963) *J. Pediat.* **63**, 1167–78.
PARKER, J. E. and MCSPADDEN, B. J. (1943) *Poult. Sci.* **22**, 170–7.
PARKER, J. E. and ARSCOTT, G. H. (1964) *J. Nutr.* **82**, 183–7.
PIERCE, A. W. (1954) *Aust. J. Agric. Res.* **5**, 470–83.
PLUMLEE, M. P., THRASHER, D. M., BEESON, W. M., ANDREWS, F. N. and PARKER, H. E. (1956) *J. Anim. Sci.* **15**, 352–67.
POMEROY, R. W. (1954) *J. Agric. Sci.* **45**, 327–30.
QUINLAN, J. (1929) *15th Ann. Rep. Dir. Vet. Serv. Un. S.Af.* p. 83.
RAY, D. E. and MCCARTY, J. W. (1965) *J. Anim. Sci.* **24**, 660–3.
REID, J. T., TRIMBERGER, G. W., ASDELL, S. A., TURK, K. L. and SMITH, S. E. (1951) *J. Dairy Sci.* **34**, 510 (Abstr.).
REID, J. T. (1960) *J. Dairy Sci.* **43**, 103–22 (Supplement).

REID, J. T., LOOSLI, J. K., TRIMBERGER, G. W., TURK, K. L., ASDELL, S. A. and SMITH, S. E. (1964) *Bull. Cornell Univ. Agric. Exp. Stn.* No. 987. 31 pages.
RICHARDSON, L. R., GODWIN, J., WILKES, S. and CANNON, M. (1964) *J. Nutr.* **82**, 257–61.
RINALDINI, L. M. (1949) *J. Endocr.* **6**, 54–62.
RIPPLE, P. H., RAMUSSEN, O. G., JENSEN, A. H., NORTON, H. W. and BECKER, D. E. (1964) *J. Anim. Sci.* **24**, 203–8.
RIVERO-FONTAN, J., PASCHKIS, K. E., WEST, E. and CANTAROW, A. (1952) *Endocrinology*, **51**, 100–9.
ROBERTS, E. J., DAVIES, G. M. and WILLIAMS, G. L. (1949) *Agric. Prog.* **24**, 30–8.
ROBERTSON, G. L., CASIDA, L. E., GRUMMER, R. H. and CHAPMAN, A. B. (1951) *J. Anim. Sci.* **10**, 841–65.
ROWLANDS, T. W. and SINGER, E. (1936) *J. Physiol.* **86**, 323.
RUSHOFF, L. L. (1950) *J. Anim. Sci.* **9**, 666.
SACHS, B. D. (1965) *J. Comp. Physiol. Psychol.* **60**, 144–6.
SAMUELS, L. T. (1948) *Nutrition and Hormones*. Charles C. Thomas, Springfield, Illinois.
SCHEER, B. T., SOULE, D. F., FIELDS, M. and DEUEL, H. J. (1947) *J. Nutr.* **33**, 583–92.
SCHNEIDER, D. L., KURNICK, A. A., VAVICH, M. G. and KEMMERER, A. R. (1962) *J. Nutr.* **77**, 403–7.
SCOTT, E. B. (1955) *Am. J. Path.* **31**, 1111–29.
SCOTT, M. G. and SCOTT, P. P. (1958) *Proc. Soc. Study Fert.* **9**, 72.
SELF, H. L., GRUMMER, R. H. and CASIDA, L. E. (1953) *J. Anim. Sci.* **12**, 954.
SELF, H. L., GRUMMER, R. H. and CASIDA, L. E. (1955) *J. Anim. Sci.* **14**, 573–91.
SERFATY, A. and OLIVEREAU, M. (1955) *J. Physiol. Paris.* **47**, 829–34.
SHAW, J. H. and GRIFFITHS, D. (1963) *J. Nutr.* **80**, 123–41.
SHILS, J. H. and MCCOLLUM, E. V. (1943) *J. Nutr.* **26**, 1–19.
SORENSON, A. H., JNR., BRATTON, R. W., HANSEL, W. and HOUGH, W. H. (1954) *J. Anim. Sci.* **13**, 1031–2.
SORENSON, A. H., JNR., THOMAS, W. B. and GOSSET, J. W. (1961) *J. Anim. Sci.* **20**, 347–9.
SQUIRES, C. D., DICKERSON, G. E. and MAYER, D. T. (1950) *J. Anim. Sci.* **9**, 683.
SREBNIK, H. H. (1964) *Endocrinology*, **75**, 716–20.
SREBNIK, H. H., NELSON, M. M. and SIMPSON, M. E. (1958) *Proc. Soc. Exp. Biol. Med.* **99**, 57–61.
SREBNIK, H. H., NELSON, M. M. and SIMPSON, M. E. (1961) *Endocrinology*, **68**, 317–26.
SREBNIK, H. H. and NELSON, M. M. (1962) *Endocrinology*, **70**, 723–30.
SREBNIK, H. H. and NELSON, M. M. (1963) *Proc. 6th Int. Congr. Nutr., Edinb.* 375–86.
STEPHENS, D. J. and ALLEN, W. M. (1941) *Endocrinology*, **28**, 580–5.
STEVERMER, E. J., KOVACS, M. F., JNR., HOEKSTRA, W. G. and SELF, H. L. (1961) *J. Anim. Sci.* **20**, 858–65.
STEWART, E. L. and WELT, L. G. (1961) *Am. J. Physiol.* **61**, 824–6.
SUOMALAINEN, P. (1950) *Nature (Lond.)* **165**, 364.
SUTTON, T. S., KRAUSE, W. E. and HANSARD, S. L. (1940) *J. Dairy Sci.* **23**, 574.
TAPLIN, D. E. and EVERITT, G. C. (1964) *Proc. Aust. Soc. Anim. Prod.* **5**, 72–81.
TAYLOR, T. G., MORRIS, T. R. and HERTELENDY, F. (1962) *Vet. Rec.* **74**, 123–5.
TAYLOR, T. G. (1963) *J. Sci. Fd. Agric.* **14**, 611–3.
THOMSON, A. M. and THOMSON, W. (1948) *Br. J. Nutr.* **2**, 290–305.
TRAINA, V. (1950) *Nature (Lond.)* **166**, 310.
TUCHMANN-DUPLESSIS, H. and ASCHKENASY-LELU, P. (1947) *C.R. Séanc. Soc. Biol.* **141**, 689–90.
ULLREY, D. E., BECKER, D. E., TERRILL, S. W. and NOTZOLD, R. A. (1955) *J. Nutr.* **57**, 401–14.
UNDERWOOD, E. J. and SHIER, F. L. (1941) *J. Dep. Agric. West Aust.* **18**, 13.
UNDERWOOD, E. J. (1962) *Trace Elements in Human and Animal Nutrition*. 2nd edn. Academic Press Inc., N.Y. and London.
VAN DEMARK, N. L., FRITZ, G. R., EWING, L. L. and MAUGER, R. E. (1960) *J. Anim. Sci.* **19**, 1338.
VAN DEMARK, N. L. and MAUGER, R. E. (1964) *J. Dairy Sci.* **47**, 798–802.
VAN DEMARK, N. L., FRITZ, G. R. and MAUGER, R. E. (1964) *J. Dairy Sci.* **47**, 898–904.

VENKATACHALAM, P. S. and RAMANATHAN, K. S. (1964) *J. Nutr.* **84**, 38–42.
WADDELL, J., STEENBOCK, H. and HART, E. B. (1931) *J. Nutr.* **4**, 53–65.
WALLACE, L. R. (1946a) *J. Physiol.* **104**, 33–4.
WALLACE, L. R. (1946b) *J. Physiol.* **104**, 34–5.
WALLACE, L. R. (1948a) *J. Agric. Sci.* **38**, 93–152.
WALLACE, L. R. (1948b) *J. Agric. Sci.* **38**, 243–302.
WALLACE, L. R. (1948c) *J. Agric. Sci.* **38**, 367–401.
WALLACE, L. R. (1958) *N.Z. J. Agric.* **97**, 545–50.
WARKANY, J. and ROTH, C. B. (1948) *J. Nutr.* **35**, 1–11.
WERNER, S. C. (1939) *Proc. Soc. Exp. Biol. Med.* **41**, 101–5.
WHITE, P. G. and ROBERTS, J. A. F. (1922) *Welsh J. Agric.* **3**, 70–9.
WHITE, F. R., WHITE, J., MIDER, G. B., KELLEY, M. G. and HESTON, W. E. (1944) *J. Natn. Cancer Inst.* **5**, 43.
WIDDOWSON, E. M., MAVOR, W. O. and MCCANCE, R. A. (1964) *J. Endocr.* **29**, 119–26.
WILLIAMS, S. M. (1954) *J. Agric. Sci.* **45**, 202–7.
WILSON, J. G., ROTH, C. B. and WARKANY, J. (1953) *Am. J. Anat.* **92**, 189–217.
WILSON, J. L., THOMAS, B. H. and CANNON, C. Y. (1935) *J. Dairy Sci.* **18**, 431–2.
WILTBANK, J. N., ROWDEN, W. W., INGALLS, J. E., GREGORY, K. E. and KOCH, R. M. (1962) *J. Anim. Sci.* **21**, 219–25.
WILTBANK, J. N., ROWDEN, W. W., INGALLS, J. E. and ZIMMERMAN, D. R. (1964) *J. Anim. Sci.* **23**, 1049–53.
WITT, H. G., WARNICK, A. C., KOGER, M. and CUNHA, T. J. (1958) *J. Anim. Sci.* **17**, 1211 (Abstr.).
WOLFE, J. M. and SALTER, H. P., JNR. (1931) *J. Nutr.* **4**, 185–92.
WOOTEN, E., NELSON, M. M., SIMPSON, M. E. and EVANS, H. M. (1955) *Endocrinology*, **56**, 59–66.

CHAPTER 13

Endocrine Control of Metabolism and Growth

A. D. CARE

Rowett Research Institute, Bucksburn, Aberdeen

GROWTH may be defined as an integrated increase with time in the mass of the body in a species characteristic way. It is considered that the maximum size and development of an individual of a particular species are fixed by the genetic constitution. Adequate nutrition is then an essential factor for the achievement of the natural size. Given an adequate level of nutrition and a favourable environment for a particular species, its inherited growth potential is fulfilled through the integrated effects of the hormones secreted by the endocrine glands.

A distinction has been made between true growth and fat deposition (Maynard and Loosli, 1962). True growth is considered to imply an increase in weight of the organs and structural tissues so that a distinction can be made between growth and fattening. However, although it may often prove convenient to use this notation in an agricultural context, there is no logical reason for not regarding an increase in depot fat as part of the growth process since such an increase is an integral part of the metabolic processes which make up the living body.

Growth in the young depends upon net protein anabolism coupled with the development of the bony skeleton. Protein anabolism, which itself involves nitrogen retention by the body, is brought about by the combined actions of growth hormone from the anterior pituitary, thyroxine and tri-iodothyronine from the thyroid, insulin from the pancreas, androgens from the adrenal cortex and the testes, and oestrogens from the adrenal cortex and ovaries. On the other hand, protein catabolism is promoted by adrenal glucocorticoids, such as cortisol. The anterior pituitary exerts further influences on growth by way of thyrotrophin and corticotrophin secretion. The growth of the skeleton is also subject to a number of the above hormones and its degree of calcification is partly controlled by parathyroid hormone and calcitonin.

Before any discussion of growth regulation and control, a distinction must first be made between growth and differentiation. By growth, is meant an increase in the number and size of the cells, whereas differentiation implies a transformation of a homogenous group of cells into two or more cell groups which differ with respect to histological or physiological properties. For example, the initial increase in the size of the fertilized ovum as a result of simple cell multiplication is growth, whereas the later formation of ectoderm, mesoderm and endoderm represents differentiation. During the first third of intra-uterine life the human foetus grows but little, whereas cell differentiation is intense. However, growth is greatly accelerated during the later stages of pregnancy. After birth, four phases of general growth may be distinguished. These are, an initial rapid increase followed by a slower period of growth which then increases again about the time of sexual maturity. After that a slow terminal increment follows. However, it would appear that the adolescent spurt in growth rate is a perquisite of primates only.

GENERAL HORMONAL EFFECT ON GROWTH AND METABOLISM

Growth Hormone

The exact role of growth hormone in the regulation of normal growth remains somewhat enigmatic since it is impossible to induce in animals an experimental state in which a specific deficiency of growth hormone exists. Experiments in which growth was shown to be impaired following hypophysectomy cannot be interpreted as proof for the indispensability of growth hormone for normal growth since removal of the adenohypophysis excludes the secretion of other hormones in addition to growth hormone. However, it has been well established that the administration of growth hormone to hypophysectomized animals does result in growth stimulation. However, with the advent of the very sensitive radioimmunoassay for growth hormone (Hunter and Greenwood, 1964) it should be possible to correlate a low plasma concentration of growth hormone with hypopituitism, manifested by dwarfism. The possibility of any simultaneous depression of either thyrotrophin or corticotrophin secretion could be eliminated by similar radioimmunoassay for these two hormones. In this way, it should be possible to demonstrate, unequivocably, the requirement of a normal secretion of growth hormone for normal growth.

Growth is a complicated process and growth hormone influences a number of biochemical pathways of metabolism (Fig. 1). It stimulates the growth of bone and cartilage, the oxidation of fat, the retention of nitrogen and the synthesis of protein, blood glucose concentration is raised and the weight of the body increased. Among the tissues where anabolism proceeds most actively are the epiphyseal plates of cartilage. Thus, the administration of growth hormone to an animal, the epiphyses of which have not yet fused, is followed by a longitudinal growth of bone associated with the formation of new cartilage. Increased production of cartilage is not associated with epiphyseal maturation, which occurs as one of the anabolic effects seen after the

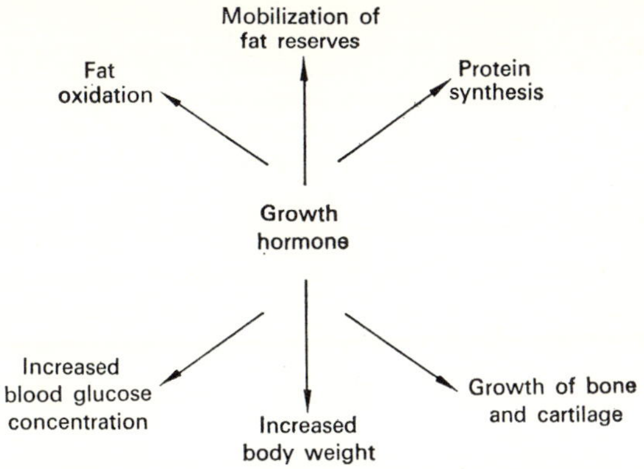

FIG. 1.—The principal effects of growth hormone on growth and metabolism.

administration of either androgenic or oestrogenic hormones (Escamilla *et al.*, 1961). It is the ability of growth hormone to increase the synthesis of both protein and mucopolysaccharide in the promotion of cartilage anabolism without leading to epiphyseal fusion that enables growth hormone to stimulate longitudinal bone growth unaccompanied by stunting.

After hypophysectomy, a sort of barrier to further calcification appears which prevents normal calcifying activity at the epiphyseal line and brings about a cessation of growth. This barrier disappears after the injection of growth hormone so that the cartilage cells at the epiphyseal line again begin to proliferate and differentiate into bone. Membrane bones (which have no epiphyses) are unaffected by growth hormone administration or hypophysectomy. Any bony tissue that has been already laid down is thickened by a sustained increase in growth hormone activity.

The increase in body growth as a whole produced by growth hormone is due to its direct action on the tissues and is not mediated through the secretion of other endocrines, although these others may exert independent effects of their own. The muscle and viscera grow as rapidly as the skeleton although in cases of excessive growth hormone secretion there is excessive development of epiphyseal bone resulting in gigantism if secreted prior to epiphyseal union, and acromegaly (overgrowth of parts of the skeleton) if the epiphyses have already joined.

Along with nitrogen retention and a positive balance of the minerals and water required for increased growth, stimulated by the administration of growth hormone, there is an increase in structural fat, i.e. in those lipids which form an integral part of the cell architecture. The fat reserves, however, are decreased and a reduction in the respiratory quotient is observed. This increased mobilization of storage fat is also associated with an increase in the plasma concentration of non-esterified fatty acids, but the increase is, to some extent, a species-related phenomenon as, although it is clearly seen in dogs and man, it is less obvious in rats. The effect of insulin in decreasing the plasma concentration of free fatty acids is not prevented by growth hormone and there is evidence that the effect of growth hormone on plasma-free fatty acid concentration is counteracted by the ingestion of food.

The mechanisms by which growth hormone produces its effects in the body remain obscure. It has been suggested that growth hormone promotes tissue anabolism by the promotion of amino-acid transport into tissues (Kostyo and Engel, 1960). However, some of the evidence which has been presented in support of this theory is insufficient to eliminate increased protein synthesis within the cell as the prime action, since this would create a concentration gradient of amino acids favouring their increased transfer into the cells as a secondary action. In lipid metabolism, it has been shown *in vitro* that the incorporation of ^{14}C acetate and palmitate into liver phospholipids is greatly reduced by the addition of growth hormone (Leal and Greenbaum, 1961).

THYROTROPHIN

In response to the secretion of thyrotrophin from the anterior pituitary not only is iodide uptake by the thyroid regulated but also the formation of enzymes involved in the conversion of iodotyrosines to the thyroid metabolic hormones, tri-iodothyronine and thyroxine. The primary action of these hormones is to stimulate metabolism in the tissues, thus increasing their oxygen consumption (Fig. 2). However, hypothyroid animals are still capable

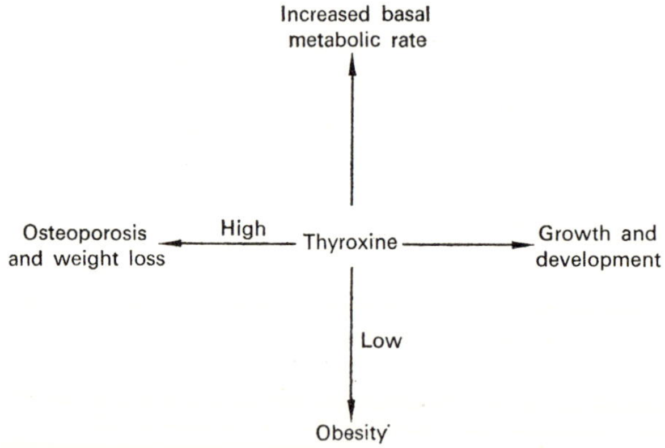

FIG. 2.—The effects of thyroxine and tri-iodothyronine on growth. HIGH and LOW indicate hyperthyroidism and hypothyroidism, respectively.

of oxidizing glucose as rapidly as usual (Mirsky and Broh-Khan, 1936). The exact way in which this stimulation of oxidation by thyroxine takes place is unknown except for the fact that the tissues are acted upon directly. The growth and development of organs in general are also promoted by the thyroid metabolic hormones.

Deficient secretion of these thyroid hormones does not result in well recognized effects on fat metabolism other than those which result from decreased utilization of proteins, fats and carbohydrates. With normal consumption of food, the hypothyroid animal becomes obese, the excess food being stored as fat (Scow, 1951). Simultaneous with this is an impairment of growth rate and a reduction in the carcass protein content. This is particularly noticeable in the epiphyseal regions of bone so that failure in the longitudinal growth of such bone and in the development of the epiphyseal centres are characteristic signs of thyroid deficiency. In the hyperthyroid animal, the excessive secretion of thyroxine leads to a removal of calcium and phosphate from bones and a progressive osteoporosis. The mode of action is unknown, but the effect is prevented by the administration of vitamin D.

As already indicated, the main regulator of thyroid hormone secretion is the anterior pituitary via thyrotrophin. Other factors which influence the secretion of the thyroid hormone are the plasma iodide concentration, changes in the ambient temperature and impulses reaching the thyroid along its sympathetic nerve (Wright, 1961). The control of thyrotrophin, on the other hand, is partly nervous via the hypothalamus and partly humoral via the plasma thyroid hormone concentration, according to a negative feedback. Actually, tri-iodothyronine is more potent than thyroxine in the suppression of thyrotrophin secretion. In some, but not all, species, if corticotrophin secretion is stimulated simultaneously the adrenocortical hormones, released in their turn, tend to depress the secretion of thyrotrophin. Thus, only in these animals, following adrenalectomy, does hypothalamic stimulation lead to a marked increase in the secretion rate of thyroid hormones.

Androgens and Oestrogens

Since androgenic and oestrogenic hormones are not elaborated in significant amounts before the onset of sexual maturity it is assumed that these hormones play no part in the regulation of growth during this early period. At adolescence, the adrenal cortex increases its secretion of sex hormones just as do the gonads. Once secretion of these hormones starts, then an anabolic effect may be observed. For example, the administration of testosterone, the principal androgen, brings about retention of nitrogen and increased anabolism (Fig. 3). In association with this, there is also a positive balance of other substances required for general growth, such as phosphorus, potassium and sulphur (Deakins, Friedgood and Ferrebee, 1944). There are, of course, also specific effects in connection with the development of primary sex organs and of secondary sex characteristics. In addition, testosterone stimulates bone growth, especially of the ossification centres. That is, increased nitrogen retention is associated with increased longitudinal growth of bones and increased development of the secondary ossification centres which are stimu-

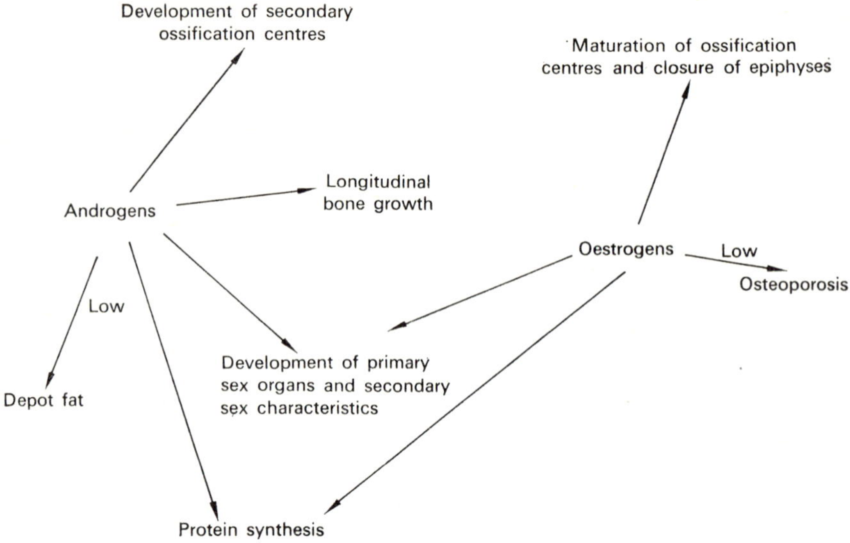

FIG. 3.—The principal effects of the sex hormones on growth. LOW indicates decreased hormone secretion.

lated to grow until fusion with the metaphyses occurs (Sobel *et al.*, 1956). Once fusion of an ossification centre with the corresponding metaphysis occurs, longitudinal growth of the bone virtually ceases. Castration of the adult is associated with an increase in weight, much of which is depot fat.

Oestrogens also exert marked effects on the osseous system. Apart from the development of the characteristic female skeleton, they too have profound effects on centres of ossification and linear growth. Ossification centres mature more rapidly and epiphyses tend to close earlier under the influence of oestrogen. The effect on bone metabolism is exemplified by the fact that generalized osteoporosis occurs in ovarian insufficiency (Albright, 1947) but the mechanisms by which oestrogens affect skeletal growth are not clearly understood, although a decrease in the excretion of nitrogen, sodium and phosphate has been demonstrated.

Oestrogen administration to ruminants induces increases in body weight but this phenomenon is not observed in non-ruminants. Clegg and Cole (1954) found that oestrogen-implanted heifers all had larger pituitary and adrenal glands than similar control animals and that the anterior pituitaries contained twice as much growth hormone than the controls. There was also increased nitrogen retention but no change in the amount digested. Other workers have noted an increased retention of both calcium and phosphorus. The extent and reasons for the species variation remain obscure.

In contra-distinction to the anabolic effects which follow the administration of oestrogens and androgens to intact adult animals, progesterone usually appears to exert a catabolic effect on tissues (Landau and Lugibihl, 1961).

However, it has been reported that if progesterone is given to normal female rats an increase in live body weight follows, of which nearly half is fat (Hervey and Hervey, 1964). This effect is not seen in male rats. There is some evidence that the increase in weight is associated with the presence of functional ovarian tissue, but not merely to the secretion of oestrogens. It is thought that the change in weight is brought about by an alteration in the central regulation of energy balance.

PARATHYROIDS

The role of the parathyroids in the control of normal growth is not clearly defined. It is well known that parathyroid hormone is intimately concerned with skeletal osteoclastic processes and in deficiency states an impairment of growth is usually seen. This may be associated with the unusual Ca:P ratio then observed in the plasma.

CORTICOTROPHIN

The main regulator of the secretion of the adrenocorticoids is the anterior pituitary through the hormone, corticotrophin. In the resting state, a negative feedback exists between plasma cortisol concentration and the secretion rate of corticotrophin. Of the adrenal cortical hormones, cortisol is probably the most important one with respect to growth processes. It is a catabolic hormone, stimulating the breakdown of tissue protein to amino acids (Fig. 4).

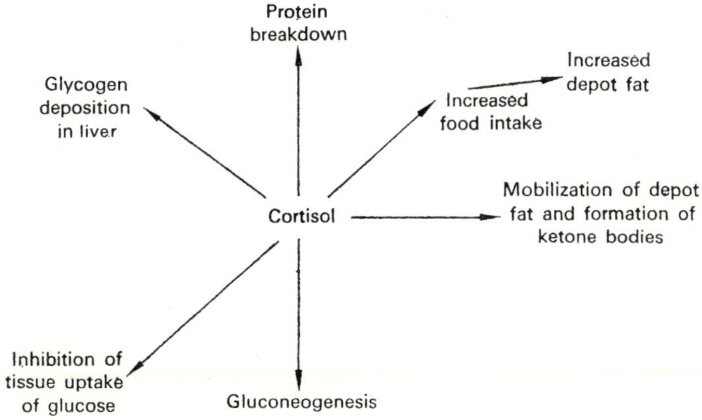

FIG. 4.—The principal metabolic functions of cortisol.

Fat is mobilized from storage depots and broken down in the liver to form ketone bodies. Thus the administration of cortisone in large doses over prolonged periods often results in retardation of growth. However, this is not an invariable finding as actual weight gain in animals has been recorded by a number of workers. This weight gain has usually been attributed to increase in appetite and thus increased food intake (Hausberger and Ramsay, 1953) whereas instances of weight loss have been explained in terms of the negative nitrogen balance observed (Rupp and Paschkis, 1957).

This paradox can be resolved when some consideration is paid to the dose rate employed (Kekwick and Pawan, 1965). At a high dose level, cortisone produces a negative nitrogen balance and loss of weight when food intake is maintained constant. Body fat is redistributed with an increase in liver fat, but the effect of cortisone on the metabolism of adipose tissue is still very obscure. Large doses of cortisol bring about an increase in food intake without gain in weight, since the loss of protein and water is balanced by an increase in carcass fat. In other experiments, using cortisone, its administration resulted in a net loss in weight but, nevertheless, an increase in body fat is also noted. At a lower dosage of cortisone and a constant food intake, it is possible to maintain nitrogen equilibrium, although total body fat is still increased. It is assumed that this increase in body fat is achieved by an alteration, either quantitative or qualitative, in the metabolic pathways used by the animal in response to cortisone or by an alteration in the rate of heat exchange. It has been suggested that the effect of cortisone on fat metabolism is an indirect one which follows an increase in the production of insulin. As the body approaches maturity there is a sudden increase in the weight of the adrenal cortex and it has been suggested that this increased adrenocortical activity may have some connection with the retardation of growth observed.

Hormonal Control of Appetite

The above account has dealt with the endocrine control of the metabolic processes involved in normal growth of the body. Such processes are only sufficient when they are provided with sufficient substrate. To achieve this it must be assumed that an adequate diet is available and that the reaction of the animal to such a diet is normal. The endocrine control of the appetite of an animal for water, minerals and food will now be discussed.

a. Thirst

There are specific osmoreceptors near the supra-optic nucleus which respond to changes in plasma crystalloid osmotic pressure (Verney, 1947). Stimulation of these osmoreceptors sets up nerve impulses which reflexly stimulate the hypothalamus and so, the neurohypophysis, to secrete antidiuretic hormone. If the plasma becomes hypertonic, as a result of water deprivation, both the receptor discharge and the secretion of antidiuretic hormone are increased, thereby increasing water retention by the kidney.

Thirst itself is not a simple sensation; there is some evidence that a thirst centre is located in the hypothalamus. For example, the injection of minute amounts of hypertonic saline directly into the dorsal and ventral hypothalami of unaesthetized goats is followed by polydipsia; electrical stimulation of a small area of the hypothalamus also causes polydipsia.

Animals have the ability to regulate their voluntary dietary intake of saline very accurately in response to previous salt depletion so that precise repletion is soon achieved (Beilharz, Denton and Sabine, 1962). The mechanism by which this is accomplished is not known.

b. Appetite

Growth in the sexually mature adult is primarily a matter of fat deposition although net changes in protein metabolism may still occur, e.g. the development of certain muscles in response to a particular exercise. Although the regulation of depot fat deposition is still by no means clear, it is thought that appetite and food intake play a prominent part. It has been shown that centres exist in the hypothalamus which are concerned with the regulation of appetite (Larsson, 1954). Evidence has been forthcoming that glucoreceptors are present in the ventromedial nuclei of the hypothalamus, since these have been shown to take up radioactive glucose from the blood more actively than do adjacent parts of the brain. If these ventromedial nuclei are destroyed, an excessive appetite is produced and obesity results. Stimulation of the lateral hypothalamic region also increases appetite, so that it has been suggested that there are two hypothalamic centres connected with the regulation of food intake (a) a ventromedial satiety centre, and (b) a lateral feeding centre. If the glucoreceptors in the satiety centre are stimulated by a low blood glucose concentration it is thought that stimuli are passed to the feeding centre which in turn influences the activity of the dorsal motor nucleus of the vagus to cause hunger contraction of the stomach and gastric secretion. Thus, it may be seen that the control of blood glucose concentration is important in the control of food intake.

c. Glucose Homeostasis

The endocrine control of blood glucose concentration is achieved by the integration of the actions of insulin, growth hormone, glucagon, thyroxine, adrenocorticoids and adrenaline (Fig. 5). This hormonal control of blood glucose concentration depends only partly on the action of these hormones on the liver; other tissues, such as muscle, are also affected. The above hormones may be divided into two groups according to their effect. In the one (A) is insulin, a hypoglycaemic hormone, whilst in the other (B) are the remaining hormones which all tend to be diabetogenic.

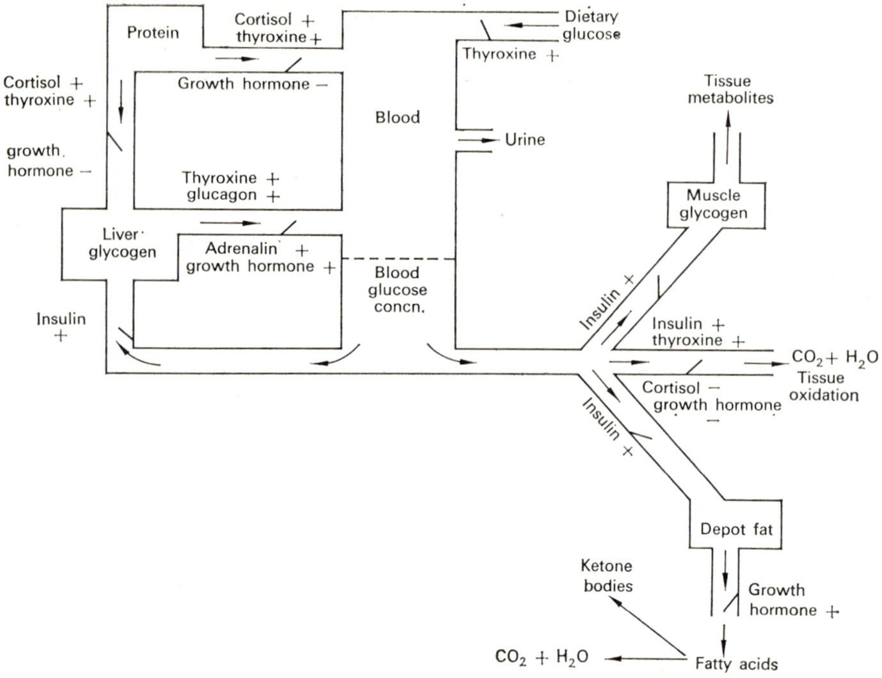

FIG. 5.—The hormonal regulation of blood glucose concentration. + or — represents stimulation or inhibition by the hormone of glucose transport in the direction of the arrow.

A. Insulin increases the deposition of glycogen, depresses gluconeogenesis and increases lipogenesis. It stimulates the rate of complete oxidation of glucose in the tissues and may also depress the conversion rate of liver glycogen to glucose.

The similarity between the effect of administering insulin and that of increasing the extracellular glucose concentration supports the suggestion that a major property of insulin is to assist in the transfer of free glucose through cell membranes for subsequent utilization by the intracellular systems, particularly the hexokinase catalysed intracellular phosphorylation of glucose to glucose-6-phosphate (Ross, 1956).

Insulin activity in the body is effected by a combination of controlled insulin secretion from the pancreas and a group of insulin antagonists. The primary regulation of insulin secretion is achieved by way of a negative feedback through the blood glucose concentration. A fine adjustment of secretion is thought to operate via branches of the vagus to the pancreatic Islets of Langerhans. However, insulin secretion can still occur despite sectioning of the pancreatic nerve supply.

B. The principal antagonist of insulin is probably the anterior pituitary factor, so that an animal, from which both the pancreas and the anterior pituitary have been removed, is not diabetic. On the other hand, lacking hormonal control, the liver reacts imperfectly and with a longer than usual time lag. It would seem that the ratio of insulin to the anterior pituitary diabetic factor seems in a large measure to control the carbohydrate metabolism of the body.

This diabetogenic factor appears to be identical with growth hormone. The effects of growth hormone would seem to be related to the existing metabolic status of the animal. That is, the same preparation may promote growth in young, or lactating, animals but be diabetogenic in adults. This action is associated with an increase in the output of glucose by the liver from glycogen as well as non-carbohydrate sources. Glucose oxidation is suppressed whilst gluconeogenesis is inhibited in favour of accelerated protein production.

With the aid of the recently developed radioimmunoassay of growth hormone in plasma, it has been shown that there is a negative feedback which operates through the blood glucose concentration (Roth *et al.*, 1963). Since growth hormone also causes a rapid release of fatty acid from reserve adipose tissue (but not from the structural fat which forms an integral part of cells), it has been suggested that growth hormone secretion may really be regulated by the demand for a non-carbohydrate source of energy, such as fatty acid, when available carbohydrate sources appear to be insufficient (Berson *et al.*, 1964).

It is not yet clear whether obesity is a cause or a result of deficient pituitary responsiveness to normal stimuli. It is well known that fasting produces a state of insulin resistance and it has recently been demonstrated that the plasma glucagon concentration increases during the fasting period. However, it seems likely that the increase in growth hormone observed is the more important factor by virtue of its essential role in the mobilization of fatty acids. Also, excessive nitrogen loss is prevented by the direct protein anabolic action of this hormone as well as by a protein-sparing effect secondary to increased depot fat utilization. In short, growth hormone suppresses the oxidation of both carbohydrate and protein but stimulates the catabolism of fat.

The glucocorticoids, e.g. cortisol, from the adrenal cortex are also diabetogenic in that they depress the peripheral utilization of glucose and increase gluconeogenesis by the liver. In the former respect, the glucocorticoids enhance the effectiveness of growth hormone but they also increase the deposition of liver glycogen, in this respect simulating the action of insulin. The diabetogenic effect, however, is usually the more important.

The thyroid hormones, thyroxine and tri-iodothyronine, stimulate tissue metabolism, in general, presumably by catalysing the enzyme systems responsible for oxidation reactions. In the liver, there is a stimulation of glycogen conversion to glucose and of gluconeogenesis. The absorption of glucose from the small intestine is increased.

The role of glucagon, another pancreatic hormone, in the regulation of blood glucose concentration may be to maintain glucose distribution to glucose-dependent tissues during glucose deprivation. This is achieved by stimulation of glycogen breakdown in liver and by increased gluconeogenesis. The control of glucagon secretion appears to be by a negative feedback mechanism operated through the blood glucose concentration.

Adrenaline is also glycogenolytic as a result of the activation of the phosphorylase used in the breakdown of liver glycogen to glucose phosphate, and later to glucose. The secretion of adrenaline from the adrenal medulla is stimulated by hypoglycaemia.

It is only too evident from the above that the hormonal interrelationships involved in the control of live weight gain, be it in the immature young or in the mature adult, are extremely complicated. Far too little experimental material is so far available for many firm conclusions to be drawn as to the way in which genetic influences may be imposed on growth processes by way of the endocrine regulated metabolic pathways.

The Application of Hormonal Relationships to the Growth of Farm Animals

In general, the rate at which an animal grows is of greater interest agriculturally than its mature weight. Few farm animals, apart from dairy cattle and some sows, ever live long enough to reach their mature weight, but the rate at which they grow has a profound influence on their profitability. The reason for this is that the food eaten by a mature animal is largely devoted to the satisfaction of maintenance requirements, the remainder leading to the deposition of fat. In the growing animal, on the other hand, a larger proportion of its absorbed food is used for growth so that the ratio of weight gain to feed intake is higher. This is a somewhat simplified approach since maximum growth rate at all ages may not always produce the best possible carcass conformation, so that it may be desirable to maximize growth rate at certain stages of an animal's life and deliberately to retard it at others (Pomeroy, 1955).

The growth-promoting effects of certain hormones have already been described. However, only certain of these hormones can be manufactured or extracted sufficiently cheaply for their large scale use in farm animals. Thus, only oestrogens, androgens, progestogens and thyroxine or thyroprotein are employed as positive agents for growth promotion. Goitrogens have also been tried for certain types of fattening.

a. The Use of Sex Hormones for Growth Promotion

Bogart, Lasley and Mayer (1944) showed that injected oestrogen would inhibit growth in normal and ovariectomized rats and that, conversely, progesterone stimulated growth. The same effect of progesterone could be obtained by the injection of luteinizing hormone, by pseudo-pregnancy or by true pregnancy. The effect of ingested oestrogens in other species has led to somewhat variable results. Thus, Dinusson, Klosterman and Buchanan (1951) found no appreciable stimulus to growth rate, only an improved efficiency of food conversion, following the subcutaneous implantation of 12 mg stilboestrol into pigs, whereas Woehling *et al.* (1951) found no effect on either growth or fattening. On the other hand, stilboestrol administration to fattening lambs was found to improve both the rate of liveweight gain and the efficiency of food conversion (Means, Andrews and Beeson, 1953; Thomson and Gill, 1963). Similar results have also been obtained with fattening bullocks (Burroughs, Culbertson, Kastelie, Cheng and Hale (1954).

However, these reported advantages of oestrogen administration have been offset by several reports of a reduction in the grade or quality of the carcasses (Bell, Smith and Erhart, 1954; Gill, Thomson and Crichton, 1956). Whiting, Clark and Allan (1954) suggested that the increase in liveweight of treated lambs might be due entirely to an increase in moisture content of the tissues and to a high percentage of offal. To some extent, this is supported by Gee and Preston (1957) who showed that the carcasses of treated lambs contained more bone and flesh but less subcutaneous fat and that these tissues had a greater percentage of protein and moisture but a lower percentage of fat than the carcasses of control lambs. Similarly, Aitken and Crichton (1956) found that the carcasses of treated steers contained more muscle and bone but less fat than carcasses of untreated animals. It would thus appear that the principal nutritional advantage of oestrogen administration stems from the increase in ability of treated animals to convert food protein to muscle protein.

The oral administration of oestrogen to young ruminants has also been found to increase the rate of growth but this is seldom a practical method of administration. In addition, it has been demonstrated that the simultaneous implantation of both oestrogen and progestogen brings about a growth response in fattening cattle.

Undesirable side effects have been noted in animals given large doses of oestrogens, e.g. mammary development in bullocks and wethers, urethral calculi and vaginal and rectal prolapse. Such changes have also been obtained following grazing on pastures containing a high proportion of subterranean clover. More important than this is the possibility of carcinogenicity in man associated with the ingestion of oestrogenic residues over a long period.

It is now well established that androgens, such as testosterone, stimulate protein anabolism in cattle and that females show a greater response than males. The results obtained with pigs and lambs, however, have been inconsistent.

b. The Use of Thyroprotein and Goitrogens for Growth Promotion

The stimulatory effect of thyroxine on the basal metabolic rate has led to the use of thyroxine and thyroprotein (iodinated casein) to increase the growth rate of young farm animals. For each animal there is an optimal dosage level of thyroxine; this varies with the species, age, sex and ambient temperature. Such dependence on experimental conditions may explain the variability in response to the administration of thyroxine which has been observed. For example, the oral administration to pigs of synthetic thyroprotein, as a constant percentage of the ration, led to a significant increase in growth rate, whereas, when it was given as a constant percentage of the body weight, it had no effect or even resulted in a smaller growth rate. These results suggest that stimulation of growth rate may be obtained if small doses of thyroid metabolic hormones are fed to young growing pigs and if the dose per unit of body weight decreases with increasing body weight. After the pigs have reached the fattening stage, the use of thyroprotein becomes detrimental. With broilers, results have been variable as have those with calves and lambs. Thus, in spite of its potential usefulness, synthetic thyroprotein is only used to a very limited extent because of this variation in response and difficulty in assessment of the correct dose.

If thyroprotein is administered orally to lactating cows with the object of stimulating lactation, additional feeding is required or else a decrease in body weight will result. Even with this increased ration, there is evidence to suggest that the total lactational effect of thyroprotein feeding is small or nil, even though its short term effect may be pronounced (Leech and Bailey, 1953).

Hypothyroidism is associated with an increased deposition of fat (Fig. 2). This has led to a number of experiments involving either partial thyroidectomy or the oral administration of goitrogens, such as thiouracil and thiourea, aimed at increasing the rate of fattening of farm animals. In this connection, a good deal of work has been done with pigs but the results obtained have been very variable. If the dose is too high or the period of administration too long, both growth rate and feed efficiency are markedly reduced. However, when an increase in liveweight gain is produced, it consists largely of fat. In poultry, on the other hand, the administration of thiouracil in combination with diethylstilboestrol does improve finish and market quality without depression of growth rate (Maynard and Loosli, 1962).

REFERENCES

Aitken, J. N. and Crichton, J. A. (1956) *Brit. J. Nutr.* **10**, 220.
Albright, F. (1947) *Recent Prog. in Hormone Res.* **1**, 293.
Beilharz, S., Denton, D. A. and Sabine, J. R. (1962) *J. Physiol.* **163**, 378.
Bell, T. D., Smith, W. H. and Erhart, A. B. (1954) *J. Anim. Sci.* **13**, 425.

Berson, S. A., Yalow, R. S., Glick, S. M. and Roth, J. (1964) *Metabolism*, **13**, 1135.
Bogart, R., Lasley, J. F. and Mayer, D. T. (1944) *Endocrinology*, **35**, 173.
Burorughs, H., Culbertson, C. C., Kastelic, J., Cheng, E. and Hale, W. H. (1954) *Science*, **120**, 66.
Clegg, M. T. and Cole, H. H. (1954) *J. Anim. Sci.* **13**, 108.
Deakins, M. L., Friedgood, H. B. and Ferrebee, J. W. (1944) *J. Clin. Endocrin.* **4**, 376.
Dinusson, W. E., Klosterman, E. W. and Buchanan, M. L. (1951) *J. Anim. Sci.* **10**, 885.
Escamilla, R. F., Hutchings, J. J., Deamer, W. C., Li, C. H. and Forsham, P. H. (1961) *J. Clin. Endocrin.* **21**, 721.
Gee, I. and Preston, T. R. (1957) *Brit. J. Nutr.* **11**, 329.
Gill, J. C., Thomson, W. and Crichton, J. A. (1956) *Brit. J. Nutr.* **10**, 226.
Hausberger, F. X. and Ramsay, A. J. (1953) *Endocrinology*, **53**, 423.
Hervey, G. R. and Hervey, E. (1964) *J. Endocrin.* **30**, vii.
Hunter, W. M. and Greenwood, F. C. (1964) *Biochem. J.* **91**, 43.
Kekwick, A. and Pawan, G. L. S. (1965) *J. Endocrin.* **31**, 265.
Kostyo, J. L. and Engel, F. L. (1960) *Endocrinology*, **67**, 708.
Larrson, S. (1954) *Acta Physiol. Scand.* **32**, Suppl. 115.
Landau, R. L. and Lugibihl, K. (1961) *J. Clin. Endocrin.* **21**, 1345.
Leal, R. S. and Greenbaum, A. L. (1961) *Biochem, J.* **80**, 27.
Leech, F. G. and Bailey, G. L. (1953) *J. Agr. Sci.* **43**, 236.
Maynard, L. A. and Loosli, J. K. (1962) *Animal Nutrition*, 5th ed. McGraw-Hill, New York.
Means, T. M., Andrews, F. N. and Beeson, W. M. (1953) *J. Anim. Sci.* **12**, 176.
Mirsky, I. A. and Broh-Khan, R. K. (1936) *Amer. J. Physiol.* **117**, 6.
Pomeroy, R. W. (1955) *Progress in the Physiology of Farm Animals*, vol. 2. Ed. J. Hammond. Butterworths, London.
Ross, E. J. (1956) *Medicine*, **35**, 355.
Roth, J., Glick, S. M., Yalow, R. S. and Berson, S. A. (1963) *Science*, **140**, 987.
Rupp, J. J. and Paschkis, K. E. (1957) *Proc. Soc. Exp. Biol., N.Y.* **95**, 477.
Scow, R. O. (1951) *Endocrinology*, **49**, 522.
Sobel, E. H., Raymond, C. S., Quinn, K. V. and Talbot, N. B. (1956) *J. Clin. Endocrin. and Metab.* **16**, 241.
Thomson, W. and Gill, J. C. (1963) *Progress in Nutrition and Allied Sciences*. Ed. D. P. Cuthbertson. Oliver Boyd, Edinburgh.
Verney, E. B. (1947) *Proc. Roy. Soc. B*, **135**, 59.
Whiting, F., Clark, R. D. and Allen, C. E. (1954) *Canad. J. Agric. Sci.* **34**, 288.
Woehling, H. L., Wilson, G. D., Grummer, C. H., Bray, R. W. and Casida, W. E. (1951) *J. Anim. Sci.* **10**, 889.
Wright, S. (1961) *Applied Physiology*, p. 496. Oxford Univ. Press, London.

CHAPTER 14

Methods of Calorimetry (A) Direct

J. D. PULLAR
Rowett Research Institute, Bucksburn, Aberdeen

INTRODUCTION

The history of direct calorimetry began with the experiments of Lavoisier and Laplace,[1] who, in 1793, constructed a calorimeter suitable for use with small animals. A measure of the amount of heat given off by a guinea pig was obtained from the quantity of water running off a pack of ice surrounding the animal. This was an adiabatic instrument since the exterior of the ice pack was protected from heat exchange with the surroundings by means of a jacket of ice and water. Despite the primitive nature of his equipment Lavoisier obtained surprisingly accurate results.

The latent heat of evaporation of volatile liquids was used in a calorimeter made by d'Arsonval[2] in 1881. This removed one major disadvantage of the ice calorimeter, the need to operate at a fixed temperature of 0°C.

In the late nineteenth century many instruments were devised which depended on the measurement of the rise in temperature of a limited quantity of water. Into this category come the calorimeters of Desplats, Laulanié and Dulong, in all of which the animal was surrounded by but not in contact with the water. Immersion of the subject in the water was necessary in the bath calorimeters of Liebermeister and Lefèvre, and also in an instrument designed by Burton and Bazett[3] as recently as 1936.

Several direct calorimeters were constructed during the period 1880–1900 which were not of the adiabatic type. Instead the heat flow through the walls was measured. Probably the most famous of these are the instruments of Rubner, Richet, d'Arsonval and Rosenthal. In general these calorimeters depended on measurement of the expansion and contraction of water or air in a jacket surrounding the heat-emitting subject.

The term "compensation calorimetry" has been applied to two fundamentally different methods. One method, also known as "differential calorimetry" employs two similar chambers. The heat production of the subject in one chamber is continuously balanced by a controlled electrical heat production in the other chamber. The second approach may preferably be termed "circulating medium calorimetry". In this method the heat produced by the animal is removed by a circulating current of air or water. The rise in temperature of the cooling medium in relation to its rate of flow gives a measure of the heat emission of the subject. This system dates back to the air-cooled calorimeter of Lefèvre.[4] It was also used in the famous classical respiration calorimeter of Atwater and Rosa,[5] designed for human studies, and that of Armsby and Fries which is large enough to accommodate cattle. The latter instrument has been described by Braman.[6]

In these classical calorimeters heat flow through the chamber walls was minimized by the use of adiabatic jackets. No heat transfer is possible to or from a double-walled chamber if the two walls are kept at the same temperature. Under these circumstances the sensible heat lost by the animal must be transferred to the stream of cooling medium which usually flowed through a complicated system of pipes within the chamber. In other calorimeters different methods were employed in an attempt to achieve the same end. Capstick[7] relied on very heavy insulation alone, while Swietoslawski[8] devised an ingenious system of labyrinth flow in which the cooling water flowed within the chamber walls in such a fashion as to give a close approximation to an adiabatic jacket.

These older calorimeters, however, had many disadvantages. Several experienced operators were continuously on duty during an experiment, and the results were obtained in a form which required considerable processing. The speed of response to changes in heat output was very slow; a long preparatory period was therefore required for stabilization. Further, the capital cost of the instruments in terms of materials and labour was high enough to impose a severe restriction on the calorimetric facilities which any single research centre could afford to provide. For these reasons, indirect calorimetry which required much less complex and costly equipment, had largely supplanted direct methods by the early years of the present century. A very detailed review of the development of direct calorimetry up to the year 1905 has been published by Lefèvre.[9] More recent instruments have been described by Kleiber.[10]

In the last fifty years the trend of direct calorimeter design has been towards reduced thermal inertia and more automatic operation. One step in this direction was taken by Murlin and Burton[11] when they constructed a semi-automatic calorimeter for human subjects at Rochester University in the early 1930's. Water, flowing in pipes within the chamber, was again used to measure most of the heat emission, but the need for adiabatic jackets was eliminated by the inclusion of gradient thermometers which measured the heat leakage through the chamber walls. Hardy and Day,[12] in 1942, described an infant calorimeter which dispensed completely with the cooling medium. Heat flow from the subject was calculated from the average temperature of twelve copper plates mounted close to the walls of the chamber. A more flexible instrument based on the gradient principle was devised by Prouty,

Barrett and Hardy[13] at Cornell University in 1948. This calorimeter was large enough to hold a full-sized cat.

A major advance in the technique of direct calorimetry came in 1949 when Benzinger and Kitzinger[14,15] first described their design for a gradient layer calorimeter. Adiabatic principles were completely abandoned in favour of direct measurement of the instantaneous flow of heat through the walls of the chamber. Not only the sensible heat loss (the sum of the radiative, convective and conductive components) but also the evaporative loss of the subject could be measured continuously and automatically. For radiant heat transfer the speed of response of the calorimeter was stated to be 42 sec for full deflection.

Calorimeters of the Benzinger type were constructed at the U.S. Naval Medical Research Institute and at the Laboratories of the American Society of Heating and Ventilating Engineers.[16] Others, for use with animals, were built by Lawton, Prouty and Hardy[17] at the Russell Sage Institute, and by Stewart[18] at the University of Missouri. A gradient layer calorimcter large enough to accommodate cattle is currently being developed by McLean[19] at the Hannah Dairy Research Institute.

At the Rowett Research Institute work commenced in 1952 to build a gradient calorimeter of the Benzinger pattern for animals up to about 50 lb (23 kg) live weight. This instrument was completed in 1955. A larger calorimeter for adult sheep or for pigs up to bacon weight was started in 1956 and completed in 1959. The remainder of this chapter is mainly based on the experience obtained in building these two calorimeters.

THEORY OF THE GRADIENT LAYER CALORIMETER

In his design for the gradient layer calorimeter Benzinger developed a technique which had previously been used by Tian[20] in his microcalorimeters. A mathematical analysis of the method has been published by Bothe.[15]

If a heat-emitting subject is completely surrounded by a uniform layer of insulating material, the average difference in temperature between the inner and outer surface of this layer, in steady-state conditions, will be proportional to the rate of sensible heat loss of the subject. The thicker the layer of material the greater will be the mean temperature gradient across it for a given heat flow, but the less quickly will this temperature difference follow changes in the thermal output of the subject. Provided the temperature of the outside surface of the layer (the reference temperature) is maintained constant, the average temperature drop across the insulating shell will be a measure of the non-evaporative heat loss of the subject, and, if thin, will follow with relatively little inertia the minute-to-minute changes in heat emission which occur.

If the animal were emitting heat equally in all directions, and were equidistant from all points of the layer, it would simply be necessary to measure the temperature of the inner and outer surfaces. Since these conditions do not hold the temperature difference must be measured at many points and the mean value established. For this purpose the layer is divided up into a large number of equal units of area, each of them small in comparison with the size of the subject. To measure the temperature drop across these elements of area a thermopile is constructed with junctions attached alternately to the inner and outer surfaces. Each pair of thermo-junctions then produces an electric output which is proportional to the heat flow through the particular element of area to which the junctions are attached. When the thermocouples are all connected up in series the electrical outputs are effectively integrated over the entire layer, and the total e.m.f. will then be proportional to the sensible heat output of the subject. Provided the heat measuring shell is uniform throughout as regards thickness and thermal conductivity, the electrical response will be independent of the position of the heat source within the layer, and of the nature and direction of the heat emitted; it is also independent of the shape of the measuring shell. The electrical output of the chamber walls, which is of sufficient magnitude for direct application to commercially available potentiometric recorders, gives a continuous and automatic measure of the sensible heat emission of the subject.

Evaporation of water, both from the skin and from the respiratory tract, forms an important fraction of the total heat loss of a live subject. The ability to measure this evaporative loss in similar fashion to the sensible fraction is one of the major advantages of the Benzinger type of direct calorimeter. Most previous instruments relied on the absorption in sulphuric acid or other drying agent of the water transferred by the animal to the ventilatory airstream. This procedure does not readily lend itself to continuous or to automatic recording.

The arrangement of the air circuit used in the adult sheep sized gradient layer calorimeter at the Rowett Institute is shown in Fig. 1. Four heat exchange units (referred to by Benzinger as 'plate-meters') are used. In these units the ventilatory airstream passes for a considerable path length between two water jackets. The efficiency is sufficiently good for the temperature of the air to approach closely that of the water before the exit of each unit is reached.

Heat exchange units 1 and 2, immediately preceding the animal chamber, are used to condition the ingoing air to the desired values of temperature and humidity. Further heat exchangers, 3 and 4, extract and measure respectively the sensible and the evaporative transfer of heat from the animal to the airstream. Two large capacity water circuits are used to supply the water jackets and at the same time to set the reference temperatures of the gradient layers. One tank supplies the water jackets of heat exchangers 1 and 4: water from the other is circulated round the chamber walls and also supplies heat exchangers 2 and 3. The water in circuit 1 is invariably colder than that in circuit 2.

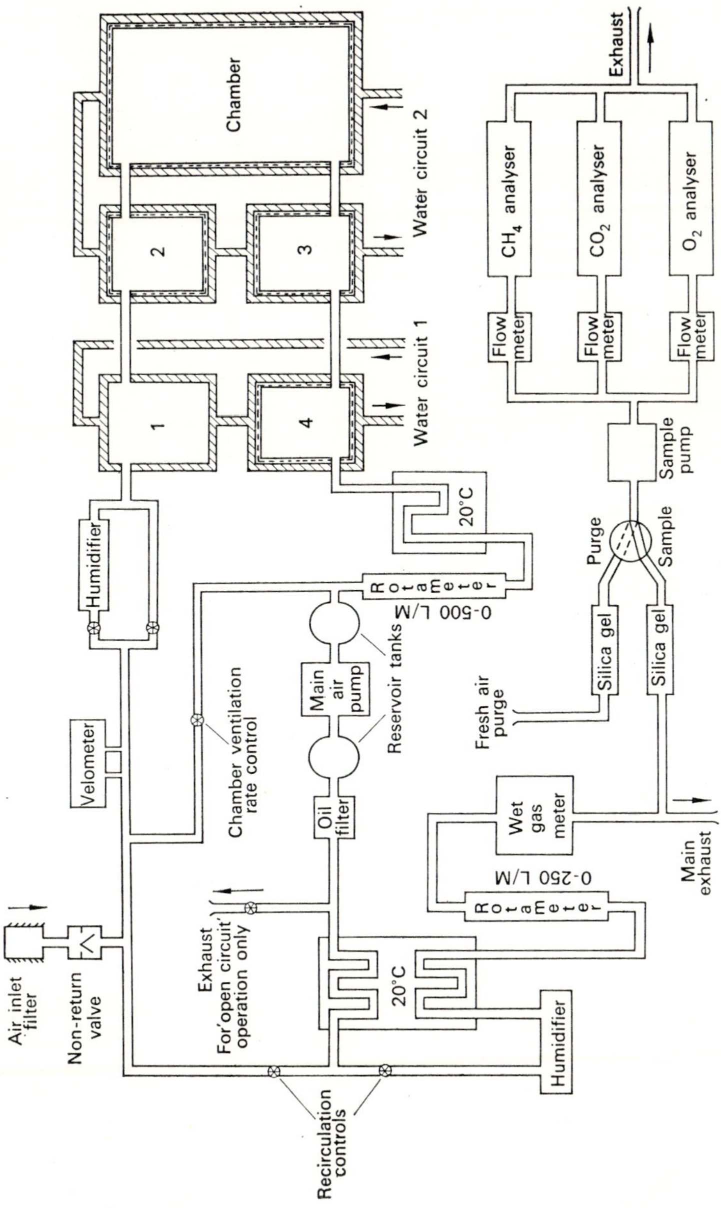

FIG. 1. Ventilatory air circuit for a gradient layer calorimeter

Outside air, entering through the inlet filter, is cooled to temperature t_1 in heat exchanger 1. This temperature is generally in the range 2–15°C. During the cooling process the air must reach and pass its saturation point. Atmospheric air will not necessarily contain sufficient water vapour for this to occur. Provision is therefore made to pass any desired proportion of the airstream through a humidifier before it is cooled.

Saturated air leaving heat exchanger 1 passes into heat exchanger 2 where it is warmed to the temperature, t_2, of the water in circuit 2. This is the temperature at which the animal chamber is being operated during any particular experiment and may be anywhere in the range 5–40°C. The relative humidity of the airstream falls as it is warmed in heat exchanger 2, and the air enters the chamber at a humidity determined by the relative values of t_1 and t_2. Ambient temperature in the chamber is fixed by the temperature of the water in circuit 2. Relative humidity of the incoming air is controlled by proper choice of the temperature of the water in circuit 1.

Outgoing air from the chamber is slightly warmer than incoming air when an animal is inside. This fraction of the total heat loss is extracted from the airstream and measured in heat exchange unit 3. The electrical output of this unit is added to the e.m.f.s from the chamber walls, the total representing the sensible heat loss of the animal. Unit 3 is not essential for the accurate measurement of total heat loss. Outgoing air from the chamber could be routed for a considerable path close to the chamber walls in order to extract and measure the sensible heat transfer to the airstream using the gradient layers in the chamber. This technique, however, makes the partitioning of heat loss less accurate, and reduces the speed of response of the calorimeter to radiant and convective transfer.

In heat exchange unit 4 the airstream is again cooled to the temperature t_1 at which it was previously saturated. With no source of water vapour in the chamber the air on cooling will just reach saturation point; when an animal is present the additional water vapour it loses to the airstream condenses out in heat exchanger 4 and gives up its latent heat which is measured by the surrounding gradient layer. Unit 4 also gives an output, constant under any one set of conditions, due to the cooling of the airstream from t_2 to t_1. This output is eliminated by fitting a similar gradient layer to heat exchange unit 2. Unit 2 then gives an e.m.f. due to the heating of the same airstream from t_1 to t_2, which can be subtracted from the output of heat exchanger 4 by connecting the e.m.f.s electrically in opposition. The difference in outputs is then proportional to the evaporative heat loss of the animal.

The gradient layer calorimeter thus automatically partitions the heat loss of the subject into its evaporative and sensible components. Benzinger's original design for a human calorimeter took this facility a step further. Provision was made for the additional partitioning of the evaporative loss into the respiratory and cutaneous contributions. In constructing the present calorimeters

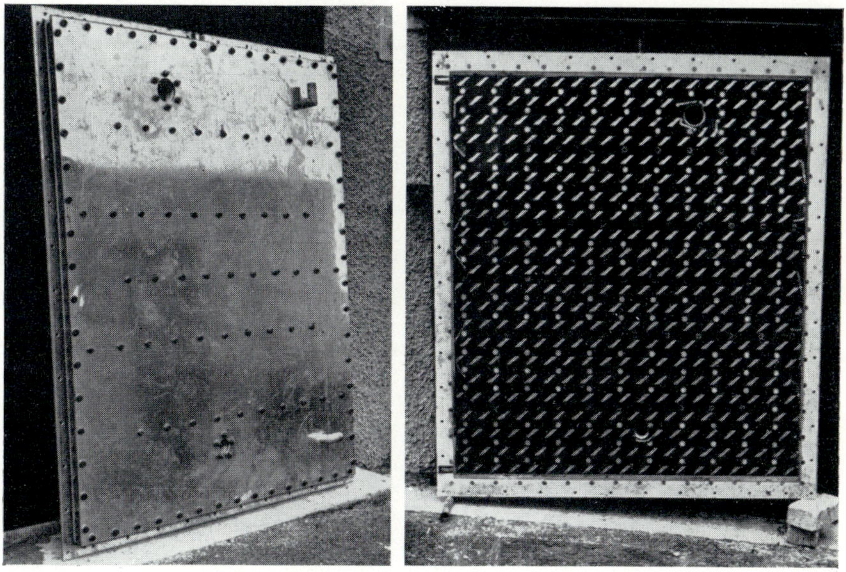

FIG. 2. Front and back views of a chamber wall.

FIG. 3. The chamber with a specialized design of animal cage inside.

for animals it was not considered that this additional facility would be of much value. Recent work[21] with trained sheep, however, would have been greatly expedited had a separate respiratory air circuit been available.

The heat measuring shell now extends over the entire inner surface area of the chamber walls, and also over the surfaces of three out of the four heat exchange units. In addition, a gradient layer lined container can be provided below the chamber to collect and measure the heat content in voided urine. It is necessary to maintain the same sensitivity in terms of electrical output for a given rate of heat flow per unit area throughout. Heat leakage in the connecting tubes between units must be minimized. These connections should therefore be short and well insulated. It will be seen that the gradient layer calorimeter must give a correct measurement of total heat loss provided the ventilatory airstream is at the same conditions of temperature, pressure and humidity at the exits of heat exchange units 1 and 4, and provided that all heat transfer with the exterior is properly measured between those points.

DESIGN AND CONSTRUCTION

The Animal Chamber

Front and back views of one wall of the sheep-sized calorimeter are shown in Fig. 2, from which many of the constructional details will be apparent. The walls of the chamber are designed as integral units fairly readily detachable from the strong angle-iron framework on which they are mounted. Each wall consists of a $\frac{1}{4}$ in. (6·35 mm) thick aluminium plate carrying the gradient layer on the inner face and the water jacket on the outside. Copper walls were used in the smaller calorimeter mentioned above. No adverse effect due to reduced thermal conductivity through the use of aluminium has been noticed in the sheep calorimeter, while the reduction in weight and cost is considerable. The internal size of the chamber is approximately 4 ft (1·22 m) × 5 ft 6 in. (1·67 m) × 4 ft 6 in. (1·37 m) (high), dimensions which have proved suitable for experiments on pigs up to 200 lb (91 kg) liveweight, though additional height would have been advantageous in some work with sheep.

Ideally, the floor of a large animal chamber should be close to ground level. This was unfortunately not possible in the present case mainly because of the urine disposal arrangements. Various designs of animal container are used in different experiments, and it is necessary to raise these cages with the animal inside through a height of about 27 in. (69 cm) to reach chamber floor level. Container and animal can then be pushed into the chamber, the cage moving ou castors which run in aluminium channels mounted just above the chamber floor.

One rather specialized design of animal cage is shown in Fig. 3, which also illustrates the door closing arrangements. The entire front wall of the chamber is hinged on a vertical axis. This door carries on its inner face, around the area of the gradient layer, a rectangular steel strip, which, on shutting the door, compresses an expanded rubber gasket. With this type of calorimeter the requirement for airtightness is not particularly stringent, and sufficiently effective door-sealing is obtained by means of four screw clamps distributed round the periphery of the door.

Figure 4 shows part of a vertical cross-section through the animal chamber. On the external side of the $\frac{1}{4}$ in. (6·35 mm) aluminium plates rubberized cork

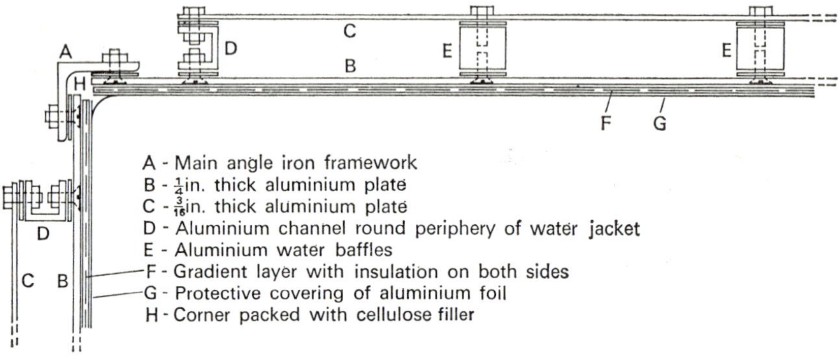

FIG. 4. Part of a vertical section through the animal chamber

gaskets are used to seal the chamber walls. This is supplemented on the inside by packing the junctions of the walls with cellulose filler. The chamber is insulated externally with a layer of fibre-glass about 3 in. (7·6 cm) thick and the whole enclosed within a plywood outer cover.

The design of the water jackets will be apparent from Fig. 4 and from the rows of bolt-heads in Fig. 2. Complete coverage of each side by the water flow is ensured by the use of square section aluminium baffles. Where it is necessary for a service, such as the ventilatory air holes in Fig. 2, to pass through the water jacket, aluminium castings with gaskets are used to provide the water seal. With a complex system such as this it is important to provide purge points so that all air can be expelled from the water circuit without difficulty. It is also advisable to cover all walls of the water channels with an epoxy coating resin to minimize the risk of corrosion.

The water jackets are approximately $1\frac{1}{2}$ in. (3·8 cm) wide on all walls except the door. Here, without adverse effect, the width was reduced to just over $\frac{1}{2}$ in. (1·27 cm) in order to lessen the strain on the hinges. The chamber water circuit is arranged in two parallel sets of three jackets in each. The rate of water flow is about 6 gal/min through each chain, which is sufficient to avoid any significant change in water temperature in passage through the chamber walls.

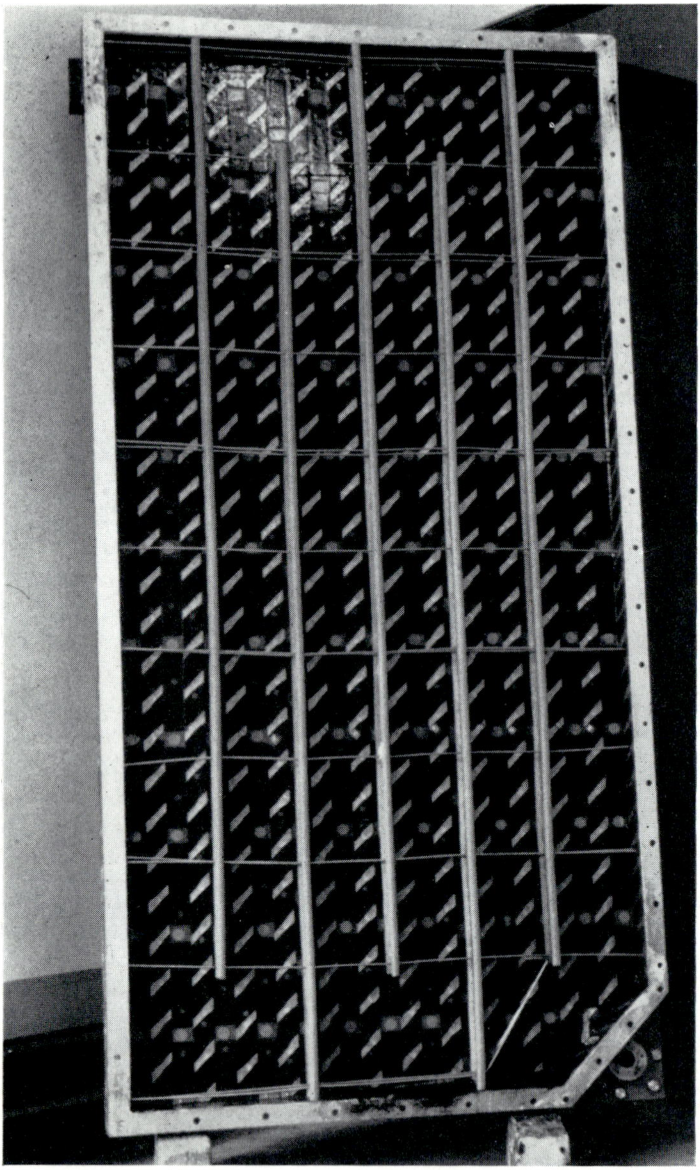

FIG. 5. The interior of a heat exchange unit.

The Heat Exchange Units

The heat exchange units are designed to operate at a maximum rate of air flow of 500 l./min. Under most conditions this is suitable for animals up to 200 lb (91 kg) live weight. The chief difficulty of design is the need to achieve the necessary efficiency (i.e. outgoing air temperature sufficiently close to water temperature) without the air pressure drop through the unit becoming excessive. Overmuch resistance to air flow would increase the pressure differential to the exterior and complicate the sealing problems throughout the air circuit.

Figure 5 shows part of a heat exchange unit, of size approximately 6 ft (1·82 m) × 3 ft (0·91 m), carrying the gradient layer on the inner face: the water jacket is mounted on the reverse. A second similar section, without the ring of aluminium channel, is bolted to the illustrated portion of the heat exchanger and renders the whole airtight. Again the main plates of these units are of ¼ in. (6·35 mm) thick aluminium. Air enters at the top left-hand corner and leaves at the top right-hand corner as shown. Baffles of light aluminium edged with expanded rubber direct the airstream over a path length of about 35 ft (10·7 m) through the unit in a channel of cross-sectional area about 1¾ in. (4·5 cm) × 5½ in. (14 cm). Not shown in Fig. 5 are aluminium deflector vanes which are positioned in the air channels so as to break up the central core of the airstream and throw the air preferentially against the surfaces of the gradient layers. A suitable compromise between efficiency and pressure drop through the unit was obtained with thirty-six of these deflectors in each heat exchanger. At maximum air flow rate and under the most severe conditions of ingoing temperature difference the outgoing air approaches within 0·2°C of the circulating water temperature for an air pressure change of under 1 in. (2·54 cm) water gauge. The four heat exchangers are all very similar in design. Unit 1 does not require gradient layers. However, dummy layers were fitted to this unit to keep the efficiency similar to that of the others.

Water vapour is continuously condensing in heat exchangers 1 and 4 and under conditions of misadjustment condensation could occur in exchangers 2 and 3. All these units have therefore suitable drainage arrangements at the lower end of the air channels, and units 1 and 4 are fitted with water traps so that the condensate can flow out without permitting the entry of air.

A major difficulty in the design of a gradient layer calorimeter is to obtain sufficient stability of the temperature of the water circulating in the jackets of the chamber and the heat exchangers. Any variation in water temperature will result in a spurious electrical output from the thermocouples, the magnitude of which is a function of the rate of change of temperature, and of the water equivalent of the material enclosed within the gradient layers. As mentioned later it is not normally possible to control the water temperature by conventional thermostatic methods. However, by reducing the amount of material within the gradient layers of the heat exchangers to a minimum it has been possible to use 'on-off' control of low-power immersion heaters in water circuit 1 (Fig. 1) without producing an intolerable saw-tooth waveform in the outputs of the gradient layers.

It is particularly necessary that heat exchangers 1 and 4 should be enclosed within an insulated container which is effectively vapour sealed. These units run for long periods at temperatures down to nearly 0°C and condensation will readily occur on the outer surfaces.

In the Benzinger design heat exchange unit 1 and the humidifier (Fig. 1) were amalgamated, the ventilatory airstream passing through a water spray within the exchanger. This system was not adopted here, partly because of the difficulty of achieving 100 per cent relative humidity and partly because of the dangers of water droplets being carried over into the next unit. Certain difficulties were originally encountered due to droplet formation and supersaturation. In addition to improving efficiency the deflector vanes within the heat exchangers play a most important role in removing droplets and providing surfaces on which condensation can occur.

To ensure that the air temperatures are maintained equal at the exits of heat exchangers 1 and 4 a thermopile is constructed with the junctions at these two points. Using a centre-zero galvanometer the operator occasionally has to vary the temperature of the humidifier or the proportion of the airstream passing through it to achieve temperature equality at those two points.

The Urine Container

Normally the urine is allowed to drain into a 70 l. container beneath the animal chamber. While the sensitivity of the gradient layer round this container must be identical with that of the remainder of the calorimeter, it is of benefit considerably to reduce the speed of response. The gradient layer is of normal thickness but additional insulating material is used to reduce the amplitude of the spikes of heat loss which occur at each urination. These tend to send the recorder off-scale. Urinary heat loss as measured by direct calorimetry is in any case of little significance since, firstly, the heat loss will actually have occurred a considerable time previously when the animal consumed food or water at ambient and heated it to body temperature, and secondly, because only part of the original heat content remains by the time the urine reaches the container.

The Gradient Layers

The material used to form the insulating shell should have a low value of thermal conductivity, be uniform in thickness and composition, and should be readily machineable. Laminated paper-board of thickness 1/32 in. (0·8 mm) was used in the present instance. In the gradient layers shown in Figs. 2 and 5 strips of this material of width 1¼ in. (3·175 cm) were used to carry the thermo-junctions. Each element of area of the measuring shell is actually 2½ in. (6·35 cm) square, but only every second strip is wound. This procedure makes it simpler to pass services through the gradient layer without loss of junctions and also reduces the total electrical resistance.

The dimensions of the various gradient layers are not all precise multiples of 2½ in. (6.35 cm). It is necessary, therefore, to use strips of non-standard width in certain places. The spacing of the thermocouples along these strips is calculated to give one pair of junctions per 2½ in. (6·35 cm) square in all cases.

Using materials similar to those mentioned above, and assuming that the signals are to be applied to a recorder of full scale 5 mV, it has been found that about 5000 pairs of copper-constantan thermo-junctions are required in the animal chamber. This figure does not vary greatly with the size of the chamber. A further 4000 or so thermocouples are necessary for the heat exchangers and the urine container. To manufacture a total of about 9000 pairs of junctions is obviously a fairly formidable task, and a mass-production technique must be employed. Various methods are available.

The simple approach used here is to wind the constantan spiral-fashion along the paper-board strip, and hand solder it to copper links which extend half-way across each face. Ribbons rather than wires of each metal are used to minimize the electrical resistance. The dimensions of the copper are ½ in. (1·27 cm)×0·002 in. (0·05 mm) and of the constantan ⅜ in. (9·52 mm)× 0·003 in. (0·075 mm). An alternative method is to wind the constantan on a former, and to deposit copper electrolytically only on the desired areas of constantan by the use of a masking lacquer which is later removed. This system has some advantages, particularly in the construction of small calorimeters, but in practice many difficulties are encountered. In both these methods the constantan ribbon is continuous along the length of each strip, effectively short-circuiting the copper links. This entails a drop of about 4 per cent compared with conventional thermocouples, but this disadvantage is well out-weighed by the reduction in possibility of the entire gradient layer becoming open-circuit through a single faulty soldered connection. Probably the simplest and most elegant method[18] of gradient layer construction is to use edge-welded copper-constantan ribbon. This material is commercially available in the U.S., and, although rather expensive, greatly reduces the labour involved. The ribbon is merely notched at suitable intervals and wound in a spiral round the former.

The gradient layers are mounted on the ¼ in. (6·35 mm) thick aluminium plates after completion of the assembly of the water jackets on the reverse side. Adhesive and casting epoxy resins are used to attach a thin layer of insulating material to the aluminium, to mount the gradient layer on this insulation, and then to encapsulate the entire layer to the exclusion of as much air as possible. Figures 2 and 5 show gradient layers at this stage of production. A further layer of insulation is then added, and finally the whole surface is protected from damage by urine or condensing water by a covering of thin aluminium foil. This final protection is particularly necessary on the floor of the animal chamber.

It is of major importance in constructing a gradient layer calorimeter to ensure that the chamber and the heat exchangers give zero electrical output under steady-state conditions when there is no heat emitting source within the chamber. Zero errors occur primarily because the gradient layers do not cover the entire surface area. If, for example, chamber and ambient laboratory temperatures are different, heat transfer will occur at any point where services pass through the chamber walls. This will not be recorded by the gradient layers. To maintain temperature stability within the chamber, however, this heat flow will be balanced by a flow of similar magnitude from the water jackets. The latter transfer will pass through the gradient layers and give rise to an error signal. Encapsulation of the gradient layer was the most important single measure taken to minimize the zero errors. It is also important to insulate windows, electrical and light entry points and other such gaps in the layers.

The Water Circuits

Acceptable stability of the base-lines of the electrical outputs of a gradient layer calorimeter demands accurate control of the temperature of the circulating water. As a first step towards achieving this, large capacity (about 100 gal) (455 l.) galvanized iron water tanks are used to supply circuits 1 and 2 (Fig. 1). Four centrifugal water pumps, two in each circuit, pass a total of 48 gal (218 l.)/ min round the water jackets. The large thermal inertia of the system and the high rate of flow minimize temperature changes due to variation in ambient conditions or in the heat loss of the animal. This, unfortunately, is achieved at the expense of flexibility as regards the speed with which a desired alteration in operating temperature can be effected during an experiment.

The maximum rate of change of water temperature which can be tolerated (i.e. without generating a significant spurious e.m.f. from the gradient layers) is largely governed by the water equivalent of the material within the gradient layers. In the case of the chamber not much can be done to reduce the internal thermal capacity. Of necessity, there is an animal and an animal container inside. Water circuit 2 (Fig. 1) therefore requires exceptionally close temperature control.

Benzinger solved this problem by a complex system involving successive approximations to the required stability. In the present calorimeters a different approach was adopted. The tanks supplying the water circuits are each fitted with refrigerating coils and several immersion heaters. The refrigerators run continuously, eliminating more heat than is necessary under all circumstances, and the excess cooling power is balanced off by the immersion heaters. Two specially designed[22] 'measurement units' also immersed in the water tanks give a continuous indication of the instantaneous rate of change of water temperature, and assist in the balancing process. In the case of water circuit 1, reduction to a minimum of the amount of materials within the gradient layers of the heat exchangers has made it possible to achieve temperature balance by using immersion heaters of up to 600 W rating switched by a conventional high-quality thermostat. This means that the main heaters, controlled by variable transformers, must be set with an accuracy of $\pm$300 W which is quite readily achieved.

For water circuit 2 a specialized form of temperature controller was devised by Cairnie,[22] and is used to reach the final balance between heat load and refrigerator cooling power. The heat output of the controller is proportional to the magnitude and sign of the out-of-balance signal of the 'measurement unit' and the action is such as to reduce the rate of change of water temperature to zero. This system has proved quite satisfactory. The rate of change of temperature during an experiment rarely exceeds 0·0002°C/min, and the absolute change in temperature is normally better than $\pm$0·1°C/day.

The refrigerator power required in circuit 1 (about 7000 Btu/hr) considerably exceeds that in circuit 2 (about 3000 Btu/hr). It is important to select the minimum acceptable value of heat elimination since the excess cooling power requires to be balanced by electrical heating and so represents the major part of the running costs of the calorimeter.

The galvanized iron storage tanks are connected to mains and hot water supplies, and are fitted with drain cocks and overflows. These all help to expedite temperature changes. Care must be taken to avoid corrosion. A mixture of zinc and aluminium surfaces with rubber hose connections has been satisfactory, but copper and brass fittings must be kept to a minimum.

The Ventilatory Air Circuit

The air circuit was originally of a straight-forward open-circuit type. A rotary pump, of positive characteristics, sucked air round the system, with a by-pass so that smooth control of ventilation rate through the chamber over the range 50–500 l./min could be obtained. With this simple circuit, however, it was not possible to do complete energy balance work on ruminants. For experiments of this type it is necessary to be able to measure methane production. Commercially existing analysers are not sufficiently sensitive to give a continuous record of the methane content of the air leaving the chamber without the rate of air flow being reduced to an undesirably low value. The effective humidity in the animal chamber would then always approach saturation level and condensation would occur on the chamber walls.

The air circuit was therefore modified to the form shown in Fig. 1 in which any desired proportion of the air flow from the pump can be recirculated round the system. By adjustment of the recirculation controls it is normal to direct up to 90 per cent of the airstream back towards the chamber. In this way a very substantial increase in methane concentration can be achieved without rise in water vapour level; heat exchanger 4 maintains the humidity at the pre-selected value. One incidental advantage of this system is that the pressure differential between the chamber and the exterior is greatly reduced when a large proportion of the airstream is recirculated. This eases the problem of keeping the chamber sufficiently airtight.

In addition to analysis for methane, it will be seen from Fig. 1 that continuous measurements are also made of the consumption of oxygen and production of carbon dioxide. Infrared analysers are used for methane and carbon dioxide, giving full-scale deflections for concentrations of 0·1 and 1·0 per cent respectively, and a paramagnetic analyser of range 20–21 per cent is used for oxygen measurements. Thus the facilities of the gradient layer calorimeter can be extended to match those of the classical respiration calorimeters. The capital cost of constructing the modern instrument remains fairly high (around £5000 for a sheep-sized calorimeter complete with three gas analysers) but the running costs are low, and the results are automatically presented in the form of a continuous permanent record which needs little in the way of correction or processing.

Filtered outside air entering the system first passes through a non-return valve. The function of this valve is to permit access from the exterior to the food trough in the chamber without major upset to the gas concentrations inside. The physical characteristics of the system are such that on opening the food admission pipe, air enters at that point and is expelled through the inlet filter. The non-return valve prevents this happening. During the few minutes it takes to introduce food, air enters the system via the food pipe at approximately the normal rate through the inlet filter. The effect of this on the calorimetric and gas analysis data is very transient, and it has not been necessary to provide any form of air-lock for the feeding trough.

Air flow is measured before the chamber by a vane anemometer (Velometer) and after the chamber by a Rotameter. The reading of the former is used routinely to ensure that the chamber is correctly sealed.

In the humidifier prior to heat exchange unit 1 the air has to travel between wet cotton wicks over a path length of about 5 ft (1·52 m) in passages 5 in. (12·7 cm) square. The efficiency of heat exchanger 1 is high enough to make the setting of the humidifier controls non-critical. By experience, the air flow through the humidifier and the temperature of the water in it are adjusted so that exchangers 1 and 4 are working at the same rate. The thermopile, mentioned above, between the air exits of these units helps in this adjustment, the real purpose of which is to cancel out any inefficiency in the functioning of the heat exchangers.

The temperature of the air leaving heat exchanger 4 is normally well below 20°C. A network of pipes in a thermostatted water tank is therefore sufficient to ensure that measurement of air flow by the Rotameter is always made at the calibration temperature.

A second water tank fitted with refrigeration coils and immersion heaters and thermostatted at 20°C is included in the air circuit after the main pump. This has two functions. Firstly, in passage through the pump the air is heated to about 50°C. Cooling the airstream to 20°C prevents the balanced working of exchangers 1 and 4 being upset by the high temperature of the recirculated air. Secondly, a wet gas meter is used to measure the flow of non-recirculated air. It is convenient to saturate the airstream at 20°C before the meter. This is accomplished by a second humidifier and a cooling coil in the thermostatted water tank.

Chamber Facilities

In designing a gradient layer calorimeter it is necessary to make provision in advance for all the services which may in the future be required to pass through the walls of the chamber. These include windows, lights, air entry and exit points, urine drain, food and water entries and electrical connections. In addition to these permanent services provision was also made in the present calorimeter to gain access to the interior of the chamber at several other places. The fittings at these entry points are modified as necessary to suit the particular experiment in progress.

Lighting of the interior is by four 100 W tungsten filament bulbs, the light from which travels down Perspex guides just projecting through the chamber roof. Thermal effects due to the transmission of infrared radiation or conduction down the light guides are barely perceptible. Urine is normally allowed to drain into the container below the chamber, but no provision is made for the removal of faeces during an experiment.

Drinking water is stored in a shallow container clamped to the outside of a chamber water jacket and covered by the insulation. When the animal drinks, water at approximately the correct temperature automatically flows from the storage container to replenish the water trough.

The chamber is ventilated in the downward direction. Air enters through distributing tubes along the top corners, and is extracted near the foot of the chamber through a flexible tube normally attached directly to the animal cage. It is necessary to fit a filter to this air outlet to prevent dust being carried through from the chamber to the rest of the system.

Instrumentation

The sheep calorimeter occupies two adjacent rooms in a small laboratory. One houses the animal chamber, heat exchangers and immediately associated equipment, the other contains most of the instrumentation. Apart from thermostatting both rooms at 20°C no exceptional care of laboratory ambient conditions has been found necessary.

Figure 6 shows part of the instrumentation. The outputs of the gradient layers and the data from the gas analysers are fed to the multi-point potentiometric recorder seen centrally. A minimum of eight points is required, five for calorimetric data and three for the outputs of the gas analysers. Sufficiently faithful recording is obtained with a plotting interval of about 1 min per variable. An event recorder, not in the photograph, is used to monitor various thermostatted devices, and to record such activities of the animal as drinking, standing up and lying down. The animal's posture is determined by a photocell and low-power light beam within the chamber. To the right of Fig. 6 is the control equipment associated with the temperature regulation of the water circulating in circuit 2.

A calorimeter designed to accommodate animals up to a maximum of 200 lb (91 kg) live weight will give quite acceptable accuracy with animals weighing as little as 20 lb (9·1 kg). Between those extremes of weight heat loss will vary by a factor of at least five. It is therefore necessary to switch the recorder input to cover several ranges if the potential accuracy of the instru-

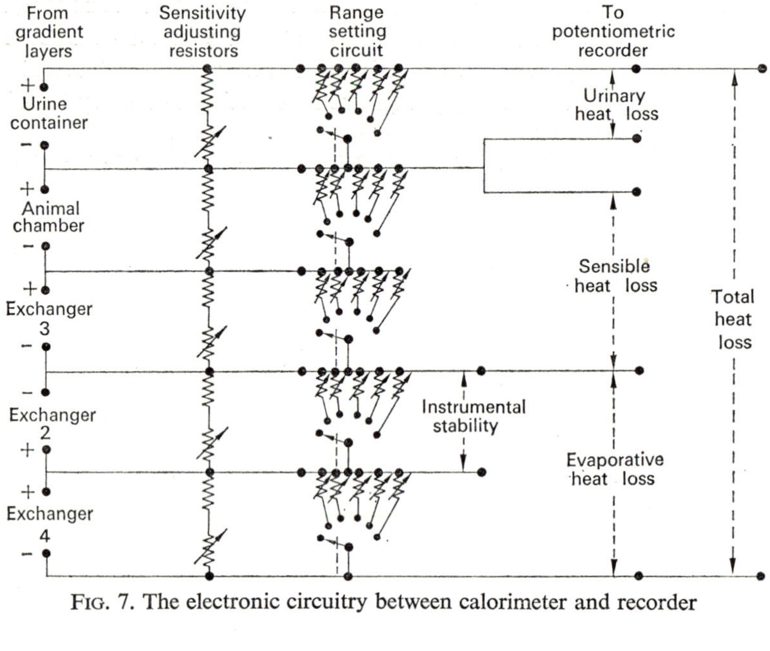

Fig. 7. The electronic circuitry between calorimeter and recorder

FIG. 6. Some of the instrumentation for the calorimetric measurements.

ment is to be realized. The circuitry used for this purpose is shown in Fig. 7. After adjustment of the sensitivity resistors, which is described later, the range-setting resistors are set so that the recorder will give full-scale deflection for 2×, 3×, 4×, . . . the basic millivolt range. It is most important that the switches and resistors used in this circuit should be of the highest quality.

Calibration Procedure

The main objective in the manufacture of the gradient layers is the production of a heat measuring shell which will at all points respond with a uniform electrical output to a given rate of heat flow per unit area. By careful design a close approach can be made to this ideal state. There will inevitably, however, be small but significant differences in the sensitivities to heat flow of the various heat exchangers, the animal chamber and the urine container. Each unit must therefore be calibrated individually and the sensitivities equalized by means of the adjusting resistors shown in Fig. 7.

Calibration of the animal chamber presents no problem. The millivolt response is measured without ventilatory air flow using a dry source of electrical heat connected to a good quality wattmeter. Movement of the source within the chamber should produce no change in output. If this does happen it will be necessary to shunt the more sensitive walls with suitable individual resistors. It should not matter whether a mainly radiative source such as an electric bulb or a mainly convective source such as a resistance winding of large physical dimensions is used.

When the ventilatory airstream is switched on, a small portion of the sensible heat generated within the chamber is transferred to heat exchanger 3. This proportion can be greatly increased by placing the source of heat within a container directly connected to the chamber air exit. The sensitivity adjusting resistors can then be set to give constant response no matter what fraction of the generated heat is carried through to the heat exchanger.

The correct adjustment of the sensitivities of heat exchangers 2 and 4 is very much a matter of trial and error. A long shallow water container fitted with an immersion heater is very useful. Such a source, placed within the chamber, can be arranged to produce various proportions of sensible and evaporative heat loss. This simulates fairly closely the heat loss of an animal under different conditions. By variation of the power input to the immersion heater, the ventilation rate and the temperatures of the circulating water, a series of successive approximations can be made towards the correct adjustment of the outputs of heat exchangers 2 and 4. It is convenient always to adjust exchanger 2 for minimum zero error without a heat source, and exchanger 4 for correct millivolt response to a given evaporative heat transfer.

The urine container is simply calibrated and adjusted by adding a measured volume of water at known temperature and allowing it to heat or cool to the reference temperature. The output of the gradient layer is recorded and the sensitivity adjusted until the desired response is obtained.

Volatile liquids such as various alcohols or acetone can be used to check the calorimeter calibration. This furnishes a measurement of total response which is independent of the accuracy of any instruments used in the electrical method. A volumetric technique can be used in which several hundred millilitres of the liquid are burned in the chamber over a period of many hours, the rate of burning being controlled by a metering pump. Alcohol checks are of great value for calibrating the measurements of total heat loss, oxygen consumption and carbon dioxide production, but are of less benefit in connection with the partitioning into the sensible and evaporative heat loss fractions. This is due to the relatively small amount of water vapour produced.

Performance

A portion of a typical record of the heat loss from a sheep is shown in Fig. 8. In this instance the spikes of urinary heat loss have not been included in the recording of total heat loss. The 'square-wave' pattern seen in this tracing is quite normal for sheep and other ruminants, and corresponds to the animal standing up and lying down. The effective speed of response of the instrument may be assessed from this recording. Virtually complete

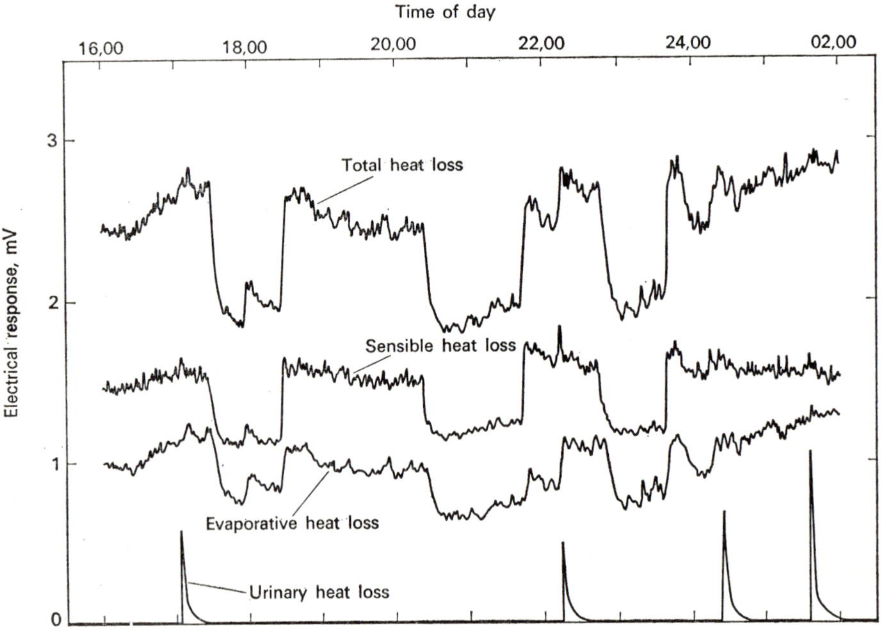

FIG. 8. The response of the calorimeter to the heat loss of a sheep

response to a variation of around 30 per cent in total heat loss on change of posture takes place within 3 min. Under different conditions change of posture of a sheep may result in as much as a 70 per cent apparently instantaneous variation in heat loss. The calorimeter follows these variations, which are partly sensible and partly evaporative, with little time lag.

The volume of the ventilatory air circuit including the animal chamber is just over 3000 l. This means that under normal experimental conditions the speed of response to a change in the level of carbon dioxide or methane production or in the rate of oxygen consumption is relatively slow. It may take up to 3 hr before acceptable gas analysis data are obtained at the beginning of an experiment. This speed of response may be accelerated with human subjects or with trained animals by the use of masks connected directly to the chamber air outlet. Recirculation of the air is not then possible and difficulties with humidity levels are encountered. An alternative solution with animals has been used,[21] and may be envisaged from Fig. 3. The groundsheet seen in this photograph can be opened out and inserted under the door seal so that the head of the sheep and its body are in separate compartments. The relatively small head compartment is then ventilated directly to the chamber air exit.

The millivolt response of the gradient layer calorimeter is completely linear over the range of heat flows which will be met in practice. It is advisable to check the sensitivity figure occasionally, a process which can conveniently be combined with calibration of the gas analysers during an alcohol check. In the present case there has been no variation in calorimeter sensitivity over a period of 10 years.

Automatic partitioning of total heat loss into the sensible and evaporative components is one of the main advantages of gradient layer calorimetry. With animals of reasonable size in relation to the volume of the chamber, or with a group of smaller animals, there is little difficulty in attaining an accuracy of measurement better than ± 1 per cent. The same accuracy is not, however, normally realized in the partitioned components. This is due to the evaporation of water from any wet material, chiefly urine and faeces, which may be lying inside the chamber. Sensible heat, which should be measured as such within the chamber, provides the latent heat of evaporation and is subsequently extracted and measured in heat exchange unit 4. There is therefore always a tendency to over-estimate evaporative measurements at the expense of sensible heat loss. This partitioning error may reach as much as 10 per cent of the total heat emission. It is thus necessary in some experiments to minimize evaporation by collecting urine and faeces under mineral oil, and in others to estimate the extent of the partitioning error involved and make suitable correction.

A major advantage of gradient layer calorimetry as opposed to indirect methods is the automatic production of continuous permanent records showing the minute-to-minute fluctuations in the heat loss of the subject. It is not necessary to decide beforehand the interval over which the heat emission is to be measured. Records, such as that shown in Fig. 8, reflect the animal's activity in a manner which is considerably more interesting and instructive than is the data from any other existing type of calorimeter.

These records must be integrated to measure the heat loss over the intervals selected. The most simple, though rather tedious, method is to integrate manually using a form of planimeter which will allow a 24-hr length of strip chart to be processed without pause. The tedium of this operation will be greatly relieved by electronic equipment now in production by several commercial firms. This equipment will either integrate the calorimeter outputs automatically over pre-selected intervals or present them on punched tape in digital form suitable for further processing by computer. Only by such methods is it possible to make full use of the very large amount of information made available by the gradient layer calorimeter.

REFERENCES

1. OEUVRES de LAVOISIER. *Mémoires de l'Académie des Sciences*, Paris (1862).
2. d'ARSONVAL. *Soc. de biol.* (1881).
3. BURTON, A. C. and BAZETT, H. C. *Am. J. Physiol*, **117**, 36 (1936).
4. LEFÈVRE, J. *Arch. de physiol.* (1895).
5. ATWATER, W. O. and ROSA, E. B. *U.S. Dept. Agr., Office Expt. Sta. Bull.*, No. 63 (1889).
6. BRAMAN, W. W. *Pennsylvania State College Bull.* 302 (1933).
7. CAPSTICK, J. W. *Handbuch der Biologischen Arbeitsmethoden.* Abt. 4, Teil 10. Urban and Schwartzenberg, Berlin (1926).
8. SWIETOSLAWSKI, W. *Microcalorimetry.* Reinhold, N.Y. (1946).
9. LEFÈVRE, J. *Chaleur Animale et Bioenergetique.* Masson et Cie, Paris (1911).
10. KLEIBER, M. *Biophysical Research Methods.* Interscience Publishers Inc., N.Y. (1950).
11. MURLIN, J. R. and BURTON, A. C. *J. Nutr.* **9**, 233 (1935).
12. HARDY, J. D. and DAY, R. *Am. J. Dis. Child*, **63**, 1086 (1942).
13. PROUTY, L. R., BARRETT, M. J. and HARDY, J. D. *Rev. Sci. Instrum.* **20**, 357 (1949).
14. BENZINGER, T. H. and KITZINGER, C. *Rev. Sci. Instrum.* **20**, 849 (1949).
15. BENZINGER, T. H., HUEBSCHER, R. G., MINARD, D. and KITZINGER, C. *J. App. Physiol.* **12** (1958), Supplement.
16. TASKER, C., HYMPHREYS, C. M., PARMELEE, G. V. and SCHUTRUM, L. F. *Trans. Am. Soc, Heating Ventilating Engrs.* **58**, 139 (1952).
17. LAWTON, R. W., PROUTY, L. R. and HARDY, J. D. *Rev. Sci. Instrum.* **25**, 370, (1954).
18. STEWART, R. E. *Agric. Eng.* **38**, 734 (1957).
19. McLEAN, J. A. Personal communication (1965).
20. TIAN, A. *Bull. Soc. Chim. Fr.* **33**, 427 (1923).
21. BROCKWAY, J. M., McDONALD, J. D. and PULLAR, J. D. *J. Physiol.* **179**, 554 (1965).
22. CAIRNIE, A. B. and PULLAR, J. D. *J. Sci. Instrum.* **36**, 249 (1959).

CHAPTER 15

Methods of Calorimetry (B) Indirect

WILLIAM P. FLATT*

United States Department of Agriculture, Beltsville, Maryland

INDIRECT calorimetry, the measurement of heat production of organisms by indirect means, has been an invaluable research tool used by physiologists, chemists and nutritionists to investigate many biological problems. Even prior to the accumulation of experimental evidence bearing out the validity of the law of conservation of energy as applied to animals, indirect calorimetric methods were being used extensively by scientists to study the effects of various dietary and physiological variables on the energy metabolism of animals.

In contrast to direct calorimetry (Chapter 14), in which physical methods of measuring heat emmission (by radiation, conduction, convection and as latent heat of water vaporization) are used, indirect calorimetric methods are dependent upon measurements of chemical changes which occur when different nutrients are catabolized or stored in the body of the organism.

The indirect calorimetric methods can be categorized as those in which energy retention is measured directly (slaughter tests, body composition studies) and those in which energy retention is measured indirectly by deducting the energy losses (faeces, urine, combustible gases, heat production) from the dietary energy intake. It is possible to calculate the heat production of an animal if the oxygen consumption and carbon dioxide production are known (R.Q. method) or if the total dietary energy intake, energy retention and energy losses in excreta are known (carbon and nitrogen balance and slaughter methods). The theoretical basis and the experimental evidence of the validity of the indirect methods will be discussed in this chapter. Several textbooks (Krogh, 1916; Lusk, 1928; Brody, 1945; Swift and French, 1954; Kleiber, 1961; Blaxter, 1962) and numerous technical and review articles (Lusk, 1922; Murlin, 1922; Paechtner, 1931; Kleiber, 1950) have been written which include detailed discussions of indirect calorimetric methods. Therefore, each method will be discussed briefly, with key references being given to provide additional details. The major emphasis is on methods which have been and are being used in studies with man and domestic

* Animal Husbandry Research Division, Agricultural Research Service.

animals, particularly those species which are of agricultural importance as sources of meat, milk, eggs or hair fibre and hides.

Within the past decade, research workers in several laboratories in Europe, Australia, Japan and America have been actively studying the energy metabolism of farm and laboratory animals. Some of the data obtained and descriptions of the methods being used in the different laboratories have been reported in the proceedings of symposia on energy metabolism sponsored by the European Association of Animal Production (Thorbek and Aersoe, 1958; Brouwer and Van Es, 1961; Blaxter, 1965).

HISTORICAL DEVELOPMENT AND EXPERIMENTAL BASIS FOR INDIRECT CALORIMETRY

Vague concepts of the relationships between life, heat and fire date back to antiquity, but these relationships were not understood and no quantitative measurements of the heat production of animals were made until late in the eighteenth century. A. Crawford (1777) and Antoine Laurent Lavoisier (1780) used animal calorimeters to measure heat production, and they also noted the similarity between the combustion of materials and animal respiration. They observed that oxygen was consumed and carbon dioxide was produced both by the burning of materials and by living animals. Lavoisier, frequently called the 'father of modern chemistry', conducted calorimetric experiments and respiration trials with guinea pigs, and he and his co-workers were the first to clearly enunciate the principles of combustion both outside and within the body. Lavoisier's theory of combustion and metabolism, that fire and animals produce the same amount of heat per unit of CO_2 produced, is still valid today, and it forms the basis for indirect calorimetric methods. Although intricate systems are involved in the intermediary metabolism and oxidation of food nutrients in the animal organism, the net oxygen consumption and carbon dioxide and heat production are the same as they would have been if these materials had been burned directly.

In order for indirect calorimetry to be a valid means of measuring the heat production of animals, the heat change in a chemical reaction must be the same whether it takes place in one or in several stages. This *law of constant heat sums* was demonstrated experimentally by G. H. Hess in 1840, and it is a direct consequence of the *law of conservation of energy*, which states that, 'although energy may be converted from one form to another it cannot be created or destroyed' (Glasstone, 1946).

Rubner (1894) obtained experimental evidence that the law of constant heat sums was applicable to living animals by measuring in a respiration calorimeter the total heat production of dogs and comparing this with the estimated potential energy of food materials calculated to have been oxidized. The average heat production obtained by the two methods agreed within less than 1 per cent. Further evidence of the validity of indirect calorimetric methods was obtained in experiments with humans by Atwater and Rosa (1899). Their direct-calorimetry results agreed so closely with those calculated from carbon and nitrogen balance experiments that they concluded that the law of conservation of energy applies to the living organism.

Armsby and his associates at Pennsylvania constructed and used extensively an adiabatic respiration calorimeter to study the energy metabolism of cattle. These workers accumulated data with cattle being fed diets differing widely in chemical composition and in quantities varying from the fasting level to more than twice the amount required to maintain energy equilibrium. In studies with steers and dry cows, as well as with lactating dairy cows, they obtained agreement (within about 1 per cent) between direct and indirect methods of measuring the total heat production of the animals. These studies are of particular importance because as much as 10 per cent of the dietary energy may be lost as methane which has not been oxidized in the animal's body. These combustible gases must be accounted for in determining the energy balance of ruminants.

Numerous other studies with many species of animals have shown the reliability of indirect calorimetry and its close agreement with results obtained with direct calorimetry. For most nutritional studies, the indirect methods have been used in preference to direct calorimetry because of the greater initial costs for equipment and increased labour requirements for operating direct calorimeters. However, direct calorimetry provides only information about heat emission of the animal, and additional measurements are required to determine energy retention and the materials oxidized as the source of heat. In many studies it is desirable to know the nutrient sources of energy and the change in body substance (fat and protein); therefore, indirect methods are used with or instead of direct methods. In nutritional studies with ruminants, the energy losses in the form of combustible gases must be considered, so that part of the advantage of direct calorimetry is lost. Automatic-recording, rapid-response, gradient-layer calorimeters (Benzinger and Kitzinger, 1949, 1957; Pullar, 1957, 1958) offer many advantages (Chapter 14), particularly for short-term studies or those in which there is doubt about factors used in indirect methods, but indirect methods have been used and will probably continue to be used extensively for most experiments with farm animals.

PRINCIPLES OF INDIRECT CALORIMETRY AND THE DERIVATION OF FACTORS FOR CALCULATING HEAT PRODUCTION

Indirect calorimetry is based primarily on the fact that, under most circumstances, the heat production is closely correlated with the oxygen consumption and the carbon dioxide production when organic compounds are oxidized. The relative amounts of oxygen used and the carbon dioxide and heat produced in the metabolism of carbohydrates, fat and protein will vary because of the different carbon and oxygen content of each of the three classes of nutrients. The total heat production of an animal can be calculated by assuming that the oxygen consumption and the carbon dioxide and nitrogen excretion during a given interval result from the breakdown of carbohydrates, fats and proteins.

CARBOHYDRATE OXIDATION

For example, if glucose were oxidized in the animal's body or in a bomb calorimeter, the result would be:

$C_6H_{12}O_6$	+	$6\ O_2 \rightarrow$	$6\ CO_2$	+	$6\ H_2O$	+ 673 kcal
1 mol glucose		134·46 l.*	134·46 l.			Heat of
180·16 g		191·99 g	264·05 g			combustion

Thus, for each mol of glucose (180·16 g) oxidized, 6 mols of oxygen (134·46 l. or 191·99 g) would be consumed and 6 mols (134·46 l. or 264·05 g) of carbon dioxide would be produced. The heat production would be 5·005 kcal/l. (673/134·46) of oxygen consumed or carbon dioxide produced. In this case, the *respiratory quotient* (mols or volume of CO_2 produced/mols or volume of O_2 consumed) is unity, and there is no difference in the factors for calculating heat production from either oxygen or carbon dioxide. This relationship is true for all carbohydrates, since 1 mol of oxygen is consumed and 1 mol of carbon dioxide is produced for each carbon atom in the molecule of a carbohydrate.

FAT OXIDATION

In the metabolism of fat or fatty acids the relative proportion of intramolecular oxygen is less and the proportions of carbon and hydrogen are greater than for carbohydrates. Consequently, the caloric equivalents of oxygen used and carbon dioxide produced will not be the same. The oxidation of palmitic acid may be used as an example:

$CH_3(CH_2)_{14}COOH$ +	$23\ O_2 \rightarrow$	$16\ CO_2$	+	$16\ H_2O$	+ 2398 kcal
1 mol palmitic acid	515·43 l.	358·56 l.			Heat of
256·43 g	735·95 g	704·14 g			combustion

For each mol of palmitic acid (256·43 g), 23 mols (515·43 l. or 735·95 g) of oxygen would be consumed and 16 mols (358·56 l. or 704·14 g) of carbon dioxide would be produced. The thermal equivalents would thus be 4·653 kcal/l. O_2 (2398/515·43) or 6·689 kcal/l. CO_2 (2398/358·56). The respiratory quotient, $16/23 = 0{\cdot}70$, is typical of the values obtained when long-chain fats

* Litres corrected to standard conditions of O°C, 760 mm Hg, dry gas.

are oxidized, but each fat has its distinctive R.Q., and the thermal equivalents of oxygen and carbon dioxide are different at each R.Q. Tables of thermal equivalents of O_2 and CO_2 and the corresponding percentages of fat and carbohydrates oxidized for different respiratory quotients have been prepared (Lusk, 1928; Zuntz and Schumburg, 1901) and reproduced in several textbooks (Brody, 1945; Swift and French, 1954; Carpenter, 1948). These values were calculated on the basis of average values for the heat of combustion of foods (4·18 kcal/g of carbohydrate and 9·46 kcal/g fat) resulting in R.Q.'s ranging from 0·707 to 1·00.

PROTEIN OXIDATION

Fats and carbohydrates are oxidized to carbon dioxide and water, but when proteins are metabolized in the body, some nitrogen is excreted as incompletely oxidized organic compounds such as urea, uric acid, hippuric acid and glycine conjugates. The most important incompletely oxidized nitrogen compound in animals is urea, which is excreted in the urine. The amount of protein oxidized may be calculated from the urinary nitrogen excretion. It is usually assumed that protein contains 16 per cent nitrogen and that all the urinary nitrogen is derived from protein oxidation, but the carbon dioxide, oxygen and heat equivalents of urinary nitrogen actually vary somewhat with the kind of protein fed. The commonly used factors were based on experiments in which meat was fed to dogs. It was found that for each gram of urinary nitrogen derived from protein metabolism, 5·94 l. of oxygen were consumed, 4·76 l. of carbon dioxide were produced, and 26·51 kcal of heat were produced. By subtracting the carbon dioxide and oxygen concerned with protein oxidation from the total amounts of each, and thereby adjusting for the heat produced from protein, the remaining amounts of carbon dioxide, oxygen and heat involved in the oxidation of fat and carbohydrate can be obtained.

The oxidation of the amino acid, alanine, in the body may be illustrated by the following equation:

$2\,CH_3CH(NH_2)\,COOH$ +	$6\,O_2 \rightarrow$	$5\,CO_2$	+ $5\,H_2O$ +	$CO(NH_2)_2$ +	623·8 kcal
2 mols alanine	134·46 l.	112·05 l.		1 mol urea	Heat of
178·19 g	191·99 g	220·04 g		60·06 g	combustion

The R.Q. resulting from the oxidation of alanine is 0·83, and the thermal equivalents are 4·539 kcal/l. O_2 and 5·567 kcal/l. CO_2 produced. Similar calculations could be made for each amino acid.

Although the errors can be minimized in a particular experiment by calculating thermal equivalents based on the specific compounds representative of the materials being metabolized (Blaxter, 1962), for most biological experiments it is satisfactory to use factors based on mixed fats, proteins and carbohydrates such as starch and cellulose. Table 1 summarizes the average constants for protein, fat and carbohydrate when oxidized in the animal body. These values were recommended by a sub-committee (E. Brouwer, K. Blaxter, K. Nehring and W. Wohlbier) on constants and factors (Brouwer, 1965) to be used in calculations of energy metabolism and they were used as the basis for preparing formulae for calculating heat production from oxygen consumption, carbon dioxide and methane production, and urinary nitrogen excretion.

TABLE 1

Constants for Protein, Fat and Carbohydrate when Oxidized in the Animal Body (Brouwer, 1965)

Compound	Carbon	O_2 consumed on oxidation of 1 g		CO_2 produced on oxidation of 1 g		Heat produced	Respiratory quotient
	(%)	g	l.	g	l.	kcal	
Protein*	52·00	1·366	0·957	1·520	0·774	4·40	0·809
Fats	76·70	2·875	2·013	2·810	1·431	9·50	0·711
Starch	44·45	1·184	0·829	1·629	0·829	4·20	1·00
Saccharose	42·11	1·122	0·786	1·543	0·786	3·96	1·00
Glucose	40·00	1·066	0·746	1·746	0·746	3·74	1·00

* Assuming the composition is 16 per cent N, 52 per cent C and 5·7 kcal/g. The physical constants used were: Atomic weights: C = 12·011; O = 15·999; H = 1·008; N = 14·007. Molar volume O_2, N_2, CO_2, CH_2, H_2 = 22·41 l.

INCOMPLETE OXIDATION

In some herbivora, especially cattle, sheep, goats and other ruminants, considerable quantities of combustible gases are formed in the digestive tract as a result of anaerobic microbial fermentation. Most of the combustible gas formed is methane, but under some circumstances hydrogen and ethane (Moustgaard and Thorbek, 1949) have been produced. Carbon dioxide is also produced by bacterial metabolism, and this is indistinguishable from that produced in the body by oxidative processes. The combustible gases represent the incomplete oxidation of organic materials in the animal's body, so they must be considered in deriving formulae for the computation of heat production based on indirect calorimetry. Empirical computation methods based on different assumptions have been derived and used (Krogh and Schmit-Jensen, 1920; Andersen, 1922), and have been discussed in detail by Swift and French (1954). The sub-committee on constants and factors (Brouwer, 1965) recommended the use of a multiple regression equation for calculating the heat production of ruminants, and a correction is included for methane production. The formula, which was adopted by general consent of the participants of the third symposium on energy metabolism is as follows:

$$HP = 3{\cdot}866\ O_2 + 1{\cdot}200\ CO_2 - 0{\cdot}518\ CH_4 - 1{\cdot}431\ N$$

where HP = kcal; O_2 = litres oxygen consumed; CO_2 = litres CO_2 produced; CH_4 = litres methane produced; N = grams urinary nitrogen excreted.

A more complete discussion of the derivation of factors in similar equations to correct for other compounds resulting from incomplete oxidation in the body was presented by Brouwer (1958).

CALCULATIONS BASED ON OXYGEN CONSUMPTION ONLY

In some experiments it is convenient to measure only the oxygen consumption of subjects, without obtaining carbon dioxide and urinary nitrogen values. This is true in most basal metabolism studies with humans, where the rate of oxygen consumption is measured under standardized conditions, usually about 12 hr after the preceding meal. The R.Q. is usually assumed to be 0·82, and the thermal equivalent of oxygen is assumed to be 4·825 kcal/l. This procedure has been used in studies with farm aminals—cattle, horses, sheep, swine and goats (Brody, 1930) and in energy metabolism studies with elephants (Brody and Procter, 1934). With a range in R.Q. from 0·707 to 1·00 the thermal equivalent of oxygen ranges from 4·686 to 5·047 kcal/litre, or a range of about 7 per cent. By assuming an R.Q. of 0·82, the computed heat production based on oxygen consumption would ordinarily be within 2 or 3 per cent of the correct value.

Weir (1949) has proposed a method of computation of heat production which requires only a knowledge of the oxygen consumption and the per cent protein in the diet. Weir's equation is:

$$HP = \frac{(O_i - O_e)\ 0{\cdot}0504 \times \text{litres expired air}}{1 + 0{\cdot}082\,p}$$

where O_i = per cent oxygen in inspired air
O_e = per cent oxygen in expired air
p = decimal fraction of protein in diet

This is only a minor improvement over using the constant of 4·825 kcal/litre oxygen consumed to calculate heat production, but it does make it possible to partially account for differences due to nitrogen excretion when diets differing in protein content are being studied.

Benedict and Talbot (1914) concluded that the best measurement of the heat production of infants was the amount of oxygen consumed rather than the amount of carbon dioxide produced. However, there are many circumstances where oxygen consumption alone does not give sufficient information, such as when there is interconversion of carbohydrates in the body.

ENERGY RETENTION METHODS

In the indirect calorimetric methods discussed above, it was necessary to know the oxygen consumption of the organism. However, in long-term nutritional studies with farm animals, it is often more desirable to know the storage or loss of body fat or protein than to have an exact measurement of the heat production. It is possible to measure the carbon and nitrogen balance of animals and to calculate the heat production by difference. The difference between the dietary energy intake and the losses of energy in the faeces, urine, combustible gases and products stored (body fat, protein, milk, eggs, etc.) is equal to the total heat produced.

$$HP = GE_i - (FE + UE + GPD) - NE_p$$

where HP = total heat production (kcal)
GE_i = gross energy intake (kcal)
FE = faecal energy (kcal)
UE = urinary energy (kcal)
GPD = gaseous products of digestion. Primarily CH_4 (kcal)
NE_p = net energy for production (kcal) including body tissue gains, milk, eggs or wool, etc. formed during the experimental periods

The nitrogen and carbon balance method has been used in nutritional studies with animals for more than a century. As early as 1839, the French chemist-farmer J. B. Boussingault measured the amounts of carbon, hydrogen, nitrogen and oxygen consumed in the feed by a lactating dairy cow and compared this with the amounts of these elements in the faeces, urine and milk. The gaseous exchange was not measured, but from the above information he calculated the amount of oxygen which must have been derived from the air to oxidize the carbon and some of the hydrogen in the feed. Later workers adopted similar methods and also measured the excretion of carbon dioxide and, in some cases, methane. Pettenkofer and Voit, Henneberg and Stohmann, Grouven, Liebig, Rubner, Kühn, and Kellner, are among the research workers who used this method extensively. The basic assumption of the method is that the materials stored in the animal's body consist of either fat or protein, with negligible changes in carbohydrate stores occurring. The protein stored is estimated from the nitrogen retained, and the fat stored is calculated from the carbon retention in excess of that stored in the protein. The values used to compute the energy retained from the carbon and nitrogen retained have usually been based on analyses of muscles (16 per cent N, 52 per cent C, 5·7 kcal/g) and depot fats (76·5 per cent C, 9·5 kcal/g). Only slight interspecies differences in the chemical composition of dried muscle were noted by Franke and Weniger (1958). They analysed dried muscle from cattle, horses, sheep, goats, swine, rabbits and chickens for their carbon, nitrogen and heat of combustion and found 16·0 per cent N, 51·1 per cent C and 5·554 kcal/g for cattle as compared to 15·0 per cent N, 50·9 per cent C and 5·527 kcal/g for horses and 16·3 per cent N, 51·4 per cent C and 5·614 kcal/g for muscle tissue of chickens.

The average caloric value of body fat from these animals was 9·463 kcal/g for cattle, 9·455 kcal/g for horses, 9·466 kcal/g for swine, 9·449 kcal/g for sheep and 9·469 kcal/g for goats, with the mean for all species being 9·461 kcal/g body fat. On the other hand, Cathcart and Cuthbertson (1931) and Cuthbertson (1933) showed that in human beings, the caloric value of fat can vary, depending on the source of the material. Blaxter and Rook (1953) adopted a statistical approach based on the carbon, nitrogen and heat of combustion of different cattle tissues. Their results indicated that the composition of protein was 51·2 per cent C, 16 per cent N and 5·322 kcal/g, while the composition of fat was 74·84 per cent C, and 9·367 kcal/g. Energy retention, based on these values, could be calculated by the use of the equation (Blaxter, 1962):

$$\text{kcal energy retained} = 12{\cdot}55 \times \text{g C retained} - 6{\cdot}90 \times \text{g N retained}$$

The above equation is slightly different from the one recommended by the subcommittee on constants and factors (Brouwer, 1965), based on the constants in Table 1. The energy balance, based on using these constants, would be:

$$\text{kcal energy retained} = 12{\cdot}388 \times \text{g C retained} - 4{\cdot}636 \times \text{g N retained}$$

COMPARATIVE SLAUGHTER METHODS

An alternative to measuring the carbon and nitrogen balance and calculating the storage of protein and fat in the body is to measure body composition changes by slaughtering animals and analysing the complete carcasses. Lawes and Gilbert (1859, 1861) employed this method in experiments with fattening cattle, sheep and pigs. Although there have been a few workers (Trowbridge, Moulton, Armsby and Moulton, Wellington, 1956) who have since analysed entire carcasses of animals, it has usually been considered to be too laborious and expensive to apply on a large scale to nutritional studies with cattle and other large farm animals. With small animals, such as rats, the whole animal can be processed and the method can be very accurate (Forbes *et al.*, 1940). The comparative slaughter technique involves slaughtering a representative group of animals (control group) at the beginning of a trial and another group of animals (test group) at the end of a trial after they have been fed a certain ration for a period of time sufficient to obtain the desired response.

NE_p = (kcal in body test group − kcal in body of control group) + kcal products formed

By analysing the feed, excreta and products formed during this time and calculating the change in body composition based on the difference between the chemical composition of the carcasses of the test and control animals, the energy retention as well as total heat production during this time can be determined. In the case of ruminants, combustible gas production would also have to be considered, but in studies with non-ruminants these losses are usually considered to be negligible.

INDIRECT METHODS OF DETERMINING BODY COMPOSITION

Comparative slaughter trials offer a number of advantages when used with small animals, but the unsuitability for studies with humans and the practical difficulties involved in experiments with large animals have led to attempts to estimate the body composition of living animals by indirect means. The basis for most of the indirect methods is that there is an inverse relationship between the fat content and the water content of the body and that the proportions of protein, mineral matter and water of the fat-free body can be predicted. By determining the body water of a living animal, it is possible to calculate the body fat, protein and ash by the use of prediction equations based on carcass analyses (Armsby and Moulton, 1925; Murray, 1922; Wellington *et al.*, 1956; Reid *et al.*, 1955; Bensadoun *et al.*, 1963).

The equations proposed for use with cattle (Reid *et al.*, 1955) were as follows:

$$F = 355{\cdot}88 + 0{\cdot}355\ W - 202{\cdot}91 \log W$$
$$P = 80{\cdot}93 - 0{\cdot}00101\ Z$$
$$A = 19{\cdot}07 + 0{\cdot}00101\ Z$$

where F = per cent fat in empty body (body less contents of digestive tract)
W = per cent body water
P = per cent protein in fat-free dry matter of the body
A = per cent ash in fat-free dry matter of the body
Z = age of the animal (days)

One of the problems in applying the above equations in studies with living animals is in measuring the 'empty body weight' with accuracy. For instance, the gastro-intestinal tract fill can vary from 5 to 30 per cent of the total body weight of sheep (Reid *et al.*, 1963). Various attempts have been made to overcome this difficulty by standardization of procedures, such as withholding feed and water from the animals for varying intervals. Bath *et al.* (1965) found that the empty body weights of cattle (heifers) was 91·6 per cent of the mean of the 'shrunk body weight' made after withholding feed and water overnight. A prediction equation for calculating the empty body weight from the shrunk body weight of sheep is as follows (Bensadoun *et al.*, 1963):

$$EBW = 0{\cdot}943\ SBW - 4{\cdot}076$$

where EBW = kg empty body weight
SBW = kg shrunk body weight

DILUTION TECHNIQUES FOR DETERMINING BODY WATER

There are a number of substances which, when injected into living animals, become evenly distributed in the body water. The total body water may be determined by measuring the concentration of these materials in body fluids such as blood or urine. The dilution of a known amount of the reference substances by the body water makes it possible to calculate the total body water. Some of the materials which have been used in determining the total body water of laboratory animals, farm animals and humans include deuterium, tritium, antipyrine (1-phenyl-2,3-dimethylpyrazolone-5-one), *N*-acetyl-4-aminoantipyrine, ^{131}I labelled 4-iodoantipyrine, urea and thiourea (Hansard, 1956, 1963; Reid *et al.*, 1958, 1963; Bensadoun *et al.*, 1963; Faller *et al.*, 1955; Kraybill *et al.*, 1951, 1953; Siri, 1956; Pace *et al.*, 1947; Panaertto, 1963; Kirton, 1964; Hevesy and Hofer, 1934).

Once the amount of water in the body is known, the amounts of fat, protein and ash may be calculated with a fair degree of accuracy. This procedure has been used advantageously in a number of nutritional and physiological experiments. However, the reference substances become distributed in the water in the gastro-intestinal tract as well as in the body water. In the case of ruminants, this can lead to considerable error unless the amount of water in the contents of the gastro-intestinal tract can be determined. The use of two or more reference substances which diffuse into the gastro-intestinal tract at different rates has been proposed as a means of estimating the gastro-intestinal tract water but the results have only been partially successful (Reid *et al.*, 1958; Blaxter, 1962).

Other techniques for estimating the body composition of living animals (sodium and potassium retention, ^{40}K, cyclopropane, body density, etc.) (Forbes *et al.*, 1961; Blaxter and Rook, 1956; Lesser *et al.*, 1956) have also been used, but most of the results obtained to date have not been sufficiently accurate to use routinely as indirect calorimetric methods. Many research workers are actively seeking improvements in the methods, and several techniques offer promising possibilities for use in the future. Descriptions of several of the techniques being used to measure body composition are summarized in the proceedings of conferences on body composition, edited by Brozek and Henschel (1961) and Brozek, Whipple and Silverzweig (1963).

CARCASS SPECIFIC GRAVITY

Another method of determining body composition is to slaughter the animal and measure the specific gravity of the carcass. This method does not require the analysis of the complete carcass, so it is applicable to studies with large farm animals. It has been used routinely in extensive studies with beef cattle and sheep at the California Agricultural Experiment Station (Garrett *et al.*, 1959; Lofgreen, 1965). The procedure they use is the same as the comparative slaughter technique discussed earlier, except that the estimation of body composition is based on the specific gravity of the carcasses. Carcass specific gravity is determined by weighing the chilled carcass in air and also when immersed in water. The specific gravity and the water content of the whole empty body may be predicted from the equations:

$$CSG = \frac{\text{weight in air}}{\text{weight in air} - \text{weight under water}}$$

$$BSG = 0{\cdot}9955\ CSG - 0{\cdot}0013$$

$$W = 100\left(4{\cdot}008 - \frac{3{\cdot}620}{BSG}\right)$$

where BSG = body specific gravity (body less contents of digestive tract)

CSG = carcass specific gravity

W = per cent body water

Once the body water is determined, the other components may be calculated and the heat production estimated in the same manner discussed previously.

METHODS OF MEASURING RESPIRATORY EXCHANGE

The methods of indirect calorimetry which have been used most widely in the past and which are still most commonly used in energy metabolism studies with humans, laboratory animals, and farm livestock, are those involving the measurement of the respiratory exchange. The techniques used to measure the oxygen consumption and carbon dioxide production of organisms have been numerous. The methods chosen have depended on the goals of the experiments, the facilities available and the type of organism being studied.

Microrespiration apparatus has been used to measure the heat production of mitochondria, tissue slices, tissue homogenates, enzyme preparations, insect eggs and microorganisms (Smith and Kleiber, 1950; Kleiber, 1961; Linderstrom-Lang, 1937). Other types of apparatus have been designed for studies with rats, mice, guinea pigs, rabbits, cats, dogs, sheep, swine, humans, cattle, horses and elephants. Even the rate of heat production of birds, fish, aquatic animals and whales have been estimated from indirect calorimetric measurements based on respiratory exchange (Scholander and Irving, 1941; Irving *et al.*, 1941; Brett, 1965). Thus, by the proper choice of apparatus, oxygen consumption ranging from micromillilitres to cubic meters and heat production from microcalories to megacalories can be measured. The ratio from the largest to the smallest measured quantity amounts to 10^{12} (Kleiber, 1961). Krogh (1916), Murlin (1922), Paechtner (1931), Brody (1945), Kleiber (1961) and Blaxter (1962) have described the different types of apparatus used and Kleiber (1961) has prepared an excellent graphical summary of the evolution of different types of respiration apparatus from 1784 to 1949.

The major methods used to measure the respiratory exchange of animals are those involving confinement of the subject in a respiration chamber and those in which a mask, tracheal cannula or similar technique is used to connect the respiratory passages directly with the measuring apparatus. In most long-term studies conducted under carefully controlled conditions and those with animals or experimental subjects which may not voluntarily cooperate, respiration chambers are usually used. The two general types of ventilation which are most commonly used are the closed circuit and open circuit systems.

CLOSED CIRCUIT (REGNAULT–REISET) SYSTEMS

The distinguishing characteristic of a closed circuit system is that the same air is recirculated through the system. The simplest system of this sort, such as was used by Lavoisier in some of his pioneering studies with guinea pigs and sparrows, is a respiration chamber of known capacity (bell jar) which contains sufficient oxygen to permit respiration of ordinary atmospheric air for a short time without discomfort to the animal. By analysis of a sample of air at the beginning and at the end of an experiment it is possible to learn from the changed composition the amount of oxygen absorbed and the amount of CO_2 given off.

A slightly modified version of this system is currently being used in energy metabolism studies with swine weighing 30 to 60 kg (Charlet-Lery, 1958). Instead of relying upon only initial and final analyses to determine oxygen consumption and carbon dioxide production, she calculates the rate of change in gas composition in the chamber from linear regression analyses of several gas samples collected during each 10–12 min experiment.

In many experiments with animals, such short intervals are not suitable, so various modifications have been made which make it possible to extend the respiration trials. Regnault and Reiset (1849) built the first closed-circuit apparatus which provided for the absorption of the carbon dioxide as rapidly as it was produced and its replacement by oxygen. They used KOH to absorb the CO_2 produced by a dog, and they were able to extend the trials until the oxygen supply was exhausted. The oxygen consumed by the animal was the amount of oxygen introduced into the chamber plus a correction for any change in concentration of oxygen in the respiration chamber during the experiment.

Closed-circuit respiration chambers which operate on the same principle as the Regnault–Reiset apparatus have been built for studies with small laboratory animals, humans, sheep, goats, cattle and horses. In order that valid results be obtained, the chamber must be absolutely air-tight, and the temperature and humidity within the chamber well defined. These conditions have been met with relatively little difficulty by workers who have used small chambers (Krogh, 1916; Kleiber, 1940). However, early attempts to use closed-circuit systems with large animals were not satisfactory, due primarily to difficulties in preventing leaks and in maintaining a constant volume in the chamber. Zuntz and Geppert (1885) constructed a closed-circuit respiration chamber large enough to conduct studies with horses and cattle, but the volume (80 m^3) was so great that an error of 1°C in temperature would cause an error of 240 l. oxygen consumption. This is equivalent to about 16 per cent of a cow's daily metabolism (Kleiber, 1935), and the possibilities of error were quite serious in experiments of short duration. Another deterrent factor which prevented research workers from using the closed circuit system in the past was the cost of absorbents for removing carbon dioxide and water vapour produced by large animals. For instance, a mature lactating dairy cow may produce as much as 40 kg water vapour and 12 kg CO_2 daily. About 250 kg silica gel and 100 kg soda lime would be required to remove this much water vapour and CO_2 (Wainman and Blaxter, 1958).

The advantages of closed circuit systems are that the total carbon dioxide and water vapour production of animals may be measured gravimetrically and that oxygen consumption may be determined by direct measurements of oxygen admitted into the chamber. Several closed circuit respiration chambers for sheep, goats and calves and one for cows and steers (9 m^3) were built by Blaxter and his co-workers at the Hannah Dairy Research Institute, Ayr, Scotland (Blaxter, 1962; Blaxter *et al.*, 1954; Wainman and Blaxter, 1958). They were successful in overcoming many of the technical difficulties, and during the past decade they have conducted extensive studies on factors affecting the energy metabolism of ruminants. Similar systems have also been built and used effectively by workers in Australia (Graham, 1962) and East Africa (Rogerson, 1960).

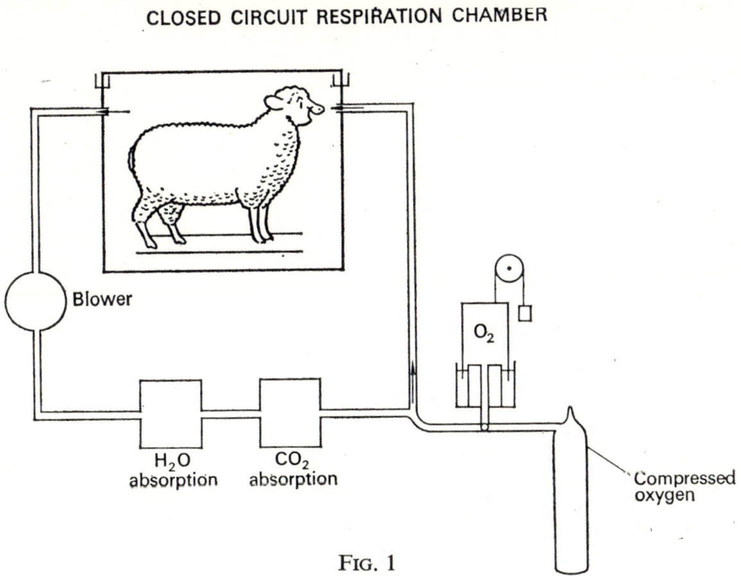

FIG. 1

Figure 1 is a schematic illustration of a closed circuit chamber. The animal is enclosed in the air-tight respiration chamber. Air is circulated by a rotary pump through the chamber, and a portion of the air stream is diverted through absorption trains for removal of water vapour and carbon dioxide, before being returned to the chamber. The consumption of oxygen by the animal results in a fall in pressure in the apparatus, and oxygen is admitted to the system from a constant-pressure reservoir (spirometer) to replace that removed by the animal. Soda lime, KOH, NaOH, $Ba(OH)_2$, Ascarite, triethanolamine, activated carbon or other material may be used to remove CO_2. Water vapour may be removed by the use of concentrated sulphuric acid, silica gel, calcium chloride, calcium sulphate (Drierite), magnesium perchlorate, phosphorous pentoxide or other chemical absorbents, or the water vapour may be condensed by the use of very low temperatures. Wainman and Blaxter (1958) found that the most suitable and economical absorbents were silica gel and soda lime due to their low costs, availability in large quantities and the ease with which silica gel could be regenerated for repeated use. The methane is allowed to accumulate in the chamber (up to concentrations of about 2 per cent) and is determined by analysis of samples of gas collected at the beginning and end of each experimental period. Compressed cylinders of oxygen of 99 per cent purity are commercially available and may be used for routine studies. The amount of oxygen introduced into the chamber may be measured volumetrically by means of a gas meter. The dimensions of the chambers and the details of the construction and operation of the equipment have been reported by Wainman and Blaxter (1958).

OPEN CIRCUIT (PETTENKOFER) SYSTEMS

The open circuit or Pettenkofer system of measuring the respiratory exchange has been the most popular method used in energy metabolism studies with large animals. In this system, illustrated by Fig. 2, the animal is enclosed in a respiration chamber through which outdoor air is passed. Outdoor air has been shown by numerous workers to be almost constant in composition (0·03 per cent CO_2, 20·94 per cent O_2). Thus, by measuring the volume of air passing through the chamber and the concentration of carbon dioxide, methane and oxygen in expired air, it is possible to calculate the carbon dioxide and methane production and oxygen consumption of the animal. The major requirements of such a system are precise means of measuring the air flow and for determining the composition of the air leaving the chamber.

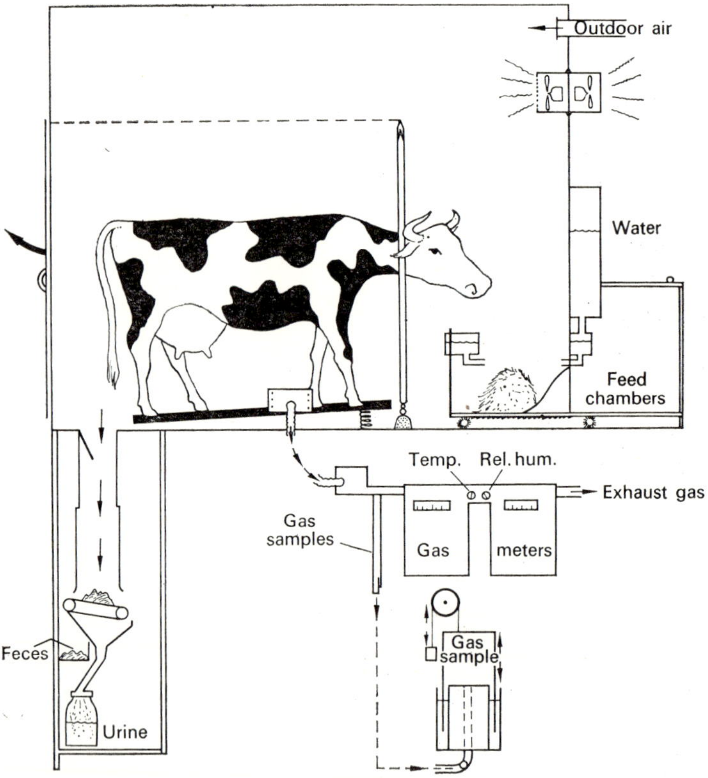

OPEN CIRCUIT RESPIRATION CHAMBER

FIG. 2

Pettenkofer (1862) built the first apparatus which operated on the open-air-current basis. The original system was used in studies with humans, and air was aspirated from the chamber by means of air pumps. The water vapour and carbon dioxide content of samples of the exhaust gas were measured gravimetrically by means of pumice stone saturated with H_2SO_4 and barium hydrate. Oxygen consumption was not measured directly, but estimates could be made by the following method:

O_2 consumption = (final weight of subject + urine + faeces + CO_2 + H_2O vapour) − (original wt. of subject + H_2O + food consumed)

This method is particularly convenient for use with small animals, where weight changes can be measured accurately by weighing the chamber plus the animal at the start and end of the experiment (Haldane, 1892; Schurch, 1958; Hoffmann and Schmitt, 1960). Sonden and Tigerstedt (1895) built a chamber (100 m^3) large enough to accommodate four or five humans, and they provided a means of collecting composite samples of the exhaust gas for oxygen analysis. A precise volumetric gas analysis apparatus was used to determine the CO_2 and O_2 concentration of the samples.

Various modifications of the Pettenkofer and Sonden–Tigerstedt open circuit systems have been built by numerous research workers throughout the world. The large open circuit chamber at the National Institute of Health, Bethesda, Maryland, being used for studies with humans (Buskirk *et al.*, 1960) is based on this principle.

GAS VOLUME MEASUREMENT

One of the distinguishing characteristics of the different types of open circuit respiration chambers for large animals is the means by which the volume of exhaust gas is measured. Most of the European laboratories (Copenhagen, Denmark, Rostock, Germany, Hochenheim-Stuttgart, Germany, Wageningen, Netherlands) use mercury sealed pumps to pipette the air and measure the volume. This is generally considered to be the most precise method available, and descriptions of such pumps have been published by Møllgaard and Andersen (1917), Møllgaard (1929), Schneider (1957), Nehring *et al.* (1958), Schiemann (1958), Wöhlbier and Schneider (1958) and Van Es (1961). The major disadvantages inherent in the mercury pump system are the high initial cost of the mercury and the potential danger to personnel due to mercury poisoning.

Kleiber (1935) designed a unique water sealed 'air pipette' system which supposedly attains the accuracy of the mercury pump method without the disadvantages inherent in the mercury pump system. One of the additional advantages of Kleiber's method is that the collecting of air over water automatically saturates the air with moisture and the high heat capacity of the water promotes a uniform temperature. In the mercury pump system a saturation tower is included in the air stream prior to measuring the volume so as to maintain a uniform degree of saturation.

It is also possible to use positive displacement dry gas meters if they are properly calibrated and are used in such a manner as to prevent condensation of moisture in the meters. The workers in New Hampshire (Ritzman and Benedict, 1929), Illinois (Mitchell and Hamilton, 1932), Beltsville, Maryland (Flatt *et al.*, 1958) and Chiba, Japan (Kaishio *et al.*, 1955) have used dry gas meters to measure the exhaust gas volume. Workers in Japan and the Netherlands and Australia (Lines *et al.*, 1938; Marston, 1948) also have used wet test meters, which theoretically are more accurate than the dry gas meters (Krogh, 1920).

Another means of measuring air flow is by the use of electric integrating flowmeters, which record the differential pressure of a gas stream when it passes through an orifice of fixed dimensions. The change in pressure is proportional to the rate of gas flow. According to Thorbek (1961) this method compared very favourably with the traditional wet test meter and mercury pump system formerly used in the Copenhagen laboratory. Both electric flowmeters and dry gas meters are used in the Beltsville, Maryland, laboratory. Hoernicke (1958) and Struss (1958) described open circuit respiration equipment in which the gas volume was measured by means of a 'ring-balance' meter ('Ringwaage') which also operates on the principle of differential pressure.

Regardless of the method used to measure the volume of gas, it is essential that the temperature, barometric pressure and moisture content of the gas be known. The gas volume must be corrected to standard temperature and pressure conditions (0°C, 760 mm Hg, dry gas) by the use of appropriate factors or equations. Another essential feature in all the systems is a means of collecting representative samples of gas for gas analysis. These samples must be true aliquots of the gas flow because the rate of carbon dioxide and methane production and oxygen consumption are not constant over a 24 hr period. Various devices are used in the different laboratories to accomplish this goal, but in each case samples are taken which are proportional to the rate of flow.

GAS ANALYSIS

The rate of gas flow through an open circuit respiration chamber is usually regulated so that the concentration of CO_2 is approximately 1 per cent by volume. According to Armstrong (1958, p. 93), there is no marked rise in the respiration rate of sheep until the CO_2 concentration of the surrounding air exceeds 2·5 to 3·0 per cent and little effect on heat production until the CO_2 concentration exceeds 5·0 to 5·5 per cent. Blaxter (1962), however, stated that small changes in respiratory frequency of calves were noted when CO_2 exceeded 0·5 per cent. The advantage of allowing the CO_2 concentration to be near 1 per cent is that errors due to gas analysis are less serious when the level of CO_2 is high, and at the 1 per cent level no physiological disturbance of the animals would usually be expected. In most laboratories, the CO_2, CH_4 and O_2 are measured with an accuracy of $\pm 0·002$ to $\pm 0·005$ vol. per cent in absolute terms. This great precision in gas analysis is necessary because the carbon dioxide production, methane production and oxygen consumption are based on the change in concentration of these components as the air flows through the chamber. Even if no error is made in measuring the gas flow, an error of $\pm 0·005$ per cent in CO_2 or O_2 analysis would result in 0·5 to 1·0 per cent error.

Several methods may be used to analyse respiratory gas with the requisite degree of accuracy, but the most commonly used method is the classical volumetric method (Petterson, 1886; Carpenter *et al.*, 1929; Krogh, 1920; Struss and Quasdorf, 1958). A fixed volume of gas is measured in a burette and the change in volume, after absorbing the CO_2 by KOH and O_2 by alkaline pyrogallol or chromous chloride, is determined. Methane and combustible gas may be analysed in such an apparatus (Carpenter *et al.*, 1930) by using heated platinum wire to burn the gas and then measuring the CO_2 formed. Such methods are slow, tedious and require considerable skill for the necessary precision to be attained. Therefore, several laboratories have investigated the possibility of using physical methods of gas analysis. Physical methods are rapid, continuous, can be automatically recorded and the sample is unchanged during analysis, thus making sequential analyses possible. Among the physical methods which have been used and seem to merit further consideration are infrared for carbon dioxide and methane, paramagnetic analysers for oxygen, and thermal conductivity, mass spectrometry, interferometry, sonic analysis, and gas chromatography for all the major components (White, 1958; Martin, 1956; Pauling *et al.*, 1946; Spoor, 1948; Romijn and Lokhorst, 1961; Lilly, 1950). A number of gas analysers which are commercially available may be used for routine energy metabolism studies. Schneider (1958) described the use of automatic infrared gas analysers for respiration trials with cattle in the Hohenheim laboratory, and similar analysers have been used routinely at the Beltsville energy metabolism laboratory for several years. Paramagnetic analysers have been used by numerous workers to determine oxygen concentration of incoming air and exhaust gas from respiration chambers. By frequently recalibrating these instruments, the necessary precision may be attained with considerably less labour and skill being required than is needed for operation of the volumetric analysers. Another advantage of the physical methods of analysis is that continuous stream analyses may be made relatively easily, thus making it possible to observe rates of change in the metabolism of animals as a result of activity, drugs, food consumption, environmental changes, and other variables.

AUTOMATIC RECORDING AND CALCULATIONS

One of the major reasons that extensive studies on the energy metabolism of large animals have not been conducted in the past was because of the excessive amount of labour required in collecting and analysing the data. The extreme slowness with which data were accumulated from respiration trials and the excessive labour costs for manually recording the gas meter readings, temperature, relative humidity and barometric pressure on a 24-hr basis discouraged many research workers. Even after the observations are made, numerous computations of a repetitive nature are required for the routine calculation of oxygen consumption and carbon dioxide, methane and heat production of experimental animals. These include corrections to reduce gas volumes to standard conditions (dry, 0°C, 760 mm Hg), adjustment of oxygen consumption and carbon dioxide and methane production for changes in respiration chamber gas composition, and corrections for gaseous losses during feeding and excreta collection. Calculations of the volume of inspired air based upon the composition and volume of exhaust gas and the composition of outdoor air, calculation of the average temperature and relative humidity of the chamber during the experiment, applications of calibration factors for gas meters and sensing elements to convert coded readings to absolute values, and various other computations are also made for each respiration trial. Logarithms and prepared tables such as those of Carpenter (1948) have been used extensively by research workers engaged in energy metabolism studies in order to reduce the labour in making these routine computations. Even then the time required to perform these calculations forms an important portion of the total labour involved in these experiments.

Automatic recording equipment which provides 24-hr records of the respiratory exchange of cattle has been used routinely with six respiration chambers at the Beltsville energy metabolism laboratory (Flatt *et al.*, 1958). The data are automatically punched into cards and recorded on a remote typewriter as illustrated schematically in Fig. 3. Medium size, digital electronic computing equipment has been used to summarize the data, resulting in considerable savings in the time required to conduct respiration trials and summarize the results (Flatt *et al.*, 1965). Speed, reliability and ease of subsequent summarization and statistical treatment of the data are among the advantages offered by the use of data processing equipment. The formulae (Flatt and Tabler, 1961) used for summarizing respiration trial data, whether manually or automatically, are listed in Tables 2 and 3.

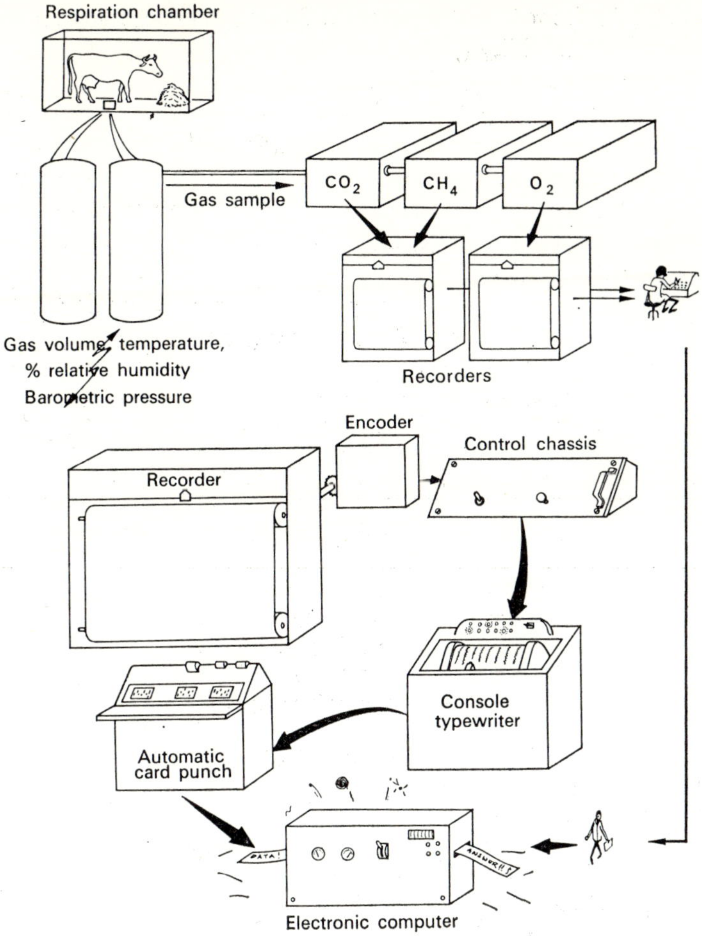

FIG. 3

MAJOR SYMBOLS

O = amount of oxygen (litres)
o = % oxygen
C = amount of carbon dioxide (litres)
c = % carbon dioxide
M = amount of methane (litres)
m = % methane
N = amount of nitrogen (litres)
n = % nitrogen
L = volume of exhaust gas (litres)
V = volume of respiration chamber and compartments (litres)

S = standard conditions correction factor (dry, 0°C, 760 mm Hg)
R = relative humidity (%)
B = barometric pressure (mm Hg)
t = temperature (°C)
E = aqueous vapour pressure at saturation (mm Hg)
e = aqueous vapour pressure under conditions measured (mm Hg)
$M_{1\text{-}4}$ = gas meter calibration factor (litres/revolution × calibration)
G = number of gas meter counts
W = animal weight (kg)
P = protein oxidized per 24 hr (grams urinary nitrogen × 6·25)
HP = total heat production of the animal (kcal)
J = number of times a designated compartment of the respiration chamber was opened.

TABLE 2

Formulæ for Correcting Gas Volumes to Standard Conditions (0°C, 760 mm Hg, dry gas)

Equation no.	Description
1	Aqueous vapour pressure of exhaust gas at saturation. $E_g = 8{\cdot}7 - 0{\cdot}25\, t_g + 0{\cdot}034\, t^2{}_g$
2	Aqueous vapour pressure of respiration chamber gas at saturation. $E_c = 8{\cdot}7 - 0{\cdot}25\, t_c + 0{\cdot}034\, t^2{}_c$
3	Aqueous vapour pressure of exhaust gas. $e_g = E_g R_g$
4	Aqueous vapour pressure of respiration chamber gas. $e_c = E_c R_c$
5	Volume correction factor for exhaust gas. $S_g = \frac{B - e_g}{760} \times \frac{273}{273 + t_g}$
6	Volume correction factor for respiration chamber gas. $S_c = \frac{B - e_c}{760} \times \frac{273}{273 + t_c}$
7	Gas meter calibration factors (examples of specific meters). $M_1 = 34{\cdot}2$ $M_2 = 35{\cdot}5$ $M_3 = 34{\cdot}1$ $M_4 = 33{\cdot}1$
8	Volume of exhaust gas, corrected to standard conditions. $L_{gs1} = \frac{G_1 M_1 + G_2 M_2}{2} \times S_{g1}$ $L_{gs2} = \frac{G_3 M_3 + G_4 M_4}{2} \times S_{g2}$
9	Respiration chamber volume corrected to standard conditions. $V_{cs1\text{-}6} = S_c\,(V_{c1\text{-}6} - W)$
10	$V_{ys1\text{-}6} = S_c\,(V_{y1\text{-}6})$
11	$V_{zs1\text{-}6} = S_c\,(V_{z1\text{-}6})$

TABLE 3
Formulae for Calculation of Respiratory Exchange and Heat Production

Equation no.	Description
12	Carbon dioxide production of the animal (litres) $C_{ps} = L_{gs}(c_d - c_a) + (c_f V_{cfs} - c_i V_{cis}) - (c_d - c_a)(J_y V_{ys} + J_z V_{zs})$
13	Methane production of the animal (litres) $M_{ps} = L_{gs} m_d + (m_f V_{cfs} - m_i V_{cis}) - m_d (J_y V_{ys} + J_z V_{zs})$
14	Change of oxygen concentration in respiration chamber, calculated from carbon dioxide change (litres) $(o_f V_{cfs} - o_i V_{cis}) = \dfrac{c_f V_{cfs} - c_i V_{cis}}{(c_d - c_a)/(o_a - o_d)}$
15	Oxygen consumption of the animal (litres) $O_{us} = L_{gs}\left(\dfrac{n_d o^a}{n_a} - o_d\right) + (o_f V_{cfs} - o_i V_{cis}) - \left(\dfrac{n_d o_a}{n_a} - o_d\right)(J_y V_{ys} + J_z V_{zs})$
16	Heat production of the animal (kcal) $HP = 3{\cdot}866\, O_{us} + 1{\cdot}200\, C_{ps} - 0{\cdot}229\, P - 0{\cdot}518\, M_{ps}$

SUBSCRIPTS

c = respiration chamber, including feed and excreta collection compartments
y = feed compartment of respiration chamber
z = excreta collection compartment of respiration chamber
g = exhaust gas
s = standard conditions (dry, 0°C, 760 mm Hg)
u = consumed or used
p = produced
a = outdoor air
d = composite or spirometer sample
i = initial conditions or residual sample from chamber at start of experiment
f = final conditions or residual sample at end of period
j = leaving
q = entering
hp = heat (kcal)
v = carbon content (g)

TRACHEAL CANNULA AND MASK TECHNIQUES

In some types of studies it is not suitable or desirable to enclose the experimental subject in a respiration chamber. Routine basal metabolism measurements with humans are usually made by connecting the pulmonary system of the subject to an oxygen spirometer, and measuring the rate of oxygen consumption by the rate of decline of the oxygen bell. Brody and his co-workers (Brody, 1945) designed masks for use with cattle and they used the closed circuit spirometer method in numerous studies. Kleiber and Edick (1952) used the mask technique to collect respiratory gas from cattle and similar systems have also been designed for use in tracer trials with radio-carbon with sheep and swine (Kleiber, 1958). Mask techniques have also been used by Orr and Magee (1923), Berman *et al.*, (1963) and numerous others. Helmets have been used in studies with humans (Buskirk *et al.*, 1960), cattle (Yousef and Johnson, 1965) and dogs (Nims and Worth, 1961).

Instead of using a mask, it is possible to use tracheostomized animals and connect the trachea directly to a light-weight portable measuring-sampling device such as a Muller-Franz (1952) meter and a Douglas (1911) bag. The advantages of such a method are that the oxygen consumption, carbon dioxide production and energy expenditure of animals may be measured for extended periods of time without restricting or confining them. The energy expenditure of animals engaged in activities such as walking, grazing, ruminating, chewing, lactating, standing and lying may be measured. The use of such a technique with grazing cattle was described by Flatt *et al.* (1958), and the results of some radioisotope and absorption studies conducted with tracheostomized cattle were reported by Hoernicke *et al.* (1965). Similar methods were used by Blaxter and Joyce (1963) to study the metabolism of tracheostomized sheep. Numerous other workers have used tracheal cannulae to study energy expenditure in short-term experiments. In the early German experiments (Zuntz and Lehmann, 1889; Lehmann *et al.*, 1894; Zuntz *et al.*, 1898) with horses it was not possible for the animal to have freedom of movement, except possibly on a treadmill because of the excessive weight of the gas meters and the elaborate, fragile aliquoting devices.

Tracheal cannula techniques have also been used successfully by Hagemann (1899), Hagemann and Karpow (1906) and Ustjanzew (1911) to study the energy expenditure of sheep. Cattle energy metabolism studies were conducted by Paechtner (1909), Dahm (1910), Heide *et al.* (1913), Klein (1912, 1915) and Møllgaard and Andersen (1917) by making use of a tracheal cannula. New Zealand workers (Cresswell, 1957; Webster and Cresswell, 1957) have maintained tracheostomized sheep over long periods to conduct climatological, nutritional and energy metabolism studies. A modification of their original technique was described by Young and Webster (1963).

The advantages to the use of tracheal cannula and mask techniques are the low initial cost and the versatility of studies which may be conducted. However, the possible effects of the tracheal transection on the health of the animal and the extra energy required for carrying the apparatus and harness may interfere with the goals of some types of experiments. Blaxter and Joyce (1963) found that the oxygen consumption of sheep measured by the tracheal

FIG. 5. Sheep with an endotracheal cannula, attached to a mobile gas metering and sampling system. (Photograph courtesy of Dr. J. L. Corbett and B. A. Young)

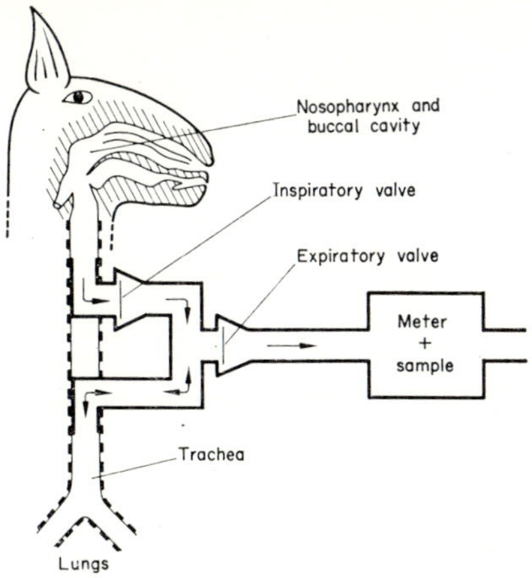

FIG. 4. Diagrammatic representation of tracheal cannula technique, showing an endotracheal divider and gas metering system.
(Courtesy Dr. J. L. Corbett and B. A. Young.)

cannula technique was the same as that measured by confinement of the same sheep given the same food in a respiration chamber. However, they experienced difficulties in maintaining tracheostomized sheep for long periods of time and they recommended the use of face masks as a superior method for collecting expired air from sheep.

INSENSIBLE WEIGHT LOSS

Another method of indirect calorimetry which has been used with large animals is the measurement of insensible weight loss. This has been suggested as a short-cut procedure for determining energy metabolism, since the only equipment required is a good set of scales and a means of accurately measuring feed and water intake and faecal and urinary excretion. Kriss (1930, 1936), Mitchell and Hamilton (1936) and Ritzman and Benedict (1938) have studied the possibility of using this method in nutritional investigations with cattle. The equation used by Mitchell and Hamilton (1936) to calculate heat production was as follows:

$$I = (W_i + FD_i + H_2O_i) - (W_f + U + F)$$

$$\frac{H_2O_i}{W} = \frac{ce}{ce + 1{\cdot}1466} r - 0{\cdot}8288$$

$$HP = \frac{ie}{1{\cdot}7241\ ce} + 1{\cdot}9769 \frac{CO_2}{O_2} - 1{\cdot}4298$$

where I = insensible weight loss
Wi = initial body weight (kg)
W_f = final body weight (kg)
FD_i = feed intake

H_2O_i = water intake
U = urine excreted
F = faeces excreted
c = fraction of total heat as vapourized water
e = heat equivalent of a litre of O_2 consumed $= 4{\cdot}688 + 1{\cdot}23 \times \dfrac{\text{litres } CO_2 \text{ prod.}}{\text{litres } O_2 \text{ prod.}}$
HP = heat produced (kcal) $= O_2 \times e$

They concluded that the method could possibly be useful under certain closely controlled conditions but that further research was needed on various aspects. The temperature, plane of nutrition, length of hair coat, animal individuality in heat loss by vapourization of water, and respiratory quotient are some of the factors which must be considered when using insensible weight loss to calculate heat production.

SUMMARY

The indirect calorimetric methods which have been commonly used in measuring the heat production of animals have been discussed. Each method has its advantages and limitations, and the method of choice will depend upon the goals of the experiment and the precision required. The theoretical basis of different methods is discussed, and brief descriptions of techniques used are given. Technological developments in the measurement and recording of physiological parameters have provided numerous tools for use in energy metabolism of animals. Further development and use of these methods in the future should be of immense aid to animal nutritionists, physiologists and biochemists in elucidating factors affecting the energy metabolism of animals.

REFERENCES

ANDERSEN, A. C. (1922) *Biochem. Ztschr.* **130**, 143.
ARMSBY, H. P. (1913) *J. Am. Chem. Soc.* **35**, 1794.
ARMSBY, H. P. (1904) *Exp. Sta. Rec.* **15**, 1037.
ARMSBY, H. P. and MOULTON, C. R. (1925) *The Animal as a Converter of Matter and Energy.* Chemical Catalog Co., Inc., New York.
ATWATER, W. O. and BENEDICT, F. G. (1905) *Carnegie Institution of Wash. Publ.* **42**, Washington, D.C.
ATWATER, W. O. and ROSA, E. B. (1899) *U.S.D.A. Exp. Sta. Bull.* **63**.
BATH, D. L., RONNING, M., MEYER, J. H. and LOFGREEN, G. P. (1965) *J. Dairy Sci.* **48**, 374.
BENEDICT, F. G. (1930) *New England J. Med.* **203**, 150.
BENEDICT, F. G., COLLINS, W. E., HENDRY, M. F. and JOHNSON, A. (1920) *N.H. Agr. Exp. Sta. Tech. Bull.* **16**.
BENEDICT, F. G. and TALBOT, F. B. (1914) *Carnegie Inst. of Wash. Publ.* **201**, 28. Washington, D.C.
BENSADOUN, A., VAN NIEKERK, B. D. H., PALADINES, O. L. and REID, J. T. (1963) *J. Animal Sci.* **22**, 604.

BENZINGER, T. H., HUEBSCHER, R. G., MINARD, D. and KITZINGER, C. (1957) *Naval Medical Institute Research Report*, **15**, 547.

BENZINGER, T. H. and KITZINGER, C. (1949) *Rev. Sci. Instr.* **20**, 849.

BERMAN, A., AMIR, S. and VOLCANI, R. (1963) *Austr. J. Agr. Res.* **14**, 874.

BLAXTER, K. L. (1965) Ed. *Energy Metabolism. Proc. 3rd Symposium on Energy Metabolism. European Assoc. of Animal Prod. Publ.* **11**. Academic Press, London.

BLAXTER, K. L. (1962) *The Energy Metabolism of Ruminants*. Charles C. Thomas, Springfield, Ill.

BLAXTER, K. L., GRAHAM, N. MCC. and ROOK, J. A. F. (1954) *J. Agr. Sci.* **45**, 10.

BLAXTER, K. L. and HOWELLS, A. (1951) *Brit. J. Nutr.* **5**, 25.

BLAXTER, K. L. and JOYCE, J. P. (1963) *Brit. J. Nutr.* **17**, 523.

BLAXTER, K. L. and ROOK, J. A. F. (1953) *Brit. J. Nutr.* **7**, 83.

BLAXTER, K. L. and ROOK, J. A. F. (1953) *Nature*, **171**, 609.

BLAXTER, K. L. and ROOK, J. A. F. (1956) *J. Agr. Sci.* **48**, 194.

BLAXTER, K. L. and ROOK, J. A. F. (1956) *J. Agr. Sci.* **48**, 210.

BOUSSINGAULT, J. B. (1839) *Ann. de chimie et de physique*, **71**, 113.

BRETT, J. R. (1965) *Sci. Amer.* **213**, 80.

BRODY, S. (1945) *Bioenergetics and Growth*. Reinhold Publ. Corp., New York.

BRODY, S., PROCTER, R. C. and ASHWORTH, U. S. (1934) *Mo. Agr. Exp. Sta. Res. Bull.* **220**.

BRODY, S., RAGSDALE, A. C., TROWBRIDGE, E. A., COMFORT, J., CHITTENDEN, D. W., MCKENZIE, F. F., HOGAN, A. G., PITCHER, R., KEMPSTER, H. L. (1930) *Mo. Agr. Exp. Sta. Res. Bull.* **143**.

BROUWER, E. (1958) *European Assoc. Animal Prod. Publ.* **8**, 182.

BROUWER, E. (1965) in *Energy Metabolism*, p. 441. Ed. K. L. Blaxter. *European Assoc. Animal Prod. Publ.* **11**. Academic Press, London.

BROUWER, E. and VAN ES, A. J. H. (1961) Ed. *Proc. 2nd Symposium on Energy Metabolism. European Assoc. Animal Prod. Publ.* **10**.

BROZEK, J. and HENSCHEL, A. (1961) Ed. *Techniques for Measuring Body Composition*. National Acad. Sci.–National Res. Council, Washington, D.C.

BROZEK, J., WHIPPLE, H. E. and SILVERZWEIG, S. (1963) Ed. *Body Composition. Annals N.Y. Acad. Sci.* **110**, 1.

BUSKIRK, E. R., THOMPSON, R. H., MOORE, R. and WHEDON, C. D. (1960) *Amer. J. Clin. Nutr.* **8**, 602.

CARPENTER, T. M. (1948) *Carnegie Inst. Wash. Publ.*, **303c**. 4th ed. Washington, D.C.

CARPENTER, T. M., FOX, E. L. and SEREQUE, A. F. (1929) *J. Biol. Chem.* **83**, 211.

CARPENTER, T. M., LEE, R. C. and FINNERTY, A. E. (1930) *Wiss. Archiv. f. Landw.* **4**, 1.

CATHCART, E. P. and CUTHBERTSON, D. P. (1931) *J. Physiol.* **72**, 349.

CHARLET–LERY, G. (1958) *European Assoc. Animal Prod. Publ.* **8**, 194.

CRAWFORD, A. (1788) *Experiments and Observations on Animal Heat and the Inflammation of Combustible Bodies*. 2nd ed. London (cited by Kleiber, 1961).

CRESSWELL, E. (1957) *Nature (London)*, **179**, 1139.

CUTHBERTSON, D. P. (1933) *Biochem. J.* **27**, 1099.

DAHM, K. (1910) *Biochem. Ztschr.* **28**, 456.

DOUGLAS, C. G. (1911) *J. Physiol.* **42**, xvii.

ES, A. J. H. VAN (1961) Between-animal variation in the amount of energy required for the maintenance of cows. Dissertation, Wageningen, Netherlands.

FALLER, I. A., PETTY, D., LAST, J. H., PASCALE, L. R. and BOND, E. E. (1955) *J.Lab.Clin. Med.* **45**, 748.

FLATT, W. P., COPPOCK, C. E., MOORE, L. A., HEATH, A., DICKINSON, F. and TABLER, K. A. (1965) in *Energy Metabolism*, p. 189. Ed. K. L. Blaxter. *European Assoc. Animal Prod. Publ.* **11**. Academic Press, London.

FLATT, W. P. and TABLER, K. A. (1961) *European Assoc. Animal Prod. Publ.* **10**, 39.

FLATT, W. P., VAN SOEST, P. J., SYKES, J. F. and MOORE, L. A. (1958) *European Assoc. Animal Prod. Publ.* **8**, 53.

FLATT, W. P., WALDO, D. R., SYKES, J. F. and MOORE, L. A. (1958) *European Assoc. Animal Prod. Publ.* **8**, 101.

FLETCHER, J. G. and WOLFF, H. S. (1954) *J. Physiol.* **123**. 67P.

FORBES, E. B., BLACK, A., THACKER, E. J. and SWIFT, R. W. (1940) *J. Nutr.* **20**, 47.
FORBES, E. B., GALLUP, J. and HURSH, J. B. (1961) *Science*, **133**, 101.
FRANKE, E. R. and WENIGER, J. H. (1958) *Archiv. Tiernahrung*, **8**, 81.
GARRETT, W. N., MEYER, J. H. and LOFGREEN, G. P. (1959) *J. Animal Sci.* **18**, 528.
GLASSTONE, S. (1946) *Textbook of Physical Chemistry*. 2nd ed. D. Van Nostrand Co., New York.
GRAHAM, N. McC. (1962) *Proc. Austr. Soc. Animal Prod.* **4**, 138.
HAGEMANN, O. (1899) *Arch. f. Anat. u. Physiol. Supp. Bd.*, p. 111.
HAGEMANN, O. and KARPOW, M. S. (1906) *Landw. Jahr.* **35** (*Suppl. Bd. 4*), 371.
HAIGH, L. D., MOULTON, C. R. and TROWBRIDGE, P. F. (1920) *Mo. Agr. Exp. Sta. Res Bull.* **38**.
HALDANE, J. S. (1892) *J. Physiol.* **13**, 419.
HANSARD, S. L. (1963) *Annals N.Y. Acad. Sci.* **110**, 229.
HANSARD, S. L. and LYKE, W. A. (1956) *Proc. Soc. Exper. Biol. Med.* **93**, 263.
HASHIZUME, T., KAISHIO, Y., MORIMOTO, H., ANBO, S., MASUBUCHI, T., ABE, B., HAMADA, T. and TANAKA, K. (1961) *Bull. National Inst. Agr. Sci. G*, **20**, 21 (Japan).
HEIDE, R. VON DER, KLEIN, W. and ZUNTZ, N. (1913) *Landw. Jahrb. Berlin*, **44**, 765.
HEVESY, G. and HOFER, E. (1934) *Nature* (*London*), **134**, 879.
HOERNICKE, H. (1958) *European Assoc. Animal Prod. Publ.* **8**, 70.
HOERNICKE, H., WILLIAMS, W. F., WALDO, D. R. and FLATT, W. P. (1965) in *Energy Metabolism*, p. 165. Ed. K. L. Blaxter. *European Assoc. Animal Prod. Publ.* **11**. Academic Press, London.
HOFFMANN, L. and SCHMIDT, M. (1960) *Archiv. f. Tierern.* **10**, 472.
IRVING, L., SCHOLANDER, P. F. and GRINNELL, S. W. (1941) *J. Cell. Comp. Physiol.* **17**, 145.
JORDAN, W. H. (1895) *Ann. Rept. Maine State College, Part II*, p. 36.
KAISHIO, Y., HASHIZUME, T., AMBO, S., ABE, G., MASUBUCHI, T., ABE, M., MORIMOTO, H., HORII, S., KURATA, Y., ISHIHARA, M. and TSUCHIYA, H. (1955) *Bull. National Inst. Agr. Sci. G*, **11**, 157.
KELLNER, O. and KOEHLER, A. (1900) *Landw. Vers. Sta.* **53**, 1.
KIHLEN, G. (1951) *Ann. Roy. Agr. College of Sweden*, **18**, 29.
KIRTON, A. H. (1964) *Proc. New Zealand Soc. Animal Prod.* **24**, 77.
KLEIBER, M. (1940) *Univ. Calif. Publ. Physiol.* **8**, 207.
KLEIBER, M. (1950) in *Biophysical Research Methods*, p. 175. Ed. F. M. Uber. Interscience Publ., New York.
KLEIBER, M. (1935) *Hilgardia* (*California Agr. Exp. Sta.*), **9**, 1.
KLEIBER, M. (1961) *The Fire of Life—An Introduction to Animal Energetics*. John Wiley, New York.
KLEIBER, M. and EDICK, M. (1952) *J. Animal Sci.* **11**, 61.
KLEIN, W. (1912) *Zentralblatt f. Physiol.* **26**, 722.
KLEIN, W. (1915) *Biochem. Ztschr.* **72**, 169.
KRAYBILL, H. F., GOODE, E. R., ROBERTSON, R. S. B. and SLOANE, H. A. (1953) *J. Appl. Physiol.* **6**, 27.
KRAYBILL, H. F., HANKINS, O. G. and BITTER, H. L. (1951) *J. Appl. Physiol.* **3**, 681.
KRISS, M. (1925) *J. Agr. Res.* **30**, 393.
KRISS, M. (1930) *J. Agr. Res.* **40**, 283.
KRISS, M. (1930) *J. Agr. Res.* **40**, 271.
KRISS, M. (1936) *Proc. Amer. Soc. Animal Prod.* **1936**, 265.
KROGH, A. (1920) *Biochem. J.* **14**, 267.
KROGH, A. (1920) *Biochem. J.* **14**, 282.
KROGH, A. (1916) *The Respiratory Exchange of Animals and Man*. Longmans, Green and Co., London.
KROGH, A. and SCHMIT-JENSEN, H. O. (1920) *Biochem. J.* **14**, 686.
KUEHN, G., THOMAS, A., MARTIN, E. *et al.* (1894) *Landw. Vers. Sta.* **44**, 257.
LAVOISIER, A. L. (1777) *Mem. de l'Academie des sciences*, 1777, 185 (*Oeuvres*, **2**, 174, 1862).
LAVOISIER, A. L. and LAPLACE (1780) *Mem. de l'Academie des sciences*, 1780, 355 (*Oeuvres*, **2**, 283, 1862).
LAWES, J. B. and GILBERT, J. H. (1859) *Philos. Trans. Roy. Soc. London*, **149**, 493.

LAWES, J. B. and GILBERT, J. H. (1860) *J. Roy. Agr. Soc. England*, **21**, 433.
LAWES, J. B. and GILBERT, J. H. (1861) *J. Roy. Agr. Soc. England*, **22**, 189.
LEHMANN, F., HAGEMANN, O. and ZUNTZ, N. (1894) *Landw. Jahrb.* **23**, 125.
LESSER, G. T., BLUMBERG, A. G. and STEELE, J. M. (1956) in *Biochemical Problems of Lipids*, p. 381. Ed. G. Popjak and E. Le Breton. Butterworths, London.
LINDERSTRØM-LANG, K. (1937) *Nature (London)*, **140**, 180.
LINES, E. W. L. (1938) *J. Agr. Sci.* **28**, 663.
LOFGREEN, G. P. (1965) in *Energy Metabolism*, p. 309. Ed. K. L. Blaxter. *European Assoc. Animal Prod. Publ.* **11**. Academic Press, London.
LUSK, G. (1922) in *Endocrinology and Metabolism*, **3**, 3. D. Appleton, Co., London.
LUSK, G. (1928) *The Science of Nutrition*. 4th ed. W. B. Saunders Co., Philadelphia.
MARSTON, H. R. (1948) *Austr. J. Sci. Res. B*, **1**, 93.
MARTIN, A. E. (1956) *Nature (London)*, **178**, 407.
MITCHELL, H. H. and HAMILTON, T. S. (1932) *J. Agr. Res.* **45**, 163.
MITCHELL, H. H. and HAMILTON, T. S. (1936) *J. Agr. Res.* **52**, 837.
MØLLGAARD, H. (1929) *Futterungslehre des Milchviehs*. M. & H. Schaper, Hannover.
MØLLGAARD, H. and ANDERSEN, A. C. (1917) *Beret. Forsogslab. K. Vet og Landb.* **94**, 1. (Denmark).
MOUSTGAARD, J. and THORBEK, G. (1949) *Beret. fra Forsogslab.* **240**, 1. (Denmark).
MUELLER, E. A. and FRANZ, H. (1952) *Arbeitsphysiologie*, **14**, 499.
MOULTON, C. R., TROWBRIDGE, P. F. and HAIGH, L. D. (1922) *Mo. Agr. Exp. Sta. Res. Bull.* **55**.
MURLIN, J. R. (1922) in *Endocrinology and Metabolism*, **3**, 514. D. Appleton Co., New York.
MURRAY, J. A. (1922) *J. Agr. Sci.* **12**, 103.
NRC COMMITTEE ON ANIMAL NUTRITION (1962) *Glossary of Energy Terms. N.R.C. National Academy of Sciences Publ.* **1040**. Washington, D.C.
NEHRING, K. (1956) *Festchr., 100 Jahre Moeckern, Deut. Akad. Landwirtsch. Berlin*, **3**, 131.
NEHRING, K., SCHIEMANN, R., HOFFMANN, L. and SCHMIDT, M. (1958) *Wiss. Abh. Deut. Akad. Landwirtsch. Berlin*, **37**, 34.
NIMS, R. M. and WORTH, W. S. (1961) *J. Appl. Physiol.* **16**, 1139.
ORR, J. B. and MAGEE, H. E. (1923) *J. Agr. Sci.* **13**, 447.
PACE, N., KLINE, L., SCHACHMAN, H. K. and HARFENIST, M. (1947) *J. Biol. Chem.* **168**, 459.
PAECHTNER, J. (1931) in *Handbuch der Ernahrung und des Stoffwechsels der Landwirtschaftlichen Nutztiere*, p. 365. Ed. E. Mangold. Julius Springer, Berlin.
PAECHTNER, J. (1909) *Respiratorische Stoffwechselforschung u. ihre Bedeutung f. Nutztierhaltung u. Tierheilkunde*. Dissertation, R. Schoetz, Berlin.
PANARETTO, B. A. (1963) *Austr. J. Agr. Res.* **14**, 594.
PAULING, L., WOOD, R. E. and STURDIVANT, J. H. (1946) *Science*, **103**, 338.
PAULING, L., WOOD, R. E. and STURDIVANT, J. H. (1946) *J. Amer. Chem. Soc.* **68**, 795.
PETTENKOFER, M. (1862) *Ann. Chem. u. Pharm.* (*Supp. Bd. 2*), p. 1.
PETTERSON, O. (1886) *Ztschr. f. anal. Chem.* **25**, 467.
PULLAR, J. D. (1958) *European Assoc. Animal Prod. Publ.* **8**, 95.
PULLAR, J. D. (1957) *Advancement of Sci.* **13**, 491.
REGNAULT, V. and REISET, J. (1849) *Annal. de Chemie et de Physique*, 3e *Serie*, **26**, 299.
REID, J. T., BALCH, C. C. and GLASCOCK, R. F. (1958) *Brit. J. Nutr.* **12**, 43.
REID, J. T., BENSADOUN, A., PALADINES, O. L. and VAN NIEKERK, B. D. H. (1963) *Annals N.Y. Acad. Sci.* **110**, 327.
REID, J. T., WELLINGTON, G. H. and DUNN, H. O. (1955) *J. Dairy Sci.* **38**, 1344.
RITZMAN, E. G. and BENEDICT, F. G. (1929) *Univ. N. H. Agr. Exp. Sta. Bull.* **240**.
ROGERSON, A. (1960) *J. Agric. Sci.* **55**, 359.
ROMIJN, C. and LOKHORST, W. (1961) *Tijd. v. Diergeneeskunde*, **86**, 153 (Netherlands).
RUBNER, M. (1885) *Ztschr. f. Biol.* **21**, 250.
RUBNER, M. (1894) *Ztschr. Biol.* **30**, 73.
SCHMIT-JENSEN, H. O. (1920) *Biochem. J.* **14**, 4.
SCHNEIDER, W. (1957) *Der Aufbau, die Arbeitsweise und die apparative Sicherheit der Hohenheimer Respirationsanlage fur landwirtschaftliche Grossnutztiere*. Dissertation, Erlau Hohenheim.

SCHNEIDER, W. (1958) *European Assoc. Animal Prod.* **8**, 127.
SCHIEMANN, R. (1958) *European Assoc. Animal Prod. Publ.* **8**, 29.
SCHOLANDER, P. F., HAUGAARD, N. and IRVING, L. (1943) *Rev. Sci. Instr.* **14**, 48.
SCHOLANDER, P. F. and IRVING, L. (1941) *J. Cell. Comp. Physiol.* **17**, 169.
SCHURCH, A. (1958) *European Assoc. Animal Prod. Publ.* **8**, 64.
SIRI, W. E. (1956) in *Advances in Biological and Medical Physics*, Vol. IV, p. 239, Ed. J. H. Lawrence and C. A. Tobias. Academic Press, New York.
SMITH, A. H. and KLEIBER, M. (1950) *J. Cell. Comp. Physiol.* **35**, 131.
SONDEN, K. and TIGERSTEDT, R. (1895) *Skand. Arch. Physiol.* **6**, 1.
SPOOR, H. J. (1948) *J. Appl. Physiol.* **1**, 369.
STRUSS, F. (1958) *European Assoc. Animal Prod. Publ.* **8**, 137.
STRUSS, F. and QUASDORF, T. (1958) *Archiv. f. Tiernahrung*, **8**, 433.
SWIFT, R. W. and FRENCH, C. (1954) *Energy Metabolism and Nutrition*. The Scarecrow Press, Washington, D.C.
THORBEK, G. (1961) *European Assoc. Animal Prod. Publ.* **10**, 73.
THORBEK, G. and AERSOE, H. (1958) Ed. *Proc. 1st Symposium on Energy Metabolism. European Assoc. Animal Prod. Publ.* **8**.
THORBEK, G. and NEERGAARD, L. (1965) in *Energy Metabolism*, p. 179. Ed. K. L. Blaxter. *European Assoc. Animal Prod. Publ.* **11**. Academic Press, London.
TROWBRIDGE, P. F., MOULTON, C. R. and HAIGH, L. D. (1915) *Mo. Agr. Exp. Sta. Res. Bull.* **18**.
USTJANZEW, W. (1911) *Biochem. Ztschr.* **37**, 457.
WAINMAN, F. W. and BLAXTER, K. L. (1958) *European Assoc. Animal Prod. Publ.* **8**, 80.
WAINMAN, F. W. and BLAXTER, K. L. (1958) *European Assoc. Animal Prod. Publ.* **8**, 85.
WEAVER, E. R. (1951) in *Physical Methods in Chemical Analysis*, Vol. II, p. 387. Ed. W. C. Berl. Academic Press, New York.
WEBSTER, W. M. and CRESSWELL, E. (1957) *Vet. Rec.* **69**, 526.
WEIR, J. B. DE V., (1949) *J. Physiol.* **109**, 1.
WELLINGTON, G. H., REID, J. T., BRATZLER, L. J. and MILLER, J. I. (1956) *J. Animal Sci.* **15**, 76.
WHITE, C. S. (1958) *The Analysis of Respiratory Gases. Aviation Medicine Selected Reviews.* AGARDograph No. 24, p. 125. Pergamon Press, New York.
WÖHLBIER, W. and SCHNEIDER, W. (1958) *European Assoc. Animal Prod. Publ.* **8**, 45.
YOUNG, B. A. and WEBSTER, M. E. D. (1963) *Austr. J. Agr. Res.* **14**, 867.
YOUSEF, M. K. and JOHNSON, H. D. (1965) *J. Dairy Sci.* **48**, 104.
ZIMMERMAN, O. T. and LAVINE, I. (1945) *Industrial Research Service's Psychrometric Tables and Charts.* Waverly Press, Baltimore.
ZUNTZ, N. (1891) *Landw. Vers. Sta.* **38**, 338.
ZUNTZ, N. and GEPPERT, J. (1885) *Pflugers Archiv.* **42**, 196.
ZUNTZ, N., HAGEMANN, O., LEHMANN, C. and FRENTZEL, J. (1898) *Landw. Jahrb.* **27** (Suppl. Bd. 3), 1.
ZUNTZ, N. and LEHMANN, C. (1889) *Landw. Jahrb.* **18**, 1.
ZUNTZ, N. and SCHUMBURG (1901) *Studien zu einer Physiologie des Marsches.* A. Hirschwald, Berlin.

CHAPTER 16

Processing and Preservation of Animal Products for Human Consumption

E. ROLFE

*Department of Food Science, University of Strathclyde, Glasgow**

INTRODUCTION

PRESENT civilized society is dependent on man's ability to preserve and distribute foodstuffs. Industrialization following the introduction of power driven machinery has resulted in the massing of populations into small areas, and has brought with it the need for stable foods capable of transportation, often over vast distances, and storage without suffering loss in palatability or nutritive value.

The traditional methods of food preservation, e.g. preparation of pemmican by the drying of meat, were practised in order to extend supplies from times of glut to those of scarcity and to provide rations for expeditions. The demands made on food preservation today are more exacting, and it is expected that the foodstuff after storage and distribution will at the time of consumption be scarcely distinguishable from its fresh counterpart. Other objectives of food processing are (i) the conversion of staple foodstuffs into new or more attractive products as in the preparation of bacon from pork and bread from wheat; (ii) the provision of convenience foods by the transfer of the preparatory work from the kitchen to the factory, e.g. the canned ham is cooked and ready to eat.

Our knowledge of the properties and behaviour of the basic foodstuffs, and of the fundamental principles of food processing methods is growing steadily but much practice is still empirical. One important realization is that good quality preserved foods cannot, in general, be prepared from surplus fresh produce. Often we can define the essential characteristics of foodstuffs that will make them particularly suited to processing, and it is not uncommon for a producer to grow specifically for this market.

Fresh foods, like meat, spoil quickly due to attack by microorganisms. Methods of food preservation seek either to destroy the microorganisms contaminating the food in a manner which also prevents re-infection, or to create an environment in and around the food so that they are unable to grow. Success in the above respect is not the complete solution to the problem of long-term storage, but may make apparent slower forms of deterioration due

*Present address: National College of Food Technology, Weybridge.

to enzymic activity or to chemical interaction between food components. A particularly difficult example of the latter is carbonyl-amine browning which causes off-flavours, loss of texture and nutritive value in many preserved foods. Procedures are, therefore, introduced into the preservation process in order to defeat or minimize these additional hazards.

MICROBIAL SPOILAGE OF FRESH MEAT, MILK AND EGGS

The tissues when associated with the healthy living animal are sterile or nearly so, but a flora of spoilage microorganisms is gained immediately during, e.g. preparation of the dead carcass. Sources of infection include (i) the large number of microbes on the skin, hoofs and in the gut of the farm animal;[1] (ii) contact with dust, water or soil;[2] (iii) animal vectors of bacteria such as rats and flies though human manipulators probably represent the greatest hazard, and (iv) contact with other spoiled food.

Heavy mould infections can be transmitted in air and dust, but bacterial and yeast infections usually are trifling unless transmitted by contact.

Factors Controlling Microbial Activity

Water. The amount of water in a food, or rather its availability, is one of the most important factors influencing microbial growth. Microorganisms are unable to multiply without water, and it must be free or available and not bound by solutes or adsorption. The availability of water in a foodstuff may be expressed as water activity A_w.[3,4]

$$A_w = \frac{P}{P_0}$$

where P = equilibrium vapour pressure of the foodstuff,

P_0 = vapour pressure of pure water at the same temperature.

Provided the water remains in the liquid phase, A_w is almost independent of temperature within the range of microbial growth.

Microorganisms differ in their minimum A_w requirements as shown in Table 1[2]

Table 1
Lowest A_w Values Permitting Development of Spoilage Organisms

Group of microorganisms	Minimum A_w value
Normal bacteria	0·91
Normal yeasts	0·88
Normal moulds	0·80
Halophilic bacteria	0·75
Xerophilic fungi	0·65
Osmophilic yeasts	0·60

A low moisture food, e.g. cheese, is most likely to spoil by mould growth, whilst on moist foods yeasts and bacteria rapidly become the dominant organisms as they generally develop much faster than moulds. Hence wet non-acid foods such as meat, milk and eggs are ordinarily spoiled by bacteria.

The tolerance to a low A_w is reduced when other environmental conditions are made less favourable to the growth of the microorganism, e.g. circumstances where pH and/or temperature are unfavourable, or less than optimal quantities of essential nutritients are available.[5,6,7]

Temperature. There is likelihood of microbial food spoilage at any temperature between −10° and +70°C. Temperature influences the composition of the spoilage flora as each microorganism has an optimum, minimum, and maximum growth temperature. Bacteria may be divided into three main groups, mesophiles and thermophiles being characterized by their optimum growth temperature, and psychrophiles by their ability to grow at refrigerator temperatures.

All so-called psychrophiles are not particularly 'cold loving' as their name would imply; rather they are cold tolerant and most would be considered as mesophiles on the basis of their optimum growth temperatures. The term psychrotrophs has been coined to describe bacteria which are able to multiply at 5°C and below irrespective of their optimum growth temperature.

Growth has been reported down to about −10°C though months may be required to produce visible colonies.[8] Very large populations can arise within a week at 0°C. It appears that—10°C is the maximum storage temperature which will prevent spoilage, but microbial enzymes elaborated during previous growth at higher temperatures may persist and cause deterioration even if subsequently the temperature is reduced to —10°C or less.[9,10,11]

Mesophiles have an optimum growth temperature between 20° and 45°C. As the majority of food spoilage bacteria are included in this group it is evident why foods spoil quickly if held at ordinary temperatures. Bacteria may not only make the food unpalatable or unattractive, but in some cases also unsafe to eat. Some mesophiles produce a toxin in the food and subsequent consumption causes food poisoning, whilst others elaborate a toxin after ingestion with the food and cause food infections.

Botulism is food poisoning caused by eating food in which *Clostridium botulinum* has grown and produced its toxin. The botulinum toxin is a protein and is believed to be the most potent poison known to man.[12] Spores of *Clostridium botulinum* are widely distributed in nature, e.g. in soil and intestinal contents of animals, so that the risk of food contamination exists almost everywhere.[13] The bacterium grows and produces toxin under anaerobic conditions only; hence the hazard with inadequately processed canned foods. Fortunately the toxin is readily destroyed by heat. Exposure to 80°C for 30 min or for a much shorter period at 100°C will inactivate it.[14] Incidence of botulism is low but mortality rate is high.

Staphylococcus aureus produces a toxin which gives rise to a common but mild form of food poisoning. The organism is widely distributed and is found on our skin and nasal passages. Food is readily contaminated by human contact, and cream-filled bakery goods and cold meats commonly provide the cases of food poisoning observed during the summer when conditions favour the growth of bacteria. The organism is readily killed by boiling in water, but the toxin can withstand boiling for 20 to 60 min although it gradually loses its potency during such heating.[15]

Clostridium perfringens has been implicated in relatively mild cases of food poisoning.

An important problem is the food infection salmonellosis caused by bacteria of the genus *Salmonella*. The toxin is elaborated after the contaminated food is eaten, and the degree of infection depends on the resistance of the consumer, the virulence of the particular strain of *Salmonella*, and the number of organisms ingested. The most important reservoirs of human salmonellosis are livestock, particularly pigs and poultry. Foods most commonly contaminated include meat products, eggs especially duck, dried egg, and frozen egg.[16] Storage below 5°C prevents growth. Adequate cooking destroys Salmonellae, but the temperature at the centre of the food must reach 74° to 77°C.[17]

Comparatively few outbreaks of food-borne infections due to *Streptococci* are reported, and usually meat products are involved. The organisms occur in milk and survive pasteurization.[15]

Thermophiles have an optimum growth temperature above 45°C. Their spores are particularly heat resistant, and if present initially in sufficiently large numbers they may survive processing in canned foods to cause spoilage with storage at tropical temperatures. Yeasts and moulds are mostly mesophilic with optimum growth temperatures around 25° to 30°C, but some are able to grow at freezing temperatures.

Acidity. pH has a marked effect on the microbial flora of a foodstuff. The failure of *Clostridium botulinum* to grow in substrates with pH below 4·5 is one factor which permits the milder heat treatments given to acid canned foods. Most bacteria grow best around neutral pH, yeasts are favoured by slight acidity, and moulds are most tolerant of acidity. Meat of pH near 6 is readily attacked by putrefactive bacteria, but if the pH is near 5·5 these bacteria are inhibited and replaced by yeasts.[2]

Availability of oxygen. Bacteria may be aerobic, anaerobic or facultative in which case they are indifferent to whether oxygen is present or not. Moulds and yeasts grow best under aerobic conditions. A slice of fresh meat exposed to air becomes coated with a slime of aerobic bacteria. Oxygen can diffuse only 1mm or so below the surface so that the interior could support the growth of putrefactive anaerobes.

Inhibitory substances. Some foods contain natural protective substances that inhibit the growth of microorganisms, for example lactenin in milk and lysozyme in eggs. The inhibitors are often specific in their action, lysozyme is particularly active against *Micrococci.*

Level of initial infection. Spoilage may be delayed by keeping initial infection to a low level. Strict attention to cleanliness in the slaughterhouse will limit psychrophile infection and extend storage life of the chilled beef. Disintegration by mincing or chopping distributes the microorganisms and increases available surface for growth so that more rapid spoilage results.

Spoilage of Fresh Meat

The flesh of healthy animals is almost if not entirely sterile due to natural defensive factors such as phagocytosis and serum reactions. It is suggested that small numbers of bacteria are constantly passing through the gut wall into the blood stream and thence into the tissues, whilst at the same time an adequate removal system operates. If the animal has been recently fed, or is fatigued, or is frightened, the number of bacteria escaping into the blood stream is increased and subsequent slaughter will give a carcass more susceptible to spoilage by anaerobes.[1] Animals should therefore be properly rested and watered before slaughter.

Bacteria found in ischiatic lymph nodes taken from freshly killed cattle resembled those found in tainted beef. It seems therefore that the infection causing bone taint may be present in the lymph nodes of the live animal and spreads into the surrounding tissues after death.[18,19] Microorganisms may gain entry to the bloodstream and thence to the muscular tissues at the time of slaughter by the use of a dirty sticking knife.[20]

The resting of animals provides additional benefits conducive to better keeping properties of the meat. Bleeding is more efficient and muscle glycogen is not depleted but remains to be converted by glycolysis to lactic acid in the muscle of the slaughtered animal. The pH of the meat consequently falls from an initial value of about 7·2 to 5·7 and the acidity exerts a pronounced inhibition on the growth of spoilage bacteria. A classical example of this effect was observed in the spoilage of hams.[21] Dry salt cured hams from farm killed pigs gave only 5 per cent slightly tainted hams, whereas with pigs killed in abattoirs or bacon factories the proportion of tainted hams shot up to 50 per cent. The incidence of taint was found to be related to muscle pH. The ultimate pH of muscles from farm killed pigs ranged from 5·5 to 5·9 but rose to 6·9 in factory killed pigs. The high pH was due to the fatigue to which the animals had been subjected during transport from farm to the slaughterhouse.

Subsequent resting and feeding decreased the ultimate pH and diminished the incidence of taint.

Major sources of infection in abattoirs are the hide, hoofs, etc., of the animals and bacteria from the open gut. The deleterious effect of the superficial contamination acquired by the carcasses is reduced by attention to cleanliness which minimizes the extent of the infection at the outset, followed by prompt cooling in the chill room.[11]

The bone taint which occasionally occurs in the deep tissues is due to *Clostridia*.[22] The optimum growth temperature is around 37°C so that the trouble is minimized by rapid chilling and subsequent refrigerated storage.[23]

The growth of contaminating microorganisms is inhibited at any dry surface. A dry unbroken surface to a carcass or a layer of fat reduces the rate and extent of microbial growth. Conversely the commonly observed film of condensed moisture on the surface of chilled meat when removed from cold store encourages spoilage, particularly by the spread of motile species such as *Pseudomonas* and *Proteus*.[24] The spoilage of fresh meat is mainly a surface phenomena with the formation of slime, and it has been shown that the storage life is inversely proportional to the logarithm of the initial number of bacteria per unit area of surface.[25]

The primary spoilage of poultry is also due to growth of bacteria on the surface. Within a few days after processing, psychrophiles predominate and cause development of off odours and slime.[26] The organisms are mainly *Pseudomonas* and the slime becomes visible at a population of about 100 million/cm^2. Bacteria do not reach the muscle unless the skin is broken,[27] but once gaining access they multiply more rapidly.

Fresh meat is being retailed increasingly in prepacked form. The physical properties of the packaging material, particularly permeability to oxygen and moisture will influence storage life and spoilage pattern. A wrap allowing access of oxygen whilst maintaining damp conditions at the surface of the meat favours spoilage. It is doubtful that prepacking serves to reduce contamination as the meat acquires spoilage bacteria during butchering and cutting operations, except perhaps in the case of pathogens which may be transmitted by e.g. sneezing at the point of sale.[28] Access of oxygen appears essential to preserve the red colour, and spoilage in oxygen permeable wraps appears to proceed normally. Oxygen is less important for poultry and storage life has been extended from 12 to 16 days at 1°C in an impermeable wrap[29] as growth of aerobic *Pseudomonads* is limited.[30] In addition the accumulated carbon dioxide appears to exert an inhibitory effect.

Spoilage of Milk

Freshly drawn milk can be obtained in a sterile condition from goats and ewes if due precautions are taken.[31] Sterile samples of cows' milk have been obtained by teat wall puncture.[32] Contamination occurs during milking and subsequent handling. Storage temperature will determine the spoilage observed. If the milk is kept cool proteolysis by psychrophiles will probably occur, whilst at room temperatures an acid fermentation is usually caused by lactic streptococci and coliforms, followed by acid tolerant lactobacilli. Bacterial counts must reach about 2×10^7/ml before a detectable change in flavour or appearance is evident. A high count is indicative of spoilage provided the initial infection was small, but milk may acquire a massive inoculum from a dirty surface, and if followed by proper cooling the population may remain dormant without causing any detectable change in flavour or appearance.

Complete exclusion of pathogens from milk is difficult, and their destruction by pasteurization followed by rigid controls on handling is the accepted way of assuring a safe milk supply.[33] Thermoduric bacteria (mesophiles which have sufficient heat resistance to survive pasteurization and subsequently cause souring) picked up at the farm, must be kept to a minimum as excessive numbers will affect the keeping quality of pasteurized milk stored above 45°F (7°C).[34]

Coliforms are usually present in raw milk and the most common source of infection is improperly cleaned utensils. The presence of coliforms in pasteurized milk indicates post pasteurizing contamination. Normally they do not grow at temperatures below 45°F, but they can give rise to a variety of defects, e.g. 'gassiness', off-odours and bitterness.

Psychrophiles have become increasingly important with greater use being made of refrigeration and increased regard for keeping quality. They are present in all raw milk supplies, *Pseudomonads* being most common.[35] Holding milk at low temperatures favours the psychrophiles which attack fats and proteins. The lipase of *Pseudomonas fragi* is produced preferentially at low temperatures, but the enzyme has optimum activity at 40°C, is specific for caprylic ester, and is heat stable.[36,37,38] This microorganism is of considerable significance as butter containing residual lipase from *Pseudomonas fragi* will undergo considerable deterioration during storage, especially if the temperature is 5°C or more. The majority of psychrophiles isolated from milk are proteolytic.[39] The liberation of extracellular proteolytic enzymes from *Pseudomonas fluorescens* is greatest at 0°C and decreases with increasing temperature[40] but a rise in temperature increases the enzymic activity.[41,42]

Psychrophiles can cause 'ropiness' in milk.[43,44] The milk becomes so viscous due to slimy capsular material from the cells that it can be pulled out in strings. Definite ropiness can occur after about 20 hr storage at 20°C or after 48 hr at 6°C when milk is inoculated with *Alcaligenes viscolactis*.[45]

Spoilage of Eggs

The contents of roughly 90 per cent or more of newly laid eggs are sterile, and infection is mainly derived from external sources and from the oviduct of the hen.[46] Eggs dirty with e.g. faecal material are invaded more quickly by bacteria than are uncontaminated eggs,[47] and spoilage is more rapid during storage.[48] Before spoilage can occur the bacteria must penetrate the shell, the two shell membranes which separate the white from the shell, and overcome the inhibitory mechanisms in the white. The membranes are more effective than the shell in hindering the entrance of bacteria into the egg.[47,49] The inner membrane appears to be the most important single barrier, and is at least as effective as both the shell and outer membrane together.[50] It has been suggested that the membranes possess antibacterial activity[51] but in fact the membranes provide a more favourable medium for growth than the white,[52] probably due to the lack of iron and other trace elements in the latter. The behaviour of the membranes appears to be that of an efficient mechanical barrier.[53]

Wet eggs are more susceptible to bacterial invasion than dry eggs,[54] thus cleaning eggs by wet washing gives rise to more spoilage than does dry abrasion.[55] Some of the results obtained by Lorenz[56] are given in Table 2.

TABLE 2
Percentage of Rots in Eggs after Cold Storage and a 10-day Marketing Period

Treatment	Percentage of rots
Unwashed clean eggs	0·6
Unwashed slightly soiled eggs	1·3
Unwashed heavily soiled eggs	7·4
Farm washed eggs	10·6
Dry-cleaned eggs	1·0

Water on farms is frequently contaminated with *Pseudomonads* and similar organisms which give rise to egg spoilage.[57] If the temperature of the water is lower than that of the eggs or the eggs cool whilst the shell is wet, penetration may be encouraged by drawing bacteria through the shell and accelerate spoilage.[58]

Even after the microorganisms have penetrated the membranes, the egg still possesses natural defences in the white. The defences have been identified as (i) high pH of the white, about 9·3; (ii) presence of anti-bacterial substances such as lysozyme; (iii) the native proteins are unavailable for bacterial nutrition, and (iv) sequestering of nutrients by avidin, ovomucoid and conalbumin. Biotin and iron are rendered unavailable by the binding action of avidin and conalbumin respectively. Conalbumin prevents the growth of Gram-negative bacteria, but ovomucoid apparently has no effect on them.[59] Gram-positive bacteria are sensitive to lysozyme. The effect of high pH may lie in its ability to increase the binding power of conalbumin for iron.[52,59] The growth of *Pseudomonads* in egg white is encouraged by adding 20 μg iron per ml of white or by reducing the pH from its normal 9·3 to 6·5.[52]

The natural defences of the egg become less effective during storage. The membranes undergo a change which enables microorganisms to multiply faster on them and penetrate them more easily. The membranes appear to become more susceptible to attack at about the twelfth day. The escape of nutrients from the yolk into the white has been shown to have no effect on microbial growth.[52]

FOOD PRESERVATION

PRINCIPLES

When attempts are made to store meat, milk or eggs, in general the cause of spoilage is attack by microorganisms. The first requirement of a preservation method is therefore either to destroy the microorganisms present in the food and prevent re-infection, or to create an environment unsuited to their growth in and around the tissue. The first process is one of sterilization, the usual agency being heat. In fact a few microorganisms may survive but remain quiescent so that the condition is often termed 'commercial sterility' or 'commercial stability'.[60] As an alternative a milder processing treatment may be applied to kill many of the microorganisms present, in particular the pathogens. The reduction in the microbial flora improves the keeping quality and the process is one of pasteurization.

Preservation methods that rely on an unfavourable environment include:

- (i) Storage below the minimum growth temperatures of spoilage microorganisms.
- (ii) Reduction of the water activity A_w in the food below the minimum level necessary to support the growth of microorganisms. There are two methods of achieving this objective; either physical removal of water from the tissue by dehydration, or by the addition of sufficient solutes, e.g. salt or sugar, to reduce A_w by colligative effects.
- (iii) Addition of chemical preservatives. Their nature varies widely, e.g. antibiotics (nisin, aureomycin), acids (acetic and benzoic acids), gases (carbon dioxide, ozone), etc.

Under certain conditions preservatives can produce sterility.

The above procedures may be employed singly, but if used in combination their effect can be additive.

Having prevented spoilage by microorganisms, still we may not have a stable food, but rather we have cleared the way to allow slower forms of deterioration to appear. First of them is enzymic activity naturally present in the food and is readily destroyed by heat. The second is a group of chemical interactions (i) between the food constituents themselves, e.g. carbonyl-amine browning, and (ii) between food constituents and oxygen e.g. autoxidation of fats. The chemical interactions are much more difficult to deal with.

Having obtained a stable food, it must be adequately packed to prevent damage by animals, insects, and mechanical causes. It is axiomatic that preservation should not detract from the wholesomeness, palatability, and nutritive value of the foodstuff.

A summary of the preservation methods may be made as follows:

For short-term storage, usually measured in days:

Pasteurization techniques—using heat.
—using ionizing rays.

Chilling—cooling to about 0°C with little or no ice formation in the tissue.

Chemical preservatives—including traditional methods of salting and smoking.

For extended storage, possibly years.

Sterilizing techniques—using heat, canning.
—using ionizing rays.

Adverse environment—freezing.
—dehydration.

PRESERVATION METHODS

Application of Heat

The objective is to supply just sufficient heat to destroy the spoilage microorganisms in the food. Excessive heat has an adverse effect on texture, flavour, and nutritive value. The vegetative cells of microorganisms are readily susceptible to heat, but two genera of bacteria, *Bacillus* and *Clostridium* can change from vegetative cells into spores which are far more resistant to unfavourable external conditions. The heat requirement for their destruction becomes the limiting factor in the canning process.

The heat resistance of bacteria varies not only with the species present but also with the environment. The following are the chief considerations.

The number of microorganisms initially present. Their destruction follows a logarithmic law,[61] hence the larger the initial number of microbes the longer the time or the higher the temperature of processing must be.

Acidity. This has two important influences. Heat resistance of cells and spores is greatest around neutral pH and is reduced as the pH is lowered. Also a low pH is inhibitory to the growth of spoilage organisms of canned foods. A contributory factor permitting short process treatments for acid canned foods is the failure of *Clostridium botulinum* and similar putrefactive anaerobes to grow below pH 4·5. Bacterial spoilage has never been reported in canned food with pH below 3·7.[62]

Canned foods have been divided into four groups based on pH.[63]

1. Low acid foods, pH above 5·3. Includes fish, meats, poultry, milk, and some vegetables e.g. peas.

2. Medium acid foods, pH between 5·3 and 4·5. Includes vegetables such as spinach and asparagus.
3. Acid foods, pH between 4·5 and 3·7. Includes fruits, e.g. tomatoes, pears and pineapple.
4. High acid foods, pH between 3·7 and 2·5. Includes the berry fruits.

It is obvious that the foods of animal origin require the most stringent treatments.

Effect of substrate. Colloidal materials such as proteins and fats can protect cells against heat damage, possibly by encapsulation. Sugars also appear able to protect some microorganisms. Curing salts, e.g. salt and sodium nitrite are believed to aid preservation by preventing germination of spores damaged by the mild heat treatment given to canned hams.[64] Antibiotics, e.g. nisin[65] and tylosin, are able to inhibit the development of spoilage thermophiles in canned foods. The spores are not destroyed by the antibiotic but their germination and development are retarded. For this reason the antibiotic must persist in the product at effective levels throughout the expected shelf life.[66,67]

Temperature. A rise in temperature accelerates chemical reactions both in the microbial cell where the reactions can cause its death and between food components where the result is a loss of quality. The temperature coefficient of a chemical reaction is conveniently expressed as a $Q_{10°C}$ value.

$$Q_{10°C} = \frac{\text{reaction velocity at } (T + 10)°C}{\text{reaction velocity at } T°C}$$

$Q_{10°C}$ for destruction of bacterial spores is between 6·6 to 15·8, the value being 10 for *Clostridium botulinum*. In marked contrast, $Q_{10°C}$ for the reactions involved in the cooking of foods is approximately 3·5, and for the destruction of aneurin and riboflavin the values are 2·1 and 2·3 respectively. Therefore a high temperature short time process will cause less overcooking and less vitamin destruction than a process of equal sterilizing value carried out at a lower temperature.[62]

Heat Processing

Pasteurization. A mild heat treatment designed to kill a part of the microbial population, mainly vegetative cells, is termed a pasteurizing process. The temperatures employed are usually below 100°C, and pasteurization is generally used to meet two specific needs (i) where a more severe heat treatment would affect adversely the quality and a short storage life is acceptable, or (ii) for the destruction of pathogens as *Salmonellae* in e.g. liquid egg.

Canning. The process requires sufficient heat to be applied to the foodstuff after being hermetically sealed in a container so that both vegetative cells and microbial spores will be destroyed. The severity of heat treatment is dependent on the heat resistance of the spores, and the time required for the heat to penetrate to the centre of the can.

A pure culture of spores heated to a constant temperature shows an exponential order of destruction. The logarithm of the number of survivors plotted against time gives a straight line. The heat resistance of the spores can be read off from the graph in terms of the decimal reduction time, which is written as the *D* value as shown in Fig. 1. The *D* value is the time required to

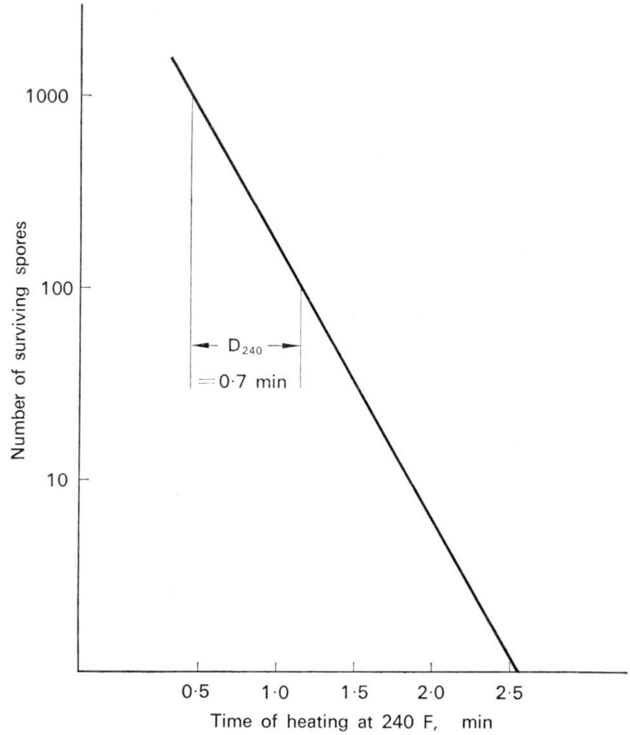

FIG. 1. Rate of destruction curve for microbial spores. The D_{240} value is 0·7.

bring about a 90 per cent reduction in the spore number at a given constant temperature, usually 250°F (121°C). The effect of temperature on spore destruction is expressed by means of the *z* value which is the number of Fahrenheit degrees which will produce a tenfold change in the *D* value. From the data of Esty and Meyer[68] it can be derived that the spores of *Clostridium botulinum* types A and B have D_{250} value of 0·2 and a *z* value of 17·6.

The food contained in the can heats up gradually to approach the retort temperature during the heat processing. The rate of heat penetration varies with the commodity, and is greater for liquid or liquid containing products, e.g. canned peas, where heating can occur by convection as compared with solid packs, e.g. corned beef, in which heat penetration is predominantly by the slower process of conduction. It is obvious that the can contents do not receive a uniform heat treatment, and the least heat treatment is given to the food located at the slowest heating point which is located near the geometric centre of the can. The process value given to the can of food must therefore be assessed on the time-temperature history recorded at the location of the slowest heating point. The heat treatment is integrated, and by means of the z value is expressed as an equivalent time in minutes at 250°F (assuming instantaneous heating and cooling) which would produce an identical lethal effect. This is termed the F value of the process, and where a z value of 18 is used in the calculation it is denoted by adding a subscript 0, i.e. F_0 value.

Commercial sterility in canned foods. In a classic study on the heat resistance of *Bacillus botulinum* spores, the "no survival" level was interpreted as the reduction of an initial population of approximately 10^{11} spores to 10^{-1} which is a destruction to the extent of 12 log cycles.[68] This work has given rise to the 12D concept in canning. The number of minutes at 250°F to achieve the 12D destruction level was found to be 2·78, but was subsequently corrected to 3·0 to allow for the protective effect of food constituents.[69] Therefore a heat treatment adequate to destroy *Clostridium botulinum* spores has a value $F_0 = 3{\cdot}0$ in a non-acid pack.

It has emerged that the critical resistance index in canning from the point of view of spoilage is not the heat resistance of *Clostridium botulinum* spores, but depends rather on the more resistant spores of mesophillic saprophytic types, particularly putrefactive anaerobes. Occasionally the resistant spores of thermophillic anaerobes constitute the spoilage hazard. A process value $F_0=3$ provides only a 5D to 7D treatment against mesophillic and thermophillic spoilage, but the satisfactory record of canning shows it gives a sufficient margin of safety.[70]

Retorting. The heat processing is usually done in retorts operating a batch system, but continuous methods using hydrostatic sterilizers are being introduced.[71,72] Rapid cooling in water immediately after processing is necessary to avoid over processing. Careful handling during cooling is essential as a small deformation in the can seam may provide temporary access for bacteria into the can. This is the explanation for the contaminated corned beef which gave rise to the outbreak of typhoid in Aberdeen during 1964.[73]

Methods for measuring temperature gradients in canned foods during processing,[74] and the application of this data to evaluate the applied heat treatment is available in the literature.[75—80]

Heat penetration into convection heated packs can be accelerated by use of rotating cookers.[81] The 'end-over-end' agitation increases convection through movement of headspace gas in the can. Developments in canning procedure may follow aseptic canning,[82] and the Smith-Ball process.[83] In the latter case the use of retorts is avoided by carrying out the heating operation in a pressurized room.

Preservation by Use of Ionizing Rays

There are perhaps two major advantages inherent in this new and potentially useful process. The resultant rise in temperature and the amount of chemical change are both small, making the method suited to the preservation of raw or lightly cooked foods. Absorption of the radiation by water raises the temperature by 2·4°C per megarad (Mrad). The second advantage is that appreciable thicknesses of material can be treated. The major obstacle to its adoption as a sterilizing process is the induced unpleasant flavours and odours. Another difficulty is to demonstrate that the irradiation products in the food are not toxic.

The sensitivity of different organisms to radiation varies widely, but in general the larger and more complex are the more susceptible. Man and higher animals are about one thousand times more sensitive than bacteria. Vegetative bacteria are more sensitive than bacterial spores.[84] The loss of the various cell functions occurs at different rates. Of greatest significance is the prevention of cell division as a sequence to interference with DNA synthesis.

Effects of environment. The undesirable side reactions in food irradiation are due to the formation and activity of free radicals and activated molecules in the aqueous environment. Manipulating the environment provides a means of modifying their effect.

A reduction in moisture content appears to reduce the effectiveness of irradiation, and possibly the resistance of bacterial spores is due to their low moisture content.

The radio sensitivity of microorganisms is reduced at temperatures below the freezing point, but at the same time the foodstuff also becomes less sensitive, so much so that it becomes advantageous to irradiate at subzero temperatures. Possibly the freezing immobilizes the free radicals.

Irradiation at low oxygen tensions also reduces undesirable side effects possibly by depressing the formation of oxidative free radicals and activated molecules.

Bacteria are more resistant to irradiation in foodstuffs than in buffer solutions, and is probably due to the presence of protective substances such as sugars, amino acids, etc. Compounds containing a sulphydryl group exert a protective action.

Interesting work is in progress on the sensitizing of bacteria and the use of spore germinating agents as vitamin K_5 (2-methyl-4-amino-1-naphthol hydrochloride). Success could mean significant reductions in sterilizing doses.

Processing treatments. The unit of radiation is the 'rad' which is equivalent to an energy absorption of 100 ergs per gram of foodstuff. The application of ionising rays to food processing may be classified conveniently according to dose.[85,86]

1. Low dose treatments: 0·01 to 0·1 Mrad (1 megarad=1 Mrad=10^6 rad).

Sprout inhibition of vegetables	0·01 Mrad
Insect disinfestation	0·02 Mrad
Destruction of parasites in fresh meat	0·02 Mrad

2. Medium dose treatment: 0·1 to 1·0 Mrad.

Destruction of *Salmonellae*	0·5 Mrad
'Pasteurization' of meats	0·6 to 0·8 Mrad
Destruction of superficial mould spores on fruit	0·2 to 0·5 Mrad

3. High dose treatments: 2 to 5 Mrad, required to achieve commercial sterility in foods. A dose of 4·8 Mrad is required to inactivate spores of *Clostridium botulinum*, but the irradiation causes simultaneously an extremely adverse effect on quality.[84,87]

Recently a new classification and terminology has been proposed.[88]

Type I. 'Radappertization.' Ionizing dose sufficient to produce commercial sterility.

Type II. 'Radicidation.' Ionizing dose sufficient to reduce the number of viable specific non-spore forming pathogenic microorganisms, e.g. destroy *Salmonellae.*

Type III. 'Radurization.' Ionizing dose sufficient to enhance keeping quality by causing a substantial reduction in the number of viable specific spoilage organisms.

Present efforts are directed more to the exploitation of low dose methods.[89] Two examples are (i) radurization—the destruction of the bulk of non-spore formers that cause food spoilage, e.g. *Pseudomonads* in meat, followed by storage at chill temperatures, and (ii) combination processes where ionizing rays are used to supplement the effect of another form of processing, e.g. application of ionizing and thermal energies to the canning of hams.

Sources of radiation. β - and γ-radiations (including X-rays) are most suited for food preservation. γ-radiation derived from radioisotopes has the greatest penetrating power. Electrons from machine sources such as the resonant transformer, Van de Graaff generator and the linear accelerator have a limited penetrating power. The penetration is approximately 0·2 in. per MeV. The most suitable radiation source for a particular purpose is dictated by considerations as dosage, food thickness, and penetration.

A source of 100,000 curies of ^{60}Co can treat 500 lb (227 kg) of material an hour with a dose of 1 Mrad assuming an efficiency of 40 per cent, but an exposure of 5 to 10 hr will be required. Spent fuel elements from reactors provide intense sources of γ-radiation, but because of a rapid rate of decay they are inconvenient to use. ^{137}Caesium, which has a long half life, can be separated from the elements and used as a source of γ energy.

The most suitable energy range for electron beams is 4 to 10 MeV. Theoretically radio nuclides may be formed, but in immeasurable amounts, when food is irradiated at energies above 1·65 MeV.[90] The magnitude of the hazard is put into perspective by appreciating that if all of one's diet were given a dose of 5 Mrad with 24 MeV electron beam the result would be equivalent to an exposure of the person to 0·26 Mrad per year. This is very small compared with one's annual exposure of 150 Mrad of which 5 Mrad is contributed by fallout.[91] Induced radio-activity cannot be detected in foods irradiated with energies of less than 10 MeV.

Of the machine sources, the linear accelerator is the most versatile and capable of high power outputs in the required energy range.[92]

Application of Refrigeration

Storage at low temperatures inhibits spoilage by slowing down or preventing the growth of microorganisms, the activity of enzymes, and chemical reactions involving the food. Chilling employs temperatures at which little or no freezing occurs and is useful for short-term storage or where it is essential to preserve the natural characteristics of the fresh food. With progressive reduction in temperature the storage life is extended, but the separation of ice causes disruption and damage to the tissue.

Chilling. At chill temperatures psychrophiles will increase in numbers and lead to spoilage. Spoilage is more rapid if the initial contamination is high, which accords with observations that cultures grown from small inocula have long lag periods.

Progressively lowering the storage temperature has two marked effects on the growth of the microbial population, (i) the lag period (where the microorganisms are getting used to their environment and there is no reproduction) increases, and (ii) the slope of the logarithmic phase (when active multiplication occurs) is much steeper at the higher temperatures. This is apparent from a study of Fig. 2. In addition it will be observed that the effect of a temperature increase from 32° to 40°F. (0° to 4·4°C) is much more marked than from 40° to 50°F (4·4° to 10·0°C). The practical implication of some importance is that a few degrees of difference at a higher temperature have little influence on keeping time, but the same temperature difference in the lower temperature range has a pronounced effect.

Bacteria responsible for food poisoning are unable to grow or grow only very slowly at temperatures below 50°F (10°C), and none have been observed to grow below 40°F (4·4°C). Chilled foods should be stored therefore at or below this temperature.

Injury to frozen cells. Freezing a suspension of microorganisms kills a proportion of the cells. The percentage of survivors varies both with the environ-

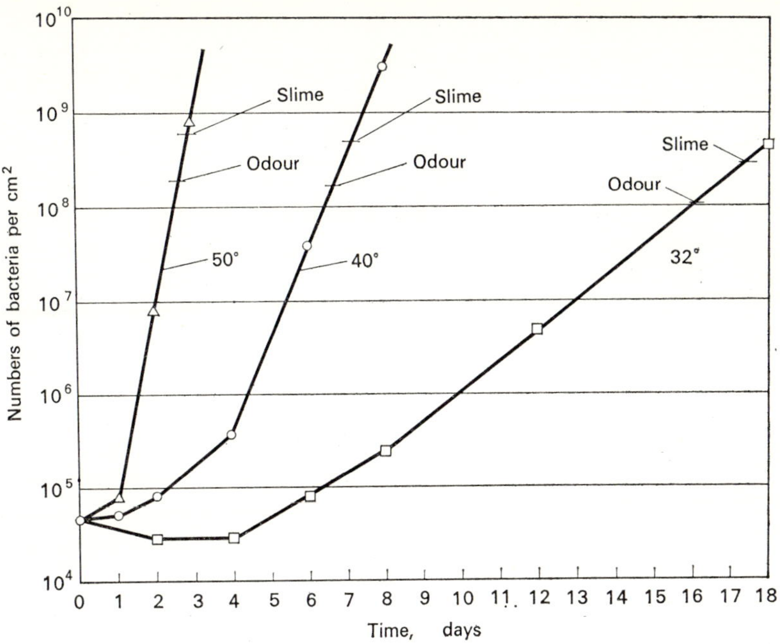

FIG. 2. Growth curves showing relationships between bacterial counts, off-odour, and slime formation on chicken stored at 32°, 40° and 50°F (0°, 4·4° and 10°C). Note particularly, (i) increased initial lag period of growth at the lower temperatures, and (ii) the increase in slope of the logarithmic phase as the temperature is raised. Ayres, J. C., Ogilvy, W. S. and Stewart, G. F., *Food Technol.* **4**, 199 (1950). Reproduced by kind permission of authors and the Institute of Food Technologists.

ment and the organism. Solutes, e.g. sugar and gelatin can exert a protective action. Generally survival is less in acid media, and as a rule yeasts and moulds are less liable to damage by freezing than most bacteria.[93] Death is due to freezing rather than merely exposure to low temperatures. 97 per cent of cells have been observed to survive in a medium supercooled to 26.6°F (−3°C), but when freezing intervened only 2 per cent survived.[94] Most of the freezing injury in yeast and red blood cells is apparently due to the combined effects of (i) exposure to concentrated solutions of electrolytes which may damage the molecular constituents of living cells, e.g. the lipoproteins of membranes,[95] and (ii) formation of large intracellular ice crystals.

During cooling, ice usually forms first in the liquid between and around the cells.[96,97] The cell contents tend to remain supercooled above temperatures of about −10°C. The vapour pressure inside supercooled cells is higher than in the phase containing ice outside the cell. Equilibrium is achieved either by diffusion of water out of the cell to freeze externally, or the water can freeze within the cell. The deciding factor is the rate of cooling. With slow cooling the cells will dehydrate, but if cooled sufficiently rapidly they will dehydrate less and will freeze intracellularly. The cooling velocity required to induce intracellular ice formation varies very considerably with the kind of cell. Yeast cells freeze at cooling velocities above 1 to 10 Centigrade degrees per minute.[98] One theory is that intracellular freezing occurs as a result of inoculation by extracellular ice, the seeding occurring through water filled pores in the cell membrane.

Very rapid cooling of cells produces intracellular ice crystals which are so small as to be innocuous. Small ice crystals have a higher surface energy than large crystals, and hence with time some of the small crystals will increase in size while others will vanish. The speed of the change is temperature dependent and is more rapid at higher temperatures. The process is termed recrystallization and has been observed in cells and tissues.[102,103]

The principal characteristics of cells which influence their behaviour towards freezing and thawing include (i) size, (ii) shape, (iii) water content, (iv) composition (nature and concentration of solutes, and nature of insoluble components), (v) type and nature of cell membranes, (vi) physical structure of protoplasm and subcellular components, and (vii) mode of aggregation of the cells (a tissue or a suspension).[104]

Freezing of foodstuffs. Frozen foods are not sterile and have been found to contain considerable numbers of viable organisms after 11 years' storage. Survival appears to increase with low storage temperatures. Some bacteria die rapidly at temperatures near —2°C, but only slowly at —20°C.[1] Microbial growth may occur in frozen foods stored at temperatures above 20°F (—6·7°C), but danger of microbial spoilage is negligible below 10°F (—12·2°C).[94]

Two steps are involved in freezing; the formation of crystal nuclei, and the growth of the nuclei into crystals. The number of crystals, and hence also their size, in a frozen tissue is therefore dependent on the extent of nucleation. Nucleation is not extensive until some appreciable degree of supercooling of the tissue is produced. Hence it is observed that ice crystals are small and numerous in rapidly frozen foods. In exceedingly pure water homogeneous nucleation occurs where purely by chance a suitable number of water molecules orient themselves into the ice-like structure of nuclei. Heterogeneous nucleation occurs when water molecules crystallize around minute solid particles, a process which takes place more easily and at much higher temperatures than homogeneous nucleation.[96] Ice nucleation particles are present in all but highly purified aqueous materials, and freezing will occur by heterogeneous nucleation in foodstuffs.[105]

The ice crystals may increase in size. Recrystallization has been observed in frozen cod stored for extended periods where the temperature fluctuated between —7° and —14°C. No recrystallization occurred with storage at a constant temperature of either —14°C or —29°C.[106]

During freezing, nucleation and crystal growth occurs preferentially between the cells of the tissue. Resultant osmotic effects induce intracellular water to diffuse into and crystallize in the intercellular spaces. The cells shrink and become compressed because of the expansion which occurs during the phase change of water into ice With sufficiently rapid freezing rates nucleation and crystal growth will occur inside the cell also. Defining the rate of freezing as the time required for the centre of a cod fillet to fall from 0° to −5°C, freezing in less than 80 min results in intracellular ice, otherwise freezing is intercellular. From the point of view of eating quality no difference has been detected between fish frozen in 2 min and fish frozen in 1000 min, and extent of cell damage also has no effect. However, after thawing, slowly frozen fish has a poorer appearance due[107] to damage to tissue proteins.

The cooling of pure water will show a sharp thermal arrest at 0°C as shown in Fig. 3. Subsequently at any temperature below 0°C the water can exist only as ice. The behaviour of aqueous solutions to freezing is quite different.

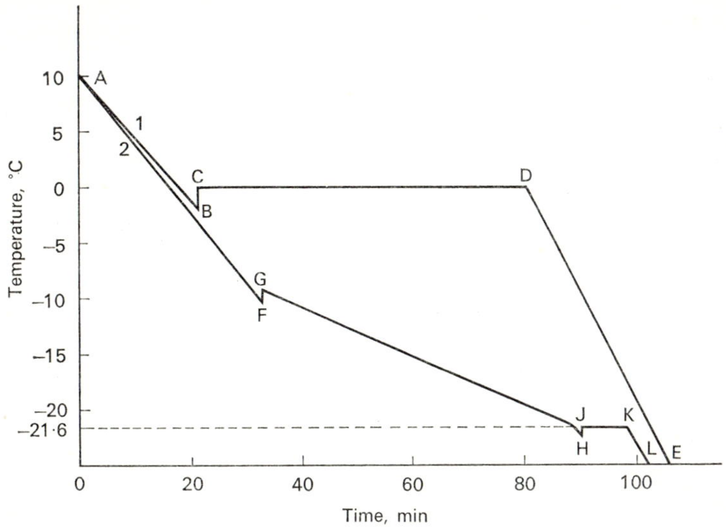

FIG. 3. Cooling curves for pure water and a dilute aqueous solution of salt. Curve 1 is for pure water. *AB*, cooling of water. *B*, slight supercooling and the temperature rises rapidly to the freezing point, *C*. *CD*, thermal arrest, and the temperature remains constant until all water is changed into ice. *DE*, cooling of the ice. *DE* is steeper than *AB* because the specific heat of ice is a little less than half that of water.

Curve 2 is for a dilute aqueous solution of salt. *AF*, cooling of the solution. *F*, slight supercooling and the temperature rises rapidly to the freezing point *G*, of the solution. *GH*, cooling of the solution with continuous separation of ice until the eutectic composition is reached. *H*, slight supercooling, and the temperature rises rapidly to the freezing point of the eutectic. *JK*, freezing at constant temperature of the residual solution of eutectic composition at the eutectic temperature. *KL*, cooling of solid eutectic plus ice.

Referring to Fig. 4 the cooling of a dilute salt solution of composition *E* will result in the onset of freezing at a sub-zero temperature *F*. However, the liberation of a small amount of ice will concentrate the solution and therefore depress the freezing point a little further. Hence with continuous cooling is obtained a gradual separation of ice accompanied by a continuous fall in temperature and the production of an increasingly concentrated solution. In fact the temperature and composition of the solution follows the conditions specified by line *AB*. Conditions at *B* represent the eutectic temperature and eutectic composition, where the remaining liquid solidifies as though a pure compound. Liquid water will always be present unless the temperature is at or below the eutectic point. The changes observed during cooling will be apparent from the cooling curve in Fig. 3. The water of foodstuffs is present as a solution of a complex mixture of solutes, and cooling results in an analogous behaviour. Apparently hard frozen foodstuffs still

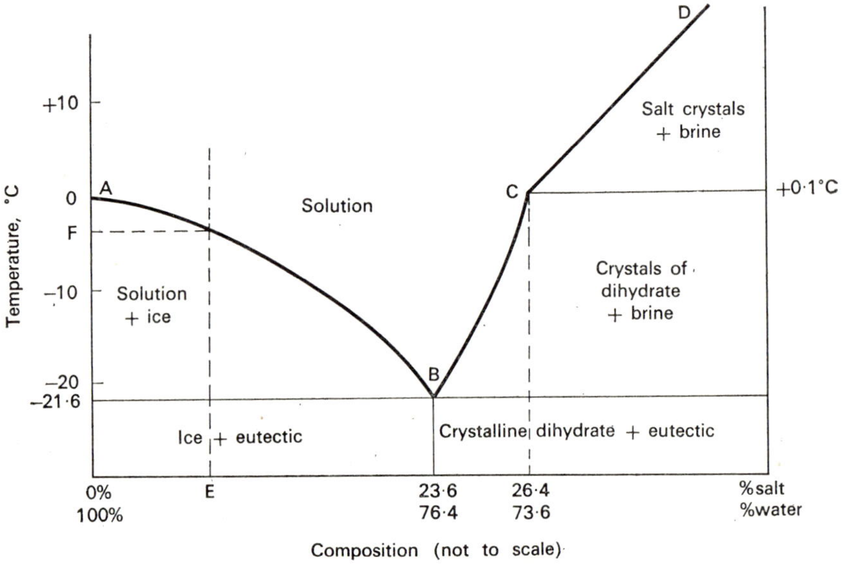

FIG. 4. Phase diagram for the system common salt and water. *A*, pure water, freezing point 0°C. *B*, eutectic of composition 23·6 per cent salt and 76·4 per cent water at the eutectic temperature of −21·6°C. *C*, transition temperature of salt crystals to crystals of dihydrate. *E*, a solution of composition *E* when cooled remains completely liquid until temperature *F* is reached and ice commences to separate. *AF*, freezing point depression of the solution.

contain appreciable amounts of liquid water. The extent of freezing in relation to temperature is shown in Fig. 5 drawn from data of Riedel.[108]

Freezing methods. The equipment and procedures for the freezing of foodstuffs have been fully described.[109] The following are the commonly employed techniques. In air blast freezing a stream of refrigerated air, temperature about

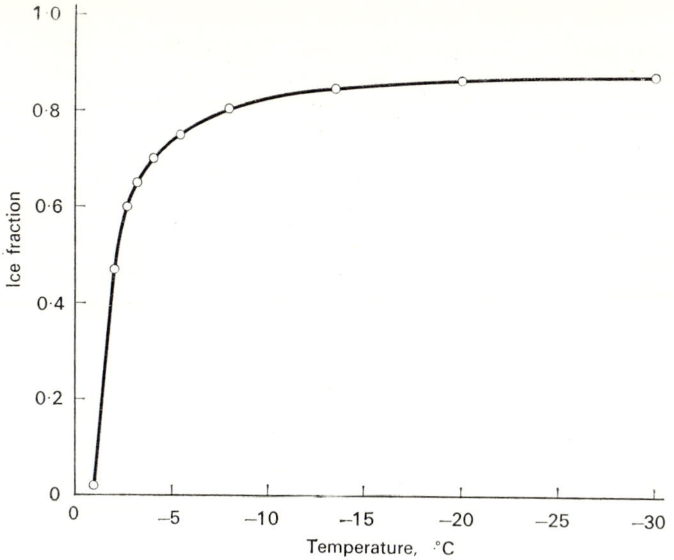

FIG. 5. The proportion of water separating as ice when lean beef of 74 per cent moisture content is cooled below its freezing point.

$$\text{ice fraction} = \frac{\text{quantity of ice formed}}{\text{total water present}}$$

−40°C, is circulated around the foodstuff. The temperature represents a compromise between throughput of the plant as controlled by the freezing time, and amount of heat extraction available from the refrigerating system. A rate controlling factor in freezing is the coefficient of heat transfer between the product and the air stream. The coefficient is proportional to $V^{0 \cdot 6}$ where V = air velocity, but the energy consumed to produce the air velocity is proportional to V^3. It becomes uneconomic therefore to use air velocities outside the range 500 to 1000 ft/min (152 to 305 m/min).[110] The procedure is useful for freezing large or irregular shaped products, while for particulate foods the process may be modified to the fluidized bed principle.[111]

The heat transfer coefficient of foods packed in cartons is small, but rapid freezing may be achieved by using contact plate freezers. The cartons, maximum thickness usually 2 in., are squeezed between horizontal refrigerated plates. The limiting factor in the rate of freezing is the thermal conductivity of the food and wraps. Freezing requires approximately 1 hr per inch of thickness of package, and the process is mainly used for consumer packs.

Provided the foodstuff is adequately protected it may be frozen by immersion in a refrigerated brine or organic liquid.[112] Contamination is avoided by immersion in a liquefied gas such as nitrous oxide[113] or nitrogen.[114] Increasing attention is being devoted to use of liquid nitrogen as a refrigerant, both for freezing and for refrigerated transport. Typical of freezing techniques is the conveyance of the food under a series of nozzles spraying liquid nitrogen in a suitably insulated chamber. On leaving, the product may have a core temperature of +30°F (—1·1°C) and external temperature of —320°F (—145°C,) but subsequently equilibrium is established.[115]

Dehydration

Considerable interest is attached to dehydration because of the peculiar advantages it possesses as a means of food preservation. Long storage life is achieved without the need of either refrigerated storage or a refrigerated distribution chain. Processing does not require the application of excessive heat as the destruction of bacterial spores is not required. Finally, there are the logistical advantages of reduction in weight and bulk, the latter being very considerable indeed if compression is also permissible.

The process. In order to retain the natural characteristics of the foodstuff, dehydration must entail more than controlled drying by artificial thermal means. The process must include other operations, e.g. the selection of raw material, and in addition, knowledge of the biochemistry of the foodstuff and of the chemistry of spoilage reactions will suggest appropriate treatments for maximum quality retention.[116]

The driers are basically of three types, air convector driers, vacuum driers, and roller driers. In air convector driers a stream of heated air supplies directly to the food the necessary heat for evaporation, and also entrains the evaporated moisture. Examples include cabinet, pneumatic, air lift, belt trough, and conveyor driers for use with piece form or particulate foods, and spray driers for the drying of liquids and purees in the form of droplets. Roller or drum driers are applicable to liquids and purees, which are dried by spreading as a thin layer on a heated rotating surface.

Vacuum drying offers two main advantages, the ability to dry foods (i) in large and recognizable pieces, and (ii) at low temperatures which is of importance with heat sensitive foods as raw meats. The methods of vacuum drying fall into two main groups.[117] One group utilizes a moderate vacuum of 5–10 mm mercury absolute pressure. During drying, water exists in the tissue as the liquid state, and its removal results in a shrivelled, shrunken product. The second group comprises freeze driers operating at about 1 mm mercury absolute pressure. Water is removed by sublimation of ice crystals from the frozen tissue, and the product is a porous, low density solid retaining the original size and shape. The porosity confers rapid reconstitution properties.

Hot air drying. [118–121] For a considerable period at the beginning of the drying cycle the surface of the foodstuff remains completely wet, and the rate of evaporation corresponds to that of pure water. This period is characterized by an unchanging rate of evaporation per unit of surface area, and is usually termed the 'constant rate period'. The rate of evaporation is a function of air velocity, wet bulb depression, shape and area of the food surface, and direction of the air stream. Subsequently during the 'falling rate period', the drying rate rapidly declines, and eventually becomes negligibly small as the residual moisture in the food approaches vapour pressure equilibrium with the surrounding air.

The drying time is approximately proportional to the square of the thickness of the tissue. The drying time therefore becomes very short if the particle dimensions are small, e.g. as in spray drying. For piece form foods the size adopted is a compromise between the conflicting needs for rapid drying and for large pieces. During the early stages the rate of drying is nearly proportional to the wet bulb depression of the surrounding air, but during the 'falling rate' period this factor gradually becomes of lesser importance. Instead, air temperature assumes increasing significance. A rise in temperature increases the rate of diffusion of moisture from the interior of the tissue to the outside surface.

During the constant rate period, the rate of drying varies with the nth power of air velocity where n equals 0·6 to 0·8. Later stages of drying are substantially independent of air velocity.

Freeze drying. Commercial interest in the freeze drying of foodstuffs was stimulated markedly by the development of the Accelerated Freeze Drying (AFD) process, which demonstrated that batches of many hundreds of pounds of meat steaks 1·5 cm in thickness could be freeze dried in periods of about 10 hours. The development of AFD stemmed from a failure to obtain a product with satisfactory quality and ease of reconstitution characteristics by using the Vacuum Contact-plate Dehydration (VCD) technique.[121,122] A description of the basic AFD technique first appeared in 1956[123] but the name 'AFD' did not appear in the literature until 1958 when the process was discussed in some detail.[117] Rapid freeze drying was achieved by drying at somewhat higher temperatures than usual, e.g. −8° to −10°C at the drying front, and also by applying an efficient heating system to the upper and lower surfaces of the foodstuff. Contact plates are not essential for rapid freeze drying, any system which can provide heat uniformly over the entire drying area will suffice; but it must also be susceptible to sensitive temperature control if overheating of the external dry surface of the partially dried product is to be avoided.

Freeze drying proceeds through the sublimation of ice crystals in the exterior of the frozen food, and the production of a dry porous outer crust. As drying progresses we are left with a diminishing frozen core surrounded by an outer dry crust of increasing thickness. The conditions are illustrated in Fig. 6, together with the temperature gradients. It is evident that as drying

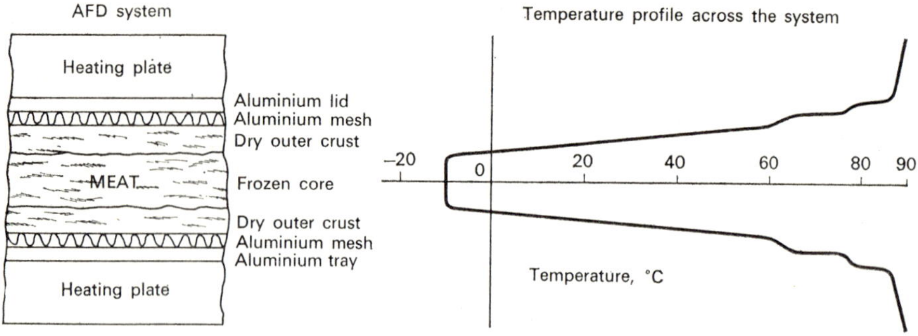

FIG. 6. Typical temperature profile to show the conditions at an intermediate stage during the accelerated freeze drying of raw beef steaks 1·5 cm in thickness. The aluminium mesh provides escape channels for the water vapour evolved from ice, whilst at the same time conducting heat to the tissue from the heating plates.

proceeds the critical factor in the rate of drying is the poor thermal conductivity of the outer layer of freeze dried material through which all the heat required to support sublimation at the ice front must be conducted. Under the conditions of freeze drying, the thermal conductivity of freeze dried meat has a value of 0·487 cal $cm^{-1}hr^{-1}$ $°C^{-1}$[124] which is little more than that of efficient thermal insulators, e.g. 0·348 cal $cm^{-1}hr^{-1}$ $°C^{-1}$ for polyurethane foam. Similar results have been reported subsequently.[125]

Studies on freeze dried tissues have shown that their thermal conductivities decrease from an observed value at atmospheric pressure to a lower constant value at pressures below 0·1 mm. Hg absolute. The higher thermal conductivity at atmospheric pressure is due to the contribution made by the air in the pores of the material.[126,127,128] The use of gas to increase the thermal conductivity of freeze dried tissues could perhaps lead to a more rapid freeze drying technique.

The carrier gas sublimation process, i.e. freeze drying at atmospheric pressure in a stream of desiccated air, must always be very slow, and be restricted to thin layers because of the low diffusivity of water vapour at such pressures. The products have been described as shrunken[129,130] which is an immediate indication that some drying from the liquid phase has intervened.

COMMODITIES

MEAT

The dressed carcass of the dead animal mainly consists of muscle, fatty tissue and bone. The relative proportions are dependent on the age of the animal and its nutrition.

Carcass composition. The percentage of edible material in a carcass increases as the animal approaches maturity. A new-born lamb is all head, skin and shanks. For every 100 lb (45·6 kg) of live weight it will yield 53 lb (24 kg) of carcass weight containing 31 lb (14 kg) of edible meat and 17 lb (8 kg) of bone. As the animal grows up the body first lengthens and then thickens consequently the proportion of head and shanks become less. Every 100 lb of live weight of the adult ram yields 67 lb (31 kg) of carcass meat containing 62 lb (28 kg) of edible meat (twice as much as from the lamb) and only 4 lb of bone.[131]

The body tissues mature in the following order, nerve, bone, muscle and fat. The early maturing tissues have priority of supply of nutrients from the blood stream. A young animal maintained initially on a low plane of nutrition until the rate of bone growth has sunk to a low level and then placed on a higher plane of nutrition will make little further bone growth but will deposit fat on a small skeletal framework. Conversely, animals fed on a high plane of nutrition whilst young and on a low plane later will have a large skeletal framework and little fat.[132–135]

Drug administration can induce desirable physiological changes in meat animals. Synthetic oestrogens, administered orally or implanted as pellets, enhance the rate of weight gain in cattle and sheep, and decrease the amount of subcutaneous, perinephritic and intermuscular fat.[136,137] The edible meat appears to have a higher moisture content than normal, but is linked with the reduction of intramuscular fat. The moisture content expressed on a fat-free basis is normal.[138]

Post-mortem changes. In living respiring muscle a high level of adenosine triphosphate (ATP) is maintained by energy derived from the oxidation of organic compounds in the tissue. Subsequent to death, oxygen is no longer transported to the cells but the ATP level remains high initially due to re-synthesis using energy released by decomposition of creatine phosphate (CP). Therefore CP in muscle disappears rapidly from the moment of death and thereafter ATP begins to disappear in spite of limited resynthesis using energy derived from glycolysis. *In vivo* actin and myosin, the two predominant contractile proteins of muscle fibres, are prevented from associating by the magnesium complex of ATP. In the absence of ATP the two proteins unite to form an inelastic complex and is made apparent by the stiffening of rigor mortis. The first sign of loss of extensibility in dead muscle is evident when ATP concentration begins to fall, becoming very marked when half the ATP has been destroyed and proceeds to completion as the pH of the muscle reaches a final value of about 5·4. The fall in pH is caused by the formation of lactic acid, the end product of glycolysis, and at around pH 5·4 the glycolytic enzymes are usually inactivated. In spite of reserves of glycogen remaining, glycolysis cannot continue.[139–145]

The extent and rate of fall in pH both have a pronounced effect on the properties of the meat. Meat of normal ultimate pH (about 5·6) is resistant to bacterial attack, lighter in colour, and more rapidly penetrated by salt and sugar, e.g. during curing. It has a wet surface when cut, and has a tendency to exude fluid.[146] Meats of higher pH are less tender, reaching a minimum at about pH 6·0, and then increase markedly in tenderness up to pH 7·0 but at the same time become jelly-like and unpalatable in texture. Meat of high pH is darker in colour. Instead of the cut surface appearing red with presence of oxymyoglobin, it is coloured purple by the predominance of reduced myoglobin, and hence the term 'dark cutting beef' which has been applied to carcasses of high pH.[147] Meat of high pH is sticky, and drip and cooking losses are reduced due to the improved water holding properties of the proteins. At pH 5·5 the actomyosin complex is near its isoelectric point and its capacity to bind water is minimal.[148]

A low pH in meat is achieved by careful slaughter of rested animals. Exercise readily depletes the glycogen reserves of pigs and the resultant pork tends to have a high ultimate pH. Steers are not susceptible to such treatments, but some have excitable temperaments which result in a low level of muscle glycogen. Selection of pigs for a high food conversion efficiency has led to the incidence of pale soft exudative (PSE) muscle. The pigs are very susceptible to stress, and a muscle characteristic is that the fall in pH post mortem is very rapid (i.e. in 0·5 to 1·5 hr) whilst the carcass is still warm. The rapid glycolysis and consequent production of lactic acid (pH may be as low as 5·1) causes a marked decrease in water binding and a pale colour to the tissue. The effect is not especially detrimental to flavour though the odour of the fresh pork may be sour. Its appearance is unpleasant and there is heavy shrinkage with cooking.[149]

Chickens 10 to 16 weeks old pass into rigor mortis in 2 to 4·5 hr when held at room temperature and reach an ultimate pH of 5·8 to 5·9. A treatment that results in a more rapid loss of ATP (i.e. a more rapid development of rigor mortis) or a more rapid loss of glycogen, as may arise in inappropriate processing, can induce toughness in the muscle.[150,151] Ante mortem factors, e.g. genetic characteristics, feeding and management can also be important.[152]

Tenderness. This is perhaps the most important attribute of meat quality and is usually associated with the age of the animal and the activity of the muscle,[153] both of which reflect the amount and nature of the connective tissue present.[154,155] The marbling fat in a joint has been accepted as indicative of tenderness, but its validity is doubtful.[156,157,158,159]

Conditioning or ageing. Holding beef for about 14 days at 1·5°C (or shorter periods at higher temperatures) has a beneficial effect on its texture and flavour.[160,161] The cause is unknown and is not due to proteolysis.[162] Ageing is the greatest single factor affecting tenderness of poultry meat, but unfortunately cannot be fitted into a continuous process. Restriction of ageing time to 40 min in continuous chillers gives rise to tough meat.[151]

Tenderized meat. The vascular system of live animals provides an excellent distribution system for the dispersal of proteolytic enzymes into carcass meat. Papain shows optimum activity in the temperature range 140° to 160°F (60° to 71°C) so that little change occurs in the meat from treated animals during distribution, and tenderizing occurs only during cooking. The level of enzyme injection is a compromise between the optimum amount in quick cooking steaks whilst avoiding the excessive tenderizing of slow cooking roasts.[163]

Curing of meat.[164] Historically curing arose through the addition of salt to meat for the purpose of preservation. Nitrate was subsequently added, which after reduction to nitrite in the pickling brine yields nitric oxide. The latter enters into reaction with myoglobin to form the typical and attractive cured meat pigment. Internal injection of curing agents into the pork sides gives a more rapid and uniform distribution of cure, and thereafter the meat is allowed to lie in pickle for 4 to 7 days. The brined pork is then stacked for about 7 days to allow the meat to mature and develop flavour. Processing is carried out in a curing cellar maintained at about 3°C to keep bacterial growth in check. The flavour may be modified by smoking. The compounds deposited from the smoke together with the surface desiccation of the meat exert a preservative action.

The conversion of pork into bacon can be accomplished rapidly if the pork is first cut into slices 2 to 8 mm thick. The slices are brined for 2 to 15 min and mature in a few hours in warm temperature conditions. A uniform product can be obtained in less than one day instead of the 10 to 14 days when treating whole sides.[165]

Canning. Packs such as corned beef must receive a sufficiently severe heat treatment to produce commercial sterility in a non-acid pack. Hams would be ruined by such treatment, and instead F_0 values of 0·3 to 0·4 are frequently used. At $F_0 = 0{\cdot}4$, a hundred-fold reduction in the number of *Cl. botulinum* spores would be obtained, and this does not appear to be sufficiently drastic to account for the stability of the product. In spite of the mild cook the products have an excellent public health record, possibly because a canned ham receiving a treatment of $F_0 = 0{\cdot}4$ at the centre may be receiving $F_0 = 4{\cdot}0$ in the outer layers where most of the spores are likely to exist,[104] and the curing salts create an environment unfavourable for the germination of bacterial spores. From presently available data it appears that canned cured meats, e.g. hams, tongue, and pork luncheon meats, will be stable with a heat treatment $F_0 = 0{\cdot}2$ to 0·4 when the salt concentration is 3·5 to 4·0 per cent in the water phase, initial nitrite is 75 to 150 parts per million (ppm) and final figure around 20 ppm, and a low incidence (less than 1/g) of clostridial spores[166].

The canned hams should be distributed and held under chill conditions. If kept in cold store for some time, and then brought out into warm conditions it is observed that the keeping time is very short. Apparently microbial growth is initiated during cold storage.[167]

Chilling. A reduction in the storage temperature markedly increases the shelf life of meat (see Fig. 2). The optimum storage temperature is 29·5°F (−1·4°C) when about 25 to 30 per cent of the water present in the meat is frozen. Careful control is necessary as lower storage temperatures will rapidly cause irreversible changes.[168] Storage life is about 30 days, but may be extended to approximately 50 days by the introduction of 10 to 12 per cent of carbon dioxide into the surrounding atmosphere. The beneficial effect of CO_2 in extending storage life of poultry meat is shown in Fig. 7. Apparently CO_2 inhibits bacterial growth by interfering with the utilization of several organic acids of the citric acid cycle.[169] Moisture is lost by evaporation from the meat during storage and surface drying is a major factor in controlling bacterial development. Excessive evaporation occurs in storage atmospheres below 85

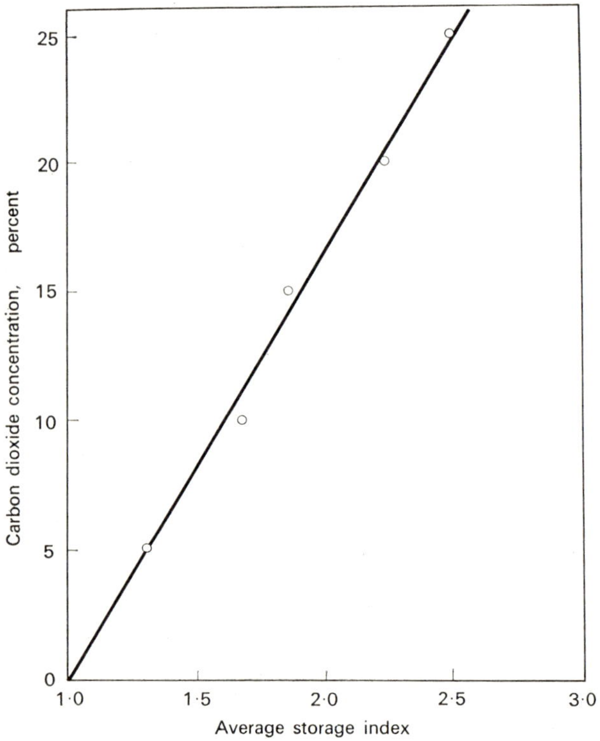

FIG. 7. The increase in storage life at 4·4°C of chilled chicken thighs relative to the surrounding concentration of CO_2.

$$\text{storage index} = \frac{\text{keeping time in } CO_2}{\text{keeping time in air}}$$

Ogilvy, W. S. and Ayres, J. C., *Food Technol.* **5**, 97 (1951). Reproduced by kind permission of authors and the Institute of Food Technologists.

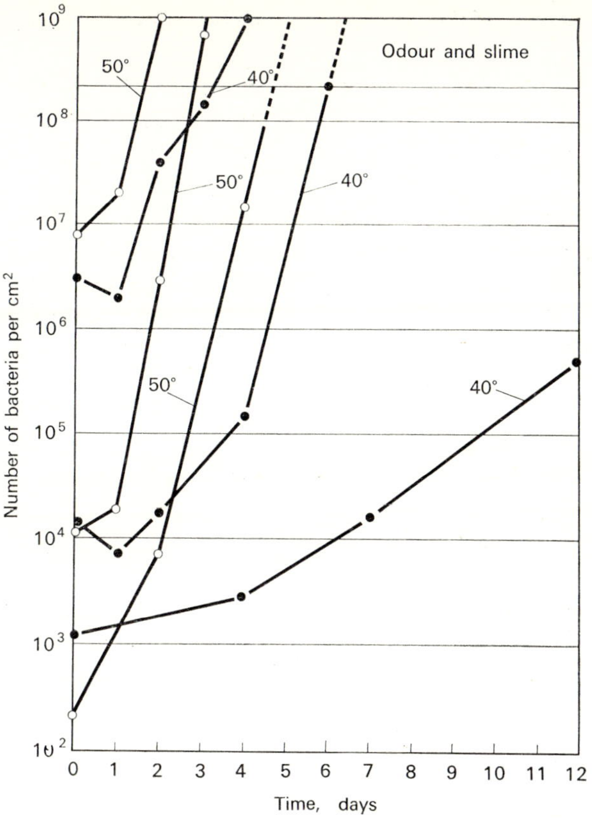

FIG. 8. The effect of a high, average, or low initial bacterial contamination on the storage life of chicken held at 40°F (4·4°C) and 50°F (10°C). Ayres, J. D., Ogilvy, W. S. and Stewart, G. F., *Food Technol.* **4**, 199 (1950). Reproduced by kind permission of authors and the Institute of Food Technologists.

per cent relative humidity (RH), and the commercial compromise is 85 to 90 per cent RH where mould growth becomes the major problem.

Meat with the highest initial microbial contamination reaches the spoiled stage most quickly as may be seen from Fig. 8. The flora on meat from recently killed animals is largely mesophilic, and during chill storage their numbers do not affect the spoilage rate; the significant organisms are the psychrophiles from e.g. soil contamination.[170] At cold room temperatures of +4° to +6°C psychrotrophic bacteria multiply rapidly on meat. In the case of pork the count on the skin has been observed to rise from $275/m^2$ to $1305/m^2$ after 2 days, $73500/m^2$ after 4 days, and 41×10^6 in 8 days. The pork is slowly invaded by the microorganisms. The muscular fluid exerts some inhibiting effect for at least 4 days but disappears by the 8th day. The bacteria rapidly invade the surfaces of cold rooms and equipment.

Bacterial decomposition is slower in undrawn poultry than in eviscerated, as the tissues of the former are protected by uncut skin.[171] Diffusion of decomposition products from the gut of undrawn poultry causes off-flavours,[172] and can lead to a shorter keeping time though bacterial growth is less.[173] Some preference has been shown in Britain for the stronger flavour of undrawn birds.[174]

Chickens are scalded to facilitate removal of feathers, but the heated skin is a better medium for the growth of microorganisms. Keeping quality is reduced, though scalding at 53°C is less detrimental than scalding at 60°C.[175]

When meat is prepacked into wraps which are relatively impermeable to oxygen, the oxygen inside the pack is rapidly consumed and growth of the main spoilage organisms, the *Pseudomonads*, is inhibited. Instead lactic acid bacteria begin to grow[176] and storage life becomes a function of the number initially present.[177,178] The prepacked fresh meats discolour during storage due to the formation of metmyoglobin in the decreased partial pressure of oxygen. Rate of discoloration is dependent on temperature as the two processes which consume the oxygen, respiration of bacteria and of the meat, are also temperature dependent. However, the pigment of cured meat is more stable under anaerobic than aerobic conditions.

Freezing. The freezing of meat offers an extended storage life but the product suffers from a serious defect, the exudation of 'drip', which arises on thawing[183]. Actomyosin has a lower water binding capacity than its components actin and myosin, and it could be expected that 'drip' would be reduced by freezing meat pre-rigor. However, the reverse effect is observed when isolated muscle strips are frozen pre-rigor. On thawing the muscle contracts and exudes large volumes of fluid.[184] The phenomenon is termed 'thaw rigor', and is caused by an excessively high rate of breakdown of ATP in the tissue as it thaws[185]. The effect has been observed on a large scale during the thawing of frozen whale meat,[186] and could be of considerable significance in meat works where faster freezing techniques are being introduced. Within a few hours of death, lamb carcasses are frozen in an air stream at −20°C, but in spite of this thaw rigor is not observed; possibly because (i) thaw rigor is diminished when contraction of the muscle is prevented as by attachment to the skeleton, and (ii) the slow thawing of carcasses in still air reduces drip formation.[187] A striking reduction in drip is obtainable by freezing meat of high ultimate pH, but the eating quality tends to be inferior above pH 6·3.[188] Drip and cooking losses are greatest in poultry frozen during rigor mortis as compared with being frozen pre- and post-rigor. The reduced water holding capacity of muscle in rigor mortis may increase the amount of water available in the tissue for freezing, and consequently lead to higher solute concentrations which would be more damaging to muscle proteins.[189]

A common belief is that the quickest freezing possible is desirable to obtain a first class product, but provided freezing is reasonably rapid, it is difficult to discover organoleptically any improvement with shorter freezing time.[190] The rate of freezing can become important in special circumstances, e.g. to produce light coloured poultry carcasses. The scalding treatment produces reddish birds which darken appreciably on exposure to air, but the desired degree of lightness can be obtained by immersion freezing in a methanol bath at —20° F (—29°C). The same effect can be produced by freezing in air blast at —100°F (—73°C).[112]

In general, frozen meat deteriorates in quality more rapidly as the temperature is raised.[191] Poultry stored at 10° F (—12° C) spoiled within 6 months, but remained palatable when held at —10° F (—23° C). A temperature of (—18°C) 0°F was suggested as the upper limit for the frozen storage of poultry.[192]

Comprehensive studies of the influence of time and temperature of storage on the deterioration of frozen foods have been carried out (e.g. the time-temperature-tolerance-TTT studies), and two important relationships have been established.[193–197]

(i) For any given product there is a certain daily loss of quality corresponding to the temperature of storage.
(ii) The quality loss arising from a given set of time-temperature influences is cumulative over the storage period and is independent of their sequence.

The time-temperature tolerance of a frozen food is expressed in terms of its 'high quality life', i.e. the storage time required to produce at any particular temperature a quality difference detectable by a taste panel. Examples are given in Fig. 9. Knowledge of this keeping quality characteristic of a frozen product allows the loss in quality which has occurred during storage to be calculated (provided the time-temperature history is known), and also an assessment to be made of its remaining shelf life.

Deterioration from many causes may occur during frozen storage, e.g. enzymic changes as lipolysis,[198] and chemical reactions as autoxidation of lipids. Turkey is more susceptible to oxidation than beef, lamb or pork,[199] probably because of its fatty acid composition, and its low level of natural anti-oxidants.[200,201] It is seen from Fig. 10 that oxidation of lipids is more marked in the red as compared with the white muscle[202] which suggests haem catalysis.[203] To determine the exact nature of many forms of deterioration in frozen foods we require a knowledge of the chemical and physico-chemical changes which occur in the unfrozen phase. Studies in model systems have shown that freezing can lead to changes in pH.[204,205] pH changes observed in frozen poultry during storage are shown in Fig. 11, and can be caused by the precipitation of salts and interaction of proteins with ionic substances.[206] The removal of calcium and magnesium ions from solution may affect stability of the proteins and exert a pronounced effect on texture. Enzymic activity is not eliminated by freezing and slow production of lactic acid

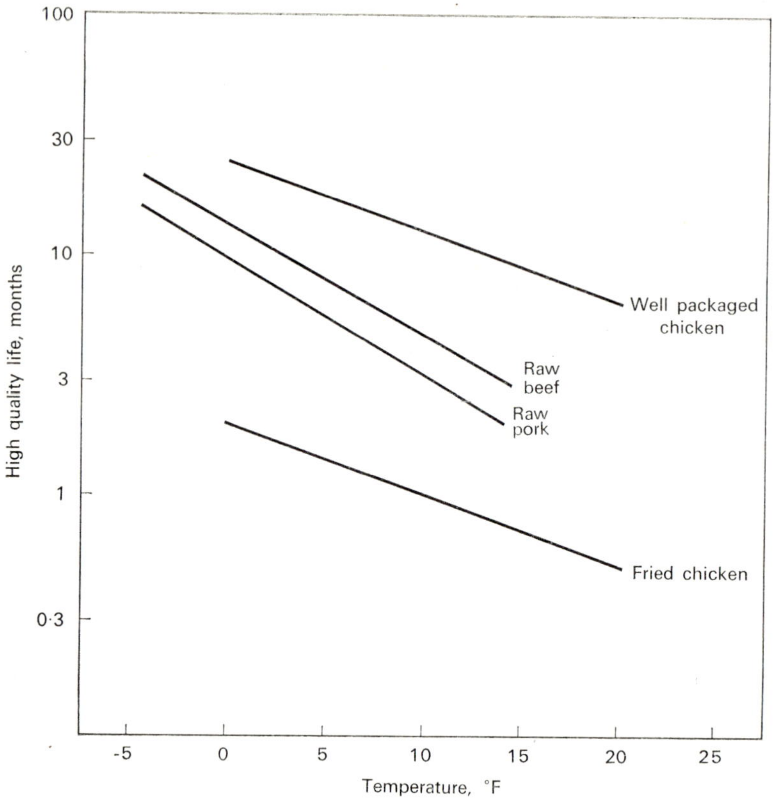

FIG. 9. The dependence of stability on the storage temperature of frozen meats. The steeper the curve, the greater is the extension in storage life obtained by reducing the storage temperature. Drawn from data contained in the Report of the Conference of the Western Regional Research Laboratory, Albany, November 1960.

during the early stages of storage of meat frozen pre-rigor is clearly evident in Fig. 11.[207]

Frozen foods must be protected during protracted storage by wraps impermeable to moisture vapour in order to avoid surface desiccation, a condition termed 'freezer burn'. The colour, flavour and texture of the desiccated tissue is irreversibly altered and is associated with the formation of a condensed layer of protein.[208,209]

Contrary to the belief that thawed foods spoil more rapidly than the fresh counterpart, available evidence shows the difference is negligible, if apparent at all.[210]

Dehydration. Traditional dried meats such as biltong and pemmican were nutritious, but texture and flavour were modified by the treatment. Dehydration aims to retain all the characteristics.

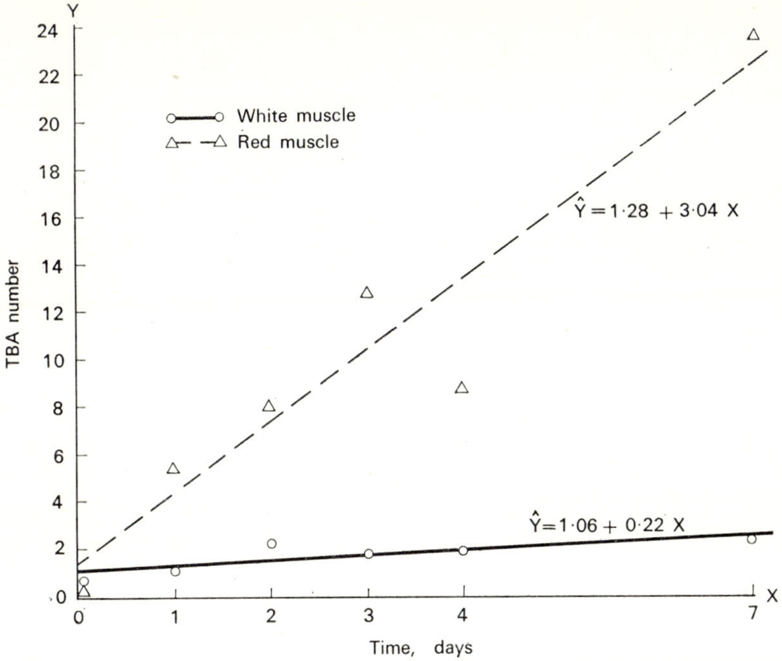

FIG. 10. Autoxidation of lipids in red and white turkey muscle stored at 4°C. The TBA number is a measure of the amount of oxidation which has occurred during storage. Marion, W. W. and Forsythe, R. H., *J. Food Sci.* **29**, 530 (1964). Reproduced by kind permission of authors and the Institute of Food Technologists.

Hot air drying. The Second World War saw the development of stable palatable products in the form of dehydrated minced cooked meats, particularly beef and pork.[211,212] Details of the process are summarized in Fig. 12. Selection of the pork carcasses is necessary to avoid the development of incipient rancidity during the drying process, and subsequent instability of the product.[213]

Attempts to dry larger pieces of cooked meat (e.g. ½ in. cubes) were not encouraging. Even with careful selection of lean meat, the final product was very irregular in water content. Drying was protracted requiring up to about 22 hr to reach an average 8 per cent moisture content, where the comparable time for a cooked mince was only 4 to 5 hr.

Products from raw meats, whether as mince or in piece form, are unpalatable—they are tough, reconstitute slowly and have a cured flavour. In addition drying is slow and uneven.

Vacuum drying. The VCD process has been applied to the dehydration of meat. Vacuum drying offers three main advantages over hot air drying.

(i) Drying is effected at low temperatures, thus avoiding the complications which arise from rendering of fat from adipose tissue. Rendering is avoided

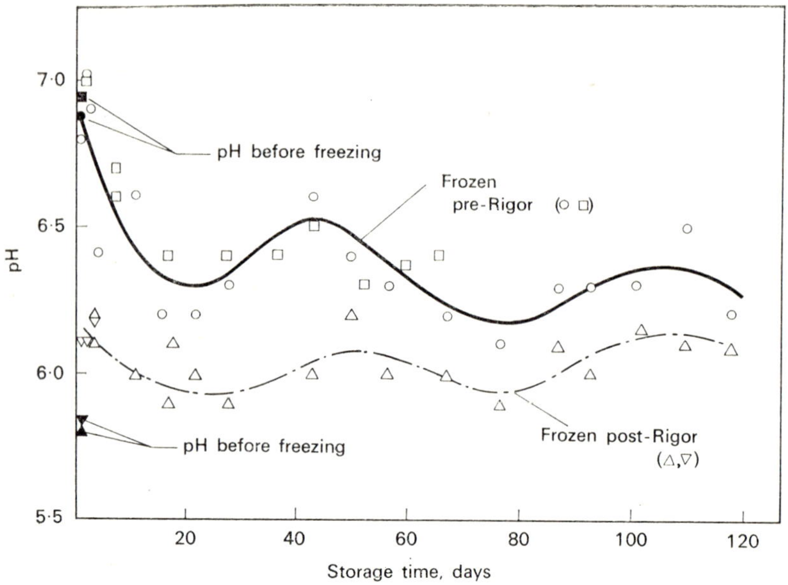

FIG. 11. Changes in pH of poultry leg meat frozen pre- and post-rigor, and subsequently stored at —10°C. Note the characteristic and marked fall in pH of meat frozen pre-rigor during the early stages of frozen storage due to enzymic production of lactic acid and possibly also CO_2. van den Berg, L., *J. Food Sci.* **29**, 540 (1964). Reproduced by kind permission of author and the Institute of Food Technologists.

in the hot air drying of minces by trimming the meat to a fat content not exceeding 30 per cent on a dry weight basis, whereas meat containing 45 per cent fat is dried without difficulty by the VCD process.[214]

(ii) Incipient rancidity has not been observed in a product prepared by the VCD process. A batch of minced cooked pork was divided into two parts. One part dried in heated air contained fat of peroxide value 19·8 m-equiv. oxygen/kg, whilst the VCD material contained fat of peroxide value 2 m-equiv. oxygen/kg.[215]

(iii) The vacuum dried cooked minces have a sweeter, fresher flavour than the air dried products. The reason is the avoidance of oxidation and incipient non-enzymic browning due to the low oxygen tension and low drying temperature respectively.

The VCD process will dry raw meats quite quickly, e.g. 1·5 cm cubes of raw beef can be dried to 3 per cent moisture content in $7\frac{3}{4}$ hr. The product is shrunken, dark in colour, and covered with an external glossy pellicle. After reconstitution, which occupies 2 to 4 hr the product may be cooked only by wet methods. Cooking in hot fat evaporates excessive moisture from the tissue. Improved drying and reconstitution characteristics are obtained if the meat is initially frozen.

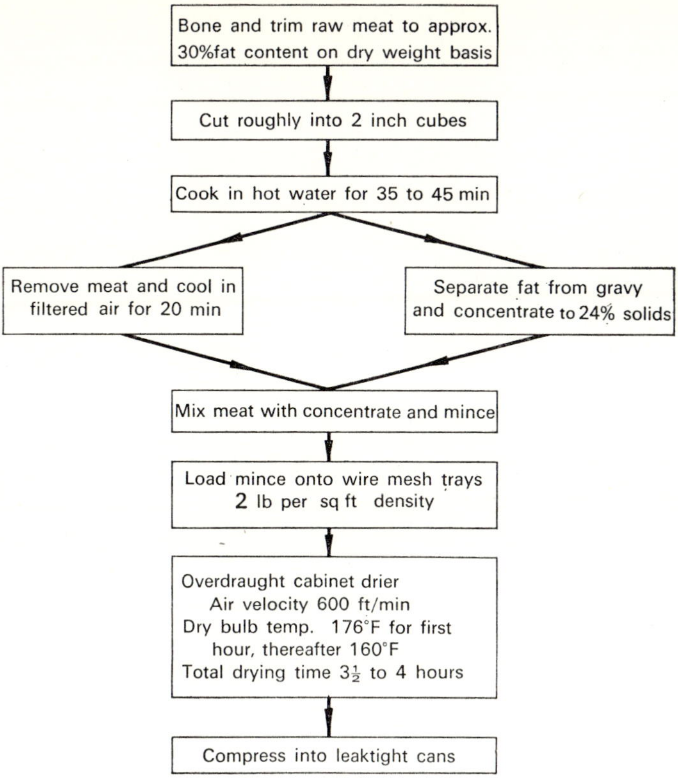

FIG. 12. Process for the preparation of hot air dried minced cooked meats.

Freeze drying. Freeze dried meat is a very porous solid, pink in colour and retains the original shape and dimensions of the starting material. The location of ice crystals, intra- or inter-cellular, and their size at the time of freeze drying, influence both the rate of freeze drying and ease of reconstitution. The freezing of raw sliced meats 1·5 cm thick in 2 hr gives fairly large extra cellular ice crystals well suited to freeze drying. The product reconstitutes rapidly as water quickly permeates the dry tissue through the pores. However if the pores are small from sublimation of small ice crystals, water penetration is inhibited by occluded air. On immersion in water the outer layer quickly reconstitutes to a gel, through which all further water must migrate to reach the dry core. Not surprisingly, reconstitution may occupy 2 hr[117] but can be avoided by reconstituting in a vacuum or by filling the pores with a water-soluble gas.[116]

Inhibiting *post-mortem* glycolysis by pre-slaughter injection of adrenalin gives meat of high ultimate pH[217] and markedly increased water holding capacity. For this reason adrenalin treated meat yields a freeze dried product of superior texture.[218,219] In addition the glucose content is depleted which confers added stability by limiting deterioration due to non-enzymic browning.

Deterioration of dehydrated meats. Dehydrated meats gradually deteriorate during storage because of slow spoilage reactions. The condition is aggravated in freeze dried products as the porosity creates a large surface area, 1·1 to 2·2 m^2/cc[220] which increases hygroscopicity and susceptibility to oxidation. The following are the major spoilage reactions.

(i) Non-enzymic browning, often termed carbonyl-amino browning.[221] In dehydrated meat it is due to interaction between reducing sugars and amino groups of e.g. proteins. It induces a wide range of defects, e.g. darkening in colour, development of scorched off-flavours and odours, a loss in nutritive value particularly by destruction of lysine, a toughening of texture, and reduced reconstitution. The reaction is inhibited in foods of low moisture content, and increases rapidly with rise in pH or an increase in storage temperature.[222,223,224] $Q_{10°C}$ lies between 3·2 and 4·3 over the temperature range 15° to 50°C, which explains the reduction in shelf life observed when dehydrated meat is stored under tropical conditions.[222] Dehydrated meats of enhanced stability can be prepared by removing the glucose at some stage either by a yeast fermentation or by use of glucose oxidase. The method is not successful with raw meats due to diffusional barriers, but is very effective with cooked meats.[225,226]

(ii) Oxidative deterioration arises from compounds such as lipids, pigments, vitamins and flavour constituents, and is particularly aggravated by a low moisture content. This has been interpreted in terms of water providing a protective monolayer.[227,228] Oxidation of the protein fraction accounts for at least half of the total oxygen absorbed by freeze dried beef.[229]

(iii) Enzymic activity. Dehydrated raw meat is less stable than the dehydrated cooked product.[213] Constituent enzymes survive dehydration and protracted storage, and taste panel assessment of the quality of freeze dried raw meat follows closely the loss of ATPase activity of the tissue.[230] Enzymes may themselves induce undesirable changes, or form reactive compounds which could accelerate spoilage.[231] The low moisture content of dehydrated foods slows down enzymic activity but does not prevent it.[232]

Irradiation. A major deterrent of sterilizing meat with ionizing rays is the development of off-flavours and odours, aptly described as 'wet dog' odour. The ionizing rays form chemically active radicals or ions (1 Mrad produces between 1 to 10 mM/kg of irradiated tissue) which attack and damage radiation sensitive substances, e.g. glutathione and sulphur containing amino acids. The end products are mercaptans which contribute largely to the irradiation odour. Carbonyl compounds are formed from the side chains of other amino acids.[233,234] The effect varies with species of meat, and in general the order of increasing sensitivity is (1) pork, chicken, (2) lamb, veal and (3) beef. No adequate reason has been given but there may be some significance in the fact that the order of increasing sensitivity follows roughly the order of increasing myoglobin content of the muscular tissue. For pork it has been shown that the flavour and odour sensitivity is not related to the intramuscular fat content of the tissue.[235]

Sterilizing doses of radiation also tenderize meat,[236] possibly by degradation of collagen,[237] but with storage irradiated meats may suffer complete loss of texture and develop unpleasant bitter off-flavours. The deterioration is greatly accelerated at warm temperatures,[236] and is largely caused by autolysis. Irradiation has only a small and variable effect on enzymes,[238] but autolysis is prevented by pre-cooking.[239]

Protection against irradiation defects is afforded by treatment of the meat at low temperatures. There is little advantage in reducing the temperature from ambient to 0°C, but a rapidly increasing protection is obtained by using subzero temperatures as shown in Table 3.

TABLE 3
Effect of Irradiation Temperature on Beef Steak Flavour
Dose 6 Mrad

Irradiation temperature °C	Flavour score
25	5·07
—25	4·37
—80	4·08
—140	3·52
—196	2·35

Flavour score recorded on a six-point scale, where 1 = no irradiated flavour and 6 = very much. 95 per cent confidence level $\pm$ 0·35.
(Data of F. L. Kaffman, J. W. Harlan and C. E. Rasmussen (1964). U.S. Army Natick Lab. Contract No. DA-19-129-QM-2000, Swift & Co. Final Report, and quoted by W. M. Urbain in *Radiation Preservation of Foods*, Publ. 1273, Natl. Acad. Sci., Natl. Res. Council, Washington D.C., 1965.)

An important part of the protective action of freezing appears to be in the limitation of free water in the tissue. Microorganisms gain similar protection but fortunately not so the spores. Hence with sterilizing procedures designed to destroy spores the benefits from freezing should be considerable.[240]

Radurization of fresh meats and poultry in moisture-proof, oxygen-permeable wraps gives a life of 1 to 3 weeks at 2° to 12°C with a dose in the range of 100,000 to 250,000 rad. At the higher levels, beef irradiated at normal temperatures exhibits a noticeable irradiation flavour.[241] Spoilage in chilled beef may be delayed by 25,000 to 100,000 rad, but as fat oxidation is accelerated the process is of doubtful application to the importing of chilled beef.[242]

For radappertization the problem of enzyme inactivation must be solved. The bitter flavour developed during storage may be due to autolytic increase of tyrosine.[243] Only heat has been found effective and practical to inactivate enzymes, an adequate treatment must be defined as a time–temperature relationship[244] as shown in Fig. 13. Hence radappertization is applied most satisfactorily to moderately well cooked meats.

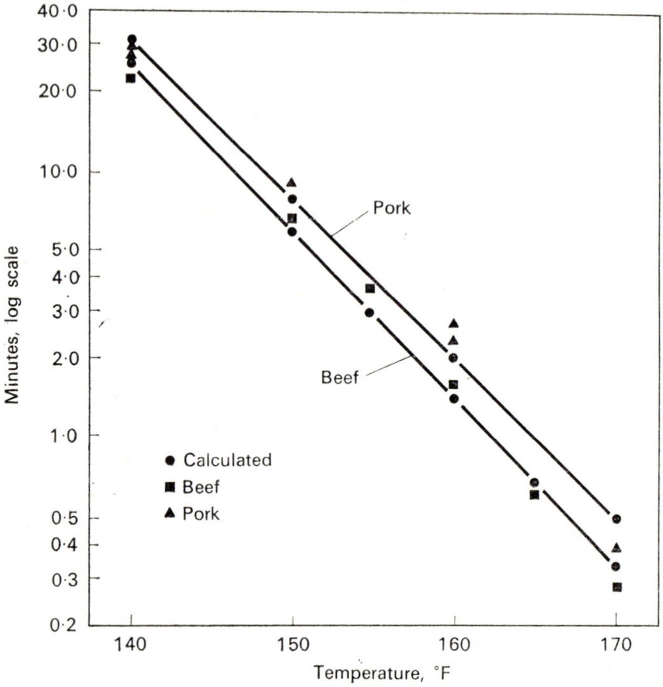

FIG. 13. A plot of time on a log. scale *vs.* temperature gives a straight line in agreement with the generally accepted theory that heat inactivation of enzyme systems follows a first order reaction curve. The above graph shows the relationship of time and temperature which would be required to inactivate enzymes in radiation sterilized beef and pork products. Chiambalero, C. J., Johnson, D. A. and Drake, M. P., *J. Agr. Food Chem.* 7, 782 (1959). Reproduced by kind permission of authors and the American Chemical Society.

An important potential in low dose treatments is the application of the complementary effect of ionizing rays and heat treatment, a so-called combination process. Applied to canned hams (where a heat treatment sufficient to produce commercial sterility will destroy the palatability of the product) the heat treatment is given first to kill vegetative microorganisms and to activate spores, and the latter are destroyed by subsequent irradiation.[245] A pasteurizing heat treatment to a centre temperature of 65°C followed by 0·3 Mrad reduces the number of micro-organisms to less than 10/g in 97 per cent of can contents, and meat shrinkage is only 10 per cent instead of 30 per cent if heat alone was used to produce the lethal effect. The heat treatment corresponding to a centre temperature of 75°C has an F_O value of 0·0003. Where the centre temperature is only 65°C the heat treatment has practically no lethal effect on bacterial spores and is too small to be expressed as an F_O value. The use of about 0·5 Mrad after the heat treatment has been reported to give rise to some off-flavours and off-odour. The irradiated hams were scored 1½ to 2 points lower on an 11 point hedonic scale than the normally heat treated controls.

Bovine Milk

Interest in the composition of cows' milk stemmed initially from its nutritional importance. In the United Kingdom the 'Sale of Milk Regulations, 1901' introduced legal presumptive minimum standards of 3·0 per cent fat and 8·5 per cent solids-not-fat (SNF) to prevent the adulteration of milk. These standards are still in force. In 1958 the Cook Committee enquired into the chemical composition of milk sold off farms in the United Kingdom and recommended that the dairy industry should attempt to improve the SNF content of milk. It advised the introduction of differential payment schemes in the form of price penalties for ex-farm supplies with an SNF content below 8·4 per cent. The recommendation was implemented by the Milk Marketing Boards in the Compositional Quality Schemes introduced in 1962. The Hygiene Quality Scheme was introduced in 1964 under which penalties are applied to producers who persistently supply milk of poor keeping quality.[246, 247]

Composition. The composition of cows' milk can vary considerably, but in general it contains less protein than fat. The proximate analysis of average mixed milk from herds of non-Channel Isle breeds is given in Table 4.

Table 4[248]

Proximate Composition of Bovine Milk

Component	g/l.
Water	873
Total solids	127
Lipid	37
Lactose	48
Protein	33
Ash	7

The average fat and SNF content of milk supplied to the Milk Marketing Board during recent years appear in Table 5.

TABLE 5[247]
Trends in the Compositional Quality of Milk (England and Wales)
All herds except Channel Island herds.

Twelve months ending	Fat per cent	SNF per cent	Total solids per cent
September 1960	3·68	8·70	12·38
September 1961	3·71	8·70	12·41
September 1962	3·73	8·72	12·45
September 1963	3·73	8·70	12·43
September 1964	3·74	8·67	12·41

Reported values are based on the simple averages of the monthly results for a sample of one out of every five supplies up to September 1961, and one out of every ten supplies since October 1961.

The composition of milk is influenced by many factors.[249,250] Seasonal effects may stem from many causes, e.g. stage of lactation, weather conditions and feeding practices. Many of the constituents of milk show significant seasonal changes, the minimum content generally occurring from May to July and highest from November to January.[251] The rise in SNF content which has been reported when cows go on spring pasture may be due to an increased energy intake by the animals.[252] An indication of the variation in milk composition with season may be derived from data in Table 6.

TABLE 6[247]
Seasonal Trends in the Composition of Milk (England and Wales)
All herds except Channel Island herds

Month	1959–60		1960–61		1961–62		1962–63		1963–64	
	Fat per cent	SNF per cent	Fat per cent	SNF per cent	Fat per cent	SNF per cent	Fat per cent	SNF per cent	Fat per cent	SNF per cent
October	3·84	8·69	3·90	8·79	3·89	8·78	3·83	8·76	3·88	8·77
November	3·89	8·71	3·97	8·75	3·94	8·75	3·94	8·76	3·98	8·72
December	3·84	8·71	3·92	8·70	3·91	8·73	3·90	8·71	3·96	8·67
January	3·72	8·69	3·81	8·67	3·81	8·70	3·86	8·70	3·81	8·61
February	3·66	8·68	3·70	8·64	3·71	8·68	3·79	8·62	3·73	8·59
March	3·61	8·66	3·60	8·62	3·69	8·68	3·70	8·67	3·69	8·58
April	3·56	8·67	3·56	8·67	3·64	8·65	3·64	8·58	3·63	8·59
May	3·46	8·76	3·51	8·75	3·54	8·72	3·54	8·69	3·49	8·74
June	3·55	8·72	3·52	8·74	3·52	8·77	3·52	8·73	3·57	8·76
July	3·65	8·66	3·63	8·67	3·61	8·70	3·62	8·69	3·63	8·69
August	3·65	8·68	3·68	8·63	3·70	8·69	3·68	8·68	3·69	8·66
September	3·76	8·75	3·77	8·73	3·76	8·75	3·78	8·75	3·80	8·71

Values are derived in the same manner as those in Table 5.

The effect of climatic conditions has been studied, and it has been observed that an increase in the temperature variables depresses secretion. There is a much greater effect on milk volume than on composition of solids.[253]

The fat and SNF contents of milk decrease with the age of the cow. The decline is consistent, and amounts to approximately 0·12 per cent SNF for each successive lactation.[254] Within any given lactation, the SNF content is relatively high for the first month, drops to a minimum during the second, and subsequently rises as lactation progresses.[254] The protein content follows essentially the same pattern of change as that of SNF with respect to age and stage of lactation.[255,256]

Infection of the mammary gland modifies milk composition by altering the permeability of udder tissue and by impairing the ability of the secretory tissue to synthesize milk constituents.[249] The result is a decrease in lactose and potassium contents which is compensated by an increase in sodium and chloride contents, and also an increase in globulin content with a decrease in casein content, the net effect being a decline in SNF content. Changes in fat content are irregular.

In recent years much knowledge has been obtained concerning the influence of feeding on milk composition, and on the relationship between rumen metabolism and milk composition.[257,258] The fat content of milk can be depressed drastically by feeding diets lacking in physical fibrousness or rich in readily soluble carbohydrates.[259] Milk yield and composition is affected by the plane of nutrition of the cow, the SNF content increasing (due largely to an increase in protein) as the level of feeding is raised.

The influence of inherited factors, for example breed, on the composition of milk is well known, and is illustrated by the data in Table 7.

TABLE 7

Composition of Milk from Single Breed Herds

Breed	Fat		SNF		Total solids per cent
	Average per cent	No. of herds	Average per cent	No. of herds	
Ayrshire	3·77	550	8·71	496	12·48
Fresian	3·58	1,997	8·64	1,789	12·22
Guernsey	4·53	432	8·93	404	13·46
Jersey	4·89	194	9·13	163	14·02
Shorthorn	3·60	527	8·73	483	12·33

Data are for period October 1959 to September 1960 and are taken from *Dairy Facts and Figures 1962*, Milk Marketing Board, Surrey.

In general, breeds which yield milk rich in fat, also give higher protein and SNF contents. The mineral content, approximately 0·68–0·70g per cent, and lactose content of about 4·9g per cent show little variation. The milk composition varies considerably between individual cows of a breed.[260] The heritabilities of milk fat, protein, SNF and lactose contents all appear to be of the order of 0·50, and should therefore respond to selection.

Composition and Properties of Milk of Importance in Processing

Milk fat. Milk fat occurs as globules about 4μ in diameter dispersed in the aqueous phase as an emulsion. The emulsion is stabilized by a layer of interface protein[261] and phospholipids in association with other lipid components[262,263] around the globule. The compounds are organized into complex structural units[264] principally by hydrophobic bonds together with significant contributions from covalent disulphide bonds.[265]

Milk contains a variety of enzymes, but only lipase causes readily observed changes, i.e. hydrolytic rancidity, which produces a characteristic off-flavour due to an accumulation of short chain fatty acids.[266] At least two milk lipases exist (i) membrane lipase normally present only in small amounts and is irreversibly adsorbed onto the fat globule membrane when fresh milk is cooled, and (ii) plasma lipase which is associated with casein until some treatment causes transfer to the globule.[267] The fat in the globules is liquid in milk as withdrawn from the cow, and the various lipases appear inactive. Membrane lipase is inactive unless it is adsorbed onto the globule as caused by prompt cooling of the fresh raw milk. This has obvious importance in respect of the rapid cooling of milk in bulk tanks on farms. Plasma lipase is activated by treatment such as foaming (e.g. in a pipe line milker[268]), homogenization, and temperature fluctuations.[269] The adverse effects of homogenization may be due to (i) coating of newly formed fat globules with casein containing lipase, and (ii) an increase in substrate surface area. Fortunately milk lipases are sensitive to heat pasteurization.[270]

Milk proteins. Milk contains a complex mixture of proteins which fall into three main groups.

Caseins account for about 80 per cent of total milk nitrogen and are unique to milk. They are relatively heat stable, and bulk milk requires about 20 min heating at 135°C to become denatured. The stability of the colloidal casein-calcium-phosphate phase of milk is important to milk processors. In the instance of cheese manufacture separation of the colloid by acid or enzymic activity is essential, whereas retention of colloid stability is required in beverage milk products. The stabilizing factors present in normal milk are sufficient to maintain the casein-calcium-phosphate in suspension in pasteurized milks, but concentrated milk products often undergo undesirable changes in fluidity during storage. Evaporated milks are stabilized by forewarming and addition of chemicals but our knowledge of the changes involved is very limited.[271] Difficulties in the development of a stable frozen concentrated milk are due principally to the instability of the casein-calcium-phosphate phase during frozen storage. The dispersibility of milk powders is dependent on the stability of the colloidal phase.

Albumins are much more heat sensitive, and represent 13–15 per cent of the total nitrogen content of milk.

Globulins make up only 1–2 per cent of the total nitrogen content and are also heat sensitive. Albumins and globulins when denatured may lessen the heat stability of casein.

The influence of heredity on milk proteins is under study and genetically controlled variations have been found. The latter may become important in the development of stable concentrated milks.[272]

Light induced changes in milk. Riboflavin is very susceptible to exposure to sunlight with up to 30 per cent destruction occurring after 30 min. Ascorbic acid is also photosensitive and up to 80 per cent may be destroyed by 2 hr exposure. Prolonged exposure is required to cause significant loss of vitamin A activity and bleaching.[273]

Exposure to sunlight also causes an off-flavour, and two different reactions are thought to be involved. In the presence of riboflavin, methionine is converted to 3–methyl thio-propanol which can produce an off-flavour at a concentration of 0·05 ppm in milk.[274] In addition occurs the photoxidation of lipids and degradation of the hydroperoxides to carbonyl compounds.[275]

Processing of Milk Using Heat

Nearly all milk sold by retail in Great Britain must be specially designated milk. There are four categories in England and Wales prescribed by the Government, viz. untreated, pasteurized, sterilized, and ultra heat treated. The milks must pass appropriate tests and licences to use the designations are subject to the conditions in the Milk and Dairies (General) Regulations 1959, the Milk (Special Designation) Regulations 1963, and the Milk (Special Designations) (Amendment) Regulations 1965, relating to the production, handling, storage, treatment and distribution of milk. 'Untreated' is used in respect of raw milk which has been bottled on the farm or at the dairy, and it replaced 'Tuberculin Tested' as the special designation for raw milk as from 1 October 1964. The Government's Area Eradication Plan to stamp out bovine tuberculosis in Great Britain was completed on 1 October 1960 and in Northern Ireland on 25 November 1960.[276]

Pasteurization. Pasteurization is a simple and efficient method of making milk safe to drink. All common pathogens likely to occur in milk are destroyed by a relatively mild heat treatment. The most resistant pathogen present is the tubercle bacillus, which is killed by either of the two prescribed treatments, i.e. (i)—the holder process—the heating of the milk to a temperature between 145°F (62·8°C) and 150°F (65·6°C) for at least 30 min and cooling immediately to less than 50°F (10°C), (ii) high temperature—short time process (HTST)—the heating of the milk to at least 161°F (71·7°C) for a minimum period of 15 sec followed by immediate cooling to a temperature of not more than 50°F (10°C).

The classical holder method may be considered obsolete. Rapid heating and cooling in the HTST process is achieved by passing the milk as a thin film through a plate type heat exchanger. The equipment is usually made of stainless steel, and the process is a reliable and economic method of obtaining a safe and palatable product. The minerals, proteins, vitamins A, B_2 and D, and fat are nutritionally unaffected by pasteurization. A small proportion of vitamins B_1 and C is destroyed but this is relatively unimportant as our main source of them is fruit and vegetables.

The keeping quality of pasteurized milk is only 2 to 3 days, and where a daily delivery service cannot be maintained a more stable product is required.

Sterilized milk. All microorganisms in this product, except a few heat resistant spores, are killed, and the consequent keeping quality is 1 to 3 weeks or longer. Sterilized milk suffers from the defects of a slight brown colour and a characteristic, marked flavour change. It is prepared by homogenizing fresh milk at about 150°F (65·6°C), filling into narrow necked bottles, sealing with a crown closure, and heating at 224° to 230°F (106·7° to 110·0°C) for 20 to 40 minutes.

Ultra heat treated (UHT) milk. The chemical changes, which give rise to the browning in milk have $Q_{10°C}$ value of 3·1 in the range of 95 to 120°C.[277] The destruction of the spores of one strain of *Bacillus stearothermophilus*, possibly the most heat resistant microorganism likely to be found in milk, has a $Q_{10°C}$ of about 11, whilst the spores of a strain of *Bacillus subtilis*, an organism which is the most common cause of spoilage in sterilized milk, has a $Q_{10°C}$ of about 30.[278] It is readily apparent that a rise in processing temperature (accompanied by a shorter contact time) will yield a pronounced improvement in sporicidal effect relative to browning as shown in Fig. 14. This principle is the basis of the UHT process, and a minimum operating temperature of 135°C held for a very short period of time should be used to ensure a survival rate not exceeding 1 spore in 5675 litres of treated milk.[279,280] The bactericidal effectiveness of a given process can be deduced from a knowledge of the temperature distribution and the flow time distribution of the milk in the plant.[281]

The plants used for UHT treatment of milk fall into two general classes, indirect and direct steam heating. Flow diagrams of the processes are shown in Fig. 15. In the indirect heating method, the heat is transferred from the steam to the milk through the stainless steel surface of a plate or tubular heat exchanger, and this procedure is essential where it is illegal to add water to milk. The recorded time temperature relationships of milk in an experimental UHT plate exchanger are shown in Fig. 16. The rate of temperature rise is relatively slow and probably 30 to 60 per cent of total sporicidal and chemical effects take place during the heating period. Formation of deposits from the milk on the heating surfaces (which would inhibit heat transfer) is

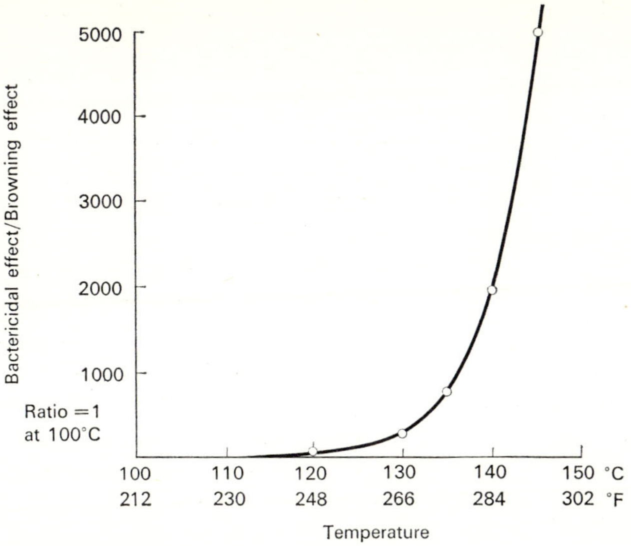

FIG. 14. The effect of increasing temperature on the ratio of bactericidal effect to browning during milk sterilization. $Q_{10°C}$ for spore destruction = 20, $Q_{10°C}$ for browning = 3. Burton, H., *J. Soc. Dairy Technol.* **18**, 58 (1965). Reproduced by kind permission of author and the Society of Dairy Technology.

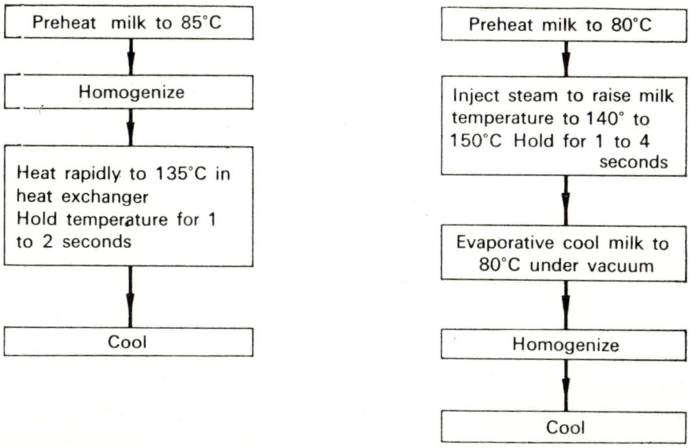

FIG. 15. Flow diagrams to illustrate the two types of processing used in the preparation of UHT sterilized milk.

reduced by preheating the milk for several minutes at 85°C, but as flavour is adversely affected this procedure is undesirable when no other heat treatment is to be given after the UHT process.

Direct heating methods employ steam under pressure, injected directly into the milk, or the milk may be sprayed into a vessel containing steam under

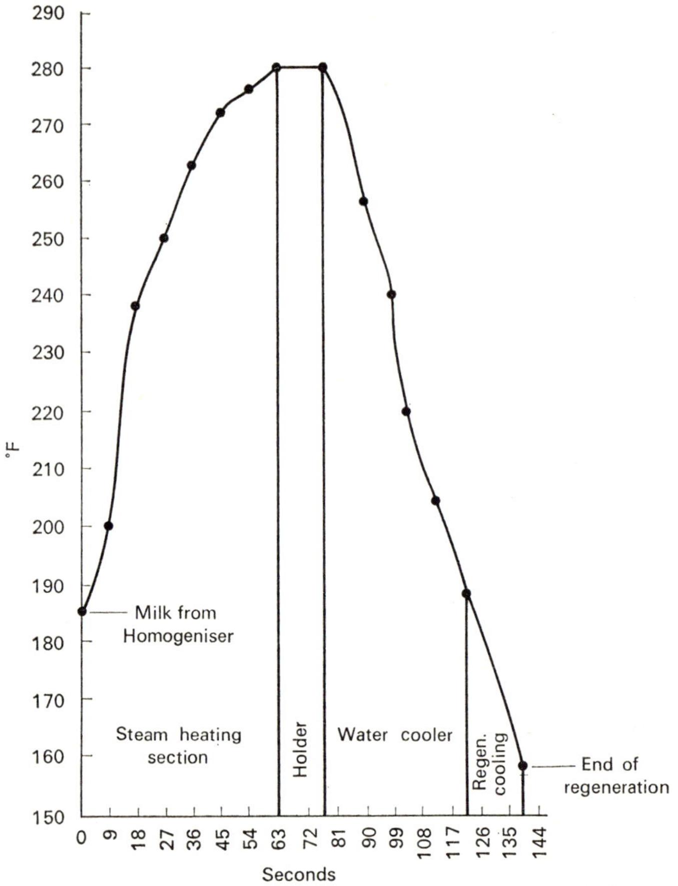

FIG. 16. Time-temperature sequence of milk passing through the APV experimental UHT plant; the latter employs indirect heating in a plate type heat exchanger. Ashton, T. R., *J. Soc. Dairy Technol.* **18**, 65 (1965). Reproduced by kind permission of the author, the Society of Dairy Technology and the A.P.V. Company Ltd.

pressure. The milk is diluted by the condensed steam, but the latter is removed subsequently by evaporation under vacuum during the cooling process. Heating to and cooling from the sterilizing temperature are almost instantaneous, and the time–temperature relationships illustrated in Fig. 17 though not obtained from actual temperature measurements, indicate the processing conditions.

The UHT process can easily produce a sterilizing effect (defined as the number of log cycles reduction in spore count) of 6 or more, and since the maximum number of resistant mesophilic spores in British raw milk is about one per ml a sterilizing effect of 6 corresponds to a spore reduction of one in

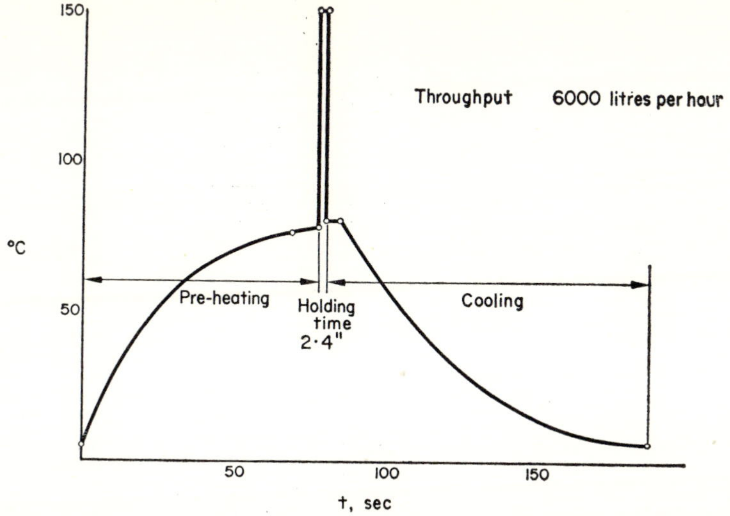

FIG. 17. Sequence of time-temperature conditions experienced by milk in the Alpura Uperization Process for UHT sterilization of milk. The milk is heated to sterilizing temperature by direct contact with steam, and the condensate is removed during the evaporative cooling which follows. Reproduced by kind permission of Alpura Ltd., Berne, Switzerland.

200 gal (909 l.) of processed milk.[282] Milk leaving the plant is whiter than fresh milk due to protein denaturation.[283] Phosphatase activity re-appears on storage.[284] The keeping quality of UHT milk is dependent on the sterility of the container, the maintenance of sterility during the filling process, and the filled container being an effective barrier against re-infection.[285] Successful aseptic bottling has been achieved, but not on a commercial scale.[286] Sterile UHT milk filled into bottles under present commercial conditions offers little or no advantage in keeping quality over pasteurized milk, possibly because of rapid growth of microbial contaminants introduced into the milk during bottling.[287] Tetra pak cartoning, which has been used commercially in Switzerland since 1962,[288] is at the moment the only available economic process for aseptic filling of UHT milk. The carton is formed from polythene coated paper, with or without an aluminium foil laminate, which is sterilized immediately before filling by the action of heat and hydrogen peroxide.[289]

Immediately after processing, UHT milk may be mildly or markedly unpleasant to taste and smell; within 2–3 days at 4·4° to 21°C the taste and smell become comparable with pasteurized milk and is at its best between 5 to 12 days. Thereafter a chalky texture (due to protein aggregation) or residual cooked flavour may be detected, and a cardboardy flavour may develop with protracted storage. In Tetra paks of aluminium foil laminate the same sequence of changes occurs, but each phase is of longer duration.[285] There is a satisfactory process for filling UHT milk aseptically into cans.[290]

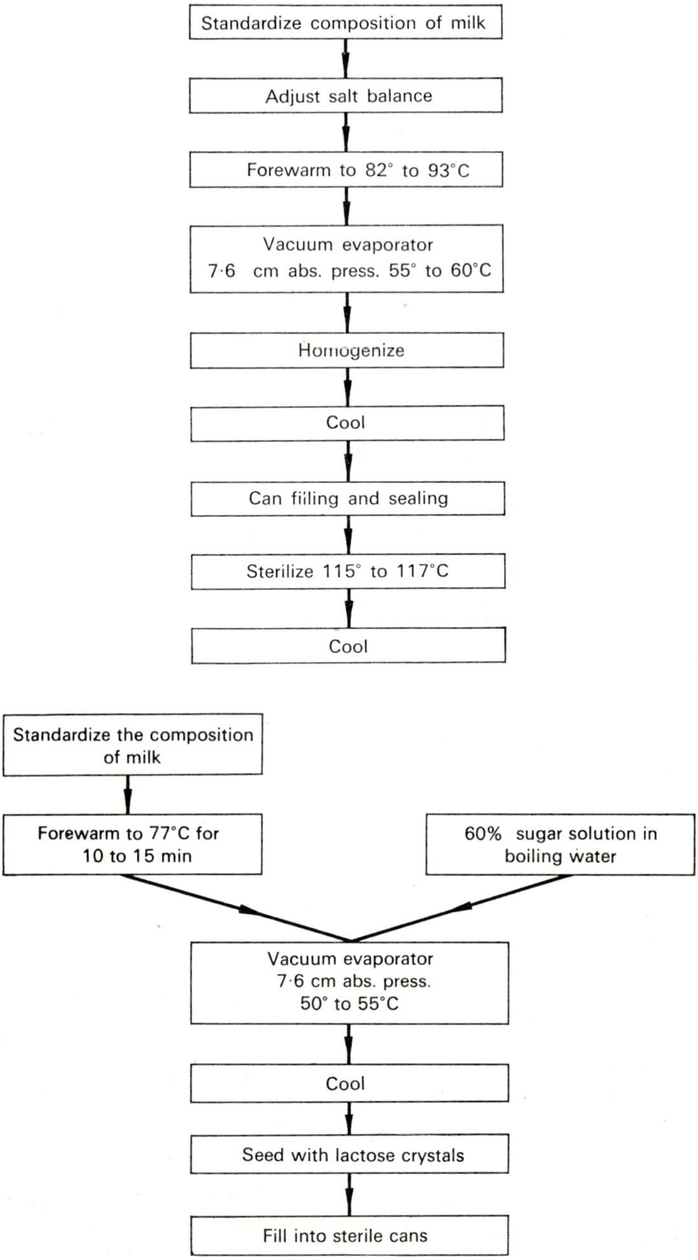

FIG. 18. Flow diagrams to show the processes employed in the preparation of evaporated milk, and in the preparation of sweetened condensed milk.

Sweetened condensed milk. A flow diagram of the operations in the manufacture of sweetened condensed milk is given in Fig. 18. The first step is adjusting the composition of the fresh milk by the addition of cream or skim milk to produce a fat : SNF ratio of 1 :2·44. The product will then satisfy the British requirements of a minimum milk fat content of 9 per cent and 22 per cent SNF.[291] Fore-warming has two objectives, viz. (i) to destroy bacteria, mould spores and enzymes, and (ii) increase the heat stability of the milk and minimize age thickening in the finished product during storage. An approximately 60 per cent solution of pure cane sugar is prepared in boiling water and is drawn into the evaporator with the milk. The mixture is evaporated under vacuum to the correct concentration. Sucrose is very soluble in water and shows no tendency to crystallize out with cooling, but the product is supersaturated with lactose. The separation of large crystals would give the product a sandy texture, and is avoided by seeding with finely crushed lactose, fine lactose crystals prepared from lactose glass,[292,293] or with previously prepared condensed milk containing fine lactose crystals. Seeding is carried out at about 30°C as the increased viscosity of the product at lower temperature would inhibit the formation of small crystals. The high concentration of sugar prevents spoilage for a considerable period of time after the can has been opened.

Evaporated milk. This name is synonymous with unsweetened condensed milk, and in the United Kingdom its composition is controlled by legislation. A flow diagram of its preparation appears in Fig. 18. The fore-warming conditions need careful selection so that the concentrated milk is prevented from coagulating during sterilizing whilst allowing sufficient thickening to occur to produce the creamy body and prevent separation of fat. The stability given by fore-warming may be supplemented by addition of salts. The balance between the natural acidic (citrate and phosphate) and basic (calcium and magnesium) salt components of milk is critical. In most cases inadequate resistance to coagulation is related to an excess of calcium and magnesium and appreciable improvement results from the addition of phosphates and citrates.

The sterilizing treatment gives the product a long shelf life but at the expense of brownish discoloration and characteristic cooked flavour which reduce its palatability as a beverage. The severity of the heat treatment may be reduced by agitation of the filled cans during sterilizing, but minimum flavour change is obtained by UHT sterilization of the concentrated milk followed by aseptic homogenization and aseptic canning.[294]

Frozen milk. Preservation of milk by frozen storage has been of interest for many years.[295] Freezing destroys the fat emulsion, and the fat globules aggregate in the thawed product. Reduction of the size of the fat globules to 1–2μ by homogenization before freezing protects the emulsion, but during frozen storage the calcium caseinate-phosphate complex is affected, and after 4 to 5 weeks the thawed product contains the precipitated proteins. It is claimed that treatment with ultrasonics (radiation dosage equivalent to 8 W of acoustical energy per pint of milk at a frequency of 1 Mc/s) followed immediately by freezing will give a product that stores indefinitely at temperatures below 10°F (−12·2°C).[296]

Frozen concentrated milk. Thawing within a few weeks of freezing a 3:1 concentrated milk, followed by reconstitution, yields a beverage indistinguishable from normal milk. During protracted frozen storage instability of the calcium caseinate-phosphate micelles occurs and protein separates on thawing.[297] A 3:1 concentrated milk is nearly saturated with lactose and protein coagulation during frozen storage is associated with the crystallization of the lactose. When water is removed from concentrated milk as ice the concentration of the dissolved salts increases, but provided the lactose does not crystallize an equilibrium is established at a salts concentration tolerable to the protein micelles. During prolonged frozen storage lactose gradually crystallizes from solution, thereby allowing more water to freeze and in consequence the salts concentration increases sufficiently to damage the micelles and induce coagulation. Analysis of unfrozen liquid from newly frozen concentrated milk showed that salt concentrations had increased by between 6 to 13 times, and pH had fallen from an initial value of 6·7 to 5·8. Subsequent progressive crystallization of lactose could be expected to raise calcium and hydrogen ion concentrations to values capable of precipitating casein.[298,299] The supersaturated solution of lactose is more stable at lower temperatures of storage where the lactose is present as an amorphous glass unable to crystallize because of the high viscosity and lack of molecular mobility. At −18°C or lower, lactose remains in solution for several months, but at −12·5°C lactose nucleation and crystallization will slowly take place and coincide chronologically with coagulation of the casein.[300]

There are other treatments which tend to retard lactose crystallization and extend storage life, e.g. the heating of concentrated milk to about 68°C for 25 min before freezing to destroy lactose nuclei, or quiescent cooling to a low temperature prior to freezing.[301]

Soluble calcium exerts a destabilizing role on casein in frozen milk, and removal of about 10 per cent of the total calcium in the milk by ion exchange improves stability.[302] It has been suggested that a loose combination between soluble lactose and ionic calcium inhibits the harmful effects of the latter.[303]

Dehydration of milk. The products, commonly known as milk powder or dried milk, can be prepared with almost 100 per cent solubility, and will reconstitute into a fluid milk of almost unchanged flavour and appearance. A high quality milk powder could provide an ideal solution to the problems of preservation and distribution of milk supplies. The three main steps in the dehydration of milk are (a) fore-warming, (b) condensing, and (c) drying. The advantages of fore-warming have been discussed above. Condensing is introduced as it is more economical to remove moisture in a vacuum evaporator than in, e.g. a spray drier. The drying of the concentrate may be effected in numerous ways, each method yielding a product with its own characteristic properties.

For roller drying the concentrated milk is fed into the trough formed between two closely set rotating rollers. The rollers are hollow and are heated by steam. The adhering film of milk is dried as the rollers rotate, and the dry product is removed as a thin crinkled sheet by a 'doctor' knife. When ground into a powder it consists of irregular platelets without air cells.

Spray drying is accomplished by spraying the homogenized concentrated milk into a drying chamber against a counter current stream of warm air (temperature up to 150°C). The system must be designed to ensure that the nearly dry particles are not overheated and thereby undergo deterioration from non-enzymic browning. The dry products are difficult to disperse in water, but in the case of non fat milk powders the difficulty may be overcome by an instantizing process, developed originally by Peebles.[304]

The particles of spray dried milk powders are generally spherical in shape and contain a variable number of air bubbles. Particle size is dependent on the spraying technique employed and the solids content of the milk concentrate, and the range is from 10μ to over 100μ. Instantized powders consist of larger particles which 'fall into' water and disintegrate instantly. Their methods of manufacture fall into three basic techniques.

1. Agglomerated powders.[305] The starting material is a normal non fat spray dried milk. A stream of the powder is moistened by exposure to wet steam or droplets of skim milk. The particles become sticky and agglomerate to larger porous masses. The aggregates are subsequently dried in a stream of warm air.
2. Single pass instantizer prepares a product of rather large individual particles. The result is achieved by spraying a high viscosity evaporated milk (such as a cool milk of high solids content, 55 per cent) into conventional drying equipment, but the larger size of droplets require extra 'flight time' in the drying chamber. Re-driers are necessary to obtain a powder of low moisture content.
3. Foam spray drying [306,307] is a new and perhaps the simplest way of preparing an instant spray dried product, and is claimed to be applicable to whole milk. The milk concentrate is converted into a foam immediately prior to atomization in the drying chamber by injecting nitrogen under pressure into the feed line between pump and spray nozzle. The product is of relatively large spheres, 60 to 100μ diameter, of dried foam, some being fused into aggregates up to 430μ.

A special application of spray drying is the Birs process,[308] in which flavour retention is improved by using a coarse spray (to reduce loss of volatiles) and an air temperature of only 25° to 30°C. The low air temperature necessitates a long contact time and the drying tower must be large, e.g. 235 ft (71·6 m) high and 50 ft to 120 ft (15·2 m—36·6 m) diameter to provide adequate capacity and the required heat transfer. The spray falls through the tower in 50 to 200 sec, and meets air which has been dried to below 3 per cent relative humidity by mechanical refrigeration and contact with silica gel. After removal from the tower the powder is dried for another 20 to 30 min in the same dry air.[309] It is said that there is no denaturation of protein and the taste is indistinguishable from fresh milk.

Preheating the liquid milk stabilizes the fat in spray dried whole milk against oxidation.[310] Heating to 180° to 190°F (82° to 88°C) for 5 to 20 min is still common practice but the present trend is towards an HTST treatment of 220°F (104°C) for 15 to 60 sec. The heat treatment is equally effective as the addition of antioxidants.[311] Adjustment of the ratio of fat: SNF to meet the Dried Milk Regulations[312] is done by the addition of cream or skim milk to the raw milk before condensing. An instant milk powder is more difficult to prepare from whole milk than skim due to the complication of milk fat in the system, but an easily reconstituted porous product may be obtained from whole milk by vacuum foam drying.[313] A milk concentrate of 43 per cent solids is homogenized, cooled to 13° C, and nitrogen dispersed into it. The gassed concentrate is fed into a drying chamber maintained at a reduced pressure where it expands into a foam which is applied and dried on a continuous belt.[314] The flavour stability is alleged to be superior to conventional spray dried milks.[315]

There is a close relationship between drying process and the shape, size and appearance of the particles of milk powder, which has been reviewed in detail.[316] The structural elements of the particle are lactose (the major component), casein micelles and whey protein particles, fat, and spherical cells of air. The proteins, fat and air are presumably dispersed in a continuous phase of amorphous lactose. During rapid drying, the viscosity of the milk droplet increases so rapidly that crystallization of the lactose cannot occur. Hence the lactose in spray and other dried powders exists as a glass with the α and β forms present in equilibrium ratio.[317] During the instantizing process, the surfaces of the milk powder particles become moistened, and partial crystallization occurs before the particles are redried.[304,318] The lactose equilibrium is thereby shifted from about 2 parts α-hydrate: 3 parts β-anhydride to 3 parts α-hydrate: 2 parts β-anhydride.[319] The conditions of instantizing must be carefully controlled to avoid loss of solubility of the powder, as crystallization of the lactose may be accompanied by insolubilization of the casein (cf frozen milk).

Deteriorative changes in milk powders. Deterioration may become evident in four characteristic forms.

1. Rancidity arising from the oxidation of fat causes a tallowy flavour. The oxidation is accelerated in powders of low moisture content, e.g. 1·7 to

2·4 per cent[320] and is a general phenomena observed with all dehydrated foods. Where a low moisture content is essential to inhibit non enzymic browning (*vide infra*) rancidity may be avoided by packing in hermetically sealed containers containing an inert gas. Residual oxygen in sealed cans of milk powder is effectively and rapidly removed by packing in nitrogen containing about 10 per cent hydrogen in the presence of a palladium catalyst.[321] This procedure may also improve flavour by the catalytic reduction of volatile carbonyls.[322] The stabilizing effect of moisture in milk powder against oxidation is reduced with storage at low tem peratures as 0°F (−17·8°C) and may be related to the physical state of the water in the milk powder.[323]

2. Caking of the powder occurs when the moisture content is too high, usually caused by absorption of watersubsequent to manufacture. Caking is due to the conversion of amorphous lactose to the crystalline state.[324] Spray dried skim milk powder of moisture content 3 to 5 per cent may be stored at 37°C for 600 days without crystallization, but at 7·6 per cent crystallization commenced after one day.[325] The hard α-hydrate crystals which gradually form, cement the milk particles together and simultaneously the caseinate system of the powder gradually becomes insoluble.

3. Non-enzymic browning is caused primarily by chemical reactions initially between the free amino groups of the milk proteins and the aldehyde group of lactose but interaction between phospholipids and lactose may play a significant part.[326] It is accelerated by a rise in storage temperature, and an increase in pH or relative humidity (the latter may be equated to moisture content) of the milk powder. The powder is not only discoloured brown, but develops an overcooked or scorched off-flavour, the dispersability of the powder is decreased, and there is a loss of nutritive value. Lysine is principally involved but arginine, histidine, methionine and tyrosine are also attacked, and with the binding of these essential amino acids the biological value of the milk powder is impaired.

4. Loss of solubility accompanies non-enzymic browning and is similarly a function of moisture content and storage temperature. Like non-enzymic browning the Q_{10} of solubility loss is high, and the solubility of moist powders falls rapidly at 37°C. A nonfat milk powder of 7·6 per cent moisture content after storage for 100 days at 37° C showed a nearly total loss of protein solubility and destruction of 70 per cent of the original amino nitrogen.[328] The deterioration was coincident to crystallization of the lactose which increased the equilibrium relative humidity from 42 to 55 per cent. The change in water activity could accelerate the deteriorative reactions.

Irradiated milk. A dose of approximately 2 Mrad is required to sterilize raw whole milk, but the treatment causes stronger undesirable off-flavours than is produced in any other food,[329] and in the case of concentrated milks browning and gelation develops in addition on storage. Free radical acceptors such as ascorbyl palmitate added to the milk or irradiating in the frozen state reduce off-flavour development.[330] Alternatively, almost all of the off-flavour can be removed by vacuum distillation of the milk soon after irradiation, but if 1–2 days elapses before distillation, increasing amounts of off-flavour remain in the milk. A rotary disc apparatus has been devised for continuous production of a sterile irradiated milk concentrate free of off-flavour based on the simultaneous removal of volatiles by vacuum distillation whilst irradiation is in progress.[331] An irradiated concentrate 4:1 gelled in 3 days at 30°C but no gelation was observed after several months at 3°C. Gelation is inhibited by the addition of 0·2 per cent sodium tripolyphosphate,[332] in which case the gelation of 4:1 concentrate is deferred for 10 days at 30°C, and for a 3:1 concentrate at least 6 months.

Ionizing radiations affect the casein, probably producing an unfolding of the molecule, and results in an increase in the coagulation time by rennet enzyme and a reduction in heat stability. The effect is probably due to an alteration of the calcium balance of casein where the available calcium is reduced by combining with the modified protein molecule.

Hen's Egg

The structure of the raw hen's egg can be seen with the unaided eye and is shown in Fig. 19.

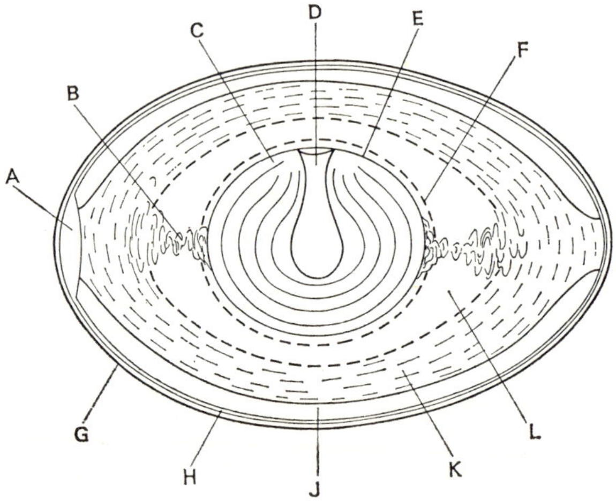

Fig. 19. The structure of the hen's egg, as shown by a section through the long axis. *A*, air cell; *B*, chalazae; *C*, yolk; *D*, germinal disc and white yolk; *E*, vitelline membrane; *F*, film of mucin; *G*, shell; *H*, shell membranes; *J*, outer thin white; *K*, thick white; *L*, inner thin white. Brooks, J. and Hales, H. P., *Biochim. Biophys. Acta* **32**, 237 (1959). Reproduced by kind permission of the publishers and of Her Majesty's Stationery Office.

The colour of the egg shell varies from white to deep brown, the pigment being a porphyrin. The intensity of colour is dependent largely on the breed of hen, and is no guide to the quality of the egg.

The shell has an inner lining composed of two parchment-like membranes which are separated at the blunt end of the egg to form an air cell. The membranes are composed of interlacing fibres of a keratin-like protein. The inner membrane is coated on the side adjacent to the white with a mucin-like substance which fills the interstices of the fibre network and forms an apparently non porous structure.[53] The inner shell membrane is the most important barrier to the penetration of microorganisms into the egg.[50]

The white (also termed albumen) is made up of three distinct concentric layers surrounding the yolk. Twenty per cent of the white consists of an outer layer of fluid thin white, a middle gel-like layer of thick white forms 50 per cent, and immediately adjacent to the yolk lies a layer of thin white which accounts for the remaining 30 per cent. (all approx).

The yolk, which lies at the centre of the egg, is enclosed within the vitelline membrane, and the latter is surrounded in turn by a film of mucin. The yolk is held in position by spiral fibrous structures known as chalazae which extend lengthwise into the white. On the surface of the yolk the germinal disc is observed as a small spot, and from it extends the flask-shaped white yolk to the centre,[333] but in the main the yolk consists of a yellow stratified material. The yolk is a complex mixture of microscopic granules suspended in a soluble phase termed the plasma. The granules represent about 23 per cent of the yolk solids, and the plasma contains about 77 per cent lipid, both being expressed on a dry weight basis.[334]

The new laid egg consists, on the average of 10·3 per cent shell, 57·6 per cent white and 32·1 per cent yolk. The yolk contains approximately 49·1 per cent water, 16·6 per cent protein and 31·7 per cent fat, whilst the white contains 87·8 per cent, 10·4 per cent and 0·1 per cent respectively. The yolk is the more complex system, and is the source of nutrients for the growing embryo during the incubation of a fertile egg.

Composition of egg white. The egg white contains a number of proteins of quite interesting biochemical properties.

Ovalbumin is present in the greatest amount and the coagulation of egg white is due primarily to the relative ease of denaturation of this protein.

Conalbumin behaves as an anti-bacterial agent by reason of its ability to sequester ferric ions in solution. Denatured conalbumin does not possess the ability to bind iron. The iron complex of the protein is very resistant both to denaturation by heat and to attack by enzymes.

Ovomucoid, the antitrypsin factor, is a glyco-protein which is quite resistant to denaturation and is not coagulated by heat.

Lysozyme is an enzyme and possesses bacteriolytic activity. It is somewhat exceptional in respect of the ease with which it may be obtained in crystalline form.

Avidin is present in egg white in only a very small amount, but has the important property of combining with biotin, and rendering the latter vitmin unavailable to microorganisms.

The thick white is not a homogeneous gel but is composed of bands of a continuous phase approximately 1 mm wide separated by bands of fibres. The rigidity may be due to the cross-linking of chains of an ovomucin-lysozyme complex into a network.[335]

Composition of egg yolk. The proteins of egg yolk may be subdivided into two main classes, the simple proteins termed livetins and the phosphoproteins. The latter contain a phosphoric acid residue and are usually present in a loose combination with phospholipids as lipoproteins.

The lipids of the egg are confined almost entirely to the yolk and make up about 32 per cent of the fresh egg yolk. The lipids consist mainly of triglycerides, phospholipids and sterols.

Factors affecting the composition of the hen's egg. The breed, age, and diet of the hen can affect the appearance and composition of the egg.

The proportion of yolk to white increases with the age of the hen during its first year of lay, and also the size of egg increases as the hen approaches maturity. Eggs from the more mature birds contain a higher proportion of yolk, but for birds of the same age the yolk proportion decreases with increasing size of egg.[336] Increasing age, particularly for birds of 18 months or older, leads to an increase in the spread of internal quality of the eggs (as measured by Haugh units[337]), and an increase in shell defects.[338] Defects of this kind limit the usefulness of the eggs for table purposes but not for manufacture.

The strain of the bird has been reported to have a significant effect on the proportion of the yolk to white, and also on the solids content of the egg white.[339]

The colour of the egg yolk is controlled by the diet of the hen, and though it appears that the pigments are of little or no nutritive value (consisting of about 98 per cent xanthophylls and 2 per cent carotenes), the colour is important in respect of appearance. The colour intensity of the yolk depends on the xanthophyll content of the feed, and some 11 per cent of the ingested pigment is deposited in the yolk.[340] Light coloured yolks are produced by hens on a diet containing about 8 mg xanthophyll/lb (17·6 mg/kg), and dark coloured yolks can be obtained by increasing the xanthophyll content to 25 mg/lb (55 mg/kg). About 10 days on a high xanthophyll diet are required to obtain maximum yolk colour. Barley, wheat, oats and maize are cereals commonly used for the feeding of hens, but only maize contains a significant amount of xanthophyll (up to 20 mg/lb (44 mg/kg)). Dried green crops (containing 100–200 mg/lb (220–440 mg/kg)) are a convenient means of providing the necessary xanthophyll to give adequate yolk colour.[341]

Fats added at 10 per cent level to the hens' diet do not alter the total lipid content of the yolk, but modify the fatty acid composition of the lipids, the acids of the neutral glyceride fraction being more susceptible to dietary manipulation than the phospholipid fraction.[342]

For a considerable number of years it has been known that egg white will develop a pink colour during storage if the hens have been fed a diet containing cotton seed or cottonseed oil. The active principles have been identified as two cyclopropenoid compounds, malvalic[343] and sterculic[344] acids.

$$CH_3 - (CH_2)_7 - \underset{\diagdown\;\; CH_2 \;\;\diagup}{C{=}C} - (CH_2)_7 - COOH$$

sterculic acid

$$CH_3 - (CH_2)_7 - \underset{\diagdown\;\; CH_2 \;\;\diagup}{C{=}C} - (CH_2)_6 - COOH$$

malvalic acid

Ingestion of the acids by hens can produce marked effects on the eggs and egg quality. In particular the hen's metabolism of fatty acids is modified, leading to an increased proportion of stearic acid and a lesser amount of oleic acid in the lipids of the yolk (and body organs of the hen) which produces the striking effect of a pasty texture in affected yolks.[345] There are many other additional effects. The pH values in the yolk and white of normal eggs after storage for a few days are about 6·4 and 9·0 respectively. In affected eggs the pH values converge rather rapidly and approach about 7·4 in yolks and 7·6 in whites.[346] A considerable transfer of water from the white to yolk occurs. The water content in normal yolks remains constant at about 50 per cent, but in eggs from hens fed with cyclopropenoid compounds it may rise to 57 per cent[346] producing a marked reduction in yolk viscosity. At the same time a major transfer of non protein nitrogen compounds, chiefly amino acids, occurs from yolk to white.[346] It has been suggested that these effects arise from the reaction of the cyclopropene ring with the sulphydryl groups in physiologically active proteins.[347]

Deterioration of Shell Eggs during Storage

Immediately after the egg has been laid, a series of deteriorative changes set in.

1. A shift in pH of the white occurs and is accompanied by a loss of carbon dioxide through the porous shell. The pH of the white of the newly

laid egg is approximately 7·9 but rises to about 9·3 after 3 days, and thereafter remains practically constant.[348] The pH of the yolk remains constant at about 6·2.

2. The thick white gradually thins and decreases in amount.
3. Water passing from the white to the yolk decreases the viscosity of the latter.
4. The vitelline membrane becomes weaker, and consequently the yolk sags when the egg is broken out. Eggs stored for too long or at too high temperatures become almost impossible to break out without rupturing the yolk.
5. Water slowly evaporates from the egg thereby increasing the size of the air cell.
6. The flavour deteriorates.

A fresh egg when broken out shows a large proportion of thick white surrounding a compact hemispherical yolk. By contrast the white of a stored egg spreads out to cover a much larger area due to thinning of the thick white and the yolk has a flattened appearance.

The $Q_{10°C}$ of both the storage thinning of the yolk membrane and of the thick white is approximately 3, and may have a similar basic mechanism.[349] One theory supposes the chemical reduction of ovomucin with sylphydryl groups from ovalbumin.[350] Since reactivity has been observed to increase on denaturation, it is possible that thinning is initiated by denaturation of a small amount of ovalbumin. Another theory suggests that the thinning is caused by the high pH in the white, either by giving rise to dissolution of ovomucin, or to products of protein– protein interaction, particularly ovomucin complexing with lysozyme.[351,352] Chicken and turkey eggs show deterioration by thinning in a similar fashion, but duck eggs are very resistant. This could be due either to the apparently low amount of sulphydryl in duck egg white, or to the very low amount of lysozyme.[353]

Storage of Shell Eggs

The optimum storage conditions for fresh eggs is in the temperature range 31°±1°F, and a humidity between 80 to 85 per cent. The latter keeps evaporation to a minimum whilst retaining an adequate margin of safety against spoilage by moulds. The internal quality of an egg after 6 months storage at 32°F (0°C) is rather better than after 3 weeks spent at 70°F (21·1°C).

Gas storage in an atmosphere containing as little as 0·5 to 0·6 per cent carbon dioxide improves storage efficiency.[354] Using an atmosphere containing 60 per cent or more of carbon dioxide prevents mould growth, and the relative humidity (RH) in the store can be maintained above 96 per cent. Evaporation is greatly reduced but is accompanied by an increase in the rate of thinning of the thick white. The yolk does not become flabby when eggs are stored in low or high concentrations of carbon dioxide because the rate of weakening of the vitelline membrane is decreased. In addition development of storage flavour is reduced.[348]

Some advantage is gained by dipping eggs for a few seconds in a light mineral oil before cold storage. The oil reduces evaporation from the egg during storage, and consequently the RH of the store may be reduced to 80 per cent so that spoilage by moulds is avoided. Carbon dioxide is produced slowly within the egg and is retained inside the oiled shell to give an internal concentration similar to that in an untreated egg stored in 1 per cent carbon dioxide. To some extent oiling reproduces the conditions of storage in a low concentration of carbon dioxide, and similar effects on the rate of weakening of the vitelline membrane and rate of thinning of thick white are observed.[355] The protective effect on internal quality is proportionately much greater at normal temperatures and humidities than at 32°F (0°C) and 80 per cent RH.

Thermostabilization[356,357] is a process in which eggs are dipped in a bath of hot oil or water held at a temperature below the normal coagulation point of egg albumen (and therefore 136°F (57·8°C) is the upper temperature limit) for an extended period of time. The net effect is an improvement in bacterial quality and the thick white is stabilized during storage.[358] A comparison of eggs stored (i) untreated, (ii) oiled, and (iii) thermostabilized by dipping in oil at 130°F (54·4°C) for 16 min, over a period of 8 months at 34°F (1·1°C) has been made. The broken out appearance showed that the untreated eggs had deteriorated to a greater extent (as measured by Haugh units),[337] but off-flavours were more evident in the oiled and thermostabilized eggs.[359]

Shell eggs have been preserved in the past by immersion in water-glass or lime water. The solutions prevent evaporation, and their alkalinity prevents bacterial invasion by sealing off the pores of the shell.

Processed Egg Products

Liquid whole egg. Large quantities of egg are used in the bakery trade, and for convenience it is distributed as liquid egg. During manufacture special care must be maintained always against infection by microorganisms. Once the white and yolk are mixed, the natural protective mechanisms of the white are destroyed and bacteria can grow rapidly in the mixture. A separate department in the factory should be used for the unpacking of the eggs to prevent contamination from the packing materials reaching the processing area. The leaky eggs are rejected, and the sound eggs are passed to the breaking room. Each egg is individually broken by hand into a small pan and the contents rejected when of unsatisfactory odour or appearance. A pan containing a rejected egg must be removed and immediately washed and sterilized. The egg whites and yolks are bulked, strained to remove pieces of shell and chalazae, and pasteurized. Raw liquid egg is liable to contain Salmonellae and consequently there is always a risk that bakery products may cause outbreaks of food poisoning. During 1963 there was some concern in the U.K. over a series of salmonella and paratyphoid outbreaks which were attributed to infected egg in bakery products. Legislation was introduced to make pasteurization of liquid egg compulsory as from 1 January 1964.[360] The requirement is that liquid egg must be retained at a temperature not lower than 148°F (64·4° C) for at least 2½ min and immediately thereafter cooled to below 38°F (3·3°C). This heat treatment is adequate to kill the most resistant strains of *Salmonella senftenberg*. The critical heat treatment for the destruction of the activity of α-amylase is 2½ min exposure to a temperature of 64·4°C, and therefore absence of α-amylase activity in liquid egg has been developed into a simple routine test for the adequacy of pasteurization. The official test will detect a fall of 1°C in the pasteurizing conditions.[361] It is essential that the heat treatment should not affect adversely the desirable baking properties of the product, and it has been shown that the prescribed heat treatment meets this requirement.[362] After pasteurization the egg is filled into rectangular tinplate containers for freezing.

There is now a growing practice to convey large quantities of chilled pasteurized liquid whole egg in insulated road tankers direct to large bakeries. Cake production is thereby made more flexible as delays involved in the thawing of frozen egg (18–48 hr) is avoided.

Liquid egg white. Egg white is more sensitive to coagulation by heat than is either whole egg or yolk, and changes in physical and functional properties are observed after heating for several minutes at temperatures above 57°C. The two reasons for this behaviour are (i) the heat lability of conalbumin which in whole egg is converted into the more stable iron-conalbumin complex by iron from the yolk, and (ii) the heating of the egg while at pH 9·0 causes a marked increase in viscosity due to the instability of one or more of the proteins. The heat stability of the major protein ovalbumin varies sharply with pH, and reaches a maximum between pH 6·5 and 7·0 when denaturation is less than 1 per cent after heating for 30 min at 65°C.

Both of the above factors can be corrected by adjusting the pH to about 7 by addition of lactic acid, and by the addition of certain metal salts, e.g. aluminium. Pasteurization can then be carried out at a temperature of 60° to 61°C for 3½ to 4 min without adversely affecting the functional properties of the egg white. Used in angel cakes it produces an excellent texture and volume.[363]

Frozen whole egg. Cans of liquid egg are frozen in a blast freezer for about 36 hr at a temperature between −10° and −40°F (−23·3° and −40°C).

Some constituents of yolk undergo an irreversible change when the temperature of freezing is below 21°F (−6·1°C), and the thawed product exhibits a greatly increased viscosity termed gelation. A similar but smaller increase is evident in thawed whole egg for the same reason. Storage time and temperature, and rate of thawing also influence the degree of gelation. Gelation is inhibited by mechanical treatment, e.g. colloid milling, or by the addition of 10 per cent salt or 10 per cent sugar prior to freezing. Salted yolk may be used in the manufacture of mayonnaise and sugared yolk in the bakery trade. Proteins in the soluble phase of egg yolk are presumably more susceptible to processing than the granule proteins which are aggregated and exist as large insoluble particles. It seems probable that lipoproteins in the soluble phase are altered by the freezing and thawing of yolk as gelation is less in crotoxin treated yolk (lipoproteins are degraded by crotoxin to lysophospholipoproteins) than in untreated yolk.[364]

Dried whole egg and egg yolk. Dried whole egg and egg yolk are usually prepared by spray drying, whilst the drying of albumen is usually carried out in trays.

The whole egg and yolk are prepared for spray drying in the same manner as for freezing and observing the same precautions against microbiological contamination. Multiplication of microorganisms is avoided in liquid egg awaiting drying by rapid cooling and holding at a temperature below 45°F (7·2°C) until required. The spray drying process is similar to that for milk. The high solids content of liquid egg obviates any need for pre-condensation, and as it is easily damaged by heat, rapid drying of a fine spray by contacting with a large volume of hot air is employed.

Long term storage of spray dried whole egg, particularly at elevated temperatures, leads to deterioration by the interaction of the glucose naturally present with proteins and with phospholipids, e.g. cephalins. The reactions are of the Maillard type and spoilage is apparent by loss of solubility, colour darkening, and development of objectionable flavours and odours. It was shown that stability is increased by reducing the moisture content of the dry powder from about 5 to 2 per cent. Additional advantage is gained by reducing the pH of the powder. The pH of liquid whole egg is adjusted to 5·5 by the addition of hydrochloric acid, and the mixture is spray dried. A sufficient quantity of powdered sodium bicarbonate to restore the natural pH on reconstitution (about 8·5) is mixed with the powder before gas-packing. Current procedure is to remove the natural glucose (approx. 0·4 per cent) in the liquid egg. The desugared egg may be prepared by two procedures. Bacteria-free bakers' yeast[365,366] is added to the acidified warm egg and fermented for 3 hr. The glucose is converted mainly to carbon dioxide and alcohol, both of which are removed during the drying. It is difficult, however, to avoid a yeasty flavour. The alternative enzyme process is more expensive, and employs a mixture of catalase and glucose oxidase.[367,368] The enzyme mixture is added to liquid egg together with hydrogen peroxide which provides a source of oxygen to convert glucose to gluconic acid. The excess of peroxide is decomposed by catalase. Desugared dried egg is stable and resistant to non-enzymic browning during prolonged storage.

The loss of palatability during lengthy storage is not completely prevented by removal of glucose. Storage in an inert atmosphere of nitrogen prevents some part of the residual deterioration. To obtain maximum stability the glucose-free spray dried powder should have a maximum moisture content of 2 per cent, and should be gas packed. Under such conditions the product shows no change in palatability after 2 months' storage at 100°F (37·8°C).

Spray dried whole egg, even when freshly prepared, has only half the aerating power of fresh or frozen egg, and this loss in functional property is not prevented by a glucose removal treatment.

The loss in this essential property may be prevented by dissolving 10 to 15 per cent of certain sugars, sucrose or lactose, in the liquid egg before spray drying, and in addition Maillard type reactions are retarded.[369,370] The latter are inhibited progressively as the level of added carbohydrate is increased.[371] Sugar dried egg is particularly suited for baking purposes because of its superior aerating power, but there is a serious disadvantage in that it rapidly develops intense off-flavours when exposed to air even at refrigerated temperatures. The spoilage reactions are first progressively inhibited by increasing the amount of added sugar to a point of maximum stability. Further addition of sugar leads to an abrupt transition to marked flavour instability. The transition corresponds to a change in the physical state of the egg lipids from one of coalescence due to emulsion breakdown during drying to a finely dispersed state at higher sugar levels. The fine dispersion may represent a separation of the surface active phospholipids from their protein complexes, perhaps at interfaces of other lipid components, and thus give rise to instability.[372] Another possibility is the effect of the added sugar on the activity of the water in the egg powder. Moisture has a marked effect on the oxidative deterioration of dehydrated foods. It has been postulated that maximum stability occurs when the water content is sufficient to form a mono layer, the latter providing a physical barrier to attack by oxygen.[228] The effect of added sugar will be to compete for adsorptive sites and influence stability.[373]

Dried albumen. Dried albumen has been manufactured in China for many years. Egg whites are allowed to ferment in wooden casks for 2 to 3 days at about 21°C (70°F), the precise time depending on the ambient temperature. The fermentation is initiated by bacteria derived from the egg shells and from the casks. At the end of the fermentation the acid produced is neutralized and the pH of the egg white corrected by the addition of a little ammonia. The liquid is dried in shallow metal trays in a warm room for several days, and allowed to harden at ordinary temperatures. The final product contains about 8 to 12 per cent water. It is broken into fragments, and packed into containers which are reasonably water vapour proof. The product is heavily infected with bacteria, and Salmonellae are frequently present.

Initial attempts to produce dried albumen more hygienically gave an unstable product which quickly became discoloured and insoluble. The instability was caused by chemical reactions involving glucose present in the egg white, and glucose is removed during the fermentation of the Chinese process. Desugaring using a yeast fermentation prior to drying gave a product of good quality and stability.[366,374,375] Fermentation is rapid (less than 10 hr at 10°C) and the product possesses superior bacteriological quality. The objectionable yeasty smell and taste is avoided by removing the yeast cells by centrifuging or filtering the fermented white before drying.[376] Glucose may be removed from egg white also by bacterial fermentation with pure cultures, e.g. *Aerobacter aerogenes*,[377] or by the use of enzymes. Drying is normally carried out in shallow trays exposed in a warm room for 24 hr. Spray drying is often used, but there is some evidence that the very high sheer rates required for pressure jets in spray driers have a damaging effect on the functional properties of the dried product. It is thought that less damage occurs with centrifugal atomizers.

Use of dried albumen infected with Salmonellae is particularly hazardous, since the dry material can become dispersed in the bakery atmosphere to be deposited on fillings where proliferation of the microorganisms may take place with serious results.

Irradiated egg. Perhaps the most interesting and potentially useful application of irradiation to egg products is the destruction of Salmonellae in frozen egg. The process has the unique advantage that the frozen egg may be treated still in the sealed cans and without thawing. A dose of 0·5 Mrad produces a reduction of at least 10^7 fold in the Salmonellae population, and causes little or no change in the flavour or odour of the sponge cakes made from the treated egg. The total cost of the treatment, including capital cost of the equipment, is approximately £5 per ton (1020 kg). In contrast, the alternative process of thawing suspected frozen egg, removing it from the cans, pasteurizing by heat treatment, repacking, and re-freezing is believed to cost approximately 6*d*. per lb, i.e. £56 per ton.[85]

REFERENCES

1. Haines, R. B., *Microbiology in the Preservation of Animal Tissues*, (Gt. Brit.) Dept. Sci. Ind. Res., Food Invest. Bd. Spec. Rept. No. 45, 1937.
2. Mossel, D. A. A. and Ingram, M., *J. Appl. Bacteriol.* **18**, 232 (1955).
3. Scott, W. J., In Mrak E. M. and Stewart, G. F., ed. *Advan. Food Res.* Vol. 7, Academic Press, New York, 1957.
4. Christian, J. H. B., In Leitch, J. M. and Rhodes, D. N., ed. *Recent Advan. in Food Sci.* Vol. 3, Butterworths, London, 1962.
5. Wodzinski, R. J. and Frazier, W. C., *J. Bacteriol.* **79**, 572 (1960).
6. Wodzinski, R. J. and Frazier, W. C., *J. Bacteriol.* **81**, 353 (1961).
7. Wodzinski, R. J. and Frazier, W. C., *J. Bacteriol.* **81**, 359 (1961).
8. Ingraham, J. L. and Stokes, J. L., *Bacteriol. Rev.* **23**, 97 (1959).
9. Michener, H. D. and Elliott, R. P., In Chichester, C. O. Mrak, E. M. an Stewart, G. F., ed. *Advan. in Food Res.* Vol. 13, Academic Press, New York, 1964.
10. Peterson, A. C. and Gunderson, M. F., *Food Technol.* **14**, 413 (1960).
11. Ayres, J. C., In Mrak, E. M. and Stewart, G. F., ed. *Advan. in Food Res.* Vol. 6 Academic Press, New York, 1955.

12. LAMANNA, C., *Science*, **130**, 763 (1959).
13. FOSTER, E. M., DEFFNER, J. S., BOTT, T. L. and MCCOY, E., *J. Milk and Food Technol.* **28**(3), 86 (1965).
14. RIEMANN, H., In AYRES, J. C., KRAFT, A. A., SNYDER, H. E. and WALKER, H. W. eds. *Chemical and Biological Hazards in Foods*, Iowa State Univ. Press, Ames, Iowa, 1962.
15. FRAZIER, W. C., *Food Microbiology*, McGraw-Hill, New York, 1958.
16. BOWMER, E. J., *J. Milk and Food Technol.* **28**(3), 74 (1965).
17. MILLER, A. A. and RAMSDEN, F., *J. Appl. Bacteriol.* **18**, 565 (1955).
18. NOTTINGHAM, P. M., *J. Sci. Food Agr.* **11**, 436 (1960).
19. LEPOVETSKY, B. C., WEISER, H. H. and DEATHERAGE, F. E., *Appl. Microbiol.* **1**, 57 (1953).
20. JENSEN, L. B., *Microbiology of Meat*, 3rd Ed., Garrard Press, Ill., 1954.
21. CALLOW, E. H., In BATE-SMITH, E. C. and MORRIS, T. N., ed., *Food Science*, Cambridge Univ. Press, 1952.
22. HAINES, R. B., *Chem. and Ind.* **19**, 413 (1941).
23. CALLOW, E. H. and INGRAM, M., *Food*, **24**, 52 (1955).
24. INGRAM, M., *J. Roy. Sanit. Inst.* **69**, 39 (1949).
25. HAINES, R. B. and BATE-SMITH, E. C. (Gt. Brit.) *Dept. Sci. and Ind. Res., Rept. Food Invest. Bd.* 1932, p. 18.
26. AYRES, J. C., OGILVY, W. S. and STEWART, G. F., *Food Technol.* **4**, 199 (1950).
27. MUNDT, J. O., STOKES, R. L. and GOFF, O. E., *Poultry Sci.* **33**, 799 (1954).
28. INGRAM, M., Low Temperature Research Station, Cambridge. *An. Rpt.* 1960, p. 17.
29. SHRIMPTON, D. H. and BARNES, E. M., *Chem. and Ind.* 1492 (1960).
30. BARNES, E. M., *Proc. 10th Int. Congr. Refrig.* **3**, 97 (1959).
31. AGENJO, *Proc. XIIIth Int. Dairy Congr.* **2**, 153 (1953).
32. MURPHY and STUART, *Cornell Vet.* **54**, 501 (1952).
33. FOSTER, E. M., NELSON, F. E., SPECK, M. L., DOETSCH, R. N. and OLSEN, J. C., *Dairy Microbiology*, Prentice-Hall Inc., New Jersey, 1957.
34. THOMAS, S. B., EGDELL, J. W., CLEGG, L. F. L. and CUTHBERT, W. A., *Proc. Soc. Appl. Bacteriol.* **13**(1), 27 (1950).
35. WITTER, L. D., *J. Dairy Sci.* **44**, 983 (1961).
36. NASHIF, S. A. and NELSON, F. E., *J. Dairy Sci.* **36**, 459 (1953).
37. NASHIF, S. A. and NELSON, F. E., *J. Dairy Sci.* **36**, 471 (1953).
38. NASHIF, S. A. and NELSON, F. E., *J. Dairy Sci.* **36**, 481 (1953).
39. THOMAS, S. B. and THOMAS, B. F., *Proc. Soc. Appl. Bacteriol.* **2**, 65 (1947).
40. PETERSON, A. C. and GUNDERSON, M. F., *Appl. Mcirobiol.* **8**, 98 (1960).
41. VAN DER ZANT, W. C., *Food Res.* **22**, 151 (1957).
42. CAMP, B. J. and VAN DER ZANT, W. C., *Food Res.* **22**, 158 (1957).
43. GAINOR, C. and WEGEMER, D. E., *Appl. Microbiol.* **2**, 95 (1954).
44. THOMAS, S. B., DRUCE, R. G. and ELSON, K., *Dairy Ind.* **25**, 202 (1960).
45. PUNCH, J. D., OLSEN, J. C. and THOMAS, E. L., *J. Dairy Sci.* **48**, 1179 (1965).
46. BROOKS, J. and TAYLOR, D. J. (Gt. Brit.) *Dept. Sci. Ind. Res., Food Invest. Bd. Spec. Rept.* No. 60, 1955.
47. FROMM, D. and MARGOLF, P. H., *Poultry Sci.* **37**, 1273 (1958).
48. TRUSSELL, P. C., TRIGGS, R. E. and GREER, B. A., *Food Technol.* **9**, 134 (1955).
49. KRAFT, A. A., ELLIOTT, L. E. and BRANT, A. W., *Poultry Sci.* **37**, 238 (1958).
50. LIFSHITZ, A., BAKER, R. C. and NAYLER, H. S., *J. Food Sci.* **29**, 94 (1964).
51. STOKES, J. L. and OSBORNE, W. W., *Food Res.* **21**, 264 (1956).
52. BROOKS, J., *J. Appl. Bacteriol.* **23** (3), 499 (1960).
53. GARIBALDI, J. A. and STOKES, J. L., *Food Res.* **23**, 283 (1958).
54. HAINES, R. B., *Chem. Ind.* **18**, 391 (1940).
55. TRUSSELL, P. C., FULTON, C. O. and CAMERON, C. J., *Food Technol.* **9**, 130 (1955).
56. LORENZ, F. W., *Calif. Agr.* **4**(4), 15 (1950).
57. BOARD, R. G., *J. Appl. Bacteriol.* **28**, 197 (1965).
58. HAINES, R. B. and MORAN, T., *J. Hyg. Camb.* **40**, 453 (1940).
59. GARIBALDI, J. A., *Food Res.* **25**, 337 (1960).
60. RIEMANN, H., *Food Manuf.* **32**, 265 and 333 (1957).

61. BIGELOW, W. D., *J. Infect. Diseases* **29**, 528 (1921).
62. GILLESPY, T. G., In HAWTHORN, J. and LEITCH, J. M., ed. *Recent Advan. Food Sci.* **2**, 93 (1962).
63. CAMERON, E. J., *J. Assoc. Offic. Agr. Chemists* **23**, 607 (1940).
64. RIEMANN, H., *Food Technol.* **17** (1), 39 (1963).
65. HAWLEY, H. B., In WOODBINE, M., ed. *Antibiotics in Agriculture*, Butterworths, London, 1962.
66. GREENBERG, R. A. and SILLIKER, J. H., *J. Food Sci.* **27**, 64 (1962).
67. DENNY, C. B., SHARPE, L. E. and BOHRER, C. W., *Appl. Microbiol.* **9**, 108 (1961).
68. ESTY, J. R. and MEYER, K. F., *J. Infect. Diseases* **31**, 650 (1922).
69. TOWNSEND, C. T., ESTY, T. R. and BASELT, F. C., *Food Res.* **3**, 323 (1938).
70. EL-BISI, H. M., In *Rad. Preserv. Foods* Publ. 1273 Natl. Acad. Sci.—Natl. Res. Coun. Washington D.C. 1965.
71. BASHFORD, T. E., Canned Foods In *Rept. Progr. Appl. Chem.*, Soc. Chem. Ind. Lond. 1959.
72. ROBINSON, L. E., *Food Manuf.* **36**(3), 110 (1961).
73. *Aberdeen Typhoid Outbreak 1964*. Cmnd. 2542 H.M. Stationery Office. Edinburgh 1964.
74. ALSTRAND, O. V. and ECKLUND, O. F., *Food Technol.* **6**(5), 185 (1952).
75. POWERS, J. J., PRATT, D. E., CARMEN, J. L., SAMANTMADJA, D. and FORTSON, J. C., *Food Technol.* **16**, 80 (1962).
76. HICKS, E. W., *Food Technol.* **5**(4), 134 (1951).
77. HICKS, E. W., *Food Technol.* **6**(5), 175 (1952).
78. HICKS, E. W., *J. Food Sci.* **26**, 218 (1961).
79. STUMBO, C. R., *Food Technol.* **3**, 126 (1949).
80. STUMBO, C. R., In MRAK, E. M. and STEWART, G. F., ed. *Advan. Food Res.* Vol. 2, Academic Press, New York, 1949.
81. ZIEMBA, J. V., *Food Eng.* **33**(6), 42 (1961).
82. ANON. CANNING REPT. *Food Eng.* **32**(1), 110 (1960).
83. SMITH, H. L. and BALL, C. O., *Food Manuf.* **30**, 139 (1955).
84. HANNON, R. S. (Gt. Brit.) *Dept. Sci. Ind. Res., Food Invest. Bd. Spec. Rept.* No. 61. 1955.
85. ROLFE, E., *Trends and Developments in Food Technol. in O.E.C.D. Countries*. O.E.C.D. Directorate Agr. Food, Div. Tech. Action. AGR/T(62) 29. Dec. 1962.
86. INGRAM, M. and RHODES, D. N., *Food Manuf.* **37**, 318 (1962).
87. SHEA, K. G., *Food Technol.* **12**(8), 6 (1958).
88. GORESLINE, H. E., INGRAM, M., MACUCH, P., MOCQUOT, G., MOSSEL, D. A. A., NIVEN, C. F. and THATCHER, F. S., *Nature* (*Lond.*) **204**, 237 (1964).
89. GOLDBLITH, S. A., *Food Technol.* **18**, 1384 (1964).
90. GOLDBLITH, S. A., In *Rad. Preserv. Foods* Publ. 1273, Natl. Acad. Sci.—Natl. Res. Coun., Washington D.C. 1965.
91. RAICA, N., In *Rad. Preserv. Foods* Publ. 1273, Natl. Acad. Sci.—Natl. Res. Coun., Washington D.C. 1965.
92. COLEBY, B. and SHEWAN, J. M., In BORGSTROM, G., ed., *Fish as Food* Vol. 4, Acad. Press, New York, 1965.
93. BORGSTROM, G., In *Proc. Low Temp. Microbiol. Symp.* Campbell Soup Co., New Jersey 1961.
94. ELLIOTT, R. P. and MICHENER, H. D., In *Conf. Frozen Food Quality*, U.S. Dept. Agr., ARS-74-21, 1960.
95. LOVELOCK, J. E., *Biochim. Biophys. Acta* **10**, 414 (1953).
96. MERYMAN, H. T., *Science* **124**, 515 (1956).
97. MERYMAN, H. T., In PARKES, A. S. and SMITH, A. U., ed. *Recent Research in Freezing and Drying* Blackwell Sci. Publ. Oxford 1960.
98. MAZUR, P., *Federation Proc.* **24**, Suppl. 15:S-175 (1965).
99. ASAHINA, E., *Nature* (*Lond.*) **191**, 1263 (1961).
100. MAZUR, P., *Ann. N.Y. Acad. Sci.* **85** (Art. 2), 610 (1960).
101. MAZUR, P., *J. Gen. Physiol.* **47**, 347 (1963).

102. MENZ, L. J. and LUYET, B. J., *Biodynamica* **8**, 261 (1961).
103. RAPATZ, G. and LUYET, B. J., *Biodynamica* **8**, 295 (1961).
104. LUYET, B. J., *Federation Proc.* **24**, Suppl. 15:S-315 (1965).
105. LUSENA, C. V., *Arch. Biochem. Biophys.* **57**, 277 (1955).
106. LOVE, R. M., *J. Sci. Food Agr.* **13**, 269 (1962).
107. LOVE, R. M., In HAWTHORN, J. and LEITCH, J. M., ed. *Recent Advan. Food Sci.* **1**, 147 (1962).
108. RIEDEL, L., *Kaltetechnik* **9**, 38 (1957).
109. TRESSLER, D. K. and EVERS, C. F., *The Freezing Preservation of Foods*, Vol. 1, Avi Publishing Co., Westport, 1957.
110. GANE, R., *Res.*, **13**, 207 (1960).
111. ANON., *Food Eng.* **34** (11), 60 (1962).
112. LENTZ, C. P. and VAN DER BERG, L., *Food Technol.* **11**, 247 (1957).
113. COOK, W. H., *Food in Can.* **6** (7), 22 (1946).
114. ANON., *Food Eng.* **34**(9), 70 (1962).
115. ANON., *Food Manuf.* **40**(10), 43 (1965).
116. ROLFE, E., *Food Manuf.* **40**(7), 46 (1965).
117. ROLFE, E., In *Fundamental Aspects of the Dehydration of Foodstuffs*, Soc. Chem. Ind. (Lond.) 1958.
118. EDE, A. J. and HALES, K. C. (Gt. Brit.) *Dept. Sci. Ind. Res., Food Invest. Bd. Spec. Rept.* No. 53, 1948.
119. JASON, A. C., In *Fundamental Aspects of the Dehydration of Foodstuffs*, Soc. Chem. Ind. (Lond.) 1958.
120. VAN ARSDEL, W. B. and COPLEY, M. J., *Food Dehydration* Vols. 1 and 2, Avi Publishing Co. Inc., Westport, 1963.
121. HAY, J. M., *J. Sci. Food Agr.* **6**, 433 (1955).
122. GOODING, E. G. B. and ROLFE, E., *J. Sci. Food Agr.* **6**, 427 (1955).
123. ROLFE, E., *Food*, **25**, 199 (1956).
124. ROLFE, E., In COTSON, S. and SMITH, D. B., ed. *Freeze Drying of Foodstuffs*, Columbine Press, Lond., 1963.
125. LUSK, G., KAREL, M. and GOLDBLITH, S. A., *Food Technol.* **18**, 1625 (1964).
126. HARPER, J. C., *Assoc. Inst. Chem. Eng. J.* **8**, 298 (1962).
127. HARPER, J. C. and EL SAHRIGI, A. F., *Ind. Eng. Chem. Fundamentals* **3**, 318 (1964).
128. SARAVACOS, G. D. and PILSWORTH, M. N., *J. Food Sci.* **30**, 773 (1965).
129. LEWIN, L. M. and MATELES, R. I., *Food Technol.* **16**(1), 94 (1962).
130. WOODWARD, H. T., *Food Eng.* **35**(6), 96 (1963).
131. HAMMOND, J., *Growth and Development of Mutton Quality in the Sheep*, Oliver & Boyd, Lond., 1932.
132. MCMEEKAN, C. P., *J. Agr. Sci.* **30**, 276, 387, 511 (1940).
133. MCMEEKAN, C. P., *J. Agr. Sci.* **31**, 1 (1941).
134. CALLOW, E. H., *J. Animal Sci.* **37**, 113 (1947).
135. PALSON, H. and VERGES, J. B., *J. Agr. Sci.* **42**, 1 (1952).
136. WILSON, L. L., DINKEL, C. A., RAY, D. E. and MINYARD, J. A., *J. Animal Sci.* **22**, 699 (1963).
137. GEE, I. and PRESTON, T. R., *Brit. J. Nutr.* **11**, 329 (1957).
138. LAWRIE, R. A., *Brit. J. Nutr.* **14**, 255 (1960).
139. BODWELL, C. E., PEARSON, A. M. and SPOON, M. E., *J. Food Sci.* **30**, 766 (1965).
140. BENDALL, J. R., In *Proceedings Meat Tenderness Symposium* Campbell Soup Co., Camden, N.J., 1963.
141. HOWARD, A. and LAWRIE, R. A. (Gt. Brit.) *Dept. Sci. Ind. Tes., Food Invest. Bd. Spec. Rept.* No. 63, 1956.
142. BATE-SMITH, E. C. and BENDALL, J. R., *Brit. Med. Bull.* **12**(3), 230 (1956).
143. BENDALL, J. R., In BOURNE, G. H., ed. *The Structure and Function of Muscle*, Volume III. Academic Press 1960.
144. HUXLFY, H. E. and HANSON, J., In BOURNE, G. H., ed. *The Structure and Function of Muscle*, Volume I. Academic Press 1960.

145. NEEDHAM, D. N., In BOURNE, G. H., ed. *The Structure and Function of Muscle*, Volume II. Academic Press 1960.
146. CALLOW, E. H., *Rept. Food Invest. Bd.* Lond. 1937.
147. HEDRICK, H. B., BOILLET, J. B., BRADY, D. E. and NAUMANN, H. D., *Univ. Mo. Agr. Expt. Sta. Res. Bull.* No. 717, 1959.
148. HAMM, R., In CHICHESTER, C. O., MRAK, E. M. and STEWART, G. F., ed. *Advan. Food Res.* Vol. 10, Academic Press, New York, 1964.
149. BRISKEY, E. J., In CHICHESTER, C. O., MRAK, E. M. and STEWART, G. F., ed. *Advan. Food Res.* Vol. 13, Academic Press, New York, 1964.
150. POOL, M. F., DE FREMERY, D., CAMPBELL, A. A. and KLOSE, A. A., *Food Technol.* **13**, 29 (1959).
151. MITCHELL, J. D. In *Proceedings Meat Tenderness Symposium* Campbell Soup Co., Camden, N.J., 1963.
152. RANKEN, M. O., In *Rept. Progr. Appl. Chem.* **48**, 495 (1963). Soc. Chem. Ind. Lond.
153. BATCHER, O. M., DAWSON, E. H., POINTER, M. T. and GILPIN, E. L., *Food Technol.* **16** (1), 102 (1962).
154. LOWE, B. and KASTELIC, J., *Iowa Agr. Expt. Sta. Bull.* No. 495, 1961.
155. PARTRIDGE, S. M. (Gt. Brit.) *Agr. Res. Council, Low Temp. Res. Sta. Cambridge, Ann. Rept.*, 1962.
156. BLUMER, T. N., *J. Animal Sci.* **22**, 771 (1963).
157. GILPIN, G., BATCHER, O. M. and DEARY, P., *Food Technol.* **19**, 834 (1965).
158. CARPENTER, Z. L. and KING, G. T., *Food Technol.* **19**, 1706 (1965).
159. WELLINGTON, G. H. and STOUFFER, J. R., *Cornell Univ. Agr. Expt. Sta. Bull.* No. 941, 1959.
160. SLEETH, R. B., HENRICKSON, R. L. and BRADY, D. E., *Food Technol.* **11**, 205 (1957).
161. DOTY, D. M. and PIERCE, J. C., U.S. Dept. Agr., *Agr. Market Serv. Tech. Bull.* No. 1231, 1961.
162. LOCKER, R. H., *J. Sci. Food Agr.* **11**, 520 (1960).
163. GOESER, P. A., *Proc. 13th Res. Conf.*, Am. Meat Inst. Found., 1961.
164. WILSON, G. D. In *Science of Meat and Meat Products*, Am. Meat Inst. Found., 1960.
165. UNILEVER LTD., Brit. Pat. 848,014.
166. SPENCER, R., *Food Manuf.* **41**(3), 39 (1966).
167. HALVORSON, H. O., WOLF, J. and SRINIVASAN, V. R., In *Proceedings Low Temperature Microbiology Symposium, 1961*, Campbell Soup Co., Camden, N.J., 1962.
168. GANE, R., In BATE-SMITH, E. C. and MORRIS, T. N., ed. *Food Science*, Cambridge Univ. Press, 1952.
169. *Div. Food Preserv. Ann. Rept. 1962-64*. Commonwealth Sci. Ind. Res. Organisation, Sydney, Australia, 1964.
170. BUTTIAUX, R. and CATSARAS, M., *Ann. Inst. Pasteur (Lille)*. **15**, 165 (1964).
170. AYRES, J. C., *Food Res.* **25**, 1 (1960).
171. NEWELL, G. W., GWINN, J. M. and JULL, M. A., *Poultry Sci.* **27**, 251 (1948).
172. STEWART, G. F., LOWE, B. and MORR, M., *U.S. Egg Poultry Mag.* **47**, 542, 571 (1941).
173. BAKER, R. C., NAYLOR, H. B., PFUND, M. C., *Poultry Sci.* **35**, 398, 1956.
174. SHRIMPTON, D. H. (Gt. Brit.) *Dept. Sci. Ind. Res., Food Invest. Bd. Tech.* Paper No. 7, 1957.
175. ESSARY, E. O., MOORE, W. E. C. and KRAMER, C. Y., *Food Technol.* **12**, 684 (1958).
176. INGRAM, M. (Gt. Brit.) *Agr. Res. Council, Low Temp. Res. Sta. Cambridge Ann. Rept.*, 1960.
177. JAYE, M., KITTAKA, R. S. and ORDAL, Z. J., *Food Technol.* **16**(4), 95 (1962).
178. SHANK, J. L., LUNDQUIST, B. R., *Food Technol.* **17**, 1163 (1963).
179. BROOKS, J., *Proc. Roy. Soc.* **118B**, 560 (1935).
180. ROBACH, D. L. and COSTILOW, R. N., *Appl. Microbiol.* **9**, 529 (1961).
181. SNYDER, H. E., *J. Food Sci.* **29**, 525 (1964).
182. FOX, J. B., *J. Agr. Food Chem.* **14**, 207 (1966).
183. LAWRIE, R. A., *J. Refrig.* **2**, 87 (1959).
184. MORAN, T., *Proc. Roy. Soc.* **105B**, 177 (1930).
185. BENDALL, J. R. and MARSH, B. B., *Proc. 8th Int. Congr. Refrig.* Lond., 1951.

186. Sharp, J. G. and Marsh, B. B. (Gt. Brit.) *Dept. Sci. Ind. Res., Food Invest. Bd. Spec. Rept.* No. 58, 1953.
187. Marsh, B. B. and Thompson, J. F., *J. Sci. Food Agr.* **9**, 417 (1958).
188. Bouton, P. E., Howard, A. and Lawrie, R. A. (Gt. Brit.) *Dept. Sci. Ind. Res., Food Invest. Bd. Spec. Rept.* No. 66, 1957.
189. Khan, A. W. and Lentz, C. P., *J. Food Sci.* **30**, 787 (1965).
190. Lorentzen, G., *Bull. Inst. Inter. Froid*, Annexe 1 (1964).
191. Klose, A. A., Hanson, H. L. and Lineweaver, H., *Food Technol.* **4**, 71 (1950).
192. Wills, R., Lowe, B. and Stewart, G. F., *Refrig. Eng.* **56**, 237 (1948).
193. *Conference on Frozen Food Quality* U.S. Dept. Agr. Publ. ARS-74-21, 1960.
194. Hanson, H. L. and Fletcher, L. R., *Food Technol.* **12**, 40 (1958).
195. Hanson, H. L., Fletcher, L. R. and Lineweaver, H., *Food Technol.* **13**, 221 (1959).
196. Klose, A. A., Pool, M. F., Campbell, A. A. and Hanson, H. L., *Food Technol.* **13**, 477 (1959).
197. *Recommendations for the Processing and Handling of Frozen Foods.* Bull. Inst. Inter. Froid, Annexe 1964.
198. Alford, J. A. and Pierce, D. A., *J. Food Sci.* **26**, 518 (1961).
199. Keskinel, A., Ayres, J. C. and Snyder, H. E., *Food Technol.* **18**, 101 (1964).
200. Mecchi, E. P., Pool, M. F., Behman, G. A., Hamachi, M. and Klose, A. A., *Poultry Sci.* **35**, 1328 (1956).
201. Mecchi, E. P., Pool, M. F., Nonaka, M., Klose, A. A., Marsden, J. J. and Lillie, R. J., *Poultry Sci.* **35**, 1246 (1956).
202. Marion, W. W. and Forsythe, R. H., *J. Food Sci.* **29**, 530 (1964).
203. Tappel, A. L., *Arch. Biochem. Biophys.* **44**, 378 (1953).
204. Van der Berg, L., and Rose, D., *Arch. Biochem. Biophys.* **81**, 319 (1959).
205. Van der Berg, L., *Arch. Biochem. Biophys.* **84**, 305 (1959).
206. Van der Berg, L., *Food Technol.* **15**(10), 434 (1961).
207. Van der Berg, L., *J. Food Sci.* **29**, 540 (1964).
208. Kaess, G., *Food Technol.* **15**, 122 (1961).
209. Kaess, G. and Weidemann, J. F., *Food Technol.* **15**, 129 (1961).
210. Kitchell, A. G. and Ingram, M., *Ann. Inst. Pasteur.* (*Lille*) **8**, 121 (1956).
211. *Committee on Dehydration of Meat*, U.S. Dept. Agr., Agr. Res. Admin. Circ. 706, 1944.
212. Sharp, J. G. (Gt. Brit.) *Dept. Sci. Ind. Res., Food Invest. Bd. Spec. Rept.* No. 57, 1953.
213. Gooding, E. G. B. and Rolfe, E., *Food Technol.* **11**, 302 (1957).
214. Rolfe, E., unpublished data.
215. Rolfe, E., unpublished data.
216. Rolfe, E., *Food Trade Rev.* **24**, 12 (1954).
217. Radouco-Thomas, C., Lataste-Dorolle, C., Zender, R., Busset, R., Meyer, H. M. and Mouton, R. F., *J. Food Sci.* **24**, 453 (1959).
218. Penny, I. F., Voyle, C. A. and Lawrie, R. A., *J. Sci. Food Agr.* **14**, 535 (1963).
219. Penny, I. F., Voyle, C. A. and Lawrie, R. A., *J. Sci. Food Agr.* **15**, 559 (1964).
220. *Am. Meat Inst. Found., Ann. Rept. 1959-60.*
221. Hodge, J. E., *J. Agr. Food Chem.* **1**, 928 (1953).
222. Sharp, J. G. and Rolfe, E., In *Fundamental Aspects of the Dehydration of Foodstuffs.* Soc. Chem. Ind. (Lond.) 1958.
223. Thomson, J. S., Fox, J. B. and Landman, W. A., *Food Technol.* **16**(9), 131 (1962).
224. Lea, C. H., In *Fundamental Aspects of the Dehydration of Foodstuffs.* Soc. Chem. Ind. (Lond.) 1958.
225. Henrickson, R. L., Brady, D. E., Gehrke, C. W. and Brooks, R. F., Univ. Missouri College of Agriculture, *Agr. Exp. Sta. Res. Bull.* No. 587, 1955.
226. Sharp, J. G., Brit. Pat. No. 767, 509.
227. Salwin, H., *Food Technol.* **13**, 594 (1959).
228. Salwin, H., *Food Technol.* **17**, 1114 (1963).
229. Tappel, A. L., *Food Res.* **21**, 195 (1956).
230. Hunt, S. M. V. and Matheson, N. A., *Food Res.* **24**, 262 (1959).
231. Matheson, N. A., *J. Sci. Food Agr.* **13**, 248 (1962).

232. ACKER, L., In CHICHESTER, C. O., MRAK, E. M., and STEWART, G. F. ed. *Advan. Food Res.* **11**, 1962.
233. RHODES, D. N., SHARP, J. G. and INGRAM, M., In *Biochemical Principles of the Food Industry*, **8**, 217. Pergamon Press, 1963.
234. DOTY, D. M., In *Radiation Preservation of Foods*, Publ. 1273, Natl. Acad. Sci., Natl. Res. Council, Washington, D.C., 1965.
235. WHITEHAIR, L. A., BRAY, R. W., WECKEL, K. G., EVANS, G. W. and HEILIGMAN, F., *Food Technol.* **18**, 230 (1964).
236. COLEBY, B., INGRAM, M. and SHEPHERD, H. J., *J. Sci. Food Agr.* **12**, 417 (1961).
237. BAILEY, A. J., BENDALL, J. R. and RHODES, D. N., *Int. J. Appl. Rad. Isotopes* **13**, 137 (1962).
238. DRAKE, M. P., GIFFEE, J. W., RYER, R. and HARRIMAN, H., *Science* **125**, 23 (1957).
239. RHODES, D. N. and MEEGUNGWAN, C., *J. Sci. Food Agr.* **13**, 279 (1962).
240. COLEBY, B., INGRAM, M., SHEPHERD, H. J., THORNLEY, M. J. and WILSON, G. M., *J. Sci. Food Agr.* **12**, 483 (1961).
241. URBAIN, W. M., In *Radiation Preservation of Foods*, Publ. 1273 Natl. Acad. Sci., Natl. Res. Council, Washington, D.C., 1965.
242. LEA, C. H., MCFARLANE, J. J. and PARR, L. J., *J. Sci. Food Agr.* **11**, 690 (1960).
243. WEST, R. G., PEARSON, A. M. and MCARDLE, J. F., *J. Food Sci.* **26**, 79 (1961).
244. CHIAMBALERO, C. J., JOHNSON, D. A. and DRAKE, M. P., *J. Agr. Food Chem.* **7**, 782 (1959).
245. HANSEN, N. H., In *Radiation Preservation of Foods*. Publ. 1273 Natl. Acad. Sci., Natl. Res. Council, Washington, D.C., 1965.
246. PROGR. REPT. 1964, *Milk Compositional and Hygienic Quality Control, England and Wales*, Joint Milk Quality Committee, Milk Marketing Board, Surrey.
247. PROGR. REPT. 1965, *Milk Compositional and Hygienic Quality Control, England and Wales*, Joint Milk Quality Committee, Milk Marketing Board, Surrey.
248. MACY, I. G., KELLY, H. J. and SLOAN, R. E., In Publ. 254. Natl. Acad. Sci., Natl. Res. Council, Washington, D.C., 1953.
249. ROOK, J. A. F., *Dairy Sci. Abstr.* **23**(6), 251 (1961).
250. LABEN, R. C., *J. Dairy Sci.* **46**, 1293 (1963).
251. NICKERSON, T. A., *J. Dairy Sci.* **43**, 598 (1960).
252. ROOK, J. A. F., LINE, C. and ROWLAND, S. J., *J. Dairy Res.* **27**, 427 (1960).
253. JOHNSON, J. C. and GWINS, R. L., *J. Dairy Sci.* **44**, 980 (1961).
254. JOHNSON, K. R., FOWT, D. L., HIBBS, R. A. and ROSS, R. H., *J. Dairy Sci.* **44**, 658 (1961).
255. TREECE, J. M., GILMORE, L. O., FECHHEIMER, N. S. and WASHBURN, R. G., *J. Dairy Sci.* **44**, 1191 (1961).
256. SLACK, N. H., MATHER, R. E. and PFAU, K. O., *J. Dairy Sci.* **45**, 144 (1962).
257. MCCARTHY, R. D., *J. Agr. Food Chem.* **10**(2), 126 (1962).
258. VAN SOEST, P. J., *J. Dairy Sci.* **46**, 204 (1963).
259. JORGENSEN, N. A., SCHULTZ, L. H. and BARR, G. H., *J. Dairy Sci.* **48**, 1031 (1965).
260. THOELE, H. W., *Holstein-Friesian World* **58**, 1721 (1961).
261. JACKSON, R. H., COULSON, E. J. and CLARK, W. R., *Arch. Biochem. Biophys.* **97**, 373 (1962).
262. HERALD, C. T. and BRUNNER, J. R., *J. Dairy Sci.* **40**, 948 (1957).
263. THOMPSON, M. P., BRUNNER, J. R., STINE, C. M. and LINDQUIST, K., *J. Dairy Sci.* **44**, 1589 (1961).
264. BRUNNER, J. R., *J. Dairy Sci.* **45**, 943 (1962).
265. HARWALKER, V. R. and BRUNNER, J. R., *J. Dairy Sci.* **48**, 1139 (1965).
266. JENSEN, R. G., *J. Dairy Sci.* **47**, 210 (1964).
267. TARASSUK, N. P. and FRANKEL, E. N., *J. Dairy Sci.* **40**, 418 (1957).
268. TARASSUK, N. P. and FRANKEL, E. N., *J. Dairy Sci.* **38**, 438 (1955).
269. HARRINGTON, B. L., *J. Dairy Sci.* **37**, 775 (1954).
270. FRANKEL, E. N. and TARASSUK, N. P., *J. Dairy Sci.* **42**, 409 (1959).
271. ROSE, D., *J. Dairy Sci.* **48**, 139 (1965).
272. ANON., *Food Technol.* **17**, 765 (1963).

273. WISHNER, L. A., *J. Dairy Sci.* **47**, 216 (1964).
274. PATTON, S., *J. Dairy Sci.* **37**, 446 (1954).
275. WISHNER, L. A. and KEENEY, M., *J. Dairy Sci.* **46**, 785 (1963).
276. *U.K. Dairy Facts and Figures*. The Federation of U.K. Milk Marketing Boards, 1965.
277. BURTON, H., *J. Dairy Res.* **21**, 194 (1954).
278. BURTON, H., FRANKLIN, J. G., WILLIAMS, D. J., CHAPMAN, H. R., HARRISON, A. J. and CLEGG, L. F., *J. Dairy Res.* **25**, 338 (1958).
279. WILLIAMS, D. J., FRANKLIN, J. G., CHAPMAN, H. R. and CLEGG, L. F., *J. Appl. Bacteriol.* **20**, 43 (1957).
280. FRANKLIN, J. G., WILLIAMS, D. J., CHAPMAN, H. R. and CLEGG, L. F., *J. Appl. Bacteriol.* **21**, 47 (1958).
281. BURTON, H., *J. Dairy Res.* **25**, 75, 324 (1958).
282. BURTON, H., *J. Soc. Dairy Technol.* **18**(2), 58 (1965).
283. BURTON, H., *Proc. XVth Intern. Dairy Congr.* **4**, 2045 (1959).
284. LYSTER, R. L. and ASCHAFFENBURG, R., *J. Dairy Res.* **29**, 21 (1962).
285. ASHTON, T. R., *J. Soc. Dairy Technol.* **18**(2), 65 (1965).
286. *J. Soc. Dairy Technol.* **19**, 31 (1966).
287. WILLIAMS, D. J., FRANKLIN, J. G., BURTON, H., KNIGHT, H. C. and BRACEGIRDLE, R. J., *J. Soc. Dairy Technol.* **13**, 46 (1960).
288. SIEGENTHALER, E., *Milchwissenschaft* **19**, 281 (1964).
289. ANON., *Dairy Eng.* **78**, 402 (1961).
290. ANON., *Food Eng.* **37**(6), 104 (1965).
291. *Condensed Milk Regulations*, S.I. 1959 No. 1098 dated 19 June, 1959.
292. BUYZE, H. G., *Neth. Milk and Dairy J.* **6**, 218 (1952).
293. EVENHUIS, N. and DE VRIES, TH.R., *Neth. Milk and Dairy J.* **11**, 184 (1957).
294. SWANSON, A. M. and CALBERT, H. E., *Food Technol.* **17**, 721 (1963).
295. SAMUELSON, E. E., THOME, K. E., BORGSTROM, G. and HJALMDAHL, M., *Dairy Sci. Abstr.* **19**, 875 (1957)
296. WEARMOUTH, W. G., *J. Refrig.* **3**, 56 (1960).
297. ROSE, D., *J. Dairy Sci.* **48**, 139 (1965).
298. TESSIER, H. and ROSE, D., *Can. J. Technol.* **34**, 211 (1956).
299. PYNE, G. T., *J. Dairy Res.* **29**, 101 (1962).
300. TESSIER, H., ROSE, D., and LUSENA, C. V., *Can. J. Technol.* **34**, 131 (1956).
301. NICKERSON, T. A., *J. Food Sci.* **29**, 443 (1964).
302. WILDASIN, H. L. and DOAN, F. J., *J. Dairy Sci.* **34**, 438 (1951).
303. EL-NEGOUMY, A. M. and BRYD, J. C., *J. Dairy Sci.* **48**, 23 (1965).
304. PEEBLES, D. D., *Food Technol.* **10**, 64 (1956).
305. HALL, C. W. and HEDRICK, T. I., *Dairy Eng.* **78**, 7 (1961).
306. HANRAHAN, F. P., TAMSMA, A., FOX, K. K. and PALLANSCH, M. J., *J. Dairy Sci.* **45**, 27 (1962).
307. BELL, R. W., HANRAHAN, F. P. and WEBB, B. H., *J. Dairy Sci.* **46**, 1352 (1963).
308. ANDERSON, T. D. C., *Food Trade Rev.* **33**(12), 52 (1963).
309. WAITE, R., *J. Soc. Dairy Technol.* **17**(3), 122 (1964).
310. TAMSMA, A., MUCHA, J. J. and PALLANSCH, M. J., *J. Dairy Sci.* **45**, 1435 (1962).
311. TAMSMA, A., MUCHA, J. J. and PALLANSCH, M. J., *J. Dairy Sci.* **46**, 114 (1963).
312. *Dried Milk Regulations*, S.I. 1964 No. 857 dated 11 June, 1964.
313. SINNAMON, H. I., ACETO, N. C., ESKEW, R. K. and SCHOPPET, E. F., *J. Dairy Sci.* **40**, 1036 (1957).
314. ACETO, N. C., SINNAMON, H. I., SCHOPPET, E. F. and ESKEW, R. K., *J. Dairy Sci.* **45**, 501 (1962).
315. HOWARD, N. and TAMSMA, A., *J. Dairy Sci.* **45**, 658 (1962).
316. KING, N., *Dairy Sci. Abstr.* **27**(3), 91 (1965).
317. CHOI, R. P., TATTER, C. W., O'MALLEY, C. M. and FAIRBANKS, B. W., *J. Dairy Sci.* **32**, 391 (1949).
318. GOULDEN, J. D. S. and WHITE, J. W., *Nature* **181**, 266 (1958).
319. BOCKIAN, A. H., STEWART, G. F. and TAPPEL, A. L., *Food Res.* **22**, 69 (1957).
320. RADEMA, L., *Neth. Milk Dairy J.* **8**, 125 (1954).

321. ABBOTT, J., WAITE, R. and HEARNE, J. F., *J. Dairy Res.* **28**, 285 (1961).
322. TAMSMA, A. and PALLANSCH, M. J., *J. Dairy Sci.* **45**, 1573 (1962).
323. TAMSMA, A. and PALLANSCH, M. J., *J. Dairy Sci.* **47**, 970 (1964).
324. TROY, H. C. and SHARP, P. F., *J. Dairy Sci.* **13**, 140 (1930).
325. LEA, C. H. and WHITE, J. C., *J. Dairy Res.* **15**, 298 (1948).
326. LEA, C. H., *J. Sci. Food Agr.* **8**, 1 (1957).
327. PATTON, S., *J. Dairy Sci.* **38**, 457 (1955).
328. HENRY, K. M., KON, S. K., LEA, C. H. and WHITE, J. C. D., *J. Dairy Res.* **15**, 292 (1948).
329. WERTHEIM, J. H., ROYCHOUDHURY, R. N., HOFF, J. E., GOLDBLITH, S. A. and PROCTOR, B. E., *J. Agr. Food Chem.* **5**, 944 (1957).
330. PROCTOR, B. E., GOLDBLITH, S. A., BATES, C. J. and HAMMERLE, O. A., *Food Technol.* **6**, 237 (1952).
331. SARAVACOS, E. D., HOFF, J. E., GOLDBLITH, S. A. and WICK, E. L., *Food Technol.* **15**, 239 (1961).
332. HOFF, J. E., SUNYACH, J., PROCTOR, B. E. and GOLDBLITH, S. A., *Food Technol.* **14**, 27 (1960).
333. BROOKS, J. and TAYLOR, D. J. (Gt. Brit.) *Dept. Sci. Ind. Res., Food Invest. Bd. Spec. Rept.* No. 60, 1955.
334. BURLEY, R. W. and COOK, W. H., *Can. J. Biochem. Physiol.* **39**, 1295 (1961).
335. BROOKS, J. and HALE, H. P., *Biochim. Biophys. Acta* **46**, 289 (1961).
336. OLSSON, N., *Proc. 6th World Poultry Congress* 310 (1936).
337. BRANT, A. W., OTTE, A. W. and NORRIS, K. H., *Food Technol.* **5**, 356 (1951).
338. KLINE, L., MEEHAN, J. J. and SUGIHARA, T. F., *Food Technol.* **19**, 1296 (1965).
339. COTTERILL, O. J., STEPHENSON, A. B. and FUNK, E. M., *Proc. 12th World Poultry Congr.* 443 (1962).
340. ANJANEYLU, K., *Poultry Sci.* **40**, 1372 (1961).
341. WARD, A. H., *Milling* **142**, 297 (1964).
342. CHEN, P. H., COMMON, R. H., NIKOLAICZUK, N. and MACRAE, H. F., *J. Food Sci.* **30**, 838 (1965).
343. SHENSTONE, F. S. and VICKERY, J. R., *Nature* **177**, 94 (1956).
344. MASSON, J. C., VAVICH, M. G., HEYWANG, B. W. and KEMMERER, A. R., *Science* **126**, 751 (1957).
345. EVANS, R. J., BANDEMER, S. L., ANDERSON, M. and DAVIDSON, J. A., *J. Nutr.* **76**, 314 (1962).
346. SHENSTONE, F. S. and VICKERY, J. R., *Poultry Sci.* **38**, 1055 (1959).
347. KIRCHER, H. W., *J. Am. Oil Chem. Soc.* **41**, 4 (1964).
348. MORAN, T., *J. Soc. Chem. Ind. Lond.* **56**, 96T (1937).
349. FEENEY, R. E., WEAVER, J. M., JONES, J. R. and RHODES, M. B., *Poultry Sci.* **35**, 1061 (1956).
350. MACDONNELL, L. R., LINEWEAVER, H. and FEENEY, R. E., *Poultry Sci.* **30**, 856 (1951).
351. HAWTHORNE, J. R., *Biochim. Biophys. Acta* **6**, 28 (1950).
352. COTTERILL, O. J. and WINTER, A. R., *Poultry Sci.* **34**, 679 (1955).
353. RHODES, M. B. and FEENEY R. E., *Poultry Sci.* **36**, 891 (1957).
354. SHARP, P. F. and STEWART, G. F., *U.S. Egg Poultry Mag.* **37**, No. 6, 30 (1931).
355. BROOKS, J., In BATE-SMITH, E. C. and MORRIS, T. N., ed. *Food Science.* University Press, Cambridge 1952.
356. FEENEY, R. E., MACDONNELL, L. R. and LORENZ, F. W., *Food Technol.* **8**, 242 (1954).
357. FUNK, E. M., FORWARD, J. and LORAH, M., *Poultry Sci.* **33**, 532 (1954).
358. EISEN, E. J., BOHREN, B. B. and MCKEAN, H. E., *Poultry Sci.* **41**, 1461 (1962).
359. BANWART, S. F., CARLIN, A. F. and COTTERILL, O. J., *Food Technol.* **11**, 200 (1957).
360. *The Liquid Egg (Pasteurisation) Regulations* 1963 (S.I. 1963 No. 1503) dated 28 August, 1963.
361. SHRIMPTON, D. H., MONSEY, J. B., HOBBS, B. C. and SMITH, M. E., *J. Hyg. Camb.* **60**, 153 (1962).
362. HELLER, C. L., ROBERTS, B. C., AMOS, A. J., SMITH, M. E. and HOBBS, B. C., *J. Hyg. Camb.* **60**, 135 (1962).

363. CUNNINGHAM, F. E. and LINEWEAVER, H., *Food Technol.* **19**, 1442 (1965).
364. LIGHTBODY, H. D. and FEVOLD, H. L., In MRAK, E. M. and STEWART, G. F., ed. *Advan. Food Res.* Vol. 1, Academic Press, New York, 1948.
365. POWRIE, W. D., LITTLE, H. and LOPEZ, A., *J. Food Sci.* **28**, 38 (1963).
366. HAWTHORNE, J. R. and BROOKS, J., *J. Soc. Chem. Ind. Lond.* **63**, 232 (1944).
367. KLINE, L. and SONODA, T. T., *Food Technol.* **5**, 90 (1951).
368. BALDWIN, R. H., CAMPBELL, H. A., THIESSEN, R. and LORANT, G. J., *Food Technol.* **7**, 275 (1953).
369. CARLIN, A. F. and AYRES, J. C., *Food Technol.* **7**, 268 (1953).
370. BROOKS, J. and HAWTHORNE, J. R., *J. Soc. Chem. Ind. Lond.* **62**, 165 (1943).
371. CONRAD, R. M., VAIL, G. E., OLSEN, A. L., TINKLIN, G. T., GREENE, J. W. and WAGONER, C., *Kansas State Agr. Expt. Sta. Tech. Bull.* **64**, 1948.
372. KLINE, L., SUGIHARA, T. F., and MEEHAN, J. J., *J. Food Sci.* **29**, 693 (1964).
373. DAVIS, J. G. and KLINE, L., *J. Food Sci.* **30**, 673 (1965).
374. AYRES, J. C. and STEWART, G. F., *Food Technol.* **1**, 519 (1947).
375. CARLIN, A. F. and AYRES, J. C., *Food Technol.* **5**, 172 (1951).
376. HAWTHORNE, J. R., *J. Sci. Food Agr.* **1**, 199 (1950).
377. HELLER, C. L., *Food Trade Rev.* **34**(3), 43 (1964).